Specific Intermolecular Interactions of Nitrogenated and Bioorganic Compounds

Alexei K. Baev

Specific Intermolecular Interactions of Nitrogenated and Bioorganic Compounds

Springer

Alexei K. Baev
Institute of Solution Chemistry
Russian Academy of Sciences
Ivanovo
Russia

ISBN 978-3-642-37471-5 ISBN 978-3-642-37472-2 (eBook)
DOI 10.1007/978-3-642-37472-2
Springer Heidelberg New York Dordrecht London

Library of Congress Control Number: 2013955906

Printed on acid-free paper

Springer is part of Springer Science+Business Media (www.springer.com)

Preface

The development of the molecular theory of solid and liquid conditions of organic and bioorganic compounds and their solutions is based on fundamental investigations of their structure, molecular interactions, and the dynamics of the process in the condensed phase. Of particular significance for the investigations of this development is that we have found interdisciplinary supramolecular chemistry interrelation in the molecular properties of condensed systems, as stipulated by intermolecular interactions. The solutions with specific interactions realized, analogous to hydrogen bonds, are related to associated solutions, and these are of great interest for fundamental and applied aspects. At present, scientific publications show a lot of completed systematization and we see a generalization dedicated to biophysics, and elucidation of the rules in the biological processes. However, the role of specific interactions in all identified knowledge fields is assigned to a scanty range, perhaps too restricted to mention. At the same time the absence of information on the energy of H-bonds allows us to consider only a network of hydrogen bonds, which is imperfect in physical chemistry. For organic and bioorganic compounds and their solutions the thermodynamic purposefulness of investigations is firmly established and developed. One of the basic values of this method is the possibility of its application when knowledge of the molecular structure of compounds is not sufficient. At the same time, because of this information and the thermodynamic properties of the vaporization process my well-founded principle about intercommunity of the enthalpy and entropy of evaporation and sublimation with the number and energy of dissociation bonds complete the invaluable role of the determination of the energies of specific intermolecular interactions and hydrogen bonds, which has been confirmed successfully for elementorganic and other classes of compounds. Thus, the key moment was the implementation of the critical thermodynamic analysis of the great quantity of alkyl compounds' main group elements (as I determined) that has accumulated. This analysis touches the notions of the sp^3-hybridization model of the carbon atom valiant orbital, which had perfect attractiveness. The results of thermodynamic analysis led to the correct conclusion, a new idea about the reverse dative bond, realized between the carbon atom of an alkyl ligand and contact with the central atom of the molecule and the carbon atom, and the ability of the end methyl ligand group to take part in the formation of specific intermolecular interactions displaying the five co-ordination stages. The fulfilled thermodynamic investigations of alkyl's aluminum substantiated the dimer molecules at the vapor phase with a stable bond (44.2 kJ mol^{-1}) formed by the five-coordinated carbon atom. The fundamental notions about the five-coordinated carbon atom, confirmed by thermodynamic investigations and, on the other side, the existence of the reverse dative bond, substantiated by the quantum-chemical calculations of methyl compounds boron, aluminum, tin, nitrogen, and oxygen resulted in a motivated refusal of the sp^3-hybridization model of the carbon atom.

The methodological principles of substantiation of specific interaction types, formed by a five-coordinated carbon atom and hydrogen bonds and determination of its energies ensured the creation of the system of these properties coordinated by each other and reflected nature-specific interactions of different identifiable oxygenated organic compounds.

A symptomatic fact of the non-stereotypic ideas for application by the specialist, dealing with the disciplines of organic and bioorganic chemistry, resulted in the development of theoretical notions on the specific interactions and information, widening the energy data for the nitrogenated and bioorganic compounds, which is the subject treated in this monograph.

The wide spectrum of notions in the theory of hydrogen bonds is analyzed in Chap. 1, where we substantiated the specific intermolecular interactions. The term "specific interaction" unites the antiparticles' interactions with participation of the five-coordinated carbon atom with the nitrogen and oxygen atoms, respectively, without formation of stable intramolecular bonds. The network of hydrogen bonds includes only this kind of intermolecular interactions, realized in solid and liquid ammonium, for aqueous matter and hydrides of nitrogen and oxygen group elements. The nontraditional approaches of the development of theory and the methodology of the determination of the energy of various types of hydrogen bonds and specific interactions with the five-coordinated carbon atom are discussed.

The specific interactions of amines, hydrazones, hydrozines, cyanide, nitrides, amine nitrites of acyclic and aliphatic series with saturated and unsaturated alkyl ligands and the different numbers of carbon atoms at the chain were studied and we determined the energies of various types of interactions, formed by the nitrogen atom and of the hydrogen bonds. We proved the law-governed nature of stability and exposed its nature (Chap. 2).

The different types of specific intermolecular interactions and hydrogen bonds of cyclic and aromatic compounds with saturated and unsaturated hydrocarbon cycles and nitrogen atoms of the functional group are discussed in Chap. 3. The values of the energies of the hydrogen bonds and different types of specific interactions obtained accurately reflect the nature of the interactions and illustrate the natural characteristic of changing its stability at the compounds of different series depending on the influence of the intermolecular reverse dative bond.

The structural-thermodynamic analysis for a wide range of compounds of aziridines, azetidines, pentamerous, and hexamerous heterocycle compounds with the nitrogen atom and saturated and unsaturated hydrocarbon cycles led to the discovery of the types of specific interactions and allowed us to reveal their nature (Chaps. 4 and 5). The idea of reverse dative bonds and the theoretical notions of specific interactions are analyzed with reference to the nitrogen atom. Located inside a cycle of availability in its one, two and three nitrogen atoms, different types of specific interactions are substantiated, their energies determined, and the levels of stability analyzed.

In Chap. 6 we consider methoxydimethylamines and *N*-(ethoxymethyl) diethylamines, methoxyacetonitriles, amides and their derivatives with hetero atoms of oxygen and nitrogen in the chain, with the network of specific intermolecular interactions and hydrogen bonds in liquid and crystal conditions and the various types of these interactions are given a basis. In the results of the thermodynamic analysis, the nature of the interactions was revealed, based on the formation of the peptide hydrogen bond by the carbonyl oxygen atom and the hydrogen atom of the amino group =C=O•••H–N and hydrogen bond O•••H–N corresponding to the amides and methoxymethylamines. The energies of these bonds and the different types of specific interactions for the crystalline and liquid compounds of various series were obtained and the range of values for their stabilization for compounds with the chain structure and benzene cycle was established.

The various types of specific intermolecular interactions of compounds with nitrogenated and oxygenated hydrocarbon cycles, pentamerous heterocycles, oxazoles, isoxazoles and thyazoles, oxazolines, azetidines and aminoisoxazole compounds with hexamerous heterocycles and their derivatives as well as the compounds with larger cycle 6-caprolactam with one and two carbonyl groups are analyzed in Chap. 7. In the results of the thermodynamic analysis, the nature of each type of the formed specific interaction and hydrogen bonds was revealed, obtaining its energies and the integral values of stabilization, which led to a significant development in the theory of intermolecular interactions.

The different types of hydrogen bonds formed and the specific interactions at the solid and liquid states of amine alcohols and their derivatives with the hydroxyl group are substantiated and the energies, illustrating the wide range of the change in stability were determined. The problems discussed in Chap. 8 have a straightforward relation with supramolecular chemistry and biochemistry.

The results of the analysis of thermodynamic characteristics of the low-molecular biological amino acids merit particular attention. The results of the investigations carried into biophysics the unknown and non-stereotypic notions of the types of specific intermolecular interactions, formed by the five-coordinated carbon atom with the nitrogen and oxygen atoms, respectively, and the formed structures of liquid and solid states with the network of specific interactions and participation of hydrogen bonds. The substantiated types of specific interactions with the five-coordinated carbon atom, hydrogen bonds, including the peptide hydrogen bonds, the determined values of the energies of different interactions, and the orders of their stabilization have enriched the molecular biophysics and biochemistry (Chap. 9).

All the preceding results of research in the field of specificity of intermolecular interactions, including organic, nitrogenous molecules and compounds with two hetero atoms of nitrogen and oxygen, provided success in the solving of problems on the energy value of peptide hydrogen bonds and all specific interactions in the formed networks in the crystalline and liquid structure of peptides. The significant results in this direction are the determination of the roles of the methylene group and the fragment with the larger number of these groups. These are located between strong acceptors of electron density by the nitrogen atoms and carbonyl oxygen atoms. They express the reduced ability in the formation of specific interaction compared with a similar fragment at the terminal part of the chain. We also see participation of the isostructural methyl group, connected with the mentioned methylene group, in the distribution of the electron density in the molecule and its extra stabilizing effect to the fragment, fringing these isostructural groups, leading to an energy contribution to the enthalpy characteristics independent of its number in the peptide and protein. The obtained values of the energies of the hydrogen bonds and specific interactions with the five-coordinated carbon atom of peptides and dipeptides led to the determination of the energy contribution by the conformer of the chain and the value of sublimation enthalpies of dipeptide, tripeptide, and tetrapeptide chains of derivative aminoacetic acid, glycine and L-(D)-α-alanine, aminopropionic acid. In the results, we first presented various protein secondary structures with parallel β-sheet and antiparallel β-sheet with all the specific interactions (Chap. 10).

The application of the theory of the hydrogen bond, specific interactions, and principles of the thermodynamic analysis to urea and its derivatives, belonging to one form of the main product of a living nature metabolism, easily extracted from the organism and playing an important role in the study of the conformation stability of globular protein, illustrated the phenomenological significance of the investigations (Chap. 11). Urea with eight hydrogen bonds of two types, formed by nitrogen and oxygen atoms with combinations of their derivatives and various functional substitutes of the hydrogen atom, formed in the amine groups, allow us to trace the influence of the substitutes on the stability of hydrogen bonds and alkyl fragments on specific interactions and the arrangement of substitutes in the amine groups with regard to the stability of the intermolecular interactions formed. The new concept of the extra stabilizing effect of the isomeric methyl group for the structure and stability of molecules of organic, nitrogenated, and bioorganic compounds, amino alcohols and acids, peptides, proteins and derivatives of urea and the value of the energy of the interactions is considered in this chapter.

The research eliminated the gap created by the absence of energy values of the hydrogen bonds and specific interactions even for the simplest and most frequently occurring compounds of water and spirits in the entire field of chemistry. The volume of energy values of different types of hydrogen bonds obtained, with most of them already known and based for the first time in different types of specific interactions, which do not even include any reference, led to the creation of energy values consistent and coherent with each other, reflecting their nature accurately. On the basis of the results obtained, also for the first time, the theory of specific intermolecular interactions was developed with the five-coordinated carbon atom; we have also undertaken the further development of the theory of hydrogen bonds, which has enriched physical chemistry, organic chemistry, biochemistry, and biophysics with fresh new ideas. The results of thermodynamic analysis of amides, compounds with nitrogenated and oxygenated hydrocarbon cycle with pentamerous and hexamerous heterocycles, amino acids, and peptides allow us to base the existence of several types of peptides' hydrogen bonds, explain the reasons causing the differences in their energies, and solve the problems by turning to the analysis of proteins on the basis of the values of the energies of hydrogen bonds and specific interactions. In this connection one should note the new perspective of quantum-chemical research of all the classes and series of compounds, first, of proteins, peptides, and amino acids and the necessity of obtaining the thermodynamic properties of these compounds. Special role is given to the research of alcohols and amino acids for determination of energies of hydrogen bonds of DNA.

The present book is intended to be a useful source of information for scientists, teachers, post–doctoral fellows, post graduate students, and other specialists working in the area of Chemistry, Biochemistry, Biophysics, and Chemical Technology.

In writing the book I have tried to reveal the depth of the analyzed problem and to provide its clear description remembering that "formulas are not the essence of the theory, and a real theoretician uses them sparingly expressing with words all that is possible to express in words" (L. Boltzmann).

I am deeply grateful to the corresponding-member of the Russian Academy of Sciences G.A. Krestov, founder of the Institute of Solution Chemistry of RAN. Ivanovo. R.F., where I worked for 25 years as the member of the Council for the doctoral thesis defense, and to my friends and colleagues of this institute and Ivanovo Chemical-Technological University for scientific activity.

Minsk, Belarus Alexei K. Baev

Synopsis

This book presents the development of the theory of specific intermolecular interactions of non-stereotypical ideas and approaches to the theory of H-bonding, peptide hydrogen bonds, and specific interactions providing inspirational ideas about the five-coordinated carbon atom with a wide spectrum of nitrogenated and bioorganic compounds, amino alcohols and acids, peptides, and proteins. New types of hydrogen bonds and specific interactions are substantiated and on the basis of a worked-out methodology, their energies are determined and a system of interconnected quantitative characteristics of specific interactions stability, exactly reflecting their nature and law-governed changes, is created. The values obtained for the energies of the hydrogen bonds and specific interactions lead to the determination of the contribution of the energy of successively iterated configurations of dipeptide, tripeptide, and tetrapeptide chains of derivative aminoacetic acid, glycine, and L-(D)-α-alanine, aminopropionic acid and to the presentation of various proteins of secondary structures with parallel β-sheet and antiparallel β-sheet with all specific interactions.

Contents

Abbreviations

Chapter 1

DA	Donor–acceptor bond
DE-Me	Dissociation energies of bond
Ddis.dim	Dissociation energies of dimmer molecule
Cn	Number of carbon atoms at chine
E.E.C.	Endoeffect coordination
Ddest. E	Destabilized effect
	Reverse dative bond on metal carbonyl
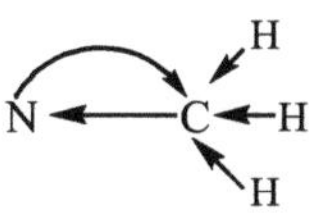	Reverse dative bond $N(CH_3)_3$
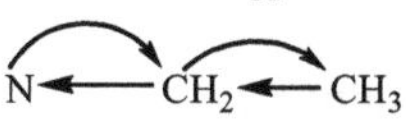	Reverse dative bond $N(C_2H_5)_3$
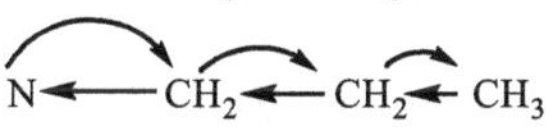	Reverse dative bond $N(C_3H_7)_3$
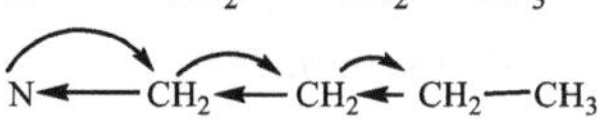	Reverse dative bond $N(C_4H_9)_3$
T m.p.	Melting point
T b.p.	Boiling point
(Iv)	Vertical Ionization Potential
$\Delta_{vap}H°(298\ K)$	Vaporization enthalpy at Standard Conditions
$\Delta_{vap}H°(T\ K)$	Vaporization enthalpy at T K
$\Delta_{sub}H°(298\ K)$	Sublimation enthalpy at Standard Conditions
$\Delta_{sub}H°(T\ K)$	Sublimation enthalpy at T K
$\Delta_{evap}H°(298\ K)$	Vaporization enthalpy at Standard Conditions

Chapter 2

$\Delta_{vap}H°(298\ K)$*iso*.pr.a	Vaporization enthalpy of isopropylamine
$\Delta_{vap}H°(298\ K)$et.a	Vaporization enthalpy of ethylamine
$\Delta_{vap}H°(298\ K)$*iso*.bu.a	Vaporization enthalpy of isobuthylamine

$\Delta_{vap}H^{\circ}$(298 K)*iso*.pr.a	Vaporization enthalpy of propylamine
$\Delta_{vap}H^{\circ}$(298 K)eda	Vaporization enthalpy of ethylenediamine
$\Delta_{vap}H^{\circ}$(298 K)bda	Vaporization enthalpy of butyldiamine
$\Delta_{vap}H^{\circ}$(298 K) dtbda	Vaporization enthalpy of di-*tert*-butyldiazene
$\Delta_{vap}H^{\circ}$(298 K) mbda	Vaporization enthalpy of *N*-Methylbutyldiazene
$\Delta_{vap}H^{\circ}$(298 K) deda	Vaporization enthalpy of diethyldiazene
$\Delta_{vap}H^{\circ}$(298 K)bdbd	Vaporization enthalpy of butylhydrazone butyraldehyde
$\Delta_{vap}H^{\circ}$(298 K)phz	Vaporization enthalpy of propylhydrazone acetone
$\Delta_{vap}H^{\circ}$(298 K)iphz	Vaporization enthalpy of isopropylhydrazone
$\Delta_{vap}H^{\circ}$(298 K)dehze	*N*,*N*-Diethyl-2-(1-methylhydrazino)ethanamine
$\Delta_{vap}H^{\circ}$(298 K)azet	Vaporization enthalpy of azidoethane
$\Delta_{vap}H^{\circ}$(298 K)azp	Vaporization enthalpy of azidopropane
$\Delta_{vap}H^{\circ}$(298 K)azp	Vaporization enthalpy of 1-azidopropane
$\Delta_{vap}H^{\circ}$(298 K)cha	Vaporization enthalpy of cyclohexylamine
$\Delta_{vap}H^{\circ}$(298 K)cac	Vaporization enthalpy of cyanoacetylene
$\Delta_{vap}H^{\circ}$(298 K)hcad	Vaporization enthalpy of hydrogencyanide
$\Delta_{vap}H^{\circ}$(298 K)acnl	Vaporization enthalpy of acetonitryle
$\Delta_{vap}H^{\circ}$(298 K)prnl	Vaporization enthalpy of propionitrile
$\Delta_{vap}H^{\circ}$(298 K)ibnl	Vaporization enthalpy of Isobutironitrile
$\Delta_{vap}H^{\circ}$(298 K)prn	Vaporization enthalpy of Propylonitrile
$\Delta_{vap}H^{\circ}$(298 K)dm.mn	Vaporization enthalpy of Dimethylmalononitrile
$\Delta_{vap}H^{\circ}$(298 K)mn	Vaporization enthalpy of Malononitrile
$\Delta_{vap}H^{\circ}$(298 K)bdn	Vaporization enthalpy of Butyldinitrile
$\Delta_{vap}H^{\circ}$(298 K)gn	Vaporization enthalpy of Glutaronitrile
$\Delta_{vap}H^{\circ}$(298 K)dea.mp	Vaporization enthalpy of Diethylamino-methylpropionitrile
ΔH_{vap}°(298 K)deac	Vaporization enthalpy of (Diethylamino) acetonitrile
$\Delta_{vap}H^{\circ}$(298 K)amc	Vaporization enthalpy of Ammoniumcyanide
$\Delta_{vap}H^{\circ}$(98 K)mac	Vaporization enthalpy of Dimethylammonium

Chapter 3

$\Delta_{vap}H^{\circ}$(298 K)cha	Vaporization enthalpy of Cyclohexylamine
$\Delta_{vap}H^{\circ}$(298 K)ch	Vaporization enthalpy of Cyclohexane
$\Delta_{vap}H^{\circ}$(298 K)cba	Vaporization enthalpy of Cyclobutylamine
$\Delta_{vap}H^{\circ}$(298 K)cb	Vaporization enthalpy of Cyclobutyle
$\Delta_{vap}H^{\circ}$(298 k)dcha	Vaporization enthalpy of N,α-Dimethylcyclohexanethylamine
$\Delta_{vap}H^{\circ}$(298 K) ech	Vaporization enthalpy of Ethylcyclohexane
$\Delta_{vap}H^{\circ}$(298 k)ccn	Vaporization enthalpy of Cyclocarbonitrile
$\Delta_{vap}H^{\circ}$(298 k)ccn	Vaporization enthalpy of Cyclohydrocarbon
$\Delta_{vap}H^{\circ}$(298 K)abz	Vaporization enthalpy of Azobenzene
$\Delta_{vap}H^{\circ}$(298 K)abcn	Vaporization enthalpy of 3-Azabicyclo[3,2,2] nonane

$\Delta_{vap}H°(298\ K)abco$	Vaporization enthalpy of 1-Azobicyclooctane
$\Delta_{vap}H°(298\ K)an$	Vaporization enthalpy of Aniline
$\Delta_{vap}H°(298\ K)dman$	Vaporization enthalpy of *N,N*-Dimethylaniline
$\Delta_{vap}H°(298\ K)man$	Vaporization enthalpy of *N*-Methylaniline
$\Delta_{vap}H°(298\ K)\ bz$	Vaporization enthalpy of Benzene
$\Delta_{vap}H°(298\ K)\ der.bz$	Vaporization enthalpy of Methylbenzene
$\Delta_{vap}H°(298\ K)etam$	Vaporization enthalpy of Ethylamines
$\Delta_{vap}H°(298\ K)etbz$	Vaporization enthalpy of Ethylbenzene
$\Delta_{vap}H°(298\ K)\textit{iso}an$	Vaporization enthalpy of isoAlkylaniline
$\Delta_{vap}H°(298\ K)\textit{iso}.bz$	Vaporization enthalpy of isoAlkylbenzene
$\Delta_{vap}H°(298\ K)damb$	Vaporization enthalpy of Diaminebenzenes
$\Delta_{vap}H°(298\ K)napha$	Vaporization enthalpy of Naphthylamine
$\Delta_{vap}H°(298\ K)naphl$	Vaporization enthalpy of Naphthalene
$\Delta_{vap}H°(298\ K)\ phpa$	Vaporization enthalpy of Phenyl-2-Propylamine
$\Delta_{vap}H°(298\ K)pbz$	Vaporization enthalpy of Propylbenzene
$\Delta_{vap}H°(298\ K)\ mphba$	Vaporization enthalpy of *N,N*-Dimethyl-3-methyl-3-phenyl-2-butaneamine
$\Delta_{vap}H°(298\ K)bam$	Vaporization enthalpy of Benzylamine
$\Delta_{vap}H°(298\ K)toe$	Vaporization enthalpy of Toluene
$\Delta_{vap}H°(298\ K)mbam$	Vaporization enthalpy of (−)α-Methylbenzylamine
$\Delta_{vap}H°(298\ K)dmba$	Vaporization enthalpy of α, α-Dimethylbenzylamine
$\Delta_{vap}H°(298\ K)dmbz$	Vaporization enthalpy of *N,N*-Dimethylbenzyl
$\Delta_{vap}H°(298\ K)tli$	Vaporization enthalpy of *o*-Toluidine, *p*-Toluidine
$\Delta_{vap}H°(298\ K)ispb$	Vaporization enthalpy of *iso*-Propylbenzene
$\Delta_{vap}H°(298\ K)mbd$	Vaporization enthalpy of *N*-Methylbenzaldehydeimine
$\Delta_{vap}H°(298\ K)bn$	Vaporization enthalpy of Benzonitrile
$\Delta_{vap}H°(298\ K)bc$	Vaporization enthalpy of Benzylcyanide
$\Delta_{vap}H°(298\ K)tn$	Vaporization enthalpy of Tolunitriles
$\Delta_{vap}H°(298\ K)dcb$	Vaporization enthalpy of Dicyanobenzines

Chapter 4

$\Delta_{vap}H°(298\ K)azd$	Vaporization enthalpy of azetidine
$\Delta_{vap}H°(298\ K)cpr$	Vaporization enthalpy of cyclopropane
$\Delta_{vap}H°(298\ K)epzl$	Vaporization enthalpy of 1-ethylpyrazole
$\Delta_{sub}H°(298\ K)\ prz$	Sublimation enthalpy of Pyrazole
$\Delta_{sub}H°(T)tz$	Sublimation enthalpy of Tetrazole
$\Delta_{sub}H°(T)mtz$	Sublimation enthalpy of 5-Methyltetrazole

Chapter 5

$\Delta_{vap}H°(298\ K)ppr$	Vaporization enthalpy of piperidine
$\Delta_{vap}H°(298\ K)appr$	Vaporization enthalpy of alkylpiperidine
$\Delta_{vap}H°(298\ K)mpp$	Vaporization enthalpy of methylpiperidine

$\Delta_{vap}H°$(2,987 K)mvp	Vaporization enthalpy of 2-methyl-5-vinylpyridine
$\Delta_{vap}H°$(298 K)mpz	Vaporization enthalpy of methylpyrazine
$\Delta_{vap}H°$(298 K)pz	Vaporization enthalpy of pyrazine
$\Delta_{vap}H°$(298 K) pprz	Vaporization enthalpy of piperazine
$\Delta_{vap}H°$(298 K)pr	Vaporization enthalpy of pyridine
$\Delta_{vap}H°$(298 K)mpr	Vaporization enthalpy of methylpyridine
$\Delta_{sub}H°$(298 K)bp	Sublimation enthalpy of bipyridine
$\Delta_{sub}H°$(298 K)ql	Sublimation enthalpy of quinoline
$\Delta_{sub}H°$(298 K)aql	Sublimation enthalpy of 3-aminoquinoline
$\Delta_{sub}H°$(298 K)hql	Sublimation enthalpy of hydroxyquinolines
$\Delta_{sub}H°$(298 K)hqxl	Sublimation enthalpy of 2-hydroxyquinoxaline
$\Delta_{sub}H°$(298 K)qxl	Sublimation enthalpy of quinoxaline
$\Delta_{sub}H°$(298 K)phbph	Sublimation enthalpy of *N*-phenylbenzophenoneimine
$\Delta_{sub}H°$(298 K)tphe	Sublimation enthalpy of triphenylmethane
$\Delta_{sub}H°$(298 K)acr	Sublimation enthalpy of acridone
$\Delta_{sub}H°$(298 K)pzo	Sublimation enthalpy of phenazine-*N*-oxide
$\Delta_{sub}H°$(298 K)qox	Sublimation enthalpy of quinoxaline-1,4-dioxide

Chapter 6

$\Delta_{vap}H°$(298 K)mdma	Vaporization enthalpy of *N*-Methoxydimethyl amine
$\Delta_{vap}H°$(298 K)meam	Vaporization enthalpy of 2-Methoxyethylamine
$\Delta_{vap}H°$(298 K)mmde	Vaporization enthalpy of *N*-(Methoxymethyl) diethylamine
$\Delta_{vap}H°$(298 K)emde	Vaporization enthalpy of *N*-(Ethoxymethyl) diethylamine
$\Delta_{vap}H°$(298 K)mph	Vaporization enthalpy of Morpholine
$\Delta_{vap}H°$(298 K)mac	Vaporization enthalpy of Methoxyacetonitrile
$\Delta_{vap}H°$(298 K)ef	Vaporization enthalpy of *N*-Ethylformamide
$\Delta_{vap}H°$(298 K)dam	Vaporization enthalpy of Vaporization enthalpy of
$\Delta_{vap}H°$(298 K)acam	Vaporization enthalpy of Acetamide
$\Delta_{sub}H°$(298 K)ban	Sublimation enthalpy of A-oxo-benseneacetonitrile
$\Delta_{sub}H°$(298 K)caph	Sublimation enthalpy of Cyanoacetophenone
$\Delta_{sub}H°$(298 K)cpph	Sublimation enthalpy of β-Cyanopropiophenone
$\Delta_{sub}H°$(298 K)aaph	Sublimation enthalpy of 4-Aminoacetophenone

Chapter 7

$\Delta_{vap}H°$(298 K)oz	Vaporization enthalpy of Oxazole
$\Delta_{vap}H°$(298 K)mos	Vaporization enthalpy of 2-Methyl-2-oxazoline
$\Delta_{sub}H°$(298 K)amioz	Vaporization enthalpy of 3-Amino-5-methylisoxazole
$\Delta_{vap}H°$(298 K) mprd	Vaporization enthalpy of *N*-Methyl-2-Pyrrolidone
$\Delta_{vap}H°$(298 K)pd	Vaporization enthalpy of 2-Pyrrolidone Pyrazole
$\Delta_{vap}H°$(298 K)tpr	Vaporization enthalpy of Tetrahydro-2-pyrinidone
$\Delta_{vap}H°$(298 K)dpao	Vaporization enthalpy of Dipropyldiazene *N*-oxide

$\Delta_{vap}H°(298\ K)dpa$	Vaporization enthalpy of Dipropyldiazene
$\Delta_{sub}H°(298\ K)hpp$	Sublimation enthalpy of *N*-(Hydroxylethyl) piperazine
$\Delta_{sub}H°(298\ K)ur$	Sublimation enthalpy of Uracil
$\Delta_{sub}H°(298\ K)cts$	Sublimation enthalpy of Cytosine
$\Delta_{sub}H°(298\ K)mhxc$	Sublimation enthalpy of 1-Methyl-*N*-Hydroxycytozin
$\Delta_{sub}H°(298\ K)mmc$	Sublimation enthalpy of 1-Methyl-4N-methoxycytosine
$\Delta_{sub}H°(298\ K)\ bfo$	Sublimation enthalpy of Benzofurazan *N*-oxide
$\Delta_{sub}H°(298\ K)mbfo$	Sublimation enthalpy of 5-Methoxybenzofurazan-1-oxide

Chapter 8

$\Delta_{vap}H°(298\ K)\ dmha$	Vaporization enthalpy of *N,N*-dimethylhydroxylamine
$\Delta_{vap}H°(298\ K)mha$	Vaporization enthalpy of *N*-Methylhydroxylamine.
$\Delta_{vap}H°(298\ K)amet$	Vaporization enthalpy of 2-Aminoethanol
$\Delta_{vap}H°(298\ K)maet$	Vaporization enthalpy of 2-(Methylamino)ethanol
$\Delta_{vap}H°(298\ K)heea$	Vaporization enthalpy of *N,N-bis*(2Hydroxyethyl) ethylenediamine
$\Delta_{vap}H°(298\ K)cel$	Vaporization enthalpy of 2-Cyanoethanol
$\Delta_{vap}H°(298\ K)map$	Vaporization enthalpy of 2-Methyl-2-amino-1-propanol
$\Delta_{vap}H°(298\ K)dea$	Vaporization enthalpy of diethanolamine
$\Delta_{sub}H°(298\ K)aph$	Vaporization enthalpy of aminophenol
$\Delta_{sub}H°(298\ K)phl$	Vaporization enthalpy of phenol
$\Delta_{vap}H°(298\ K)fdho$	Vaporization enthalpy of formaldehyde oxime
$\Delta_{vap}H°(298\ K)hbox$	Vaporization enthalpy of 2-hydroxybenzaldoxime
$\Delta_{vap}H°(298\ K)ao$	Vaporization enthalpy of Acetone oxime
$\Delta_{vap}H°(298\ K)bo$	Vaporization enthalpy of 2-Butanone oxime
$\Delta_{sub}H°(298\ K)cho$	Sublimation enthalpy of Cyclohexanone oxime
$\Delta_{sub}H°(298\ K)dmap$	Sublimation enthalpy of 1-(Dimethylamino)-2-propanone
$\Delta_{sub}H°(298\ K)dpdz$	Sublimation enthalpy of dipropyldiazene *N*-oxide

Chapter 9

$\Delta_{sub}H°(298\ K)af$	Sublimation enthalpy of Aminoformic acid
$\Delta_{sub}H°(298\ K)saa$	Sublimation enthalpy of Amino acetic acid
$\Delta_{sub}H°(298\ K)ag$	Sublimation enthalpy of *N*-Acetylglycine
$\Delta_{sub}H°(298\ K)phg$	Sublimation enthalpy of *N*-Phenylglicine
$\Delta_{sub}H°(298\ K)ba$	Sublimation enthalpy of Benzoic Acid
$\Delta_{sub}H°(298\ K)mba$	Sublimation enthalpy of 2-Methylbenzoic acid

$\Delta_{sub}H°(298\ K)$amba	Sublimation enthalpy of Aminomethylbenzoic acids
$\Delta_{sub}H°(298\ K)$pca	Sublimation enthalpy of Pyridinecarboxylic acids
$\Delta_{sub}H°(298\ K)$pcao	Sublimation enthalpy of Pyridinecarboxylic acid *N*-oxides
$\Delta_{sub}H°(298\ K)$mpzc	Sublimation enthalpy of 2-Methyl-5-pyrazine carboxylic acid
$\Delta_{vap}H°(298\ K)$mc	Vaporization enthalpy of Methyl carbamate
$\Delta_{vap}H°(298\ K)$amc	Vaporization enthalpy of Ammonium carbamate
$\Delta_{vap}H°(298\ K)$dmap	Vaporization enthalpy of 1-(Diethylamino)-2-propanone

Chapter 10

$\Delta_{vap}H°(298\ K)$mca	Vaporization enthalpy of *N*-Methyl Carbamic acid
$\Delta_{vap}H°(298\ K)$aba	Vaporization enthalpy of *N*-Acetyl-*N*-butylacetamide
D=C=O•••H–N–CH_2-*iso*	Vaporization enthalpy of (2-Pyridyl)acetamide
$\Delta_{vap}H°(298\ K)NAA_2A$	Vaporization enthalpy of *N*-Acetyl-L-alanyl-L-alaninamide
$\Delta_{vap}H°(298\ K)$paa	Vaporization enthalpy of (2-Pyridyl)acetamide

Chapter 11

$\Delta_{sub}H°(T)$et	Sublimation enthalpy of 1-Ethylurea
$\Delta_{sub}H°(T)$pr	Sublimation enthalpy of 1-Propylurea
$\Delta_{sub}H°(T)$is.pr	Sublimation enthalpy of 1-*iso* Propylurea
$\Delta_{sub}H°(T)$is bu	Sublimation enthalpy of 1-*iso*-Buthylurea
$\Delta_{sub}H°(T)$*tert* bu	Sublimation enthalpy of 1-*tert*-Buthylurea
$\Delta_{sub}H°(298\ K)$dmu	Sublimation enthalpy of Dimethyluracil
$\Delta_{sub}H°(298\ K)$ur	Sublimation enthalpy of Uracyl
ΔsubH°(298 K)amur	Sublimation enthalpy of Aminouracil
$\Delta_{sub}H°(298\ K)$tmu	Sublimation enthalpy of Trimethyluracil
$\Delta_{sub}H°(298\ K)$mcy	Sublimation enthalpy of Methylcytosine

Types of Specific Intermolecular Interactions with Pentacoordinated Carbon Atom of Nitrogenated Organic and Bioorganic Compounds

D–N → CH_3–N	CH_3 – Methyl group
D–N → CH_3–CH_2–N	C_2H_5 – Ethyl group
D–N → CH_3–CH_2–CH_2–N	C_3H_7 – Propyl group
D–N → CH_2–N	CH_2-Mehelene group
D–N → CH_2–CH_2–N	CH_2–CH_2–N
D–N → CH_2–CH_2–CH_2–N	CH_2–CH_2–CH_2–N fragment
D–N → CH_2=CH–N	CH_2=CH–N fragment
D–N → CH_3–CH=CH–N	CH_3–CH=CH–N fragment
D–N → CH_2–(CH_2)n–N	CH_2–(CH_2)n–N fragment

D–N → CH_2–	CH_2– group
D–N → CH=	CH= group
D–N → C=	>C= group
D=N → CH_3–N=	CH_3 – Methyl group
D=N → CH_3–CH_2–N=.	C_2H_5 – Ethyl group
D=N → CH_3-CH_2-CH_2-N=	C_3H_7 – Propyl group
D=N → CH_2=CH–N=.	CH_2=CH–N=. group
D=N → CH_3–CH=CH–N=.	CH_3–CH=CH–N=. fragment
D=N → CH_2–N=	CH_2–N= group
D=N → CH–N=	CH–N= group
D=N → CH	CH group
D≡N → C≡	–C≡N cyanide group
D≡N → CH_2–C≡	Malononitrile
D≡N → CH_2–CH_2–C≡	Malononitrile
D≡N → CH_2–CH_2–C≡	Butyldinitrile
D=N → C=N=.	=N=C= group
D=N → N=	Nitrogen atoms of Tetrazoles and Aminotetrazoles

Specific Intermolecular Interactions formed by *iso*-structural groups

DH_3C → H–CH_2	*iso*-CH_3 group
—CH_2—H_2C<(H···H)>CH_2—CH_2—	of *iso*-CH_3 group
DH_2C → H–CH	DH_2C group
D=HC → H–C	=HC group
D≡C → H–C≡	H–C≡ group
DCH_2–CH_3 → CH_2–CH_3	*iso*-CH_2–CH_3
DCH_2–CH_2–CH_3 → CH_2–CH_2–CH_3	*iso*-CH_2–CH_2–CH_3

Specific Interactions formed by CH_2 and CH groups of saturated and unsaturated hydrocarbon cycle of Nitrogenated and Oxygenated Organic and Bioorganic Compounds

D–H_2C → CH_2	=CH_2 group
D=HC → CH	=CH group
D–H_2C → HC	HC group
D=HC → C	HC group
D=C → C	C atoms of Carbazole, Indole and et ce.

Contribution's Energy at Vaporization Enthalpy

DCH_3	*iso*-CH_3 group
DCH_2	–CH_2-group
DCH	–CH–
DCH_2–CH_2–	–CH_2–CH_2– group

DCH_2–CH_2–CH_2–	–CH_2–CH_2–CH_2– group
D(H)	Substituted hydrogen atom (H) of CH_3, CH_2, and CH

Hydrogen Bonds formed by

D–N•••H–N	NH_3, $NHAlk_2$, NH_2Alk, –NH
D–N•••H–CH	Hydrogen bonds of *bis*(Dimethylamino)methane
D=N•••H–N=	Hydrogen bonds of HN=NH, HN=NAlk, =NH
D≡N•••H–C≡	Hydrogen bonds of H–C≡ group
D–O•••H–N	NH_3,HN=NH, HN=NAlk, =NH
D–O•••H–O	–O–H group
D=C=O•••H–C=O	Aldehyde group
D=C=O•••H–N	NH_3,NH_2, NH, HN=NH, HN=NAlk, =NH
D–N → CH_2–O	CH_2–O fragment

Hydrogen Bonds of low stability

D–N•••H–C	CH– group
D–N•••H–CH	CH_2– group
D–N•••H–CH_2	CH_3– group

Types of Specific Intermolecular Interactions of Oxygenated Bioorganic Compounds

D=C=O → CH_3–CH_2–CH_2–	Propyl group
D–O → CH_3–CH_2–	Ethyl group
D–O → CH_3	Methyl group
D–O → CH_2	Methylene group
D–O → CH	CH group
D–O → C=	Carbon atom
D–O → C=O	Carbonyl group
D=O → C=	Methyl isocyanate
D–O → CH_3–N	NCH_3
D–O → CH_3–CH_2–N	NC_2H_5
D–O → CH_3–CH_2–CH_2–N	NC_3H_7
D–N–O → N	*N*-oxides group
DO=C → CH–C=	CH–C= group
D=C=O → C=	Carbon atom
D=C=O → CH	CH group
D=C=O → CH_2	Methylene group
D=C=O → CH_3	Methyl group
D=C=O → C=O	Carbon atom of C=O group
D=C=O → H_3C–N	NCH_3
D=C=O → CH_3–CH_2–N	NC_2H_5
D=C=O → CH_3–CH_2–CH_2–N	NC_3H_7

D–N → CH–N	N–CH
D–O → CH_2–O	O–CH_2
D–O → CH_3–O	O–CH_3
D–O → CH_3–CH_2–O	O–C_2H_5
D–O → CH_3–CH_2–CH_2–O	O–C_3H_5
D=C=O → CH_3–C=	Acetic group
D=C=O → CH_2–C=	Acetal group
D=C=O → CH_3–CH_2–C=O	Propenal group
D=C=O → CH_2=CH–C	Alkenes ligand
D=C=O → CH_3–CH=CH–CH_2–C	Alkenes ligand

Hydrogen Bonds of Oxygenated Organic and Bioorganic Compounds

D–O•••H–O	–O–H group
D=C=O•••H–C=O	Aldehyd group
D=C=O•••H–N	Amine group

Hydrogen Bonds of low stability

D=C=O•••H–C
D=C=O•••H–CH
D=C=O•••H–CH_2

Strength Hydrogen Bonds of Amides, Amino acids, Peptide, and Proteins

D=C=O•••H–N	H–N
D=C=O•••H–N–CH_2-*iso*	H–N–CH_2-*iso*
D=C=O•••H–N–CH_2	H–N–CH_2
D–O•••H–O	–O–H group

Chapter 1
Hydrogen Bonds and Specific Intermolecular Interactions of Nitrogenated and Bio-organic Compounds and the Methodology of their Investigation

1.1 Some of the Problems of Hydrogen Bonds

The hydrogen bond is caused by an additional interaction between a covalently bonded hydrogen atom and a more electronegative atom of the same group or any other molecule. Thus, the purely covalent bond leads to a situation of equal sharing of electrons by the interacting hydrogen atoms in an H_2 molecule; only in this case the degree of socialization is 50 % [1]. Preferential shifting of the electron density to the atom of the element interacting with the hydrogen atom, i.e. polarization, provides the molecule with an ionic component. Acting in the covalent bond, the electric forces of the nuclear electron interaction and the magnetic forces, together with electrostatic forces of atoms attractingwith opposite charges, correlate with interatomic distances of atoms in the molecule (in the X–H bond distance), and this is observed in the fundamental X–H stretching vibration frequencies [2]. A practically similar definition of the hydrogen bond is given in Pauling's monograph titled "The Nature of the Chemical Bond" [3].

The use of various quantum-chemical approaches in the research of H-bonded complexes in the paper by Scheiner [4] has reached the area of research, entailing the structure, energetics, spectral, and electronics properties. Development of computers and equipment for experiment created opportunities for thorough research; Bader's theory of atoms in molecules [5] was developed by expanding information and providing a deeper understanding of H-bonded interactions. The hydrogen bonds formed with an electronegative atom of O, N, or F,

O-H•••O, N-H•••N, F-H•••F, O-H•••N, O-H•••F, N-H•••O, N-H•••F, F-H•••O and F-H•••N

are classified by the values of the energetic parameters. The most stable hydrogen bonds are defined by the energy interval of 83–167 kJ mol^{-1} [4, 6–8], the moderately stable ones by energy values of 16.7–62.7 kJ mol^{-1}, and weak bonds by 5–20 kJ mol^{-1} [7, 9–11]. Taking into account this classification, however, it should

A.K. Baev, *Specific Intermolecular Interactions of Nitrogenated and Bioorganic Compounds*, DOI 10.1007/978-3-642-37472-2_1,

Table 1.1 General parameters of H-bond's types [9]

H-Bond characteristics	Strong	Moderate	Weak
Type of interaction	Strongly covalent	Mostly electrostatic	Electrostatic/dispersed
Bond length (H–Y[Å])	1.2–1.5	1.5–2.2	>2.2
Lengthening of X–H(Å)	0.08–0.25	0.02–0.08	<0.02
H-Bond length (X–Y[Å])	2.2–2.5	2.5–3.2	>3.2
H-Bond angle (°)	170–180	>130	>90
H-Bond strength (kJ mol^{-1})	83–167	16.7–62.7	5–16.7

be noted that, in addition to the weak bonds with energy of 5–20 kJ mol^{-1} the authors of this work [11] classify electrostatic and dispersive interactions of liquid water as exceeding the energy value of the hydrogen bond (10.99 kJ mol^{-}) [12]. Table 1.1, taken from Steiner [9] displays the general characteristics of three different types of hydrogen bonds with strong, moderate, and weak classifications, respectively, with contributions from electrostatic interactions (acid/base), polarization (hard/soft) effect, and dispersion (repulsion) interactions, respectively. One should mention that, being attributable to the range of weak dispersive interactions, estimated by energy values of 5–20 kJ mol^{-1} as terminal, and taking into account the combination of two equivalent hydrogen bonds of a dimeric molecule, the limit values are almost two-fold higher than the energy value of the hydrogen bond of liquid water (10.99 kJ mol^{-1}), obtained with the use of vaporization enthalpy [12].

Data given in the Table 1.1 illustrate the following.

First, the traditional use of physical parameters of bond lengths, lengthening, and H-bond lengths for the development of the theory of the hydrogen bond and the search for ways of obtaining energetic parameters;

Second, development of quantum-chemical research and use of thermodynamic properties for obtaining the correct energy values of the hydrogen bonds;

Third, it does not attract attention to the participation of carbon atoms and the functional groups formed to the contributing energy of the intermolecular interactions formed by these groups.

A water molecule, which is the classic example of hydrogen bond formation, a cluster structure, in accordance with the results of research [13], is the dimeric form of the molecule. Research on molecular electric resonance [14, 15] and diverse research approaches to this homodimer [15–20] and quantum-chemical calculations [4, 6–8] are grounded in the gas phase and allow us to obtain the structure and energetic parameters of the water dimer. The results of the calculations allow us to estimate the energy value of the dimer hydrogen bond on the condition of the formation of

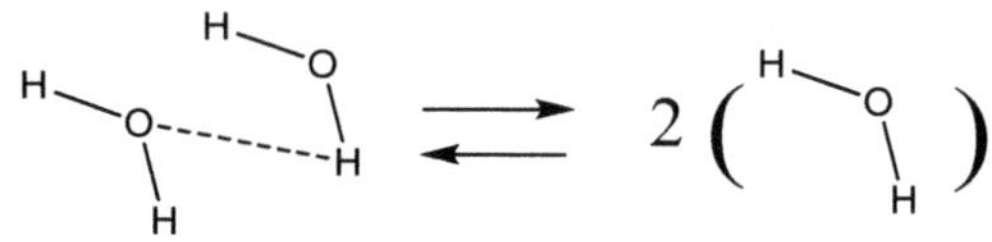

a single bond of this type equal to 20.9 ± 0.4 kJ mol^{-1} [20], which significantly exceeds the experimentally estimated energy value of the same type in a similar calculation of, 15.0 ± 2.0 kJ mol^{-1} at 375 K [21]. However, the real energy value of the hydrogen bond of the dimer form of a water molecule with two such bonds is 10.45 ± 0.4 kJ mol^{-1} [12]. Nevertheless, the single value of the energy bond, obtained from the vaporization enthalpy of water, is 10.99 kJ mol^{-1}. It follows that the energy of the hydrogen bond of liquid water and the dimer molecules ranges within the experimental errors of the estimated value on the basis of quantum-chemical calculations [20]. It is symptomatic that the energy of dissociation of the heterodimer of methyl spirit with water (10.8 kJ mol^{-1}) lies within error of the energy value of the hydrogen bond of liquid water (Table 1.2). It means that a molecule of methyl alcohol has a clear ability for dimerization in vapors, and the increased energy value of the hydrogen bond in liquid conditions (13.27 kJ mol^{-1}) expresses the reduced tendency to the formation of heterodimers with the water molecule (10.8 kJ mol^{-1}) in comparison with phenol (13.2 kJ mol^{-1}). Thus, the degree of changes in the energy of the hydrogen bond in combination with liquid water, heterodimer molecule, and liquid methanol (Fig. 1.1) is caused by the varied influences of the oxygen atom of the material, connected directly to the carbon atom of the methyl group, which is able to shift the electron density more than the oxygen atom of a water molecule, connected to a hydrogen atom. It follows that the benzene ring of a phenol shifts the electron density more to the oxygen atom, which in turn acts as the donor and supersedes the acceptor of hydrogen atom with regard to increased electron density rather than the carbon atom of the methyl group of methanol.

Consequently, the observed tendency of the hydrogen bonds to stabilize is an appropriate one.

E H-O•••H-O: Homodimeh H_2O (9.45) < Geterodimer CH_3OH•••H_2O (10.8) ≈ Homodimeh CH_3OH•••CH_3OH (10.9) < C_6H_5OH•••H_2O (13.2 kJ mol^{-1})

The small difference in the energy values of hydrogen bonds in liquid water and its dimer form points to the low partial component of dimer water in vapors, and is caused by the influence of vapor pressure, which is sharply reduced with the increase in temperature [22].

Using the various variants of quantum-chemical methods, the stable existence of the dimers methanol, formaldehyde, formamide, formic acid, acetic acid, and *N*-methylacetamide in the gas phase was substantiated and the dissociation energies determined [23–29]. From the results obtained above, vibration frequencies, electronic, geometric, and energy properties were later analyzed [9], on the basis of which the natural sequence of change of strengths of O–H•••O–H, O–H•••N–H,

Table 1.2 The energies of hydrogen bonds (kJ mol^{-1}) of homodimer and heterodimer of complexes [11]

H-Bonded complexes	Stabilization energy	Compounds	Hydrogen bond energy [12]
$H_2O{\cdots}H_2O$	10.45	Water	10.99
$C_6H_5OH{\cdots}H_2O$	26.2:2 = 13.2	Phenol	12.40
$CH_3OH{\cdots}CH_3OH$	21.8:2 = 10.9	Methanol	13.27
$CH_3OH{\cdots}H_2O$	21.6:2 = 10.8	Methanol	13.27

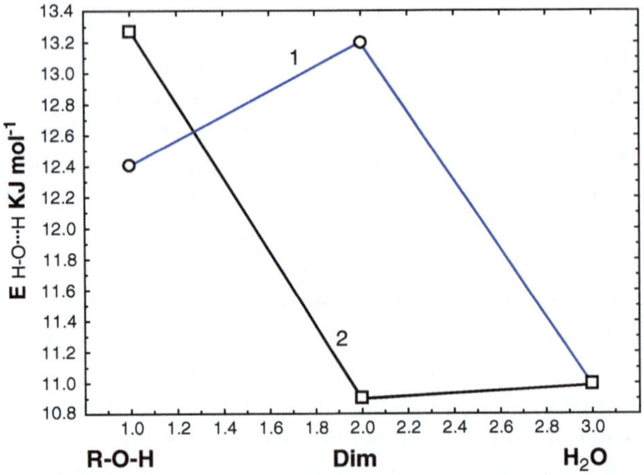

Fig. 1.1 Changing of the hydrogen-bonding energy of phenol–water (*1*) and methanol–water (*2*) dimers

N–H•••O–C, O–H•••O–C, π•••O–H, π••• H–N, π•••H–C, and π•••π interactions was proposed.

The carbonyl group =C==O in formaldehyde, acetaldehyde,

Formaldehyde	Acetaldehyde	Formamide	Acetamide	*N*-Methyl-acetamide
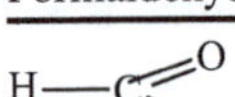			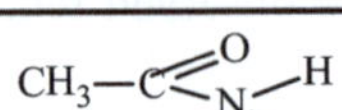	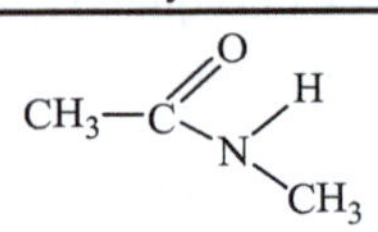

formamide, and acetamide plays a special role in the formation of hydrogen bonds. In a molecule of *N*-methyl-acetamide with a saturated hydrogen atom an amino group is converted, practically, to a peptide group. The presence of carboxyl groups in amino acids leads to the formation of a hydroxyl group of the second type, with its inherent stability as for example in the case of carboxylic acids [12].

The absence of this functional group at the peptide in the repeating configuration of a peptide chain leads to the peptide group with increased stability of the hydrogen bond =C==O•••H–N, which is the most important interaction, governing protein structure, folding, binding, and enzyme catalysis [11].

Dixon et al. [30] declared the energy value between the hydrogen atom of the amino group and the carbonyl oxygen atom of the carboxyl group

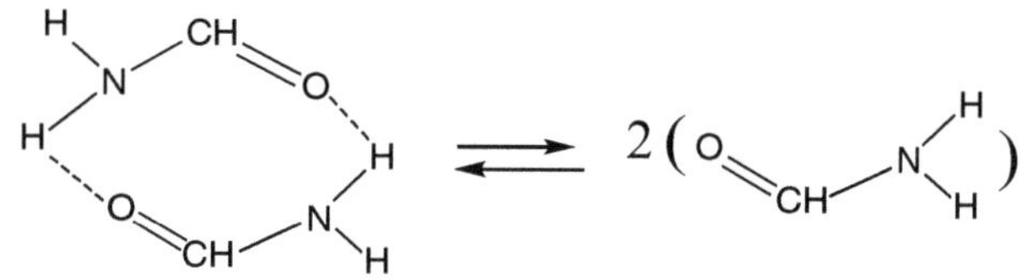

Fig. 1.2 Dimer of formamide with two hydrogen-bonded interactions [33]

N-methylacetamide to be equal to 21 kJ mol^{-1}. In the paper [31] the value of the hydrogen-bonding energy of 20.2 kJ mol^{-1} was obtained. In the paper, the approach of a direct quantum-chemical method for estimation of the intramolecular hydrogen-bonding energies in polypeptide [32] estimated the hydrogen-bonding energies of the intramolecular seven-membered ring =C==O•••H–N in glycine and alanine dipeptides. For the selected five models the calculations were implemented with the help of the Gaussian 03 program [33], using second-order Moller-Plesset perturbation (MP2) theory. Optimized structures were obtained at the MP2 level with 6-31G(d) and 6-31G(d,p) basis sets.

The authors [34] successfully applied this method to the hydrogen-bonding energies of the intramolecular 10-membered ring =C==O•••H–N in peptides and the individual intermolecular hydrogen-bonded dimers of amino-acetaldehyde, 2-aminoacetamide, formamide, and oxalamide with two identical hydrogen bonds (Fig. 1.2).

Reduction of the number of coordinating molecules and interacting functional groups at the transition from condensed condition to vapor leads to the stabilization of the remaining hydrogen bonds. As a result the obtained energy values of the last two hydrogen bonds in a dimer molecule of formamide are the maximum possible in comparison with the energy values of the same types, realized in the condensed condition of this compound. The energy values, obtained by the authors by the quantum-chemical method are presented in the Table 1.3 ; they are well in compliance with the results of the assessment of energy values of the same type of bond obtained with the help of a supramolecular method. Significantly reduced values of energies of hydrogen bonds in dimer molecules of amino-acetaldehyde and 2-aminoacetamide point to its low stability, which excludes the possibility of its existence in vapor form. In the case of formamide and oxalamide the assumption of the existence of dimer molecules in the vapor form is theoretically realistic.

The results of calculations from three optimized models of structures that conformed to the 10-membered ring =C==O•••H–N pointed to the energy values of 28.9, 18.5, and 29.1 with an average value of 25.4 kJ mol^{-1}. Assessments of the intramolecular 10-membered ring =C==O•••H–N hydrogen-bonded energies in peptides and =C==O•••H–N intermolecular hydrogen-bonded energies allowed the authors [34] to conclude that the intermolecular =C==O•••H–N hydrogen bonds are more stable in comparison with intramolecular hydrogen bonds. It follows that the intramolecular hydrogen-bonded interaction is not realized in liquid and under crystalline conditions of these compounds and their peptides. The authors [35] assume that intermolecular hydrogen bonds have an energy value of 36.1 kJ mol^{-1} for the *N*-methylacytamide dimer, which could be of

Table 1.3 The hydrogen – bonding =C==O•••H–N energies (kJ mol^{-1}) of dimers [34]

Compounds	The MP2/6-311 + +G(d,p) level [32]	Supramolecular method
Amino-acetaldehyde	7.4:2 = 3.7	7.7
2-Aminoacetamide	7.0:2 = 3.5	7.2
Formamide	26.5:2 = 13.25	26.1
Oxalamide	20.2:2 = 10.1	20.6

particular relevance for the existence of the dimer form of the molecule of this compound in the vapour form with an energy value of 18.1 kJ mol^{-1}. Schirmeister et al. [36] point out that the intermolecular hydrogen bond interaction of an aziridiny peptide in crystalline condition is more stable than the intramolecular interaction.

It is evident that the formation of stable hydrogen bonds is accompanied by the weakening of the covalent O–H•••O bond, leading to"blue shifting" stretching frequency [4, 6–10, 37]. In various theoretical approaches [38] these O–H•••O interactions may be fairly characterized as a true H-bond [39]. Study of the acetic acid dimer allowed us to estimate the energy of each C–H•••O hydrogen bond, ranging in the interval 3.3–31.3 kJ mol^{-1} [40]. By later calculations of the energy of C–H•••O bond aldehyde with variations of proton acceptor energy, ranging in the interval of 3.3–10.8 kJ mol^{-1} was obtained [41]. Calculations of dimer formamide pointed to the big strength of hydrogen bonds of the same type, ranging in the interval of 10.4–16.7 kJ mol^{-1} [42–44].

On the basis of an analysis of the experimentally observed difference of spectral bands, describing the stretching of bonds of different natures in gaseous and liquid methanol, the authors of [45] have drawn the conclusion of the possibility of formation of this compound in the condensed state by one bond C–H•••O for each methanol molecule. Calculations by the method of density functionals [46] indicated that the bond C–H•••O, like the hydrogen bond of O–H•••O, has special significance in the formation of a trimeric molecule of methanol. In this connection the authors [47] conducted an additional analysis by quantum-chemical calculation of monomer CH_3OH by different methods, including the approximation of Hartree-Fock (HF), considering electron correlation on the basis of the method of Moller-Plesset (MP2, MP4), connected clusters (CC), and configuration interactions (CI), and on the basis of the method of using the functional density (B3LYP, B3PW91, MPW1PW91). A comparison of the results of the calculation of the structure parameters of monomers, lengths of intermolecular bonds, valence and torsion angles, electron densities at the atoms, and spectral characteristics led the authors of [47] to the conclusion of a disparity of protons in the methyl group of the methanol molecule. According to the authors [47] mentioned, the characteristics of the proton, located in *trans*-position relative to the hydroxyl hydrogen atom, differ from similar characteristics of the two other hydrogen atoms of the methyl group. On the basis of the value of the energetic barrier of the inner rotation of CH_3-группы, the value of energy of the hydrogen atom of the methylene group formed in gaseous phase was estimated be equal to 4 kJ mol^{-1} and it was assumed that, possibly, that was the reason pointed out in [45] for the possibility of formation of

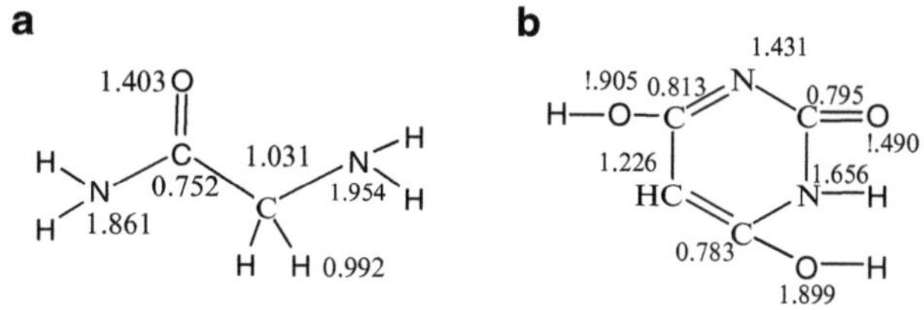

Fig. 1.3 Distribution of charges on atoms of glycinamide (**a**) and barbituric acid (lactim-lactam) (**b**) molecules [49]

one bond C–H•••O to each molecule. In this connection it should be noted that as implemented in the papers [12, 48] research directly shows the formation in liquid methanol of the specific interaction H–O → CH_3 with an energy of 5.63 kJ mol^{-1}.

We should draw attention to [40–42] such significant differences of stability of 3.3 and 31.3 kJ mol^{-1} in one type of hydrogen bonds C–H•••O, formed by the hydrogen atom of CH group, whose energies differ practically by more than an order of magnitude with each other. The charge of a hydrogen atom of 0.992e of a methylene group of C–H molecule glycinamide (Fig. 1.3) with low acceptor properties is not able to form hydrogen bonds of so high a stability with the carbonyl oxygen atom, whose donor properties at a negative charge of 1.403e are incomparably high, which is why it forms a very strong peptide hydrogen bond, =C==O•••H–N with the oxygen atom of the amino group. At the same time the carbon atom of the carbonyl group with its high positive charge is not also indifferent and forms a specific interaction with a less positively or more negatively charged atom chain, of which the free electron pair remains unoccupied.

The differences between the charges of oxygen atoms of hydroxyl groups with carbonyl oxygen atom and nitrogen atom of the chain and amide group on one side and the carbon atom and CH group cycle of barbituric acid on the other side use the principle of the formation of more stable specific interactions for reaching the minimum of the energy of the system and forming hydrogen bonds and for specific interactions with the carbon atom as well, leading to a pentacoordinated condition [12, 48].

Certainly the development of quantum-chemical methods of calculations has significantly advanced the theory of the hydrogen bond; nevertheless, we have significant restrictions, precluding the consideration of coherence of specific intermolecular interactions of carbon atoms in a chain and cycles of nitrogenous and bio-organic compounds. As for organic compounds, we may create a unified system of consistent energies of specific interactions for them, including hydrogen bonds, correctly revealing their nature [12].

1.2 Inconsistency of the sp³-Hybridization Model with Respect to Carbon Atom

Conducted in parallel with theoretical research on intramolecular interactions, the study of physical-chemical properties and the establishment of their thermodynamic characteristics led to the necessity for further analysis of the nature of intermolecular interactions in the phases of various aggregate conditions of these compounds. Perhaps the first in this direction has been the research on carbonyls of metals [50–53]. Detection of strong intermolecular interactions in solid and liquid conditions and the existence of dimer molecules in the vapors of carbonyls of nickel, iron, chrome, molybdenum, and tungsten

Compounds	$Ni(CO)_4$	$Fe(CO)_5$	$Cr(CO)_6$	$Mo(CO)_6$	$W(CO)_6$
$\Delta_{dis}H°(T)$dimer $M_2(CO)_{2n}$ kJ mol^{-1}	23.9	32.3	25.9	29.7	33.9

disproved the idea of Sidgwik that compounds with 18-electron configuration of metal atoms are not able to form associations. The high value of the energy of dissociation of the dimer molecule of carbonyl metals points to the fact that the reverse dative bond formed in the molecule of carbonyl metal, $M \rightleftarrows CO$, returning part of the electron density to the metal atom, provides it with the ability to form a dimer molecule with the metal-metal bond with an energy of 23.9–33.2 kJ mol^{-1}.

Obtained by thermochemical methods, the average values of the enthalpy of dissociation of organoelement compounds remain with singular physical-chemical characteristics of molecules, which allow us to gain information on the peculiarities in molecular interactions. However, for example, at alkyl compounds of nontransition elements it remains unclear what reasons cause inadequate character of change in the average energy at the first representatives in a number of carbon atoms as regards growth in alkyl ligand [12].

Critical analysis of the whole set of thermochemical and thermodynamic properties of alkyl compounds, obtained with high reliability, inevitably touches on the ideas of the classic model of sp^3-hybridization of valent orbitals of the carbon atom, which is, beyond all doubt, very attractive. One should mention the principal work of V.L. Talroze and A.L. Luybimova, where for the first time the cation CH^{5+} was described with a hypercoordinated carbon structure [54]. The principally important research result is the fundamental for the dimerization [55–57] in vapors of triethylaluminum and almost full dimerization of triethylaluminum (Fig. 1.4), in which the carbon atom of two methyl groups of methyl and ethyl ligands is in a pentacoordinated condition.

It follows that the carbon atom of one methyl group of the ethyl ligand acts as a donor of electron density, partly obtained from aluminum through the methylene group, to the 3pz-orbital of the aluminum atom of the second molecule of Al $(CH_2–CH_3)_3$, and forms specific intermolecular interaction $DAl–CH_2–CH_3 \rightarrow Al–CH_2–CH_3$ of high stability with an energy of 33.1 kJ mol^{-1}. Stability of dimer molecule significantly increases with transition from liquid condition to vapor, accompanying the reduction of the number of coordinate molecules. In this connection one should mention that the fact of the existence of a pentacoordinated carbon atom does not match the ideas of sp^3-hybridization or,

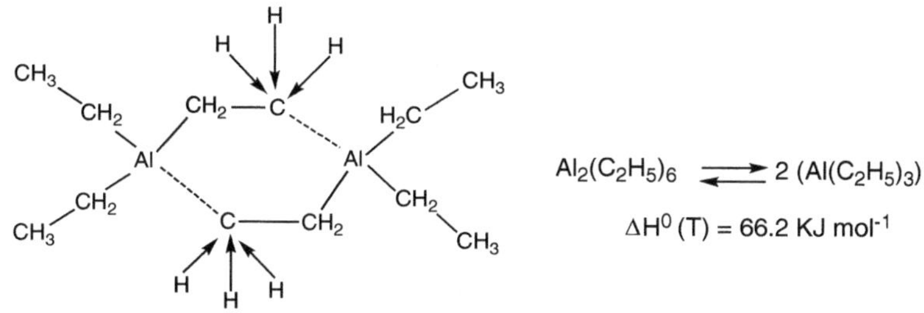

Fig. 1.4 Dimer of the triethylaluminum molecule

in other words, these ideas do not correspond with the objective fact of the existence of a fifth coordinated carbon atom. Thus, in the dimers $Al_2(CH_3)_6$ and $Al_2(C_2H_5)_6$ two carbon atoms of two terminal methyl groups, forming donor-acceptor bonds with aluminum are in the pentacoordinated condition. Similar specific intermolecular interactions are realized in liquid conditions of alkyl elements of boron, germanium, and zinc. The irrefutable fact of the existence of a fifth coordinated carbon atom became indicative and a stimulating beginning in the development of the concept of specific intermolecular interactions, revealing the reasons for the inconsistency in the natural change of vaporization enthalpies in the number of elements and periods of elementorganic compounds of the II–VI groups with properties required by the model of sp^3-hybridization [57–60]. Understanding the established patterns of change in the thermodynamic series with the shifting of the electron density became possible on the basis of the idea of the intermolecular reverse dative bond, realized between the contacting carbon atom of the alkyl ligand with metal atom donor or element-acceptor of electrons, respectively in II–IV and V–VI groups [61–64]. This idea is analyzed in detail in [12], therefore, we represent it briefly here.

The maximum ability for the formation of reverse dative bonds is expressed in alkyl compounds by a methyl ligand. Therefore it is characterized by a higher value of the dissociation energy. Replacement of the methyl ligand by the ethyl one with the C–C bond changes the ability of the carbon atom to contact an atom of a non-transition element in the electron transition from $2s^2$ orbital of the carbon atom to the p_z orbital of atoms of the third and fourth groups and from the p_z orbital of atoms of the fifth and sixth groups to the $2s^2$ orbitals of carbon. Thus in the considered molecules the average values of the energies of dissociation of the element-ethyl bonds are always lower than the dissociation energies of $DE–CH_3$. It is clear that the difference $DE–CH_3 – DE–C_2H_5$ could act as a measure of the value of the energy, contributed by the dative bond, stabilizing the dissociation energy of $E–CH_3$. This energy value of stabilization of the bond $E–CH_3$ could be considered the capacity of the relative value of the contribution of the dative bond.

Between shifting of the electron density in alkyl compounds from atoms of the elements of the II and IV groups to the dative bond of carbon atom and from the carbon atom to the atoms of the elements of the V–VI groups and to the dative bond of the last, it is difficult to draw a clear line; rather these effects complement each other. Nevertheless, the values of the difference in the energies of the E–CH_3 and E–C_2H_5 dissociation bonds reflect the influence of the second carbon atom of the ligand on the participation of the $2s^2$ electron pair of the carbon atom, in contacting with the atom of the element.

Because the difference in the energies of the 2s carbon atoms and the p-orbitals of P-As-Sb group grows, there appears a reduction in the role of the dative bond at the transition from arsenic to antimony. Thus the value of the difference of dissociation energies of DS-Me and DS-Et reflects the influence of two opposing effects. Its increase in the number from NR_3 to PR_3 testifies to the dominating influence of dative bonds, while the reduction of this value points to the all the more significant reduction in the role of the dative bond because of the increasing difference in the energies of the p-orbitals of P, As, Sb with the energy of the $2s^2$ orbital of carbon.

Furthermore it is important to note that at the transition from methyl to ethyl and further to propyl, butyl compounds the average energy of dissociation M-R (E-R) is reduced to slightly different values at the concrete value of index "n" (number of carbon atoms in the ligand). This is a direct indication that the increase in the number of carbon atoms after three in the ligand practically does not influence the distribution of the electron density between the central atom and the contacting carbon atoms, and also does not appear in the energy of dissociation and the contribution of the dative bond. In other words the manifestation of the influence

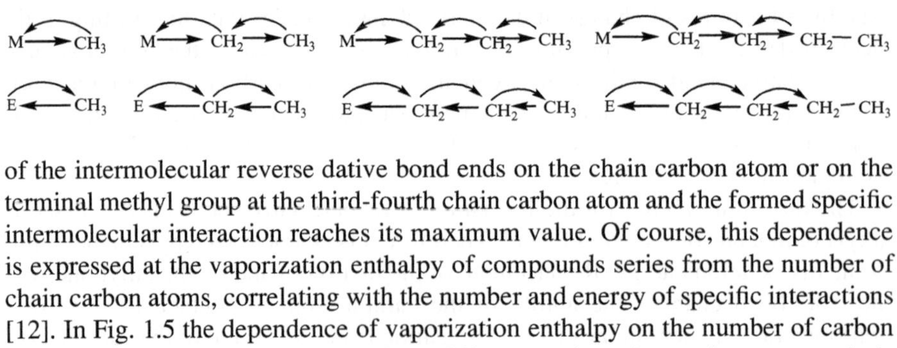

of the intermolecular reverse dative bond ends on the chain carbon atom or on the terminal methyl group at the third-fourth chain carbon atom and the formed specific intermolecular interaction reaches its maximum value. Of course, this dependence is expressed at the vaporization enthalpy of compounds series from the number of chain carbon atoms, correlating with the number and energy of specific interactions [12]. In Fig. 1.5 the dependence of vaporization enthalpy on the number of carbon atoms of the alkyl chain for the four different series of organic compounds is presented, described by two parallel lines. Its crossing at the third chain carbon atom of the alkyl ligand reflects the completion of stabilization of specific interactions, and further increase in the enthalpy characteristics is owing to the contribution of the energy of an increasing number of methylene groups in the chain. Consequently the process of dimerization of triethylaluminum is carried out without reaching the maximum possible value of the electron density of the carbon atom of the methyl group of the ethyl ligand (Fig. 1.6).

On the basis of these data one should draw the conclusion that an intermolecular dative bond exists in the molecules of alkyl compounds with saturated hydrocarbon ligands as well as in molecules of the other series of compounds and that it has a general character. In modern scientific literature the idea of an additional orbital

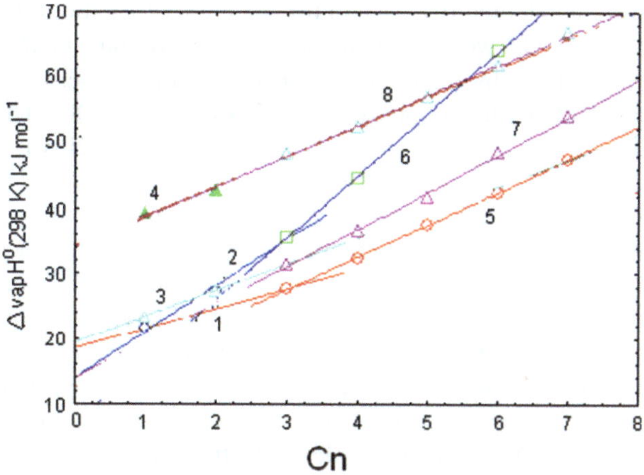

Fig. 1.5 Dependence of vaporization enthalpy of compounds versus the number of carbon atoms in alkyl chain (*1*, *5*) – aldehydes, (*2*, *6*) – symmetrical ethers, (*3*, *7*) – asymmetrical methyl ethers and (*4*, *8*) – 1-alcohol [12]

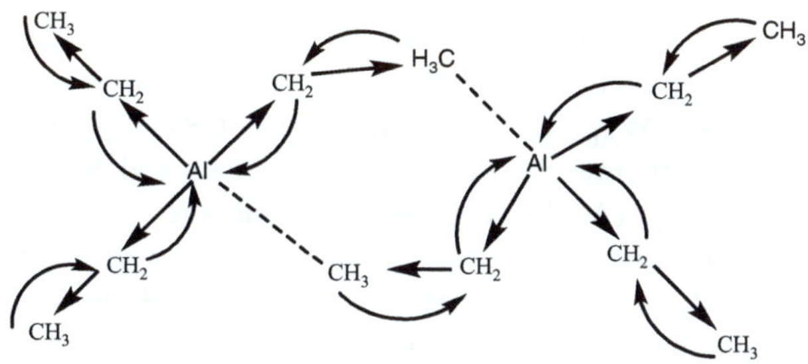

Fig. 1.6 Reverse dative bond in dimer molecule of triethylaluminum

interaction of reverse dative type in the electron structure of molecules of saturated and unsaturated hydrocarbons and its coordination derivatives, which demands the implementation of quantum-chemical research, has not been developed [12].

In this connection we should mention the work of [65], which states that in a molecule of CH_4 with sp^3-hybridization it is not as simple because special calculations conducted showed that 2s-level of carbon atom lies significantly lower than the 1s-level of hydrogen. As a result in the methane molecule the $2s^2$(c)-electron pair remains significantly undivided because MO symmetry a_1 remains significantly unbonding. For this reason the bonds in the CH_4 molecule are formed mainly as a result of interactions 2p(c) -1s(H) owing to the filling of bonding MO symmetry t_1.. As a result the bonds C–H are orbital-deficit hypervalent bonds, rather than covalent, as usual. Owing to the fact that interactions of σ-orbitals for four equivalent ligands with three "Cartesian" p-orbitals of the central atom best is taken to be the tetrahedral configuration of ligands, the geometric structure of the molecule of CH_4 does not connect with sp^3-hybridization [65].

By limitation method of Hartree-Fock (RHF) and by the method of B3LYP (variant of the density function DFT) with the use of three different atomic bases (MINI, 6-31G*, LanL2DZ) quantum-chemical calculations of the electron structures $Al(CH_3)_3$, $Ga(CH_3)_3$, $In(CH_3)_3$, $Ge(CH_3)_4$, dimer $Al_2(CH_3)_6$, $[Tl(CH_3)_3]_4$

were conducted [61]. Ab initio calculations by the RHF method in the bases MINI and 6-31G* were implemented by the program GAMESS [66, 67] and ab initio calculations by the method of B3LYP in the bases of 6-31G* and LanL2DZ were conducted on the program GAUSSIAN-94 [68]. The results of the quantum-chemical research were analyzed in detail in [12], which is why we consider only the basic points. Directly calculated characteristics of the valent electron structure of trimethyl $A(CH_3)_3$ point to the presence of a configuration reflecting hypervalent interactions in the fragment A–CH_3, along with the configuration of covalent interactions, corresponding to the scheme of sp^3 hybridization of valent orbitals of the carbon atom in the composition of CH_3. Consequently in the valent electron structure of $A(CH_3)_k$ one of the electron pairs of each fragment of A–CH_3 is significantly localized at the atom C and, thereby, there are preconditions for the formation in the further intermolecular interactions of the dative bond C(2e) → A(0e). In the valent electron structure of $[A(CH_3)_k]_n$ complexes with bridging CH_3–groups on the example of dimer $Al_2(CH_3)_6$ and tetramer $[Tl(CH_3)_3]_4$ the indices of the multiplicities of two-center bonds A–C_b of central atoms A with bridging carbon atoms C_b almost coincide with the indices of the multiplicities of two-center bonds A–C_t, including terminal carbon atoms C_t, in particular, $W_{Al–C(b)} = 0.87$, $W_{Al–C(t)} = 0.89$. It points to the fact that in the three-centered fragment A–C_b–A, one two-centered bond as in the case of terminal CH_3–group (A–C_t) (Fig. 1.2), and another dative bond C(2e) → A(0e) were formed. Thus, the obtained results show that in the electron structure of di- and tetrameric complexes $[A(CH_3)_k]_n$ (Figs. 1.4 and 1.6), including the bridging CH_3–groups (C_2H_5), the dative bonds discussed really exist. The dative bond between the carbon atom C of the alkyl ligand R and central atom A in complexes AR_k in its electron structure such that the dative bond C(2e) → A(0e) can be formed by participation, from one side, of a valent electron pair of the carbon atom (in particular, $2s^2$(C)–electron pair), or precisely, an electron pair of alkyl ligand R, localized mainly at the atom C, and, from the other side, a vacant orbital, localized mainly on the atom A. The latter may be, first, an unoccupied one-center valent orbital of atom A or, second, a free delocalized multi-centered orbital of complex AR_k, for example, free valent MO, nonbonding or weakly anti-bonding with respect to the bonds A–R, with quite low energy. The basis for such a representation are the following features. First, even in the molecule of CH_4 the valent $2s^2$-electron pair remains significantly localized at the atom C, and only two electrons (from the fourth valent electrons) of the C atom at the third of its valent AO $2p_x$, $2p_y$, $2p_z$ effectively participate in hyper-valent (electron-deficient) C–H bonds with all four hydrogen atoms (from the point of view of the symmetry of the tetrahedron this fully conforms to the demands of group theory). It is therefore likely that the same feature (reservation of pseudo-atomic $2s^2$(C)–electron pair) is a characteristic feature. For the electron structure of the methyl group, coordinated at atom A (A–CH_3), where the carbon atom forms four bonds (one C–A and three C–H) owing to only two electrons at the third of its own valent AO $2p_x$, $2p_y$, $2p_z$, there is orbital electronegativity of atom A, which is not higher than the orbital electronegativity of atom H. Second, the electron structure of the carbonium cation

$[CH_5]^+$ can most simply be described by the collection of the same as in the molecule of methane CH_4, hypervalent bonds of carbon atom with four atoms of H atoms and a fifth dative bond of one atom with hydrogen cation (formally) owing to the valent electron pair of atom C and vacant orbital of the hydrogen cation. Similar electron structures possess other complexes of the pentacoordinated carbon that were detected in the past decade. Third, the existence of methyl complexes with bridging (μ_2) CH_3-groups in the classic example $Al_2(CH_3)_6$ and $Al_2(C_2H_5)_6$ predominates the availability in its structures, as well as in the structure of the cation carbonium, the pentacoordinated carbon atom. In the electron structure of the bridging of the μ_2-group CH_3, connected simultaneously with two atoms A, the carbon atom forms four bonds (with three of their atoms H and one atom A). In the case of terminal CH_3-groups, connected only with one atom A, because of two electrons at the third of their valent AO $2p_x$, $2p_y$, $2p_z$; the fifth bond with the second atom A, the carbon atom bridging the μ_2-group CH_3 forms, apparently, by the dative mechanism owing to its valent $2s^2$-electron pair and the vacant orbital of this atom A. As the tetrahedral configuration is the one most preferred for interaction of σ-orbitals of four equivalent ligands with three Cartesian p-orbitals of the central atom, the considered geometry of CH_4 does not connect with sp^3-hybridization in general [65]. Finally, in the fourth, effective intermolecular interactions are consistently found, in alkyl (in particular, in methyl) compounds of non-transition elements, to be based on the same phenomena [69].

The experimentally based fact of the pentacoordinated carbon atom, its participation in the formation of specific interactions, and the existence of an intermolecular reverse dative bond, providing the explanation for the contradiction of thermodynamic properties of organometallic compounds with predictions by the model of sp^3-hybridization are sound arguments pointing to its inconsistency. Rejection of the model of sp^3 hybridization leads to the grounding of new types of specific interactions, formed by the essentially unshared $2s^2$(c) electron pair of the pentacoordinated carbon atom and hydrogen bonds, determination of its energies, and development of theories of these interactions and bonds, and of the theory of the reverse dative bond in organic, organometallic compounds, nitrogenous compounds, peptides, proteins, and molecular solutions. In general, the problems to be solved are: energy determination of peptide hydrogen bonds and energies of conformers necessary for determination of energies of peptides and proteins.

1.3 Methodology of Investigation of Specific Interactions of Nitrogenous and Bio-organic Compounds

From the analysis of hydrogen bond problems presented in the first chapter, which provides the data of the energy parameters of the hydrogen bonds of quantum-chemical calculations made per single bond (for example formed by the hydrogen and oxygen atom), the possibility of the formation of two hydrogen bonds is not considered. In the case of calculation of the energy of the hydrogen bond of a dimer of formamide, the resulting value is the energy of dissociation per two monomer forms. Since all implemented calculations are given for the gaseous phase in nonequilibrium conditions, it is almost impossible to reduce the accumulated information to standard conditions.

The second approach of obtaining the energy of hydrogen bonds and specific interactions, developed in the work [12], was prepared on the basis of the elaborated ideas on the chemical nature of phase transformations [70–72], from which we have the following.

There is a natural (community) affinity for polymorphic transformation, melting, sublimation, vaporization, dimerization, trimarization, formation, or destruction of tetrameric molecules, i.e. all these processes are the chemical processes of chemical dissociation.

The enthalpy of the melting process and vaporization correlate with the number and the energy of dissociation of specific intermolecular interactions.

The entropy of these processes is interrelated with the number and energy of the bursting specific intermolecular interactions [71, 72].

Energies of nonspecific interactions are laid within the error limits of calorimeters and other methods of determination of thermodynamic characteristic and therefore not included in the calculations of specific interactions. The energy values of cooperative interactions depend on the energies of the specific interactions formed, or to be precise, on the lack of such information and the possibility of its correct assessment. Therefore, it is part of the evaluation margins of the values of enthalpy characteristics of vaporization.

The implementation of structural-energetic analysis depends on the information of the structural parameters of the crystal and liquid conditions as the enthalpy and entropy of vaporization are interrelated with the energy contribution of each functional group or, more precisely, with the bond vacancy forming the specific interaction with pseudo-atomic essentially unshared $2s^2$-electron pair of carbon atoms, lying in a pentacoordinated condition; with hydrogen atoms, forming a hydrogen bond; and with atoms of oxygen and nitrogen, forming hydrogen or specific interactions dependending on the second participant of this interaction. Considering the structure of the liquid, being a unique system of where the elementary volume acquires a definite sense containing extremely few particles on condition of its subordination to the law of statistics [73], knowledge of the numbers and types of bond vacancies and the interacting particles lead to the possibility of determining the energies of specific interactions. The correct choice of the specific interactions in combination with other parameters (for example, with the results of the spectral method of research) predetermines the ground for the network of corresponding interactions. In the choice of a probabilistic structure of liquid it is necessary to ignore the model of sp^3-hybridization of the carbon atom and take into account the presence of the essentially unshared $2s^2$ electron pair, forming a specific interaction and providing the carbon atom with the pentacoordinated condition. In Fig. 1.7a a schematic picture is

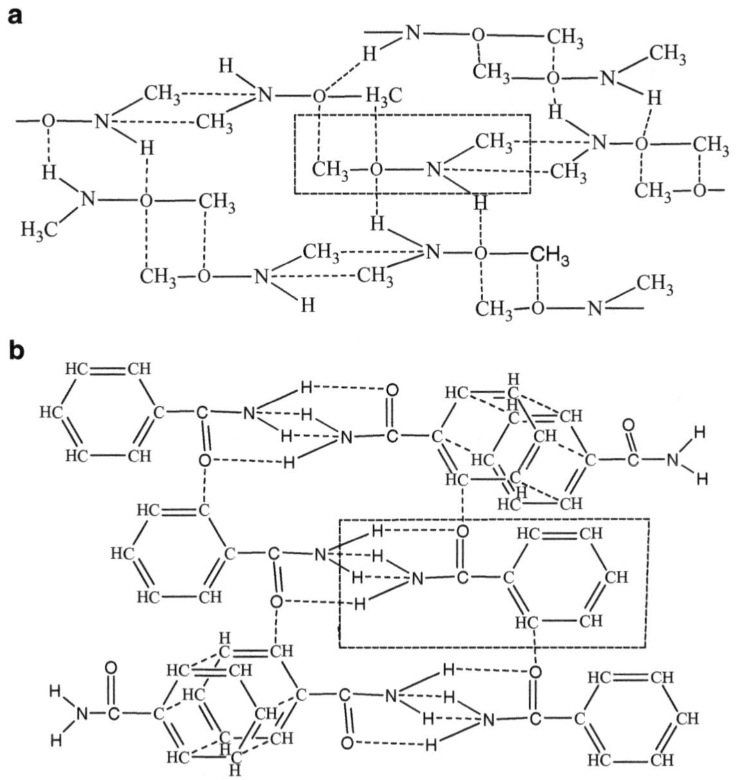

Fig. 1.7 Schematic picture of the crystalline structure of 2-methoxyethylamine (**a**) and benzamide (**b**) with the network of hydrogen bonds and specific interactions

presented of the structure of the crystalline and liquid condition of 2-methoxyethylamine with the network of specific interactions and hydrogen bonds, illustrating the employment of all bond vacancies of a molecule with two free electron pairs of the oxygen atom and one free electron pair of the nitrogen atom. The molecule of benzamide with free bond vacancies of essentially unshared $2s^2$(c) electron pairs of carbon atoms of the CH group of the ring, two free electron pairs of oxygen atoms of the carbonyl group and one free electron pair of the nitrogen atom forms the typical network of hydrogen bonds and specific interactions. Note that at the melting temperature of the crystalline and liquid condition the compounds have insignificant differences.

Innovative approaches of thermodynamic analysis of liquid and solid saturated and unsaturated organic compounds, elementorganic nitrogenous compounds, peptides, and proteins with various numbers of bond vacancies in the determination of energies of specific interactions and hydrogen bonds are based on the following principles [12]:

1. Reasonable ignoring of the model sp^3 hybridization of the carbon atom.

2. The presence of an essentially unshared $2s^2$(c)-electron pair at the carbon atom providing it with the ability to transit to the fifth coordinated condition and forming of a specific interaction
3. All free bond vacancies of the molecules of organic compounds participate in the formation of specific interactions with the fifth coordinated carbon atom and hydrogen bonds, respectively
4. Enthalpies (entropies) of the processes of sublimation, vaporization, melting, and polymorphic changes are connected with the number and energies of bursting specific interactions and hydrogen bonds.

Knowledge of the molecule structure and the number of free bond vacancies is of great importance, if we are to understand the structure of a liquid organic compound and identify the different types of specific interactions and hydrogen bonds.

Using this approach and enthalpy and entropy characteristics it is possible to receive the energies and entropies of specific interactions and hydrogen bonds with the error of experimentally received value of thermodynamic property of compound.

References

1. Krestov GA, Beresin BD (1986) Based notions of contemporary chemistry. Chemistry, Leningrad, p 102
2. Berezin BD, Krestov GA (1999) Basic laws of chemistry. Chemistry, Moscow, p 95
3. Pauling L (1960) The nature of the chemical bond. Carnell University Press, Ithaca
4. Scheiner S (1997) Hydrogen bonding. A theoretical perspective. Oxford University Press, Oxford
5. Bader RFW (1990) Atoms in molecules. A quantum theory. Clarendon, Oxford
6. Jeffrey GA, Saenger W (1991) Hydrogen bonding in biology and chemistry. Springer, Berlin
7. Jeffrey GA (1997) An introduction hydrogen bonding. Oxford University Press, New York
8. Desiraju GR, Steiner T (1999) The weak hydrogen bond in structural chemistry and biology. Oxford University Press, Oxford
9. Steiner T (2002) The hydrogen bond in the solid state. Angew Chem Int Ed Engl 41:48–76
10. Desiraju GR (2002) Hydrogen bridges in crystal engineering interactions without borders. Acc Chem Res 35:563–573
11. Parthosarathi R, Subramanian V (2006) Characterization of hydrogen bonding: from van der Waals interactions to covalent. In: Grabovski SR (ed) Hydrogen bonding – new insights, vol 3. Springer, Dordrecht, pp 1–50
12. Baev AK (2012) Specific intermolecular interactions of organic compounds. Springer, Hiedelberg/Dordrecht/London/New York, 434p
13. Kollman PA, Allen LC (1972) Theory of the hydrogen bond. Chem Rev 72:283–303
14. Dyke TR, Mack KM, Muenter JS (1977) The Structure of water dimer from molecular beam electric resonance spectroscopy. J Chem Phys 66:498–510
15. Odutola JA, Dyke TR (1980) Partially deuterated water dimers microwave spectra and structure. J Chem Phys 72:5062–5070
16. Smith BJ, Swanton DJ, Pople JA, Schaefer HF, Random L (1990) Transition structures for the interchange of hydrogen atoms within the water dimer. J Chem Phys 32:1240–1247
17. Xantheas SS, Dunning TH Jr (1993) Ab initio studies of cyclic water clusters $(H_2O)_n$ n = 1–6.1 optimal structure and vibrational spectra. J Chem Phys 99:8774–8792

18. Chacravorty SJ, Davidson ER (1993) The water dimer correlation energy calculation. J Phys Chem 93:6373–6383
19. Scheiner S (1994) Ab initio studies of hydrogen bonds: the water dimer paradigm. Annu Rev Phys Chem 45:23–56
20. Feyreistn MW, Feller D, Dixon DF (1996) Hydrogen bond energy of the water dimer. J Phys Chem 100:2993–2997
21. Koch N, Holthausen CN (2001) A chemistry guide to density functional theory. Wiley-VCH/Wiley-VCH Verlag. GmbH, New York
22. Baev AK (1987) Chemistry of Gas-heterogenic systems of elementorganic compounds. Science and Technology, Minsk
23. Curtiss LA, Blander M (1988) Thermodynamic properties of gas-phase hydrogen bonded complexes. Chem Rev 88:827–841
24. Sokulski WA, Harihuran PC, Raufman JJ (1983) A self-constant field interactions energy of 12 hydrogen-bonded dimers. J Phys Chem 87:2803–2810
25. Parthasarathi R, Subramaman V, Sathyamurthy N (2006) Hydrogen bonding without borders: an atoms-in- molecules perspective. J Phys Chem 110:3347–3351
26. Sine F, St. Amant A, Papai I, Salagub DR (1992) Gaussian density functional calculation on hydrogen-bonded systems. J Am Chem Soc 114:4391–4400
27. Kilningar M, Sunai S (1994) Density functional studies on hydrogen-bonded complexes. Int J Quantum Chem 51:465–478
28. Kim R, Jordan RD (1995) Comparison of density functional theory yield accurate thermodynamics of hydrogen bonding? Chem Phys Lett 297:112–119
29. Parthosarathi R, Subramanian V, Sathyanmurthy N (2006) Hydrogen bonding without borders: an atoms-in- molecules perspective. J Phys Chem Lett 110:3349–3351
30. Dixon DA, Dobbs KD, Valentune JJ (1994) Amino-water and amide-amide hydrogen bond strength. J Phys Chem 98:13435
31. Lantz MA, Jarvis SP, Tokumoto H, Martynski T, Kusumi T, Nakamura C, Miyake J (1999) Stretching the alpha-helix: a direct measure of the hydrogen-bond energy of a single-peptide molecule. Chem Phys Lett 315:61
32. Wang C-S, Zhang Y, Gao K, Yang Z-Z (2005) A new scheme for determining the intramolecular seven-membered ring N-H•••O C hydrogen-bonding energies of glycine and alanine peptides. J Chem Phys 123:024307
33. Frisch MJ, Trucks GU et al (2003) Gaussian 03, revision B 03. Gaussian, Pittsburgh
34. Zwang Y, Chang-Sheng W (2008) Estimated on the intramolecular 10-membered ring N-H•••O=C hydrogen-bonding energies in glycin and alanin peptides. Wiley Inter Science
35. Vargas R, Garza J, Friesner RA, Stern H, Hay BP, Dixon DA (2001) Strength of the N-H center dot center O=C and C-H center dot center O=C bonds in formamide and N-methylacetamide dimers. J Phys Chem A 105:4963
36. Schirmeister T, Breoning A, Murso A, Stalke D, Mladenovic M, Engels B, Szeghalmi B, Schmitt M, Kilfer W, Popp J (2004) Conformation and hydrogen bonding properties of an aziridinylpeptides: X-ray structure analysis, Raman spectroscopy and theoretical investigations. J Phys Chem A 108:11398–11408
37. Legon AC, Roberts BP, Wallwork AL (1990) Rotational spectra and geometries of the gas-phase dimers (CH4, HF) and (CH4, HCL). Chem Phys Lett 173:107
38. Scheiner S, Kar T (2002) Red-versus and blue-shifting hydrogen bonds: are there fundamental distinctions? J Phys Chem A 106:1784–1789
39. Gu Y, Kar T, Scheiner S (1999) Fundamental properties of the C-H•••O interaction; is it true hydrogen bonds? J Am Chem Soc 121:9411–9422
40. Turi L, Danninberg J (1993) Molecular orbital study of acetil acid aggregations. 1. Monomers and dimers. J Phys Chem 97:12197–12204
41. Kim K, Fresner RA (1997) Hydrogen bonding between amine acid backbone and side chain analogues. A higher-level ab initio study. J Am Chem Soc 119:12952–12961

42. Neuhenser T, Hess BA, Reuter C, Webwr E (1994) Ab initio calculations at supramolecular recognition modes cyclic versus noncyclic hydrogen bonding in the formic acid/formamide system. J Phys Chem 98:6459–6467
43. Scheiner S, Kar T, Gu Y (2001) Strengths of the CH•••O hydrogen bond of amino acid residues. J Biol Chem 276:9832–9837
44. Scheiner S, Gu Y, Kar T (2000) Evaluation of the H-bonding properties CH•••O interactions based upon NMR spectra. J Mol Struct (THEOCHEM) 500:441–452
45. Yuchnevich GV, Tarakanova EG (1998) Hydrogen bond C-H•••O in liquid methanol. J Mol Struct 447:257
46. Mandado M, Grana M (2003) On the structure of the methanol trimer and their cooperative effects. Chem Phys Lett 381:22–29
47. Ivanov EV, Puchovskie, Yu P, Abrosimov VK (2004) About formation of weak hydrogen bonds CH•••O in liquid methanol abstract. IX international conference on the problem of solvation and complex formation in solutions, Plyos, p 345
48. Baev AK (1992) Structural-energetic aspects of association interactions of alkyls elements of I-VI groups. Russ J Chem Thermodyn Thermochem 1:143
49. Kalinina FL, Lobov VP, Zidkov VA (1971) Reference on biochemistry. Naukova dumka, Kiev, p 1010
50. Baev AK (1969) Thermodynamics investigation of metal carbonyls nickel and aurum. In: General and applied chemistry, vol 2. Vysheischaia Educating, Minsk, p 146
51. Baev AK, Belozersriy NA, Krichevskaja OD (1969) Thermodynamic investigation of molibdenium carbonyl. In: General and applied chemistry, vol 2. Vysheischaia Educating, Minsk, p 161
52. Baev AK, Demjanchuk VV (1970) Thermodynamics investigation of metal carbonyls of chromium and tungsten, General and Applied chemistry, vol 2, Belarusia, Minsk, p 167
53. Baev AK (1986) Structure and energy of metals carbonyls. Vischaia Schkola, Minsk, p 144
54. Talrose VL, Lubimova AL (1952) Secondary process in the ion source of the mass spectrograph, vol 86. Doclady A.S. SSSR, Russia, p 99
55. Baev AK, Vitalev SE (1989) Vesti Akademii Navuk Belarusii 2:58–63
56. Vitalev SE, Baev AK (1993) Russ J Izvestiya Vischich Ucheb Zaved Chem Chem Technol 36 (6):36–40
57. Baev AK (1996) Pentacoordinated carbon atom at intermolecular interactions. Russ J Coord Chem 22:399–402
58. Baev AK (2001) Russ J Izvestiya Vischich Ucheb Zaved Chem Chem Technol 44:3–13
59. Baev AK (2001) Thermodynamics and specific intermolecular interactions in solutions of non-electrolytes with pentacoordinated carbon atom. Book of abstract. 27th international conference on solution chemistry, Vaals, p 36
60. Baev AK (2001) Specific interactions in solvates solutions of non-electrolytes with isostructural molecules components and pentacoordinated carbon atom. Book of abstract. VIIIth international conference on the problems of solution and complex formation in solution, Ivanovo, p 246
61. Baev AK, Korolkov DV (2003) Phenomenon of dativy bond in elementorganic compounds. Book of abstract. XVth FECHEM conference on organometallic chemistry. University of Zurich, Zurich, p 350
62. Baev AK (2003) Novel approaches to the structure of functional solvents and elementorgani compounds with saturated hydrocarbon ligands and their solvates in binary systems. Book of abstract. 27th international conference on solution chemistry, Debrecen, p 79
63. Baev AK (2004) Insolvency of sp^3-Gibrydixation model of carbon atom and chemistry of non-electrolytes solutions. Book of abstract. IXth international conference on the problems of solution and complex formation in solution, Ivanovo, pp 62–64
64. Baev AK (2004) Lack of co-ordination of thermodynamic properties of elementorganic compounds with sp^3-Gibrydixation model of carbon atom and its insolvency. Book of abstract.

Russian research symposium on thermochemistry and calorimetry, Niznie Novgorod, pp 40–41
65. Schustorovich EM (1978) Chemical bonds in coordination compounds. Knowledge, Moscow
66. Becke AD (1993) Density – functional thermochemistry II New mixing Hartree-Fock and local density – functional theory. J Chem Phys 98:5648
67. Bartolotti LJ, Parr RG, Tal Y (1998) Thomas – Fermi theory from a perturbative treatment of the hydrogenic limit energy density functional a possible link to local density functional theory. Int J Quantum Chem 34:509–515
68. Frisc MA, Trucks GW, Schigell HR, Gill PM, Johnson BG, Robbs MA, Gaussian 94, Revision G. 3
69. Baev AK (2008) The thermodynamic characteristics of diethylzink – diethylsulfur solutions. Russ J Phys Chem 82(8):1427–1434
70. Baev AK (1969) Problems of chemical nature of phase transformation. In: General and applied chemistry, vol 1. Vysheischaia Educating, Minsk, pp 197–206
71. Baev AK (1969) Phase condition and complex formation ability of halogenide metals. In: General and applied chemistry, vol 1. Vysheischaia Educating, Minsk, pp 207–218
72. Baev AK (1972) Complex formation ability halogenides of second – six groups periodical system. In: General and applied chemistry, vol 5. Vysheischaia Educating, Minsk, pp 35–51
73. Krestov GA, Vinogradov VI, Kesle YM (1986) Modern problem of solution chemistry. Nauka, Moscow

Chapter 2
Specific Intermolecular Interactions of Cyclic Alkylamines and Nitriles

2.1 The Energies of Hydrogen Bonds and Specific Interactions of Alkylamines

2.1.1 Liquid Alkylamines

Among modern ideas, controversies on the n-orbital of the ammonia molecule are reflected on the photoelectron spectra of amino-derivative compounds [1–3]. Violation of the nuclear framework of the molecules of noncyclic alkyl compounds in the range of the trigonal pyramid to plane ones with removal of the n-electron leads to a wide oscillatory circuit of the first photoelectron stripe, which appears at the intramolecular and intermolecular interactions. As a result a change in the coordinating ability of the nitrogen atom takes place and the specific interactions gain thermal stability. Despite the great number of papers on photoelectron spectra the best conformity with inclusion of the vertical ionization potential Iv has been established for the mono-substituted methylamine. It means that the maximum value in thermodynamic construction and specificity of the intermolecular interaction should be reflected precisely in this connection. It was shown at chapter 1 [12] that all linked vacancies of the molecules close to the environment in liquid ammonia are realized in a coordinated state with six nitrogen atoms with the location of the hydrogen atoms on top of the trigonal pyramid (Fig. 2.1), forming a similar structure at the interaction of the nitrogen atom of the ammonia molecule in contact. Hydrogen atoms or alkyl groups of the pyramid base of the interacting molecules are located at maximum distance, which is never reached because their locations relative to each other are not symmetrical. Regarding the productivity of this approach we note that for reliable determination of changes of H-cohesion of a molecule in condensed condition one should indicate changes in coordination when changing the structure of ammonia and all amino-derivative compounds on planar structure with the nitrogen atom, acting as coordinating center in, for example, hydrazones $>C{=}N{-}N<$. The equivalence of all six hydrogen bonds of one type

A.K. Baev, *Specific Intermolecular Interactions of Nitrogenated and Bioorganic Compounds*, DOI 10.1007/978-3-642-37472-2_2,

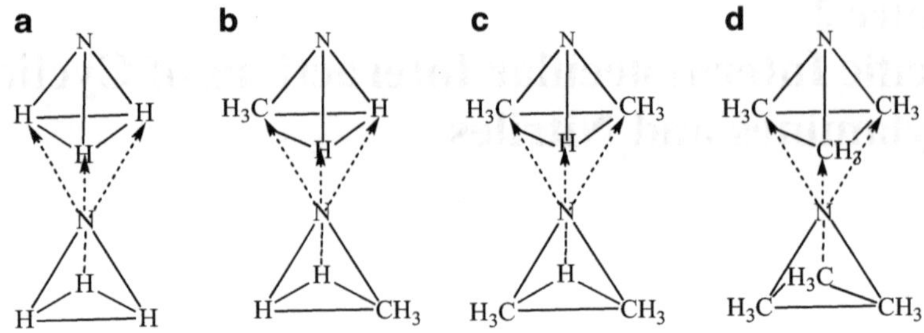

Fig. 2.1 Intermolecular interactions at the near environmental: ammonium (**a**), Methylamine (**b**), Dimethylamine (**c**) and Trimethylamine (**d**)

and its gap in the transition of the ammonia molecule to steam allows the use of enthalpy characteristics of the corresponding process to determine the energy of the hydrogen bonds in the condensed state. It follows that the energy of the H-bond N–H•••N is equal to the quotient of the sublimation enthalpy and evaporation of the number of realized hydrogen bonds [4–6]. The results of the calculation performed show reduced stability of the hydrogen bond, formed by the nitrogen atom of the ammonia molecule with the structure of a trigonal pyramid (4.23 and 3.86 kJ mol^{-1}), being less stable compared with coordination of the four hydrogen atoms (6.34 and 5.91 kJ mol^{-1}) with planar structure of the molecule in solid and liquid states, respectively.

According to [1], replacement of the pyramidal nuclear framework of the molecule of acyclic akylamine by a planar one is accompanied by an increase of the distance to n-electrons and the appearance of a wide oscillatory circuit of the first PhE-strip [7, 8]. Simultaneously there is a predominance of the contribution of delocalization of n-electrons to the 10.92-eV reduction of the vertical ionization potential Iv (n) of the size of 10.92 eV in an ammonia molecule and up to 8.66 eV in a trimethylamine molecule [7]. Replacing the hydrogen atom of the methyl group leads to a small decrease of the electron density of the nitrogen atom [9, 10]because a significantly increased part of the obtained electron density is located on the $3p_z$ orbitals of the nitrogen atom, transferred to the carbon atom, which leads to the formation of a reverse dative connection $>C\text{----}\blacktriangleright N<$ [11–15], reducing and changing the sign of the charge of the nitrogen atom to +0.024e. At the same time,

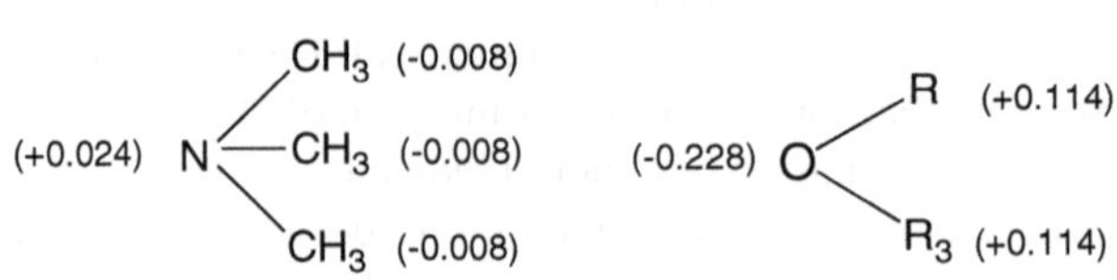

the hydrogen atom in liquid dimethyl and diethyl esters preserves the high negative value because of the high level of positive charge of the carbon atom.

This bond leads to the stability of the intermolecular bond and low energy boosts of the specific intermolecular interaction $N \rightarrow CH_3{-}N$ compared with the hydrogen bond N•••H–N. The energy of these two bonds, formed with a partial transfer of the electron density with the same p_z-orbital on the nitrogen atom or on the carbon atom of the methyl group, satisfies the inequality

$$D\,N \leftarrow\cdots H \leftarrow\cdots N < D\,N \leftarrow\cdots CH_3 \leftarrow\cdots N.$$

This type of intermolecular interaction is based on the thermodynamic analysis of organometallic compounds of the main group and the quantum chemistry calculations of their configurations [11–19]. Thus, the energy feedback is caused, on the one hand, by the electron atom configuration, electronegativity, and the capacity of the electric density obtained, and on the other hand, by transferring a part of the electron density on the hydrogen atom or the contacting carbon atom and further on the carbon chain. Certainly, this duality in the formation of charge on the hydrogen atom in the methyl group in methylamine, ethylamine, and propylamine leads to a stable hydrogen bond and specific intermolecular interaction formed by the essentially unshared $2s^2$(c) electron pair of the fifth coordination of the carbon atom. It is important to pay attention to the explained steric effect in compounds, which hides the opportunity to understand the nature of the specific interaction and the patterns of change of the energy parameters in a homologous series of compounds with two and more ligands.

Because of this, we consider the examples (Figs. 2.2a and 2.2b), reflecting basic differences in the value of the average energies of the dissociation of A-C, selecting a molecule with a smaller radius than the central atom. In this connection one can use a trimethylamine molecule with a pyramidal structure and trimethylboron molecule with a planar structure. The radii of the central atoms of these molecules are on Poling [20] rB = 0.34 A°, RN = 0.70 A°, carbon atom 0.77 A°, and interatomic spacing $r_{B\text{-}C}$ = 1.578 A° and $r_{N\text{-}C}$ = 1.451 A° [21].

Figure 2.2 shows the endo effect of coordination at the successive increase of the number of methyl groups in $B(CH_3)_3$ and $N(CH_3)_3$. The dotted line from the origin in terms of the dissociation energy of the first methyl group, located in a planar structure of an equilateral triangle, illustrates the change of dissociation energies in the order $BCH_3 > B(CH_3)_2 > B(CH_3)_3$ in the absence of a steric effect. That is why the difference between the extrapolated values (dotted lines) and experimentally established values of energies of dissociation B-CH_3 (solid lines) reflect the value of the endo effect of coordination of the carbon atoms of methyl groups. The endo effect value with the sequential addition of two or three methyl groups to the boron atom is equal to 18 and 48 kJ·mol^{-1} for $B(CH_3)_2$ and $B(CH_3)_3$, respectively. Sequential addition of the second, third, and fourth methyl group to the nitrogen atom with the increased covalent radius 0.70 A° is accompanied by an increase of the endo effect coordination, which is 19, 50, and 115 kJ mol^{-1}, respectively. Thus, a double increase of the radii of coordinated nitrogen atoms in comparison with boron atoms, and a decreased interatomic distance in the molecule of

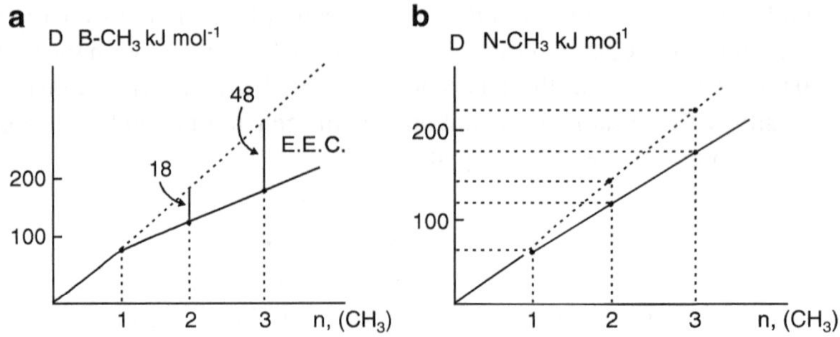

Fig. 2.2 Dependence of the dissociation energy B–CH_3 and N–CH_3 at molecules $B(CH_3)_3$, $N(CH_3)_3$ from the number of methyl groups and endo effect coordination (EEC) [22, 23]

trimathylamine with $r_{N\text{-}C} = 1.451$ A° [21] do not practically lead to the appearance of a change of the endo effect of the coordination of methyl groups in trimethylboron and trimethylamine. An important aspect of this problem is the consistency of the endo effect of the coordination of molecules with planar $B(CH_3)_3$ and trigonal pyramidal structures, with the location of methyl groups at the base of the equilateral triangle of the molecule $N(CH_3)_3$ at different interatomic distances in the molecules, with $r_{B\text{-}C} = 1.578$ A° and $r_{N\text{-}C} = 1.451$ A°. It can be concluded that the steric effect occurs between coordinated carbon atoms of methyl groups in trimethylboron and trimethylamine molecules, directly interacting with the boron or nitrogen atoms (Fig. 2.3). The increasing number of the carbon atoms in mobile alkyl chains does not appear on endo effect coordination, because they are able to locate in the space without violating the molecular structure with minimal Coulomb interaction.

The next major issue when defining the correct energy values for specific interactions is clarification of the reverse dative bond on the carbon atoms, which is dependent on the length of the alkyl chain. Valuable information flows from the analysis of the dependence of the enthalpy of vaporization of alkymines on the number of the carbon atoms in the chain connected with the number and energy of (Fig. 2.3) bursting intermolecular interactions [4–6]. The experimentally determined values of the enthalpy of vaporization are described by the two different dependencies: methylamines-propylamines and butylamines/octylamines, reflecting the manifestation of two different factors on the compounds of the homologous series. The displacement of the electron density at first number of this series to atom nitrogen is increasing from the third carbon atom of the chain to the second atom, in the same time the

transferring electron density from $3p_z$ – orbital of nitrogen atom along the chain is decreasing and comply at the same third carbon atom. The increasing role of the reverse dative bond occurs with an increase in the number of alkyl ligands in

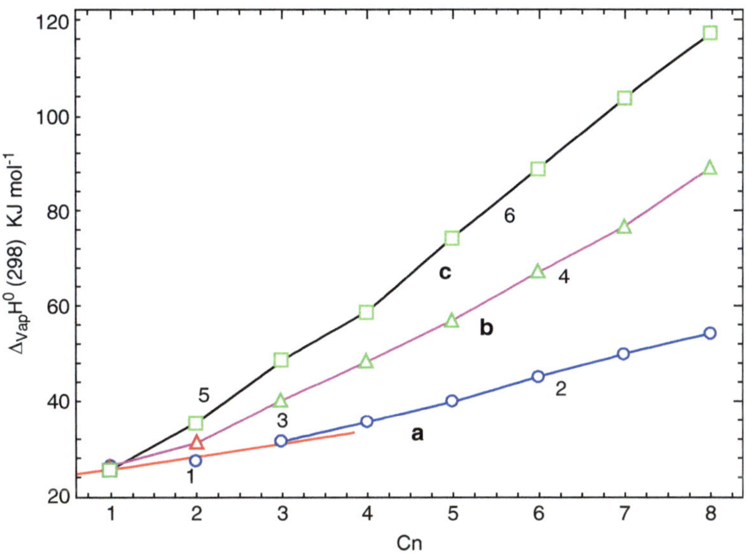

Fig. 2.3 Dependence of the vaporization enthalpy of compounds versus the number of the carbon atoms in alkyl chain: (**a**) methylamines, (**b**) dimethylamines, (**c**) trimethylamines; (*1*, *2*. *3*) influence of reverse dative bonds, (*4*, *5*, *6*) contribution of the energy by CH_2-groups

the alkylamine molecules. The greatest manifestation of the influence of the reverse dative bond is observed at tripropylamine. As a result, the carbon atom of the methyl group of a propyl ligand has higher acceptor properties compared with the ethyl and CH_3 groups in methylamines. As a consequence, the propyl fragment forms a more stable specific interaction N → CH_3–CH_2–CH_2–N–,the energy value of which is kept constant for all the compounds of the homologue series. A description of the enthalpy of vaporization of butylamines–oktylamines reflects the linear dependence of the equal contribution of the energy to the enthalpy characteristics by the increasing the number of the CH_2 groups. The subsequent increase in the number of methyl groups in a chain must make a progressively decreasing energy contribution to the enthalpy characteristics, which should be within the error limits of the experiment by the definite number of compounds of the homologous series. That is why the dependence presented in the figure allows using the method of extrapolation to be used to estimate the enthalpy of vaporization of alkymines that are not studied.

A similar dependence of the enthalpy characteristics of vaporization on the number of carbon atoms in an alkyl chain is manifest in different classes of organic compounds (Fig. 2.4), characterized by the influence of the reverse dative bond and the effect of the energy of increasing number of CH_2 groups in the alkyl chain.

Substitution of the carbon atom to the methyl group decreases the number of realized hydrogen bonds from the sixth in the solid and liquid ammonia up to the fourth in methylamine, and up to two indimethylamine (Fig. 2.1). In the crystalline

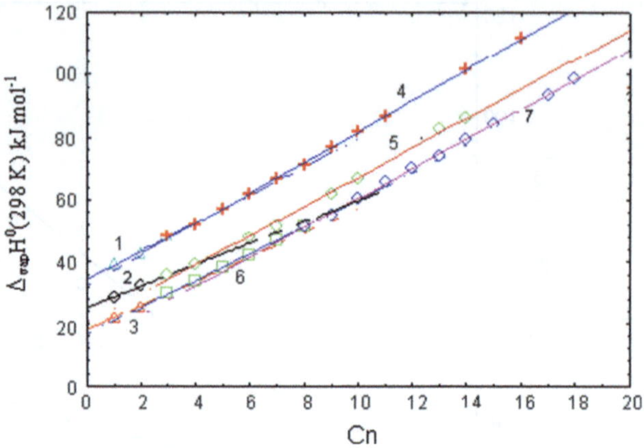

Fig. 2.4 Dependence of the vaporization enthalpy of compounds versus the number of the carbon atoms in alkyl chain (*1*, *5*) – aldehydes, (*2*, *6*) – symmetrical ethers, (*3*, *7*) – asymmetrical methyl ethers and (*4*, *8*) – 1-alcochol [24]

and liquid trialkylamine six specific interactions N → CH_3–N are formed. Thus, the energy of the hydrogen bonds of ammonia is defined by the equation

$$D\text{-}N\cdots H\text{-}N = \Delta_{vap}H°(T)/6\ (NH_3); \qquad (2.1)$$

and the energies of the specific intermolecular interactions come from

$$D\text{-}N\rightarrow CH_3\text{-}(CH_2)_n\text{-}N = (\Delta_{vap}H°(T) - 4D\text{-}N\cdots N\text{-}H)/2\ \text{(Alkylamine)},$$
$$D\text{-}N\rightarrow CH_3\text{-}(CH_2)_n\text{-}N = (\Delta_{vap}H°(T) - 2DN\cdots N\text{-}H)/4\ \text{(Dialkylamine)} \qquad (2.2)$$

We use the principle of additive contribution to the enthalpy of vaporization of ammonia and methyl or alkyl derivative compounds, and of trialkylamines from Eq. 2.3:

$$D\text{-}N\rightarrow R_3\text{-}N = (\Delta_{vap}H°(298\ K)/6\ \text{(Trialkylamine)} \qquad (2.3)$$

where R–CH_3, CH_2–CH_3, CH_2–CH_2–CH_3.

Our conclusion on the stabilization of the specific interaction N → CH_3–N compared with the hydrogen bond N•••H–N is confirmed by the results of the implemented thermodynamic calculations (Table 2.1). The difference is not great (1.5 kJ mol^{-1}); however, its value is located outside of the error of the experimentally established vaporization enthalpy of these compounds. At the same time, the energies of the specific intermolecular interactions of the liquid methylamines (Table 2.1) are described by the logical series of its stabilization

$$\text{Trimethylamine } (4.25) < \text{Dimethylamine } (4.8) < \text{Methylamine } \left(5.33\ \text{kJ mol}^{-1}\right)$$

correlated with the decline of the endo effect of the carbon atoms' coordination with a decrease in its number in the molecule. The increase of the number of carbon atoms in the alkyl chain is accompanied by a stabilization of the specific interactions

Table 2.1 The energies of the specific interactions and H-bonds (kJ mol^{-1}) of liquids methylamines-butylamines (T = 298 K)

Compounds	Formula	$\Delta_{vap}H°$(298 K) [25]	ΣDCH_2	D–N•••H–N	D–N → CH_3–N	D–N → CH_3 CH_2–N	D–N → CH_3–$(CH_2)_2$–N
Trimethylamine	C_3H_9N	25.5	–	–	4.25	–	–
Dimethylamine	C_2H_7N	26.2[a]	–	3.86	4.8	–	–
Methylamine	CH_5N	26.1[a]	–	3.86	5.33	–	–
Triethylamine	$C_6H_{15}N$	35.17 ± 0.1	–	–		5.85	–
Diethylamine	$C_4H_{11}N$	31.3	–	3.86		6.0	–
Ethylamine	C_2H_7N	27.38	–	3.86		6.0	–
Tripropylamine	$C_9H_{21}N$	48.5[b]	–	–	–	–	8.1[b]
Dipropylamine	$C_6H_{15}N$	40.0 ± 0.1	–	3.86	–	–	8.1
Propylamine	C_3H_9N	31.6[b]	–	3.86	–	–	8.1[b]
Tributylamine	$C_{12}H_{27}N$	58.7	13.8	–	–	–	7.9
Dibutylamine	$C_8H_{19}N$	48.50	9.2	3.86	–	–	7.9
Butylamine	$C_4H_{11}N$	35.6 ± 0.1	4.6	3.86	–	–	7.8

[a]T-278 K
[b]Enthalpies were calculated using energies of its specific interactions and hydrogen bonds of analogous

$$\text{Trimethylamine } (4.25) < \text{Triethylamine } (5.85) < \text{Tripropylamine } \left(8.1\ \text{kJ mol}^{-1}\right),$$
$$\text{Dimethylamine } (4.8) < \text{Diethylamine } (6.0) < \text{Dipropylamine } \left(8.1\ \text{kJ mol}^{-1}\right),$$
$$\text{Methylamine } (5.33) < \text{Ethylamine } (6.0) < \text{Propylamine } \left(8.1\ \text{kJ mol}^{-1}\right),$$

which are in direct reference to the absence of a steric effect between alkyl fragments in the molecules of the compounds considered with an increase of the chain of the alkyl ligand. The following series of the energies of the specific intermolecular interaction

$$\text{Tripropylamine}(8.1) = \text{Dipropylamine}(8.1) = \text{Propylamine}\left(8.1\ \text{kJ mol}^{-1}\right)$$

in liquid propylamine visually illustrates the absence of such an effect.

Completion of the influence of the reverse dative bond at propyl ligand on the stabilization of the specific interaction indicates the necessity to record the influence of the energy of the CH_2 group in the enthalpy of vaporization in butylamines, pentylamines, and other compounds of alkylamine derivatives. The contribution of this group is established in the capacity of the average value of the difference of enthalpy of vaporization of heptylamine and propylamine, being equal to 4.6 kJ mol^{-1}. Calculations performed of the energies of the specific interactions $D\text{–}N \rightarrow C_3H_7\text{–}N = 7.9 \pm 0.1$ kJ mol^{-1} in butylamines illustrate the invariability of the number of butyl fragments, located within the limits of experimental error with the energies of the specific interactions in liquid propylamines (Table 2.1). Thus, the energy equations of the specific interactions in liquid propylamines and butylamines is experimental proof of the feasibility of an additive contribution by the CH_2-groups of alkyl bond to the values of enthalpy vaporization and invariability of energies of these interactions in alkylamines, regardless of the number of the hydrogen atoms in the alkyl fragment. It follows that for the established energies of the specific interactions of liquid propylamines and butylamines at the implementation of identical contribution of the energies of the hydrogen bonds of liquid ammonia and contribution of the CH_2-groups there exists a correct way to calculate the energies of the specific interactions, independent of the number of the carbon atoms of the alkyl fragment and its number in the alkylamine and vaporization enthalpy of nonstudied compounds. The vaporization enthalpy of compounds of a number of CH_2 groups greater than three is equal to the sum of the contributions of the energies of the specific interactions, hydrogen bonds, and the same methylene groups

$$\Delta_{vap}H^0(298\text{ K}) = nDN{\rightarrow}CH_3\text{-}(CH_2)\text{-}N + mDN{\bullet\bullet\bullet}H\text{-}N + qDCH_2 \qquad (2.4)$$

where we have n, m specific interactions and H-bonds, and q CH_2 groups, higher than in the propyl fragment. The results of the calculations of the energies of the specific interactions for the compounds with experimentally established enthalpy of vaporization (Table 2.2) for the number of compounds

Table 2.2 The energies of the specific interactions and H-bonds (kJ mol^{-1}) of liquids pentylamines – oktylamines (T = 298 K)

Compounds	Formula	$\Delta_{vap}H°$(298 K) [25]	D–N•••H–N	D(CH_2)	D–N → CH_3–$(CH_2)_2$–N
Pentylamine	$C_5H_{13}N$	40.1 ± 0.1[a]	3.86 × 4	4.6 × 2	7.73[a]
Dipentylamine	$C_{10}H_{23}N$	57.0	3.86 × 2	4.6 × 4	7.73
Tripentylamine	$C_{15}H_{33}N$	74.0	–	4.6 × 6	7.73
Hexylamine	$C_6H_{15}N$	45.0 ± 0.1	3.86 × 4	4.6 × 3	7.88[a]
Dihexylamine	$C_{12}H_{27}N$	67.1	3.86 × 2	4.6 × 6	7.88
Trihexylamine	$C_{18}H_{39}N$	88.7	–	4.6 × 9	7.88
Heptylamine	$C_7H_{17}N$	50.0 ± 0.1[a]	3.86 × 4	4.6 × 4	8.08[a]
Diheptylamine	$C_{14}H_{31}N$	76.8	3.86 × 2	4.6 × 8	8.08
Triheptylamine	$C_{21}H_{45}N$	103.7	–	4.6 × 12	8.08
Oktylamine	$C_8H_{19}N$	54.5	3.86 × 4	4.6 × 5	8.08
Dioktylamine	$C_{16}H_{35}N$	89.0	3.86 × 2	4.6 × 10	8.08
Trioktylamine	$C_{24}H_{51}N$	117.5	–	4.6 × 15	8.08
Pentylamine	$C_5H_{13}N$	37.3[a] (435 K)	3.86 × 4	4.6 × 2	6.35[a]
Dipentylamine	$C_{10}H_{23}N$	52.4[a] (435 K)	3.86 × 2	4.6 × 4	6.35[a]
Tripentylamine	$C_{15}H_{33}N$	65.7[a] (435 K)	–	4.6 × 6	6.35[a]

[a]Experimental obtained received vaporization enthalpy

$$\text{Butylamine } (7.8) \approx \text{Pentylamine } (7.85) \approx \text{Hexylamine } (7.85)$$
$$\approx \text{Heptylamine } \left(8.08 \text{ kJ mol}^{-1}\right)$$

are within the limits of experimental error and could be equal to 7.9 kJ mol^{-1} for all alkylamines of this series. This allows one to conclude that the structural and energy effects of the specific interactions and H-bonds in all alkylamines are similar. Given the above, we conducted thermodynamic calculations of the enthalpy of vaporizations of nonstudied compounds of amines with different alkyl ligands (Table 2.2).

Using the method of extrapolation, we estimated the enthalpy of vaporization of non-amines-undekanones (Table 2.3).

The influence of the temperature on the decrease of the energies of the specific interactions in liquid alkylamines can be traced at penthylamine, of which vaporization enthalpy is defined at the temperature of 435 K. The calculations of the energies of these bonds are made with the help of the energy of H-bonds, established at lower temperatures, and thus, the values obtained may significantly differ from the real values. Nevertheless, it draws attention to the fact that for all penthylamines (Table 2.2) the value of the energy of the specific interaction D–N → CH_3–(CH_2) –N equal to 6.35 kJ mol^{-1}, which is 1.55 kJ mol^{-1} lower than the values of energies of the same bond type established at 298 K was obtained.

The established distribution of structural-energetic parameters at amines with different numbers of alkyl ligands allows us to wait for the analogue in the behavior of polyligand alkylamines. In this connection we used the value of the established energy of the specific interaction, formed by the methyl group (Table 2.1), and we

Table 2.3 Vaporization enthalpy (kJ mol^{-1}) of nonylamines – undecanylamines

Compounds	$\Delta_{vap}H°$ (T)	Compounds	$\Delta_{vap}H°$ (298 K)	Compounds	$\Delta_{vap}H°$ (298 K)
Nonanylamine	59.3	Decanylamine	64.0	Undecanylamine	68.3
Dinonanylamine	96.7	Didecanylamine	106.7	Diundecanylamine	116.0
Trinonanylamine	132.7	Tridecanylamine	147.2	Triundecanylamine	162.7

calculated the value of the energy of the formed specific interaction by ethyl and butyl fragment with a bigger number of the carbon atoms. The results of the implemented thermodynamics calculations establishing the energies of the specific interactions of amines with different alkyl ligands are presented in Table 2.4. First of all, we draw attention to the fact of coincidence of values of energies of the specific interaction D–N → CH_3–CH_2–N y *N,N*-dimethylethylamine (5.7 kJ mol^{-1}) and *N,N*-diethylmethylamine. Amines with methyl ligand methylbutylamines (8.1 kJ mol^{-1}) have several increased values of the energy of the specific N → CH_3–CH_2–CH_2 interaction, higher than at ethylbutylamines (7.9 kJ mol^{-1}), although the difference differs little from experimental error. At the same time, triethylamine (5.83 kJ mol^{-1}) with three ethyl ligands is characterized by higher stability in comparison with similar interactions of methylethylamines. The two given contradictory facts are not considered against the steric effect, because the set values of energies are within the experimental error of the identification of the enthalpy characteristics of the compounds considered. Such a similarity is also seen at other amines (Tables 2.1 and 2.2).

To establish the energies of the specific interactions of alkylamines with an isomeric methyl group it is necessary to define the value of the energetic contribution of this group to the enthalpy characteristics. Analysis of the characteristics of alkylamines of normal structure and with isomeric methyl group (Table 2.5) indicates that compounds with isostructural group have lower values of enthalpy vaporization and higher values in comparison with the values of these characteristics for compounds, for example, ethyl or propyl ones in comparison with isopropyl or isobutyl containing ligand. It follows that the contribution of the isostructural methyl group should be set by the difference of the vaporization enthalpy, in particular, isopropyl containing ligand and its ethyl analogue as well as amine with the isostructural methyl group of butylamine and compounds with propyl ligand of normal structure divided by the number of these groups.

$$D iso CH_3\text{-} = (\Delta_{vap}H°(298\ K) iso.\text{pr.a} - \Delta_{vap}H°(298\ K).\text{et.a})/n \quad (2.5)$$

$$D iso CH_3\text{-} = (\Delta_{vap}H°(298\ K) iso.\text{bu.} - \Delta_{vap}H°(298\ K).\text{pr.a})/n \quad (2.5a)$$

The calculations (Table 2.5) lead to the conclusion that in the case of triisobuthylamine and isobuthylamine the value of the contribution of the three isostructural methyl groups to the enthalpy of vaporization is equal to 6.90 kJ mol^{-1} or per contacting molecule of the local environment 1.15 kJ mol^{-1}. For the rest of

Table 2.4 The energies of the specific interactions and H-bonds (kJ mol^{-1}) of liquid of different ligand alkyl amines (T = 298 K)

Compounds	Formula	$\Delta_{vap}H°$(298 K) [25]	ΣDCH2	D–N•••H–N	D–N → CH_3–N	D–N → CH_3–CH_2–N	D–N → CH_3–$(CH_2)_2$–N
N,N-Diethylmethylamine	$C_5H_{13}N$	31.8	–	–	4.25	5.7	–
N,N-Dimethylethylamine	$C_4H_{11}N$	28.4[a]	–	–	4.25	5.7[a]	–
N-Methylbutylamine	$C_5H_{13}N$	38.1	4.6	3.86	4.8	–	8.1
N,N-Dimethylbutylamine	$C_6H_{15}N$	33.5[a]	4.6	–	4.25	–	8.1[a]
Dibutylmethylamine	$C_9H_{21}N$	40.9[a]	9.2	–	4.25	–	8.1[a]
Butylethylamine	$C_6H_{15}N$	40.2 ± 0.1	4.6	3.86	–	6.0	7.9
Dibutylethylamine	$C_{10}H_{23}N$	52.8[a]	9.2	–	–	6.0[a]	7.9[a]
Diethylbutylamine	$C_8H_{19}N$	44.5[a]	4.6	–	–	6.0[a]	7.9[a]

[a]Experimental values of vaporization enthalpies

Table 2.5 The energies of the specific interactions and H-bonds (kJ mol^{-1}) of liquid isostructural alkyl amines (T = 298 K)

Compounds	Formula	$\Delta_{vap}H°(298\ K)$ [25]	D–N•••H–N	D–N → R–N	D*iso*CH$_3$–
Triisopropylamine	$C_9H_{21}N$	38.6	–	6.0	1.15, 0.60
Diisopropylamine	$C_6H_{15}N$	34.6 ± 0.1	3.86	6.0	1.65, 0.80
Isopropylamine	C_3H_9N	28.0[a]	3.86	6.0	0.62, 0.30
Triisobuthylamine	$C_{12}H_{27}N$	55.4	–	8.1	2.30, 1.15
Diisobuthylamine	$C_{12}H_{27}N$	43.1 ± 0.3	3.86	8.1	1.55, 0.80
Isobuthylamine	$C_{12}H_{27}N$	33.9 ± 0.1	3.86	8.1	2.30, 1.15
N-Methylisopropylamine	$C_4H_{11}N$	30.7 ± 0.1	3.86	4.8, 6.0	1.40, 0.70
N-Ethylisopropylamine	$C_5H_{13}N$	33.1 ± 0.1	3.86	6.0, 8.1	1.40, 0.70
Dipropylisopropylamine	$C_9H_{21}N$	41.8	–	6.0, 8.1	1.60, 0. 80
Diisopropylpropylamine	$C_9H_{21}N$	43.0	–	6.0, 8.1	1.60, 0.80
N-Isopropylpropylamine	$C_6H_{15}N$	37.3	3.86	6.0, 8.1	1.60, 0. 80
N-Butylethylamine	$C_6H_{15}N$	40.2 ± 0.1	3.86		
N-Butylisopropylamine	$C_7H_{17}N$	42.1 ± 0.1	3.86	6.0, 8.1	1.90, 0.95
N-Butylisobutylamine	$C_8H_{19}N$	41.2	328 K		
tert-Butylmethylamine	$C_5H_{13}N$	32.3 ± 1.4	3.86; 297 K	4.8, 6.0	1.40, 0.70
N,N-Dimethyl *tert*-butylamine	$C_6H_{15}N$	34.8	–	–	1.80, 0.90
tert-Butylisopropylamine	$C_7H_{17}N$	35.7 ± 1.0	287 K	6.0	1.40, 0.70

[a]Calculated from the energies of specific interactions and H-bonds

the isostructural amines the contribution is changed within 1.15–1.90 15 kJ mol^{-1} or per interacting molecule of the local environment 0.60–0.95 kJ mol^{-1}. It is remarkable that similar values of energetic contribution of isostructural methyl group to the vaporization enthalpy were obtained for the other classes of organic groups [23]. Thus, one can state that isostructural methyl groups contribute similarly in mean contributions to the enthalpy of vaporization with its location at the first and second carbon atom of alkyl bond independentof the number of these groups and ligands in a molecule. This value for the experimental error and of values given in Table 2.5 could be accepted to be equal to 0.80 ± 0.10 kJ mol^{-1}.

2.1.2 *Saturated and Unsaturated Diamines*

According to [1] the value of n-orbitals, splitting in diamines with isolated nitrogen atoms is caused by competition of interactions "through the space" and "through connections" [26, 27] and compounds with linear molecule are characterized by a lower role of σ-(CC)-interactions; it remains significantly lower in comparison with compounds with cyclic molecule structure. One can suppose that it is this property that is responsible for the values of the vaporization enthalpy of diamines (Table 2.6), changing in the range

Table 2.6 The energies of the specific interactions and H-bonds (kJ mol^{-1}) of liquid diamines

Compounds	Formula	$\Delta_{vap}H°$(298 K) [25]	T K	nD–N•••H–N	D–N → CH_3–N	D*iso*CH_3	D–N → (CH_2)n–N
bis(dimethylamino)methane	$C_5H_{14}N_2$	32.3	310	N•••H–CH 0.6:2	3.7	–	n = 1 0.9
Ethylenediamine	$C_2H_8N_2$	45.0 ± 0.1	298	8 × 3.86	–		n = 1 3.6
1,3-Propanediamine	$C_3H_{10}N_2$	50.2 ± 0.1	298	8 × 3.86	–	–	n = 1.5 4.83
1,6-Hexanediamine	$C_6H_{16}N_2$	51.3	353	8 × 3.86	–	–	n = 3 5.1
3-Hexenediamine	$C_5H_{14}N_2$	49.4	361	6 × 3.86		–	D–N → CH=CH–CH_2–N = 4.9
3-Hexynediamine	$C_6H_{14}N_2$	44.4	368	8 × 3.86	–	–	D–N → C≡C–CH_2–N = 3.4
1,7-Heptanediamine	$C_7H_{18}N_2$	(46.5)	288	8 × 3.86	–	–	n = 3.5 (3.8)
3-Heptenediamine	$C_7H_{16}N_2$	(46.5)	288	8 × 3.86	–	–	n = 3.5 (3.8)
1,2-Propanediamine	$C_3H_{10}N_2$	44.2 ± 0.2	298	8 × 3.86	–	0.30 × 2	n = 1 3.2
2-Methyl-1,2-propanediamine	$C_4H_{10}N_2$	43.6 ± 0.1	298	8 × 3.86	–	0.30 × 4	n = 1 3.0
1,2-Butanediamine	$C_4H_{12}N_2$	46.9, 46.3 ± 0.2	298	8 × 3.86	–	D*iso*C_2H_5 1.3:2	n = 1 3.5
N,N,N,N-Tetramethyl-1,3-butanediamine	$C_8H_{20}N_2$	49.2	272–363	–	3.7	0.30 × 4	n = 1.5 4.83
N,N,N,N-Tetramethylethylenediamine	$C_6H_{14}N_2$	43.6[a]	298	–	3.7	–	n = 1 3.55
N,N,N,N-Tetramethyl-1,3-propanediamine	$C_7H_{18}N_2$	(48.8[a])	(298)	–	3.7	–	n = 1.5 4.83

(continued)

Table 2.6 (continued)

Compounds	Formula	$\Delta_{vap}H°(298\ K)$ [25]	T K	nD–N•••H–N	D–N → CH_3–N	D*iso*CH_3	D–N → (CH_2)n–N
N,N-Dimethyl-1,3-propanediamine	$C_5H_{14}N_2$	45.2	318	4 × 3.86	3.7	–	n = 1.5 3.8[b]
N,N-Diethyl-1,3-propanediamine	$C_7H_{18}N_2$	46.4	344	4 × 3.86	3.95[b]	–	n = 1.5 3.8[b]
1,1,4,4-tetrame thyltetramethylenediazine	C_8H_{16}N2	50.1 ± 0.4	298	–	3.7	–	n = 2 5.12
1,4,8,11-tetraazaundecane	$C_7H_{20}N_4$	100.0 ± 2.5.	298	12 × 3.86	–	–	n = 2 7.53

[a]Elucidated from the energies of the specific interactions
[b]It is not corresponded to the standard conditions

$$bis(\text{imethylamino})\text{methane } (32.3) < \text{Ethylenediamine } (45.0) < 1,3\text{-Propanediamine } \left(50.2 \text{ kJ mol}^{-1}\right).$$

This low value of the vaporization enthalpy of *bis*(dimethylamino)methane indicates that the value of the energy of its specific interaction N → CH_3–N should be less stable than the similar interaction formed by trymethylamine (4.25 kJ mol^{-1}) and the hydrogen bond of the liquid N_2H_4 diamine not containing the methylene group.

The presence in each amino group of two ligands of the hydrogen atoms or methyl groups leads to the formation of the pyramid structure of the liquid condition and crystal. Interaction "through the space" *bis*(dimethylamino)methane is carried out with participation of the methylene group and two and three ethylenediamine and 1,3-propanediamine, respectively and is accompanied by such high difference of enthalpy in these characteristics. It means that energies of the separate types of specific interactions of the given compounds depends greatly on the number of groups in the alkyl chain and distribution of the electron density in bond vacancies at its equal number in the diamines molecules. Amino groups of contacting molecules of diamines form a pyramidal structure with molecules from the local environment, while nitrogen atoms form specific intermolecular interactions with the CH_2 groups of the chain N → CH_2–N, the essentially unshared $2s^2$(C) by electron pair (Fig. 2.5) forming a grid between molecules of the local environment. Thus, liquid and crystal diamines are characterized by the formation of volumetric-grid structure. A characteristic feature of *bis* (dimethylamino)methane with one methylene group is the formation by the second atom of the amino group of the low stability hydrogen bond with the atom of the CH_2-group N•••H–CH (Fig. 2.5a) with energy 0.60 kJ mol^{-1}. Two pyramidal structures, formed by the methyl groups or hydrogen atoms of amino groups, are generated by eight specific interactions DN → CH_3–N and by the same number of the hydrogen bonds D–N•••H–N *bis*(dimethylamino)methane (Fig. 2.5a), ethylenediamine (Fig. 2.5b), 1,3-propanediamine (Fig. 2.5c), 1,6-hexanediamine. Ethylenediamine, and 1,6-hexanediamine respectively with an even number of the methylene groups in the chain characterized by formation of four specific interactions N → CH_2–N and N → CH_2–CH_2–CH_2–N, respectively. Equivalent displacement of the electron density on the chain to the nitrogen atoms of amino groups occurs when the odd number of methylene groups in the molecule 1,3-Propanediamine is conditionally accepted formed by the specific interaction N → $(CH_2)n$–N at n = 1.5.

As it was mentioned above, alkylamines and liquid ammonia are practically unchanged when formed by amino group hydrogen bonds. This is sufficient reason to accept that the energies of the hydrogen bonds formed in liquid diamines to be equal to the energies of the corresponding bonds in liquid ammonia. Considering the enthalpy characteristics of vaporization connected with the energy of the specific interactions and their number, there is enough ground to conclude that the energy of the specific interaction D–N → CH_2–N is obtained from the

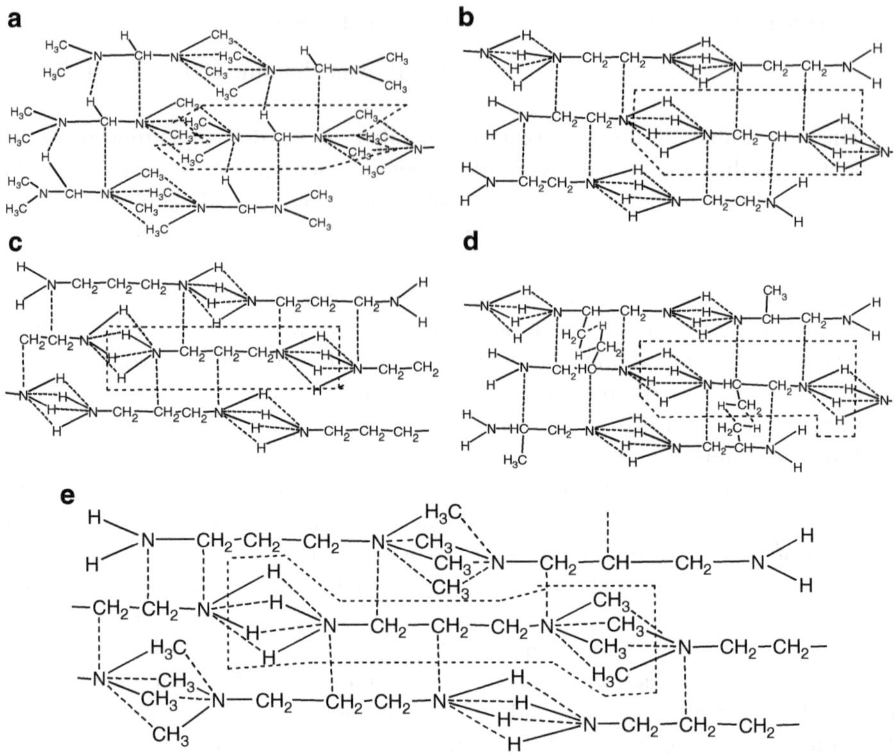

Fig. 2.5 Schematic picture with network of the specific interactions: *bis*(dimethylamino)methane (**a**), ethylenediamine (**b**), 1,3-propanediamine (**c**), 1,2-propanediamine (**d**), *N,N*-dimethyl-1,3-propanediamine (**e**)

difference between the enthalpy of vaporization and the total contribution of the energies of interactions D–N → CH_3–N and hydrogen bonds 2D–N•••H–CH.

$$\text{D-N}\rightarrow\text{CH}_2\text{-N} = (\Delta_{\text{vap}}H°(298\ \text{K}) - 8\text{D-N}\rightarrow\text{CH}_3\text{-N} - 2\text{D-N}\cdots\text{H-CH})/2 \quad (2.6)$$
(*bis*(Dimethylamino)methane

The enthalpy of vaporization and eight hydrogen bonds ethylenediamine and 1,3-Propanediamine

$$\text{D-N}\rightarrow\text{CH}_2\text{-N} = (\Delta_{\text{vap}}H°(298\ \text{K}) - 8\text{D-N}\cdots\text{H-N})/4 \quad (2.7)$$

$$\text{D-N}\rightarrow\text{CH}_2\text{-CH}_2\text{-CH}_2\text{-N} = (\Delta_{\text{vap}}H°298\ \text{K}) - 8\text{D-N}\cdots\text{H-N})/4\ \text{(1,6-Hexanediamine)} \quad (2.8)$$

The results of the calculations of the specific interactions energies formed by the nitrogen atom with methylene groups of the chain (Table 2.6)

(D-N→CH_2-N) Ethylenediamine (3.6) < (D-N→(CH_2)n-N, n=1.5) 1.3-Propanediamine (4.8) < (D-N→CH_2-CH_2-CH_2-N) 1,6-Hexanediamine (5.1 kJ mol^{-1}, 353 K)

indicate its stabilization with the increase in the number of methylene groups of the chain, owing to the increased displacement of the electron density in each nitrogen atom of diamine and weakening of the influence of the intramolecular reverse dative bonds with increasing number of methylene groups of intramolecular chains up to three. Low values of the specific interactions reflect small difference of charges with opposite signs of nitrogen and hydrogen atoms. An even smaller difference of charges of nitrogen and carbon atoms with a single CH_2-group *bis* (dimethylamino)methane allows one to expect a reduced value of the energy D–N → CH_2–N in comparison with the value of the energy from the same type of interaction in ethylenediamine. Obviously, the established value of the energy of the specific interaction 1,6-hexanediamine at 353 K should have a greater value under standard conditions. The results of the calculations of energies of the specific interactions realized in liquid 3-hexenediamine and 3-hexynediamine (Table 2.6) illustrate

D-N→CH_2-CH_2-CH_2-N (5.1) 1,6-Hexanediamine > D-N→(=CH)-CH_2-CH_2-N (4.9) 3-Hexenediamine > D-N→(≡C)-CH_2-CH_2-N (3.4 kJ•mol^{-1}) 3-Hexynediamine

the natural tendency of manifestation of the influence of double and triple bonds in the molecules of diamines to destabilize specific interactions.

In sources in the literature [25] there are enthalpies of vaporization of 1,7-heptanediamine and 3-heptenediamine (46.5 kJ mol^{-1}, 288 K) without indication of experimental errors and their values are too low compared with enthalpy characteristics of 1,3-propanediamine, 1,6-hexanediamine, and 3-hexenediamine. Their values of enthalpies of vaporization should be estimated from the established energy values of the specific interactions of derivatives 1,6-hexanediamine; however, for implementation of calculations the values contributed by the energies of one CH_2-group is lacking.

A characteristic feature of 1,2-propanediamine and 1,2-butanediamine is the presence in their molecules of methyl and ethyl groups, located outside of the chain, which significantly change the distribution of electric density in a molecule, leading them to play the role of isostructural groups and form specific interactions, H_3C → H–CH_2, of low stability (Fig 2.5d). The influence of the given methyl group is so great in diamines that it leads to a lower value of the enthalpy characteristics of 1,2-propanediamine compared with the enthalpy of vaporization of ethylenediamine, which has the identical number of the methylene groups (Table 2.5). The second isostructural methyl group 2-methyl-1,2-propanediamine decreases the vaporization enthalpy practically in equal parts (0.6 kJ mol^{-1}). This value corresponds to the contribution of one isostructural methyl group to the vaporization enthalpy of isopropylamine (0.30 kJ mol^{-1}), which participates in the interaction of the molecules of the local environment. The results of the conducted calculations of the energies of the specific interactions of D–N →

CH_2–N, formed by the nitrogen atom with CH_2-group of the chain (Table 2.6), using Eq. 2.9,

$$\text{D-N}\rightarrow\text{CH}_2\text{-N} = (\Delta_{vap}H°(298\ K)eda - 8\text{D-N}\cdots\text{H-N} - nDisoCH_3)/4 \quad (2.9)$$

illustrate the tendency for it to be destabilized with increasing number of isostructural methyl groups in the molecules 1,2-propanediamine < 2-methyl-1,2-propanediamine.

The contribution of the energies of the isostructural ethyl group to the vaporization enthalpy of 1,2-butanediamine is equal to the difference of vaporization enthalpies of diamines with the same number of the methylene groups chains

$$DisoCH_3) = (\Delta_{vap}H°(298K)bda - \Delta_{vap}H°(298\ K)\,eda)/2 \quad (2.10)$$

The values obtained for the energy of the specific interaction of D–N → CH_2–N reflect the natural tendency of its destabilization in the range of isostructural diamines

$$\text{Ethylenediamine}(3.5) \approx 1,2\text{-Butanediamine}(3.5) > 1,2\text{-Propanediamine}(3.2) > 2\text{-Methyl-1, 2-propanediamine}(3.0\ \text{kJ mol}^{-1})$$

The value obtainedfor the energy of the specific interaction of D–N → CH_2–N (4.8 kJ mol^{-1}) of liquid 1,3-propanediamine is the result of the offset of the electron density through the chain of the methylene group and from the atom of hydrogen of amino group. Taking into account the difference of 0.8 kJ mol^{-1} between the vaporization enthalpy of 1,3-Propanediamine and *N,N,N,N*-tetramethyl-1,3-butanediamine at the equal number of the hydrogen bonds of D–N•••H–N and specific interactions of D–N → CH_3–N, we have reason to assume permanence of the energy of interactions in the D–N → CH_2–N chain (4.8 kJ mol^{-1}) at the given compounds. The value obtained for the energy of the specific interaction D–N → CH_3–N by Eq. 2.11 confirms the correctness of small

$$\text{D-N}\rightarrow\text{CH}_3\text{-N} = (\Delta_{vap}H°(298\ K) - 4\text{D-N}\rightarrow\text{CH}_2\text{-N})/8 \quad (2.11)$$

differences of energies in this type of interaction and in the hydrogen bond (Table 2.6). It follows that like the hydrogen diamine bond the value of the energy of the specific interaction D–N → CH_3–N will be unchangeable at compounds *bis* (dimethylamino)methane – *N,N,N,N*-tetramethyl-1,3-propanediamine. Thus, there is a way to obtain the total value of the energy of the hydrogen bond of low stability D–N•••H–CH and D–N → CH_2–N *bis*(dimethylamino)methane and to estimate the vaporization enthalpy of *N,N,N,N*-tetramethylethylenediamine and *N,N,N,N*-tetramethyl-1,3-propanediamine. The obtained values of the energies of the hydrogen bond D–N•••H–CH (0.6) and the specific interaction of D–N → CH_2–N (0.9 kJ mol^{-1}) *bis*(dimethylamino)methane and the total number of two bonds of each type

(2.7 kJ mol^{-1}) correlate well with the stabilization of the above number of specific interactions.

D-N→CH_2-N *bis*(Dimethylamino)methane – 1,6-Hexanediamine.

It is possible to settle the influence of double bonds –CH=CH– on the energy of the specific interaction, formed by this fragment of the chain, by the example of 1,1,4,4-tetrame thyltetramethylenediazine.

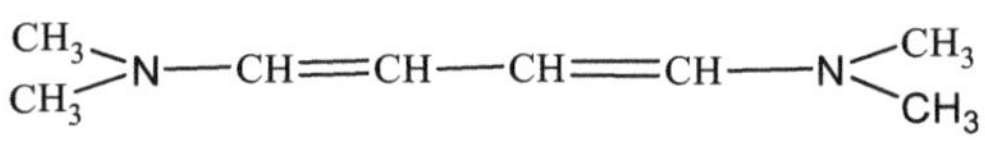

It is quite correct to make the contribution of the energy by the methyl group introduced to the vaporization enthalpy of this compound equal to the contribution of the same group *N,N,N,N*-tetramethyl-1,3-butanediamine of 3.7 kJ mol^{-1}. The obtained value of the energy of the D–N → CH=CH–N formed by this fragment is 5.1 kJ mol^{-1}, which exceeds the energy of the interaction formed by ethyl ligand *N, N,N,N*-tetramethyl-1,3-butanediamine (4.83 kJ mol^{-1}).

The thermodynamical analysis performed allows drawing a very important conclusion.

First of all, independentof the number of the methylene groups in the intramolecular chain located between the two amino groups with strong acceptors of the nitrogen atom, energies of the specific interactions formed by ethylene groups and their fragments have significantly low values compared with the values of energies of interactions, formed with the same number of hydrogen atoms with an open chain, and

Second, the energy of the specific interactions formed by the unsaturated ligand is stabilized at the same number of carbon atoms at the inner chain. With an increased number of the carbon atoms of the chain it is less stable than the one formed by a ligand with the same number of atoms of carbon of the open chain.

The position of the specific interactions of1,4,8,11-tetraazaundecane with four non-contacting nitrogen atoms in a molecule functioning as amino groups is attractive. External nitrogen atoms in positions N(1) and N(11), playing the rolc of methylamine, form hydrogen bonds and specific interactions. Likewise for methylamine, the nitrogen atom forms four hydrogen bonds and leads to more stable specific interactions with the participation of the methylene groups. The sixth coordinated condition of the nitrogen atom leads to the reduced energy of the hydrogen bond, which is practically the same as the value of the energy realized in liquid ammonia (3.86 kJ mol^{-1}). Nitrogen atoms in positions N(2) and N (3) acquire

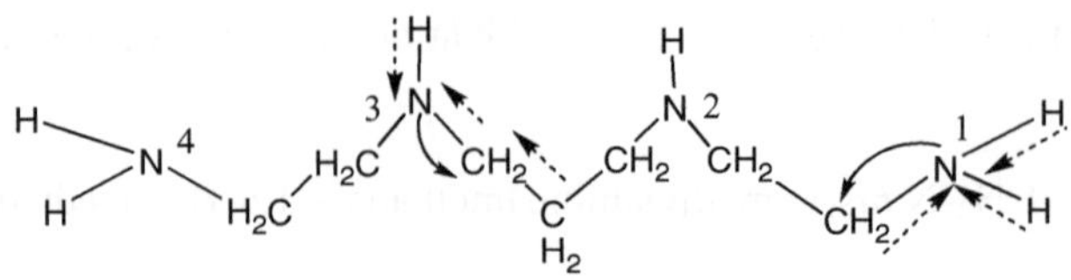

relative abundance of the electron density compared with carbon atoms in positions N(1) and N(4) from two ethylene groups, a part of which shifts to the contacting carbon atoms, forming the reverse dative bond, stabilizing the intramolecular bond, and leading to decreasing negative charge of the nitrogen atom. Thus, the energies of the hydrogen bonds formed and the specific interactions are described by the stability factor.

The 12 hydrogen atoms DN•••H–N formed, and seven interactions 2D–N → CH_2–N and 5D–N → CH_2–CH_2–N form the structure of liquid 1,4,8,11-tetraazaundecane with a grid of the specific interactions (Fig. 2.6). The complexity in identification of the energy types of the specific interactions formed by the carbon atoms of the ethylene groups is because they belong to two neighboring nitrogen atoms and obviously come with little difference in stability. Accepting them as equal we determined the energy from the difference in the enthalpy of vaporization and the total contribution to the energy of the 12 hydrogen bonds

$$\text{D-N}\rightarrow\text{CH}_2\text{-CH}_2\text{-N} = (\Delta_{\text{vap}}H^0(298\text{ K}) - 12\text{ D-N}\cdots\text{H-N}) / 7 \quad (2.12)$$

The calculations (Table 2.6) point to the fact that the obtained value of the energy of the considered specific interactions lies between the energies of the same type of interactions realized in liquid diathyl DN → CH_3–CH_2–N and dipropyl amines: DN → CH_3–$(CH_2)_2$–N

$$\text{Diethylamine }(6.0) < 1,4,8,11\text{-Tetraazaundecane }(7.35) < \text{Propylamine }\left(8.1\text{ kJ mol}^{-1}\right).$$

It follows that reducing coordination of the nitrogen atoms from the sixths to the fifths at its location of N(2) and N(3) for ethylene groups leads to the stabilization of the specific interaction up to the energy formed by the propylene fragment in liquid dipropylamine and the CH_2-group up to ethyl in diathylamine.

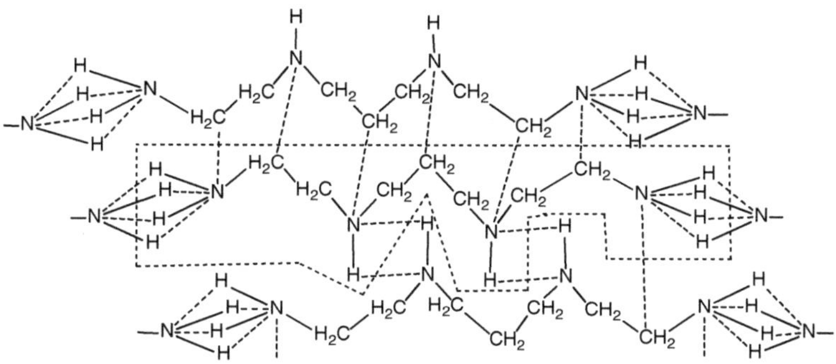

Fig. 2.6 Schematic picture of the solid structure of 1,4,8,11-tetraazaundecane

2.2 The Energies of Specific Interactions of Liquid Hydrozines

Extensive research of FE-spectrum of acyclic hydrozines led the authors of [28–32] to establish the change in the values of splitting of n-orbitals of isolated nitrogen atoms ΔI(n) with the following consequence

N_2R_4 (0.5-0.6 eV) < N_2HR_3 (0.6-.7 eV) < RHNNHR (0.7-0.80 eV) < N_2H_4 (0.8 эВ) < N_2H_3R (0.9-1.0 eV) < R_2NNH_2 (1.2-1.3 eV).

This means that depending on the number and location of R–ligands and hydrogen atoms significant changes in properties take place. The most significant difference should be expected at N_2R_4 and R_2NNH_2. Provided all four ligands in a hydrazine molecule are of the same type, equivalent displacements of the electron density take place for each nitrogen atom (Fig. 2.7). In turn, there occurs a transfer of electron density from the p_z orbital of the nitrogen atoms to the hydrogen atoms of the methyl group, forming an intramolecular reverse dative bond. For example, relocationof one hydrogen atom from an amine to a methyl group violates the uniformity in replacement of the electron density and one of thc hydrogen atoms in *trans*-position transfers a somewhat higher proportion of the electron density and gets increased positive charge compared with two other hydrogen atoms. Study of FE-spectra of a significant number of hydrozines of the acyclic series [1] showed that at room temperature strips of transposition are not indicated. At the same time, the result of research by other methods indicates the possibility of its contribution to be up to 30 %. The sequence of ionizing potentials Iv σ-levels has not been discovered [1].

Partial transfer of the relative access to the electron density in the p_z orbital of the nitrogen atom to the carbon atom of the methyl group and the hydrogen atoms leads to a reduction of its positive charge and stability of the hydrogen bonds and specific interactions, respectively. But a single methyl group of the *trans*-position shows higher propensity to the formation of more stable interactions for N → CH_3–N

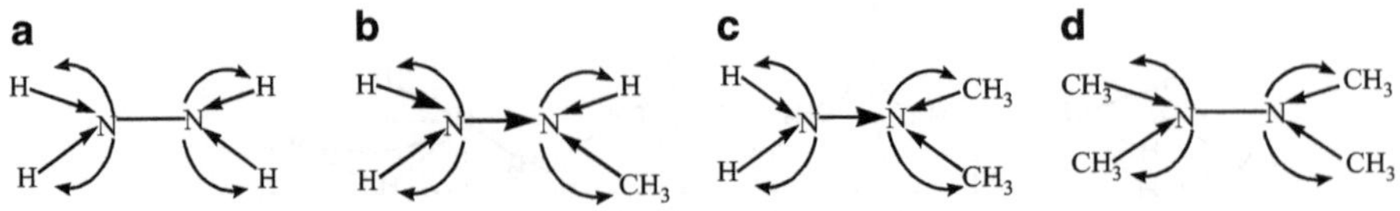

Fig. 2.7 Displacement of the electron density and reverse dative bonds in hydrozines

compared with other methyl groups, including the location of the second CH_3-group in *trans*-position. At the same time, in the *trans*-position the hydrogen atom with respect to the methyl group it shows higher acceptor ability compared with the hydrogen atom in N_2H_4. Thus, the energies of the H-bonds and specific intermolecular interactions of hydrozines change depending on their location in the hydrazine molecule. Consequently, a unique solution to the problem of placement of real values of H-bonds energies and specific interactions is not possible. However, taking into account the necessity for definite assumptions, it is expedient to combine approaches and assumptions in order to obtain more exact information on the energies of realized intermolecular interactions in liquid and solid hydrozines. It is necessary to notice that the least stable ones from the considered series of interactions of hydrozines are in the hydrogen bonds formed by hydrogen atoms. That is why the smallest error in thermodynamical calculations of energies of the specific interactions, realized in the condensed state of compounds, will be on condition that we take adequate values for the energies of the H-bond in liquid ammonia and hydrozine in the case when both have equal coordination numbers of nitrogen atoms. The structure of liquid hydrozines can be presented by a schematic picture, with the network of the specific interactions given in Fig. 2.8. Taking into account that all bond vacancies of a molecule should be occupied, compounds of this series are characterized by each nitrogen atom satisfying the five-coordination condition at its interaction with molecules of the local environment with the energy of the hydrogen bond of diamine 3.86 kJ mol^{-1}, equal to the energy of H-bond of liquid ammonia. Consequently, vaporization enthalpy of diamine is equal to 30.9 kJ mol^{-1}. The correctness of the given value is confirmed by the value of the enthalpy characteristics of trimethylhydrozine (Table 2.7). The formation in condensed hydrozines of two types of intermolecular interactions directly leads to the fact that the sum of energies of the specific interactions formed by pentacoordinated carbon atom is equal to

$$nD\text{-}N{\rightarrow}CH_3\text{-}N = \Delta_{vap}H°(298\ K) - mD\text{-}N{\cdots}H\text{-}N \text{ и } D\text{-}N{\rightarrow}CH_3\text{-}N = (\Delta_{vap}H°(298\ K) - mD\text{-}N{\cdots}H\text{-}N)/n \quad (2.13)$$

or, in general,

$$D\text{-}N{\rightarrow}CH_3\text{-}N = (\Delta_{vap}H°(298\ K) - mD\text{-}N{\cdots}H\text{-}N)/n \quad (2.14)$$

where “n” and “m” are the numbers of the corresponding bonds.

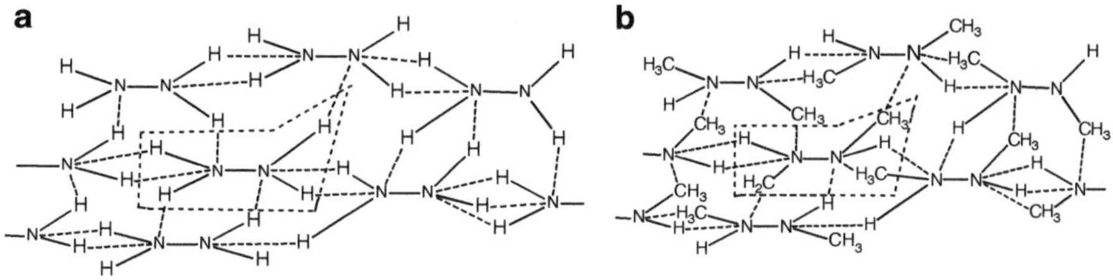

Fig. 2.8 Schematic picture of the liquid structure of hydrozines: diamine (**a**), 1,2-dimethylhydrozine (**b**)

The thermodynamic properties of acyclic hydrozines are limited by the small number of compounds; nevertheless, the thermodynamical calculations led to interesting results.

From the set of values of the energy properties (Table 2.7) it is obvious that energies of the specific interactions D–N → CH_3–N are stabilized with reducing the number

$$\text{Tetramethylhydrazine } (4.1) < \text{Trimethylhydrozine } (4.30) < 1.1-\text{Dimethylhydrozine } (4.70) < 1.2 - \text{Dimethylhydrozine } (6.40) < \text{Methylhydrozine } (9.10 \text{ kJ mol}^{-1})$$

of the methyl group in a molecule of hydrozine dependending on its location. A similar pattern is observed in ethylhydrozines

$$\text{Tetraethylenediamine } (4.30) < \text{Ethylenediamine } (10.7 \text{ kJ mol}^{-1}).$$

Note that values obtained for the energies of the specific interactions include errors, connected with the mismatching of the given values of the vaporization enthalpy of some compounds to the standard conditions. Nevertheless, the established integral changes in the values of the specific interactions energies fully agrees with the above conclusion on the change in the values of splitting of n-orbitals of isolated nitrogen atoms ΔI(n) and notions of mutual influence of *cis*-location and *trans*-location of ligands in the molecules of considered compounds [1].

Among solid ethylhydrozine the enthalpy of sublimation is known only for ethylenediamine. In calculating the energy of the specific interaction D–N → CH_3–CH_2–N, the energy of the hydrogen bond of solid ammonia D–N•••H–N, which is equal to 6.34 kJ mol^{-1}, was used. The established values of the specific interaction energies of solid and liquid ethylenediamine of 13.9 and 10.7 kJ mol^{-1} at 363 and 298 K, respectively, reflect the high impact of the crystal field on the stabilization of the specific interaction.

The peculiar feature of the 2,2-diaminodiethylamino is the difference in nitrogen atoms in formatted coordination (Fig. 2.9). In the structure of the liquid four hydrogen bonds and two more stable H-bonds are formed by pentacoordinated

Table 2.7 The energies of the specific interactions (kJ mol^{-1}) of liquid hydrozines

Compounds	Formula	The structure	$\Delta_{vap}H°(298\ K)$ [25]	T K	nD–N•••H–N	D–N → CH_3–N
Diamine	N_2H_4	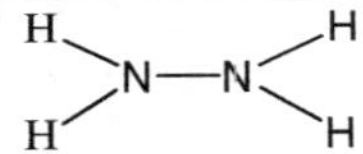	30.9		8 × 3.86	–
Tetramethylhydrazine	$C_4H_{12}N_2$	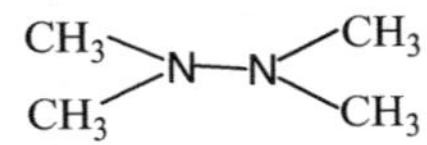	32.9	305	–	4.1[a]
Trimethylhydrozine	$C_3H_{10}N_2$	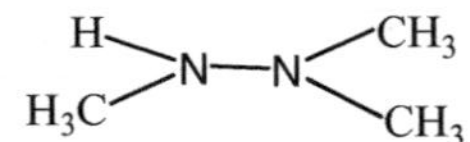	33.4 ± 0.1	292	2 × 3.86	4.30
1,1-Dimethylhydrozine	$C_2H_8N_2$	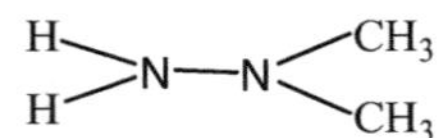	34.1	267–309	4 × 3.86	4.70
1,2-Dimethylhydrozine	$C_2H_8N_2$	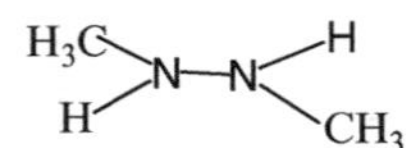	41.0	267–309	4 × 3.86	6.40
Methylhydrozine	CH_6N_2	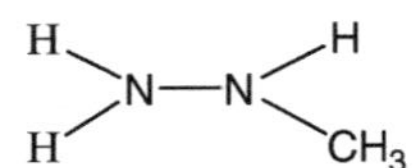	41.8	288	6 × 3.86	9.1
						D–N → CH_3–CH_2–N
Tetra-ethylenediamine	$C_2H_8N_2$	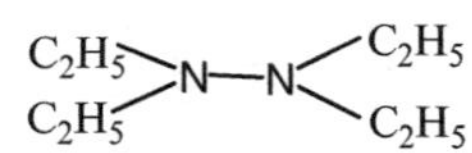	33.4	323	–	4.20[a]
Ethylenediamine	$C_2H_8N_2$	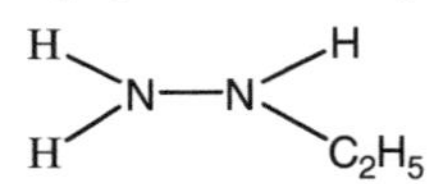	45.0 ± 0.1	298	6 × 3.86	10.7
Etylenediamine	$C_2H_8N_2$	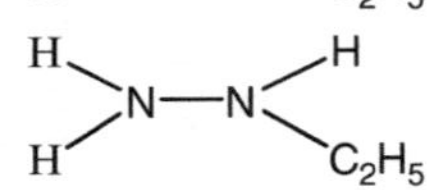	65.8[b]	263	6.34[b]	13.9[b]

2,2-diaminodiethylamino	$C_4H_{13}N_3$	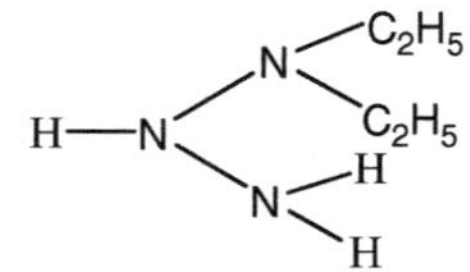	63.4 ± 0.7	298	4 × 3.86 2 × 5.9	9.0
Diethylenetriamine	$C_4H_{13}N_3$	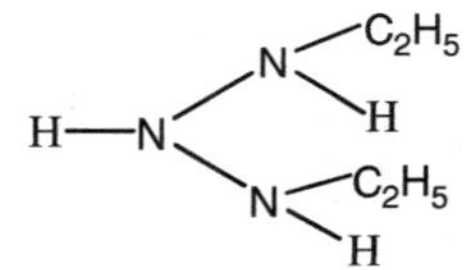	54.8	386	4 × 3.86 2 × 5.9	6.9[a]
Triethylenetriamine	$C_6H_{18}N_4$	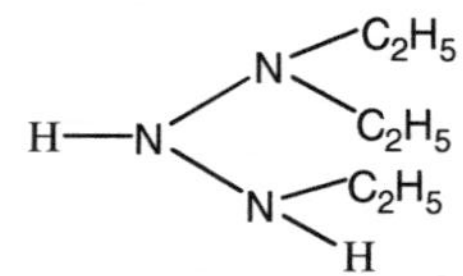	59.8	446	2 × 3.86 2 × 5.9	6.7[a]
Soled						
Ethylenediamine	$C_2H_8N_2$	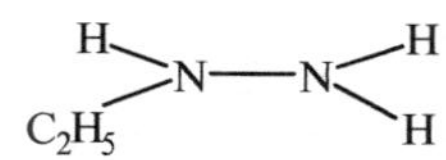	65.6	263	D–N•••H–N 6.34	13.8[a]

[a]T > 298 K
[b]Solid

Fig. 2.9 Schematic picture of liquid structure of 2,2-diaminodiethylamine

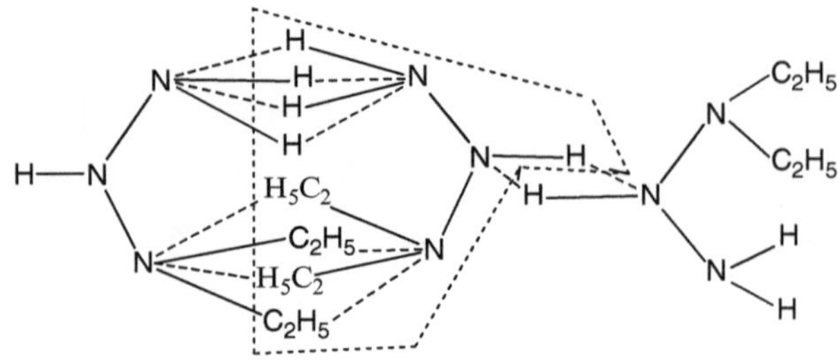

and fourth-coordinated nitrogen atoms with energies 3.86 and 5.90 kJ mol^{-1}, respectively. It follows that the specific interaction energy, formed by the ethyl fragment, is equal to

$$D\text{-}N \rightarrow CH_3\text{-}CH_2\text{-}N = (\Delta_{vap}H^0(298\ K) - (4D\text{-}N\cdots H\text{-}N + 2D\text{-}N\cdots H\text{-}N^{***}) / 4 \quad (2.15)$$

comprising the vaporization enthalpy and summary values of the hydrogen bond energies.

The high value of the energy of the specific interaction D–N → CH_3–CH_2–N with participation of the ethyl fragment points to the significant splitting of electric density from the carbon atom of the methyl group under the influence of the chain of the two nitrogen atoms. As a result, on one side, the carbon atom receives higher positive charge and, on the other, the nitrogen atom receives higher negative charge and the energy of the specific interaction D–N → CH_3–CH_2–N for liquid 2,2-diaminodiethylamino is greatly stabilized compared with the same type of interaction for liquid diethylamine (6.0 kJ mol^{-1}). The estimated energies of the specific interaction D–N → CH_3–CH_2–N for liquid diethylenetriamine and triethylenetriamine lead to the conclusion that the energies depend on the location of the ethyl groups in their molecules, reflecting the difference in charges at the nitrogen atoms of the ethyl fragments in the methyl groups.

The value of the sublimation enthalpy of ethylenediamine, available in the literature, the only one for the amines, provides an insight into the stabilization of the specific interaction D–N → CH_3–CH_2–N (6.0 kJ mol^{-1}, 298 K) under the influence of the crystal field.

The energy of this type of interaction is obtained with the help of Eq. 2.16,

$$D\text{-}N \rightarrow CH_3\text{-}CH_2\text{-}N = (\Delta_{sub}H°(298\ K) - 6D\text{-}N\cdots H\text{-}N)/2 \quad (2.16)$$

and found to be equal to 13.8 kJ mol^{-1}, 263 K, which value is increased twice with the temperature dependence compared to the standard conditions.

2.3 The Energies of Specific Interactions of Liquid Diazene, Hydrazone, and Azidoalkyles

Compounds with the nitro-group –N=N– and reduced number of ligands possess lower coordination ability and lower number of pentacoordinated carbon atoms of the specific interactions and hydrogen bonds with higher stability compared with amines. The simple compound HN=NH and its derivatives have four bond vacancies, two of which belong to nitrogen atoms and two to hydrogen atoms or, in the general case to the end methyl groups of the alkyl ligand. Compounds of this class form a chain structure of liquid and crystal condition (Fig. 2.10) with a network of the specific interactions. The formation by the azoamine molecule of the single type hydrogen bond and by the molecules azometane, diethyldiazene, and dipropyldiazene of the specific intermolecular interaction D=N → CH_3–N= or with an ethyl or a propyl fragment, respectively, directs attention to the fact that the general contribution of the energies of these types of interactions corresponds to the value of the enthalpy characteristics referred to the number of interactions. The presence of two types in the structure of liquid methyldiazene allow determination of the summary contribution of the energies of the D=N → CH_3–N= interactions from the difference of the enthalpy characteristics and energies of two H-bonds 2D–N•••H–N=

$$D{=}N{\rightarrow}CH_3\text{-}N = (\Delta_{vap}H°(298\ K) - 2D{=}N{\cdots}H\text{-}N{=})/2 \qquad (2.17)$$

In liquid *N*-methylbutyldiazene each of the fragments forms two specific interactions, D=N → CH_3–N= and D=N → CH_3–$(CH_2)_2$–N=. One of the CH_2-groups of the butyl ligand does not participate in the reverse dative bond, which is why the energy of one of the types of the specific interactions is determined from the difference of the vaporization enthalpy, and the contribution of the energies of the specific interactions and of the methylene group

$$D{=}N{\rightarrow}CH_3\text{-}CH_2\text{-}CH_2\text{-}N = (\Delta_{vap}H°(T) - 2D{=}N{\rightarrow}CH_3\text{-}N - DCH_2)/2 \qquad (2.18)$$

Each of the two ligands of the molecule dibutyldiazene contributes energy to the enthalpy characteristics of the ethylene group, which should be considered in the calculation of the energy of the specific interaction, equal to the energy of the same type of interaction, for the propyl ligand. The contribution of the energies of two isostructural methyl groups in diisopropyldiazene and four in di-*tert*-butyldiazene is equal to the difference of vaporization enthalpies of the latter and diethyldiazene with the same ethyl ligand.

$$D iso CH_3 = \left(\Delta_{vap}H°(298\ K)dtbda - \Delta_{vap}H°(298\ K)deda\right)/n \qquad (2.19)$$

The established values of energies of the hydrogen bond and specific interactions of symmetric nitro-compounds are given in Table 2.8. The energy of the specific interaction D–N → CH_3–N and the hydrogen bond of liquid methyldiazene can be

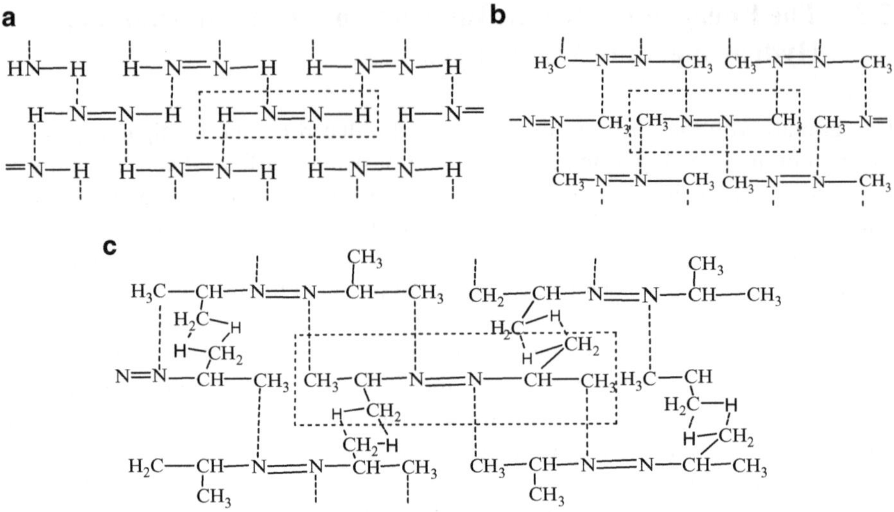

Fig. 2.10 Schematic picture of the liquid structure of azoamine (**a**), azometane (**b**), and diisopropyldiazene (**c**) with network of the specific interactions

determined using the approximation method of additive contribution and established values of corresponding interactions of liquid azoamine and azometane. In the second approach it is necessary to use Eq. 2.17 and the established value of the energy of one of the two types of bonds of liquid azoamine and azometane. Full compliance of energies of the specific interaction D=N → CH_3–N= and hydrogen bond DC–H•••N= independent of the chosen approach illustrates the rules of addictiveness and reflects the weak influence of the double bond —N=N— to the interaction of the molecules with different ligands. As a consequence, matching the energy of the specific interaction of the propyl group of butyl fragments of liquid *N*-methylbutyldiazene and the same group of dipropyldiazene (Table 2.8) is established. Introduced by two isostructural types, the quantity of energy required for the the vaporization enthalpy of diisopropyldiazene, which is equal to the difference of enthalpy characteristics with diethyldiazene, is 2.30 kJ mol^{-1} and for a one isostructural methyl group is 1.35 kJ mol^{-1}. The interaction of these groups with similar coordinating molecules doubles the number of the specific interactions of low stability, DH_3C → H–CH_2, for which the value of the energy is equal to 0.68 kJ mol^{-1} (Fig. 2.10c). The contribution of the energy of one of the isostructural methyl groups in di-*tert*-butyldiazene to the enthalpy of vaporization is 1.47 kJ mol^{-1} and the energy of the specific interaction of low stability for DH_3C → H–CH_2 is 0.73 kJ mol^{-1}. The established values of the energy of this specific interaction for diethyldiazene and di-*tert*-butyldiazene are located within the errors of experimentally established vaporization enthalpies of these compounds. The value of the energy of the specific interaction of the ethyl group, remains unchanged in liquid diisopropyldiazene, di-*tert*-butyldiazene, and diethyldiazene. The energy of the specific interaction D=N → CH_3–N= (6.6 kJ mol^{-1}) points to its higher stability in comparison with the specific interaction of

Table 2.8 The energies of the specific interactions and H-bonds (kJ mol^{-1}) of liquid derivative diamines

Compounds	Formula	The structure	$\Delta_{vap}H^\circ$ (298 K) [25]	T K	D=N•••H–N=	D=N → CH_3–N=	D=N → CH_3–$(CH_2)_n$–N=
Azoamine	N_2H_2	H—N=N—H	(28.8)	272	7.2	–	–
Methyldiazene	CH_4N_2	H_3C—N=N—H	27.5	272	7.2	n = 0 6.6	–
Azometane	$C_2H_6N_2$	H_3C—N=N—CH_3	26.4	258	–	n = 0 6.6	–
Diethyldiazene	$C_4H_{10}N_2$	CH_3—CH_2—N=N—CH_2—CH_3	(33.6)	298	–	–	n = 1 8.4
Dipropyldiazene	$C_6H_{14}N_2$	C_3H_7—N=N—C_3H_7	39.9	298	–	–	n = 2 10.0
Diisopropyldiazene	$C_6H_{14}N_2$	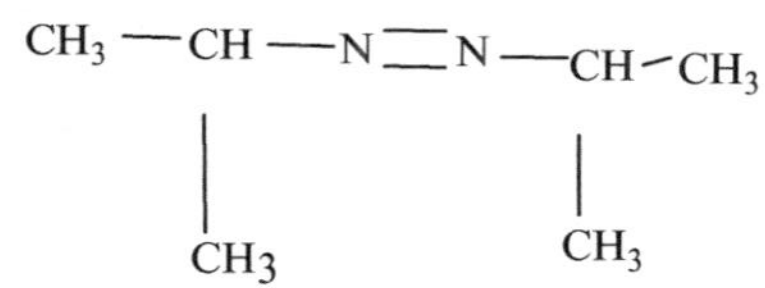	35.9 ± 0.4	298	–	Di*so*CH_3 = 1.15 × 2	n = 1 8.4
		CH_3—$(CH_2)_2$—CH_2—N=N—CH_3					
N-Methylbutyldiazene	$C_5H_{12}N_2$	CH_3—$(CH_2)_2$—CH_2—N=N—CH_3	36.4 ± 0.2	298	DCH_2 = 3.2	6.6	n = 1 10.0
Dibutyldiazene	$C_8H_{18}N_2$	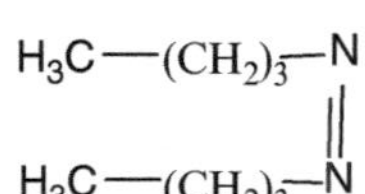	49.3 ± 0.2	298	DCH_2 = 4.7 × 2	–	n = 1 10.0
Di-*tert*-butyldiazene	$C_8H_{18}N_2$	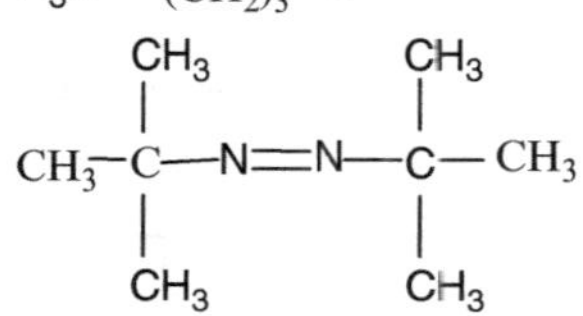	39.1 ± 0.3	298	–	Di*so*CH_3 = 1.47 × 4	n = 1 8.4
	$C_{16}H_{34}N_2$		66.5 ± 0.6	298	DCH_2 = 4.3 × 4	Di*so*CH_3 = 1.47 × 8	n = 2 10.0

(continued)

Table 2.8 (continued)

Compounds	Formula	The structure	$\Delta_{vap}H^{\circ}$ (298 K) [25]	T K	D=N•••H–N=	D=N → CH_3–N=	D=N → CH_3–$(CH_2)_n$–N=
bis(1,1,3,3-Tetramethylbutyl) diazene		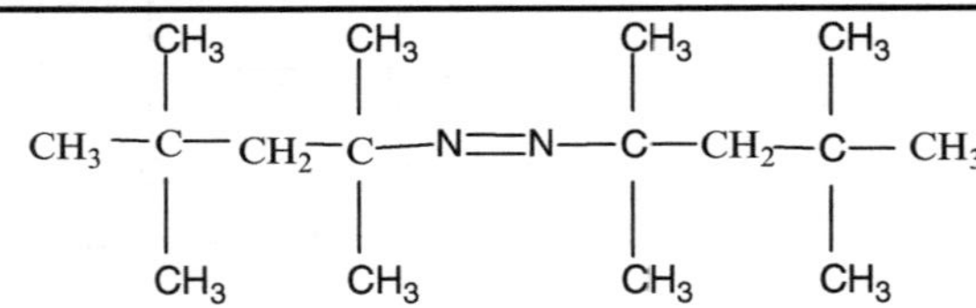					
					D–N•••H–N	DH_2C → CH_2	D=N → CH_3–$(CH_2)_n$–N=
Butylhydrazone butyraldehyde	$C_8H_{18}N_2$	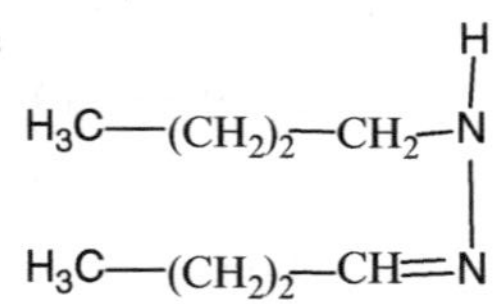	55.6	298–323	3.1	4.7 × 2	10.0 × 4
Isobutylhydrazone isobutyraldehyde	$C_8H_{18}N_2$	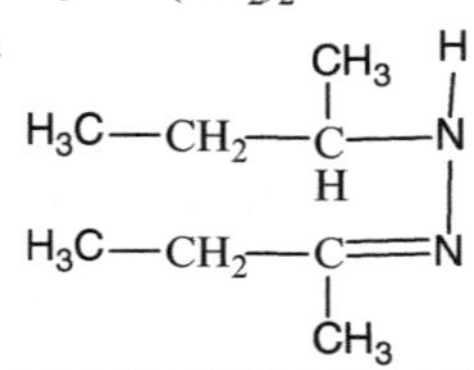	57.2	288–313	7.5	D*iso*CH_3 = 1.15 × 2	10.0 × 4

liquid dimethyl ether (5.63 kJ mol^{-1}), where the oxygen atom has a higher negative charge and higher donor properties. The energy of the specific interaction D–N → CH_3–N of nit-compounds and amino-derivatives are described by the range of stabilization:

$$\begin{aligned}&\text{Methylhydrozine}(9.10) > \text{Azometane}(6.6) \approx 1,2\text{-Dimethylhydrozine}(6.4)\\&> \text{Dimethyl ether}(5.63) < \text{Methylamine}(5.33) > \text{Dimethylamine}(4.80)\\&> \text{Trim ethylamine}\left(4.25\ \text{kJ mol}^{-1}\right)\end{aligned}$$

Increasing number of carbon atoms of the alkyl bond provide the decrease in influence of the reverse dative bond and stabilization of the specific interactions in the range of compounds

D=N→CH_3-N= (6.6) Methyldiazene = D=N→CH_3-N= (6.6) Azometane < D=N→CH_3-CH_2-N= (8.4) Diethyldiazene < D=N→CH_3-$(CH_2)_2$-N=(10.0) Dipropyldiazene = D=N→CH_3-$(CH_2)_2$-N= (10.0) N-Methylbutyldiazene = D=N→CH_3-$(CH_2)_2$-N= (10.0) Dibutyldiazene = D=N→CH_3-$(CH_2)_2$-N= (10.0 kJ•mol^{-1}) *bis*(1,1,3,3-Tetramethylbutyl)diazene

with peak value in the propyl ligand. The influence of the reverse dative bond in the fourth carbon atom of alkyl chain causes immutability of the energy of the specific interaction. Thus, subsequent increase in the vaporization enthalpy of the compounds of the gives series is determined by the contribution of the energy of an increasing number of CH_2-groups in the molecules. It allows using the methods of extrapolation and additive contribution of the energies for determination of enthalpy characteristics of unstudied compounds of the homologous series.

The conversion of the structure of a molecule of dibutyldiazene with double bond –N=N– with preservation of number of atoms of carbon and hydrogen in the butylhydrazone butyraldehyde molecule and changing the type of bonds –NH–N= and formation of fragments

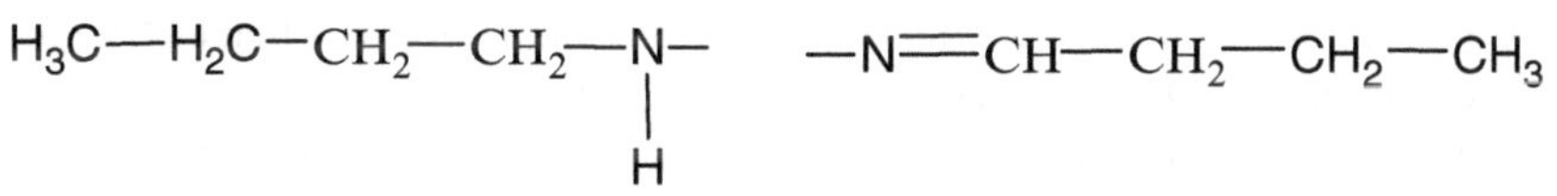

is accompanied by increase in the vaporization enthalpy of 6.1 kJ mol^{-1}, which contributes to the value of the energy, formed by the NH group. This is correct, because, first, replacement of a single bond by the double carbon atoms of a butyl fragment does not appear, in practice on the energy of the specific interaction D–N → CH_3–CH_2–CH_2–N– = DN → CH_3–CH_2–CH=N–; second, contribution of the excess methylene group of butyl ligand, which does not appear to experience the influence of the reverse dative bond, does not change. It means that the energy of the hydrogen bond formed by the NH group is necessary to define with the help of Eq. 2.18a, taking into account the contributions of the energies of the specific interactions and two methyl groups

$$DN{\bullet\bullet\bullet}H\text{-}N = (\Delta_{vap}H°(T)bdbd - 2DN \rightarrow CH_3\text{-}CH_2\text{-}CH{=}N - 2DN \rightarrow CH_3\text{-}CH_2\text{-}CH_2\text{-}N\text{-} - 2DCH_2)/2 \quad (2.18a)$$

The obtained value of the energy of the hydrogen bond (3.1 kJ mol^{-1}) is within the error of experimental value of the vaporization enthalpy of Butylhydrazone. It draws special attention to the fact of a significant increase of the vaporization enthalpy when the methylene groups are replaced in the molecule of butylhydrazone butyraldehyde compared to the isostructural methyl groups in the molecule of isobutylhydrazone. From that we can conclude that the influence of the isostructural methyl groups of the molecule isobutylhydrazone is manifested significantly on the offset of electric density in a fragment of the NH group, which in spite of a decrease of in the enthalpy of vaporization of Isobutylhydrazone takes place with an increase in 1.6 kJ mol^{-1} (Table 2.8). Taking the value of the contribution of the energy of two isostructural methyl groups, isobutylhydrazone as an equal amount of similar groups of diisopropyldiazene (2.3 kJ mol^{-1}), we calculated the energy of the hydrogen bond

$$D\text{-}N{\bullet\bullet\bullet}H\text{-}N = (\Delta_{vap}H°(T)bdbd - 2D\text{-}N \rightarrow CH_3\text{-}CH_2\text{-}CH{=}N - 2D\text{-}N \rightarrow CH_3\text{-}CH_2\text{-}CH_2\text{-}N\text{-} - 2D\mathit{iso}\,DCH_3)/2 \quad (2.18b)$$

with the help of Eq. 2.18b, the value of which is comparable to the energy of the hydrogen bond, formed by the imido group.

Molecules of methylhydrozine with two practicaly identical methylene and methyl ligands, forming specific interactions of almost the same stability,

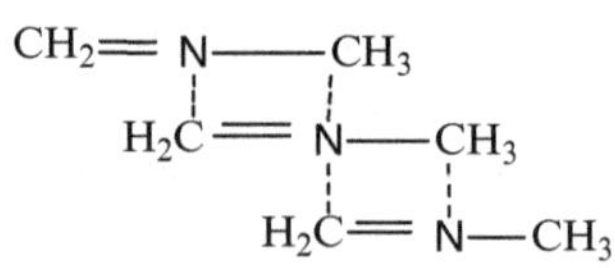

contribute similar values to the enthalpy characteristics. It follows that energy of the specific interaction D–N $\rightarrow$ CH_3–N = DN $\rightarrow$ CH_2=N should be determined by the division of enthalpy of vaporization among the number of formed interactions (Table 2.9). We obtain a similar value of the energy of the specific interaction D–N $\rightarrow$ CH_2–N taking the energy of the specific interaction D–N $\rightarrow$ CH_3–N of liquid methyldiazene and azometane (6.6 kJ mol^{-1}).

Molecules of propylhydrazone acetone and isopropylhydrazone acetone with acetone $(H_3C)_2C{=}$ fragment and two free bond vacancies of the essentially unshared $2s^2$-electron pair of the carbon atoms of the methyl groups behave like isopropyl fragment with isostructure CH_3-group at the lack of electron density of the donor's atoms. In this regard, molecule propylhydrazone acetone forms two specific interactions N $\rightarrow$ CH_3–CH_2–CH_2–N, N $\rightarrow$ CH_3–C=N, hydrogen bonds D=N•••H–N= and two specific interactions of low stability, formed by

Table 2.9 The energies of the specific interactions and H-bonds (kJ mol^{-1}) of liquid derivative hydrozines and hydrazone

Compounds	Formula	The structure	$\Delta_{vap}H^{\circ}$ (T) [25]	T K	D=N•••H–N=	D=N → CH_2–N=	D=N → CH_3–$(CH_2)_n$–N=
Methylhydrozine	C_2H_5N	CH_2=N-CH_3	26.4	288	–	6.6	n = 0 6.6
Propylhydrazone acetone	$C_6H_{14}N_2$	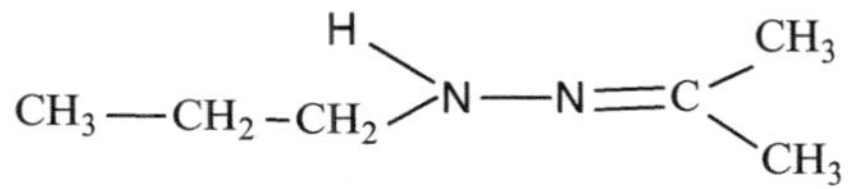	44.0	300	3.86	D*iso*CH_3 = 1.35	n = 2 10.0 n = 1 7.45
Isopropylhydrazone acetone	$C_6H_{14}N_2$	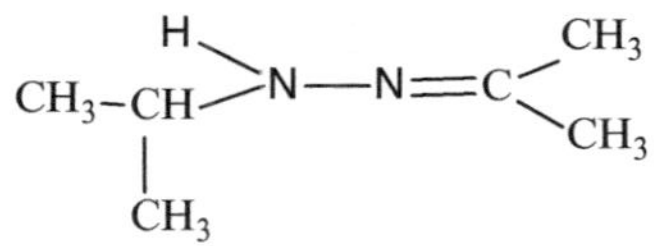	44.6	303	3.86	D*iso*CH_3 = 1.35 × 2	n = 1 8.65
N,N-Diethyl-2-(1-methylhydrazino) ethanamine	$C_7H_{18}N_3$	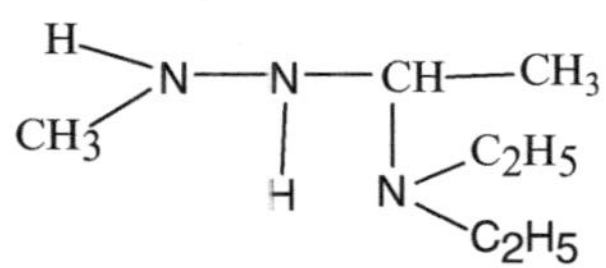	61.6	298	3.86	3.70	n = 1 5.80
Azidoethane	$C_2H_5N_3$	CH_3–CH=N–N≡N–H	31.5 28.9	303 268	2 × 5.90	4.92 4.30	n = 1 8.65
1-Azidopropane	$C_3H_7N_3$	CH_3–CH_2–CH=N–N≡N–H	31.1	268	2 × 5.90	–	n = 1.5 4.82
2-Azidopropane	$C_3H_7N_3$	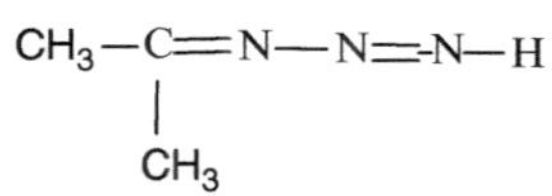	33.2	268	2 × 5.90	–	n = 1 5.35[a]

[a]Established for ketone's group 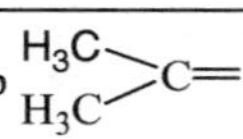

isostructural methyl group. The molecule of isopropylhydrazone acetone should be attributed to the compounds with two isostructural methyl groups and two ethyl ligands, which form four specific interactions =N → CH_3–C=N, two hydrogen bonds D=N•••H–N=, and four specific interactions of low stability. Both compounds form the structure of liquid condition with the network of the specific interactions (Fig. 2.11). The increased value of the vaporization enthalpy of isopropylhydrazone acetone (44.6) with the low number of the carbon atoms in the ethyl ligand compared with propylhydrazone acetone (44.0 kJ mol^{-1}) points to the *trans*-location of ethyl groups in the molecule of the isopropyl compound. The contribution of the energy of the isostructural groups to the enthalpy characteristics should be assumed to be the contribution of the same groups to the enthalpy characteristics of diisopropyldiazene and energies of the hydrogen bonds of liquid hydrozines (3.86 kJ mol^{-1}). The energy of the specific interaction of the propyl ligand should be assumed to be the contribution of the energy of the same fragment of dipropyldiazene (10.0 kJ mol^{-1}). It allows us to determine the energies of the specific interactions of the formed acetone and ethyl groups of liquid propylhydrazone acetone and isopropylhydrazone acetone with the help of Eqs. 2.20 and 2.21, respectively,

$$\text{D-N}\rightarrow\text{CH}_3\text{-C=N} = (\Delta_{vap}H°(298\ \text{K})\text{phz} - 2\text{D-N}\rightarrow\text{CH}_3\text{-CH}_2\text{-CH}_2\text{-N} - 2\text{D=N}\bullet\bullet\bullet\text{H-N=} - \text{D}iso\text{CH}_3)/2 \quad (2.20)$$

$$\text{D-N}\rightarrow\text{CH}_3\text{-CH}_2\text{-N} = (\Delta_{vap}H°(298\ \text{K})\text{iphz} - 2\text{D=N}\bullet\bullet\bullet\text{H-N=} - \text{D}iso\text{CH}_3)/4 \quad (2.21)$$

Presented in Table 2.9 are the results of the calculations of energies of the specific interactions described by the integral values of its stabilization by the series of compounds

D-N→CH_3-C=N: Propylhydrazone acetone (7.45) < Diisopropyldiazene (8.40) < Isopropylhydrazone acetone (8.65 kJ mol^{-1})

The molecule of *N,N*-diethyl-2-(1-methylhydrazino)ethanamine with two hydrogen atoms of the amino groups and one methyl group forms four hydrogen bonds and two specific interactions D–N → CH_3–N with energies 3.86 and 3.70 kJ mol^{-1}, respectively (Table 2.9). Specific interactions by three ethyl fragments are identified with the help of Eq. 2.22

$$\text{D-N}\rightarrow\text{CH}_3\text{-CH}_2\text{-N} = (\Delta_{vap}H°(298\ \text{K})\text{dehze} - 2\text{D-N}\rightarrow\text{CH}_3\text{-N} - 2\text{D-N}\bullet\bullet\bullet\text{H-N})/6 \quad (2.22)$$

As presented in Table 2.9, the value of the energy of the specific interaction D–N → CH_3–CH_2–N of these compounds (5.80 kJ mol^{-1}) is in good compliance with the one obtained for triethylamine (5.85 kJ mol^{-1}), which indicates the correctness of the approaches of thermodynamic calculations and the choice of

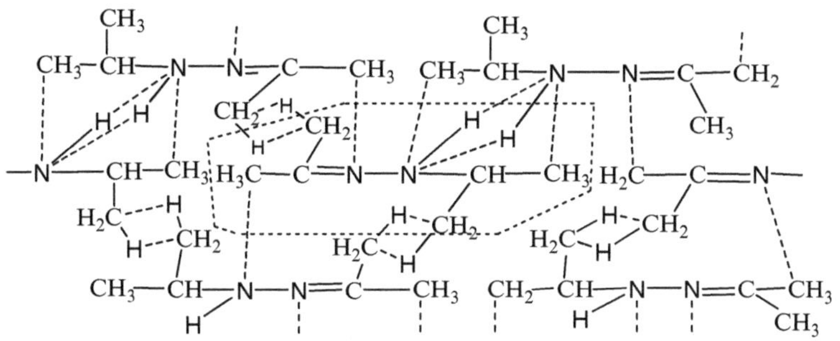

Fig. 2.11 Schematic picture of the liquid structure of isopropylhydrazone acetone

the values of energies of the specific interactions and hydrogen bonds, adequately reflecting the nature of the interactions.

Results of the implemented thermodynamic analysis and obtained energies of the interactions, formed by the methyl and ethyl fragments of hydrozines and hydrazones allow drawing the conclusion that acetone $\begin{smallmatrix}H_3C\\H_3C\end{smallmatrix}\!>\!C=$ fragment with one

D=N→CH_3-CH_2-N= (5.80) N,N-Diethyl-2-(1-methylhydrazino)ethanamine < D=N→CH_3-N= (6.6) Methylhydrozine < D=N→CH_3-CH_2-N= (7.45) Propylhydrazone acetone < D=N→CH_3-CH_2-N= (8.65 kJ mol^{-1}) *trans* Isopropylhydrazone acetone

methyl group with isostructural properties causes stabilization of the specific interaction. Molecules of azidoethane, 1-azidopropane, and 2-azidopropane with two double bonds with plate structure and six bond vacancies form two hydrogen bonds D–N•••H–N and four specific interactions and the structure of the liquid condition with the network of the specific interactions (Fig. 2.12). The presence of three nitrogen atoms in the molecule azidoethane leads to CH_3- and CH-groups and the independent participation in the formation of the specific interactions D–N → CH_3–N and D–N → CH–N. The carbon atom of the CH-group connected directly with the nitrogen atom differs by a positive charge from the nitrogen atom of the CH_3-group and, consequently, by the energy of the specific interaction. But because of the complexity in the determination of differences in the energies of these bonds one should make them equal and operate with average values. Participation of similar groups of propyl ligand of the molecule 1-azidopropane with three carbon atoms in the formation of the specific interaction allows one to consider its energies in the capacity of the average value of D–N → CH_3–CH_2–N, in which the "1.5 CH_2-group" participates fundamentally.

In the 2-azidopropane molecule the central carbon atom is connected with practically similar intermolecular interactions. The location of the nitrogen atoms and methyl groups in the same plane like ketenes causes, for these groups, as an analogue with acetone, the formation of the specific interaction N → CH_3–C–N, of similar stability. Thus, one has reason to conclude that the specific interactions of

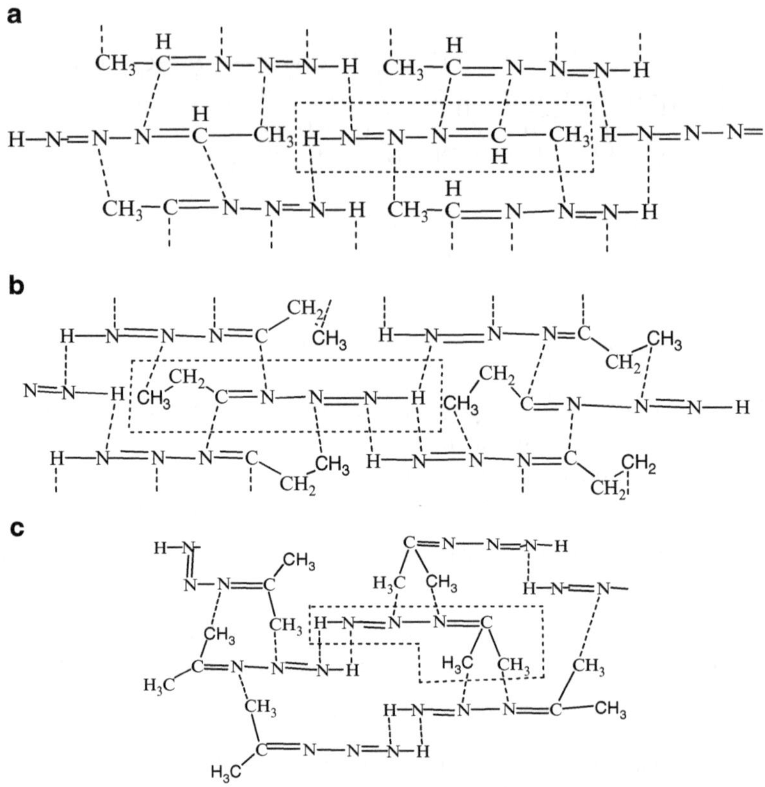

Fig. 2.12 Schematic picture of the liquid structure of azidoethane (**a**), 1-azidopropane (**b**), and 2-azidopropane (**c**) with network of the specific interactions and hydrogen bonds

the molecules azidoethane and 1-azidopropane on one side and 1-azidopropane on the other side form different series of the specific interactions.

The energies of the specific interactions formed by azidoethane,1-azidopropane and 2-azidopropane are determined with the help of Eqs. 2.23, 2.24, and 2.25, respectively,

$$D\text{-}N{\rightarrow}CH_3 = DN{\rightarrow}CH = (\Delta_{vap}H°(298\ K)azet - 2D\text{-}N{\bullet\bullet\bullet}H\text{-}N)/4 \quad (2.23)$$

$$D\text{-}N{\rightarrow}CH_3\text{-}CH_2 = DN{\rightarrow}CH\text{-}CH_2 = (\Delta_{vap}H°(298\ K)azp - 2D\text{-}N{\bullet\bullet\bullet}H\text{-}N)/4 \quad (2.24)$$

$$D\text{-}N{\rightarrow}CH_3\text{-}C{<} = = (\Delta_{vap}H°(298\ K)az2p - 2D\text{-}N{\bullet\bullet\bullet}H\text{-}N)/4 \quad (2.25)$$

The results of the calculations of the specific interaction energies (Table 2.9) point to the stabilization at the first two

D-N→CH_3-N Azidoethane (4.30) < D-N→CH_3-CH_2-N 1-Azidopropane (4.82) < D-N→CH_3-C< 2-Azidopropane (5.35 kJ mol^{-1})

compounds in connection with an increasing number of the carbon atoms in alkyl fragment and at 2-azidopropane with formation of more stable specific interactions of the ketone series by two similar fragments CH_3–C< with two carbon atoms. The established values of the energy of the specific interaction of azidoes and the estimated value of the energy of the hydrogen bond are a good combination and consistent with the research results by the method of RE-spectroscopy, and we have substantial reduction of the electron density at the nitrogen atom on the replacement of the hydrogen atom by the methyl group [9, 10].

2.4 The Energies of Hydrogen Bonds and Specific Interactions of Nitrides

2.4.1 Liquid Nitrides

The electronic structure of the CN group with the simplest compounds was analyzed in [2] and [33] on photo-electron spectra. From these studies it follows that availability of σ-π-pairing in aceton-nitryl $H_3C–C{\equiv}N$ leads to $\pi(-C{\equiv}N)$-n_N-splitting from 0.4 to 0.9 eV at HCN [33, 34]. In its turn, analysis of the qualitative difference of the vibrational structure of F-bands $1e(CH_3)$ and $2e(\pi)$-orbitals leads directly to an insignificant value of delocalization of two e-orbitals, which leads us to expect no high donor properties of the nitrogen atom and acceptor properties of the carbon atom, for the specific interaction ≡N → C≡ and ≡N → CH_3–N≡ and leading to a value of of 33.0 kJ mol^{-1} for the enthalpy of vaporization. At the same time cyanogen with two similar fragments N≡C in a molecule is characterized by a reduced value of the enthalpy of vaporization of 23.5 kJ mol^{-1}, established at a temperature of 267 K, i.e. at 30° lower than the standard one. It means that the molecule of the given compound with the specific interactions ≡N → C≡ should be less stable than liquid acetonnitryl. Cyanogen and acetylene molecules with triple bond and plane structure form analogue structures of liquid and crystal condition with a grid of the specific interactions (Fig. 2.13a) and four specific interactions ≡N → C≡ and ≡C → H, respectively, energies of which are determined from the enthalpy characteristics $\Delta H^0(T)/4$ (Table 2.10). The energies established for two different series of interactions of these compounds illustrate the excess quantities of energy of the specific interaction ≡N → C≡ compared with the energy of the ≡C → H–C≡ bond (4.18 kJ mol^{-1}) at 1.7 kJ mol^{-1}. Taking into account that the temperature of the established enthalpy of vaporization for acetylene at 210 K is lower by 57° of the enthalpy characteristics of cyanogen, it should be recognized that there is more difference in the stability of the considered specific interactions. Considering the nature of the temperature dependence of

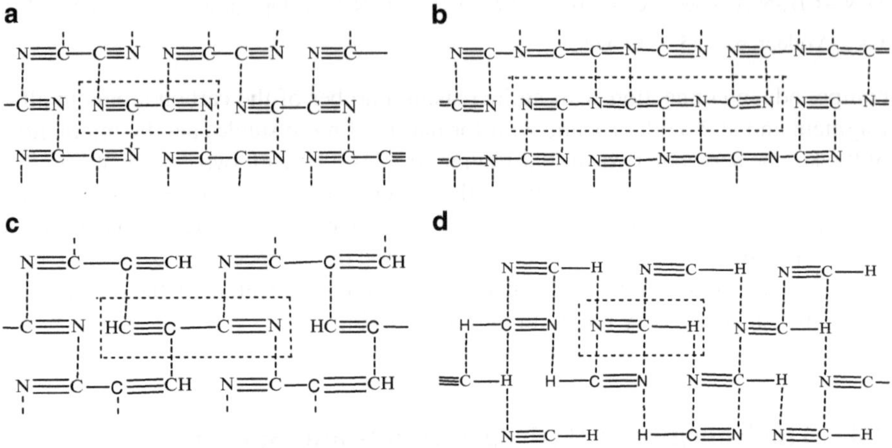

Fig. 2.13 Schematic picture of the liquid and solid structure of cyanogen (**a**), dicyanoacetylene (**b**), cyanoacetylene (**c**), and hydrogencyanide (**d**) with network of the specific interactions

energies at low value of the specific interaction ≡C → H–C≡ of liquid acetylene, one may suppose that its value at 298 K will be comparable with the energy of the hydrogen bond of liquid ammonia, 3.86 kJ mol^{-1}.

Dicyanoacetylene is of great interest because its molecule contains two –C≡N fragments (Fig. 2.13b, c), forming four specific interactions ≡N → C≡, and two fragments, =N=C=, connected by a double bond, also form four specific interactions =N → C=N=. Thus, the molecule with eight specific interactions leads to a reduced vaporization enthalpy (27.3 kJ mol^{-1}), compared to cyanogen with four specific interactions. It means that the presence in the molecule of two fragments =N=C= with three double bonds, separated from fragments N≡C– single bonds, fundamentally alters the redistribution of the electron density in a molecule. The result of this is a difference of charges between nitrogen atoms on one side and carbon atoms with triple and double bonds on the other side, and a small difference in donor and accept properties of the nitrogen atoms and carbon atoms. Such a difference of the properties mentioned create significant difficulties in experiment implementation at its definition and estimation of energie values of two different series of the specific interactions ≡N → C≡ and =N → C=N=. That is why we determined its average value from enthalpy of vaporization $\Delta_{vap}H^{\circ}$ (294)/8. The established value of the energy of the interaction D=N → C=N= (3.4 kJ mol^{-1}) is incommensurable in quantity with the value of the energy obtained of the specific interaction of liquid cyanogen (Table 2.10).

Replacement of the nitrogen atom in the central part of the molecule of dicyanoacetylene by the carbon atoms results in formation of two acetylene groups in the dicyanobutane molecule, leading to the increase of the enthalpy of vaporization up to 30.2 kJ mol^{-1}. Based on this, one should draw the conclusion that all carbon atoms with triple bonds contribute to little change in the redistribution of the electron density in the molecule dicyanobutane. Nevertheless, we see shifting of the

Table 2.10 The energies of the specific interactions and hydrogen bonds (kJ mol^{-1}) of simple cyanides

Compounds	Formula	The structure	$\Delta_{vap}H^{\circ}$ (T) [25]	T K	D≡N → C≡	D≡N → CH_3–(CH_2)n–N≡
Cyanogen	C_2N_2	N≡C—C≡N	23.5	267	5.9	–
Acetylene	C_2H_2	H—C≡C—H	16.7	210	–	D≡C → H–C = 4.18[a]
Dicyanoacetylene	C_4N_4	N≡C—N=C=C=N—C≡N	27.3	310	3.4	–
Dicyanobutadiene	C_6N_2	N≡C—C≡C—C≡C—C≡N	30.2	356	D≡N → C≡C–C≡ 7.55	–
Cyanoacetylene	C_3HN	N≡C—C≡CH	28.1	294	8.7	D–C≡N → CH≡C– = 5.35
Hydrogencyanide	HCN	H—C≡N	28.1	273–315	8.7	D≡C → H–C = 5.35
Deuterium cyanide	CDN	D—C≡N	27.6	265–393	8.45	D≡C → H = 5.35

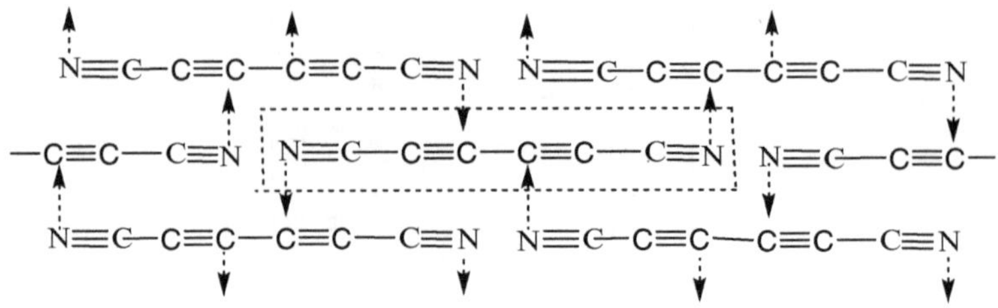

Fig. 2.14 Schematic picture of the liquid and crystalline structure of dicyanobutadiene with network of the specific interactions

electron density in the molecule from central carbon atoms along the chain to the nitrogen atom, which, in its turn, transfers

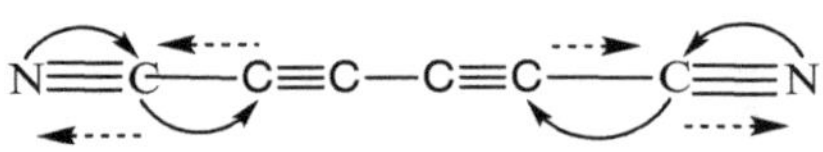

the part of the electron density with $3p_z$ orbitals to the contacting carbon atom and further to the second one, forming the reverse dative bond, leading to greater difference of charges at the nitrogen atoms and the third carbon atom of the chain. The essentially unshared $2s^2$ electron pair of the carbon atom participates in the formation of the specific intermolecular interaction D≡N → C≡C–C≡ and in the formation of the network structure of liquid and crystal condition of the considered compounds (Fig. 2.14).

The energies of four specific interactions of one series dicyanobutane are determined form the vaporization enthalpy $\Delta_{vap}H°(T)/4$. The quantity of interaction energy D≡N → C≡C–C≡ = 7.55 kJ mol^{-1} obtained at T = 356 K is lowered compared with the standard quantity. Nevertheless it greatly exceeds the value of the energy of the specific interaction D≡N → C≡ liquid cyanogen, established at 267 K.

The influence of the temperature in the narrow range on the stabilization of the hydrogen bonds and specific interactions with relatively low values is within experimental error. Thus, the equation of enthalpy characteristics of vaporization cyanoacetylene and hydrogencyanide at different temperatures 294 and 278 K are within the experimental error and, respectively, illustrate the practically similar contribution of the energies of the specific interaction formed by acetylene –C≡CH-group and the hydrogen bond D≡N•••H–C≡, formed by hydrogen atom H–C≡ group with the nitrogen atom –C≡N of the group of the molecule hydrogencyanide (Fig. 2.13d), to its enthalpy characteristics. It means that the energies of the given specific interaction and hydrogen bonds are equivalent, D–C≡N → CH≡C– = D≡N•••H–C≡ = 5.35 kJ mol^{-1}, and are within the error of the experimentally established values This value is different from the value of the energy of the hydrogen bond of liquid ammonia with four-coordinated condition

(5.9 kJ mol^{-1}) and reduced coordination of the nitrogen atom with the plane structure of the molecule. In Part 2.1 it was shown that the energy of the hydrogen bond of liquid ammonia with marked coordination of liquid amines exceeds the energy of the specific interaction, formed by methyl and CH-group, which is equal to 5.35 and the ethyl and propyl groups, of 6.0 and 7.8 kJ mol^{-1}, respectively.

Thus, taking into account the adequacy of energies of the specific interactions D≡N → CH≡ = D≡N•••H–C≡, DN → CH_3–N = DN → CH_3–C≡N and DN → CH_3–CH_2–N = DN → CH_3–CH_2–C≡N, there is a possibility to determine the value of the energies of the specific interaction D≡N → C≡ from the equations

$$D{\equiv}N{\to}C{\equiv} \; = (\Delta_{vap}H^0(298\ K)cac\ \text{-}2D\text{-}C{\equiv}N{\to}CH{\equiv}C\text{-})/2 \quad \text{(Cyanoacetylene)} \tag{2.26}$$

$$D{\equiv}N{\to}C{\equiv} \; = (\Delta_{vap}H^0(298\ K)hcad\text{-}2D\ {\equiv}N{\to}H\text{-}C{\equiv},)/2 \quad \text{(Hydrogencyanide)} \tag{2.26a}$$

$$D{\equiv}N{\to}C{\equiv} \; = (\Delta_{vap}H^0(298\ K)acnl - D\ {\equiv}N{\to}CH_3\text{-}C{\equiv}\ )/2 \quad \text{(Acetonitryle)} \tag{2.26b}$$

$$D{\equiv}N{\to}C{\equiv} \; = (\Delta_{vap}H^0(298\ K)prnl - D{\equiv}N\ {\equiv}N{\to}CH_3\text{-}CH_2\text{-}C{\equiv}\ )/2 \quad \text{(Propionitrile)} \tag{2.26c}$$

The value obtained for the energy of the specific interaction D≡N → C≡ (8.7 kJ mol^{-1}) for hydrogen cyanide has higher stability in comparison with the same series of interaction of liquid cyanogen. Results of the calculations of the energies for specific interactions given in Tables 2.10 and 2.11 are described by the integral value of stabilization,

$$\begin{aligned}&\text{Dicyanoacetylene }(3.4) < \text{Cyanogen }(5.9) < \text{Dicyanobutadiene }(7.55)\\ &\qquad < \text{Deuterium cyanide}\\ &(8.45) < \text{Hydrogencyanide }(8.7) = \text{Cyanoacetylene }(8.7\,\text{kJ mol}^{-1})\end{aligned}$$

on which basis one can conclude that the most stable interaction is formed by compounds with one cyanide group, –C≡N, in the molecule.

The thermodynamic properties of alkylnitriles allow for obtaining reliable values of the energies for specific interactions of this series of nitrogenous compounds, discovering regularities of change, adequately reflecting the nature of the interaction in liquid and crystal conditions. A description of the vaporization enthalpies of the given class of hydrogen cyanide – valeronitrile compounds of functional dependences $\Delta_{vap}H°(298\ K) = f(Cn)$ from the number of the carbon atoms in the chain by the two crossing lines (Fig. 2.15) is caused by the manifestation of two factors: First, the shifting of the electron density

Table 2.11 The energies of the specific interactions and H-bonds (kJ mol^{-1}) of liquids alkyl nitriles

Compounds	Formula	The structure	$\Delta_{vap}H°$(298 K) [25]	T K	D≡N → C≡	D≡N → CH_3–(CH_2)n–N≡
Hydrogencyanide	HCN	H—C≡N	28.1	273–315	8.7	D≡N•••H = 5.35
Acetonitryle	C_2H_3N	CH_3—C≡N	33.0	298	11.15	5.35; n = 0
Propionitrile	C_3H_5N	CH_3—CH_2—C≡N	35.3	298	11.7	n = 1 6.0
Butironitrile	C_4H_7N	CH_3—CH_2—CH_2—C≡N	39.2 ± 0.1	298	11.7	n = 2 6.0
Valeronitrile	C_5H_9N	CH_3—CH_2—CH_2—CH_2—C≡N	44.3	298	11.7	n = 2 ΣDCH_2 = 8.3
Hexanenitrile	$C_6H_{11}N$	CH_3—$(CH_2)_3$—CH_2—C≡N	49.6[a]	298	11.7	n = 3 14/.2
Heptilonitrile	$C_7H_{13}N$	CH_3—$(CH_2)_4$—CH_2—C≡N	54.8[a]	298	11.7	n = 4 D(CH_2) × 3 = 19.4
Octanenitrile	$C_8H_{15}N$	CH_3—$(CH_2)_5$—CH_2—C≡N	56.6 ± 0.3	298	11.7	–
Nonanenitrile	$C_9H_{17}N$	CH_3—$(CH_2)_6$—CH_2—C≡N	62.3[a]	298	11.7	–
Decanenitrile	$C_{10}H_{19}N$	CH_3—$(CH_2)_7$—CH_2—C≡N	66.8 ± 0.4	298	11.7	–
Undecanenitrile	$C_{11}H_{21}N$	CH_3—$(CH_2)_8$—CH_2—C≡N	71.1 ± 0.1	298	11.7	–
Lauronitrile	$C_{12}H_{23}N$	CH_3—$(CH_2)_9$—CH_2—C≡N	76.1	298	11.7	–
Tridecanonitrile	$C_{13}H_{25}N$	CH_3—$(CH_2)_{10}$—CH_2—C≡N	80.4[a]	298	11.7	–
Myristonitrile	$C_{14}H_{27}N$	CH_3—$(CH_2)_{11}$—CH_2—C≡N	85.3 ± 0.3	298	11.7	–
Pentadecanenitrile	$C_{15}H_{29}N$	CH_3—$(CH_2)_{12}$—CH_2—C≡N	89.5[a]	298	11.7	–
Palmironitrile	$C_{16}H_{31}N$	CH_3—$(CH_2)_{13}$—CH_2—C≡N	93.4[a]	298	11.7	–
Heptadecanonitrile	$C_{17}H_{33}N$	CH_3—$(CH_2)_{14}$—CH_2—C≡N	97.8[a]	298	11.7	–
Stearonitrile	$C_{18}H_{35}N$	CH_3—$(CH_2)_{15}$—CH_2—C≡N	103.5[a]	298	11.7	68.1

Isobutironitrile	C_4H_7N	$CH_3-CH(CH_3)-C\equiv N$	37.5	321	11.7 D*iso*CH_3 = 2.20	6.0 n = 1
2-Methylbutyronitrile	C_5H_9N	$CH_3-CH_2-CH(CH_3)-C\equiv N$	42.5 ± 0.3	298	11.7 D*iso*CH_3 = 3.30	6.0 n = 2 DCH_2 = 3.9
Pivalonitrile	C_5H_9N	$CH_3-C(CH_3)_2-C\equiv N$	37.8	298	11.7 D*iso*CH_3 = 1.25 × 2	6.0 n = 1

[a]Determined by inter- and extrapolation

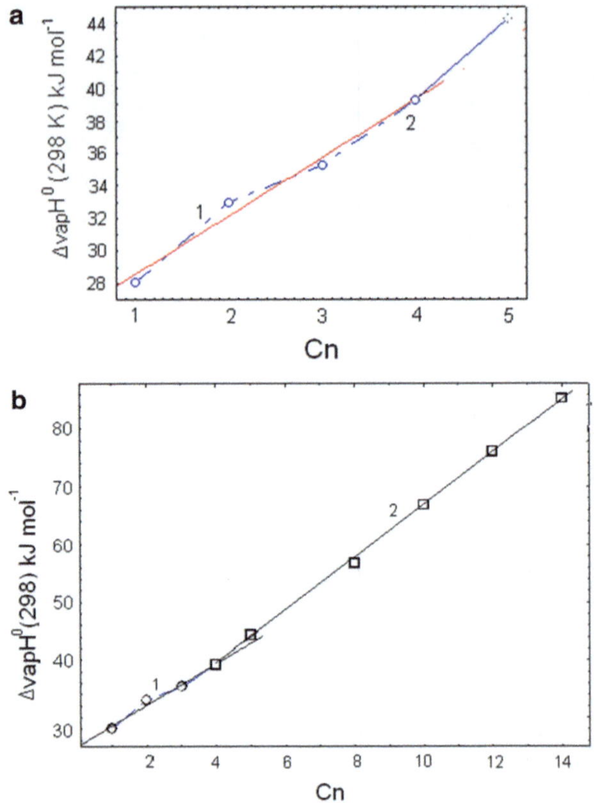

Fig. 2.15 Dependence of the vaporization enthalpy of alkylnitriles versus the number of the carbon atoms in alkyl chain: acetonitryle (**a**) – valeronitrile, acetonitryle (**b**) – myristonitrile (*1*) – (*2*)

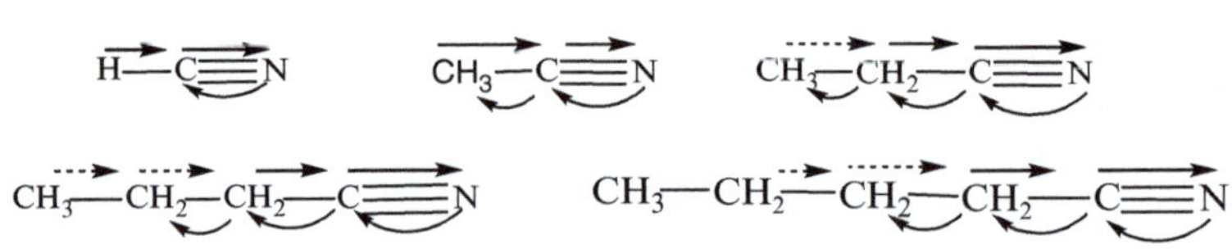

from the carbon atom to the nitrogen atom and the declining value of its shifting from the hydrogen atom of hydrogencyanide andfrom the methyl group of acetonitryle and with low shifting from the methyl group of propionitrile and even less at butironitrile, respectively. In the same sequence there occurs a decrease in the role of the reverse dative bond on the stabilization of the formed specific interaction, finally ending on the butyl fragment of butironitrile. Thus, the first part of the functional dependence $\Delta_{vap}H°(298\ K) = f(Cn)$ corresponds to the stabilization of the specific interactions formed.

$$D{\equiv}N\bullet\bullet\bullet H\text{-}C{\equiv} < D{\equiv}C{\rightarrow}CH_3\text{-}C{\equiv} < D{\equiv}C{\rightarrow}CH_3\text{-}CH_2\text{-}C{\equiv}\ .$$

The subsequent increase in the vaporization enthalpy of the compounds of the

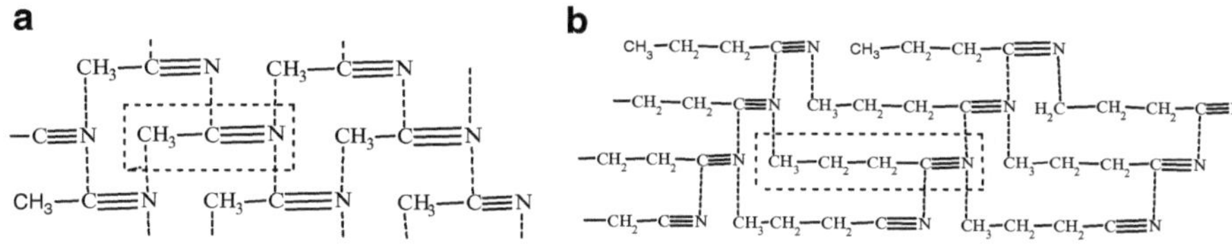

Fig. 2.16 Schematic picture of the liquid and solid structure of acetonitryle (**a**) and butironitrile (**b**) with network of the specific interactions

considered series in the range of butironitrile – myristonitrile is owing to the contribution of the energy of the increasing number of methylene groups of the alkyl chain (Fig. 2.15b). The steeper upward stroke of dependency $\Delta_{vap}H°$ (298 K) = f(Cn) is caused by the increased contribution of the energy of methylene group compared with manifestation of the reverse dative bond with increasing number of carbon atoms on the three compounds of the series.

Relocation of the hydrogen atom in hydrogen cyanide to the methyl group in acetonitryle or ethyl group is not accompanied by a change of the number of bond vacancies, forming the network structure of liquid and crystal (Fig. 2.16a, b). The number of coordinating molecules in the local environment is stable for all the compounds of the given homologous series. That is why the calculation of the energies of the specific interactions is carried out with the help of Eqs. 2.26a, 2.26b, and 2.26c (Table 2.11). The contribution of the energies of methylene group to the enthalpy characteristics is set by the difference of the vaporization enthalpy of nearby compounds or in the capacity of the average value of the range myristonitrile – valeronitrile. The results of the calculations (Table 2.11) point to the integral values of the changes of stability in the specific interaction D≡N → C≡

$$\text{Hydrogencyanide}\ (8.7) < \text{Acetonitryle}\ (11.15) < \text{Propionitrile}\ (11.7)$$
$$= \text{Butironitrile}\ \left(11.7\ \text{kJ mol}^{-1}\right)$$

Thus, our conclusion on the decreased stability of the specific interactions of alkylcianide on the whole is proved by the established values of energies of the interactions. The influence on the reverse dative bond at the third carbon atom of the propyl chain, excluding the carbon atom with triple bond, causes the immutability of the energy of the specific interaction at all following alkylnitriles. The location of the enthalpy characteristics of the compounds of the series butironitrile – myristonitrile at one plane dependence $\Delta_{vap}H°$(298 K) = f(Cn) with low error (Fig. 2.15) allows us to use with high accuracy the method of extrapolation and interpolation for determination of the enthalpy of vaporization of non-studied compounds (Table 2.11).

Alkylnitrides with isostructural group and the grid of the specific interactions (Fig. 2.17) and similar types of the specific interactions, in particular at compounds

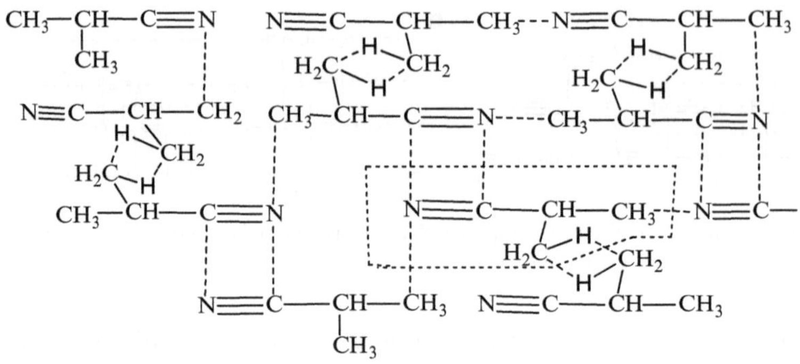

Fig. 2.17 Schematic picture of the liquid structure of isobutironitrile with the network of the specific interactions

with propyl carbon chain propylonitrile, isobutironitrile, and pivalenitrile, allow us to determine the contribution of the energy of the isostructural methyl groups to the enthalpy characteristics and energies of interactions of low stability, formed by this group from the differences of the vaporization enthalpy$\Delta_{vap}H°(298\ K)$ of isobutironitrile or pivalonitrile and propylonitrile, as well as $\Delta_{vap}H°(298\ K)$ 2-methylbutyronitrile and butironitrile, respectively.

$$DisoCH_3 = (\Delta_{vap}H°(298\ K)ibnl - \Delta_{vap}H°(298\ K)prn) \quad (2.27)$$

Participation in such interactions of two isostructural methyl groups of contact molecules, forming two specific interactions of low stability allows us to determine its energies.

$$DH_3C{\rightarrow}H\text{-}CH_2 = (\Delta_{vap}H°(298\ K) - 2\ D{\equiv}N{\rightarrow}C{\equiv}\ -\ 2D{\equiv}N{\rightarrow}CH_3\text{-}(CH_2)n\text{-}C)/2 \quad (2.28)$$

The above conclusion on the conservation of the energy of the specific interaction at the stable number of carbon atoms of the chain allows us to calculate the contribution of the energy of isostructural methyl group to the enthalpy characteristics with the help of the equations for the compounds with two or more isostructural methyl groups in the molecule.

$$DisoCH_3 = (\Delta_{vap}H°(298\ K) - 2D{\equiv}N{\rightarrow}C{\equiv} - 2D{\equiv}N{\rightarrow}CH_3\text{-}CH_2\text{-}C)/n \quad (2.29)$$

The established values of the energies of the contribution of isostructural methyl group to the enthalpy characteristics illustrate its increased dependence on itslocation in the alkyl chain.

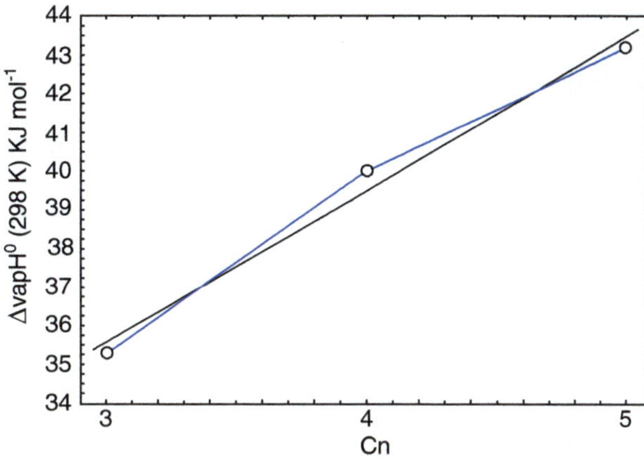

Fig. 2.18 Dependence of the vaporization enthalpy of alkenenitriles versus the number of the carbon atoms in alkenes chain of acrilonitrile – (Z) 2-pentenenitrile

$$\text{Pivalonitrile } (1.25) < \text{Isobutironitrile } (2.20) \ (321\text{ K}) < \text{2-Methylbutironitrile } \left(3.30\text{ kJ mol}^{-1}\right)$$

The calculated values of energies of the specific interactions of low stability, formed by isostructural methyl groups,

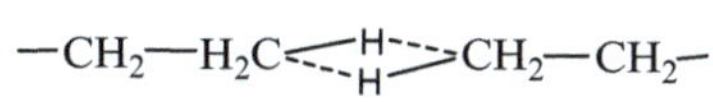

are described by the range of its stabilization with decreasing numbers of these groups in the molecule and the number of the carbon atoms of the chain at the permanent place of its location.

$$\text{Pivalonitrile } (0.62) < \text{Isobutironitrile } (1.10) < \text{2-Methylbutironitrile } \left(1.65\text{ kJ mol}^{-1}\right)$$

The discrepancy between the established enthalpy of vaporization of isobutironitrile and the standard conditions leads to some underestimation in the value of the energy of the given series of interactions.

As shown in Tables 2.10 and 2.11, replacement of the alkyl fragment of propylonitrile, butironitrile, and pentylonitrile with single bond by an alkene with a double bond is accompanied by change of the vaporization enthalpy of −0.2, +0.8, and −1.1 for Z-form and 0.5 kJ mol^{-1} for E-form, respectively. The absence in the literature of the directions for experimental error of cited values of the vaporization enthalpy complicates and often practically excludes the possibility of matching the considered enthalpy characteristics for clarifying the role of double and triple bonds for stabilization of the specific interactions. At the same time it has been shown above that displacement of the alkyl ligand by the alkene on the fragment with triple bond does not lead to a significant change of the enthalpy

Table 2.12 The energies of the specific interactions and H-bonds (kJ mol^{-1}) of liquids of unsaturated nitriles

Compounds	Formula	The structure	$\Delta_{vap}H°(T)$ [25]	T K	$D{\equiv}N \rightarrow C{\equiv}$	$D{\equiv}N \rightarrow CH_2{=}(CH)n\ C{\equiv}$
Acrilonitrile	C_3H_3N	$CH_2{=}CH{-}C{\equiv}N$	35.5	237[a]	11.7	6.05 n = 1
Methacrylonitrile	C_4H_5N	$H_2C{=}C(CH_3){-}C{\equiv}N$	36.5	288	11.7 D*iso*CH_3 = 1.0	6.05 n = 1
2-Ethylacrilonitrile	C_5H_7N	$H_2C{=}C(C_2H_5){-}C{\equiv}N$	37.1	259[a]	11.7[a] D*iso*C_2H_5 = 1.6	6.05 n = 1
2-Butenenitrile	C_4H_5N	$CH_3{-}CH{=}CH{-}C{\equiv}N$	40.0	298	11.7	6.05 n = 1
3-Butenenitrile	C_4H_5N	$CH_2{=}CH{-}CH_2{-}C{\equiv}N$	40.0	298	11.7	8.3 n = 2
2-Methylbutyronitrile	C_5H_5N	$CH_2{=}C{=}C(CH_3){-}C{\equiv}N$	42.5 ± 0.1	298	11.7 D*iso*CH_3 = 2.5	8.3 n = 2
E 2-Butenenitrile	C_4H_5N	$CH_3{-}CH{=}CH{-}C{\equiv}N$	38.9	298	11.7	7.8 n = 2
cis !-Cyano-1,3-butadiene	C_5H_5N	$CH_2{=}CH{-}CH{=}CH{-}C{\equiv}N$	40.7	348[a]	10.1[a]	8.3; n = 3 DCH_2 = 3.9
cis-crotonitrile	C_4H_5N	$H_2C{=}C(H){-}CH(H){-}C{\equiv}N$	37.1	312[a]	10.3	8.3; n = 2 DCH_2 = 3.9
trans-crotonitrile	C_4H_5N	$CH_2{=}C(H){-}CH(H){-}C{\equiv}N$	39.7	307[a]	11.5	8.3; n = 2 DCH_2 = 3.9
(Z) 2-Pentenenitrile	C_5H_7N	$CH_3{-}CH_2{-}CH{=}CH{-}C{\equiv}N$	43.2	298	11.4	8.3; n = 3 DCH_2 = 3.9
(E) 2-Pentenenitrile	C_5H_7N	$CH_3{-}CH_2{-}CH{=}CH{-}C{\equiv}N$	44.9	298	12.1	8.3; n = 3 DCH_2 = 4.2
(E) 3-Pentenenitrile	C_5H_7N	$CH_3{-}CH{=}CH{-}CH_2{-}C{\equiv}N$	44.8	298	12.0	8.3; n = 3 DCH_2 = 4.2

[a]T > 298 K

characteristics of compounds with the same number of the carbon atoms in the chain. However, the dependence of vaporization enthalpy on the number of the carbon atoms of the chain of unsaturated cyanides (Fig. 2.18) should be specified.

The functional dependence for analogous compounds with maximum carbon bond with the difference of these characteristics not exceeding 1.0 kJ mol^{-1} is shown in Fig. 2.15. It means that the contribution of the hydrogen atom to the enthalpy characteristics is equal to 0.5 kJ mol^{-1}. Taking into account that the energy of more stable specific interactions D≡N → C≡ is less subject to the fluctuation effects of the values due to experimental error, we take the value of the energy of given interactions to be unchanged on saturated and unsaturated acyclic cyanides.

Contribution of the CH_2- and CH-groups to the vaporization enthalpy in the absence of influence of the reverse dative bond on the specific interaction at the compounds with five-six carbon atoms in a chain is determined from the difference of the vaporization enthalpy with butenenitrile.

The results of the implemented thermodynamic calculations of energies of the specific interactions D≡N → CH_2=(CH)n–C≡ compounds with alkene fragment with the help of Eqs. 2.27, 2.28, and 2.29 practically coincide with the values of similar interactions of saturated alkylcyanides (Table 2.12).

The practical equality of D≡N → C≡ bonds at saturated alkylnitriles and alkenenitriles and specific interactions of the second type D≡N → CH_3–(CH_2) n–N≡, D≡N → CH_2=(CH–CH_2)n–C≡ are caused by the weak influence of saturated and unsaturated carbon-hydrogen fragments on the shifting of the electronic density. It means that replacement of the single –C–C– by the double –C=C– bond in the carbon-hydrogen chain is reflected to a small extent in the distribution of the electron density inside the molecule, which practically does not change the charge on the carbon atom of the terminal methyl group and practically leads to sustainability of the properties of the essentially unshared $2s^2$ of electron pair of the carbon atom. Unlike alkylnitriles and alkenenitriles the most significant difference in the energies of the specific interaction D≡N → C≡ of *cis*-cyano-1,3-butadiene and *cis*-crotonitrile (10.3) and *trans*-crotonitrile (11.5 kJ mol^{-1}) illustrate the influence of the location of the CH_2– and –C≡N group in the crotonitrile molecule.

Using the above procedure of calculation we identified the contribution of the energy of the isostructural methyl group by the methacrylonitrile and 2-methylbutyronitrile molecules to the vaporization enthalpy. We observed that the contrasting values (1.0 and 2.5 kJ mol^{-1}, respectively) of the energies of specific interactions should be attributed to the influence of two double bonds and an increase in the number of the carbon atoms of the chain in the 2-methylbutyronitrile molecule.

Fig. 2.19 Dependence of the vaporization enthalpy of dinitriles versus the number of the carbon atoms in alkyl chain cyanogen – glutaronitrile (*1*), glutaronitrile – sebaconitrile (*2*)

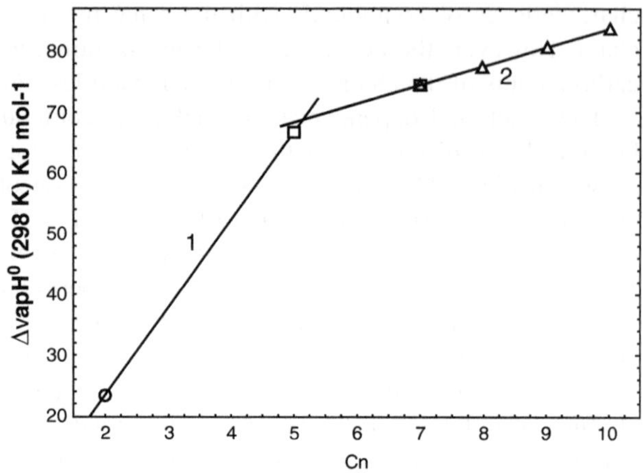

2.4.2 Dinitriles and Their Derivatives

Unlike alkylnitriles with an open molecule structure, forming two different types of specific interactions, dinitrile molecules with two end –C≡N groups form four specific interactions D≡N → C≡. The location of alkyl fragments between two nitrogen atoms in the diamine molecule is accompanied by equivalent shifting of the electron density from the central carbon atom of the chain dimethylmalononitrile or methylene group glutaronitrile, leading to negative charges of the nitrogen atom and positive charges of the central atom and carbon atoms of the nitride groups. A practically twofold increase in the vaporization enthalpy of dimethylmalononitrile (47.5 at 337 K) compared with the same characteristics of the originator of dinitriles cyanogen (23.5 at 267 K) illustrates the significant shifting of the electron density to the nitrogen atoms, leading to the formation of stable specific interactions with positive charge of the carbon atom of the same –C≡N group of the contacting molecule. The presence of two terminal nitryl groups in the molecules of dinitriles broadens the number of methylene,

We have groups of up to five, and the influence of the intermolecular reverse dative bond on these groups compared with alkylnitriles, appears in the shifting of the electron density to the nitrogen atoms in the chain. The correctness of the given presentation is confirmed by the location of the enthalpy of vaporization of compounds of the dinitriles series from the number of the carbon atoms on two crossing lines (Fig. 2.19). The integral values of the changes of the vaporization enthalpy of compounds with two nitryl groups, including cyanogen, malononitrile, butyldinitrile, and glutaronitrile, illustrates the decrease in influence of the reverse

Table 2.13 The energies of the specific interactions and H-bonds (kJ mol^{-1}) of liquids dinitriles

Compounds	Formula	The structure	$\Delta_{vap}H°$(298 K) [25]	T K	D*iso*CH_3	Σ D≡N → (CH_2)n–C≡	D≡N → C≡
Cyanogen	C_2N_2	N≡C—C≡N	23.5	267	–	–	5.9
Malononitrile (Propyldinitrile)	$C_3H_2N_2$	N≡C—CH_2—C≡N	37.7[a]		D≡N•••H–CH= 0.6 × 2	5.35 × 2 n = 1	6.45[a]
Dimethylmalononitrile	$C_5H_6N_2$	N≡C—C(CH_3)(H_3C)—C≡N	47.5	337	9.8:2 = 4.9 DH_3C → H–CH_2 = 2.45	n = 1 D≡N•••H–CH= 0.6 × 2 5.35 × 2	6.45[a]
Butyldinitrile	$C_4H_4N_2$	N≡C—$(CH_2)_2$—C≡N	52.4[a]		–	n = 2 5.35 × 4	7.70[a]
Glutaronitrile	$C_5H_6N_2$	N≡C—$(CH_2)_3$—C≡N	66.8	277–303	–	n = 2 5.35 × 2 + n = 2 6.0 × 2	11.0
					ΣDCH_2	22.7	
Hexanedinitrile	$C_6H_8N_2$	N≡C—$(CH_2)_4$—C≡N	70.0[a]	318	3.2	22.7	11.0
Adiponitrile	$C_6H_8N_2$	N≡C—$(CH_2)_4$—C≡N	58.7	367	–	22.7	9.0[b]
3-Hexenedinitrile	$C_6H_8N_2$	N≡C—CH_2—CH=CH—CH_2—C≡N	49.4	386	–	22.7	6.1[b]
1,5-Dicyanopentane	$C_7H_{10}N_2$	N≡C—$(CH_2)_5$—C≡N	74.5	318	7.7	22.7	11.0
Suberic acid dinitrile	$C_8H_{12}N_2$	N≡C—$(CH_2)_6$—C≡N	77.3	318	10.5	22.7	11.0
Azelaic acid dinitrile	$C_9H_{14}N_2$	N≡C—$(CH_2)_7$—C≡N	80.4	323	13.6	22.7	11.0
Sebaconitrile	$C_{10}H_{16}N_2$	N≡C—$(CH_2)_8$—C≡N	83.7	318	16.9	22.7	11.0
Nonanenitrile	$C_{11}H_{18}N_2$	N≡C—$(CH_2)_9$—C≡N	87.0[a]	318	20.2	22.7	11.0
Dodecanenitrile	$C_{12}H_{20}N_2$	N≡C—$(CH_2)_{10}$—C≡N	90.5[a]	318	23.9	22.7	11.0
2,2-Dicyanopropionitrile	$C_5H_3N_3$	CH_3—C(C≡N)(C≡N)—C≡N	55.2	293–333	5.0	–	8.4

[a]Determined by interpolation and extrapolation
[b]Elucidated at nonstandard conditions

dative bond for the stabilization of the specific interaction D≡N → C≡, formed by the nitryl groups. The steep turn of the given dependence $\Delta_{vap}H°(298\ K) = f(Cn)$ at this part (Fig. 2.19) compared with single bonds of saturated and non-saturated nitriles (Figs. 2.15 and 2.18) is caused by the display of the reverse dative bond of two nitrogen atoms of dinitriles. The location of the vaporization enthalpy of hexanedinitrile – sebaconitrile compounds on the second plane of the functional dependence $\Delta_{vap}H°(298\ K) = f(Cn)$ (Fig. 2.19) illustrates the sequential equal contribution of the energy by the increasing number of the methylene groups in the molecules of these compounds.

The functional dependence presented in Fig. 2.19 was used for the determination of the vaporization enthalpy of non-studied compounds of the glutaronitrile – dodecanenitrile series with allowance for error in the experimental research method and estimate of the vaporization enthalpy of malononitrile and butyldinitrile (Table 2.13).

The enthalpy characteristics of malononitrile obtained allowed us to identify the influence of the energy of two isostructural methyl groups on the vaporization enthalpy of dimethylmalononitrile from the difference of the vaporization enthalpy of

$$D\mathit{iso}CH_3 = \left(\Delta_{vap}H°(298\ K)dm.mn - \Delta_{vap}H°(298\ K)mn\right)/2 \qquad (2.30)$$

The value of the contribution of the energy of isostructural methyl group (4.9 kJ mol^{-1}) exceeds a similar contribution by the isostructural methyl group pivalonitrile (1.25 kJ mol^{-1}) by four times and the contribution of the same isobutironitrile group (2.20 kJ mol^{-1}) by two times. Such increased contribution of the isostructural methyl group of dimethylmalononitrile is consistent with the increased shifting of the electron density under the influence of two terminal nitrogen atoms and the experimentally obtained value of the vaporization enthalpy at 337 K, which was significantly lower compared with the valuc corresponding to standard conditions.

The molecule of malononitrile with four specific interactions, formed by two nitrile groups, additionally contains bond vacancy, shown clearly by the essentially unshared $2s^2$ electron pair of the carbon atom of the methylene group. The remaining unoccupied bond vacancy of the second nitrogen atom of the nitrile group forms a H-bond of low stability with the hydrogen atom, as in the CH_2-group. As a result the molecule of malononitrile takes the structure of liquid and crystal condition with the network of the specific interactions and two H-bonds D≡N•••H–CH of low stability and two specific interactions D≡N → CH_2–N≡ (Fig. 2.20a). The butyldinitrile molecule with two methylene groups forms in crystal and in liquid conditions the network of specific interactions including four interactions D≡N → C≡ and four less stable ones D≡N → CH_2–C≡ (Fig. 2.20b). Beacuse of the fact that the nitrogen atom is in three-coordinated condition when forming these bonds, we have every reason to accept that the energy of the specific interaction D≡N → CH_2–C≡ is equal to 5.35 kJ mol^{-1} and the energy of the

H-bond of low stability is comparable to the energy of the same series formed by the hydrogen atom of the CH_3 group 0.6 kJ mol^{-1} in liquid malononitrile. The glutaronitrile molecule with three CH_2 groups forms two specific interactions D≡N → CH_2–C≡ and two with longer carbon fragments D≡N → CH_2–CH_2–C≡ and energy 6.0 kJ mol^{-1} (Table 2.13, Fig. 2.20c). Thus, the energies of the specific interaction D≡N → C≡ of these compounds are determined by the equations:

$$\text{Malononitrile: D}\equiv\text{N}\rightarrow\text{C}\equiv\ = (\Delta_{vap}H°(298\text{ K})\text{mn.} - (2\text{D}\equiv\text{N}\rightarrow\text{CH}_2\text{-C}\equiv\ + 2\text{D}\equiv\text{N}\cdots\text{H-CH})/4 \quad (2.31)$$

$$\text{Butyldinitrile: D}\equiv\text{N}\rightarrow\text{C}\equiv\ = (\Delta_{vap}H°(298\text{ K})\text{bdn.} - 4\text{D}\equiv\text{N}\rightarrow\ \text{CH}_2\text{-CH}_2\text{-C}\equiv)/4 \quad (2.32)$$

$$\text{Glutaronitrile:D}\equiv\text{N}\rightarrow\text{C}\equiv\ = (\Delta_{vap}H°(298\text{ K})\text{gn.} - (2\text{D}\equiv\text{N}\rightarrow\ (\text{CH}_2)_3\text{-C}\equiv\ + 2\text{D}\equiv\text{N}\rightarrow\text{CH}_2\text{-CH}_2\text{-C}\equiv)/4 \quad (2.33)$$

Following the procedure of calculations, we define the energy of the specific interaction D≡N → C≡ of dinitrile (Table 2.13), whose stabilization naturally rises as the numbers of the methylene groups in alkyl fragments grow,

$$\text{Cyanogen (5.9)} < \text{Malononitrile (6.45)} < \text{Butyldinitrile (7.70)} < \text{Glutaronitrile}\left(11.0\text{ kJ mol}^{-1}\right)$$

leading to the decrease in the influence of the reverse dative bond, ending at glutaronitrile. The energies of the specific interactions of all the following compounds of this series remain unchanged and the vaporization enthalpies depend on the contribution of the energy and the number of the methylene groups. The low-energy differences of the specific interactions of alkylnitriles and dinitriles

$$\text{Propionitrile (11.7)} = \text{Valeronitrile (11.7)} < \text{Glutaronitrile}\left(11.0\text{ kJ mol}^{-1}\right)$$

point to the slight influence of wo nitrile groups located at the ends of the alkyl fragments in reducing stability of the specific interactions D≡N → C≡ compared with one given group in the alkylnitriles molecules. In this connection it is indicative that we have the molecules with three carbon atoms in a pivalonitrile chain with two isostructural methyl groups and with one nitryl group, dimethylmalononitrile with two isostructural methyl and two nitrile groups and 2,2-dicyanopropionitrile with three nitrile groups and one methyl group,

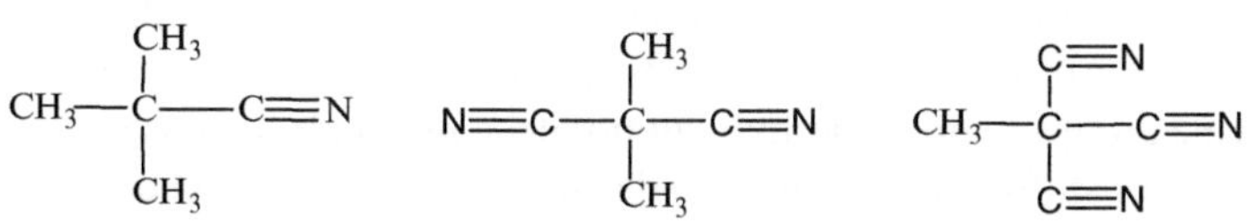

Pivalonitrile Dimethylmalononitrile 2,2-Dicyanopropionitrile

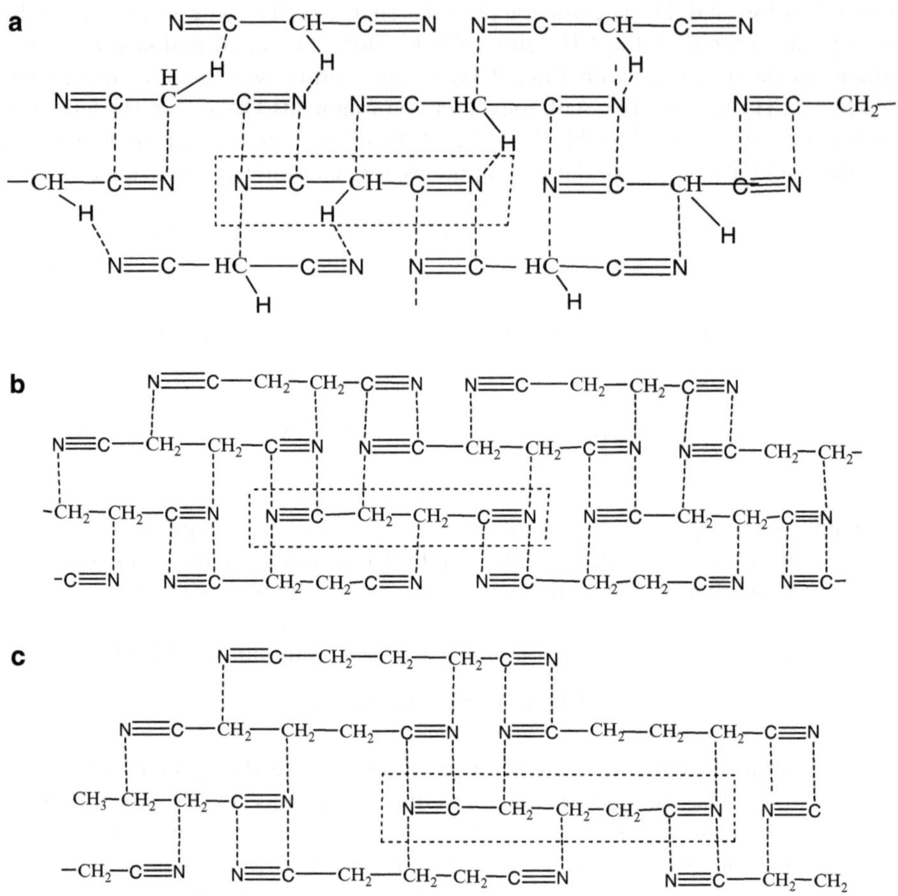

Fig. 2.20 Schematic picture of the liquid malononitrile (a), butyldinitrile (**b**), and glutaronitrile (**c**) with the network of the specific interactions

implementing the function of the isostructural group.

$$\text{Pivalonitrile}(11.7) > 2,2\text{-Dicyanopropionitrile}(8.4)$$
$$> \text{Dimethylmalononitrile}\left(6.45\ \text{kJ mol}^{-1}\right)$$

In the case of pivalonitrile (11.7 kJ mol^{-1}) in the open molecule form there is maximum stability of the specific interaction, caused by the presence of three carbon atoms in alkyl fragment, and, in this connection, the absence of the destabilizing effect of the reverse dative bond. The presence of two symmetrical terminal nitride groups in the dimethylmalononitrile molecule supposes more than a double decrease of shifting of the electron density compared with what is realized in the pivalonitrile molecule. At the same time two symmetrically located isostructural methyl groups at the single, central carbon atom do not shift their electron density. As a result, the dimethylmalononitrile molecule forms the less stable specific interactions of D≡N → C≡. In the case of 2,2-dicyanopropionitrile the location of the nitride group symmetric to the methyl group leads to the moving the electron density to the central carbon atom and further to the nitrogen atoms, which leads to an increased difference of charges. From this we can conclude the special role of isostructural methyl group in redistribution of the electron density in the molecule. Thermodynamic analysis of dinitriles and its derivatives leads to the conclusion that the location of the methylene group and ligands, with two and three between two nitrogen atoms of acceptors of electrons of the nitride groups, is accompanied by formation of the specific interactions with low stability D≡N → C≡ compared with the compounds formed with open alkyl chain by alkinitriles. Close to the equivalent values of energies of these series of the specific interactions of the two given compounds (11.0 and 11.7 kJ mol^{-1}, respectively) are reached at the five methylene groups chain, where every nitride group shifts the electron density practically onto three methylene chain groups.

2.4.3 (Dimethylamino)Alkyl Nitriles and (Diethylamino) Alkylnitriles

The molecules of (dimethylamino)alkylnitriles and (diethylamino)alkylnitriles have the structure of the molecules of trimethylamine and triethylamine with trigonal pyramid and equilateral triangle on the basis of a relocation of the hydrogen atom of the terminal methyl group to the fragment CH_2–C≡N. The given fragment creates definite steric density and creates some distortion in the structure of the molecules. The location of the methylene group between nitrogen atoms by the acceptors of electrons of amine and nitryl groups is accompanied by a not quite equivalent displacement of the electron density, leading to negative charges. The molecules of (dimethylamino)acetonitrile and (dimethylamino)propionitrile, and (diethylamino) acetonitrile, with replaced acetonitrile for propionitrile fragment contain eight bond vacancies, three of which belong to the nitrogen atom, two of them to methyl and ethyl groups of the amine fragment, one to the methylene group, and two bond vacancies belong to the nitrogen atom and carbon atom of nitrile –C≡N group. Eight specific interactions formed were of two types, 4D–N → CH_3–N or 4D–N → CH_3–CH_2–N and 2D–N → CH_2–C≡N with acetonitrile fragment or 2D–N → CH_2–CH_2–C≡N with propionitrile fragment forming a volume of mesh structure of crystal and liquid condition of these compounds (Fig. 2.21). Fragments of the molecule with volume structures are connected with each other by more stable interactions, formed by nitrile groups. Thus, the volume mesh structure contains a net of the specific interactions, formed by nitrogen atoms with the essentially unshared $2s^2$ electron pair of the carbon atoms in the fifth-coordinated condition. The energies of the specific interaction of liquid amonitriles, correspond to the energies of the interactions, realized in liquid trimethylamine or triethylamine and tripropilamine, and we have reason to accept the values are equal to 4.25, 5.85, and 7.9 kJ mol^{-1}, respectively (Part 2.1). The methylene CH_2 group, ethylene C_2H_4 group, connected with the nitride group, with the ligand function, formed by the specific interactions, is presented as D–N → CH_2– and D–N → CH_2–CH_2–. The values of the energies of these interactions are equal to the contribution of the energies of the specific interactions minus the energy contribution of 0.60 kJ mol^{-1} by the saturated hydrogen atom for the liquid condition of the compounds, and are equal to 3.65, 5.35, and 7.3 kJ mol^{-1}, respectively. Thus, the energies of the specific interactions formed by the nitride group D≡N → C≡ should be determined with the help of the equations:

(Dimethylamino)acetonitrile

$$D{\equiv}N{\rightarrow}C{\equiv} = (\Delta_{vap}H°(T)ma.n - 4D\text{-}N{\rightarrow}CH_3\text{-}N - 2D\text{-}N{\rightarrow} CH_2\text{-})/ 2 \quad (2.34)$$

(Dimethylamino)propionitrile

$$D{\equiv}N{\rightarrow}C{\equiv} = (\Delta_{vap}H°(T)ma.n - 4D\text{-}N{\rightarrow}CH_3\text{-}N - 2D\text{-}N{\rightarrow} CH_2\text{-}CH_2\text{-}) / 2 \quad (2.35)$$

(Diethylamino) acetonitrile

$$D{\equiv}N{\rightarrow}C{\equiv} = (\Delta_{vap}H°(T)ma.n - 4D\text{-}N{\rightarrow}CH_2\text{-}CH_2\text{-}N - 2D\text{-}N{\rightarrow} CH_2\text{-}) / 2 \quad (2.36)$$

3-(Diethylamino)propionitrile

$$D{\equiv}N{\rightarrow}C{\equiv} = (\Delta_{vap}H°(T)ma.n - 4D\text{-}N{\rightarrow}CH_2\text{-}CH_2\text{-}N - 2D\text{-}N{\rightarrow}CH_2\text{-}CH_2\text{-})/ 2 \quad (2.37)$$

2-(diethylamino)propionitrile, 2-diethylamino-2-methylpropionitrile and 2-(diethylamino)pentanenitrile

$$D{\equiv}N{\rightarrow}C{\equiv} = (\Delta_{vap}H°(T)ma.n - (4D\text{-}N{\rightarrow}CH_2\text{-}CH_2\text{-}N - D\text{-}N{\rightarrow}(CH_2)n\text{-}N) - qD\mathit{iso}CH_3)/ 2 \quad (2.36) \quad (2.38)$$

where n is 1, 2 or 3, and q the number of isostructural methyl groups.

The calculation of energies of the specific interactions D≡N → C≡ reflect low-governed stabilization at the compounds with similar errors

$$\text{(dimethylamino)propionitrile } (8.3) < \text{diethylamino) acetonitrile } (9.6)$$

$$< 3\text{-(diethylamino) Propionitrile}\left(9.9\,\text{kJ mol}^{-1}\right)$$

in the experimentally obtained enthalpies of vaporization at the replacement of methyl groups by the ethyl ones in the amino fragment and at the increasing number of the methylene groups in the alkyl chain, fringed by nitrogen atoms of an amino and a nitryl fragment at the increased influence of the first one.

Molecules of 2-(diethylamino)propionitrile and 2-(diethylamino)pentanenitrile formed by the isostructural methyl group of the propionitrile and pentanenitrile fragment, respectively, and additionally two specific interactions DH_3C → H–CH_2

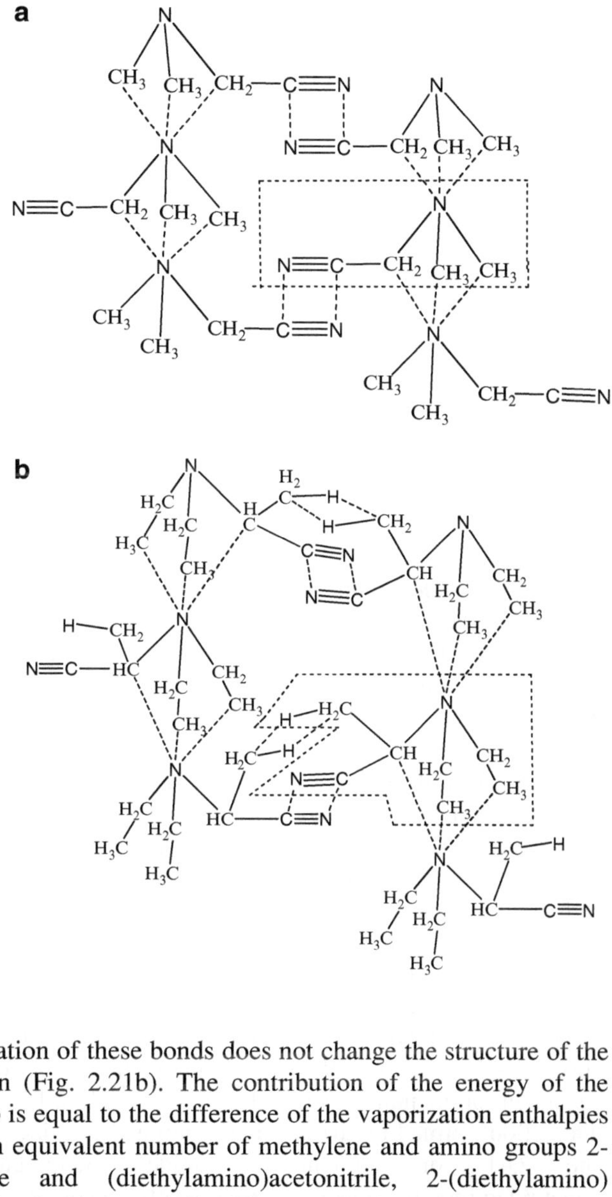

Fig. 2.21 Schematic picture of the liquid (dimethylamino)acetonitrile (a) and 2-(diethylamino) propionitrile (**b**)

of low stability. The formation of these bonds does not change the structure of the liquid or crystal condition (Fig. 2.21b). The contribution of the energy of the isostructural methyl group is equal to the difference of the vaporization enthalpies of the compounds with an equivalent number of methylene and amino groups 2-(diethylamino)propionitrile and (diethylamino)acetonitrile, 2-(diethylamino) pentanenitrile and (diethylamino)butanenitrile. The contribution of the energies of two isostructural methyl groups of 2-diethylamino-2-methylpropionitrile is determined from the difference with the vaporization enthalpy of (diethylamino) acetonitrile

$$2\text{Diso}_3 = \Delta H^\circ(298\ \text{K})\text{dea.mp} - \Delta H^\circ(298\ \text{K})\text{deac}. \qquad (2.39)$$

Presented in Table 2.14 are the results of calculations of the energies of the contribution of isostructural methyl groups pointing to the tendency of the stabilization of 2-(diethylamino)propionitrile (0.90) < 2-(diethylamino)pentanenitrile (1.30 kJ mol^{-1}) with increasing number of the methylene groups at the intermolecular carbon chain. The high contribution of the energy of two isostructural methyl groups of 6.4 kJ mol^{-1}, or for one methyl group of 3.2 kJ mol^{-1}, to the vaporization enthalpy 2-diethylamino-2-methylpropionitrile, located at the single methylene group, at the high value of the energy of the specific interaction D≡N → C≡ points to the fact that isostructural methyl groups prevent the shifting of the electron density to the carbon atom of the nitrile group. As a result, carbon and nitrogen atoms of the nitrile group keep the increased difference in charges, leading to the formation of a stable specific interaction.

The nitrogen atom in the ammonium cyanide molecule and its derivatives with four hydrogen atoms or at the definite combination of methyl groups and cyanide fragment is in a pentacoordinated condition. That is why the structure of the given type can be presented in the shape of a tetragonal pyramid, at which peak there is located a cyanide group. At the formation of the structure of liquid and crystal ammonium cyanide all four hydrogen atoms have the same propensity to the formation of the hydrogen bond. However, existing knowledge in coordination chemistry does not allow for the possibility of the eight-coordinated condition of the nitrogen atom. In a similar variant it is common to use the presentations of cooperative interaction. Nevertheless, available thermodynamic data allow making thermodynamical analysis and thermodynamical calculations of energies of the specific interactions formed by the cyanide group of ammonium cyanide and its derivatives. Shown in Fig. 2.22 is a schematic picture of the structure of liquid ammonium cyanide and methylammonium cyanide illustrating the formation of eight hydrogen bonds, whose energy at the eight-coordinated condition of the nitrogen atom could be accepted equal to the energy of the same bond of liquid ammonia with the coordination of six hydrogen atoms (3.86 kJ mol^{-1}). The replacement of the hydrogen atom in ammonia by the methyl group is accompanied by an insignificant decrease of the electron density at the nitrogen atom [9, 10]. This allows for the energy of the specific interaction to be formed by the methyl group dependent on its number in methylamine (Table 2.15). The energy of the specific interaction, formed by the cyanide group is determined with the help of Eqs. 2.40 and 2.41 for ammonium cyanide, methylammonium cyanide, and dimethylammonium cyanide, respectively, where n is the number of specific interactions

$$\text{D≡N→C≡} = (\Delta_{vap}H^\circ(298\ \text{K})\text{amc} - 8\text{D-N}\cdots\text{H-N})/2 \qquad (2.40)$$

Table 2.14 The energies of the specific interactions (kJ mol^{-1}) of aminenitriles

Compounds	Formula	The structure	$\Delta_{vap}H°(298\ K)$ [25]	T K	$D{\equiv}N \rightarrow (CH_2–)$ n	$D{\equiv}N \rightarrow CH_3–(CH_2)$ n–N	$\sum DisoCH_3$	$D{\equiv}N \rightarrow C{\equiv}$
(dimethylamino)acetonitrile	$C_4H_8N_2$	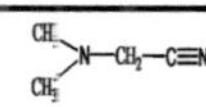	45.4	277–307	n = 1 3.65 × 2	n = 1 4.25 × 4	–	10.6
(dimethylamino) propionitrile	$C_5H_{10}N_2$	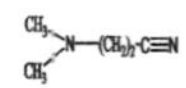	44.1 ± 0.2	290–317	n = 2 5.25 × 2	n = 1 4.25 × 4	–	8.3
(diethylamino) acetonitrile	$C_6H_{12}N_2$	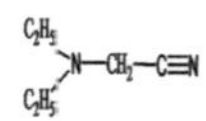	49.9 ± 0.3	283–318	n = 1 3.65 × 2	n = 2 5.85 × 4	–	9.6
3-(diethylamino) propionitrile	$C_7H_{14}N_2$	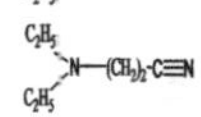	53.7	353	n = 2 5.25 × 2	n = 2 5.85 × 4	–	9.9
(diethylamino)butanenitrile	$C_9H_{18}N_2$	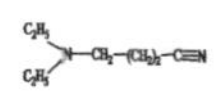	56.1 ± 0.4	298	n = 2 7.5 × 2	5.85 × 4 8.1 × 2	–	9.1
2-(diethylamino) Propionitrile	$C_7H_{14}N_2$	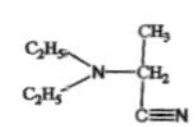	50.8 ± 0.3	278–315	n = 1 3.65 × 2	5.85 × 4	$DisoCH_3 = 0.90$ $DH_3C \rightarrow H–CH_2 = 0.45$	9.6
2-Diethylamino-2-methylpropionitrile	$C_8H_{16}N$	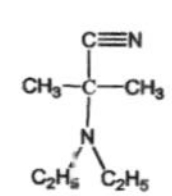	56.3 ± 0.2		n = 1 3.65 × 2	5.85 × 4	$DisoCH_3 = 6.4{:}2$ $DH_3C \rightarrow H–CH_2 = 1.6$	9.6
2-(diethylamino) pentanenitrile	$C_9H_{18}N_2$	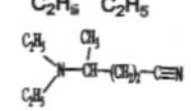	57.4 ± 0.4	298	n = 2 7.5 × 2	5.85 × 4	$DisoCH_3 = 1.30$ $DH_3C \rightarrow H–CH_2 = 0.65$	9.1

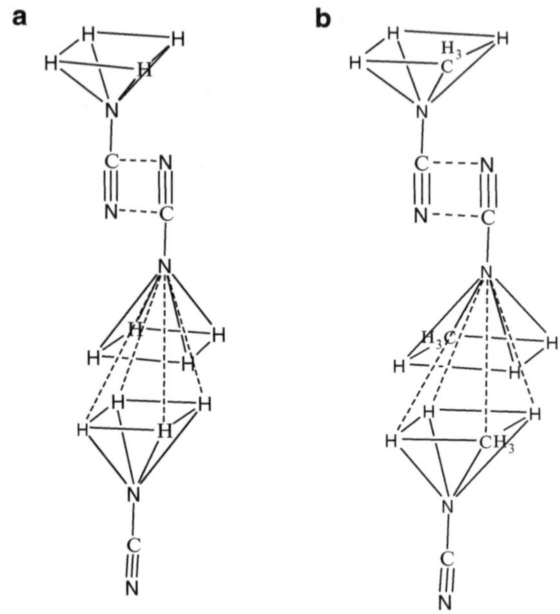

Fig. 2.22 Schematic picture of the liquid structure of ammoniumcyanide (**a**) and methylammonium cyanide (**b**)

$$D{\equiv}N{\rightarrow}C{\equiv} = (\Delta_{vap}H°(98\ K)mac - nD\text{-}N{\rightarrow}CH_3\text{-}N - (8\text{-}n)D\text{-}N{\bullet\bullet\bullet}H\text{-}N)/2 \quad (2.41)$$

The results of the calculations illustrate the logical stabilization of the specific interaction D≡N → C≡ at a number of compounds

$$\text{Ammoniumcyanide } (8.1) > \text{Methylammonium cyanide } (7.65) > \text{Dimethylammonium cyanide } \left(7.2\ \text{kJ mol}^{-1}\right),$$

caused by insignificant decreased shifting of the electron density at the replacement of the hydrogen ligand to the methyl group.

2.4.4 Crystalline Cyan Derivative Compounds

The great difference in the enthalpies of sublimation of the simple nitrogen-containing compounds (Table 2.16) reflects the significant influence of the crystal field on the stability of the specific interactions. Such high values of the sublimation enthalpies of the number of compounds and difference in the energies of different series of the specific interactions suggest the possibility of polymorphic changes in their crystals, caused by rupture of the least stable bonds in the process of heat exposure. An important feature of this class of compounds is the sharp increase of the sublimation enthalpies at the relocation of one hydrogen atom to the N≡C–

Table 2.15 The energies of the specific interactions and H-bonds (kJ mol^{-1}) of liquids ammonium cyanides

Compounds	Formula	The structure	$\Delta_{vap}H°(298\ K)$ [25]	T K	$D{\equiv}N \rightarrow CH_3{-}N$	$D{\equiv}N \rightarrow C_2H_5{-}N/DN{\cdots}H{-}N$	$D{\equiv}N \rightarrow C{\equiv}$
Ammonium cyanide	CH_4N_2	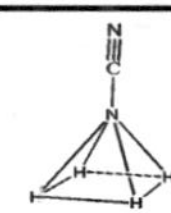	47.1	237	–	3.86 × 8	8.1
Methylammonium cyanide	$C_2H_6N_2$		49.1	280	5.33 × 2	3.86 × 6	7.65
Dimethylammonium cyanide	$C_3H_8N_2$	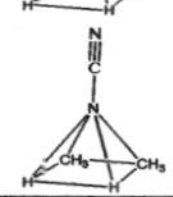	49.0	251–295	4.8 × 4	3.86 × 4	7.2

group (Table 2.16). It follows, that the appearance of the N≡C– group results in a significant change in the distribution of the electron density and charges on the hydrogen atoms and negative charge on the nitrogen atom and strengthening of the influence of the crystal field. The analogue of structures of liquid and solid condition of the specific interactions by acetylene and cyanogen molecules (Fig. 2.13) of one character type D≡C → H–C≡ and D≡C←N≡ interaction allows you to set values of energies directly from sublimation enthalpies $\Delta_{sub}H°(T)/4$ (Table 2.16). Using the above described procedure of estimation of the energy of the hydrogen bond formed in solid hydrogen cyanide, we took into account that the value of its energy should have increased the value compared with the value of the energy of solid acetylene D≡N•••H–C≡ > D≡C → H = 5.43 kJ mol^{-1} and corresponding to the energy of the hydrogen bond of solid ammonia D–N•••H–N at the same coordination of the nitrogen atom (6.34 kJ mol^{-1}). This is justified by the fact that the nitrogen atom possesses increased donor properties rather than the carbon atom with the essentially unshared $2s^2$ electron pair.

It was shown above that the contribution of the energy of the specific interaction D=HC → C≡, formed in liquid cyanoacetylene by the fragment H–C≡C–, corresponds to the similar value of the hydrogen bond in liquid ammonia. The procedure for the calculation of the energy of the specific interaction D≡N → C≡ cyanoacetylene and hydrogen cyanide was carried out with the help of Eqs. 2.26 and 2.26a. The results of the calculations are given in Table 2.16.

The energies of the mono-type specific interactions D≡N → C–C≡ and D≡N → C≡C–C≡ of crystal dicyanoacetylene and dicyanodiacetylene at the maximum number of the carbon atoms, for which the effect of the reverse dative connection does not appear in its stabilization, are determined from the enthalpy characteristics $\Delta_{sub}H°(298\ K)/4$. Discrepancy in the obtained energy values of the specific interactions from the standard conditions is difficult to establish from the patterns of its changes. Nevertheless, they provide a general view on the values of energies of the considered bonds and allow us to estimate its values at the standard conditions using the analogue of temperature dependence.

However, we note that one could not expect the decrease in enthalpy characteristics of dicyanoacetylene on 10 kJ mol^{-1} on increasing the temperature to 30°. Thus, it should be supposed that the difference of the carbon chain of dicyanodiacetylene and dicyanoacetylene on the –C≡C– fragment has little effect on the energy of the specific interactions formed D≡N → C≡C–C≡ and D≡N → C–C≡ of these compounds.

The molecules of tetracyanoethylene and tetracyanomethane with eight bond vacancies and the same number of the specific interactions in one series (Fig. 2.23a) allow us to determine the value of their energies by dividing the enthalpy of sublimation by the number of interactions. However, in this approach an error will be introduced because of the unaccounted contribution of the energy by the central carbon atom (2.5 kJ mol^{-1}) of the tetracyanomethane molecule and by two atoms of ethylene group of tetracyanoethylene. The structure of the molecules of these compounds, caused by the presence of the double =C=C< and triple -C≡C– bonds, are an important argument for the assumption of a planar network of their

Table 2.16 The energies of the specific interactions and hydrogen bonds (kJ mol^{-1}) of solid cyano and nitrile's derivatives

Compounds	Formula	The structure	$\Delta_{sub}H°$ (T) [35]	T K	D≡C → H	D–N•••H–N	D≡N → C≡
Acetylene	C_2H_2	H—C≡C—H	21.9	298	D≡C → H 5.43	–	–
Cyanogen	C_2N_2	N≡C—C≡N	34.4	223	–	–	8.3
Cyanoacetylene	C_3HN	HC≡C—C≡N	42.3	268	D≡C → H 5.43	–	15.7
Hydrogencyanide	CHN	H—C≡N	35.6	251	D≡N•••H–C≡ = 6.34	–	11.5
Dicyanodiacetylene	C_6N_2	N≡C–C≡C–C≡C–C≡N	34.4	302	–	–	D≡N → C≡C–C≡ 8.6
Dicyanoacetylene	C_4N_2	N≡C–C≡C–C≡N	44.3	268	–	–	D≡N → C–C≡ = 11.1
Tetracyanomethane	C_5N_2	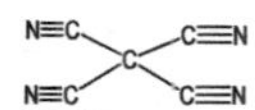	61.1 ± 8.0	298	DC ≈ 2.5	–	7.3
Tetracyanoethylene	C_6N_4	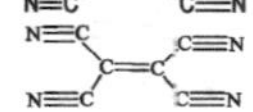	84.3	302	2DC ≈ 2.5 × 2	–	9.8
2,2-Dicyanopropionitrile	$C_5H_3N_3$	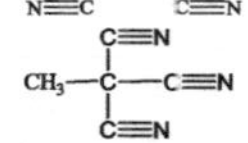	73.9 ± 0.5	293–333	D–C–CH_3– → C–CH_3 = 7.5	–	9.8
Malononitrile	$C_3H_2N_2$	N≡C—CH_2—C≡N	78.2 ± 1.0	298	1.0 × 2	6.0 × 2	16.0
Succinonitrile	$C_4H_4N_2$	N≡C–CH_2–CH_2–C≡N	70 ± 0.3	289	6.0 × 4	D≡N → CH_2–C≡N = 6.0 × 4	11.5
Fumaronitrile	$C_4H_2N_2$	N≡C–HC=CH–C≡N	68.6	298	5.65 × 4	D≡N → CH=C≡N = 5.65 × 4	11.5
Cyanamide	CH_2N_2	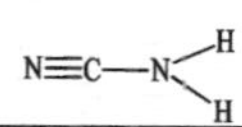	75.2	227–289	DN•••H–N = 9.7	–	18.2

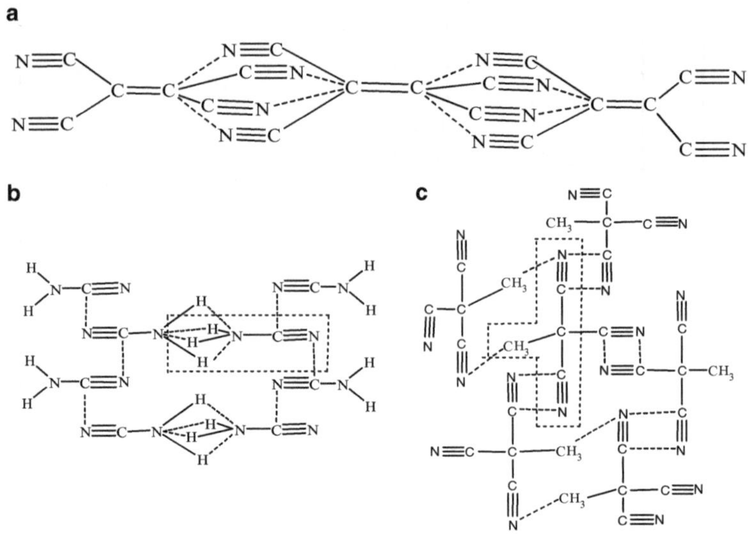

Fig. 2.23 Schematic picture of the liquid structure of tetraceanoethylene (**a**), cyanamide (**b**) with network of the specific interactions

crystal (Fig. 2.23a) formed by specific interactions. The results of the calculations of the energies of the specific interactions are given in Table 2.16.

The 2,2-dicyanopropionitrile molecule with three nitride groups and C–CH_3 fragment provide the structure of the liquid and crystal condition with the net of stable specific interactions. Formed in liquid acetonitryle by the given fragment, the specific interaction D≡N → CH_3–C contributes 5.35 kJ mol^{-1} to the vaporization enthalpy. Taking into account that the number of coordinating –C≡N groups in the molecule 2,2-dicyanopropionitrile takes on the average value compared with tetracyanomethane and tetracyanoethylene, and because of the increased stability of the methyl group to shift the electron density to the amino groups, one can suppose that nitrile groups 2,2-dicyanopropionitrile and tetracyanoethylene form specific interactions with a small difference in energy (9.8 kJ mol^{-1}). In this connection it makes sense to define the energy of the interaction D≡N → CH_3–C of crystal dicyanopropionitrile with the help of Eq. 2.42

$$D{\equiv}N{\rightarrow}CH_3\text{-}C = (\Delta_{sub}H°(T)\ dcp - 6\ D{\equiv}N{\rightarrow}\ \ C \equiv)/2. \tag{2.42}$$

The obtained value of the energy of the considered type of interaction in crystal 2,2-dicyanopropionitrile exceeds by 2.15 kJ mol^{-1} its value in liquid condition. Note that the energies of the specific interaction D≡N → C≡ of liquid and crystal hydrogen cyanide differ by 3.2 kJ mol^{-1}.

The molecule of cyanamide with six bond vacancies forms under the crystal condition four H-bonds of D–N•••H–N, corresponding to the values of the energy of the hydrogen bonds of solid ammonia with five coordination of the nitrogen atom (6.34 kJ mol^{-1}), and specific interaction D≡N → C≡ (Fig. 2.23b), which energy is determined with the help of Eq. 2.43

$$D{\equiv}N{\rightarrow}C{\equiv} = (\Delta_{sub}H°(298\ K) - 4D\text{-}N{\cdots}H\text{-}N)/2. \quad (2.43)$$

The presence of single bonds in amino fragments points to the formation of the volume structure. That is why the given class of nitrogen-containing compounds should have a three-chain structure of liquid and crystal condition with grid of the specific interactions.

The results of the thermodynamic analysis and the procedure for the calculations illustrate:

First, stabilization of the specific interaction D≡N → C≡ in the number of compounds

D≡N→C≡: Cyanogen (8.3) < Dicyanodiacetylene (8.6) < Dicyanoacetylene (11.1) < Hydrogencyanide (11.5) < Cyanoacetylene (15.7 kJ mol^{-1})

caused by the increasing capability for shifting the electron density to the carbon atom and further to the nitrogen atom of the nitrile group, accompanied by an increase in the number of the carbon atoms in a chain, following the decrease in the number of acetylene groups in the molecule dicyanoacetylene, replacement of this group on the hydrogen atom in hydrogen cyanide and on the terminal acetylene group,

Second, the tendency of stabilization at the replacement of double bonds with the single ones in

$$\text{Fumaronitrile } (9.9) \approx \text{Succinonitrile } \left(10.1\ \text{kJ mol}^{-1}\right)$$

Third, stabilization of the specific interactions in the compounds series

$$\text{Tetracyanomethane } (7.3) < \text{Dicyanopropionitrile}(9.8) \approx \text{Tetracyanoethylene } \left(9.8\ \text{kJ mol}^{-1}\right)$$

caused by the decrease in the number of coordinating cyanide groups in the molecule, and the appearance of an alkyl fragment with increased capability to shift the electron density.

As was shown above, the contribution of the energy of the methylene group, located "through bonds" between nitrogen atoms of electrons acceptors (Part 2), is equal to 5.35 kJ mol^{-1}. Taking into account that the contribution of the energy by the hydrogen atom of the given group in liquid and crystal condition made to malononitrile(propyldinitrile) and butyldinitrile is equal to 0.60 and 1.00 kJ mol^{-1}, respectively, we use the difference of these values for the estimation of the contribution of the energies of the methylene group with two hydrogen atoms

(6.00 kJ mol^{-1}) to the sublimation enthalpy. Using the above procedure for calculation of energies of the specific interactions D≡N → C≡ in crystalline malononitrile (propyldinitrile) and butyldinitrile (succinonitrile), we determined the energies of this type of interactions for these compounds (Table 2.16). The difference in the enthalpies of sublimation succinonitrile with two methylene groups and fumaronitrile with double bond –CH=CH– (1.40 kJ mol^{-1}) reflects the decreased energy of the specific interactions DN → CH= compared with the methylene group fragment $–CH_2–CH_2–$ with the limit bond. The value of the energy of the specific interaction formed by –CH= group of fragment with double bond was determined from the Eq. (2.44) to be equal to 5.65 kJ mol^{-1}.

$$DN{\rightarrow}CH == (\Delta_{sub}H°(298\ K)frn - 2D{\equiv}N{\rightarrow}{\equiv}C)/4 \tag{2.44}$$

Presented in Table 2.16 are the energies of the specific interactions DN → CH= of the considered crystal compounds,

$$\text{Malononitrile}\ (16.0) > \text{Succinonitrile}\ (11.5) = \text{Fumaronitrile}\ \left(11.5\ \text{kJ mol}^{-1}\right)$$

which are described by the number of its stabilization, which under the condition of correct experimentally measured values of the sublimation enthalpies of the compounds, points to the great difference between the charges of the carbon atoms and nitrogen atoms of the group –C≡N of malononitrile molecules, rather than succinonitrile and fumaronitrile, caused by less shifting of the electron density from the methylene group to the carbon atoms of the cyanide group compared with the given electron density from two methylene groups of the succinonitrile molecule and –CH=CH–group of the fumaronitrile molecule. The increased value of the energy of the specific interaction with participation of the methylene group

D-N→CH_2- : Malononitrile (6.00) = Succinonitrile (6.00) > DN→CH=: Fumaronitrile (5.65 kJ mol^{-1})

also reflects the increased shifting of the electron density compared with the CH= group.

References

1. Nefedov VI, Vovna VI (1989) Electronic structure of organic and elementorganic compound. Nauka, Moscow, 198 p
2. Nefedov VI (1987) Electronic structure of chemical compounds. Nauka, Moscow, 347 p
3. Vovna VI (1987) Bonded character and s-p-hybridization orbital's of hydride's molecules on data photo electronic spectroscopy. Russ J Struct Chem 28(5):166–168
4. Baev AK (1969) Problems of chemical nature of phase transformation. General and applied chemistry, vol 1. Vysheischaia Educating, Minsk, pp 197–206
5. Baev AK (1969) Phase condition and complex formation ability of galogenide metals. General and applied chemistry, vol 1. Vysheischaia Educating, Minsk, pp 207–218

6. Baev AK (1972) Complex formation ability halogenides of second – six groups periodical system. General and applied chemistry, vol 5. Vysheischaia Educating, Minsk, pp 35–51
7. Vovna VI, Vilesov FI (1974) Photoelectron specters and structure molecular orbital's methylamines. Opt Spectrosc 34(2):436–438
8. Takahashi M, Watanabe I, Ikeda S (1985) He(I) photoelectron-spectra of various alkylamines. J Electron Spectrosc 37(3):275–290
9. Karlson TA (1981) Photo electronic and OGE-spectroscopy. Mechanical Engineering, Leningrad. 431p
10. Nefedov VI (1984) X-ray spectroscopy of chemical compounds. Chemistry, Moscow, 255p
11. Baev AK (1996) Five-coordinated carbon atom at intermolecular interactions. J Coord Chem 22(5):399–402
12. Baev AK (1997) Insolvensy sp^3-hybridization and revers dative bond of alkyl compound and molecular complexes. Book of abstracts of international symposium "the research of Georg Wittig relevance to chemistry today", Heidelberg, p 21
13. Baev AK (2001) Pentacoordinated carbon atom in intermolecular interactions in solutions of alkyl compounds of non-transition elements. Abstract of sixth international symposium on carbanion chemistry, Marburg, p 75
14. Baev AK, Korolkov DV (2003) Phenomenon of dative bonds in elementorganic compounds with alkyl ligand. Abstract XVth FECHEM conference on organometallic chemistry, University of Zurich, p 350
15. Baev AK (2003) Novel approaches to the structure of functional solvents and elementorganic compounds with abstract international conference on solution chemistry, Debrecen, p 79
16. Baev AK (2004) Insolvence of model sp^3-hybridization of carbon atom and chemistry of non aqua solutions. In: IX international conference on the problem of solvation and complex formation in solutions, Plyos, pp 62–64
17. Baev AK (2004) Lack of co-ordination thermodynamic properties of elementorganic compounds with sp^3-hybridization of carbon atom and its insolence. Russian scientific symposium on thermo chemistry and calorimetry, Niznei Novgorod, pp 40–41
18. Baev AK (2002) Thermodynamics of solutions non-electrolytes with five- coordinated carbon atom at specific intermolecular interactions. Abstract XIV international conference on chemical thermodynamic, St. Petersburg, p 135
19. Baev AK (2009) Phenomenological significance of nonidentical sp3 huvridization model and a nature of interaction in chemistry of liquid non-electrolites and chemical thermodynamics. Abstracts. X international conference on physical and coordination porphyrin's chemistry and their analogous (ICPC-10), Ivanovo, p 86
20. Pauling L (1970) General chemistry. W.N. Freeman, San-Fransco, 846 p
21. Vilkov LV, Mastrukov VS, Sadova NI (1978) Determination of geometrical structure of free molecules. Chemistry, Leningrad
22. Baev AK, Leschenyc NS, Zuk A. Chemical problems of chemiluminescencec analazis of atmospheric and stratospheric ozone. Izvestia of higher education. Chem Chem Technol 53(3): 3–33
23. Baev AK (2012) Specific intermolecular interactions of organic compounds. Springer, Hiedelberg/Dordrecht/London/New York, 434p
24. Vasiliev IA, Petrov VM (1984) Thermodynamic properties of oxygen organic compounds. Chemistry, Leningrad, 238p
25. Chickens JS, Acree WE (2003) Enthalpies of sublimation of organic and organometallic compounds, 1910–2001. J Phys Chem Rev 32(2):519
26. Gleiter R (1979) Photoelectron spectra and bonding in small ring hydrocarbons. Top Curr Chem 86:197–285
27. Honegger E, Yang ZZ, Hellbronner E et al (1985) Lone-pair ionization energies of diazabicycloalkanes. J Electron Spectrosc 36:297–304
28. Lyc AE, Gaborakipov VZ, Shligina IA, Petropavlov VA (1986) Photoelectron specters of saturated nitrogen containing organic compounds. Russ J Struct Chem 27(4):146–169
29. Kirssinger M, Rademacher P (1979) Konformationsanalyse durch Photoelektronen-Spektroskopie. Angev Chem 91(11):885–896

30. Vovna VI (1974) Investigation of photo-ekectron spectroscopy phosphorated compounds. PhD thesis, Leningrad, 154 p
31. Vovna VI, Vilesov FI, Lopatin SI (1975) Photoelectron specters of hydrozines and series of the its alryl derivatides. Opt Spectrosc 38(2):259–262
32. Nelsen SF, Weisman GR, Clennan EL, Peacock VE (1976) Carbon-13 NMR studies of 9-methyl-9-azabicyclo[3.3.1]nonane and related compounds. J Am Chem Soc 98(22): 6893–6896
33. Dammel R, Bock H (1986) The ionization patterns of cyanomethanes H4-NC(CN)n and their assignment. J Electron Spectrosc 41(3/4):311–323
34. Asbrink L, Fridh C, Lindholm E (1979) Interpretation of electron-spectra. 1. Spectra of furan, pyrrole and cyclopentadiene, studied with ham-3. J Electron Spectrosc 16(3):65–76
35. Chickens JS, Acree WE (2002) Enthalpies of vaporization of organic and organometallic compounds, 1880–2002. J Phys Chem Rev 31(2):537

Chapter 3
Specific Intermolecular Interactions of Cyclic Aromatic Compounds with Nitrogen Atom in Their Functional Groups

3.1 Energies of Specific Interactions of Cyclic Compounds with Saturated and Unsaturated Hydrocarbon

The thermodynamic properties of processes of vaporization of cyclic compounds with saturated and unsaturated cycles and nitrogen atoms in the functional group, established at a standard temperature, are known for a limited number of compounds (Table 3.1). Nevertheless it allows us to find the role of the specific interactions $H_2C \rightarrow CH_2$, $H_2C \rightarrow H\text{–}CH$, $HC \rightarrow CH$ and $HC \rightarrow H\text{–}C$, formed by relevant groups CH_2, CH of the carbon-hydrogen cycles in the energy of the formed specific interaction $D{\equiv}N \rightarrow C{\equiv}$ by the cyanide group $\text{–}C{\equiv}N$ and the hydrogen bond of the amino group.

In spite of the extent of research implemented on the photo-electron spectra of amines, a correct classification of the vertical ionization potential (Iv) was obtained only for methylamine CH_3NH_2 [1]. For compounds with an active cycle $RN(CH_2)n$ the values of Iv significantly exceed the values for acyclic amine. In the case of the simple amine $HN(CH_2)_2$, the difference of these values is equal to 0.9 eV [2]. Existing stabilization of electrons in the molecule of this compound is explained by the increased contribution of the N2s-orbital [3]. However, this representation, according to Nefedov and Vovna [2], is not correct because at the similar regularity at cyclic O $(CH_2)_n$ the top n-orbital O2s do not contribute. Even more significant stabilization appears by the deep valence 1a'(N2s)-orbital, part of 1.3 eV in the molecule HN $(CH_2)_2$. In addition, we have hypothesised that the approximation of the canonical MO amine B3MO shows an antibonding effect, which destabilizes its N2s-H1s bridging [2]. In this connection it makes sense to talk about reverse dative bond, i.e., transmission of an electron from the p_z-orbital of the nitrogen atom to the 1s-orbital of the hydrogen atom or on the essentially unshared $2s^2$ electron pair of the carbon atom [4]. As a result, unusually high Iv in the molecule $HN(CH_2)_2$, rather than in $HN(CH_3)_2$, should be responsible for reduction of the electron density at hetero nitrogen atom and destabilizing interaction of n-σ-type. The mentioned factors should manifest for cyclic compounds with a small number of carbon atoms in a cycle

A.K. Baev, *Specific Intermolecular Interactions of Nitrogenated and Bioorganic Compounds*, DOI 10.1007/978-3-642-37472-2_3,

Table 3.1 Contribution of the energy (kJ mol^{-1}) of CH_2- and CH-groups to the vaporization enthalpy of the saturated and unsaturated cyclic compounds

Compounds	Formula	Structure	$\Delta_{vap}H°(T)$ [8, 9]	T K	DCH_2/ orDCH
Cyclopropane	C_3H_6	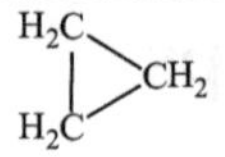	19.9	254	6.63
Cyclobutane	C_4H_8	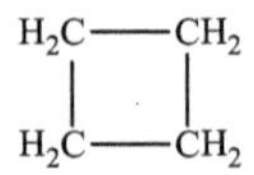	25.2	217–285	6.30
Cyclobutene	C_4H_6	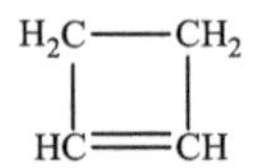	24.6	206–275	6.15
1,2-butadiene	C_4H_6	$H_3C–HC=C=CH_2$	23.9	298	5.97
1,3-butadiene	C_6H_6	$H_2C=HC–HC=CH_2$	23.0	285	5.75
Cyclopentane	C_5H_{10}	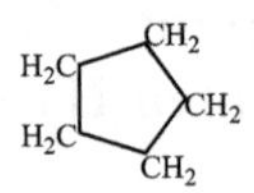	28.5 ± 0.1 29.2 29.5	298	5.70 5.84 5.90
Cyclohexane	C_6H_{12}	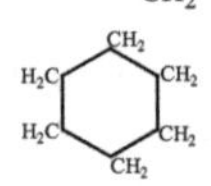	33.0 ± 0.1 33.3 ± 0.1	298	5.50 5.55
Cycloheptane	C_7H_{14}	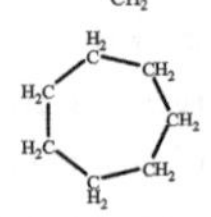	38.5 ± 0.2	298	5.50
Cyclooctane	C_8H_{16}	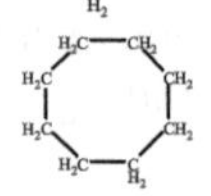	43.3 ± 0.2	298	5.41
Cyclooctene	C_8H_{14}	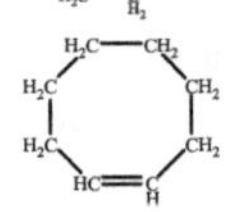	41.6	300	5.20
Cyclopentene	C_5H_6	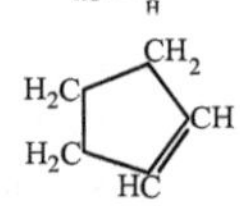	28.9	299	5.78
Cyclohexene	C_6H_{10}	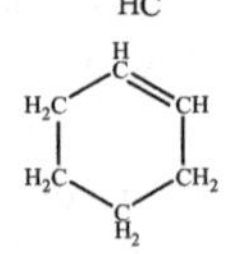	33.5 ± 0.5 33.7	298	5.58 5.62
1,3-Cyclohexadiene	C_6H_8	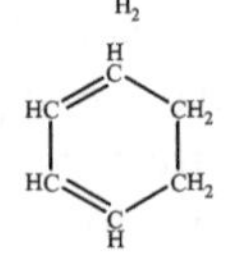	32.6	322	5.45
1,4-Cyclohexadiene	C_6H_8	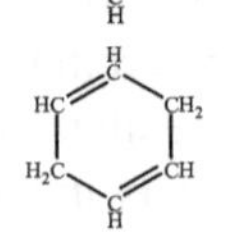	33.9	313	5.65
1,3,6-*cis* hexaditriene	C_6H_8	$H_2C=HC–HC=HC–HC=CH_2$	33.3	314	5.55
2,4-hexadiene	C_6H_{10}	$H_3C–HC=HC–HC=HC–CH_3$	33.2	308	5.53

(continued)

Table 3.1 (continued)

Compounds	Formula	Structure	$\Delta_{vap}H°(T)$ [8, 9]	T K	DCH_2/ orDCH
1-Hexyne	C_6H_{10}	$H_3C{-}(H_2C)_3{-}C{\equiv}CH$	33.4	280	5.57
Benzene	C_6H_6	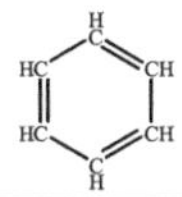	33.8 ± 0.1	298	5.65

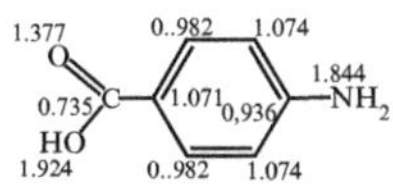

Fig. 3.1 Charges distribution on atoms of molecule of aminobenzoic acid [5]

with functional –C≡N, amine, isostructural CH_3, or another group at the energies of the specific interactions formed, in particular, D≡N → C≡.

Cyclic hydrocarbons should form liquid and crystal structures like benzene. Each molecule has two molecules close in the environment that interact with formation of the specific interactions H_2–C → H–CH and H–C → H–C, the number of which does not exceed the number of CH_2- or CH-groups in cycle. The essence of such interaction is as follows: cycle carbon atoms have no equivalent charges and as a consequence, hydrogen atoms connecting with them also differ in the positive charges [4]. This idea is based on other properties with the cycle and linear structure of molecules of saturated and unsaturated compounds. It also follows from the distribution of charges at the atoms of complex molecule aminobenzoic acid (Fig. 3.1).

The difference of charges at atoms reflects the disparity of the electron density of the essentially unshared $2s^2$-electron pair of the carbon atom, providing those donor or acceptor properties that are more often expressed and, consequently, the capacity to participate in the formation of the specific intermolecular interaction. This interaction is realized at the location of molecules that are close in the crystal and liquid environments so that carbon atoms of groups CH_2, CH, or =C= molecules have had the greatest difference of charges. Depending on the number of cycle carbon atoms, for example, cyclobutane, such a condition is reached by rotating molecules not more than 90°.

In thermodynamic calculations of the energies of the specific interactions formed by nitrogenous groups, one requires knowledge of the contribution of the energy by cycle to the enthalpy characteristics and energy of the substituted hydrogen atoms to the functional groups. In this connection it is necessary to establish the contribution of the energy by the cycles and hydrogen atoms of the CH_2, CH groups that are participating in the formation of the specific interactions H_2C → H–CH, and H–C → H–C with low stability of the enthalpy characteristics. A correlation between weak specific interactions and H-bonds, formed by hydrogen atoms of H–CH– and H–C-groups, in particular, DC=O•••H–C– is confirmed by a small difference in vaporization enthalpies and energies of the hydrogen bonds of saturated and unsaturated acids and spirits [6]. A similar conclusion follows from the analysis of the contribution of the CH_2 groups of the initial members of a homologous series of cycloalkanes to the vaporization enthalpy (Table 3.1). The difference in the conditions of experimentally measured vaporization enthalpies of these compounds are not compensated for by the contribution of its temperature dependence and the tendency of change is maintained.

$$\text{Cyclopropane (6.63)} > \text{Cyclobutane (6.30)} > \text{cyclobutene} \left(6.15\ \text{kJ mol}^{-1}\right)$$

A similar tendency to change in the values of the CH_2 group's contribution to the vaporization enthalpy, obtained in standard conditions with error ±0.1–0.2 kJ mol^{-1}, is kept for the compounds with a growing number of carbon atoms in the molecular cycle.

$$\text{Cyclopentane (5.70)} > \text{Cyclohexane (5.50)} = \text{Cycloheptane (5.50)} < \text{Cyclooctane} \left(5.41\ \text{kJ mol}^{-1}\right).$$

If the energies of the contributions of these groups for two compounds, differing by one CH_2-group, are located within the limits of the experimental error of determination of vaporization enthalpies, the last representative of this series illustrates the increased difference of the values of the similar contribution.

Replacing the single bond in cyclopentane and cyclohexane with the double bond in 1,3-cyclopentadiene reflects the tendency of change of the average value

$$\text{1,3-Cyclopentadiene (5.67)} \approx \text{Cyclopentane (5.70)} < \text{Cyclopentene} \left(5.78\ \text{kJ mol}^{-1}\right)$$

$$\text{1,3-Cyclohexadiene (5.43)} < \text{Cyclohexane(5.50)} < \text{Cyclohexene (5.58; 5.62)} \approx \text{1,4-Cyclohexadiene} \left(5.65\ \text{kJ mol}^{-1}\right)$$

of the contribution of the CH_2-groups to the enthalpy characteristics at the minimum value at 1,3-cyclopentadiene and 1,3-cyclohexadiene, respectively, having values practically within experimental errors. It follows that the contribution of the energy of the CH and CH_2– groups is located within the same experimental error as enthalpy characteristics measured using calorimetric and other methods. The contribution of the CH group of benzole is also 5.65 kJ mol^{-1}, thus, forming the specific interactions $H_2C \rightarrow$ H–CH and HC → H–C by these groups have practically constant values. This leads to the conclusion that there are grounds for taking the average value of the contribution of the energies of these specific interactions for compounds

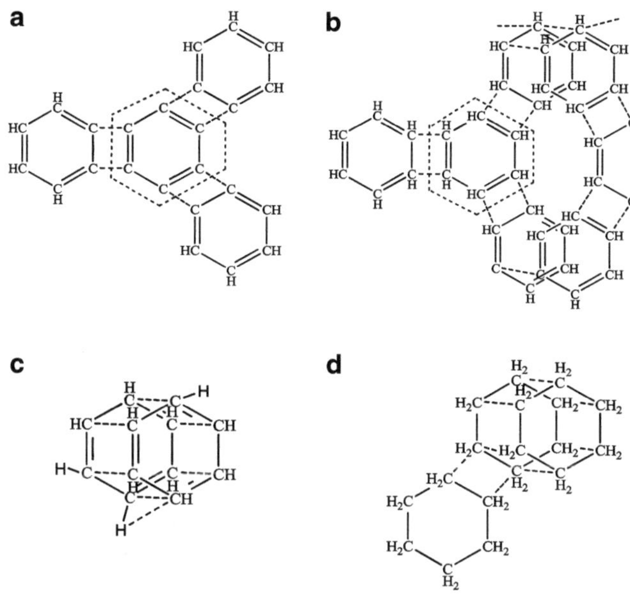

Fig. 3.2 Schematic picture of the liquid and crystalline structure of benzene (**a–c**) and cyclohexane (**d**) with the increased (HC → C–H) and low (HC → H–C) stability of the specific interactions

$$\text{Cyclohexene } (5.58;\ 5.62) \approx 1,4\text{-Cyclohexadiene } (5.65) = \text{Benzene } \left(5.65\ \text{kJ mol}^{-1}\right)$$

independent of the number CH_2 and CH groups in molecules of the given series. The energies of the formed interactions CH_2 and CH groups of cyclopentane and its derivatives are also within the error with the value of the energy of interaction, formed by the CH group of benzene. It follows that contacting CH_2 groups of molecules that are close in the environment form two pairs of interactions $H_2C \rightarrow H\text{–}CH$, which do not have stability. That is why the value of the energy obtained by the division of its contribution to the enthalpy characteristics by the number of interactions (four), gives an average value (1.41 kJ mol^{-1}). The given value can be used in calculations of the energies of the specific interactions and thermodynamic properties of compounds with the use of the principle of additivity. In its turn, interacting CH groups of benzene molecules that are close in the environment and form two specific interactions HC → H–C with an average energy value of 2.82 kJ mol^{-1} could not be considered as real characteristics. This is because CH and CH_2 groups of contacting molecules are able to form specific interactions at the difference of charges at the carbon atoms with participation of the essentially unshared $2s^2$-electron pairs. Thus, the CH group of benzene or unsaturated cyclic hydrocarbon can form two types of specific interactions, one of which (H-C → C-H) has increased stability and the other (HC → H–C) has low stability (Fig. 3.2). In turn, the carbon atom

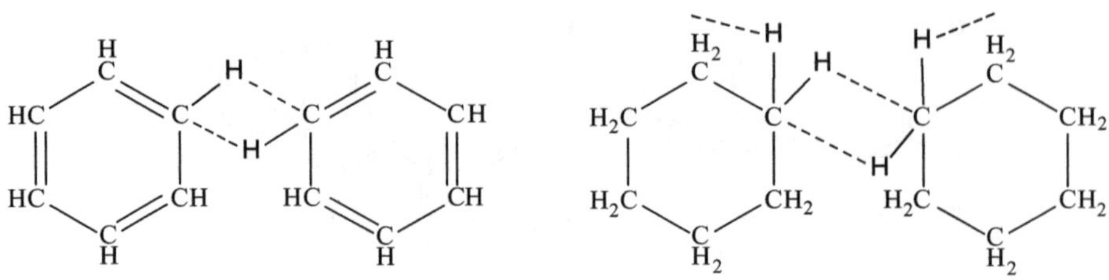

of the interacting methylene groups of cyclohexane that are close in the environment form the specific interaction $H_2C \rightarrow CH_2$ by the essentially unshared $2s^2$ electron pairs of the carbon atoms with differing charges but the same stability and specific interaction of low stability $H_2C \rightarrow$ H–C H. The second hydrogen atom of the given methylene group participates in stitching chains of hydrogen atoms that differ by the positive charge of similar groups, which are practically cooperative interactions. A benzene molecule with a planar structure, for which double bonds are responsible, coordinates three molecules, leads to employment of all bond vacancies of the carbon atoms, which are responsible for the essentially unshared $2s^2$ electron pairs, leading to the formation of the specific interaction and the fifth coordination. This result of the intermolecular interaction in the formation of a network structure of liquid and crystal conditions (Fig. 3.2a). Nevertheless the possibility of formation of the specific interaction HC → C–H between carbon atoms of benzene molecules, located in neighboring grids (Fig. 3.2b, c) cannot be excluded. Similar structures in liquid and crystal conditions form cyclohexane and other compounds (Fig. 3.2d).

Ivanova et al. [7] showed that the hydrogen atom located in the CH_3 group in the *trans*-position to the oxygen atom of the hydroxyl group of methyl alcohol forms a H-bond O•••H–C with an estimated value of the energy 4 kJ mol^{-1}. But justifications for the adopted value have not been discovered. That the comparability of the given value with the energy values of hydrogen bond of liquid ammonia (3.86 kJ mol^{-1}) and the specific interaction of liquid dymethyl ether (5.63 kJ mol^{-1}) is a strong argument but it is not correct. The estimation of the energy of the specific interaction of low stability corresponds to DH–C → C–H ≈ $DH_2C \rightarrow$ H–CH ≈ 0.6 and 1.0 kJ mol^{-1}, for liquid and crystal conditions, respectively. Considering these values, the contribution of the specific interactions DH–C → C–H and D–$H_2C \rightarrow CH_2$ of cyclic saturated and unsaturated hydrocarbons to enthalpy characteristics is equal to 5.05 kJ mol^{-1}. The energy of this interaction has a decreased value compared with the specific interaction DO → CH_3 (5.63 kJ mol^{-1}) of dimethyl ether [6].

Thus, the contribution of the energy of each CH_2 group of the considered compounds of different series is slightly dependent on its number in a molecule of cyclic alkanes and alkenes. It is also important that the contributions of the CH_2 and CH groups to the vaporization enthalpy of derivatives of cyclopentane and cyclohexane, unsaturated 1,3,6-*cis*-hexaditriene, 2,4-hexadiene with linear molecule structure with one, two or three double bonds –CH=CH– and with one triple bond –C≡C– at 1-hexyne remain within the same experimental error.

Table 3.2 Contribution of the energy of CH_2- and CH-groups (kJ mol^{-1}) at sublimation enthalpy of saturated and unsaturated cyclic compounds

Compounds	Formula	Structure	$\Delta_{sub}H°(T)$ [8, 10]	T K	DCH_2/orDCH
Cyclopropane	C_3H_6	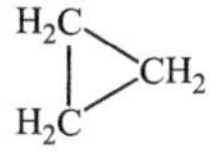	27.6	298	9.20
Cyclobutane	C_4H_8	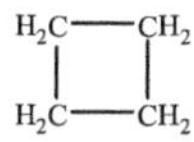	36.4	(145)	9.13
Cyclopentane	C_5H_{10}	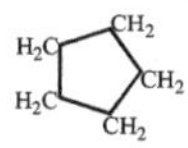	42.6	(122)	8.52
Cyclohexane	C_6H_{12}	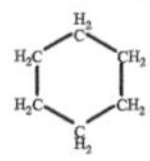	37.6	265	6.25
Ceclooctane	C_6H_{16}	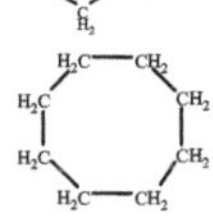	58.7	160	7.40
Benzene	C_6H_6	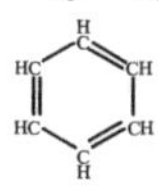	44.4	298	7.40
Cyclodeca-1,2,6,7-tetraene	$C_{10}H_{12}$	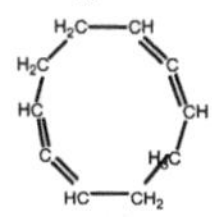	73.0 ± 0.4	298	7.30
EEE-1,5,9-Cyclododecatriene	$C_{12}H_{18}$	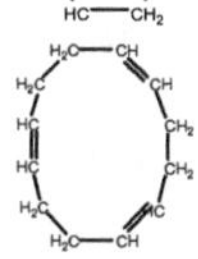	74.7 ± 0.8	298	6.22
Cyclododecahexadiene	$C_{12}H_{12}$	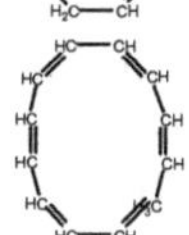	75.6	298	6.30
Cyclododecane	$C_{12}H_{24}$	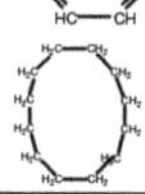	76.4 ± 1.7	298	6.37

The available data on the sublimation enthalpies of cyclic hydrocarbons and the contribution of the energy to characteristics of similar groups (Table 3.2) points to the fact that its tendency to change is also maintained at the maximum and minimum values of the crystal compounds of cyclopropane (9.2 kJ mol^{-1}) and cyclohexane (6.25 kJ mol^{-1} at 265 K), respectively. High differences in contributions of the energies of these groups of crystal and liquid conditions of cyclopropane (2.6) and benzene (1.75 kJ mol^{-1}) point to the significant stabilization by the crystal field, which is significantly decreased with increasing number of methylene groups in saturated and unsaturated cyclic hydrocarbons.

$$\text{Cyclopropane} > \text{Cyclohexane} > \text{Cyclododecane}$$

$$\text{Benzene} > \text{Cyclodeca-1,2,6,7-tetraene} > \text{Cyclododecahexadiene} \geq \text{1,5,9-Cyclododecatriene.}$$

The given tendency is greatly disturbed at cyclododecanes when it reaches the cycle of 12 methylene groups, which reflects a significant change in the charges of the carbon atoms.

The high values of the energies contributed by the CH_2 groups of cyclopropane to the vaporization and sublimation enthalpies point to the fact that carbon and hydrogen atoms have the biggest difference between negative and positive charges, respectively, rather than molecules of other cyclic hydrocarbons. As a result, the specific interaction $H_2C \rightarrow H–C$, formed by the essentially unshared $2s^2$ electron pairs of the donor carbon atom with the acceptor hydrogen atom, has the biggest stability in cyclic compounds with a large number of methylene groups. On the other side, difference between the charges at the carbon atoms leads to formation of specific interaction ($H_2C \rightarrow CH_2$) of high stability. Since it is difficult to estimate the values of the energies of the given two types, to ensure the error in the calculations of stable specific interactions is a minimum it is necessary to accept as universal the value of interaction of low stability $DH_2C \rightarrow H–CH$ as equal to 0.60 and 1.00 kJ mol^{-1}, respectively, for liquid and crystal conditions. Such an assumption is quite correct because the error at the given values is lower than the experimental error in the definition of enthalpy characteristics.

Taking into account that the contribution of the carbon atom to the enthalpy characteristics is the difference between the energies contributed by the H-group and the hydrogen atom (0.60, 1.00 kJ mol^{-1}), in the case of benzene we obtain the energy contributed by the carbon atom for liquid and crystal conditions as equal to 5.05 and 6.40 kJ mol^{-1}, respectively. Detailed analysis of enthalpy characteristics of vaporization and sublimation of cyclic saturated and unsaturated hydrocarbons solves the problem of finding the energies of the specific interactions, formed by different functional groups including heteroatoms.

3.2 Energies of Specific Intermolecular Interactions of Saturated and Unsaturated Cyclic Amines

A relatively large number of values of vaporization enthalpies of the amines with saturated and unsaturated hydrocarbon cycles available in the literature do not allow for a detailed thermodynamic analysis because of their limited number, established at the standard conditions. The planar structure of molecules of cyclic amines with single bonds of the nitrogen atom implies the formation of a complex structure of liquid and crystal conditions, in which two coordinating molecules interact by cycles on one hand and, on the other, amine NH_2 groups forms four hydrogen bonds D–N•••H–N with location in different planes (Fig. 3.3a, b).

a
b
c

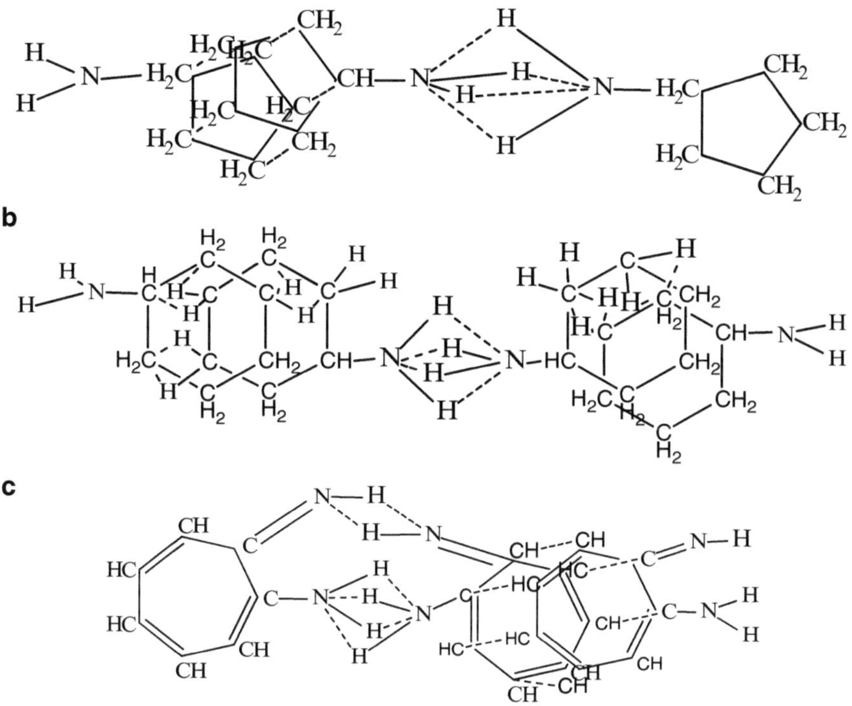

Fig. 3.3 Schematic picture of the liquid and crystalline structure of cyclopentylamine (a), cyclohexylamine (**b**), and 1-amino-7-imino-1,3,5-cyclohepta-triene (c)

Such chains are connected to the network of the specific interactions with the help of low stable interactions formed CH_2 and CH group in saturated and unsaturated cyclic amines. Methylene groups form specific interactions, whose energies are taken into account by enthalpy characteristics of cyclic hydrocarbon. The energies of the hydrogen bonds are established from the difference of vaporization enthalpies of cyclopentylamine and cyclopentane, cyclohexylamine and cyclohexane, reduced by the contribution (0.60 kJ mol^{-1}) of the saturated hydrogen atoms of the CH_2 groups on the amine functional groups, and divided by the total value of the number of formed H-bonds (Eq. 3.1).

$$D\text{-}N\cdots H\text{-}N = (\Delta_{vap}H°(298\ K)cha\ -\ \Delta_{vap}H°(298\ K)ch. + D\text{-}H_2C{\rightarrow}H)/n \quad (3.1)$$

The results of the thermodynamic calculations (Table 3.3) allow one to conclude that the hydrogen bonds D–N•••H–N formed in liquid cyclopentylamine and cyclohexylamine lie within the error of enthalpy characteristic measurements. This is because at four or more number of CH_2 groups in the cycle of influence of the reverse dative bond in a molecule does not appear to stabilize the specific interaction and, consequently, the energy of interaction remains unchanged for all subsequent

Table 3.3 Energies of the hydrogen bonds (kJ mol^{-1}) of the amines with hydrocarbon cycle

Compounds	Formula	Structure	$\Delta_{vap}H°(298\ K)$ [9]	T K	$\Delta_{vap}H°(T)cyc\ \text{-}DH_2C \rightarrow H\text{–}CH$	DC–N•••H–N≡
Cyclopentylamine	$C_5H_{11}N$	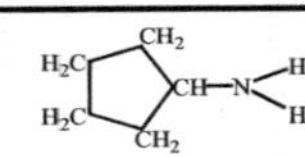	40.2 ± 0.4	298	27.9	3.10
Cyclopentane	C_5H_{10}	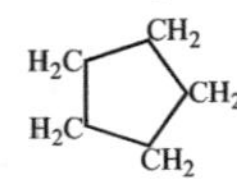	28.5 ± 0.1	298	27.9	–
Cyclohexylamine	$C_{6.}H_{13}N$	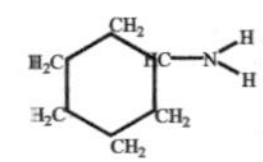	42.8 ± 0.1	298	32.4	2.60
Cyclohexane	C_6H_{12}	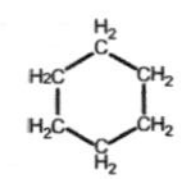	33.0	298	32.4	–
1,4-Diaminocyclohexane	$C_{6.}H_{14}N_2$	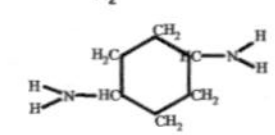	48.2	398	31.8	2.0[b]
Cycloheptylamine	$C_7H_{15}N$	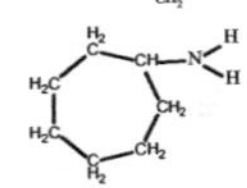	49.1[a]	298	37.9	2.8
Cyclooctylamine	$C_8H_{17}N$	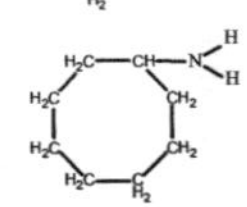	53.9[a]	298	42.7	2.8
Cyclononylamine	$C_9H_{19}N$	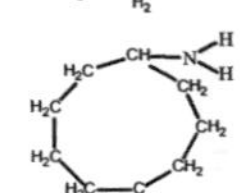	58.8[a]	298	47.5[a]	2.8
Cyclodecylamine	$C_{10}H_{21}N$	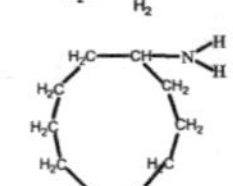	63.5[a]	298	52.3	2.8

1-Amino-7-imino-1,3,5-cyclohepta-triene	$C_7H_8N_2$	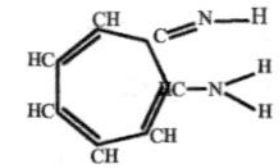	49.4 ± 0.4	298	37.6	2.0[a]
1,3,5-cyclohepta-triene	C_7H_8	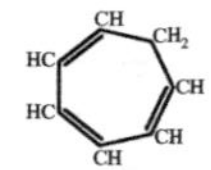	38.7	298	37.6	–
N, α-dimethylcyclohexane thylamine	$C_{10}H_{21}N_2$	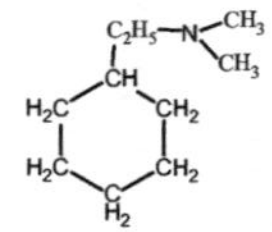	50.2	285	39.9	2.6
Ethylcyclohexane	C_8H_{16}	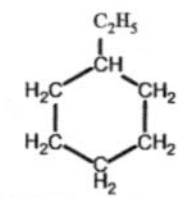	40.5 ± 0.1		39.9	

[a]Estimated with using of the hydrogen bond energy
[b]Determined at 398 K

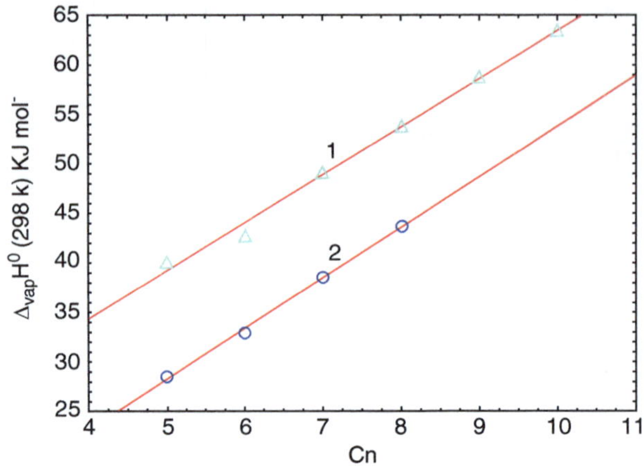

Fig. 3.4 Dependence of the vaporization enthalpy versus the number of carbon atoms in the cycle of the amines with saturated hydrocarbon cycle (*1*) and saturated cyclic hydrocarbons (*2*)

connections series. This is why the more correct value of H-bond will be the average value 2.8 ± 0.5 kJ mol^{-1}. In conclusion, the consistent molecular interconnection by hydrogen bonds, independent of the number of methylene groups in a cycle, ensures the establishment of reliable values of vaporization enthalpies of unstudied compounds of the series cyclobutylamine – cyclodecaneamine with the help of Eq. 3.2.

$$\Delta_{vap}H°(298\,K)cba = \Delta_{vap}H°(298\,K)\ cb. - DH_2C{\rightarrow}H + 4D\text{-}N{\bullet\bullet\bullet}H\text{-}N \qquad (3.2)$$

Estimated vaporization enthalpies of the considered amines are well described by the functional dependence $\Delta_{vap}H°(298\ K) = f(Cn)$ (Fig. 3.4).

The established value of the hydrogen bond has decreased lower value compared with that realized in liquid ammonia (3.86 kJ mol^{-1}). We can then conclude that the hydrocarbon cycle significantly decreases the shifting of the electron density to the methylene group, located at the nitrogen atom providing it with a low negative charge and donor properties. In view of the fact that temperature dependence of vaporization enthalpies is less affected by energies of the specific interactions of low stability, we determined the energy of the hydrogen bond of liquid 1,4-Diamimocyclohexane with the help of Eq. 3.1. In calculations the vaporization enthalpy cyclohexane was used and the number of hydrogen bonds increased. Using a similar procedure of calculation and vaporization enthalpy of 1,3,5-cycloheptatriene, with the help of Eq. 3.1 we identified the average values of the energies of hydrogen bonds of 1-amino-7-imino-1,3,5-cycloheptatriene, formed by the amino and imino groups. From the given in Table 3.5 the values of the energies of the hydrogen bonds of 1,4-diamimocyclohexane and 1-amino-7-imino-1,3,5-cycloheptariene follows the value (2.0 kJ mol^{-1}), established at 398 and 298 K, respectively, and points to the special influence of the three double bonds of 1,3,5-cycloheptatriene on the destabilization of the hydrogen bond N•••H–N. This conclusion is proved by the energies of the hydrogen bonds of saturated liquid cyclopentylamine (Eq. 3.10) and cyclohexylamine (2.6 kJ mol^{-1}).

The energies of the specific interactions formed by methyl groups of the amine fragment N,α-dimethylcyclohexanethylamine are determined from the difference of vaporization enthalpies with ethylcyclohexane reduced by the energy introduced by the hydrogen atom substituted

$$DN{\rightarrow}CH_3\text{-}N = (\Delta_{vap}H°(298\ K)dcha\ -\ (\Delta_{vap}H°(298\ K)ech. + DH_2C{\rightarrow}H))/4 \quad (3.1a)$$

by the isostructural ethyl group of N,α-dimethylcyclohexanethylamine, and divided by the hydrogen cycle from the amino group. The established value of the energy (2.6 kJ mol^{-1}) of this interaction is practically no different from the value of the hydrogen bond of the amines with saturated and unsaturated hydrocarbon cycles. It follows that the ethyl group shows little effect on the change of the nitrogen atom charge, forming a specific interaction of the understated stability, which energy value is lower than that realized in liquid methylamines.

Thus, thermodynamic analysis of vaporization enthalpies of saturated and unsaturated cyclic amines allows us to determine the energy of the hydrogen bonds and clarify the influence of cycles on the energies of this bond D–N•••H–N. This conclusion is consistent with the values of vertical ionizing potentials, causing weak donor activity of NH_2 group [2, 11].

3.3 Energies of Specific Interactions of Compounds with Cyanide Functional Group at Saturated and Unsaturated Hydrocarbon Cycles

The cyclic molecule of saturated and unsaturated hydrocarbons with a heteroatom functional group undergoes great changes in the distribution of the electron density. The presence of the nitrogen atom in cyclic carbonitrile leads to the shifting of the electron density to the carbon atom of the nitrile group and further to the p_z-orbital of the nitrogen atom,

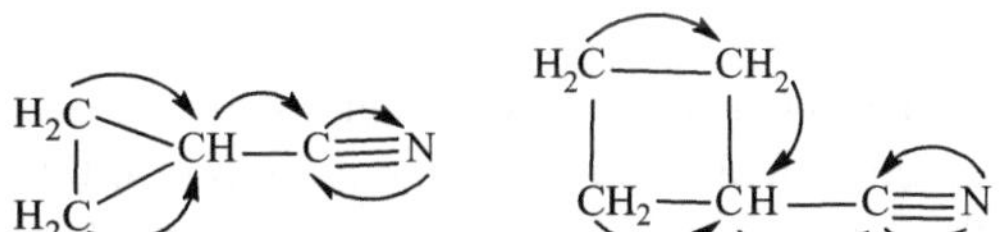

part of which is then transferred to the essentially unshared $2s^2$- electron pair of the contacting carbon atom. Like planar molecules the presence of the three carbon atoms in cyclopropylcyanide brings the molecule to the maximum shift of the electron density of the cycle to the nitrogen atom. This is why increase from three to four CH_2 groups in the cyclobutanecarbonytrile molecule is not accompanied by further enrichment of the negative charge of the nitrogen atom, which should be reflected in the energy of the specific interaction formed by the

Fig. 3.5 Dependence of the vaporization enthalpy of compounds versus the number of carbon atoms in cycle of cyclopropylcyanide – cyclobutanecarbonytrile (*1*) and cyclobutanecarbonytrile – cyclohexanecarbonitrile (*2*)

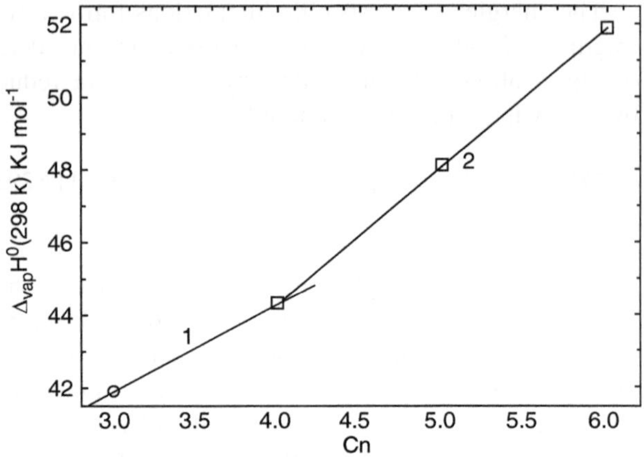

cyanide –C≡N group. In this connection the further increase of enthalpy characteristics of vaporization is determined by the increasing number of CH_2 groups of saturated hydrocarbon cycles, CH_2 and CH groups of unsaturated hydrocarbons or CH group benzene, cyclooctanetetraene. In Fig. 3.5 the dependence of the vaporization enthalpy on the number of carbons in the cycle of compounds of the series cyclopropylcyanide – cyclohexanecarbonitrile is presented and illustrates the location, on one hand, of cyclopropylcyanide and the final manifestation of the reverse dative bond in the stabilization of the specific interaction, and, on the other, compounds of the series cyclobutanecarbonytrile – cyclohexanecarbonitrile at the correlative line, causing the change of the energy contribution of the increasing number of CH_2 groups to the enthalpy characteristics. The good description of the experimentally established vaporization enthalpies ($\pm$0.1 kJ mol^{-1}) of cyclocarbonitriles allows using this dependence for establishing by the method of extrapolation the enthalpy characteristics of cycloheptanitrile – cyclononanenitrile (Table 3.4), which have not been researched.

The molecule of cyclopropylcyanide with two bond vacancies of the C≡N group and three vacancies of the CH_2 groups of the propyl cycle form five stable specific interactions of two types D≡N → C≡ and $DH_2C \rightarrow CH_2$ with molecules that are close in the environment. In similar molecules of the series of cyclocarbonitriles compounds, the number of the specific interactions formed by the cycle rises with increasing number of methylene groups. The planar structure of the cyclocarbonitrile molecule at the interaction with molecules that are close in the environment forms chains sewn into the grid by weak specific interactions (Fig. 3.6). The specific interaction CH_2-groups $H_2C \rightarrow CH_2$ and low stability $H_2C \rightarrow$ H–CH are included in the enthalpy characteristic cycle.

The energy of the specific interaction D≡N → C≡ should be determined from the difference of the vaporization enthalpy of cyclocarbonitrile and saturated and unsaturated cyclic compounds with the same number of methylene or CH-groups in the cycle (Tables 3.1 and 3.2) reduced by the energy, introducing the saturated hydrogen atom (0.60) of the corresponding cycle on the ≡N → C≡ group. Thus, the energy of the specific interaction D≡C ← N≡ is determined with the help of Eq. 3.3

Table 3.4 Energies of the specific interactions (kJ mol^{-1}) of cyanides with hydrocarbon cycle

Compounds	Formula	Structure	$\Delta_{vap}H°(T)$ [9]	T K	$\Delta_{vap}H°(298\ K)$ –$DH_2C \rightarrow H–CH$	$D{\equiv}C \leftarrow N{\equiv}$
Cyclopropylcyanide	C_4H_5N	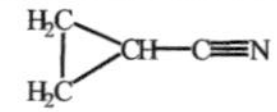	41.9 ± 0.1	298	19.3	11.0
Cyclobutanecarbonytrile	C_5H_7N	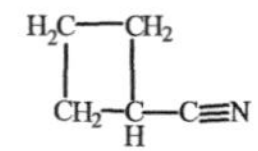	44.3	298	24.6	9.85[a]
Cyclopentane-carbonitrile	C_6H_9N	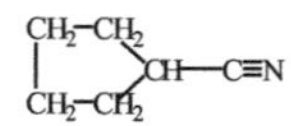	48.1 ± 0.1	298	27.9	10.1
Cyclohexanecarbonitrile	$C_7H_{11}N$	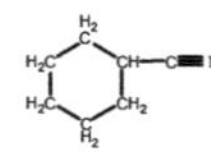	51.9 ± 0.1	298	32.4	9.80
Cycloheptanenitrile	$C_8H_{13}N$	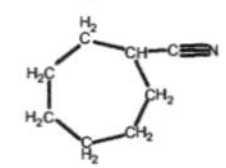	56.3[a]	298	37.9	9.7
Cyclooctanenitrile	$C_9H_{15}N$	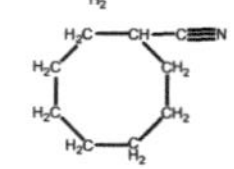	60.4[a]	298	41.4	9.7[a]
Cyclononanenitrile	$C_{10}H_{17}N$	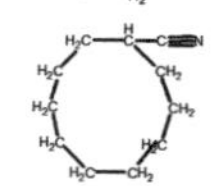	64.3[a]	298	45.9[a]	9.7
3-Metilenecyclo-butanecarbonitrile	C_6H_7N	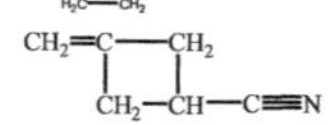	41.6	345	28.5	9.15
Methylene cyclobutane	C_5H_8	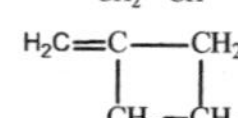	29.1	298	28.5	–
Bicyclo[2.1.0]pentane-1-carbonitrile	C_6H_7N	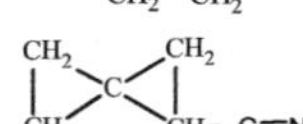	41.8	343	27.4	7.2
Bicyclo[2.1.0]pentane-	C_5H_8	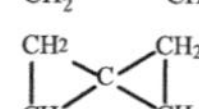	28.0 ± 0.5	298	27.4	–

(continued)

Table 3.4 (continued)

Compounds	Formula	Structure	$\Delta_{vap}H°(T)$ [9]	T K	$\Delta_{vap}H°(298\ K)$ –$DH_2C \rightarrow H–CH$	$D{\equiv}C \leftarrow N{\equiv}$
Bicyclo[1,1,0]butane-1-carbonitrile	C_5H_5N	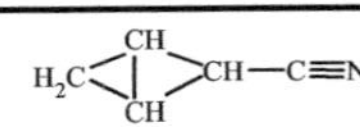	48.0	333	23.3	12.9
Bicyclo[1,1,0]butane	C_4H_4	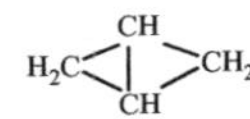	23.9	298	23.3	–

[a]Estimated by the extrapolation method of the dependence $\Delta_{vap}H°(298\ K) = f(Cn)$

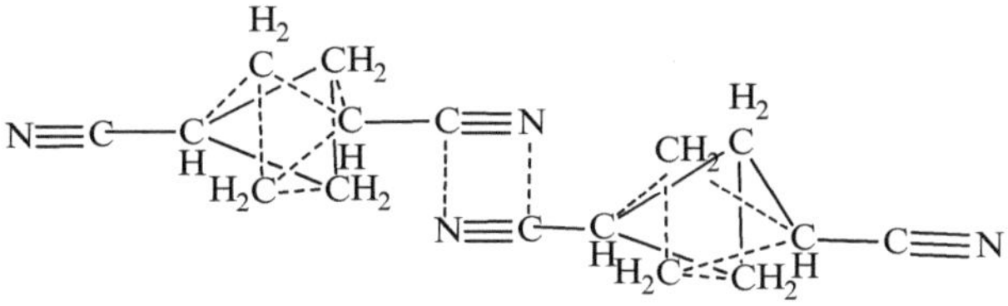

Fig. 3.6 Schematic picture of the liquid structure of cyclopropylcyanide

$$D\equiv N\rightarrow C\equiv\ = (\Delta_{vap}H°(298\ K)ccn.\ -\ \Delta_{vap}H°(298\ K)ch.. + DH_2C\rightarrow H)/2 \quad (3.3)$$

which can be used for all cyclocarbonitriles, because the number of specific interactions formed by bond vacancies of saturated and unsaturated hydrocarbon cycle of molecules of these compounds are taken into account by enthalpy characteristics reduced by the energies introduced by saturated hydrogen atoms.

In the case of 3-metilenecyclobutane carbonytrile one should use the vaporization enthalpy 3-methyl-1-butane, consisting of a molecule of the methyl group and double bonds at the same atoms as on the 3-metilenecyclobutane molecule. A similar procedure should be applied to the calculation of the energy of the specific interaction D≡C ← N≡ of liquid bicyclo[2.1.0]pentene-1-carbonitrile, using the vaporization enthalpy bicyclo[2.1.0]pentene. The thermodynamic calculations of the specific interaction energy D≡N → C≡ (Tables 3.4 and 3.5) of different cyclic nitriles illustrate the sharp decrease in the energy of the specific interaction D≡N → C≡ at the first representatives of the cyclopropylcyanide and cyclobutanecarbonytrile series,

$$\begin{aligned}\text{Cyclopropylcyanide }(11.0) &< \text{Cyclobutanecarbonytrile }(9.85)\\ &\approx \text{Cyclopentanecarbonitrile }(10.1)\\ &\approx \text{Cyclohexanecarbonitrile }(9.80\ \text{kJ mol}^{-1})\end{aligned}$$

practically unchanged at the further compounds of this series. In conclusion, the consistent energy in the coupling of molecules for the cyclobutanecarbonytrile – cyclohexanecarbonitrile compounds in crystal and liquid conditions is proved by the dependence of the vaporization enthalpies on the number of methylene groups of a cycle (Fig. 3.2). It leads to the conclusion that the influence of the reverse dative bond on the energy of the specific interactions of the molecules of cyanides ends at cyclopropylcyanide and therefore the further invariability or the tendency to small changes in the energy of the specific interactions D≡N → C≡ at the considered compounds series is because of the increasing number of methylene groups.

We draw attention to the inequality in the energies of the specific interactions cyclobutanecarbonytrile (9.85) > 3-metilenecyclo-butanecarbonitrile (9.15 kJ mol^{-1}) formed by the cyanide group, illustrating the effect of the methylene group on shifting the electron density in a cycle and of the decrease in positive and negative charges at the atoms of carbon and nitrogen of the cyanide group, respectively.

Table 3.5 Energies of the specific interactions (kJ mol^{-1}) of cyclic carbonitriles of unsaturated series

Compounds	Formula	Structure	$\Delta_{vap}H°(298\ K)$ [9]	T K	$\Delta_{vap}H°(T)cyc - DH_2C \rightarrow H–CH$	$D{\equiv}N \rightarrow C{\equiv}$
Cyclobutenecarbonitrile	C_5H_5N	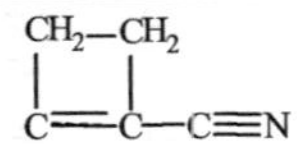	43.7[a]	298	23.0	10.35
Cyclobutene	C_4H_6		23.6[a]	298	–	–
Cyclopentenecarbonitrile	C_6H_7N	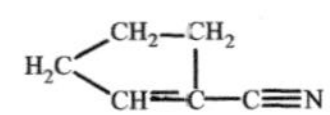	48.5[a]	298	27.8	10.35
Cyclopentene	C_5H_8		28.4	298		
2-Cyclopentene-1-carbonitrile	C_6H_7N	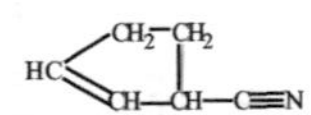	44.9 ± 0.1	298	27.6	8.85
1-Cyclohexene-1-carbonitrile	C_7H_9N	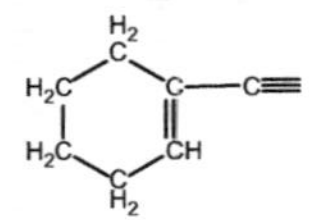	53.6	298	32.9	10.35
Cyclohexene	C_6H_{10}		33.5 ± 0.5	300	32.9	
Cycloheptene	C_7H_{12}	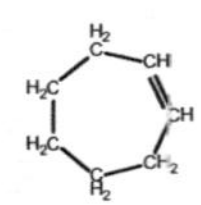	36.7	300	36.1	–
Cycloheptenenitrile	$C_8H_{11}N$	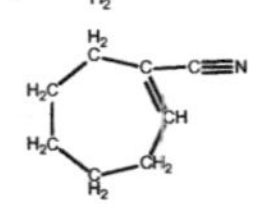	56.8[a]	298	36.1	10.35[a]
Cyclooctenenitrile	$C_9H_{13}N$	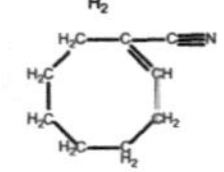	62.3	298	41.6	10.35
Cyclononenenitrile	$C_{10}H_{15}N$	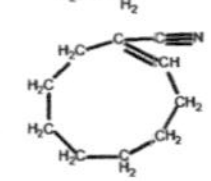	67.4[a]	298	46.7	10.35[a]

Cyclodecenenitrile	$C_{11}H_{17}N$	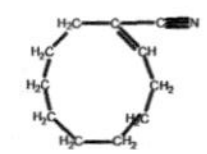	72.8	298	52.1	10.35
Cycloundecenenitrile	$C_{12}H_{19}N$	–	77.7[a]	298	57.0	10.35
Cyclododecenenitrile	$C_{13}H_{21}N$	–	82.6[a]	298	61.9	10.35

[a]Estimated by extrapolation and interpolation

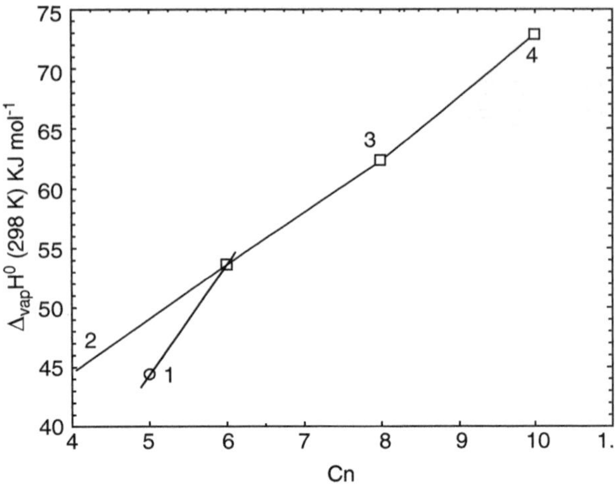

Fig. 3.7 The dependence of the vaporization enthalpy of cyclocarbonitriles of unsaturated series on the number of carbon atoms in cycle of 2-cyclopentene-1-carbonitrile (*1*), cyclobutenecarbonitrile – cyclooctenenitrile (*2, 3*), and cyclononenenitrile – cyclodecenenitrile (*3, 4*)

At the same time, considering that such calculations using thermodynamic characteristics do not meet standard conditions, it seems advisable to use a reasonable combination of the considered approaches to obtain reliable information on specific of interactions and the energies of biocyclic compounds. As evidence of this we have bicyclo[2.1.0]pentane-1-carbonitrile, for which a reasonable combination of calculations of the energy of the specific interactions, formed by the cyanide group, was used. The value obtained is slightly understated because of the inappropriate conditions under which the experimental research was conducted. In the calculations of the energy of the same type of interactions, formed in liquid bicyclo[1,1,0]butane-1-carbonitrile, the maximum possible value of interaction energy was obtained. However, considering destabilization of this type of interaction with the increase in the methylene groups in the ring cyclopropylcyanide and cyclobutanecarbonytrile, the value of the energy D≡N → C≡ (12.9 kJ mol^{-1}) of liquid bicyclo[1,1,0]butane-1-carbonitrile obtained definitely has reliability.

The enthalpy characteristics of nitriles with a saturated hydrocarbon cycle of four-six carbon atoms at the linear dependence chain, which is a consequence of the influence of the reverse dative bond on the stabilization of the specific interactions by the butyl cycle leads to the conclusion that compounds with one double C=C bond in the hydrocarbon cycle of cyclocarbonitriles with the nitrile group located at the first carbon atom should be described by a similar regularity. It follows that the dependence ΔH°(298 K) on the number of carbon atoms in a cycle, given in Fig. 3.7, should be extrapolated at 1-cyclopentene and 1-cyclobutene (Table 3.5). The marked increase in the value of the vaporization enthalpy for cyclopentene-carbonitrile at 3.6 kJ mol^{-1}, obtained by the method of extrapolation compared with the experimental value of the enthalpy characteristics 2-cyclopentene-1-carbonitrile, reflects the influence of double C=C bond to reduce the bias of the electron density from the second carbon atom CH- to the nitryl group –C≡N, located at the first carbon atom.

From the data in Table 3.5 the results of the calculations of the energies of the specific interactions, D≡N → C≡, of unsaturated cyclic nitryles established with the help of the enthalpies of vaporization in Eq. 3.1,, it follows that the values at the compounds

D≡N→C≡ Cyclobutenecarbonitrile = D≡N→C≡ Cyclopentenecarbonitrile = D≡N→C≡ (1-Cyclohexene-1- carbonitrile) = D≡N→C≡ (Cyclooctenenitrile) = D≡N→C≡ Cyclodecenenitrile =10.35 kJ mol^{-1}

remain invariable and vaporization enthalpies are determined by the introduction of the energies of the methylene and CH groups and their number in a cycle. It follows from this conclusion that vaporization enthalpies of the unstudied cyclocarbonitriles are easily determined from the average value of the energy contributed by ethylene and CH group, its number in a cycle, and the established energy of the specific interaction D≡N → C≡ or enthalpy of vaporization of unsaturated cyclic hydrocarbon and energies of its bonds reduced by the contribution of the energy of the substituted hydrogen atom of a cycle (Table 3.5). The reduced stability of the interaction D≡N → C≡ of unsaturated cyclocarbonitriles at 0.60 kJ mol^{-1} reflects the influence of the double bond in the cycle at the carbonitrile group on the decrease in the difference of charge at the carbon and nitrogen atoms of the nitryl group in shifting the electron density to the carbonitrile group. The symptomatic fact of the reduced stability of the specific interaction D≡N → C≡ 2-cyclopentene-1-carbonitrile, the foundation of the series of cyclic carbonitriles with double C=C bond at the second carbon atom, allows us to take the amount of energy (8.85 kJ mol^{-1}) of the given interaction for corresponding compounds 2-cyclopentene-1-carbonitrile and 2-cyclododece-1-nenitrile and provides us with calculation procedures for vaporization enthalpies.

3.4 Energies of Specific Interactions of Azobenzenes and Azabicycloalkanes

Enthalpies of sublimation *cis*- and *trans*-azobenzenes, differing at 1.2 kJ mol^{-1}, points to the low difference in the energies of interactions, formed by the nitrogen atom with CH groups of benzene. Azobenzene molecules that are in close contact in the environment in crystal and liquid conditions form four specific interactions and the next four CH groups of each benzene fragment form specific interactions H–C → CH, using the differences in its charges at the location of the carbon atoms, for example, at position C(3) and C(4). Carbon atoms of the CH groups of planar structure of the benzene ring and nitro-group form the grid structure of

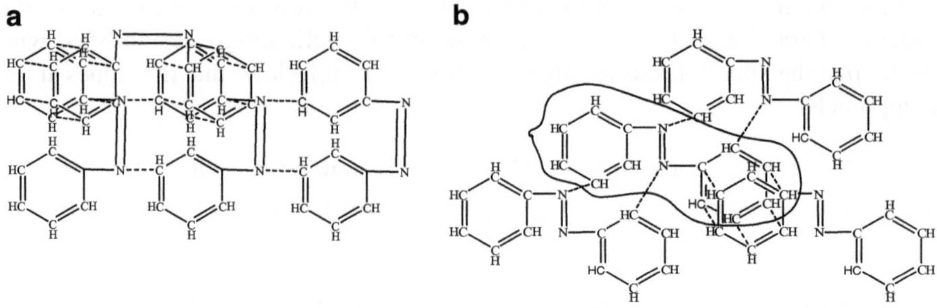

Fig. 3.8 Schematic picture of the liquid structure of *cis*- (**a**) and *trans*-azobenzenes (**b**)

crystal and liquid conditions with strong specific interactions, in which HC → CH bonds contribute 6.80 kJ mol^{-1} to the enthalpy of sublimation for *cis*-azobenzene and *trans*-azobenzene (Fig. 3.8). It means that the energy of the specific interaction N → CH is calculated with the help of Eq. 3.4.

$$D\text{-}N{\rightarrow}CH = (\Delta_{vap}H^0(298\ K)abz - 8D{=}HC{\rightarrow}CH)/4 \tag{3.4}$$

The given equation is right for the liquid *trans*-azobenzene, whose energies contributed by the CH groups to the enthalpy of vaporization are equal to 3.38 kJ mol^{-1}. The results of the calculations (Table 3.6) illustrate the high stability of the specific interactions in azobenzene crystals under the influence of the crystal field. The energy of the same type of specific interactions with participation of the nitrogen atom in liquid *trans*-azobenzene is equal to 7.50 kJ mol^{-1}. Taking the difference in the energies of crystal and liquid *cis*-azobenzene forms as invariable (0.30 kJ mol^{-1}), we have an opportunity to establish the vaporization enthalpy of *cis*-form as equal to 72.2 kJ mol^{-1}.

The 3-azabicyclo[3, 2, 2]nonane molecule with two planar (pentagon and square) fragments forms the crystal and the liquid, in which each nitrogen atom with three ligands from the two methylene groups is located in the plane.

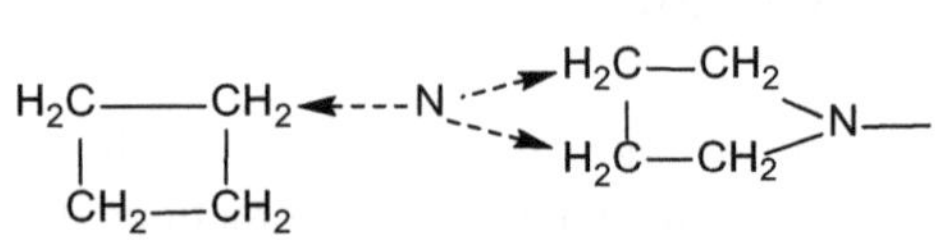

One of these groups remains free and has to contribute the energy to the enthalpy characteristics equal to the contribution of the same groups of crystal or liquid *cis*-bicyclo[4.2.0]octane 5.90 and 5.34 kJ mol^{-1}. Each molecule forming six specific interactions N → CH_2–CH_2–N, constitutes the structure with the network of stable specific interactions (Fig. 3.9). The energy of this type of specific interaction is equal to the difference between the enthalpy characteristics and the contribution of the energies of the two methylene groups, divided by the number of specific interactions.

Table 3.6 Energies of the specific interactions (kJ mol^{-1}) of the liquid and solid azobenzenes and azabicycloalkanes

Compounds	Formula	Structure	$\Delta_{evap}H°(T)$ [9, 10]	T K	D=CH	D=N → CH
Crystal [10]						
cis-Azobenzene	$C_{12}H_{10}N_2$	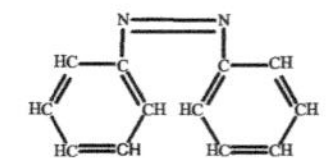	92.9	288	6.80	9.60
trans-Azobenzene	$C_{12}H_{10}N_2$	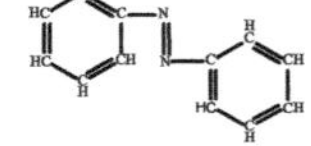	94.1 ± 0.8	298	6.80	9.90
Biphenyl	$C_{12}H_{10}$	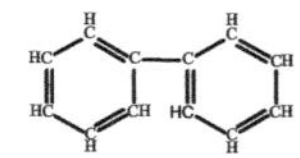	81.8 ± 0.4	298	6.80	–
Liquid [9]						
trans-Azobensene	$C_{12}H_{10}N_2$	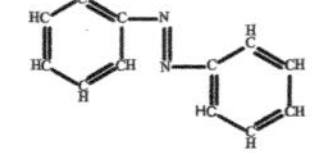	72.8 ± 0.7	298	5.38	7.50
cis-Azobenzene	$C_{12}H_{10}N_2$	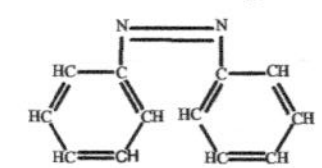	72.2[a]	298		7.20[a]
Biphenyl	$C_{12}H_{10}$	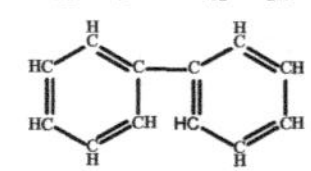	64.52.2	298	5.38	–
Crystal [10]						
					DCH_2–CH_2–	DN → CH_2–$(CH_2)n$–N
3-Azabicyclo[3, 2, 2]nonane	$C_8H_{15}N$	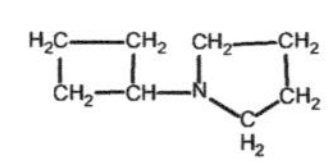	57.8 ± 1.3	298	5.95	7.65 n = 1
Bicyclo[2.2.2]octane	C_8H_{14} s	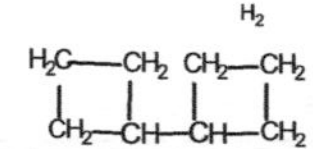	47.7 ± 0.8	298	5.95	–

(continued)

Table 3.6 (continued)

Compounds	Formula	Structure	$\Delta_{evap}H°(T)$ [9, 10]	T K	D=CH	D=N → CH
1-Azobicyclo octane	$C_{7.}H_{13}N$	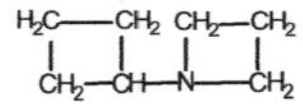	50.8 ± 0.2	298	5.62	7.65 n = 1 4.5 n = 0
Bicycle[2.2.1]heptane	C_7H_{12}	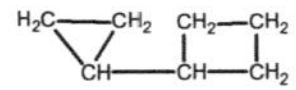	39.33 ± 0.13	298	5.62	–
Liquid [9]						
3-Azabicyclo[3,2,2] nonane	$C_{8.}H_{15}N$	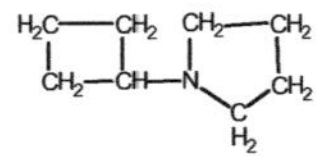	52.2	318	5.34	6.92 n = 1
cis-Bicyclo[4.2.0]octane	$C_{8.}H_{14}$	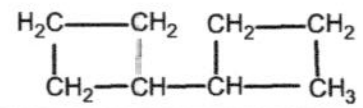	42.7 ± 1.2	298	5.34	–

[a]Estimated using D=HC → CH = 7.20 kJ mol^{-1}

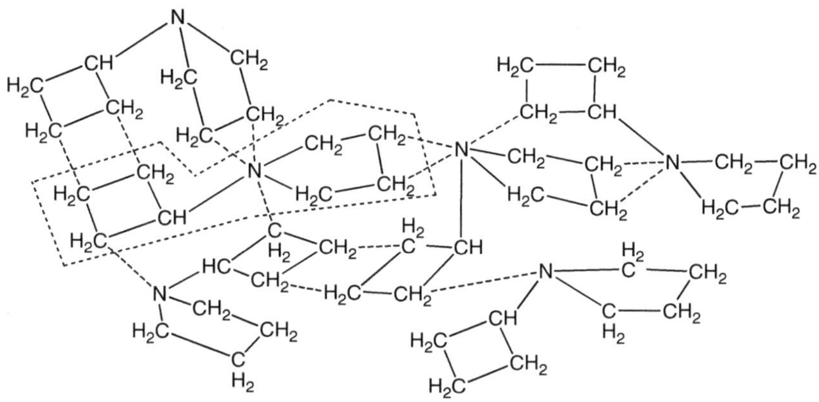

Fig. 3.9 Schematic picture of the liquid and crystalline structure of 3-azabicyclo[3,2,2] nonane with a network of the specific interactions

$$DN \rightarrow CH_2\text{-}CH_2\text{-}N = (\Delta_{vap}H^0(298\ K)abcn - 2DCH_2)/6 \quad (3.5)$$

The results of the calculations (Table 3.6) point to the low effect (0.70 kJ mol^{-1}) of stabilization of the specific interaction D–N → CH_2–CH_2–N by crystal field.

A distinctive feature of molecule 1-azobicyclooctane is the presence in one of the cycle three methylene groups, forming one specific interaction N → CH_2–CH_2–N and a second one N → CH_2–N. Thus, in crystal azobicyclooctane there are formed four specific interactions D–N → CH_2–CH_2–N and two of low stability D–N → CH_2–N. The contribution of energy from each of the two methylene groups of the second cycle to the enthalpy characteristics (5.62 kJ mol^{-1}) is equal to the contribution to the enthalpy of sublimation for bicycle [2.2.1]heptane. At the same time the energies of the specific interaction D–N → CH_2–CH_2–N of crystal 3-azabicyclo[3, 2, 2]nonane and 1-azobicyclooctane are equal, which is why the energy of the specific interaction of reduced stability, D–N → CH_2–N, is determined with the help of Eq. 3.6.

$$D\text{-}N \rightarrow CH_2\text{-}N = (\Delta_{vap}H^0(298\ K)abco - 2DCH_2 - 4D\text{-}N \rightarrow CH_2\text{-}CH_2\text{-}N)/2 \quad (3.6)$$

The data presented in Table 3.6 point to inequality between the energies of the specific interactions

DN→CH_2-CH_2-N Azabicyclo[3,2,2]nonane (6.92) > DN→CH_2-N 1-Azobicyclooctane (6.45 kJ mol^{-1})

reflecting significant differences of charges at the end carbon atoms of the fragments of one cycle –CH_2–CH_2–N and –CH_2–N. It follows that assumptions made on the equality of the energies of the specific interactions D–N → CH_2–CH_2–N in liquid azabicyclo[3,2,2]nonane and 1-azobicyclooctane do not introduce significant errors. That is why it should be used for the thermodynamic analysis and calculations of the energies of the specific interactions.

3.5 Energies of Specific Interactions of Aromatic Compounds with Nitrogen Atom in the Functional Group

3.5.1 *Aniline and Its Derivatives*

On the basis of a survey of the electron structure of compounds with aromatic benzene rings by the FES method it is seen that many intramolecular conversions are connected with high symmetry C_6H_6, which does not allow the (s)-orbital to mix with the molecular orbital of p-series [2]. The values of vertical ionization potential (Iv) and Eсв (C1s) lead to the conclusion of total acceptor nature of OH groups and weak expressed NH_2 groups, leading to energy difference in the specific interactions formed. Immiscibility of the s- and p-series orbitals of the carbon atom provides the opportunity to participate in the formation of the specific intermolecular interactions of different stability. The special role of the substituent is evident in the case of diamine molecules $C_6H_6(NH_2)_2$, in which the difference in the varieties of vertical ionizing potentials ΔI_{1-2} decreases from 1.87 eV at *para-* to 0.57 eV at *meta-* and 0.87 eV at *ortho*-diaminbenzene [12, 13]. The specificity of the intermolecular interactions in a molecule of benzene between the carbon and hydrogen atoms determines their ability in forming the specific interactions and the energies, adequately reflected at vaporization enthalpies. On the other hand, similar intermolecular interactions appear on the energies of interaction, formed by functional groups, replacing hydrogen atoms of the C_6H_6 cycle.

Aniline molecule with two bond vacancies of hydrogen and two nitrogen atoms of the amino-group form four hydrogen bonds N•••H–N with molecules that are close in the environment, forming a pyramidal structure with four hydrogen atoms in the bottom (Fig. 3.10). The planar (or rather weakly expressed having the form of an arm-chair) C_6H_6 of the benzene ring and amino groups form liquid and liquid aniline structures with strong specific interactions of the benzene ring and the hydrogen bonds of the amine fragment. The stable bonds are connected to the network by weak interactions formed by hydrogen and carbon atoms of the benzene rings in contact. As mentioned above, the specific interactions, stable and of low stability, formed by the benzene ring reduced by the contribution of the substituted hydrogen atom and amino group lead to the aniline vaporization enthalpy. That is the reason why the value of the energy of the hydrogen bond D–N•••H–N of aniline should be determined with the help of Eq. 3.7.

$$\text{D-N}\cdots\text{H-N} = (\Delta_{vap}H^0(298\ \text{K})\text{an} - \Delta_{vap}H^0(298\ \text{K})\text{bz} - \text{D (H)n}) / 4 \qquad (3.7)$$

A similar equation, including the energy contributed by the methyl groups of the amine fragment *N,N*-dimethylaniline, are used for the determination of the energy of the specific interaction N → CH_3–N.

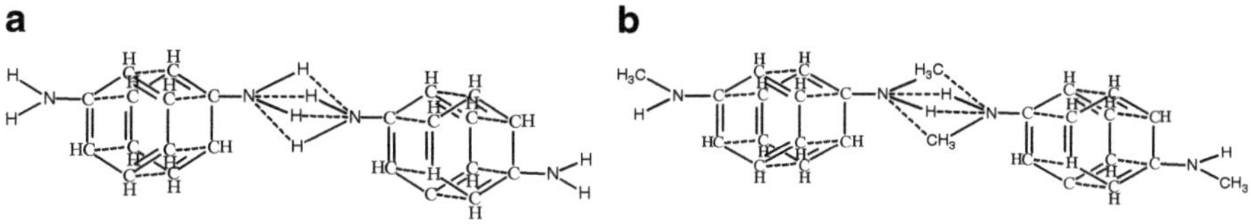

Fig. 3.10 Schematic picture of the liquid and crystalline structure of the amine derivative aniline: aniline (**a**), *N*-methylaniline (**b**)

$$DN{\rightarrow}CH_3\text{-}N = (\Delta_{vap}H^0(298\ K)dman - \Delta_{vap}H^0(298\ K)bz - D\ (H)n)\ /\ 4 \quad (3.7a)$$

The established values of the energies of these types of interactions as presented in Table 3.6 point to the increased stability of the hydrogen bond (5.10 kJ mol^{-1}). These data are consistent with the results of research conducted by the method of RE-spectroscopy, and point to the reduction of the electron density at the nitrogen atom when it replaces the hydrogen atom in the methyl group [14, 15]. This effect of stabilization of the hydrogen bond should be strengthened at the joint presence of the hydrogen atom and methyl group in the amine fragment *N*-methylaniline, whose energy should be determined with the help of the following equation.

$$D\text{-}N{\bullet\bullet\bullet}H\text{-}N = (\Delta_{vap}H^0(298\ K)man - \Delta_{vap}H^0(298\ K)bz - D(H)n - 2D\text{-}N{\rightarrow}CH_3\text{-}N)/2 \quad (3.8)$$

The summary contribution of energy of the specific interaction, formed by methyl group is also worth considering. Taking the value of the energy D–N → CH_3–N of this interaction to be unchanged in liquid *N*-methylaniline and *N,N*-dimethylaniline, we determine the energies of the hydrogen bond, and point to the tendency of its stabilization D–N•••H–N (5.10) Aniline < D–N•••H–N (5.30 kJ mol^{-1}) *N*-methylaniline.

Taking the obtained value of the energy of the hydrogen bond *N*-methylaniline as constant for the compounds with different alkyl ligands, we used Eq. 3.8 and determine the energy of the specific interaction, formed by the ethyl group ethylaniline (Table 3.7). The value of the energy of the specific interaction (7.25 kJ mol^{-1}) obtained leads to the increased stability compared with liquid ethylamine (6.00 kJ mol^{-1}), which can be explained by the simultaneous stabilization of the hydrogen bond of liquid ethylaniline. In this regard, the tendency of its stabilization should be continued in the range of compounds with increasing number of carbon atoms in the alkyl ligand.

D-N•••H-N: (5.10) Aniline < (5.30 kJ mol^{-1}) N-Methylaniline < (5.80 kJ mol^{-1}) Ethylaniline

D-N→CH_3-N: N-Methylaniline (4.90) < DN→CH_3-CH_2-N (6.75 kJ mol^{-1}) Ethylaniline

Table 3.7 Energies of the specific interactions (kJ mol^{-1}) of the liquid aromatic amines of derivative anilines at 298 K

Compounds	Formula	Structure	$\Delta_{vap}H°(298\ K)$ [9]	$\Delta_{vap}H°(298\ K)bz–D(H)n$	DN → CH_3–N	DN•••H–N
Benzene	C_6H_6	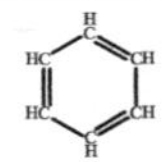	33.8	33.2	–	–
Aniline	C_6H_7N	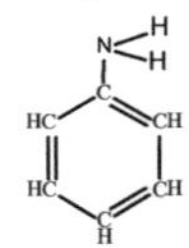	53.6	33.2	–	5.10
N-Methylaniline	C_7H_9N	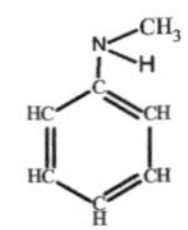	53.6	33.2	4.90	5.30
N,N-Dimethylaniline	$C_8H_{11}N$	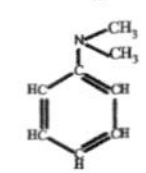	52.8 ± 0.1	33.2	4.90	–
Ethylaniline	$C_8H_{11}N$	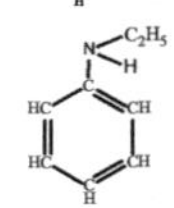	58.3 ± 0.6	33.2	(7.25) 6.75	5.30 5.80
2,4-Dimethylaniline	$C_8H_{11}N$	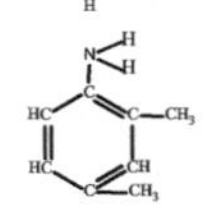	61.9 ± 0.6	42.1	–	4.95
2,6-Dimethylaniline	$C_8H_{11}N$	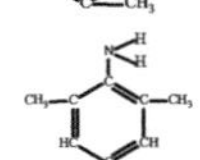	59.6 ± 0.3	42.1	–	4.90
1,3-Dimethylbenzene	C_8H_{10}		42.7 ± 0.1	42.1	–	–
2,5-Dimethylaniline	$C_8H_{11}N$	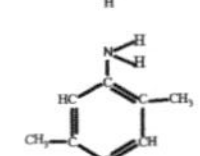	61.7 ± 0.7	41.7	–	5.25

1,4-Dimethylbenzene	C_8H_{10}		42.3 ± 0.1	41.7	–	–
2-Ethylbenzeneaniline	$C_8H_{11}N$	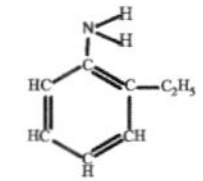	60.6 ± 0.9	41.8	–	4.70
4-Ethylaniline	$C_8H_{11}N$	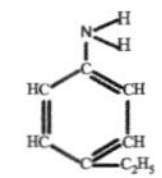	54.8 (340 K)	41.8	–	3.25
Ethylbenzene	$C_{10}H_{10}$		42.4 ± 0.1	41.8	–	–
2,6-Diethylaniline	$C_{10}H_{15}N$	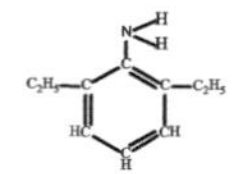	65.9 ± 0.6	51.9	–	3.50
1,3-Diethylbenzene	$C_{10}H_{14}$		52.5	51.9	–	–
2-isopropylaniline	$C_9H_{13}N$	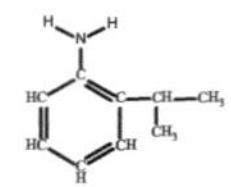	61.8 ± 0.9	44.5	–	4.30
Isopropylbenzene	C_9H_{12}	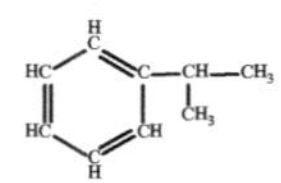	45.1 ± 0.1	44.5	–	–
2,6-Diisopropylaniline	$C_{12}H_{19}N$	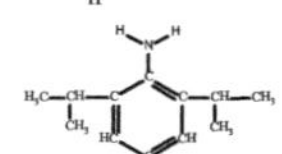	69.5 ± 0.3	55.6	–	3.50
1,3-Diisopropylbenzene	$C_{12}H_{18}$	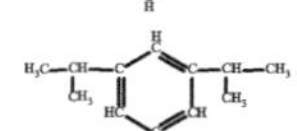	56.2 ± 0.8	55.6	–	–
2-*tert*-Butylaniline	$C_{10}H_{15}N$	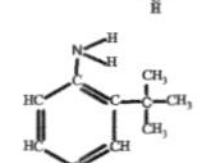	62.7 ± 0.4	46.9	–	3.95
tert-Butylbenzene	$C_{10}H_{14}$	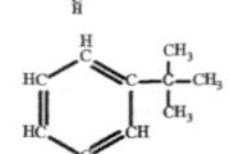	47.5 ± 0.4	46.9	–	–

(continued)

Table 3.7 (continued)

Compounds	Formula	Structure	$\Delta_{vap}H°(298\ K)$ [9]	$\Delta_{vap}H°(298\ K)bz–D(H)n$	DN → CH_3–N	DN•••H–N
N-Ethylaniline	$C_8H_{11}N$	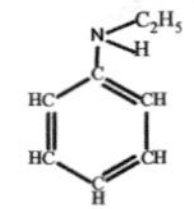	58.3 ± 0.1	33.2	DN → CH_3–CH_2–N 7.25	5.30
N-Benzylaniline	$C_{13}H_{13}N$	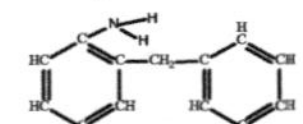	79.6 ± 1.1	67.3	–	3.10
Diphenylmethane	$C_{13}H_{12}$	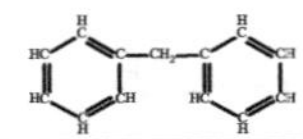	67.9	67.3	–	–

The energy of the hydrogen bond of dimethylaniline should be determined from the difference between the vaporization enthalpies of the aniline and methylbenzene derivatives with the corresponding location of the methyl groups at the carbon atoms of the benzene ring reduced by the energy contributed by the replaced hydrogen atom of the benzene derivative.

$$D\text{-}N\bullet\bullet\bullet H\text{-}N = (\Delta_{vap}H^0(298\ K)\ der.an - \Delta_{vap}H^0(298\ K)mbz - D\ (H)n\ )/4 \quad (3.9)$$

A similar equation should be used for the determination of the energies of the hydrogen bonds of ethylamines.

$$D\text{-}N\bullet\bullet\bullet H\text{-}N = (\Delta_{vap}H^0(298\ K)etam - \Delta_{vap}H^0(298\ K)etbz - D(H)n\ )/4 \quad (3.10)$$

The results of the calculations presented in Table 3.6 point to the tendency to reduce the stability of the hydrogen bonds of liquid dimethylanilinecompared with aniline and to its stabilization at a maximum distance of the relative position of the methyl groups in the benzene ring.

D-N•••H-N: 2,6-Dimethylaniline (4.90) ≈ 2,4-Dimethylaniline (4.95) < 2,5-Dimethylaniline (5.25 kJ mol^{-1})

The ethyl isostructural group displays to a greater extent the destabilizing influence on the energy of the hydrogen bond.

$$\text{Aniline } (5.10) > 2,4\text{-Dimethylaniline } (4.95) > 2\text{-Ethylaniline } \left(4.70\ \text{kJ mol}^{-1}\right).$$

The location of the two ethyl groups at the carbon atom with the amino group is accompanied

$$2,6\text{-Dimethylaniline } (4.90) > 2.6\text{-Diethylaniline } \left(3.50\ \text{kJ mol}^{-1}\right)$$

by maximum destabilization of the hydrogen bond.

Considering the above, it should be concluded that the isostructural methyl and ethyl groups participate in the distribution of the electron density in the aniline molecule, leading to the reduction of the electron density and negative charge at the nitrogen atom of the amino group. Increasing the number of carbon atoms in the ethyl group is accompanied by an increasing number of destabilizing effects, because the energy for 2,6-diethylaniline (3.50) reaches the reduced value of the same bond in liquid ammonia (3.86 kJ mol^{-1}) owing to the hydrogen bonds of the amino groups,.

The problem of the specific interactions of aerated hydrocarbons was considered earlier [6] Sect. 3.1. Nevertheless, these aspects of this interaction problem did not affect the methyl groups at isopropyl compounds. In an isostructural methyl group the carbon atom with the essentially unshared $2s^2$ electron pair gets the additional electron density from the hydrogen atoms, leading to a relatively high negative charge to the hydrogen atoms and positive charge to the hydrogen atoms, causing

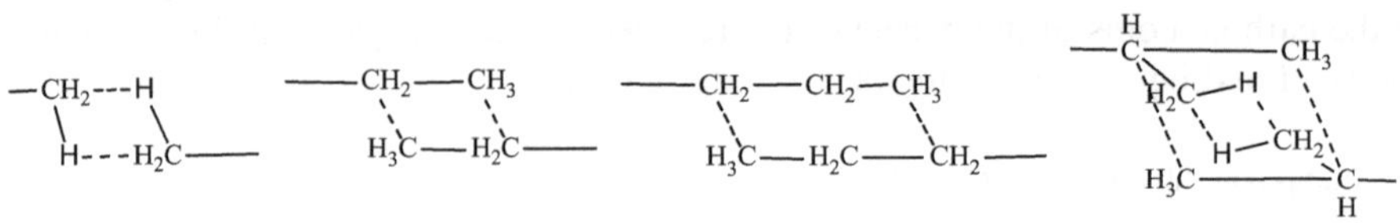

4.8:2 = 2.4 9.2 :2 = 4.6 13.0 :2 = 6.5 kJ mol^{-1} 11.9 - 9.2 = 2.7
DisoCH$_3$= 2.7: 2 = 1.35

corresponding donor and acceptor properties as also the possibility to form specific interactions of low stability. Increase in the number of carbon atoms in the ethyl isostructural group is accompanied by the formation of the biggest difference of charges at the carbon atoms CH_2 and terminal methyl groups. Hence, the principle of achievement of the minimum energy by the system leads to the interaction of the given carbon atoms of the essentially unshared $2s^2$ electron pairs, leading to more stable specific interactions and pentacoordinated condition to each carbon atom [6, 16–18]. The energy of this interaction is significantly higher than the energy of the specific interactions formed by an isostructural methyl group. Further increase of the carbon atoms in the propyl isostructural group is accompanied by increased shifting of the electron density to the terminal methyl group and demonstration of the

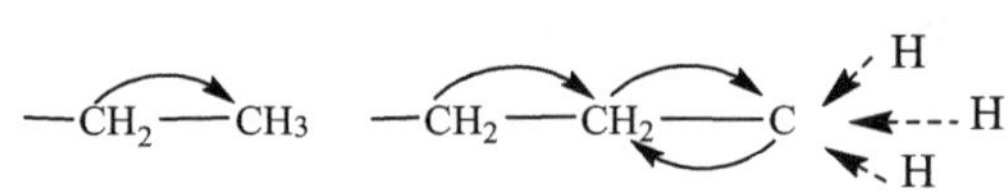

reverse dative bond, which does not preclude the formation of increased negative charge at the carbon atom of the same methyl group. As a result, the isostructural propyl group forms the most stable specific interaction. The manifestation of the reverse dative bond at the three carbon atoms provides this group with a maximum energy contribution to the enthalpy characteristics. The contribution of the isostructural groups to the vaporization enthalpy is equal to the difference of enthalpy characteristics of methylbenzene, ethylbenzene, propylbenzene or isopropylbenzene and benzene, respectively, reduced by the energy contributed by the substituted hydrogen atom. It follows that the structure of the liquid and crystal compounds are chains, connected by strong specific interactions, formed by the CH groups of the benzene ring and stable interactions, formed by ethyl, propyl isostructural groups (Fig. 3.11). Chains are connected by weak interactions of the hydrogen atoms of the CH group of the benzene ring and isostructural group in networks with specific interactions.

The results of the calculations given above point to the increased contribution of the energies in the series of isostructural methyl – propyl groups and stabilization of the specific interactions $DH_3C \rightarrow H{-}CH_2$, $D{-}CH_2{-}CH_2{-}CH_3 \rightarrow CH_2{-}CH_2{-}CH_3$,

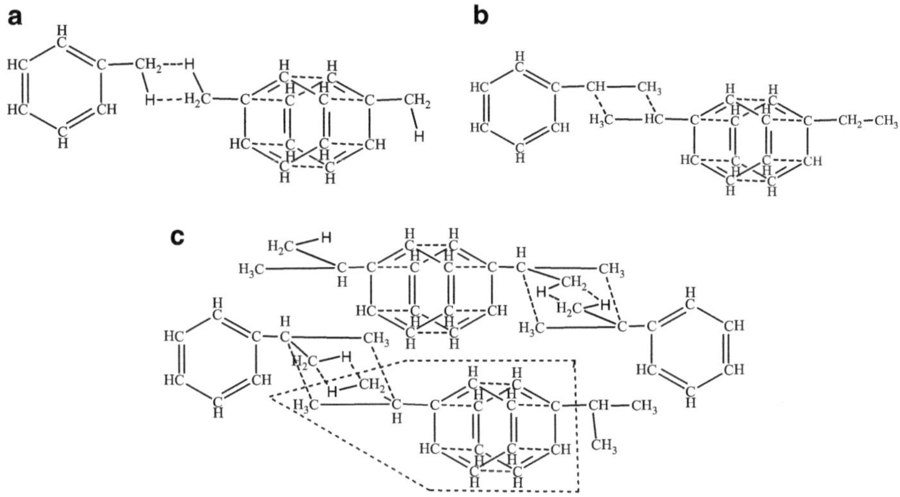

Fig. 3.11 Schematic picture of the liquid and crystalline structure of methylbenzene (**a**), ethylbenzene (**b**), and isopropylbenzene (**c**)

D–CH_2–CH_3 → CH_2–CH_3 equal to 2.4, 4.6, and 6.5 kJ mol^{-1}, respectively. Taking into account that the ethyl isostructural group contributes an equal energy to the enthalpy characteristics with a similar group like isopropylbenzene, we determined the value of the contribution of the isostructural methyl group of the isopropyl fragment from the difference between the energy contributions of the latter and ethyl ligand ethylbenzene and the energy of the specific interaction of low stability, DH_3C → H–CH_2=1.35 kJ mol^{-1}. From the established energies of the specific interactions of the same series, formed by isostructural methyl groups methylbenzene and isopropylbenzene, it follows that the interaction of the isostructural methyl group of the isopropyl fragment is approximately two times less stable.

The structure of molecules of the benzene-derivatives of compounds with the isostructural alkyl group stands practically in the position of defining the structure of liquid and crystal benzeneamines. That is why the structure of liquid and crystal 2-isopropylaniline should provide a network, containing the chains linked to the specific interactions (Fig. 3.12). The value of the energy contribution of the specific interactions of isostructural propyl groups 2-isopropylaniline and 2,6-diisopropylaniline is identical to that contributed by the interactions in liquid isopropylbenzene and 1,3-diisopropylbenzene. Hence, the energy of the hydrogen bonds of the given amines are determined with the help of a similar equation, taking the contribution of one saturated hydrogen atom by the amino-group (Fig. 3.11).

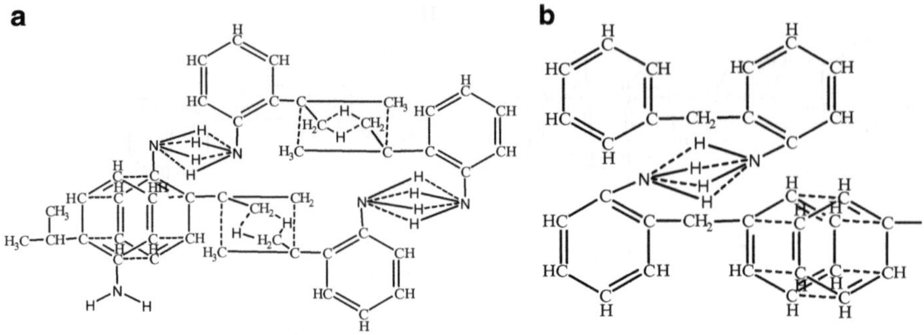

Fig. 3.12 Schematic picture of the liquid structure of 2-isopropylaniline (a) and *N*-benzylaniline (**b**) with network of the specific interactions and hydrogen bonds

$$D\text{-}N\bullet\bullet\bullet H\text{-}N = (\Delta_{vap}H^0(298\ K)iso\text{an} - \Delta_{vap}H^0(298\ K)iso.\text{bz} - D(H)n) / 4 \quad (3.11)$$

The results of the implemented thermodynamic calculations of the energies of the hydrogen bonds of iso-propyl derivatives of aniline (Table 3.7) allow us to present the integral values of reducing energies of these bonds.

$$\text{Aniline }(5.10) > \text{2-tert-Butylaniline }(3.95) > \text{2-Ethylaniline }(4.70) > \text{2-Isopropylaniline }\left(4.30\ \text{kJ mol}^{-1}\right)$$

This reflects the participation of the isostructural methyl group of isoropyl ligand in the destabilization of the hydrogen bond and more significantly the influence of the two isostructural ethyl and isopropyl groups in the destabilizing effect.

$$\text{Aniline }(5.10) > 2,6\text{-Dimethylaniline }(4.90) > 2.6\text{-Diethylaniline }(3.50) = 2,6\text{-Diisopropylaniline }\left(3.50\ \text{kJ mol}^{-1}\right) >$$

One should mention that the specific interaction, formed by ethyl group *N*-ethylaniline can be determined with the help of transformed Eq. 3.12, if we take

$$D\text{-}N\rightarrow CH_3\text{-}N = (\Delta_{vap}H^0(298\ K)\text{an} - \Delta_{vap}H^0(298\ K)\text{bz} - D(H)n - 2\ D\text{-}N\bullet\bullet\bullet H\text{-}N) /2 \quad (3.12a)$$

the energy of the hydrogen bond equal to the one realized in liquid *N*-methylaniline. The value obtained for the energy of the specific interaction $N \rightarrow CH_3\text{–}CH_2\text{–}N$ ($7.25\ \text{kJ mol}^{-1}$) is higher than the energy of the similar interaction of liquid diethyl ether, where the negative charge is significantly higher than the charge of the nitrogen atom in diethylamine. It means that the benzene cycle shifting the electron density to the nitrogen atom provides it with a negative charge higher than the charge of the oxygen atom of the diethyl ether.

Table 3.8 Contribution energy of the isostructural methyl group to the vaporization enthalpy of alkyl derivative benzene and hydrocarbons

Compounds	$\Delta_{vap}H^\circ$ (T) [9]	D*iso*CH_3/n	Compounds	$\Delta_{vap}H^\circ$ (T) [9]	D*iso* CH_3/n
Ethylbenzene	42.2 ± 0.1 42.4 ± 0.1	–	*iso*-Propane	21.6	2.8
iso-Propylbenzene	45.1 ± 0.1	2.7	2,2-Dimethylpropane	21.85	1.50
1,3-Diisopropylbenzene	56.2 ± 0.8	2.7	2-Methylbutane	25.0	2.6
tert-Butylbenzene	47.5 ± 0.4	2.65	2,2-Dimethylbutane	27.7 ± 0.1	2.65
–	–	–	2,2-Dimethylpentane	32.4 ± 0.1	2.9
–	–	–	3,3-Dimethylpentane	33.0 ± 0.1	3.3

To obtain correct values of the energies of the specific interactions, consistent with experimentally established thermodynamic characteristics of the process of vaporization of aromatic compounds with the nitrogen atom in a functional group, an isostructural group and isostructural methyl groups, it is necessary to make a more detailed thermodynamic analysis, considering the contributions of the groups and adequately reflecting the specific interaction in liquid and crystal compounds. In Table 3.8 the energies contributed by the isostructural methyl groups of alkyl compounds of benzene and hydrocarbons are given, established from the difference between the enthalpy characteristics of compounds containing an equal number of carbon atoms in the alkyl chain, for example:

$\Delta_{vap}H^0$(298 K) *iso*-Propylbenzene - $\Delta_{vap}H^0$(298 K) Ethylbenzene, $\Delta_{vap}H^0$(298 K) *tert*-Butylbenzene - $\Delta_{vap}H^0$(298 K) Ethylbenzene, $\Delta_{vap}H^0$(298 K) 2,2-Dimethylpropane - $\Delta_{vap}H^0$(298 K) Propane.

First of all, attention is drawn to the fact that the contribution of the isostructural methyl group of alkyl benzene compounds, *iso*-propane and 2-methylbutane are located within the limits of the experimental error of the vaporization enthalpies obtained. We have the minimum value of the energy contribution by an isostructural methyl group to the compounds with two isostructural methyl groups located at the one chain carbon atom, and this is typically the tendency of the growth of the energy contribution to the enthalpy characteristics and, consequently, stabilization of the specific interaction in the series of compounds,

$$\text{2,2-Dimethylpropane}\,(1.50) < \text{2-Methylbutane}\,(2.6) <$$
$$\text{2.2-Dimethylbutane}\,(2.65) < \text{2.2-Dimethylpentane}\,(2.9)$$
$$< \text{3.3-Dimethylpentane}\,(3.3\,\text{kJ}\,\text{mol}^{-1})$$

at first, with increasing chain length of molecules, and second, with removal of two isostructural CH_3 groups to the central carbon atom.

It leads to the conclusion that the minimum value of the energy contribution of the isostructural methyl group to the vaporization enthalpy of 2,2-dimethylpropane (1.50 kJ mol^{-1}) is based on the possibility of its participation in the redistribution of the electron density between two terminal methyl groups, in which the carbon atom of the isostructural CH_3-group takes minimum charge. Taking into account that the energy contribution to the vaporization enthalpy (1.50 kJ mol^{-1}) is caused by the interaction of two methyl groups of contacting molecules that are close in the environment with the formation of two specific interactions $—H_2C\overset{H}{\underset{H}{\rightleftarrows}}CH_2—$ of low stability, the real value of the energy of this interaction is equal to 0.75 kJ mol^{-1}.

3.5.2 *Crystalline Diaminebenzenes and Naphthylamine*

The most dense and energetically favorable structure of crystal diamines is reached at the location that is close to two molecules in the environment, forming stable chains, caused by interactions of the CH groups of the benzene ring and amino groups (Fig. 3.13), forming a tetragonal bi-pyramidal structure.

Chains are connected by weak interactions with participation of the hydrogen atoms and carbon atoms of the benzene ring. The contribution of the benzene ring to the sublimation enthalpy of diamine is equal to the sublimation enthalpy reduced by the contribution of two substituted hydrogen atoms (2.0 kJ mol^{-1}). Hence, the energy of the hydrogen bonds of diaminebenzenes is determined with the help of Eq. 3.13.

$$DH\text{-}N\cdots H\text{-}N = (\Delta_{vap}H^0(298\ K)damb - \Delta_{vap}H^0(298\ K)bz - 2D\ (H))\ /\ 8 \qquad (3.13)$$

The results of the calculations (Table 3.9) illustrate the stabilization of the hydrogen bonds dependent on the place of the amine groups in diamine molecules.

$$1,2\text{-Diaminebenzene}\ (5.36) < 1,3\text{-Diaminebenzene}\ (6.00) < 1,4\text{-Diaminebenzene}\ \left(6.20\ kJ\ mol^{-1}\right)$$

This sequence of the hydrogen bond stabilization is accompanied by the above mentioned dependence of mesomeric splitting of e_{1g}-orbitals on the location of derivatives in the benzene ring [2]. The minimum value of the energy of the hydrogen bonds of crystal 1,2-diaminebenzene (5.35) exceeds by 0.25 kJ mol^{-1} the value of the same bond of liquid aniline (5.10 kJ mol^{-1}). It allows us to conclude that on formation by the molecule the large number of hydrogen bonds D–N•••H–N is not accompanied by destabilization.

Naphthylamines form a crystal similar to aminobenzene. Hydrogen bond energies in the crystal condition with one amino-group in a molecule are determined from the difference of sublimation enthalpies of naphthylamine and naphthalene reduced by the contribution of the saturated hydrogen atom to the amine fragment, using Eq. 3.13a.

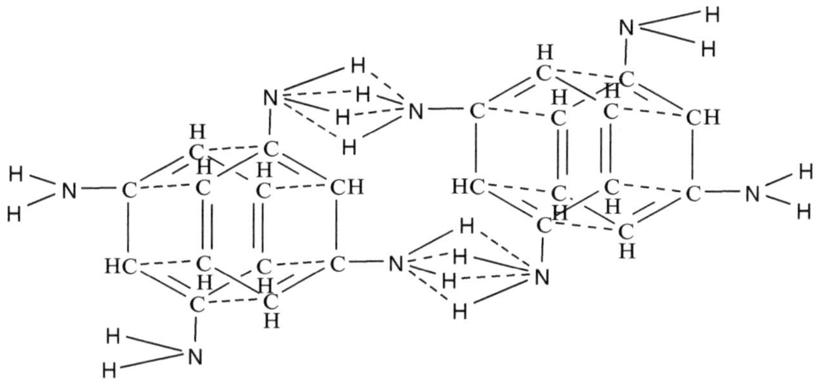

Fig. 3.13 Schematic picture of the crystalline structure of 1,3diaminebenzene

Table 3.9 Energies of the specific interactions (kJ mol^{-1}) of the crystalline diaminebenzenes at 298 K

Compounds	Formula	Structure	$\Delta_{sub}H°(298\ K)$ [9]	$\Delta_{sub}H°(T)$bz–D(H)n	D–N•••H–N
Benzene	C_6H_6		44.6	42.6	–
1,2-Diaminebenzene	$C_6H_8N_2$	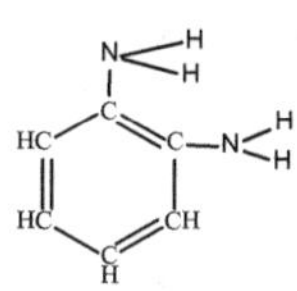	85.5 ± 0.3	42.6	5.35
1,3-Diaminebenzene	$C_6H_8N_2$	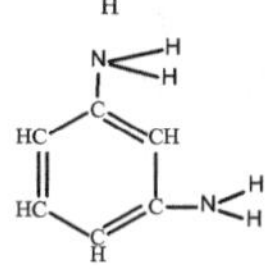	90.4 ± 0.4	42.6	6.00
1,4-Diaminebenzene	$C_6H_8N_2$	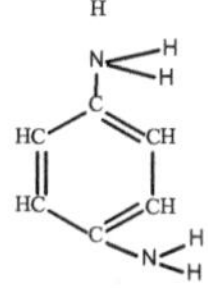	92.2 ± 0.2	42.6	6.20
Naphthalene	$C_{12}H_8$	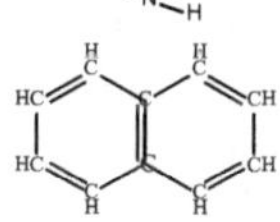	72.1 ± 0.25	71.1	–
1-Naphthylamine	$C_{10}H_9N$	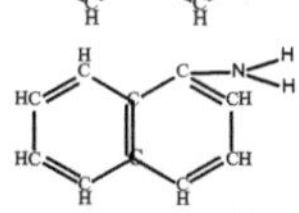	90.0 ± 4.2	71.1	4.7
2-Naphthylamine	$C_{10}H_9N$	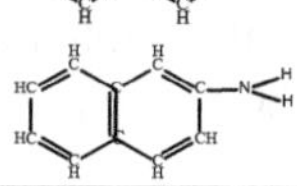	88.3 ± 4.2	71.1	4.3

$$D\text{-}N\bullet\bullet\bullet H\text{-}N = (\Delta_{sub}H^0(298\ K)napha - \Delta_{sub}H^0(298\ K)naphl - D(H)) / 4 \quad (3.13a)$$

The established energies of the hydrogen bonds (Table 3.9) have a reduced stability compared with the same bonds of diaminebenzene and as observed in crystal ammonia (6.34 kJ mol^{-1}) in the planar structure with four coordination. It follows that the presence of two cycles in the naphthylamine molecules with a reduced number of hydrogen atoms for the cycle reduces the shifting of the electron density to the nitrogen atom and its donor properties. Reducing of the hydrogen bond stability at the replacement location in diaminebenzene by 2-naphthylamine reflects the difference between charges at the carbon atoms at locations C(1) and C(2) at its reduced value in the latter.

3.5.3 Phenyl-2-Propylamine and Phenyl-2-Butaneamines

The location of the amine group at the second carbon atom of the propyl fragment leads to difficulties in the shifting of the electron density from the rigid structure of the benzene ring. Hence, the nitrogen atom obtains a negative charge at the shifting of the electron density from the hydrogen atoms or carbon atoms directly connected to it. But the high capability of the nitrogen atom to present a part of its electrons, located at p_z-orbitals, provides it with low negative charges and a low difference of charges with carbon atoms. As a consequence, the compounds of this series have hydrogen bonds and specific interactions of low stability. The following characteristic property of phenyl-2-butaneamine is by implementation by the methyl group, located in position C(1), of the function of the isostructural group. As a result, phenyl-2-propylamine and phenyl-2-butaneamine and its derivatives have an identical propyl fragment and the methyl groups located at C(2) and C(3) are also isostructural groups. The planar structure of the benzene ring with propilamine fragment and 12 bond vacancies form a network structure in which the amine groups contribute volume direction (Fig. 3.14). Six CH groups of the benzene ring form five specific interactions D=HC → CH and one D=HC → C, which energies are determined from benzene vaporization enthalpy (5.75 kJ mol^{-1}) reduced by the energy contributed to the enthalpy characteristics by the substituted hydrogen atom with energy 5.45 kJ mol^{-1}. The propyl ligand forms two specific interactions, $D\text{–}CH_2\text{–}CH_2\text{–}CH_3 \rightarrow CH_2\text{–}CH_2\text{–}CH_3$, and four hydrogen bonds, D–N•••H–N. Determination of the unknown energies of the specific interactions and the hydrogen bond can be simplified, taking into account the special location of the amino group, practically excluding the shifting of electronic density from the benzene ring.

In this connection one should use the energy contribution of the amine group, which should be determined from the differences of the vaporization enthalpies of phenyl-2-propylamine and propylbenzene reduced by the energy contributed by the saturated hydrogen atom equal to

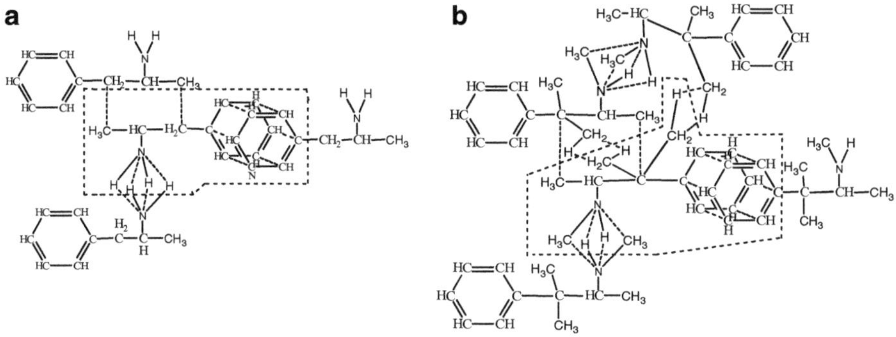

Fig. 3.14 Schematic picture of the liquid and crystalline structure of 1-phenyl-2-propylamine (**a**) and *N*-methyl-3-methyl-3-phenyl-2-butaneamine (**b**)

$$D\text{-}N\cdots H\text{-}N = (\Delta_{vap}H^0(298\ K)\ phpa - (\Delta_{vap}H^0(298\ K)pbz - D(H)n)/4 \quad (3.14)$$

0.60 kJ mol^{-1} for liquid compounds at standard temperature. The energy of the specific interaction formed by the propyl ligand is determined from the difference of vaporization enthalpies of propylbenzene and benzene reduced by the energy contributed by the substituted hydrogen of the latter,

$$D\text{-}CH_2\text{-}CH_2\text{-}CH_3\rightarrow CH_2\text{-}CH_2\text{-}CH_3 = (\Delta_{vap}H^0(298\ K)pbz - (\Delta_{vap}H^0(298\ K)bz - D(H)n)/2 \quad (3.14a)$$

The results of the calculations (Table 3.10) direct attention to the low stability of the hydrogen bond (2.55 kJ mol^{-1}), obtained at the temperature maintained in the experiment, 344 K.

Derivatives of phenyl-2-butaneamine with isostructural methyl groups form a similar structure of liquid condition with amine fragments and specific interactions; the number depends on the number of isostructural methyl groups.

Using similar approaches, the energy of the specific interactions of the methyl groups of the amine fragment is determined by the difference of the vaporization enthalpies of *N,N*-dimethyl-3-methyl-3-phenyl-2-butaneamine or *N,N*-dimethyl-3-methyl-3-phenyl-2-butaneamine, respectively, reduced by the vaporization enthalpies of propylbenzene, the energy contributed by the isostructural methyl groups, and the contribution of the substituted hydrogen atom of propylbenzene.

$$D\text{-}N\rightarrow CH_3\text{-}N = (\Delta_{vap}H^0(298\ K)\ mphba - (\Delta_{vap}H^0(298\ K)pbz - D(H)n) - mD\mathit{iso}CH_3)/4 \quad (3.14b)$$

A similar equation is used for calculation of the hydrogen bond energy, formed in liquid *N*-methyl-3-methyl-3-phenyl-2-butaneamine and *N*-methyl-2,3-dimethyl-3-phenyl-2-butaneamine

Table 3.10 Energies of the specific interactions (kJ mol^{-1}) of the liquid phenyl-2-propylamine and derivative phenyl-2-butaneamines

Compounds	Formula	Structure	$\Delta_{vap}H°(T)$ [9]	$\Delta_{vap}H°(298\ K)bz - D(H)n$	$DCH_2–CH_2–CH_3 \rightarrow CH_2–CH_2–CH_3$	DN•••H–N
Benzene	C_6H_6	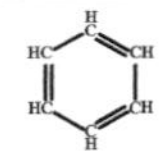	33.8 31.4 (346 K)	33.2 31.4	–	–
Propylbenzene	C_9H_{12}	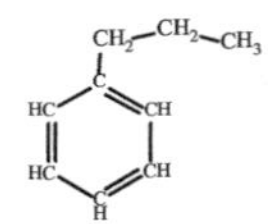	46.2 (298 K) 43.8 (355 K)	33.2 (298 K) 31.2 (355 K)	6.5 (298 K) 6.2 (355 K)	–
1-Phenyl-2-propylamine	$C_9H_{13}N$	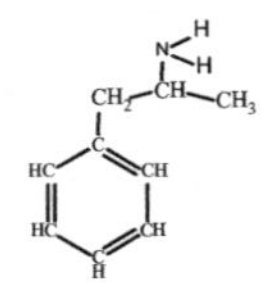	53.4 (343 K)	43.2 30.8 (355 K)	6.2 (343 K)	2.55
N,N-Dimethyl-3-methyl-3-phenyl-2- butaneamine	$C_{13}H_{21}N$	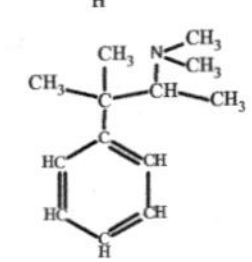	60.3 ± 0.7 (298)	45.6 33.2 2D*iso*CH_3 = 3.6	6.5 D–N → CH_3–N = 2.9	–
N,N-Dimethyl-3-methyl-3-phenyl-2- butaneamine	$C_{14}H_{23}N$	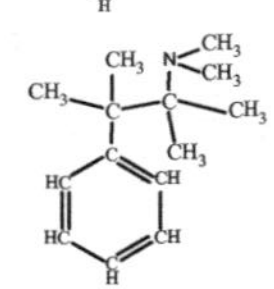	66.4 ± 1.3 (298)	45.6 33.2 2D*iso*CH_3 = 3.8	6.5 DN → CH_3–N = 3.82	–

N-Methyl-3-methyl-3-phenyl-2-butaneamine	$C_{12}H_{19}N$	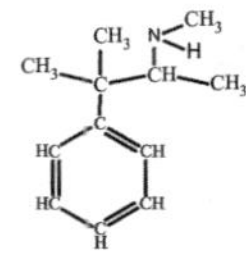	67.5 ± 0.8 (298)	45.6 33.2 2D*iso*CH_3 = 3.8	6.5 D–N → CH_3–N = 3.82	4.45
N-Methyl-2,3-dimethyl-3-phenyl-2-butaneamine	$C_{13}H_{21}N$	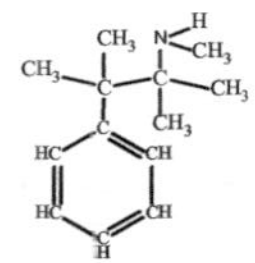	72.5 ± 1.1 (298)	45.6 33.2 3D*iso*CH_3 = 5.4	6.5 D–N → CH_3–N = 3.82	6.05

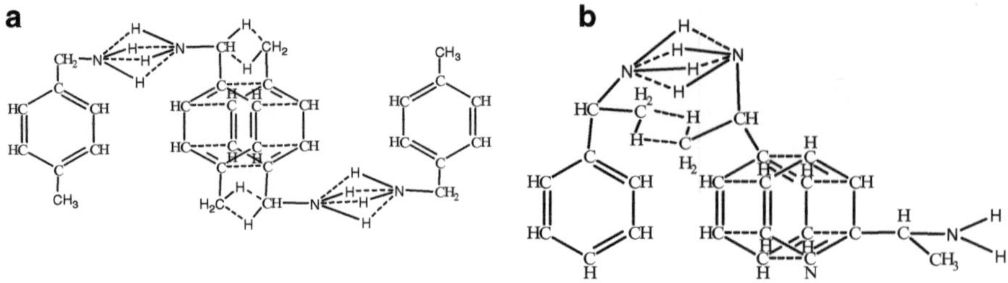

Fig. 3.15 Schematic picture of the liquid and crystalline structure of the amines: 4-methylbenzylamine (**a**), α- methylbenzylamine (**b**)

$$D\text{-}N\cdots H\text{-}N = (\Delta_{vap}H^0(298\ K)\ mphba - (\Delta_{vap}H^0(298\ K)pbz - D(H)n) - 2D\text{-}N\rightarrow CH_3\text{-}N - mD iso CH_3)/2 \quad (3.14c)$$

in which m is the number of isostructural methyl groups and n the number of saturated hydrogen atoms in the propylbenzene molecule. The energy contribution of the isostructural methyl group was obtained from the differences in vaporization enthalpies of sobuthylbenzene and propylbenzene, equal to 1.80 kJ mol^{-1}. The results of the calculations point to the stabilization of specific interactions formed by the hydrogen bonds of the methyl group and by the hydrogen atom of amino groups

$DN\rightarrow CH_3\text{-}N$: N,N-Dimethyl-3-methyl-3-phenyl-2- butaneamine (2.9) < N,N-Dimethyl-3-methyl-3-phenyl-2- butaneamine (3.82 kJ mol^{-1})

$D\text{-}N\cdots H\text{-}N$: N-Methyl-3-methyl-3-phenyl-2-butaneamine (4.45) < N-Methyl-2,3-dimethyl-3-phenyl-2-butaneamine (6.05 kJ mol^{-1})

and reach maximum values at the location of the amino and isostructural methyl group at the second carbon atom in the propyl ligand, which reflects the special role of isostructural group in the distribution of the electron density in molecules of *N*, *N*-dimethyl-3-methyl-3-phenyl-2- butaneamine and *N*-methyl-2,3-dimethyl-3-phenyl-2-butaneamine.

3.5.4 Amines: The Toluene Derivatives

Amine – derivatives of toluene form the structure of liquid and crystal condition which is similar to aniline and we have the same number of hydrogen bonds and specific interactions, formed by the CH groups of the benzene ring (Fig. 3.15a).

The total value of the hydrogen bond energies is equal to the difference in vaporization enthalpies of the amine reduced by the corresponding toluene derivative and the contributed energy of the substituted hydrogen atom at methyl group 0.60 or

1.0 kJ mol^{-1} for liquid and crystal condition, respectively. Thus, the energy of the hydrogen bond should be determined with the help of Eq. 3.15.

$$D\text{-}N\bullet\bullet\bullet H\text{-}N = (\Delta_{vap}H^0(298\ K)bam. - (\Delta_{vap}H^0(298\ K)toe - D(H)n))\ /\ 4 \quad (3.15)$$

The results of the calculations of the hydrogen bond energies of (Table 3.11) benzylamine and 4-methylbenzylamine illustrate its difference, which could be attributed to the discrepancy in the temperature conditions of experimentally measured vaporization enthalpies of these compounds (317 and 368 K, respectively). The reduced stability of the hydrogen bonds of toluene derivatives compared with the same bonds of aniline (5.1 kJ mol^{-1}) and its derivatives and amines should be attributed to the influence of the methylene group connected directly with the amino group and benzene ring, preventing shifting of the electron density to nitrogen.

A characteristic feature of α-methylbenzylamines is the presence of the methyl group, replacing the hydrogen atom in the methylene group, connected with nitrogen atom and benzene ring. It leads them to the implementation of the function of the isostructural methyl group and isomeric molecules forms of (+)α-methylbenzylamine and (−)α-methylbenzylamine with difference in the values of vaporization enthalpies 0.5 kJ mol^{-1}, forming specific interactions of low stability (Fig. 3.14b) with the values of the energies, lying practically within the experimental error. It means that the total value of the hydrogen bond energies formed by the nitrogen atom and two hydrogen atoms of the amino-group is equal to the difference in vaporization enthalpies of amino derivatives of toluene reduced by the contribution of the substituted hydrogen atoms and the isostructural methyl groups. As in the previous case, the hydrogen bond energy is established with the help of Eq. 3.16.

$$D\text{-}N\bullet\bullet\bullet H\text{-}N = (\Delta_{vap}H^0(298\ K)mbam. - (\Delta_{vap}H^0(298\ K)toe. - D(H)n) - D_{iso}CH_3)\ m)\ /\ 4 \quad (3.16)$$

A similar equation is used for the determination of the energy contributed by the specific interaction, formed by methyl groups of the amine fragment of *N*, *N*-dimethylbenzyl.

$$DN{\rightarrow}CH_3\text{-}N = (\Delta_{vap}H^0(298\ K). - (\Delta_{vap}H^0(298\ K)toe. - D(H)n\ ) - D_{iso}CH_3)\ m)\ /\ 4 \quad (3.16a)$$

The amino-group of the molecule of α, α-dimethylbenzylamine forms two hydrogen bonds and two specific interactions D–N → CH_3–N; hence, to obtain the hydrogen bond energy one should use Eq. 3.16b

Table 3.11 Energies of the specific interactions (kJ mol^{-1}) of the liquid benzylamines and methylbenzylamines

Compounds	Formula	Structure	$\Delta_{vap}H°(T)$ [9]	T K	$\Delta_{vap}H°(T)$tl–D(H)n/DN → CH_3	D–N•••H–N
Toluene	C_7H_8	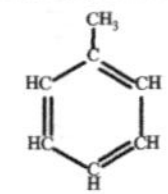	38	298	37.4	–
Benzylamine	C_7H_9N	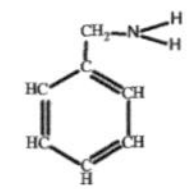	51.8	317	37.4	3.60
1,4-Dimethylbenzene	C_8H_{10}	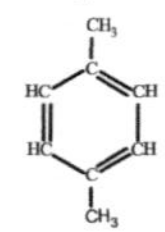	42.3 ± 0.1	298	41.7	–
4-Methylbenzylamine	$C_8H_{11}N$	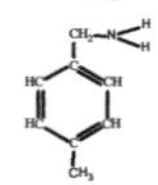	54.4	368	41.7	3.20
(+)α- Methylbenzylamine	$C_8H_{11}N$	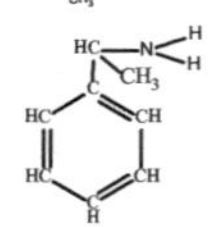	54.1 ± 0.1	298	36.8 D*iso*CH_3 = 1.50	4.00
(−)α- Methylbenzylamine	$C_8H_{11}N$	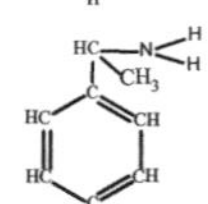	54.6 ± 0.1	298	36.8 D*iso*CH_3 = 1.50	4.10
(Dl)α- Methylbenzylamine	$C_8H_{11}N$	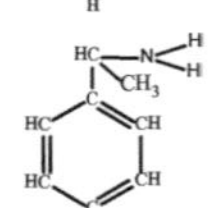	54.5 ± 0.1	298	36.8 D*iso*CH_3 = 1.50	4.05

N,N-Dimethylbenzyl amine	$C_9H_{13}N$	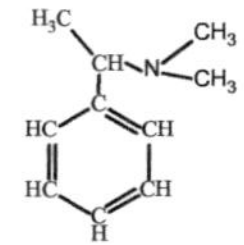	49.5 ± 0.4 50.1 ± 0.9	298	36.2 D*iso*CH_3 1.50	D–N → CH_3–N = 2.95
α,α- Dimethylbenzylamine	$C_9H_{13}N$	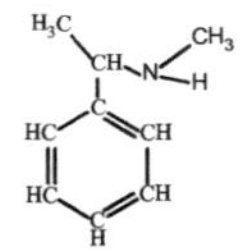	56.7 ± 0.7	298	D*iso*CH_3 = 1.50	D–N•••H–N = 6.20 D–N → CH_3–N = 2.95

$$D\text{-}N\cdots H\text{-}N = (\Delta_{vap}H^0(298\ K)dmba. - (\Delta_{vap}H^0(298\ K)dmbz. - D(H)n\ ) - 2D\text{-}N\rightarrow CH_3\text{-}N - D_{iso}CH_3)\ m) / 2 \quad (3.16b)$$

in which we have taken the energies contributed by the specific interactions D–N → CH_3–N of the methyl groups of the amine fragments *N,N*-dimethylbenzyl and α, α-dimethylbenzylamine.

The results of the calculations point to the fact that values of the hydrogen bond energies (Table 3.11) of benzylamines show a natural tendency to stabilize

$$\text{4-Methylbenzylamine (3.20)} < \text{Benzylamine} \left(3.60\right) <$$
$$\left(+\right)\alpha\text{-Methylbenzylamine} \left(4.0\right) < \left(-\right)\alpha\text{-Methylbenzylamine (4.10)}$$
$$< \alpha, \alpha\text{-Dimethylbenzylamine} \left(6.20\,\text{kJ mol}^{-1}\right)$$

except 4-methylbenzylamine, for which the value of the energy of the hydrogen bond corresponds to 368 K. The reduced stability of the hydrogen bond of α-methylbenzylamine compared with the energy of the same bond of aniline (5.10 kJ mol^{-1}) reflects the influence of the methylene group of the toluene derivative, preventing displacement of the electron density from the benzene ring to the amine group. The stabilization of the hydrogen bond of α,α-dimethylbenzylamine is effected by the increased mixing of the electron density from the remaining hydrogen atom to the nitrogen atom, ensuring the differences in charges of the nitrogen and hydrogen atoms of the amino group compared with the charges of the same atoms of α-methylbenzylamines are double the number of the latter.

3.5.5 *Toluidines*

From the relatively big number of experimentally established vaporization enthalpies of toluidines only a small part is defined at the standard or near standard conditions (Table 3.12). The toluidines presented in Table 3.12 should be divided into three groups. The first are the compounds with methyl group, i.e., amines of toluene derivatives. The second are compounds with two methyl groups dimethyl-toluidines, replaced the hydrogen atoms of the CH_3-group of toluene. However, in the given alkyl fragment only one methyl group acts as isostructural group, while the second one together with the methylene group acts as ethyl ligand and together with the CH_3 group of isostructure is a isopropyl fragment (Fig. 3.16) participating in the formation of the network of specific interactions in liquid and crystal conditions.

The third group is *N*-ethyl-3-toluidine, an ethyl group which together with the methylene group, connected directly with carbon atom of benzene, acts as a propyl

Table 3.12 Energies of the specific interactions and the hydrogen bond (kJ mol^{-1}) of the liquid and crystalline toluidines

Compounds	Formula	Structure	$\Delta_{evap}H°$ (298 K) [9, 11]	T K	$\Delta_{evap}H°$(298 K) tl –D(H)n	DN•••H–N
Liquid						
Toluene	C_7H_8	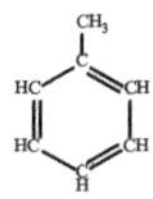	38	298	36.8	–
o-Toluidine	C_7H_9N	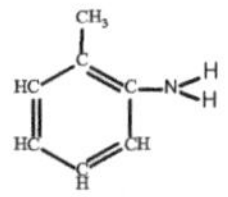	57.8	300	37.4	5.08
p-Toluidine	C_7H_9N	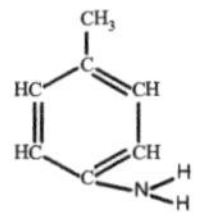	54.9	330	37.4	4.40
Iso-Propylbenzene	C_9H_{12}	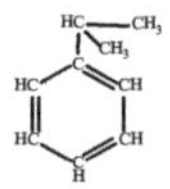	45.1 ± 0.1	298	44.5	–
N,*N*-Dimethyl-2-toluidine	$C_9H_{13}N$	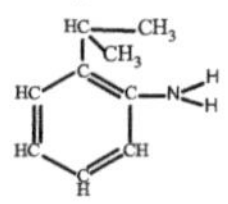	52.4	316	43.9	2.10
N,*N*-Dimethyl-3-toluidine	$C_9H_{13}N$	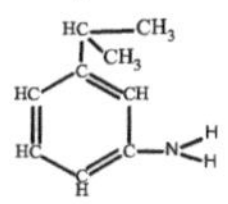	58.7 ± 6.9	298	43.9	3.70
N,*N*-Dimethyl-4-toluidine	$C_9H_{13}N$	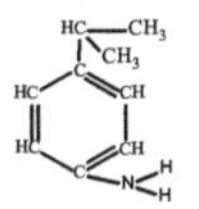	60.7	332	43.9	4.20
Propylbensene	C_9H_{12}	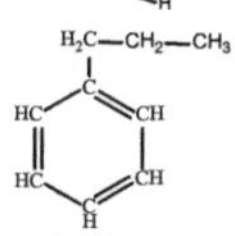	46.2	298	45.6	–
N-Ethyl-3-toluidine	$C_9H_{13}N$	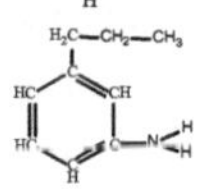	60.0 ± 3.0	298	45.0	4.00
Solid						
Toluene	C_7H_8	–	43.1	298	42.1	–
p-Toluidine	C_7H_9N	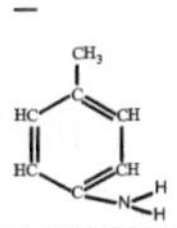	78.8 ± 0.5	298	42.1	9.18

fragment. This ligand practically plays an independent role in the complex molecule of propylbenzene. From the differences in vaporization enthalpies of toluene, isopropylbenzene, propylbenzene, andbenzene, respectively, we obtain the contribution of corresponding groups to the vaporization enthalpy of the first representatives of this series.

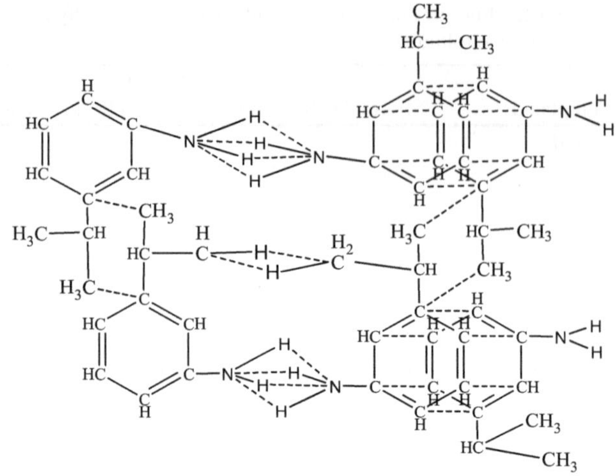

Fig. 3.16 Schematic picture of the liquid and crystalline structure of dimethyltoluidine

$-CH_3$	$-CH-(CH_3)-CH_3$	$-CH_2-CH_2-CH_3$
5.4	11.0	13.6 kJ mol^{-1}
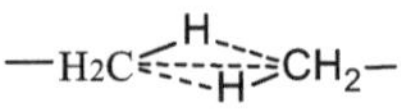	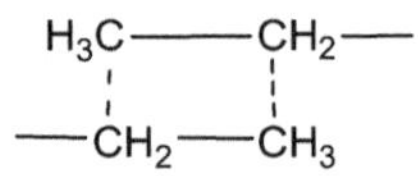	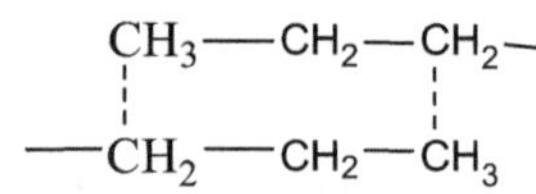
2.7	5.5	6.8 kJ mol^{-1}
D–O → CH_3	D–O → CH_3-CH_2-	D–O → $CH_3-CH_2-CH_2-$
5.63	6.78	8.92 kJ mol^{-1}

Since the isostructural methyl group of the isopropyl fragment contributes 1.50 kJ mol^{-1}, the values of the energies obtained are relevant for the two specific interactions. The sequence of its stabilization corresponds to the integral value, in particular, in liquid simple ethers. The maximum value of the specific interactions is reached, practically, for the compounds with propyl ligand, which is caused by reduction of the influence of the inter-molecular reverse dative bond, ending at the propyl fragment. The energies of the hydrogen bonds of the marked toluidines should be determined with the help of Eqs. 3.17, 3.18, and 3.19.

$$\text{D-N}\cdots\text{H-N} = (\Delta_{vap}H^0(298\text{ K})\text{tli.} - \Delta_{vap}H^0(298\text{ K})\text{to.} - \text{D(H)n)}) / 4 \qquad (3.17)$$

$$\text{D-N}\cdots\text{H-N} = (\Delta_{vap}H^0(298\text{ K})\text{tli.} - \Delta_{vap}H^0(298\text{ K})\text{ispb.} - \text{D(H)n)}) / 4 \qquad (3.18)$$

$$\text{D-N}\cdots\text{H-N} = (\Delta_{vap}H^0(298\text{ K})\text{tli.} - \Delta_{vap}H^0(298\text{ K})\text{pb.} - \text{D(H)n)}) / 4 \qquad (3.19)$$

As is evident from Table 3.12, the established energies of the hydrogen bonds are more stable at *o*-toluidine and *p*-toluidine, in which molecules there is no isostructural methyl group. Compounds with an isopropyl fragment are characterized by the change in stability of the hydrogen bonds in correspondence with the location of the amino groups in the molecule.

N,N-Dimethyl-2-toluidine $(2.10)\,316\,K <$ N,N-Dimethyl-3-toluidine $(3.70)\,298\,K <$ N,N-Dimethyl-4-toluidine $(4.20\,kJ\,mol^{-1})\,332\,K$

Replacement at *o*-toluidine (5.08 kJ mol^{-1}) of the methyl group by the isopropyl ligand leads to *N,N*-dimethyl-2-toluidine (2,10 *N,N*-dimethyl-2- toluidine, 316 K), reduces by more than double the energy of the hydrogen bond. In its turn, the replacement of the isopropyl ligand at *N,N*-dimethyl-3- toluidine (3.70 kJ mol^{-1}) by the propyl at *N*-ethyl-3-toluidine (4.00 kJ mol^{-1}) is accompanied by its stabilization at 0.30 kJ mol^{-1}. Thus, replacement of the methyl group in the molecule *o*-toluidine by the ethyl one at the isostructural CH_3 group of *N,N*-dimethyl-2-toluidine is accompanied by destabilization of the hydrogen bond; however, the further increase in the number of carbon atoms at the propyl fragment leads to the stabilization of the hydrogen bond from 2.10 to 4.00 in liquid *N*-ethyl-3-toluidine. On the basis of this result it must be concluded that the isostructural methyl group participates in the redistribution of the electron density and distribution of charges not only at the functional group but also at the benzene ring. This is evident at the value of the energy of the hydrogen bond of *N,N*-dimethyl-2-toluidine. When removing the amino group from its location in the isopropyl fragment, the charge of the carbon atom of the benzene ring changes little and the destabilizing effect of isostructural methyl group is reduced.

3.5.6 Biphenylamines

Compounds with amino group with two or more benzene rings form a similar network of the specific interactions at the structures of liquid and crystal conditions (Fig. 3.17). Reducing the enthalpy of sublimation at 5.0 and vaporization at 2.1 kJ mol^{-1} of biphenyl compared with two C_6H_5 groups introduced and change of the average value of the contribution of the CH group for benzene from 7.20 to 6.82 and for biphenyl from 5.53 to 5.37 kJ mol^{-1} in crystal and liquid condition, respectively, (Table 3.13) reflects the reduction of charges of the carbon atoms at the benzene cycles and reducing the stability of the C–C bond, respectively, connecting carbon atoms of the two cycles. On the contrary, the replacement of the C–C bond at biphenyl by the interactions by CH group at diphenylmethane is accompanied by an increase in enthalpies of sublimation at 5.8 and vaporization at 2.1 kJ mol^{-1}. It follows that in the thermodynamic analysis it is necessary to use biphenyl and diphenylmethane molecules, because the constituent molecules of benzene will lead to significant errors. Such conversions exist at the replacement of the C–C bond pointed out by the amino group *N*-diphenylamine. In this bond we refer to the fact that the replacement of the C–C bond between carbon atoms at biphenyl formed on the CH_2 group (5.8 kJ mol^{-1}) is accompanied by stabilization of the crystal field from 2.1 in liquid condition to 3.8 kJ mol^{-1} in crystal. Thus, the procedure of thermodynamic calculation of the hydrogen bond energies and

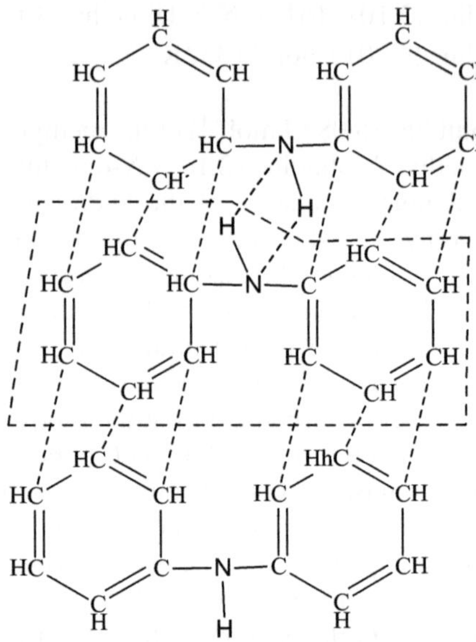

Fig. 3.17 Schematic picture of the liquid and crystalline structure of *N*-diphenylamine with the network of the specific interactions

Table 3.13 Energies of the hydrogen bond (kJ mol^{-1}) of the liquid derivative biphenylamines

Compounds	Formula	Structure	$\Delta_{evap}H^{\circ}$ (298 K) [9, 11]	T K	$\Delta_{evap}H^{\circ}$ (298 K) dhlm –D (H)n	DN•••H–N
Crystal						
Benzene	C_6H_6	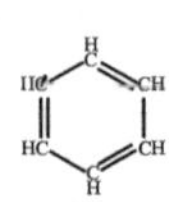	44.4 44.6	298	43.4	–
Biphenyl	$C_{12}H_{10}$	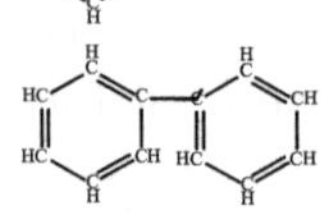	81.8 ± 0.2	298	–	–
Diphenylmethane	$C_{13}H_{12}$	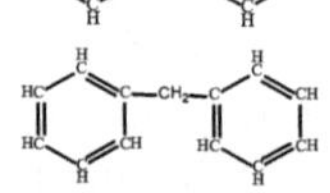	87.67 ± 0.8	298	86.6	–
N-Phenyl benzylamine	$C_{13}H_{13}N$	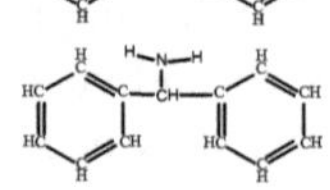	103.6 ± 1.6	303	86.6	4.25
N-Diphenylamine	$C_{12}H_{11}N$	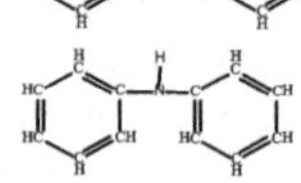	96.7 ± 2.3	310	80.8	4.5
Liquid						
Phenylnaphthalene	$C_{16}H_{12}$	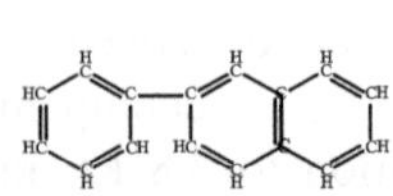	85.9	298	–	–
N-Phenyl-1-naphthylamine	$C_{16}H_{13}N$	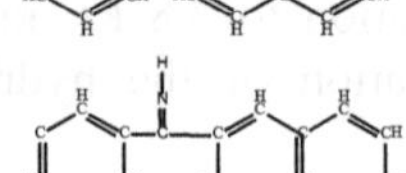	89.6	353	85.3	2.2

specific interactions , formed by the pentacoordinated carbon atom, as suggested above should take into account the contributions of the energies to the enthalpy characteristics of compound molecules reduced by the contribution of the energies of the substituted hydrogen atoms.

Of the large number of vaporization enthalpies of aminobiphenyls only a very limited part is defined at the standard conditions (Table 3.13). The energy of the hydrogen bonds of *N*-phenylbenzylamine is determined from the difference with enthalpy of sublimation of diphenylmethane

$$\text{D-N}\bullet\bullet\bullet\text{H-N} = (\Delta_{sub}H^0(298\ \text{K})\text{phba.} - \Delta_{sub}H^0(298\ \text{K})\text{dph.m}) / 4 \qquad (3.20)$$

and formed in crystal *N*-diphenylamine, from the difference

$$\text{D-N}\bullet\bullet\bullet\text{H-N} = (\Delta_{sub}H^0(298\ \text{K})\text{dpha.} - \Delta_{sub}H^0(298\ \text{K})\text{bph.}) / 2 \qquad (3.20a)$$

of sublimation enthalpies with biphenyl reduced by the contribution of the energy of the substituted hydrogen atom of the methylene group with the help of Eqs. 3.20 and 3.20a.

From the data of Table 3.13 it follows that like benzylamine (3.60 kJ mol^{-1}, Table 3.11) the energy of the hydrogen bond of crystal *N*-phenylbenzylamine (4.25 kJ mol^{-1}) also has the low value that is based on the methylene group, located between benzene rings. It means that a carbon atom of this group has a low negative charge, providing it with low donor properties. The absence of the methylene group in a molecule of *N*-biphenylamine leads to insignificant stabilization of the formed hydrogen bond. Thereby, the number of the hydrogen bond stabilization

$$\text{Benzylamine}\ \left(3.60\ \text{kJ mol}^{-1}\right) < N\text{-Phenyl benzylamine}\ (4.25) < N\text{-Diphenylamine}\ \left(4.5\ \text{kJ mol}^{-1}\right)$$

allows us to draw the conclusion that two benzene cycles of the *N*-diphenylamine molecule stabilize the hydrogen bond at 0.9 and the methylene group *N*-phenyl benzylamine at 0.65 kJ mol^{-1}, respectively, compared with benzylamine.

Taking the average value of the contribution of the energy by the CH group of liquid 5.37 kJ mol^{-1} and crystal 6.83 kJ mol^{-1} of biphenyl as the permanent value of similar compounds, in particular phenylnaphthalene, we have the possibility to use the sum of the contribution of the average values of the energies of the CH group of naphthalene with one saturated hydrogen atom for estimation of the values of vaporization enthalpies 85.9 and sublimation 109.3 kJ mol^{-1}. Using the latest, we form the differences of the enthalpy characteristics of *N*-phenyl-1-naphthylamine or *N*-phenyl-2-naphthylamine and phenylnaphthalene that have the values (4.4) and (6.5 kJ mol^{-1}) and energy values of the hydrogen bonds of liquid and crystal condition equal to 2.2 and 3.2 kJ mol^{-1}, respectively. Significantly reduced values of the hydrogen bonds are caused by the discrepancy in the vaporization enthalpies used with the standard conditions and, on the other side, the

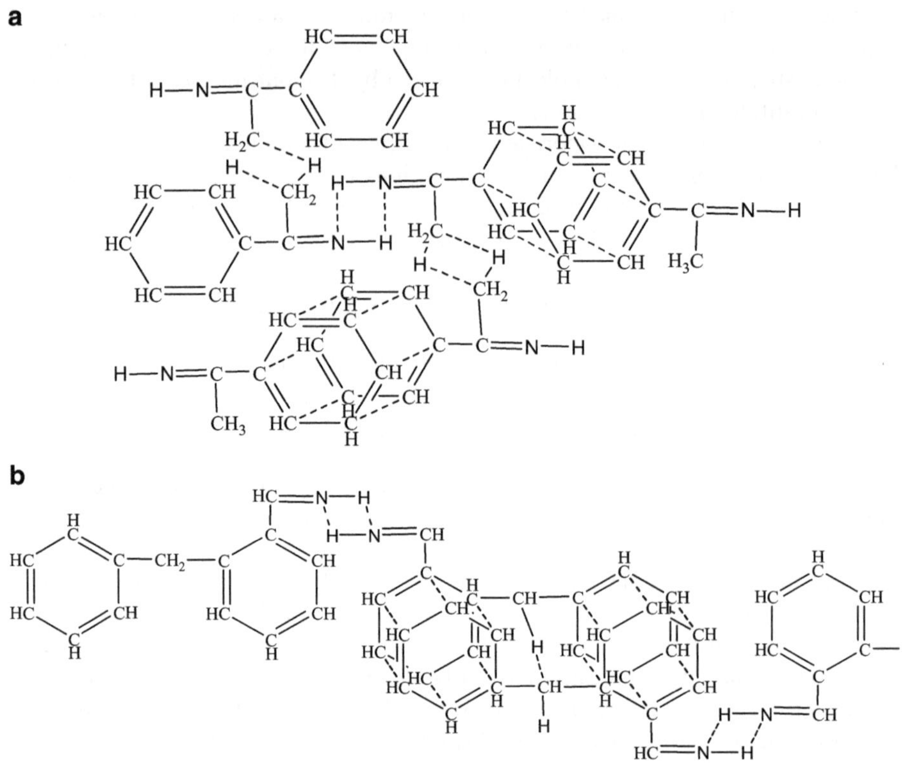

Fig. 3.18 Schematic picture of the liquid and crystalline structure of *N*-methylbenzaldehyde –imine (**a**), *N*-benzylbenzaldehydeimine (**b**) with the network of the specific interactions

unaccounted contribution of the stabilizing effect of the =C–C= bond at the formation of phenylnaphthalene.

3.5.7 Hydrogen Bonds of Benzaldehyde Imines

Imines with the benzene cycle and by cyclic structure with one =N–H group form the network of the specific interactions in liquid and crystal conditions (Fig. 3.18). In spite of the specific interactions, formed by the benzene ring molecules of imines, two hydrogen bonds D=N•••H–N=form. Formed by the methyl group molecules of *N*-methylbenzaldehydeimine the specific interactions of low stability participate together with the CH group in the sewing of chains. The methyl group of the fragment $-\underset{\mathrm{CH_3}}{\underset{|}{\mathrm{C}}}=\mathrm{N}-\mathrm{H}$ acts as an isostructural group of the isopropyl ligand, which allows us to accept the introduced energy contribution to the vaporization enthalpy equal to 1.50 kJ mol^{-1}, to draw the conclusion of similarity of the structural effects of the given group with developing molecules of *N,N*-dimethyltoluidines. In this bond the energies of the hydrogen bonds D=N•••H–N=, formed by *N*-Methylbenzaldehydeimine, (Eq. 3.20) should be obtained from the differences of vaporization enthalpies with toluene

$$D{=}N{\cdots}H\text{-}N{=}. = (\Delta_{vap}H^0(298\ K)mbd. - \Delta_{vap}H^0(298\ K)\ toe. - 3D(H) - D_{iso}CH_3)/2 \quad (3.21)$$

reduced by the contribution of three substituted hydrogen atoms and isostructural methyl group with the help of the equation of the group (Eq. 3.21).

The energy of the hydrogen bond benzophenoneimine, *N*-phenylbenzaldehydeimine should be determined from the difference in the vaporization enthalpy of diphenylmethane reduced by the energy contributed by two hydrogen atoms by methyl and methylene group and *N*-benzylbenzaldehyde-imine with enthalpy characteristics (4-methylphenyl)phenylmethane reduced by the contributions of the two substituted hydrogen atoms (Table 3.14). The results of the thermodynamic calculations (Table 3.14) basically agree with the values of the energies of the hydrogen bonds of aromatic compounds.

The relatively high value of the hydrogen bond energy of liquid *N*-methylbenzaldehyde-imine with one benzene ring is caused by the stabilizing effects with the location of the amino group in the plane of the benzene ring, the minimum number of hydrogen atoms, and the low coordinating number. Being exact in the planar structure of the ammonia molecule with the coordination of four hydrogen atoms in liquid ammonia the energy of the hydrogen bond reaches 5.91 kJ mol^{-1}, instead of 3.86 kJ mol^{-1} with coordination of six hydrogen atoms. Thus, stabilization of the hydrogen bond D=N•••H–N= of liquid *N*-methylbenzaldehydeimine (6.5 kJ mol^{-1}) is an appropriate one and reflects the increase in the nitrogen atom charge of the N–H group, resulting in the increase of its donor properties. In this connection the reduced values of the energies of the same series of the hydrogen bonds of liquid benzophenoneimine (3.65 kJ mol^{-1}) and crystal benzophenoneimine (4.05), dibenzylideneazine (3.85) and *N*-phenylbenzophenone imine (4.20 kJ mol^{-1}) seem symptomatic. This directs attention to the fact that the location of imine group between two benzene cycles which is in direct connection the carbon atom of the same imine group. It follows that the given carbon atom partly shifts its electron density to the contacting carbon atoms of the benzene rings and simultaneously reduces the nitrogen atom charge, providing it with the formation of a hydrogen bond equal to, or differing little from, the stability of the same bond in liquid ammonia (3.86 kJ mol^{-1}). In this connection anomalously high is the value of the energy of the series of the hydrogen bond of liquid *N*-benzylbenzaldehydeimine (8.35 kJ mol^{-1}), caused by the location of the imine group in the benzene ring

Table 3.14 Energies of the hydrogen bond (kJ mol^{-1}) of the liquid and crystalline aromatic imines at 298 K

Compounds	Formula	Structure	$\Delta_{evap}H°(298\ K)$ [9]	DisoN → CH_3	$\Delta_{evap}H°(298\ K)bz$ –D (H)n	D=N•••H–N=
Liquid						
Toluene	C_7H_8	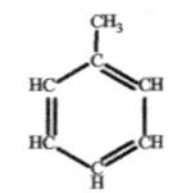	38	–	36.4	–
N-Methylbenzaldehyde-imine	C_8H_9N	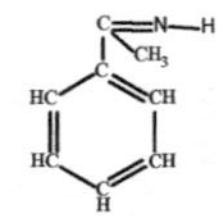	51.2 ± 0.2	1.50	36.8	6.5
Benzophenoneimine	$C_{13}H_{11}N$	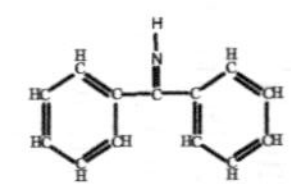	75.7 ± 1.0	–	68.4	3.65
Diphenylmethane	$C_{13}H_{12}$	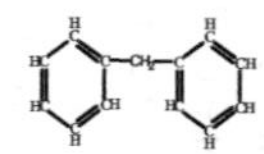	66.6 ± 0.1	–	–	–
N-Benzylbenzaldehyde-imine	$C_{14}H_{13}N$	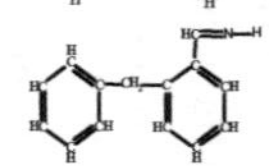	85.0 ± 1.2	–	68.3	8.35
(4-Methylphenyl)phenylmethane	$C_{14}H_{13}$	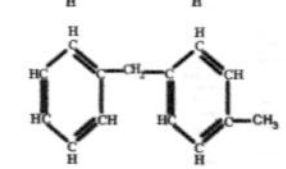	69.5 ± 0.3	–	68.3	–
Crystal [11]						
N-Cyclohexyl-(2,4,6-trimethyl) benzaldehyde imine	$C_{16}H_{23}N$	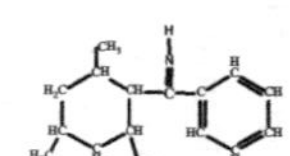	104.3±	–	–	–
N-Phenyl-benzaldehyde imine	$C_{13}H_{11}N$	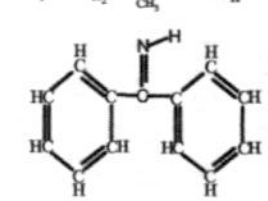	93.7 ± 0.9	–	85.6	4.05

Diphenylmethane	$C_{13}H_{12}$	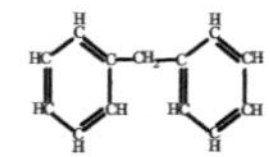	87.6 ± 0.8	–	85.6	–
Dibenzylideneazine	$C_{13}H_{12}N_2$	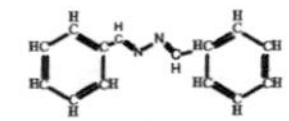	93.3 ± 2.1	–	85.6	3.85
N-Phenyl benzophenone imine	$C_{19}H_{15}N$	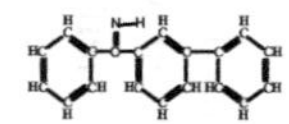	119.7 ± 1.8	–	103 ± 0.8	4.2[a]
Triphenylmethane	$C_{19}H_{16}$	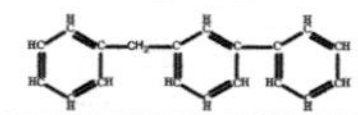	105 ± 08	–	103 ± 0.8	–

[a]Estimated

C(2) in close proximity to the methylene group (1), which shifts the electron densityand enriches the negative charge of the nitrogen atom of imine group and stabilizes the hydrogen bond formed.

The changes in the energy values of the hydrogen bonds of toluidines with two hydrogen atoms at the amino group in the range of 5.10–4.40 kJ mol^{-1}, and for dimethyltoluidines in the range 4.00–3.70 kJ mol^{-1}, are caused by the influence of the amino groups locatedn in the benzene ring. Its minimum values combined with the values of the hydrogen bonds of *N*-phenylbenzylamine and *N*-diphenylamine 4.25 and 7.50 kJ mol^{-1}, respectively, are determined by reveal; in each case this features the distribution of the electron density in the molecules of the compounds considered.

3.6 Energies of Specific Interactions of Liquid Aromatic Cyanobenzines

The double bond of the carbon atom of the benzene ring with –C–C≡ bond and –C≡N fragment leads to the formation of a planar structure of the benzonitrile molecule, which forms bonds with stable specific interactions ≡N → C≡ and CH groups of the benzene ring, connected by specific interactions of low stability D=HC → H–C, formed by hydrogen atoms with the carbon atoms of contacting molecules. The methylene group of fragment –CH_2–C≡N leads to the definite mobility and participation of the hydrogen atoms in the formation of the specific interaction of low stability, participating in linking of chains to the network system (Fig. 3.19a). The methyl groups of molecules α-methylbenzylcyanide and α,α-dimethylbenzylcyanide, replaced hydrogen atoms of the ethylene group act as the isostructure group

interacting between each other with formation of the specific interactions of low stability (Fig. 3.19b). A gradual transition from benzonitrile to benzylcyanide and further to α-methylbenzylcyanide and α,α-dimethylbenzylcyanide is accompanied by an increase in the number of specific interactions of low stability, formed by pentacoordinated carbon atom of the corresponding functional groups. Replacement of the hydrogen atoms of the CH_2 group by the methyl one is accompanied by preservation of the total number of interactions at the replacement of the hydrogen atoms by the CH_3 group. The practical invariability of the values of vaporization enthalpies α-methylbenzylcyanide (60.9 ± 0.7) and α,α-dimethylbenzylcyanide (60.6 ± 0.6 kJ mol^{-1}), respectively, with one and two isostructural methyl groups

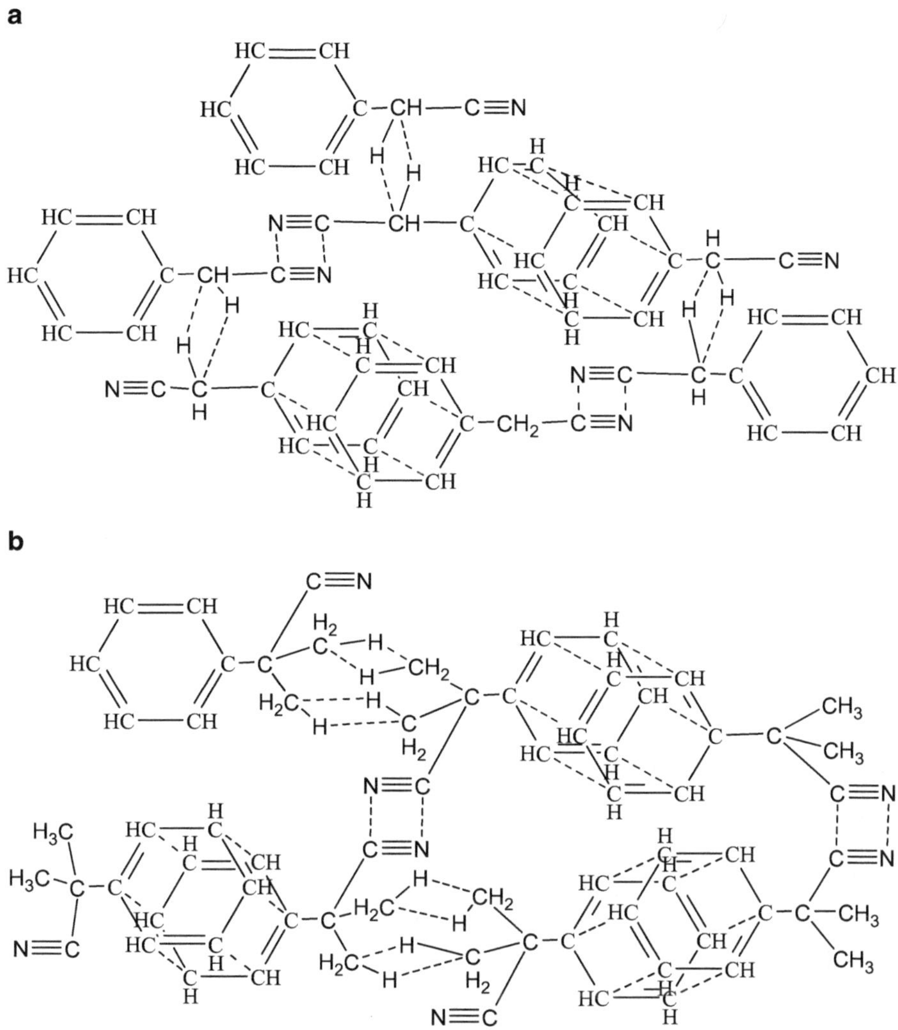

Fig. 3.19 Schematic picture of the liquid and crystalline structure of benzylcyanide (**a**), α, α-dimethylbenzylcyanide (**b**) with network of the specific interactions

in molecules direct attention to the fact that replacement of the terminal hydrogen atom in the methylene group at the first compound by the second isostructural group (Table 3.15) does not contribute additional energy to the enthalpy characteristics. One can draw the conclusion that participation of two isostructural methyl groups of molecule α,α-dimethylbenzylcyanide to the redistribution of the electron density contribute equally to the vaporization enthalpy, and is equal to the contribution by the similar series of molecule α-methylbenzylcyanide. In other words, the energy of the contribution of the isostructural methyl group α-methylbenzylcyanide is equal to the contribution of the two similar series of α,α-dimethylbenzylcyanide.

Table 3.15 Energies of the hydrogen bond (kJ mol^{-1}) of the liquid aromatic nitriles

Compounds	Formula	Structure	$\Delta_{vap}H°$(298 K) [9]	T K	$\Delta_{vap}H°$(298 K) –nD(H)/qDisoCH$_3$	D≡N → C≡
Cyclopentene	C_6H_8	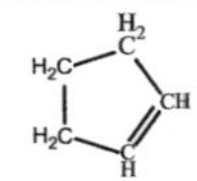	24.8 28.4	299 300	24.2 27.8	10.35 8.55
2-Cyclopentene-1-carbonitrile	C_6H_7N	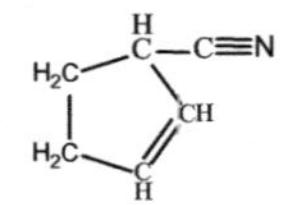	44.9 ± 0.1	298	–	–
Benzene	C_6H_6	–	33.8	298	33.2	–
Benzonitrile	C_7H_5N	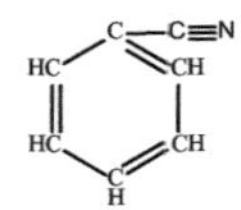	49.1	316	33.2	8.0
Toluene	C_7H_8	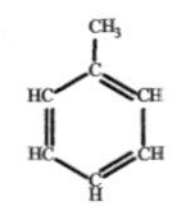	38	298	–	–
Benzylcyanide	C_8H_7N	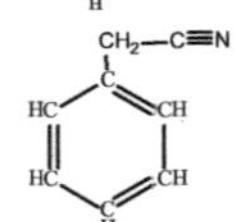	60.5 ± 0.7	298	37.4	11.6 ± 0.7
Benzylisocyanide	C_8H_7N	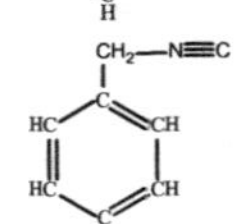	42.3 (45.8)[a]	348 298	37.4 37.4	2.5 (4.2)
α-Methylbenzylcyanide	C_9H_9N	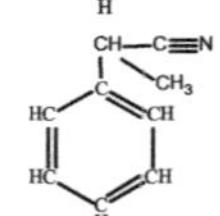	60.9 ± 0.7	298	36.8 DisoCH$_3$ = 1.50	11.3 ± 0.7

α,α-Dimethylbenzylcyanide	$C_{10}H_{11}N$	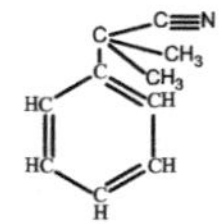	60.6 ± 0.6	298	36.2 D*iso*CH_3 × 2 = 1.50	11.45 ± 0.7
α-Ethylbenzylcyanide	$C_{10}H_{11}N$	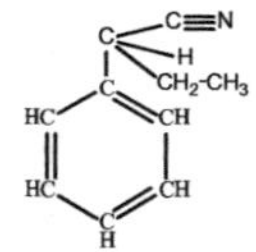	64.3 ± 0.6	298	36.2 D–CH_2CH_3 → CH_2–CH_3 = 2.45	11.45 ± 0.7

[a]Estimated

Taking into account that the fragment with the isostructural methyl group α-methylbenzylcyanide is equivalent to isopropyl we had reason to take the value of the contribution of this group in the previous chapters as equal to 1.50 kJ mol^{-1}, which is reasonable. Consequently, each of the two isostructural groups of similar fragments for α,α-dimethylbenzylcyanide has a value that equals 0.75 kJ mol^{-1}. Considering the participation in interactions of the two isostructural methyl groups of molecules that are in close contact in the environment, the energy of the contribution to the enthalpy characteristics of each such group is equal to 0.38 kJ mol^{-1}. Certainly, this conclusion has particular importance for a thermodynamic analysis of aromatic nitriles, imines, and other series of organic compounds.

As was shown above, the identity of the specific interactions and its energies, formed by the CH groups of the benzene ring, the methyl group of toluene, and with the saturated hydrogen atom of the CH_2 group, the isostructural methyl group allows the interpretation of thermodynamic characteristics and structure reorganizations of liquid and crystal conditions. Hence, using the location of relationship of the vaporization enthalpy with the number and energy of bursting specific interactions [19–21], we can state that the total value of the energy of two specific interactions D≡N → C≡ is equal to the difference of the vaporization enthalpy of benzylcyanide and benzenederivative compound reduced by the contribution of the saturated hydrogen atoms and the contribution of the isostructural groups. The energy of the specific interaction D≡N → C≡ is determined with the help of the equation:

Benzonitrile:

$$D{\equiv}N{\rightarrow}C{\equiv} = (\Delta_{vap}H^0(298\ K)bn - \Delta_{vap}H^0(298\ K)C_6H_5) / 2 \quad (3.22)$$

Benzylcyanide, Benzylisocyanide

$$D{\equiv}N{\rightarrow}C{\equiv} = (\Delta_{vap}H^0(298\ K)bc - \Delta_{vap}H^0(298\ K)tol - D(H)) / 2 \quad (3.22a)$$

α-Methylbenzylcyanide, α-Ethylbenzylcyanide, and α,α-Dimethylbenzylcyanide

$$D{\equiv}N{\rightarrow}C{\equiv} = (\Delta_{vap}H^0(298\ K)bc - \Delta_{vap}H^0(298\ K)tol - nD(H) - qD\mathit{iso}CH_3) / 2 \quad (3.22b)$$

As seen from the results of the thermodynamic calculation presented in Table 3.15, the energy of the specific interaction D≡N → C≡ does not depend on the number of substituted hydrogen atoms in the methylene group and number of isostructural groups in molecules at the α- and α,α- methylene group locations. It should be mentioned that the location of the cyanide group in the benzene ring with the CH_2 group stabilizes the specific interaction approximately at 3.5 kJ mol^{-1} compared with the similar group, located at the carbon atom of the benzene ring of benzonitrile. Taking into account the invariability of the energy of the specific interaction of D≡N → C≡ for benzylcyanides, one should define the meaning of the contribution of the isostructural ethyl group to the vaporization enthalpy, located at α-ethylbenzylcyanide, whose value is determined with the help of the equation, reorganized from Eq. 3.22c.

$$D iso CH_2\text{-}CH_3 = (\Delta_{vap}H^0(298\ K)bc - \Delta_{vap}H^0(298\ K)tol - 2D(H) - 2D{\equiv}N{\rightarrow}C{\equiv}\ )/2 \quad (3.22c)$$

The value obtained of 4.9 kJ mol^{-1} belongs to the energies of two specific interactions formed by isostructural ethyl groups of the molecules that are close in the environment in liquid

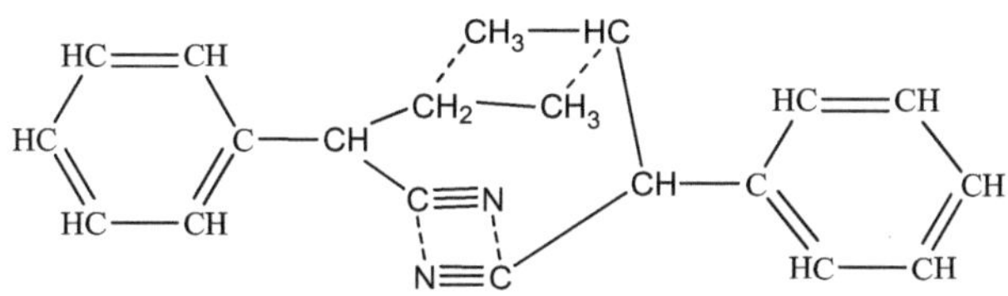

α-ethylbenzylcyanide (2.45 kJ mol^{-1}).

The calculation of the energy of the specific interaction D–N≡ → C (2.5 kJ mol^{-1}), formed by the isocyanide group –N≡C of liquid benzylisocyanide, established at 348 K, is significant because of the unaccounted temperature dependence of enthalpy characteristics. Using the temperature dependence of the vaporization enthalpy, a more correct value of the energy of this interaction can be obtained, if the vaporization enthalpies of styrene (C_8H_8), changing in the same temperature interval 43.5 ± 0.4 (298 K) and 40.2 kJ mol^{-1} (348 K) are considered. The value of differences in the vaporization enthalpy values of styrene (3.5 kJ mol^{-1}) is equal to benzylisocyanide, and hence, the value of the vaporization enthalpy at standard conditions could be taken to be equal to 45.8 kJ mol^{-1} with sufficient reliability. The results of the estimation of the vaporization enthalpy of benzylisocyanide lead to the energy of the specific interaction D–N≡ → C being equal to 4.2 kJ mol^{-1}.

The energies of the specific interactions D≡N → C≡ of liquid tolunitriles are determined from the difference of the vaporization enthalpy of tolunitriles and toluene reduced by the energy of the saturated hydrogen atom of the benzene ring on Eq. 3.22d.

$$D{\equiv}N{\rightarrow}C{\equiv}\ = (\Delta_{vap}H^0(298\ K)tn - \Delta_{vap}H^0(298\ K)tol - D(H))/2 \quad (3.22d)$$

The value of the energies of the specific interactions D≡N → C≡ tolunitriles obtained illustrates the absence of its dependence from the location of the nitrile group in the benzene cycle (Table 3.16). On the other side, a higher value of the energy of this interaction of benzonitrile (8.0), than of 2-tolunitrile and 4-tolunitrile (5.3 kJ mol^{-1}) and approximately double the understatement of the last compared with the value of the energy of the same series of interaction (11.6 kJ mol^{-1}), formed in liquid benzylcyanide, allows us to draw the conclusion of the significant influence of the methyl group of toluene on the destabilization in the specific interaction D≡N → C≡. In its turn, the methylene group of benzylcyanide, connected in the common fragment —CH_2—C≡N directly with benzene ring stabilizes the same type of the specific interaction. It follows that the methyl

Table 3.16 Energies of the specific interactions (kJ mol^{-1}) of the liquid tolunitriles

Compounds	Formula	Structure	$\Delta_{vap}H°$ (T) [9]	T K	$\Delta_{vap}H°$(T)tn –D (H)	D≡N → C≡
2-Tolunitrile	C_8H_7N	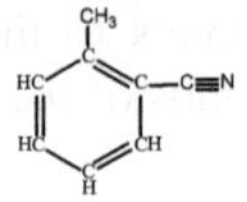	48.0	330	37.4	5.3
4-Tolunitrile	C_8H_7N	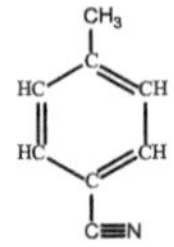	48.0	330	37.4	5.3
2-Toluizocyanide	C_8H_7N	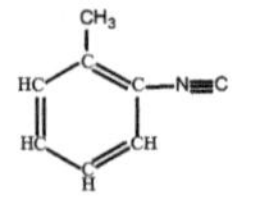	48.5	313	37.4	5.55
Phenylisocyanide	C_7H_5N	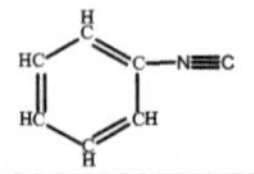	46.2	300	33.2	6.50

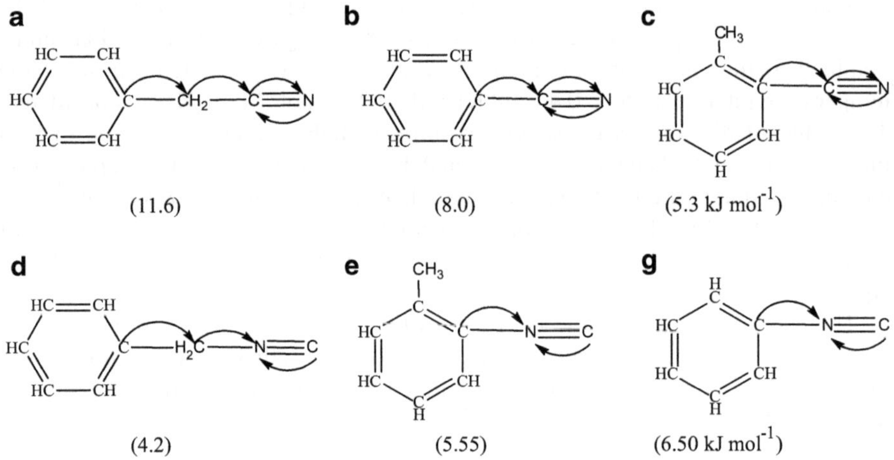

Fig. 3.20 Schematic picture of displacement electron density in molecules benzylcyanide (**a**), benzonitrile (**b**), 2-tolunitrile (c), benzylisocyanide (**d**), toluizocyanide (**e**), phenylisocyanide (**f**)

group of toluene prevents the electron density from shifting to the carbon atom of nitrile and, vice versa, the CH_2 group shifts electric density to the carbon atom of the same nitrile group and further to the nitrogen atom (Fig. 3.20).

In its turn, the electron density located at the p_z-orbital of the nitrogen atom participates lightly in the reverse dative bond. As a consequence the carbon and nitrogen atoms have a bigger difference in the values of positive and negative charges, and the nitrogen atom of benzylcyanide possesses high donor properties.

The energies of the specific interactions of D–N≡ → C≡ isocyanides are located at

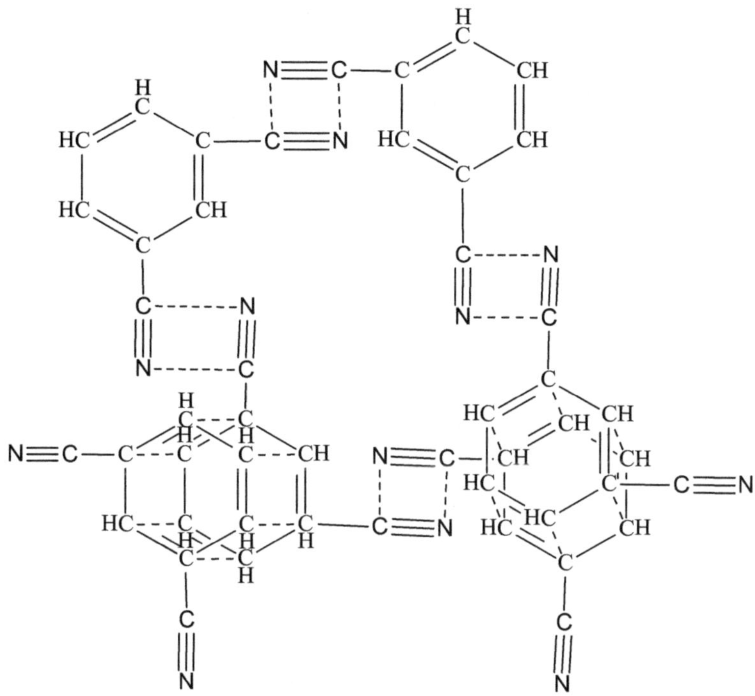

Fig. 3.21 Schematic picture of the crystalline structure of 1,3-dicyanobenzine

D-N≡→C≡ Benzylisocyanide (4.2) < D-N≡→C≡ 2-Toluizocyanide (5.55) < D-N≡→C≡ Phenylisocyanide (6.50 kJ mol^{-1})

The stabilization reflects the small difference in charges at the nitrogen and carbon atoms of the cyanide group benzylisocyanide compared with molecules of 2-toluizocyanide and phenylisocyanide. It follows that the nitrogen atom of benzylisocyanide, receiving electron density from the CH_2 group, transfers part of it to the carbon atom of N≡ → C≡, reducing its positive and negative charge. In the phenylisocyanide molecule the electron density is shifting from the carbon atoms of the benzene ring with negative charge to the nitrogen atom, which transfers its part from the $3p_z$ orbital to the same carbon atom. As a result, the carbon atom of the isocyanide group keeps its positive charge and the nitrogen atom has an increased negative charge.

The thermodynamic properties of crystalline aromatic compounds established at standard conditions with the cyanide group are known for the five compounds. Nevertheless, conducting the thermodynamic analysis is of great importance because it allows us to determine the energies of the specific interactions and develop a general representation of the structural energetic correspondence.

The molecule of 1,3-dicyanobenzine with two cyanide groups interacts in the near environment, at least, with three molecules and is able to form the cycles from the four molecules (Fig. 3.21).

Table 3.17 Energies of the specific interactions (kJ mol^{-1}) of the crystalline aromatic nitriles

Compounds	Formula	Structure	$\Delta_{vap}H°$ (T) [9]	T K	$\Delta_{vap}H°$ (T)–2D (H)	D≡N → C≡
Benzene	C_6H_6		44.6	298	42.4	–
1,2-Dicyanobenzine	$C_8H_4N_2$	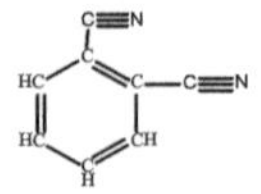	86.9 ± 1.6	298	42.6	11.1
1,3-Dicyanobenzine	$C_8H_4N_2$	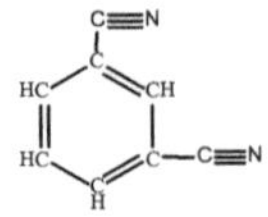	90.1 ± 1.5	298	42.6	11.9
1,4-Dicyanobenzine	$C_8H_4N_2$	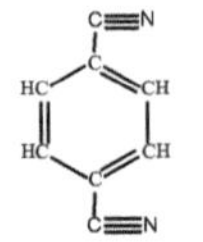	88.2 ± 2.1	298	42.4	11.4
2,6-Dimethylbenzene	C_8H_{10}	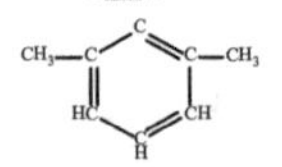	60.1	248	59.1	–
2,6-Dimethylbenzonitrile	C_9H_9N	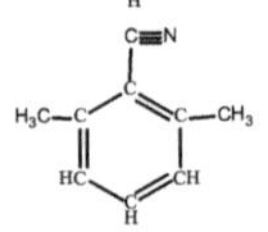	83.9 ± 2.8	298	59.1	12.4
2,4,6-Trimethylbenzene	C_9H_{12}	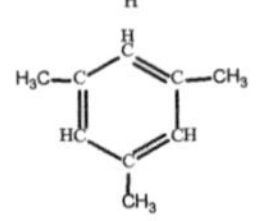	59.7	298	58.7	–
2,4,6-Trimethylbenzonitrile	$C_{10}H_{11}N$	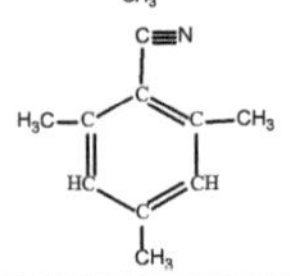	82.9 ± 1.6	298	58.7	12.1 ± 0.7

The energies of the specific interactions of D≡N → C≡ of crystal dicyanobenzines calculated with the help of Eq. 3.22e.

$$D{\equiv}N{\rightarrow}C{\equiv} = (\Delta_{vap}H^0(298K)dcb - \Delta_{vap}H^0(298K)bz - 2D(H)) / 2 \quad (3.22e)$$

Table 3.17 illustrate the small values, almost the same as the values of the same type of interaction of liquid benzenecyanides. It means that the location of the two cyanide groups at the marked carbon atoms of the benzene ring equivalently influences the shifting of the electron density from carbon atoms of the CH group to the carbon atom of –C≡N and further to the nitrogen atom, leading to the value of the energy of the specific interaction D≡N → C≡, being within the experimental error of the determination of the enthalpy characteristics.

The influence of the methyl groups, the replaced hydrogen atoms of the benzene ring, on the energies of the specific interactions can be found for example in the case of dimethylbenzonitrile and trimethylbenzonitrile. The total value of the energies of the specific interactions for dimethylbenzonitrile and trimethylbenzonitrile is equal to the difference in sublimation enthalpies with enthalpy characteristics of 2,6-dimethylbenzene and 2,4,6-trimethylbenzene, respectively, reduced by the contribution of the energies of the substituted hydrogen atoms of the benzene ring to the nitrile fragment –C≡N. The enthalpy characteristics of 2,6-dimethylbenzene used in calculations and established at a temperature of 248 K leads to some reduction in the value of the specific interaction energy of D≡N → C≡ (12.4 kJ mol^{-1}); hence, its established values, lying within experimental error, reflect the tendency of its change. Considering this fact, the values of the specific interactions energies of dicyanobenzines and methylbenzonitrile should be described as

$$\text{1,2-Dicyanobenzine } (11.1) \approx \text{1,4-Dicyanobenzine } (11.4) \leq$$
$$\text{1,3-Dicyanobenzine } (11.9) \approx \text{2,4,6-Trimethylbenzonitrile } (12.1)$$
$$< \text{2,6-Dimethylbenzonitrile } \left(12.4 \text{ kJ mol}^{-1}\right)$$

the tendency of its stabilization from the compounds with two nitrile groups to the compounds with one nitrile group. Taking into account the stabilizing influence of the toluene methyl group, the established values of the energies of the specific interactions of the methyl-derivatives of benzonitriles point to the stabilizing influence of the two methyl groups located at the second and sixth carbon atoms of the benzene ring of the 2,6-dimethylbenzonitrile molecule.

References

1. Lyc AE, Gaborakipov VZ, Shligina IA, Petropavlov VA (1986) Photoelectron specters of saturated nitrogen containing organic compounds. Russ J Struct Chem 27(4):146–169
2. Nefedov VI, Vovna VI (1989) Electronic structure of organic and elementorganic compound. M. Nauka, Russia, p 198
3. Cauletti C, Divona ML, Gargano P et al (1986) Ring-size effect on the ionization potential of N-substituted azacycloalkanes. J Chem Soc Perkin Trans Pt Ii 2(5):667–670
4. Baev AK (2004) Insolvence of model sp^3-hybridization of carbon atom and chemistry of non aqua solutions. In: IX international conference, The problem of solvation and complex formation in solutions, Plyos, pp 62–64
5. Kalinina FL, Lobov VP, Zidkov VA (1971) Reference on biochemistry. Naukova dumka, Kiev, p 1010
6. Baev AK (2012) Specific intermolecular interactions of organic compounds. Springer, Hiedelberg/Dordrecht/London/New York, 434p
7. Ivanov EV, Puchovskie Yu P, Abrosimov VK (2004) About formation of weak hydrogen bonds CH•••O in liquid methanol Abstract. In: IX international conference, The problem of solvation and complex formation in solutions, Plyos, p 345
8. Lebedev Yu A, Miroschnichenko EA (1981) Thermochemistry of the vaporization of organic compounds. The vaporization and sublimation enthalpy and saturation pressure. M. Nauka, Moscow, Russia, p 218
9. Chickens JS, Acree WE (2002) Enthalpies of vaporization of organic and organometallic compounds, 1880–2002. J Phys Chem Ref Data 31(2):537

10. Chickens JS, Acree WE (2003) Enthalpies of sublimation of organic and organometallic compounds, 1910–2001. J Phys Chem Ref Data 32(2):519
11. Vovna VI, Vilesov FI (1974) Photoelectron specters and structure molecular orbital's methylamines. Optica Spectrosc 34(2):436–438
12. Palmer MN, Moyes W, Spiers M, Ridyard IN (1979) The electronic structure substituted benzenes; a study of aniline, the toluidines, phenilenediamines and phenoloanilines by photoelectron spectroscopy and ab initio calculation. J Mol Struct 53(2):235–249
13. Palmer MH, Moyes W, Spiers M, Ridyard JNF (1979) The electronic structure of substituted: benztnes: Ab initio calculations and photoelectron spectra for phenol, the methyl- and fluoroderivatieves, and the dihydroxybenztnes. J Mol Struct 52(2):293–307
14. Karlson TA (1981) Photo electronic and OGE-spectroscopy. Mechanical Engineering, Leningrad. 431p
15. Nefedov VI (1984) X-ray spectroscopy of chemical compounds. Chemistry, Moscow, 255p
16. Baev AK (2003) Novel approaches to the structure of functional solvents and elementorganic compounds with saturated hydrocarbon ligands and their solvates in binary systems. In: Abstracts of the 28th international conference on solution chemistry, Debresen, 23–26 Aug 2003, p 79
17. Baev AK (2009) Phenomenological significance of nonidentical sp3 hybridization model and a nature of interaction in chemistry of liquid non-electrolites and chemical thermodynamics. In: Abstracts of the X international conference on physical and coordination porphyrin's chemistry and their analogous (ICPC-10), Ivanovo, 2009, p 86
18. Baev AA, Baev AK (2011) Structural-energies properties of liquid isostructural acetilene alcohols C_{15}, C_{18} and their solutions with network of specific interactions. In: Abstracts of the XI internftionfl conference "problems of salvation and complex formation", Ivanovo, 10–14 Oct 2011, p 180
19. Baev AK (1969) Problems of chemical nature of phase transformation. Gen Appl Chem 1:197–206, Vysheischaia Educating
20. Baev AK (1969) Phase condition and complex formation ability of halogenide metals. Gen Appl Chem 1:207–218, Vysheischaia Educating
21. Baev AK (1987) Chemistry of gas-heterogenic systems of elementorganic compounds. Science & Technology, Minsk, p 175

Chapter 4
Specific Intermolecular Interactions of Nitrogen Containing Five-Membered Heterocycles

4.1 Energies of Specific Interactions of Liquid Nitrogenated Hetero Compounds with Hydrocarbon Cycle

In Sect. 3.1 we mentioned that the amino B3MO shows an antibonding influence, which destabilizes its N2s-H1s overlap [1]. At the same time, transmission of the electron density from the $3p_z$-orbital of the nitrogen atom to the 1s-orbital of hydrogen atom or, for a significantly undivided $2s^2$ electron pair of carbon atoms, respectively, a corresponding reduction in its charges, and the stability of specific interactions reflects the manifestation of the reverse dative bond.

Mentioned factors should be expressed particularly clearly at cyclic compounds with small number of carbon atoms in cycle at the presence in the molecule of amino group at the energies of the formed specific interactions, hydrogen bonds and contributions of energies of the methylene and formed by them interactions $H_2C \rightarrow CH_2$ и $H_2C \rightarrow$ H–CH.

The implementation of a thermodynamic analysis requires reliable enthalpy characteristics of the vaporization process. However, it should be noted that a lack of errors casts doubt on the correctness of experimentally measured vaporization enthalpies of aziridine (34.9 kJ mol^{-1}) and pyrrolidine. Second, there is a natural increase of the contribution of the methylene group to the enthalpy characteristics in the series of compounds

$$\text{Cyclohexane } (5.50) < \text{Cyclopentane } (5.70) < \text{Cyclobutane } (6.30) < \text{Cyclopropane } \left(6.61 \text{ kJ mol}^{-1}\right) < (CH_2)_2$$

which should be continued to the hypothetic cycle from two methylene groups $(CH_2)_2$, reducing the number from six to two, respectively, in the compounds considered.

The dependence of the vaporization enthalpy of cyclic substituted hydrocarbon compounds, is presented in Fig. 4.1

A.K. Baev, *Specific Intermolecular Interactions of Nitrogenated and Bioorganic Compounds*, DOI 10.1007/978-3-642-37472-2_4,

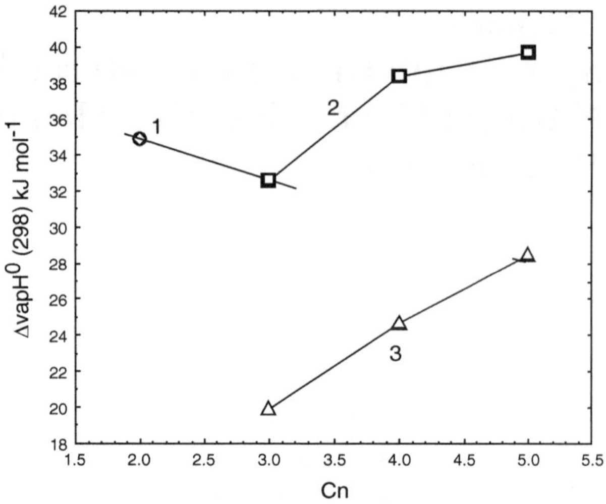

Fig. 4.1 Dependence of the vaporization enthalpy of cyclic saturated hydrocarbons and nitrogenated hetero compounds with hydrocarbon cycle versus the number of carbon atoms in alkyl chain: (*1*) aziridine (*1*), azetidine – pyrrolidine – piperidine (*2*, *3*), cyclopropane – cyclobutane – cyclopentane (*3*) [2, 3]

This and the number of carbon atoms in a cycle are well described by two straight lines, covering cyclopropane – cyclopentane and cyclohexane – cyclooctane (Chap. 3). The given character of the dependence gives the possibility to obtain the vaporization enthalpy of cyclobutane by the implemented experimental measurements of the enthalpy characteristics. The description of highly accurate experimental data by linear dependence means that at the second section of this line, covering cyclohexane – cyclooctane, each next CH_2-group contributes a similar energy (4.8 kJ mol^{-1}) to the vaporization enthalpy. In this connection, the given functional dependency allows us to use the method of extrapolation to determine the vaporization enthalpies of unstudied compounds.

At the same time, reduction of the number of methylene groups from five to three in the cycle is accompanied by a significant change in the distribution of the electron density and charges at the carbon atoms of the cycle, and, consequently, at the hydrogen atoms of this group. It follows that further reduction of the number of methylene groups in the molecule (up to three) will lead to additional changes in the distribution of the electron density in the molecule of the hypothetical compound. It will facilitate the maximum shifting of the electron density upon introduction to the structure of a heteroatom of nitrogen, providing the aziridine molecule with maximum stability of the formed hydrogen bond D–N•••H–N for all of the series of heterocyclic compounds aziridine – cyclohexylamine. Thus, the dependency of the vaporization enthalpy of compounds of the series $(CH_2)_2$ – cyclopentane on the number of methylene groups in the molecule should also be described by a straight line, allowing the use of the method of extrapolation to determine the vaporization of the hypothetical compound, which is estimated to have a value of 17.0 kJ mol^{-1} (Table 4.1).

Since the enthalpy characteristic of vaporization is related to the number of specific interactions and their energies [4–6], the values for the relation of the

Table 4.1 Vaporization enthalpy (kJ mol^{-1}) of cyclic saturated hydrocarbons and nitrogenated heterocyclic hydrocarbons

Compounds	Structure	$\Delta_{vap}H°(298\ K)$ [2, 3]	$\Delta_{vap}H°(298\ K)/n$	Compounds	Structure	$\Delta_{vap}H°(298\ K)$ [2, 3]	$\Delta_{vap}H°(298\ K)/n$
–	–	–	–	$(CH_2)_2$	–	17.0[a]	8.50
Aziridine	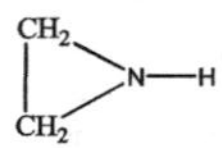	34.9 (288–293 K) 30.5[a]	7.62	Cyclopropane	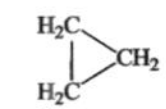	19.9 (297–359 K)	6.61
Azetidine	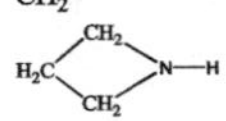	32.6 (288)	6.52	Cyclobutane	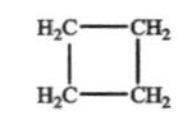	25.2 (272) 24.7 (298)	6.18
Pyrrolidine	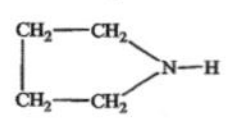	38.4 (288) 36.5[a]	6.08	Cyclopentane	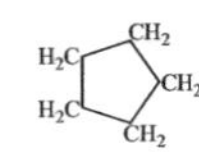	28.5 ± 0.1 (298)	5.70
Piperidine	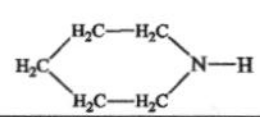	39.7 (298)	5.67	Cyclohexane, Cycloheptane	–	–	5.50

[a]Estimated by extrapolation

vaporization enthalpies to the total number of CH_2 groups and two hydrogen bonds, formed by NH-group reducing in the

Aziridine < Azetidine < Pyrrolidine < Piperidine (Table 4.1)

series, is caused by manifestation of the factors mentioned, realized in the molecule of the simplest amine $HN(CH_2)_2$ and for the more complex heterocyclic compounds. The influence of these factors is gradually reduced with the increasing number of carbon atoms in the cycle due to the redistribution of the electron density. At the same time, increased values of the relations hinted at for the given series of compounds, at the cyclic saturated hydrocarbons (Table 4.1), reflect the following.

First, the nitrogen atom influences the change of charge of the carbon atoms of all compounds of the series, providing stabilization of specific interactions $H_2C \rightarrow CH_2$ and $H_2C \rightarrow$ H–CH, formed by methylene groups of nitrogenated compounds.

Second, there is reduction of the stability of the hydrogen bond with increasing number of carbon atoms in the cycle for the series aziridine–piperidine, approaching the energy of the same bond with minimum value at piperidine and remaining constant for the compound with six or more methylene groups in the cycle. In its turn, a similar dependency of the values of the vaporization enthalpies on the number of carbon atoms in the cycle of saturated cyclic compounds reflects the reduction of the difference of the charges of close carbon atoms, in particular, $C(1)H_2$ and $C(2)H_2$ and $C(2)H_2$ and $C(3)H_2$. This factor also applies to cyclic nitrogenated compounds.

Third, the lower end of the energy of the specific interaction $H_2C \rightarrow CH_2$ reaches the saturated minimum value of the energy of the same series of interaction of liquid cyclopentane (5.70 kJ mol^{-1}).

Saturated heterocyclic amines form a chain structure for liquid and crystal conditions with possible angles of rotation of the molecules close in the environment of 45°, 60°, 90°, and 180° for reaching the location of the methylene groups contacting with maximum favorable charge difference of carbon atoms of the cycle (Fig. 4.2). Bond vacancies of the amino groups form chains with the interacting molecules, united in the network by specific interactions of $H_2C \rightarrow$ H–CH, HC $\rightarrow$ H–C (Fig. 4.2) of reduced stability.

The contribution of specific interaction energies to the enthalpy characteristics, formed by methylene groups, depends on the number of the latter and is considered by vaporization enthalpy of the corresponding hydrocarbonated cycle. Consequently, the total contribution of the energies of two hydrogen bonds to the enthalpy characteristics is equal to the difference in the vaporization enthalpies of heterocyclic amine and the corresponding saturated hydrocarbon cycle, for example, aziridine and hypothetical $(CH_2)_2$ or azetidine and cyclopropane:

$$D\text{-}N\cdots H\text{-}N = (\Delta_{vap}H^0(298\ K)azd - \Delta_{vap}H^0(298\ K)cpr)/2 \qquad (4.1)$$

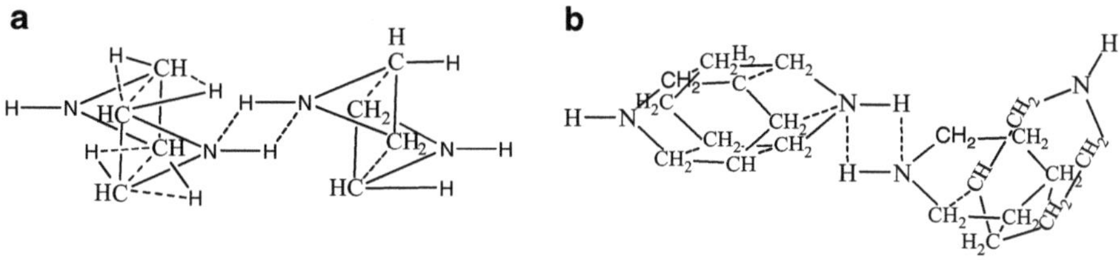

Fig. 4.2 Schematic picture of the structure of liquid saturated heterocyclic nitrogen compounds of aziridine (**a**), piperidine (**b**)

The results of the calculations of energies of the hydrogen bonds (Table 4.2) of heterocyclic amines point to the wide range of its change – from a value of 6.75 in liquid aziridine to 5.85 kJ mol^{-1} for pyrrolidine and 5.60 kJ mol^{-1} for piperidine, remaining the same without change beyond that. The hydrogen bond of aziridine is characterized by stability, which is almost the same as the stability of the same bond of liquid ammonia (5.90 kJ mol^{-1}) with a four-coordinated nitrogen atom. The reduced stability of this series at liquid piperidine (5.60 kJ mol^{-1}) means that heterocycles with five or more methylene groups shift less electron density to the nitrogen atom compared with two hydrogen atoms of ammonia. Thus, we conclude that in the aziridine molecule the carbon atoms and hydrogen atoms have the highest positive charges, more stable hydrogen bonds, and specific interactions D–H_2C → H–CH, DHC → H–C, formed by CH_2 groups, compared with other compounds of this series. It means that heterocyclic compounds with a nitrogen atom in the cycle, in contrast to linear and aromatic amines, are characterized by a strengthening of the role of the reversed dative bond with an increasing number of carbon atoms in the chain cycle,

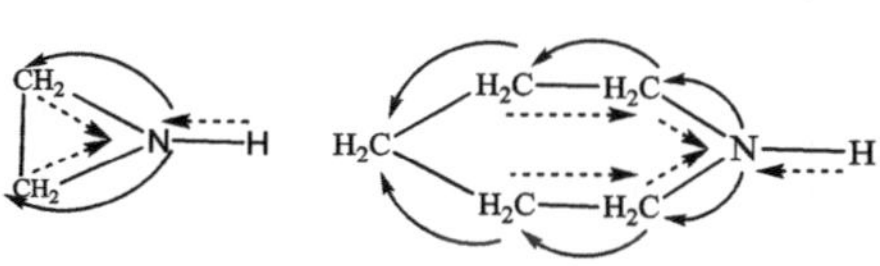

whose influence is totally ended at the six CH_2 groups in molecules with a cycle for the heterocyclic compound. It follows that the minimum possible value of the energies of H_2C → CH_2, H_2C → H–CH of specific interactions of piperidine is equal to the energies of these bonds realized in liquid cyclopentane (5.70 kJ mol^{-1}). Taking this value and considering the energy contributed by the hydrogen bond (5.60 kJ mol^{-1}) to the enthalpy characteristics, we established the vaporization enthalpy of piperidine at the standard conditions to be equal to 39.8 kJ mol^{-1}. The results of the calculations on hydrogen bonds and summary values of the energies, formed by methylene groups of cycles H_2C → CH_2, H_2C → H–CH are given in Table 4.2. Using the value of the energy of specific interaction DH_2C → H–CH of low stability (0.60 kJ mol^{-1}) established in Chap. 3, we determined the energies of interaction DH_2C → CH_2, described by the integral value of its stabilization

Table 4.2 Energies of hydrogen bonds (kJ mol^{-1}) of liquid nitrogenated hetero compounds with saturated hydrocarbon cycle

Compounds	Formula	Structure	$\Delta_{vap}H°(298\ K)$ [3]	T K	$\Delta_{vap}H°(298\ K)$ chc	D–N•••H–N	DCH_2
Aziridine	$N(CH_2)_2$	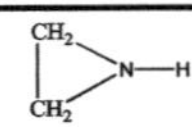	34.9	288–293	17.0	6.75	8.50
			30.5[a]				
Azetidine	C_3H_7N	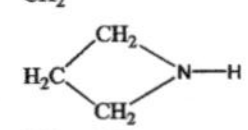	32.6	288	19.9	6.35	6.31
Pyrrolidine	C_4H_9N	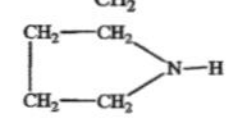	38.4	288	25.3	5.85	6.20
			36.5[a]		24.8		
Piperidine	$C_5H_{11}N$	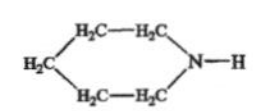	39.8	298 (338)	28.5	5.60	5.70
			36.6				
Hexahydroazocine	$C_6H_{13}N$	–	44.2[b]	298	33.1 ± 0.1	5.60	5.51
Heptahydroazocine	$C_7H_{15}N$	–	49.7[b]	298	38.5 ± 0.1	5.60	5.50
Octahydroazocine	$C_8H_{17}N$	–	54.5[b]	298	43.3 ± 0.2	5.60	5.41
1-Methyl-pirrolidine	$C_5H_{11}N$	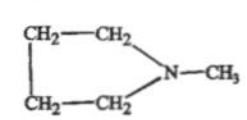	34.2 ± 0.7	298	24.8	D–N → CH_3–N = 4.70	5.90
			35.2 ± 0.7	284			

[a]Estimated by interpolation
[b]Extrapolation of dependency of the vaporization enthalpy of heterocyclic saturated hydrocarbons from the number of methylene groups

D-$H_2C \rightarrow CH_2$: Aziridine (7.90) < Azetidine (5.70) < Pyrrolidine (5.60) < Piperidine (5.10 kJ mol^{-1}).

The practically unchanged energy of the hydrogen bond of the liquid six-membered hetero compound piperidine and compounds with a large number of CH_2 groups in a cycle points to the application of this rule for other hetero cyclic compounds with content at the saturated hydrocarbon cycle the hexyl, heptyl, octyl fragment. From this we can conclude that the influence of the reverse dative bond ends practically at the compounds with six carbon atoms in the cycle. This conclusion is important for calculations of the vaporization enthalpies and energies of hydrogen bonds of the experimentally unstudied compounds.

The replacement of the hydrogen atom of the amino group at pyrrolidine by the methyl at 1-methyl-pirrolidine is accompanied by reduction of the vaporization enthalpy at 2.3 kJ mol^{-1}, which reflects the reduced stability of the specific interaction $N \rightarrow CH_3$ compared with the hydrogen bond and, consequently, the reduced ability of the carbon atom of the methyl group to shift the electron density to the nitrogen atom. The energy of the given specific interaction is also determined with the help Eq. 4.1. The value of the energy of this series of interactions (4.70 kJ mol^{-1}) practically corresponds to the energy of the same series of interactions of liquid dimethylamine (4.80 kJ mol^{-1}) and is within the experimental error of the established vaporization enthalpies of the compounds. Reduced stability of these interactions compared with the energy of the hydrogen bond is caused by the reduced shifting of the electron density from the methyl group and then from the hydrogen atom pyrrolidine (5.85 kJ mol^{-1}). Notice that research by the method of RE-spectroscope pointed to the reduced electron density at the nitrogen atom on the replacement of the hydrogen atom of the amino group by the methyl group [7, 8].

4.2 Energies of Specific Interactions of Liquid Pyrroles

In accordance with the ideas of [1], at the pentamerous hetero cycle of pyrrole C_4H_4NH, orbital $a_2(\pi_3)$ is localized at the butadiene fragment, and the molecular orbitals $1b_1(\pi_1)$ and $2b1(\pi_2)$ are the combination of bottom π-C_4H_4 fragment with the n-orbital of the nitrogen atom. The experimentally established value of vertical ionizing potential Iv is based both theoretically on [9, 10] and on patterns at photoelectron-spectra of compounds of the pyrrole analogues – C_4H_4NH, C_4H_4O, and C_4H_4S.

Experimental research of shifts of two upper levels of hetero cyclic compounds led the authors of the work [1] to the conclusion that introduction of deputies at location C(2) and C(5) leads to the largest contribution to $1a_2$-molecular orbital by 2p-orbital of the same carbon atoms, at the same time we have the contribution to the $2b_1$-orbital of carbon atoms at locations C(3) and C(4). These contributions lead to a different shifting of the electron density from atoms C(2) and C(5) compared with latter carbon atoms. As a result, there exists a difference in charges at the atoms C(2), C(5) and C(3), C(4) and donor-acceptor properties. The distinction of the considered contributions of each of the two pairs of carbon atoms C(2), C(5) and C(3), C(4) to the same molecular orbitals clearly appeared at photoelectron spectra of pyrazole and imidazole [11].

The experimentally established values of vertical ionizing potentials led to the conclusion [8], that in the series of azoles the stabilization of π-electrons occurred in the sequence of constitution of CH to the nitrogen atom, because

$$C_4H_4NH(IV = 10.0\ eV) < C_3H_3N_2H\ (Iv = 10.9\ eV) < C_2H_2N_3H\ (Iv = 11.9\ eV) < CHN_4H\ (12.3\ eV).$$

At the same sequence there occurs an increase of vertical ionizing potential Iv (n)

$$\text{diazols}\ (10.3\ eV, 10.7eV) < \text{triazols}\ (11.5\ eV) < \text{tetrazols}\ (12.3\ eV).$$

Replacement of the single bond C–C at cyclobutane by the double bond for cyclobutenes is accompanied by a reduction of the vaporization enthalpy. At the same time replacement of single bonds

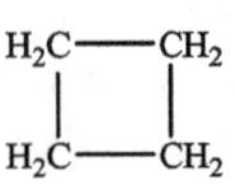

22.7

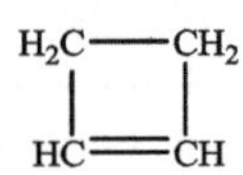

22.2

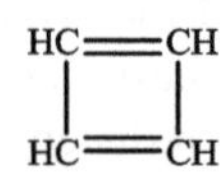

21.1 kJ mol^{-1}

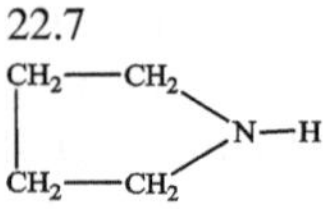

38.4

Pyrrolidine

–

–

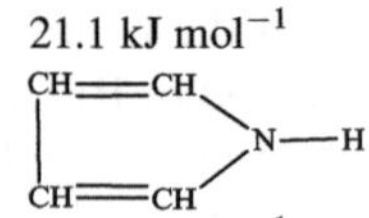

42.5 kJ mol^{-1}

Pyrrole

at pyrrolidine by the double bonds leads to the increase of the enthalpy characteristics of pyrrole. It means that at hetero compounds with saturated hydrocarbon cycle the main part of the electron density is transferred from the hydrogen atom to the carbon atoms and then its part shifts from atoms C(3) and C(4) along the chain to the nitrogen atom, providing carbon atoms of corresponding pairs with small differences in charges. Replacement of CH_2 group by CH groups at pyrrole is accompanied by the sharp reduction of the electron density shifting with a single hydrogen atom of these groups. The predominant part is shifting from carbon atoms at locations C(2) and C(5), providing the positive charges and to the nitrogen atom the negative charge. As a result, this pair of carbon atoms of pyrrole, interacting with a similar pair at locations C(3) and C(4), forms more stable specific interactions D=HC $\rightarrow$ H–C > D–H_2C $\rightarrow$ H–CH. On the other side, nitrogen atom of pyrrole has a reduced negative charge, compared to the carbon atom of pyrrolidine, and forms a hydrogen bond with reduced or equal stability.

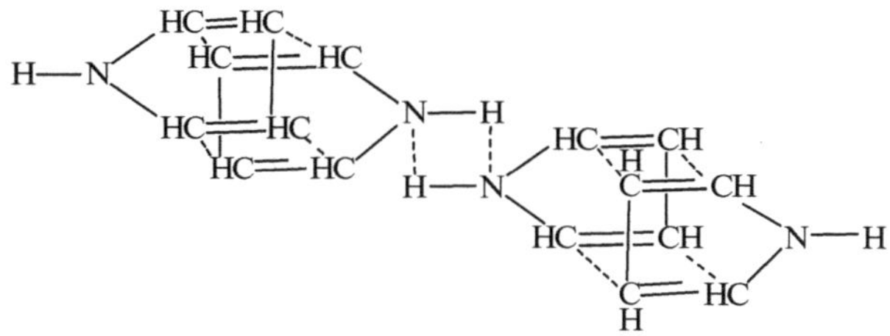

Fig. 4.3 Schematic picture of the structure of liquid and crystalline pyrrole

Similar to hetero cyclic compounds with saturated hydrocarbon cycle the planar structure of pyrrole molecule forms the chain structure by strong specific interactions D=HC → CH and hydrogen bonds N•••H–N (Fig. 4.3). Chains are crosslinked by weak specific interactions. Each molecule forms four specific interactions, two hydrogen bonds, and four interactions D=HC → H–C of low stability. Thus, the vaporization enthalpy is equal to the sum of the energy contributions of each series of interaction and its number:

$$\Delta_{vap}H^0(298\ K) = 4D{=}HC{\rightarrow}CH + 2D\text{-}N{\bullet\bullet\bullet}H\text{-}N \qquad (4.2)$$

Taking the energy of hydrogen bonds of liquid pyrrole and pyrrolidine to be equal (5.85 kJ mol^{-1}), we determine the value of the energy contribution of CH groups to the vaporization enthalpies or the total value of the energy of the CH group of specific interactions D=HC → CH and DHC → H–C, being equal to 7.70 kJ mol^{-1}. The energy estimated over and above the specific interaction D=HC → H–C (0.60 kJ mol^{-1}) leads to the value of the energy of stable specific interaction D=HC → CH, equal to 7.10 kJ mol^{-1}. This value of the energy of this series of specific interactions of liquid pyrrole (Table 4.3), formed by the essentially unshared $2s^2$-electron pair of carbon atoms of the molecules that are close in the environment, illustrates the high stability, the energy of which increases to the same energy of the specific interactions of liquid cyclopentane and cyclohexane (Table 3.5) and pyrrolidine. Thus, the difference in the charges of carbon atoms at locations C(3), C(4) and C(2), C(5) of liquid pyrrole determines the stabilizing "effect of double bond."

Replacement of the hydrogen atom of the amine group in the pyrrole molecule by the methyl group is accompanied by reduction in the vaporization enthalpy of 1-methylpyrrole at 2.9 kJ mol^{-1}. This fact illustrates the reduced donor properties of the carbon atom of the methyl group and reduced stability of the specific interactions D–N → CH_3–N compared with the hydrogen bond of liquid pyrrole. Since the difference in vaporization enthalpies of pyrrole and 1-methylpyrrole belongs to the two specific interactions D–N → CH_3–N of contacting molecules, the energy of this interaction equals, under 338 K, 4.40 kJ mol^{-1}. The value of the energy of specific interaction of D=HC → CH obtained is the most stable in the range of

Table 4.3 Energies of the specific interactions and H-bonds (kJ mol^{-1}) of liquid pentamerous heterocyclic compounds with one nitrogen atom

Compounds	Formula	Structure	$\Delta_{vap}H°(298\ K)$ [3]	T K	D–N•••H–N	D=HC → CH + D=HC → H–C
Pyrrole	C_4H_5N	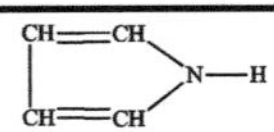	42.5	303	5.85	7.70
1-Methylpyrrole	C_5H_7N	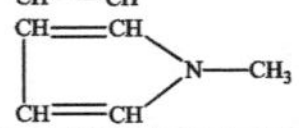	39.6	338	DN → CH_3–N 4.40	7.70

D-H_2C→CH_2 Cyclohexane (4.90) < D-H_2C→CH_2 Cyclopentane (5.10) = D-H_2C→CH_2 Piperidine 5.10 < D=HC→CH Benzene (5.53) < D-H_2C→CH_2 Pyrrolidine (5.60) < D-H_2C→CH_2 Azetidine (5.71) < D=HC→CH Pyrrole (7.10 kJ mol^{-1})

compounds including nitrogenous hetero cyclic compounds with saturated hydrocarbon cycle.

4.3 Energies of Specific Interactions of Liquid Pentamerous Heterocyclic Compounds with Two Nitrogen Atoms

On replacement of C(2)H at pyrrole by the nitrogen atom there occurs preferential stabilization of the $1a_2$-molecular orbital. At the same time the replacement of C(3)H for the nitrogen atom contributes the stabilizing effect to the $2b_1$-molecular orbital, providing the difference in vertical ionizing potentials Iv: $\Delta 1_1 = 0.92$, $\Delta 1_2 = 0.40$ eV and $\Delta 1_1 = 0.54$, $\Delta 1_2 = 0.9$ eV, respectively [8]. The pyrazole and imidazole molecules with two nitrogen atoms and three carbon atoms in a cycle have significant difference at the location of the CH groups, which should greatly influence the stabilization of the specific interactions.

At the sixth bond vacancies in pyrazole and imidazole molecules and two stable hydrogen bonds D–N•••H–N formed, the two specific interactions D–N → CH, and two D=HC → CH, the difference of these compounds consists of the distribution of the electron density and charges at the nitrogen atoms, carbon atoms, and hydrogen atoms. The pyrazole molecule has two closely located nitrogen atoms, which gives rise to

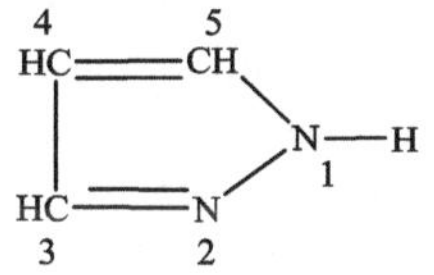

the maximum shifting of the electron density from carbon atoms at location C(3)H. Nitrogen atom N(1)H, receiving the electron density from the connected hydrogen atom, shows a reduced ability to take the additional electron density from the carbon atom C(5)H. Less susceptible to the shifting of the electron density of the carbon atom at location C(4)H, that is, by taking it from the hydrogen atom of the same CH group, it remains with negative charge. In this connection the carbon atoms of the two neighboring CH groups, C(3)H and C(4)H, also have different charges and form relatively less stable specific interactions with similar carbon atoms of the molecules that are close in the environment, the energies of which should differ little from the values of the energies of the same series of specific interaction of liquid pyrrole. The carbon atom C(3)H transfers part of the electron density to the nitrogen atom N(2), which provides it with an increased negative charge and donor properties and, as a sequence, it forms a specific interaction N → CH of increased stability. Thus, the pyrazole molecule forms three types of specific interactions, the stability of which is described by the series D–N•••H–N < D=HC → C–H < D–N → CH.

Continuing a similar analysis of the shifting of the electron density in the imidazole molecule, we can draw the conclusion that in spite of the shifting of the electron density from

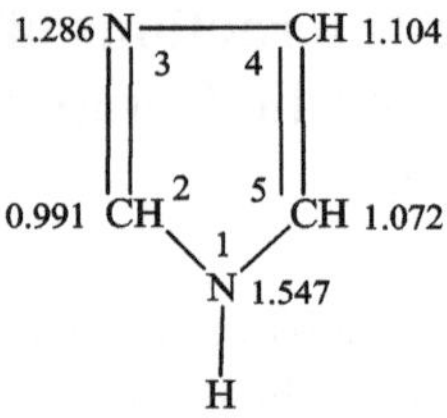

the hydrogen atom at the location of the carbon atom at location C(2)H between two nitrogen atoms it creates a condition where the loss of the highest electron density compared with two other carbon atoms provides it with positive charge. The maximum shifting of the electron density occurs at N(1)H group from the hydrogen atom and the least shift from the carbon atom at location C(5)H. Hence, the nitrogen atom at location N(1) has maximum negative charge. At the same time, the carbon atom C(5)H, obtaining electron density from the hydrogen atom and partly transferring it to the nitrogen atom located at N(1)H, remains with a negative charge. At the same time the nitrogen atom at location N(3) obtains electron density from two closely located carbon atoms with increased shifting from carbon atoms located at C(4)H. Thus, this nitrogen atom has a lower negative charge compared with the nitrogen atom located at N(1)H, but significantly exceeding the charge of the nitrogen atom N(2) in the pyrazole molecule. Consequently, the nitrogen atoms mentioned in the imidazole and pyrazole molecules differ by donor properties and we see the stability of the same type of specific interactions in imidazole DN → CH > in pyrazole DN → CH. The shifting of the electron density in the chain of the imidazole molecule cycle N ← CH ← CH → NH from the carbon atom located at C(5)H cannot be changed at the considered carbon atoms. The principle of maximum difference of the charges at the atoms determines the formation of a series of specific interactions, and its stability, described by the similar series D–N•••H–N < D=HC → C–H < D–N → CH–N and in combination with the structure of the molecule, causes the formation of the structure of liquid and crystal conditions. Hence, pyrazole and imidazole are characterized by the formation of chains, crisscrossed by weak intermolecular interactions HC → H–C (Fig. 4.4).

The thermodynamic properties of liquid pyrazoles and imidazoles are limited by two compounds, which results in some difficulties in the estimation of the energies of the specific interactions. However, the similarity in the shifting of the electron density of CH groups in pyrrole and pyrazole molecules allows us to take the energies of the specific interaction D=HC → C–H and the hydrogen

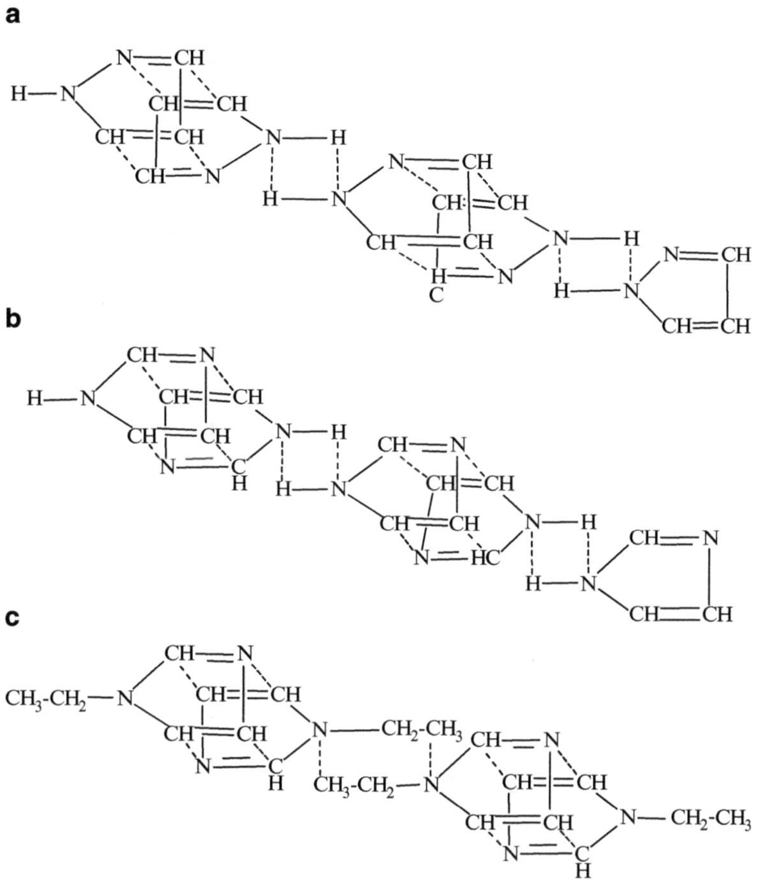

Fig. 4.4 Schematic picture of the structure of liquid and crystalline pyrazole (**a**), imidazole (**b**) and 1-ethylimidazole (**c**)

bond equal to 7.70 and 5.85 kJ mol^{-1}, respectively. The energy of the specific interaction D–N → CH–N in liquid pyrazole should be determined with the help of the equation considering the contribution to the vaporization enthalpy of the three types of interactions,

$$D\text{-}N\rightarrow CH\text{-}N = (\Delta_{vap}H^0(298\ K) - 2D{=}HC\rightarrow CH - 2D\text{-}N\cdots H\text{-}N \qquad (4.3)$$

The obtained value of the energy of the specific interaction D–N → CH–N (Table 4.4) allowed us to determine the energy of the specific interaction, formed by ethyl fragment 1-ethylpyrazole, using the equation given in Eq. 4.4

Table 4.4 Energies of the specific intermolecular interactions and hydrogen bonds (kJ mol^{-1}) of liquid heterocyclic derivative pyrazoles and imidazoles (298 K)

Compounds	Formula	Structure	$\Delta_{vap}H°$(298 K) [3]	DHC→CH	D=N → C(=)H–N	D–N•••H–N	D–N → C_2H_5–N
Pyrazole	$C_3H_4N_2$	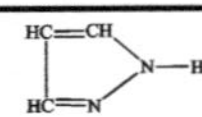	48.4[a]	7.70	10.65	5.85	–
Imidazole	$C_3H_4N_2$	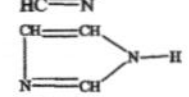	61.4[a]	11.00	13.85	5.85	–
1-Ethyl-pyrazole	$C_5H_8N_2$	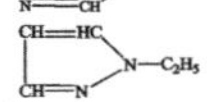	53.3 ± 2.3	7.70	10.65	–	8.30
1-Ethyl-imidazole	$C_5H_8N_2$	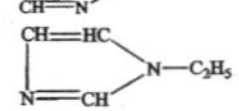	66.0 ± 3.4	11.00	13.95	–	8.30
1,3-Dimethyl-2-imidazol-idinone	$C_5H_8N_2$	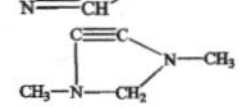	54.3[a]	DN→CH_3–N = 4.70	DH_2C–N→CH_2 = 7.70	D(N–)≡ C→C≡ = 10.0	

[a]At T = 375 K

$$D\text{-}N{\rightarrow}CH_3\text{-}CH_2\text{-}N = (\Delta_{vap}H^0(298\ K)epzl - 2D{=}HC{\rightarrow}CH - 2D\text{-}N{\rightarrow}CH\text{-}N)/2 \quad (4.4)$$

The enthalpy characteristics of 1-ethylpyrrole and 1-ethylimidazole (Table 4.4) illustrate the difference of the vaporization enthalpies at 12.7 kJ mol^{-1}. It follows that for the same type of 2D=HC → CH and 2D–N → CH–N and number of specific interactions, the nitrogen atoms of 1-ethylimidazole at locations N(1) and N(3) make the increased contribution (12.7 kJ mol^{-1}) to the vaporization enthalpy in comparison with nitrogen atoms of 1-ethylpyrrole, lying at the N(1) and N(2) positions. It means that the energies of these specific interactions of pyrazole and imidazole and the derivatives are described by the natural number of the stability,

D-N→CH-N Imidazole N(3) > D=HC→CH Imidazole N(3) > D-N→CH-N Pyrazole N(2) > D=HC→CH Pyrazole N(2)

Taking the average value of the stabilizing effect (3.3 kJ mol^{-1}) for the nitrogen atom in position N(3) of the imidazole molecule and 1-ethylimidazole for each specific interaction D–N → CH–N and D=HC → CH, we obtain the energies of these interactions; they are equal to 13.80 and 11.00 kJ mol^{-1}, respectively. Using the value of the energy of the hydrogen bond realized in liquid pyrrole and pyrazole, we determined the vaporization enthalpy of imidazole (Table 4.4).

The value of the energy of the formed specific interaction by the ethyl fragment of 1-ethylimidazole D–N → CH_3–CH_2–N is determined with the help of Eq. 4.4.

The estimated values of the energies of specific interactions and vaporization enthalpies of pentamerous heterocyclic

Pyrrole	Pyrazole	Imidazole
42.5	48.4	61.4 kJ mol^{-1}

compounds with replacement of C(2)H and C(3)H by the nitrogen atom, illustrate the stabilization of interactions with participation of the nitrogen atom (Table 4.4) and the natural growth of the enthalpy characteristics, proved by the differences given above between the vertical ionizing potentials and photoelectron spectra of diazols [8].

It is interesting to estimate the energy of specific interaction of D(N–)≡C → C≡, formed in liquid 1,3-dimethyl-2-imidazolidinone, because it allows us to obtain definite information on the energy contribution to the enthalpy characteristics of a compound with saturated carbon atom C(3) to the nitrogen atom of the derivative imidazole with triple bond –C≡C–.

The molecule of 1,3-dimethyl-2-imidazolidinone with seven vacancies forms with the molecules that are close in the environment four D–N → CH_3–N, one H_2C → CH_2 and two (N–)≡C → C≡ specific interactions. The location of two CH_3 groups, connected with nitrogen atoms in a plane, is directed to the fourth-coordinated condition of the nitrogen atom and leads to the formation of the network structure of the liquid condition (Fig. 4.5). The replacement of the carbon atom at location C(3) by the nitrogen atom is accompanied by significant

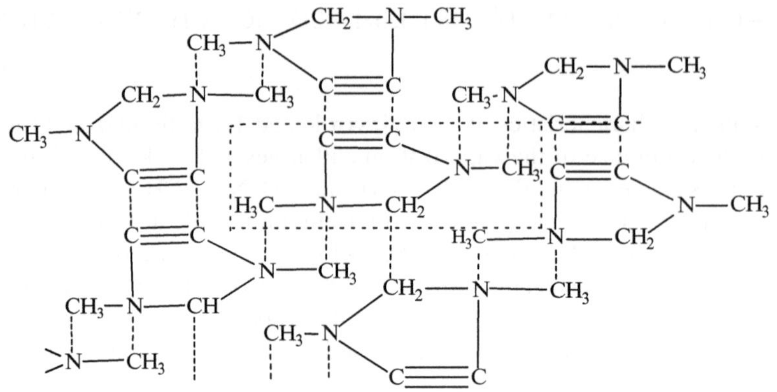

Fig. 4.5 Schematic picture of the liquid and crystalline structure of 1,3-dimethyl-2-imidazolidinone with the network of specific interactions

stabilization of the specific interaction. The presence of a methyl group at the given nitrogen atom N(3) is accompanied by additional shifting of the electron density and an increase in its negative charge and donor properties, which leads to the stabilization of specific interaction D(N–)≡C → C≡. The connection of the nitrogen atom N(3) by single bonds with the carbon atoms of the methyl and methylene groups leads to the formation of a specific interaction with the energy of interaction, realized in liquid 1-methylpirrolidine (4.70 kJ mol^{-1}). The energy of the formed specific interaction by methylene groups connected with the nitrogen atom N(3) in a cycle without the stabilizing effect of the double bond and with minimum value should be taken to be equal to 7.70 kJ mol^{-1}. The energy of the specific interaction D(N–)≡C → C≡, determined with the help of the equation

$$D(N\text{-})\equiv C \rightarrow C\equiv\, = (\Delta_{vap}H°298\ K) - 4\ D\text{-}N \rightarrow CH_3 - D\text{-}H_2C \rightarrow CH_2)\,/\,2 \qquad (4.5)$$

(10.0 kJ mol^{-1}), reasonably reflects the nature of the considered interaction.

4.4 Energies of Specific Interactions and Hydrogen Bonds of Crystalline Pentamerous Heterocyclic Compounds of Nitrogen

As was shown above, the energy of the hydrogen bond of liquid pyrrolidine, pyrazole, and imidazole and its derivatives does not exceed the values of the energy of the same bond of liquid ammonia with a tetracoordinated nitrogen atom. In this connection it is reasonable to take the energy of the hydrogen bond of the crystal compounds as equal to the energy of the H-bond of crystal ammonia (6.34 kJ mol^{-1}) with the same coordination of the nitrogen atom. The most successful compound

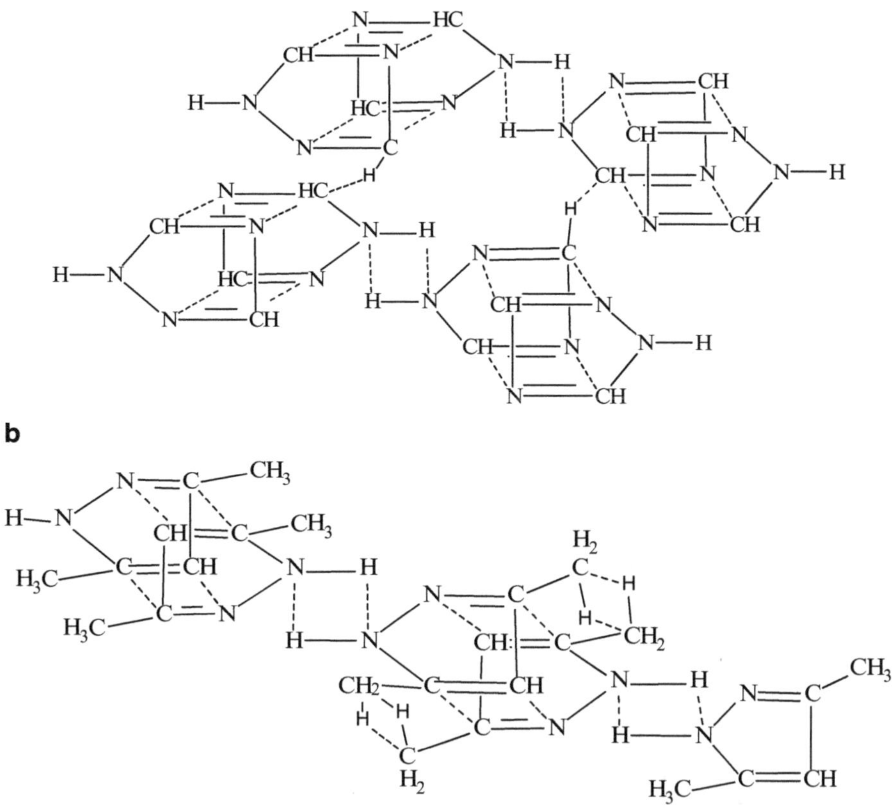

Fig. 4.6 Schematic picture of the structure of liquid and crystalline 1,2,4-triazole (**a**) and 3,5-dimethylpyrazole 3,3,5,5-tetramethyl-1-pyrazoline (**b**) with network of specific interactions

for thermodynamic analysis is 1,2,4-triazole, with a molecule with planar structure and four-coordinated nitrogen atoms of the NH group and six free bond vacancies, which forms a chain-network structure of liquid and crystal conditions by two hydrogen bonds N•••H–N and four specific interactions N → CH (Fig. 4.6a).

The energies of the specific interactions of this compound, formed by the nitrogen atoms at locations (2) and (4) with carbon atoms C(3) and C(5), are determined by the difference of the sublimation enthalpy and energies of the hydrogen bonds, using Eq. 4.6

$$\text{D-N}\rightarrow\text{CH-N} = (\Delta_{\text{sub}}H^{\circ}(298\ \text{K})\ - 2\text{D-N}\bullet\bullet\bullet\text{H-N})/4 \qquad (4.6)$$

The value of the energy established reflects the high stability of the specific interaction D–N → CH–N (Table 4.5).

From the three types of specific interactions of crystalline pyrazole the energies of the hydrogen bond and specific interaction D–N → CH–N are known, which

Table 4.5 Energies of the specific interactions and hydrogen bonds (kJ mol^{-1}) of crystalline pentamerous heterocyclic compounds with two nitrogen atoms (298 K)

Compounds	Formula	Structure	$\Delta_{sub}H°(298\ K)$ [9]	D=HC → CH	D=N→CH–N	D–N•••H–N	D*iso*CH_3
1,2,4-Triazole	$C_2H_3N_3$	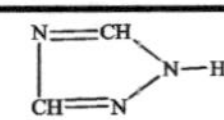	80.7 ± 0.5	–	17.0	6.34	–
Pyrazole	$C_3H_4N_2$	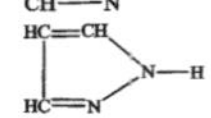	74.5 ± 0.4	13.65	17.0	6.34	–
3,5-Dimethyl pyrazole	$C_5H_6N_2$	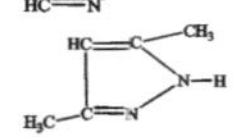	83.4 ± 2.4	13.65	17.0	6.34	5.05 × 2
Imidazole	$C_3H_4N_2$	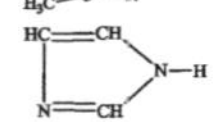	83.1 ± 0.2	15.8	19.2	6.34	–
2-Methyl-imidazole	$C_4H_6N_2$	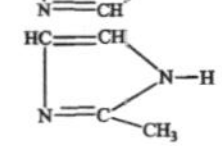	88.4 ± 0.2	15.8	17.3 ± 2.0 19.45	6.34	2.95 × 2 5.9
2-Ethyl-imidazole	$C_5H_8N_2$	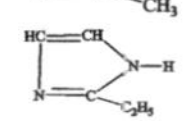	89.6 ± 0.4	15.8	17.3 ± 2.0 19.45	6.34	3.55[a] 7.1
3,3,5,5-Tetramethyl-1-pyrazoline	$C_7H_{14}N_2$	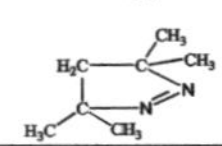	61.6 ± 0.2	–	13.0[b]	–	2.35 × 4

[a]D*iso*C_2H_5
[b]Average value of the energy of two specific interactions (N → C–N + N → CH_2–C–N)/2

is why the unknown energy value of interaction D=HC → CH is determined from the difference of sublimation enthalpy and the total value of the energies of the two types of specific interactions

$$D{=}HC{\rightarrow}CH = (\Delta_{sub}H°(298\ K)\ - 2D\text{-}N{\rightarrow}CH\text{-}N - 2D\text{-}N{\bullet\bullet\bullet}H\text{-}N)/4 \qquad (4.7)$$

It was shown above that the increased value of the sublimation enthalpy of imidazole compared with the value of the enthalpy characteristics of pyrazole is caused by the stabilizing effect as a result of the replacement of the carbon atom C(3) by the nitrogen atom. Thus, the value of the difference in sublimation enthalpies of these compounds (8.6 kJ mol^{-1}) is the result of the stabilizing effect to the energies of specific interactions of the two series D=HC → CH and –N → CH–N of crystalline imidazole or the average value of the energy contribution (2.15 kJ mol^{-1}) to each of the fourth bonds, formed by the molecules that are close in the environment (Table 4.5). The established values of the energies of the hydrogen bonds of crystalline pentamerous heterocyclic compounds with differing locations of the nitrogen atom in molecules of compounds and its number in the cycle are described by the series

D=HC→CH-N: 1,2,4-Triazole (17.0) = Pyrazole (17.0) < Imidazole (19.2 kJ mol^{-1}),

reflecting the high influence of the crystalline field on the stabilization of these bonds.

We draw attention to the established fact that the isostructural methyl group changes the value of the vaporization enthalpy with the saturated hydrocarbon cycle and practically leaves the energies of specific interactions and the hydrogen bond without changes. The enthalpy characteristics of the derivatives of pyrazole and imidazole given in Table 4.5 reflect the significant influence of the isostructural CH_3 groups and its number in the increase of the enthalpy characteristics of nitrogenous cyclic compounds with the unsaturated hydrocarbon cycle. Consequently, the difference of the sublimation enthalpies of pyrazole and 3,5-dimethylpyrazole reflects the contribution of the energy of the specific interactions of low stability, formed by the isostructural methyl groups of the molecules that are close in the environment (Fig. 4.6). The contribution of the energy of the two isostructural groups is the difference in the sublimation enthalpies of 3,5-dimethylpyrazole and pyrazole reduced by the energy contributed by two saturated hydrogen atoms (0.60 kJ mol^{-1})

$$D iso CH_3 = (\Delta_{sub}H°(T) - (\Delta_{sub}H°(298\ K)prz - 2DH))/2 \qquad (4.8)$$

The total value of the energy contribution of the two isostructural methyl groups is 10.6 or, per one CH_3 group, 5.3 kJ mol^{-1}. Each methyl group forms two specific interactions $H_3C \rightarrow H\text{–}CH_2$ and, as a result, the energy of this interaction is equal to 2.65 kJ mol^{-1}.

From the differences of the enthalpy characteristics of compounds with isostructural methyl 2-methylimidazole or ethyl group 2-ethylimidazole and imidazole, the contributions of the energies to the vaporization enthalpy of the corresponding compounds, formed by these groups of specific interactions of low stability were

Fig. 4.7 Schematic picture of the structure of liquid and crystalline 3,3,5,5-tetramethyl-1-pyrazoline with network of specific interactions

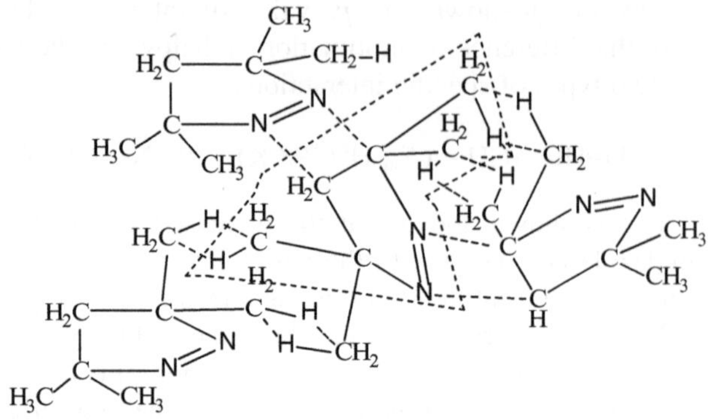

determined. The values of the energies of the isostructural methyl group for the vaporization enthalpy per one such group have a tendency to increase the contribution to the sequence of replacements of the nitrogen atom from C(2)H to C(3)H group,

$$3,5\text{-Dimethylpyrazole } D iso\mathrm{CH_3} = 5.05 < 2\text{-Methylimidazole } D iso\mathrm{CH_3}$$
$$= 5.90\ \mathrm{kJ\ mol^{-1}}$$

It is possible to estimate the tendency to increase the energy of the specific interaction of low stability from,

3,5-Dimethylpyrazole D*iso*$H_3C \rightarrow H\text{-}CH_2$ (2.50) ≤ 2 -Methylimidazole D*iso*$H_3C \rightarrow H\text{-}CH_2$ (2.95 kJ mol^{-1}).

Relatively stable specific interactions are formed by the isostructural ethyl group (7.10 kJ mol^{-1}) with the energy of specific interaction D–CH_2–CH_3 → CH_3–CH_2– being 3.55 kJ mol^{-1} .

The most notable feature for the compounds with three nitrogen atoms in the cycle is that they are characterized by the formation of one series of specific interactions independent of the location of the nitrogen atom in the cycle. Nevertheless, saturated C(2)H and C(3)H groups should express the different stability of specific interaction N → CH–N, caused by the difference in the charges of the nitrogen atoms located at N(2) and N(4). It broadens the possibilities of a thermodynamic analysis and allows us to obtain additional information to trace the influence of this cycle with the specific interactions D–N → CH–N, formed by the essentially unshared $2s^2$-electron pair of carbon atoms on the energy of hydrogen bond.

The 3,3,5,5-tetramethyl-1-pyrazoline molecule with two nitrogen atoms and three carbon atoms in the cycle form four specific interactions, two of which are formed by the carbon atom with two isostructural methyl groups N → C and two others by the CH_2–C fragment of the same cycle N → CH_2–C as also with two isostructural methyl groups at the carbon atom (Fig. 4.7). Four isostructural methylgroups form eight specific interactions H_3C → H–CH_2 of low stability. The energy contribution of the isostructural methyl group to the sublimation enthalpy is taken as equal to the contribution of the same group for 2-methylimidazole, providing the low value of the energy of the specific interaction $DH_3C \rightarrow H\text{–}CH_2$ (1.13 kJ mol^{-1}) with location of these groups at the two carbon atoms of the cycle. The presence of three carbon atoms in the cycle, shifting the electron density and fringing the nitrogen atom in the 3,3,5,5-tetramethyl-1-pyrazoline molecule, allows us to determine the average value of the energy of the specific interactions $N \rightarrow C\text{–}N$ and $N \rightarrow CH_2\text{–}C\text{–}N$, formed by $N \rightarrow C\text{–}$ and $N \rightarrow CH_2\text{–}C\text{–}$ fragments with the help of Eq. 4.6a

$$D\text{-}N{\rightarrow}CH\text{-}CH\text{-}N = (\Delta_{sub}H°(298\ K)\ - 4D iso CH_3)/4 \qquad (4.6a)$$

The value of the energy obtained points to the reduced stability of the specific interaction $(N \rightarrow C + N \rightarrow CH_2\text{–}C)/2$ compared with the unsaturated five-membered nitrogenous compounds that are formed. (Table 4.5).

4.5 Energies of Specific Interactions of Crystalline Tetrazoles

Four nitrogen atoms with one CH group in the heterocycle of the tetrazole molecule lack electron density. Its location at the 1, 2, 3, and 4 cycle of imidazole presupposes the uneven distribution of the electron density between nitrogen atoms.

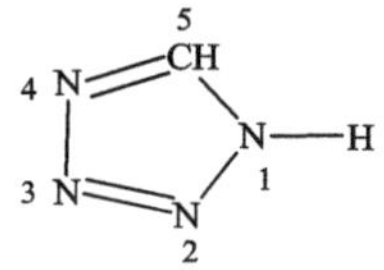

The location of the fourth nitrogen atom in the molecule is similar to the imidazole molecule and provides it with significant negative charge because of the shifting of the electron density with the essentially unshared $2s^2$-electron pair of the carbon atom of CH group and the high positive charge at the carbon atom of the same group. In its turn the shifting of the electron density from the hydrogen atom of the NH group creates a negative charge at the nitrogen atom at location N(1) and a positive charge at the hydrogen atom. The difference in the negative charges of the nitrogen atoms $N_1^{\delta-} < N_4^{\delta-}$ at locations 1 and 4 leads to the inadequate shifting of the electron density of the nitrogen atoms at locations 2 and 3 and inequality of its charges. As a result, among the atoms the charges in the cycle of the tetrazole molecule at the nitrogen atoms there should be implemented the condition of alternation of its value $N_4^{\delta-} > N_1^{\delta} > N_3^{\delta-} > N_2^{\delta-}$. Consequently, nitrogen atoms at location 2 and 3 are able to participate in intermolecular interactions with the molecules close in the environment due to the overlap of $2p_z$-electron clouds, forming the specific interaction $N_3 \rightarrow N_2$.

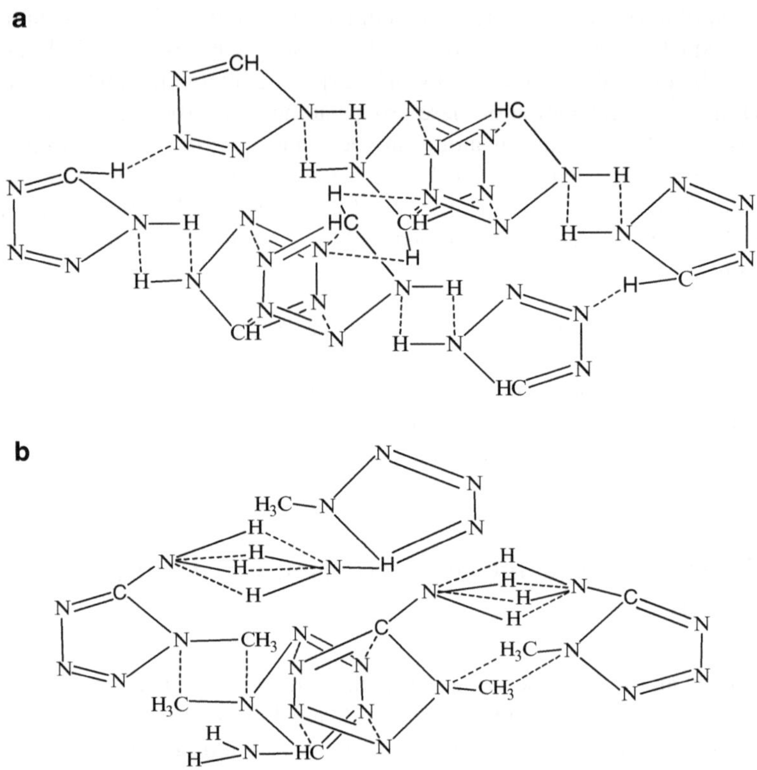

Fig. 4.8 Schematic picture of the liquid and crystalline structure of tetrazole (**a**) and 1-methyl-5-aminotetrazole (**b**)

The principle of the formation of specific interactions with maximum difference of charges of the atoms allows us to suppose that the nitrogen atoms ($N_4^{\delta-}$) and the carbon atom ($C^{\delta+}$) of CH-group lead to the increased stability of the interaction D=N → CH. The reason for taking the condition of equality of this specific interaction with the energy D=N → CH (19.2 kJ mol^{-1}) imidazole is its similarity with the structure of tetrazole (Table 4.5) and the maximum difference of negative and positive charge of the nitrogen atoms and carbon atoms.

The molecule of tetrazole with six bond vacancies form two hydrogen bonds, D–N•••H=N, two specific interactions, D=N → CH, and two interactions, D=N → N, forming with the molecules close in the environment the stable chains. The latter are crosslinked to the grid by weak bonds HC → H–C, formed by the hydrogen atoms of CH-groups of contacting molecules close in the environment (Fig. 4.8a). The presence of information on the value of the energy of hydrogen bond (6.34) of crystalline 1,2,4-triazole does not exceed the energy of the type of the solid ammonia with four-coordinated nitrogen atom and the energy of the specific interaction of the nitrogen atom at location N(4) with CH- group (19.2 kJ mol^{-1}) allows us to estimate the value of the energy of D=N → N, it being equal to the differenceof enthalpy of sublimation and energies of two interactions 2D–N → CH–N and two hydrogen bonds D–N•••H–N

$$D{=}N{\rightarrow}N = (\Delta_{sub}H°(298\ K) - (2D\text{-}N{\rightarrow}CH + 2D\text{-}N{\cdots}H\text{-}N)/2 \quad (4.9)$$

The implemented thermodynamic calculations (Table 4.6) for 353 K do not take into account the temperature dependency of the energies of the hydrogen bonds and specific interactions DN → CH–N, established at the standard conditions, and that is why the value of the energy of the specific interaction DN → N is minimal possible. It illustrates the high stability of the considered series of specific interaction.

The differences of the sublimation enthalpies of tetrazole and 1-methyltetrazole (1.5 kJ mol^{-1}) reflect the reduced stability of the specific interaction D–N → CH_3–N in comparison with the hydrogen bond formed by these compounds. Taking into account that in the crystal there participate two molecules in the interaction, and the destabilizing effect of the one of the specific interaction is equal to 0.75 kJ mol^{-1}. Thus, the energy of the specific interactions D–N → CH–N is equal to 5.60 kJ mol^{-1}. The energy of this bond one could obtain with the help of the equation

$$D\text{-}N{\rightarrow}CH_3\text{-}N = (\Delta_{sub}H°(298\ K) - (2\ D{=}N{\rightarrow}CH\text{-}N + 2\ D{=}N{\rightarrow}N)/2. \quad (4.10)$$

The increased value of the sublimation enthalpy of 5-methyltetrazole in comparison with tetrazole at the similar series of specific interactions and hydrogen bond points to the significant contribution of the energy by the isostructural methyl group, equal to the difference of the enthalpy characteristics of these compounds minus the energy, contributed by the substituted hydrogen atom (0.60 kJ mol^{-1}) in the tetrazole molecule or sublimation enthalpy of 5-methyltetrazole

$$D_{iso}CH_3) = (\Delta_{sub}H°(T)mtz - (\Delta_{sub}H°(T)tz - DH) = (\Delta_{sub}H°(T)mtz - (2D\text{-}N{\rightarrow}CH + 2D\text{-}N{\rightarrow}N + 2DN{\cdots}H\text{-}N) \quad (4.8a)$$

and the total energy of the stable specific interactions, for the hydrogen bond of the substituted hydrogen atom of the tetrazole molecule. The contribution of the isostructural methyl group to the enthalpy characteristics is equal to 6.2 kJ mol^{-1} and, consequently, the energy of the specific interaction of low stability is equal to 3.1 кJ mol^{-1}.

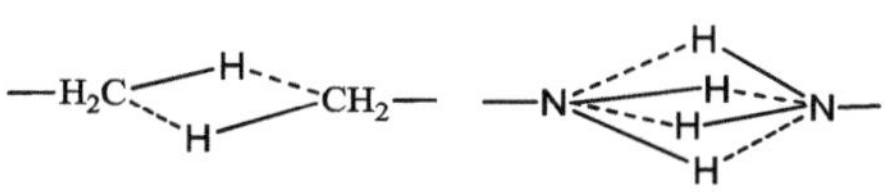

A similar procedure is used for the determination of the energy contribution of the isostructural methyl group of 1,5-dimethyltetrazole to the sublimation enthalpy of 1-methyltetrazole.

Table 4.6 Energies of the specific interactions and hydrogen bonds (kJ mol^{-1}) of crystalline tetrazoles

Compounds	Formula	Structure	$\Delta_{sub}H°$(298 K) [12]	T K	D–N•••H–N	D–N → CH–N	D=N → N
Tetrazole	CH_2N_4	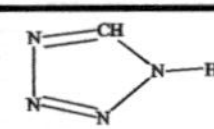	88.16	353	6.34	19.2	15.0
1-Methyl-tetrazole	$C_2H_4N_4$	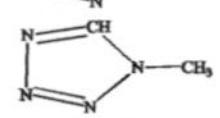	86.7 ± 1.9	282–312	D–N → CH_3–N = 5.60	19.2	15.0
5-Methyl-tetrazole	$C_2H_4N_4$	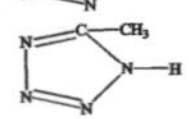	93.8 ± 0.5	323–416	6.34	19.2 D*iso*CH_3 6.20:2 = 3.1	15.0
1,5-Dimethyl-Tetrazole	$C_3H_6N_4$	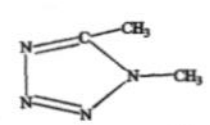	86.1 ± 1.0	303–343	D–N → CH_3–N = 5.60	19.2	15.0
1-Methyl-5-amino-tetrazole	$C_2H_5N_5$	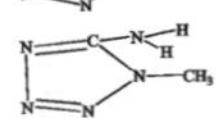	90.6 ± 1.7	298	D–N → CH_3–N = 5.63	19.2 D–N•••H–N 4.50:4 = 1.12	15.0
1-Aminol-5-methyl-tetrazole	$C_2H_5N_5$	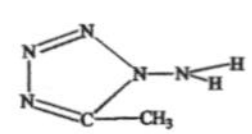	90.6 ± 1.1	310–373	D–N•••H–N 3.15:4 = 0.80	D*iso*CH_3 6.20:2 = 3.1	15.0

The replacement of the hydrogen atom at location C(5)H of 1-methyltetrazole by the amino group is accompanied by the increase of the number of bond vacancies and the formation of four hydrogen bonds of low stability (Fig. 4.8b) and an energy contribution (4.50 kJ mol^{-1}) to the sublimation enthalpy of 1-methyl-5-aminotetrazole. The energies of the formed hydrogen bonds equal 1.12 kJ mol^{-1}. It is symptomatic that replacement of hydrogen atom in the molecule 5-methyltetrazole at location N(1) by the amino group -NH_2 is accompanied by the formation of 1-aminol-5-methyltetrazole and reduction of the sublimation enthalpy at 3.15 kJ mol^{-1}. It follows that the energy contribution by the hydrogen atom, forming the hydrogen bond with energy 6.34 kJ mol^{-1}, is not compensated by the energy contribution by the amino group, equal to 3.15 kJ mol^{-1}. It means that the nitrogen atom of the amino group reduces its negative charge by the partial transmission of the electron density to the carbon atom and nitrogen atom at their location at C(5) and N(1) in tetrazole.

The practical invariability of sublimation enthalpies of 1-methyltetrazole and 1,5-dimethyltetrazole points to the significant influence of the isostructural methyl group to the redistribution of the electron density in the molecule. It follows that the location of the methyl group at location C(5) and N(1) is expressed at the reduction of negative charge of the nitrogen atom and reduction of the donor properties.

4.6 Energies of Specific Interactions of Liquid 1-Phenyl-pyrazole and 1-Phenyl-imidazole

The 1-phenyl-pyrazole and 1-phenyl-imidazole compounds with single C–C bond allows us to clarify the influence of the nitrogen atom of the pentamerous heterocycle, by the location at C(2) and C(3), respectively, to the energies of the specific interactions, formed by the carbon atoms of the benzene ring and heterocycle. The molecules of these compounds have 12 bond vacancies and an equal number of each series of specific interactions and hydrogen bonds. However, the location of the second nitrogen atom of these compounds is significant as shown by the energies of the specific interactions (Table 4.4) of the compounds given. The coordination of molecules in liquid and crystal condition of the compounds is accompanied by the implementation of the principle of minimum energy by the system. But the location of the nitrogen atoms in the 1-phenyl-pyrazole molecule does not create conditions for the formation of specific interactions, not only with CH groups of pyrazole's cycle, and that is why the CH groups of the benzene ring are forced to participate in the formation of such interactions. The pentamerous cycle forms two specific interactions D–N•••H–N with energy of 5.85 kJ mol^{-1}, two specific interactions D–N → CH of the increased stability and two specific interactions with D=HC → CH groups of coordinating molecule (Fig. 4.9a). At the same time, four CH groups of the benzene ring form similarly four specific interactions with similar groups of benzene cycle of the molecules close in the

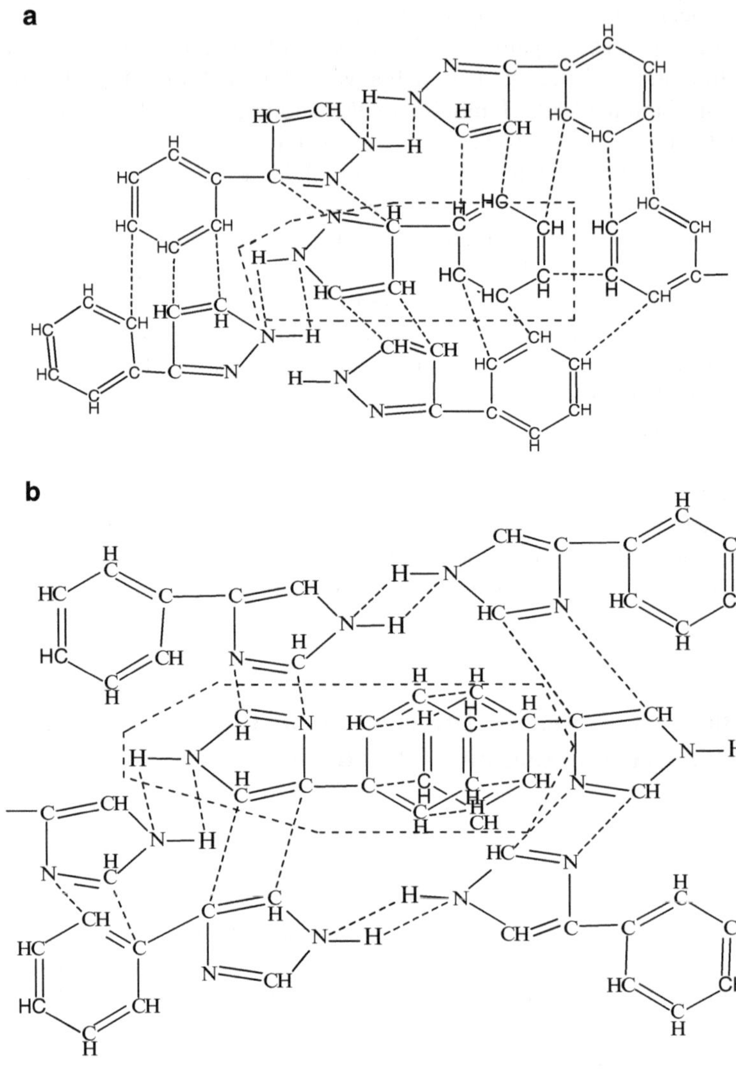

Fig. 4.9 Schematic picture of the crystalline structure of 1-phenyl-pyrazole (**a**) and 1-phenyl-imidazole (**b**) with the network of specific interactions

environment with energy of 5.7 kJ mol^{-1} and two left over with two carbon atoms of the pentamerous cycle with different energy, D=HC → CH and D=HC → C, with 5.7 and 5.1 kJ mol^{-1}, considering the energy of the saturated hydrogen atom of benzene ring; we present the contribution of the benzene cycle to the vaporization enthalpy of 1-phenylpyrazole. The total contribution of the specific interactions with known values of the energies is the difference between vaporization enthalpy for 1-phenyl-pyrazole and with energy values of the two series of specific interactions,

$$2D\text{-}N{\rightarrow}CH + 2D{=}HC{\rightarrow}CH = (\Delta_{vap}H^0(298\ K)\ - 5D{=}HC{\rightarrow}CH - 2D\text{-}N{\cdots}H\text{-}N - D{=}HC{\rightarrow}C) = 25.2\ kJ\ mol^{-1} \quad (4.11)$$

on the basis of which one can assume that the energies of the specific interactions D=HC → CH, formed by CH groups of the benzene ring and pentamerous cycle of this compound are practically equal (5.7 kJ mol^{-1}). Thus, we obtain the energy value of the more stable specific interaction, formed by pentamerous cycle of 1-phenylpyrazole, equal to 6.9 kJ mol^{-1} (Table 4.7).

The 1-phenylimidazole molecule forms the network structure of a crystal, and four specific interactions of two series 2D–N → CH and 2D=HC → CH of increased stability and two hydrogen bonds D–N•••H–N are formed by nitrogen atoms and carbon atoms of the pentamerous cycle (Fig. 4.9b). The benzene ring of this compound forms five specific interactions D=HC → CH and one D=HC → C with energies 5.7 and 5.2 kJ mol^{-1}, respectively. Using the known values of the energies of the specific interactions, hydrogen bonds and the equality (4.12), we obtained the value of the energy of the total contribution of the energies of two series of specific interactions, 2D–N → CH + 2D=HC → CH = 40.4 kJ mol^{-1}. The contribution of the energies of the same series to the vaporization enthalpy of imidazole is equal to 49.7 kJ mol^{-1}. Taking the correlation between the energies of the specific interactions for compounds of the same series one can use these values and the energy of the specific interaction D–N → CH = 13.85 kJ mol^{-1} (Table 4.4) for estimation of the energies of the same two series of interactions (Table 4.7). The energy values obtained of the specific interactions D–N → CH and D=HC → CH allow us to draw a conclusion on the significant influence of nitrogen atoms located at C(3) in the 1-phenyl-imidazole molecule on the shifting of the electron density from the benzene ring in comparison with the nitrogen atom at the location C(2) of the 1-phenylpyrazole molecule.

4.7 Energies of Specific Interactions of Crystalline Carbazoles

Being located between two benzene rings, the amino group of the molecule carbazole leads to the increased negative charge at the nitrogen atom due to the shifting of the electron density from hard cycles. As a result, the nitrogen atom changes little, however, the difference in the charges of nitrogen and hydrogen increases and the forming hydrogen bond should stabilize in comparison with crystalline pyrazole and tetrazole (Tables 4.5 and 4.6). The carbazole molecule with 14 bond vacancies forms the grid structure of a crystal by two hydrogen bonds, eight specific interactions D=HC → CH, and four interactions D=C → C (see Fig. 4.10).

Similarly for liquid 1-phenyl-pyrazole, the energy of the specific interaction of crystalline carbazole we take equal to the energy of the interaction of crystalline

Table 4.7 Energies of hydrogen bonds and specific interactions of crystalline 1-phenyl-pyrazole and 1-phenyl-imidazole at 298 K

Compounds	Formula	Structure	$\Delta_{vap}H°$(29 K) [3]	D=HC → CH/D=HC → C	D–N•••H–N	D*HC → CH	D=N → CH
1-Phenyl-pyrazole	$C_9H_8N_2$	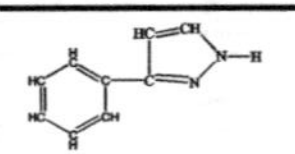	70.2 ± 3.4	5.7/5.2	5.85	–	6.9
1-Phenyl-imidazole	$C_9H_8N_2$	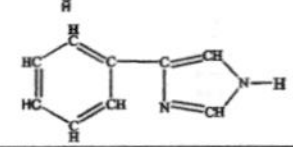	84.8 ± 3.2	5.7/5.2	5.85	8.6	11.6

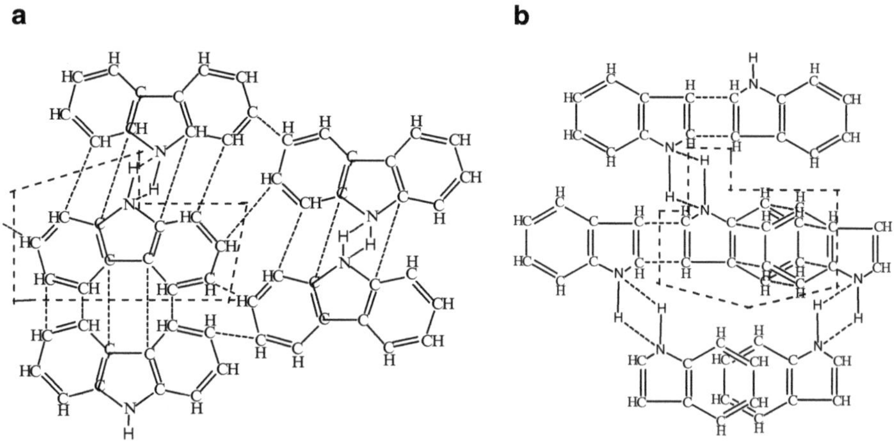

Fig. 4.10 Schematic picture of the crystalline structure of carbazole (**a**) and indole (**b**) with the network of specific interactions

benzene D=HC → CH, 7.40 kJ mol^{-1} and the energy of the interaction D=C → C equal to the energy of the first type of interaction minus the energy contribution of the two saturated hydrogen atoms (5.40 kJ mol^{-1}) to the sublimation enthalpy. Taking the equality of the enthalpy characteristics with the contributions of all realized specific interactions, we calculated the energy of the hydrogen bond with the help of Eq. 4.12 as presented in Table 4.8.

$$D\text{-}N\cdots H\text{-}N = (\Delta_{vap}H^0(298\ K) - 8D{=}HC{\to}CH - 4D{=}C{\to}C)/2 \tag{4.12}$$

The value obtained for the energy of the hydrogen bond points to its increased stability in comparison with tetrazole (6.34) < carbazole (8.0 kJ mol^{-1}) realized in crystalline derivatives of tetrazoles. It follows that the excess nitrogen atoms in the tetrazole molecule is accompanied by reduction of the charges of the nitrogen atoms and hydrogen atoms in the molecule of this compound in comparison with carbazole.

The amide group of the indole molecule is located between the benzene ring and two CH groups of the pentamerous cycle, which allows the nitrogen atom to take the increased electron density from the chain of these groups, and from the hard benzene ring. It means that CH groups of the pentamerous cycle of the indole molecule are able to form a more stable specific interaction HC → CH in comparison with the formed similar interaction CH groups of the benzene ring. The indole molecule with ten bond vacancies forms the grid structure of a crystal by the two hydrogen bonds, two specific interactions D=HC → CH, formed by CH groups of the pentamerous cycle and two ones of the same series, formed by similar groups of the benzene ring with an energy of 7.40 and four interactions D=HC → C of low stability of 6.40 kJ mol^{-1} (Fig. 4.10b). Considering the contributions of each type of specific interaction and hydrogen bond and its

Table 4.8 Energies of hydrogen bonds and specific interactions (kJ mol^{-1}) of crystalline pentamerous hetero compounds with benzyl cyclic at 298

Compounds	Formula	Structure	$\Delta_{sub}H°$(298 K) [12]	D=HC → CH	DC→C/D=HC → C	D–N•••H–N	D=HC → CH
Carbazole	$C_{12}H_9N$	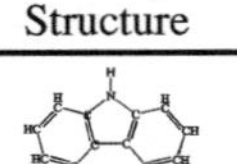	97.7 ± 0.3 103.3 ± 1.1	7.40 × 8	5.40 × 4	8.0	–
N-Ethylcarbazole	$C_{14}H_{14}N$	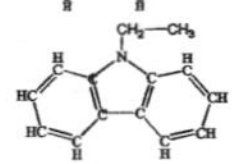	99.10.3	7.40 × 8	5.40 × 4	DN → CH_3–CH_2–N = 8.7	–
Indole	C_8H_7N	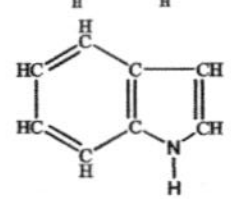	77.8 ± 1.6	7.40 × 2	6.40 × 4	8.0	10.7
3-Methylindole	C_9H_9N	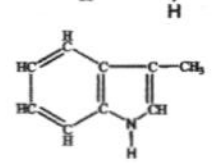	83.3[a]	7.40 × 2	5.40 × 4	8.0	10.7
Benzimidazole	$C_7H_6N_2$	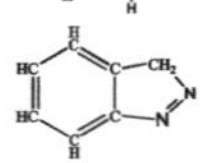	102.2 ± 0.4	7.40 × 4	5.40	D=N → C = 15.8	DN → CH 17.8
Indazole	$C_6H_7N_3$	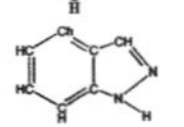	91.1 ± 01	7.40 × 4	6.40 × 4	8.0	D=N → CH 17.35
1-H-Benzotriazole	$C_6H_5N_3$	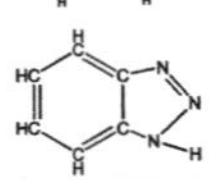	98.2 ± 0.7	7.40 × 2	5.40 × 2	8.0	D=N → C 14.15

[a]Obtained at 303 K

number, we implemented the calculations by Eq. 4.13a of the energy of the specific interaction,

$$D^*HC{\rightarrow}CH = (\Delta_{vap}H^0(298\ K)\ - 2D{=}HC{\rightarrow}CH - 4D{=}HC{\rightarrow}C - 2D\text{-}N{\bullet\bullet\bullet}\ H\text{-}N)\ /2 \quad (4.13a)$$

formed by the pentamerous cycle (Table 4.8). The energy obtained for the series of specific interaction considered reflects its significant stabilization

$$D^*HC{\rightarrow}CH\ (10.7) < D{=}HC{\rightarrow}CH\ (7.40\ kJ\ mol^{-1})$$

compared with the interaction formed by benzene ring.

The methyl group of 3-methylindole functions as an isostructural group, the value of the energy contribution of which is determined by the difference of sublimation enthalpies of this compound and indole (5.5 kJ mol^{-1}). The unusually high value of the energy contribution of this group to the enthalpy characteristics may be related to its high influence on the distribution of the electron density in the molecule and to the increase in negative charge of the carbon atom C(3), leading to the high difference between the charges of carbon and hydrogen atoms for this group. But the sublimation enthalpy of 3-methylindole is given in the sources in the literature [9] without indication of the method of research and experimental error, which impedes the interpretation of this value. Using the value obtained for the energy contribution to the enthalpy characteristics, the energy of the formed specific interaction $DH_3C \rightarrow H{-}CH_2$ should be taken to be equal to 2.75 kJ mol^{-1} and the energies of the specific interactions and hydrogen equal to the values of the corresponding interactions, realized in crystalline indole.

Increasing the number of nitrogen atoms in the pentamerous fragment of benzimidazole is accompanied by the increasing acceptor properties, which provides the increase in the difference of its charges and the charges of carbon atoms. That is why at the nine bond vacancies of the indole and benzimidazole molecules the second compound forms specific interactions with significantly increased stability. The molecule of benzimidazole forms the grid structure of the crystal (Fig. 4.11a) with two types of increased stability of the specific interactions $D{-}N \rightarrow CH_2$ and $D{-}N \rightarrow C$, four $D{=}HC \rightarrow CH$ specific interactions of CH groups of benzene ring (7.40 kJ mol^{-1}) and an interaction of reduced stability, $D{=}C \rightarrow C$ (5.40 kJ mol^{-1}). Thus, the total contribution of the two types of specific interactions with increased stability is determined by the difference of sublimation enthalpy and interactions with known energy.

$$2D\text{-}N{\rightarrow}CH_2 + 2D\text{-}N{\rightarrow}C = (\Delta_{vap}H^0(298\ K)\ - 4D{=}HC{\rightarrow}CH - D{=}C{\rightarrow}C) \quad (4.13b)$$

Taking into account that the energy contribution of the saturated hydrogen atom to the enthalpy characteristics is equal to 1.0 kJ mol^{-1} and the average

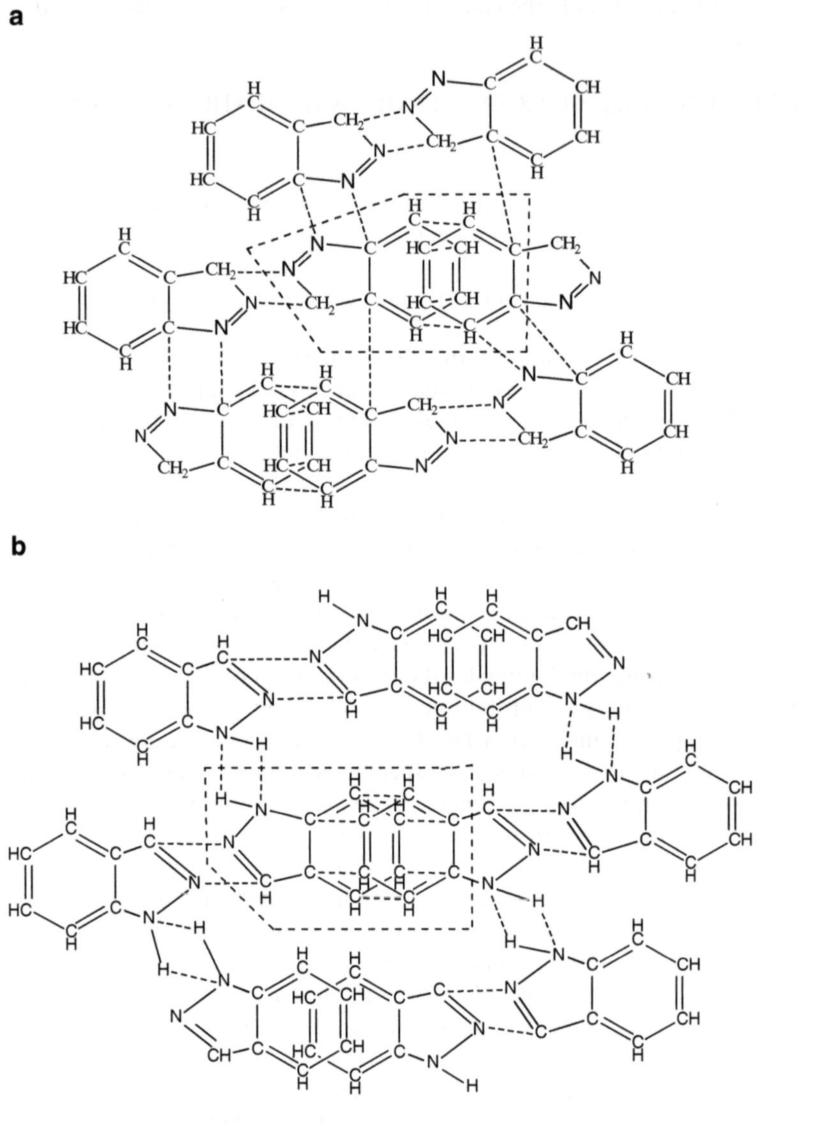

Fig. 4.11 Schematic picture of the crystalline structure of benzimidazole (**a**) and indazole (**b**) with the network of specific interactions

value of the energies of the two series is 33.6 kJ mol^{-1}, we estimated the value of the energy of each series of specific interaction D–N → CH_2 and DN → C (Table 4.8).

Replacement of the nitrogen atom in the benzimindazole molecule by the amide group in indazole is accompanied by the reduction in sublimation enthalpy of the latter to 11.1 kJ mol^{-1}. It means that the hydrogen bond formed has additional stability, approximately at the same value or at 9 kJ mol^{-1} under the

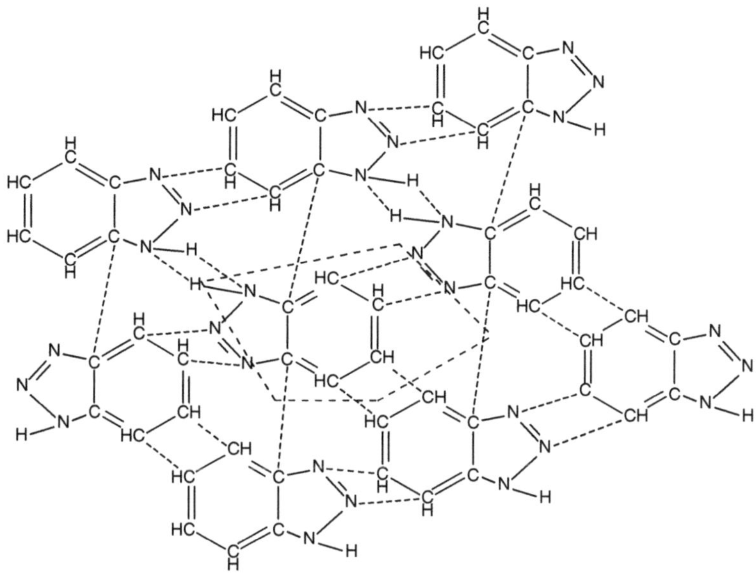

Fig. 4.12 Schematic picture of the crystalline structure of 1-H-benzotriazole with network specific interactions

condition of additional replacement of the specific interaction D=HC → CH to D=HC → C. The indazole molecule with ten bond vacancies forms a similar network structure of a crystal by two hydrogen of the amino group, two of increased stability of specific interactions D–N → C and, forming a benzene ring, the four specific interactions D=HC → C and two DHC → CH (Fig. 4.11b). The calculations of the energy of the specific interaction D=N → CH, by an equation similar to (4.13b), direct attention to its coincidence with the value of the energy of the same series of interactions of the crystalline benzimidazole (Table 4.8).

The increase in the number of nitrogen atoms in the 1-H-benzotriazole molecule and reduction of CH groups cause a reduction in the difference of the charges at the donor and acceptor atoms of the electrons and reduction of the stability of the specific interactions. In this connection benzimidazole and 1-H-benzotriazole are remarkable. The 1-H-benzotriazole molecule with three nitrogen atoms and ten bond vacancies form a similar network structure, in which the number of specific interactions formed by CH groups and carbon atoms reduces up to four 2D=HC → CH and 2D=C → C and there is an increase up to four in the number of more stable specific interactions D=N → CH (Fig. 4.12). The fourth formed series of specific interactions is the hydrogen bond. From the difference of sublimation enthalpy and total contribution of the energies of the two specific interactions and the hydrogen bonds we determined the energy of specific interaction, given in Table 4.8. From this table we know the values of the energies of the series D=HC → CH, formed by CH groups of the benzene ring (7.40 kJ mol^{-1}) and pentamerous cycle (10.7 kJ mol^{-1}), formed by indole. The energies of the most stable specific interactions, formed with nitrogen atoms, are described by the integral value of its stabilization, first, with

D-N→CH_2 (17.8) Benzimidazole > D-N→CH (17.45) Indazole > D-N→CH (14.15 kJ mol^{-1}) 1-H-Benzotriazole

reduction in the number of hydrogen atoms, connected with the carbon atom, and second, with an increasing number of nitrogen atoms in the pentamerous cycle of 1-H-benzotriazole.

References

1. Nefedov VI, Vovna VI (1989) Electronic structure of organic and elementorganic compounds. Nauka, Moscow, p 199
2. Lebedev Yu A, Miroschnichenko EA (1981) Thermochemistry of the vaporization of organic compounds. The vaporization and sublimation enthalpy and saturated pressure. Nauka, Moscow
3. Chickos JS, Acree WE (2002) Enthalpies of vaporization of organic and organometallic compounds, 1880–2002. J Phys Chem Rev Data 31:519–879
4. Baev AK (1969) Phase condition and complex formation ability of halogenide metals. General and applied chemistry, vol 1. Vysheischaia Educating, Belarusia, Minsk, pp 207–218
5. Baev AK (2001) Thermodynamics of evaporizations of alryl compounds seconds – six groups elements. Russ J Izvestiya Vischich Ucheb Zaved Chem Chem Technol 44:3–13
6. Baev AK (1987) Chemistry of Gas-heterogenic systems of elementorganic compounds. Science and Technology, Minsk, p 175
7. Karlson TA (1981) Photo electronic and OGE-spectrosopy. Mechanical Engineering, Leningrad, 431p
8. Nefedov VI (1984) X-ray spectroscopy of chemical compounds. Chemistry, Moscow, 255p
9. Bieri G, Åsbrink L, von Niessen W (1981) Photoelectron spectra of organic molecules. J Electr Spectrosc Relat Phenom 23:281–322
10. Asbrink L, Fridh C, Lindholm E (1979) Interpretation of electron-spectra. 1. Spectra of furan, pyrrole and cyclopentadiene, studied with ham-3. J Electron Spectrosc 16(3):65–76
11. Nguyen CV, Palmer GR, Singer MA, Tse MY (1987) The diffusion of small ions through phospholipid bilayers. J Chem Phys 86:7218
12. Chickos JS, Acree WE (2003) Enthalpies of sublimation of organic and organometallic compounds, 1910–2001. J Phys Chem Rev Data 32:537–698

Chapter 5
Specific Intermolecular Interactions of Nitrogen Containing Six-Membered Heterocycles

5.1 Energies of Specific Interactions of Hexamerous Heterocyclic Compounds with Saturated Hydrocarbon

5.1.1 Piperidine and Its Derivatives

Piperidine molecules and their derivatives with methylene groups and hetero cyclic nitrogen atoms, connected with the hydrogen atom, are influenced by the latter by the shifting of the electron density in comparison with pyridines. The nitrogen atom of piperidine gets electron density from the hydrogen atom of the amino group and contacting by carbon atoms $C(2)H_2$ and $C(6)H_2$, reducing its negative charge and reaching the increased electron density at its own $3p_z$-orbital [1–5]. However, the further formation by the same nitrogen atom of the reverse dative bond $\overset{\frown}{N \leftarrow CH_2}$ is accompanied by the transmission of part of the electron density from this orbital to the essentially unshared $2s^2$ electron pair of the carbon atoms. As a result, there appears a reduction of its own negative charge and relative enrichment of the electron density of the carbon atoms C(2) and C(6) of the methylene groups, of which charges differ from the charges of the carbon atoms C(3), C(5) and, of course, C(4). It follows that the nitrogen atom contributes to a definite change in the charges of all five carbon atoms of the piperidine molecule. Thus, each of the five free bond vacancies of the essentially unshared $2s^2$ electron pair of the carbon atoms of the methylene group of piperidine forms the specific interaction $D\text{–}H_2C \rightarrow CH_2$, contributing to the vaporization enthalpy being equal to the contribution of the CH_2 group, the energy value (5.70 kJ mol^{-1}) of which is identical to the energy of the realized similar specific interaction in liquid cyclopentane (Table 3.1). This value of the energy of the specific interaction formed by the methylene group is slightly higher than the energy of the specific interaction formed by the benzene CH-group (5.63 kJ mol^{-1}). It directs attention to the fact that differences of the structure of MO pyridine C_5H_5N as regards symmetry positions are insignificant compared to the electron structure of the benzene molecule and, consequently,

A.K. Baev, *Specific Intermolecular Interactions of Nitrogenated and Bioorganic Compounds*, DOI 10.1007/978-3-642-37472-2_5,

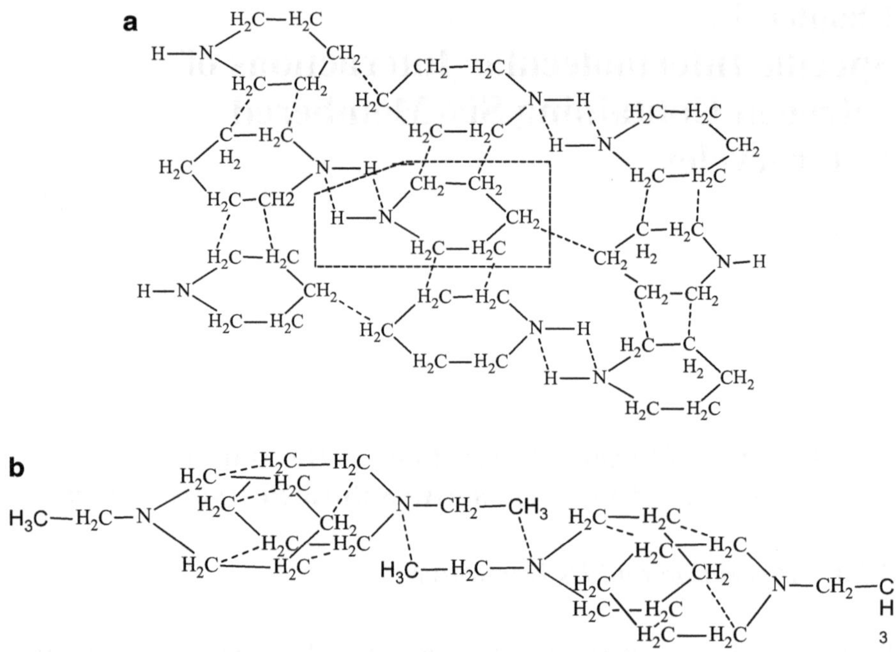

Fig. 5.1 Schematic picture of the liquid structure of Piperidine (**a**) and *N*-Ethylpiperidine (**b**) with the network of the specific interactions and hydrogen bonds

cyclohexane, splitting $e(\sigma)$ –MO at $a_1 + b_2$, $e(\pi)$ – at $a_2 + b_1$ and transforming one a_1 (σ) –orbital of radial type to the *n*-orbital [6, 7]. It follows that the accepted assumption of the energy contribution by the methylene group of cyclohexane for conducting thermodynamic calculations of the energies of the specific interactions of piperidine, piperazine, and their derivatives is correct and grounded. The rule of using the energy contributed by the CH group of benzene to the enthalpy characteristics of the vaporization process of pyridine and its derivatives is also correct.

The hydrogen atom of the NH-group forms the hydrogen bond N•••H–N, forming the chains, crisscrossed by specific interactions of D–$H_2C \rightarrow CH_2$ of molecules close in the environment, forming the grid structure of liquid and crystal (Fig. 5.1a, b).

It means that the energy of the hydrogen bond formed by the molecule of piperidine should be determined with the help of the equation

$$\text{D-N}\bullet\bullet\bullet\text{H-N} = (\Delta_{vap}H°(298\ \text{K})\text{ppr} - 5\ \text{D-}H_2C\rightarrow CH_2)/2 \tag{5.1}$$

considering the total value of the contribution of the energies of the five specific interactions D–$H_2C \rightarrow CH_2$. A similar equation is used to determine the energies of the specific interactions, formed by methyl, ethyl, and propyl groups of *N*-alkyl derivatives of piperidine

$$D\text{-}N{\rightarrow}R\text{-}N = (\Delta_{vap}H°(298\ K)appr - 5D\text{-}H_2C{\rightarrow}CH_2) / 2 \qquad (5.1a)$$

The completion of the influence of the reverse dative bond at the propyl fragment and invariability of the energy of the formed specific interaction $N \rightarrow CH_3\text{-}CH_2\text{–}CH_2\text{–}N$ at compounds of this series with a high number of methylene groups allows us to define the contribution of the remaining one, two, or three CH_2-groups by butyl – hexyl fragment to the vaporization enthalpy of the compounds of this series with the help of Eq. 5.1b,

$$DCH_2 = (\Delta_{vap}H°(298\ K)appr - 2D\text{-}N{\rightarrow}CH_3\text{-}CH_2\text{-}CH_2\text{-}N - 5D\text{-}H_2C{\rightarrow}CH_2) / n \qquad (5.1b)$$

where n is equal to the number of CH_2 groups.

The results of the calculations of the energy of the hydrogen bond, specific interactions, and vaporization enthalpies of liquid piperidines are given in Table 5.1.

The energies of the specific interactions of liquid derivatives of piperidines are naturally stabilized on the replacement of methyl by the ethyl, propyl fragment of *N*-alkylpiperidines

$D\text{-}N{\rightarrow}CH_3\text{-}N$ N-Methylpiperidine (4.10) < $D\text{-}N{\rightarrow}CH_3\text{-}CH_2\text{-}N$ N-Ethylpiperidine (6.20) < $D\text{-}N{\rightarrow}CH_3\text{-}CH_2\text{-}CH_2\text{-}N$ N-Propylpiperidine (8.20) = $D\text{-}N{\rightarrow}CH_3\text{-}CH_2\text{-}CH_2\text{-}N$ N-Butylpiperidine (8.20 kJ mol^{-1}).

A similar character of stabilization of the given type of interactions is expressed at methyl-, ethyl-, and propylamines.

$D\text{-}N{\rightarrow}CH_3\text{-}N$ Trimethylamine (4.25) < $D\text{-}N{\rightarrow}CH_3\text{-}CH_2\text{-}N$ Triethylamine (5.85) < $D\text{-}N{\rightarrow}CH_3\text{-}CH_2\text{-}CH_2\text{-}N$ Tripropylamine (8.10 kJ mol^{-1})

The difference in the energies of the specific interactions, formed by an equal number of carbon atoms of the two series of compounds with a hexamerous heterocycle with a nitrogen atom and alkylamines with a linear structure of fragments within the error (0.1 kJ mol^{-1}) of measurement of the experimentally established enthalpy characteristics points to the correctness of the accepted assumptions on the equality of total values of the energy $D\text{–}H_2C \rightarrow CH_2$ of liquid piperidines and cyclopentane. It follows that the influence of the nitrogen atom of piperidine on the stabilization of these interactions does not exceed the experimental error mentioned. Thus, one should conclude that the nitrogen atom of heterocycle *N*-methylpiperidine – *N*-hexylpiperidine expresses the influence, equivalent to the nitrogen atom of alkylamines, on the energy of the specific interactions formed by methyl – hexyl fragments. Using the correctness of the rules of the energy contribution by the CH-group of benzene to the vaporization enthalpy of piperidine, we define the energy of the specific interaction $D\text{–}N \rightarrow CH_3\text{–}N$ in its liquid condition with the help of

$$D\text{-}N{\rightarrow}CH_3\text{-}N = (\Delta_{vap}H°(298\ K)appr - 4(D\text{-}H_2C{\rightarrow}CH_2) / 4 \qquad (5.1c)$$

considering the equal number of this type of bond and formed by four CH groups.

Table 5.1 Energies of the hydrogen bonds and specific interactions (kJ mol^{-1}) of liquid piperidines at 298 K

Compounds	Formula	Structure	$\Delta_{vap}H°$(298 K) [8]	D–$H_2C \rightarrow CH_2$	D–N•••H–N	D–N → R–N	D*iso*CH_3
Piperidine	$C_5H_{11}N$	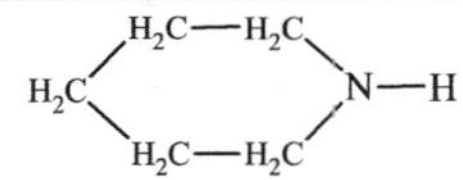	39.8 36.6[a]	5.70	5.60	–	–
N-Methyl-piperidine	$C_6H_{13}N$	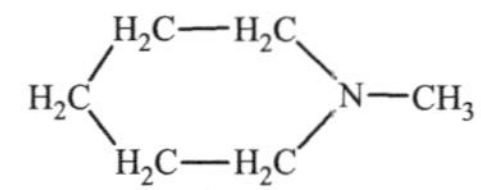	36.7 ± 0.1	5.70	–	4.10	–
N-Ethyl-piperidine	$C_7H_{15}N$	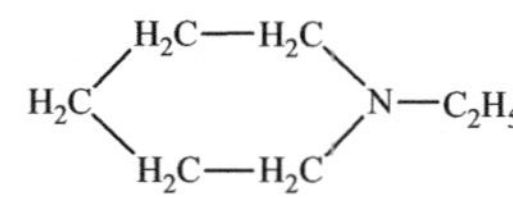	41.1 ± 0.1 40.8 ± 0.6	5.70	–	6.20 ± 0.1	–
N-Propyl-piperidine	$C_8H_{17}N$	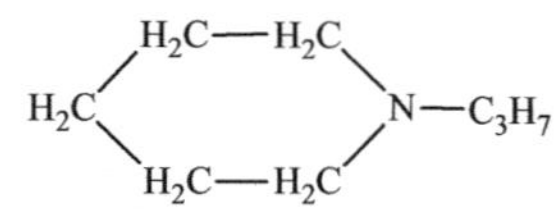	44.9 ± 0.4	5.70	–	8.10 ± 0.1	–
N-Butyl-piperidine	$C_9H_{19}N$	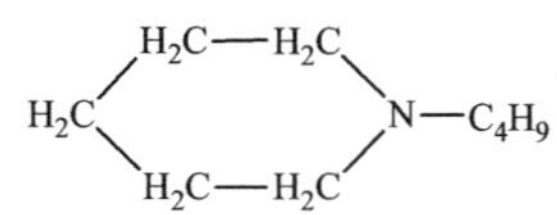	48.9 ± 0.2 49.2 ± 0.1	5.70	–	8.10 ± 0.1	DCH_2 4.00
N-Pentyll-piperidine	$C_{10}H_{21}N$	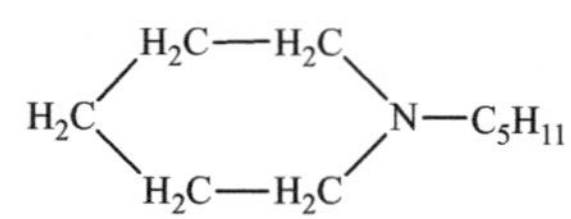	53.0[b]	5.70	–	8.10 ± 0.1	DCH_2 4.05
N-Hexyl-piperidine	$C_{11}H_{23}N$	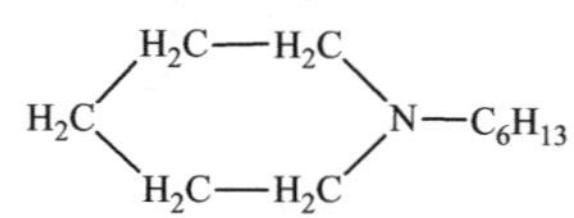	57.2[b]	5.70	–	8.10 ± 0.1	DCH_2 4.10

N-Cyclohexylpiperidine	$C_{11}H_{21}N$	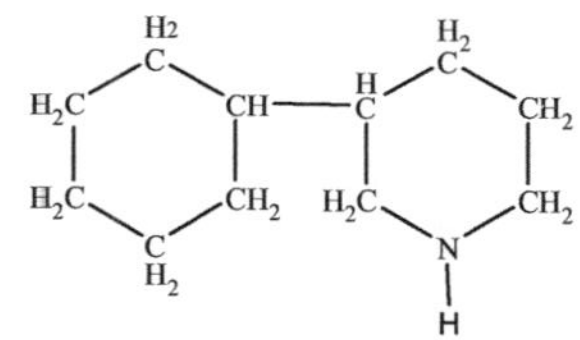	60.5 ± 0.6	5.00[b]	5.60		
Bis(piperidine)methane	$C_{11}H_{22}N_2$	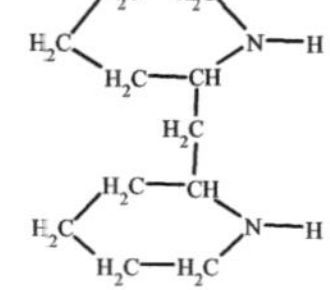	62.2 ± 0.9	4.93[b]	5.60	–	–
N-cycloheptyl-piperidine	$C_{10}H_{19}N$	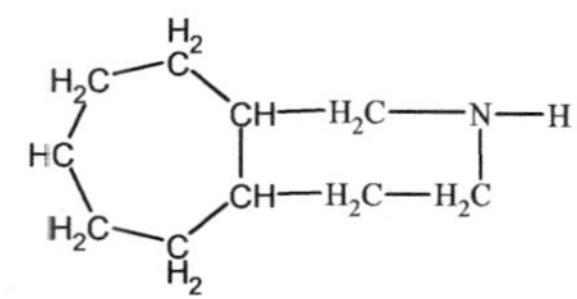	55.1 ± 0.5	4.96[b]	5.60	–	–

[a]At 338 K
[b]Received by additively method

The procedure for the calculations illustrates the increase in stabilization of this type of bond

N-methylpiperidine D-N→CH_3-N (4.10) < 1,4-dimethylpiperidine D-N→CH_3-N (5.25 kJ mol^{-1})

for 1,4-dimethylpiperidine with reduction in the molecule of the last number of the carbon atoms in the cycle and replacement of a single C–C by the double C=C bond.

The underestimated value of vaporization enthalpy of *N*-cyclohexylpiperidine with vaporization enthalpy (60.5 kJ mol^{-1}) in comparison with the total value of the enthalpy characteristics of cyclohexane (33.0 ± 0.1 kJ mol^{-1}) and piperidine (37.6 kJ mol^{-1}) points to the change of the charges of the carbon atoms of the cycles, reducing the energies of the specific interactions, formed by the methylene groups.

Bis(piperidine)methane, consisting of two molecules of piperidine, connected by C-C bonds with the methylene group, turn into vapor with the value of the vaporization enthalpy (62.2 ± 0.9 kJ mol^{-1}), which in comparison with the total value of vaporization enthalpy of toluene (38.0 kJ mol^{-1}) minus the energy contribution of the two saturated hydrogen atoms (76.6 kJ mol^{-1}) reflects the significant changes in distribution of the electron density, accompanied by reduction of the carbon atoms of the cycle. At the same time, the difference of vaporization enthalpies of 1.70 kJ mol^{-1} for (piperidine)methane and *N*-cyclohexylpiperidine with an equal number of carbon atoms points to the small difference in the energies of the specific interactions, formed by CH_2 groups of the cycles. Hence, the value of the marked difference of the enthalpy characteristics of compounds should be attributed to the energy contribution of a similar group, united by two cycles of piperidine at bis (piperidine)methane. The planar structure of the molecules of *N*-cyclohexylpiperidine and bis(piperidine)methane forms the structure of the liquid and crystal condition, peculiar to the similar compounds with the network of the specific interactions (Fig. 5.2).

Taking the energy value of the hydrogen bonds of these compounds equal to 5.60 kJ mol^{-1}, we defined the contribution of the CH_2-group to the enthalpy characteristics by the difference of vaporization enthalpies of these compounds and the energies of the hydrogen bonds or more correctly the energy contribution of the specific interactions $H_2C \rightarrow CH_2$ formed by this group. The results of the implemented calculations of the energy contribution of CH_2-groups to the enthalpy characteristics illustrate its small difference at *N*-cyclohexylpiperidine (5.00 kJ mol^{-1}) and bis(piperidine)methane (4.93 kJ mol^{-1}) and significant difference from the contribution of the same groups by molecules of piperidines (5.70 kJ mol^{-1}) and cyclohexane (5.5 kJ mol^{-1}).

Thus, hexamerous molecules of heterocyclic nitrogen compounds with saturated hydrocarbon cycle form specific interactions $H_2C \rightarrow CH_2$ with reduced energies values.

Derivatives of piperidines with isostructural methyl group are characterized by an increased value of vaporization enthalpy that is caused by additional energy to

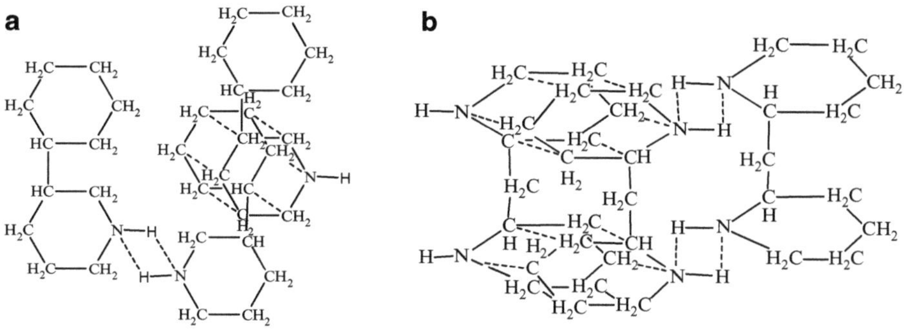

Fig. 5.2 Schematic picture of the liquid structure of *N*-cyclohexylpiperidine (**a**) and bis(piperidine)methane (**b**) with the network of the specific interactions and hydrogen bonds

the enthalpy characteristics. It follows that the energy contributed by the specific interactions of the isostructural methyl group equal to the difference of vaporization enthalpy of methylpiperidine and piperidine reduced by the energy contributed by the hydrogen atoms DH (0.60 kJ mol^{-1}), divided by the number of these groups (n) is

$$\mathrm{D}iso\mathrm{CH_3} = (\Delta_{vap}H°(298\ \mathrm{K})\mathrm{mpp} - (\Delta_{vap}H°(298\ \mathrm{K})\ \mathrm{ppr} - n\mathrm{DH})) = (\Delta_{vap}H°(298\ \mathrm{K})\mathrm{mpp} - 2\mathrm{D\text{-}N{\bullet}{\bullet}{\bullet}H\text{-}N} - 5\mathrm{D\text{-}H_2C{\rightarrow}CH_2} - n\mathrm{DH})/n \quad (5.2a)$$

or *N*-alkylpiperidines with different numbers of isostructural methyl groups and alkyl – CH_3–, C_2H_5–, C_3H_7, and C_4H_9, respectively, is

$$\mathrm{D}iso\mathrm{CH_3} = (\Delta_{vap}H°(298\ \mathrm{K})\mathrm{mpp} - (\Delta_{vap}H°(298\ \mathrm{K})\ \mathrm{ppr} - n\mathrm{DH})) = (\Delta_{vap}H°(298\ \mathrm{K})\mathrm{mpp} - 2\mathrm{D\text{-}N{\rightarrow}CH_3\text{-}N} - 5\mathrm{D\text{-}H_2C{\rightarrow}CH_2} - n\mathrm{DH})/n \quad (5.2b)$$

The results presented in Table 5.2 of the thermodynamic calculations illustrate the low levels of energies contributed by the isostructural methyl groups, described by the dependence on the number of isostructural groups at the carbon atom of the cycle and energies of the specific interactions of alkyl group, replacing the hydrogen atom of piperidine, from the location at C(2)H and C(6)H-group.

5.1.2 *Piperazines*

Hexamerous heterocyclic compounds with saturated hydrogen cycle of piperidines and piperazines with one and two nitrogen atoms, respectively, show the significant difference in thermodynamic properties. In particular, vaporization enthalpies of *N*-methylpiperidine (36.7 ± 0.1 kJ mol^{-1}) and 1,4-dimethylpiperazine

Table 5.2 Estimated value of vaporization enthalpies (kJ mol^{-1}) of methylpiperidines at 298 K, (5 DHC → CH = 28.5 kJ mol^{-1})

Compounds	Formula	Compounds	$\Delta_{vap}H°$ (298 K) [8]	D–H_2C → CH_2 DH_2C → H–CH	D–N•••H–N	D–N → CH_3	nD*iso* CH_3/DH_3C → H–CH_2=
2-Methylpiperidine	$C_6H_{13}N$	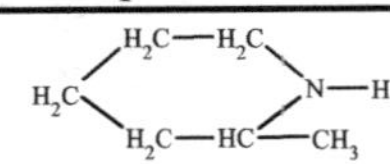	38.2 (338) 41.35	5.70	5.60	1.45	Σ 2.25/1.13
2,6-Dimethylpiperidine	$C_7H_{15}N$	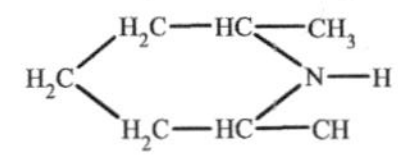	43.0	5.70	5.60	3.20:2 = 1.60	–
2,2,6,6-Tetramethyl-piperidine	$C_9H_{19}N$	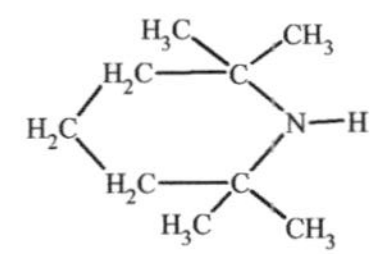	44.5 ± 0.5 (300)	5.70	5.60	4.70:4 = 1.27	Σ 7.20 DH_3C → H–CH_2 = 0.90
					D–N → R–N		
N-2-Dimethylpiperidine	$C_7H_{15}N$	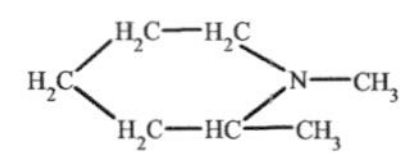	38.0 ± 0.1	5.70	4.07	1.35	n = 1 Σ 2.25/1.13
N-2,6-Trimethylpiperidine	$C_8H_{17}N$	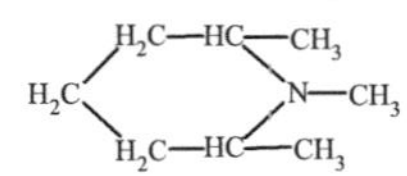	39.4	5.70	4.07	2.76:2 = 1.38	n = 2 Σ 4.50/1.13
N-2,2,6,6-Pentamethyl-piperidine	$C_{10}H_{21}N$	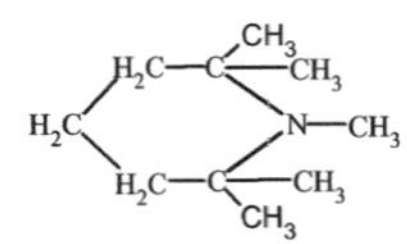	43.3	5.70	4.07	6.66:4 = 1.67	n = 4 Σ 7.20 DH_3C → H–CH_2 = 0.90
N-Ethyl-2-methyl-piperidine	$C_8H_{17}N$	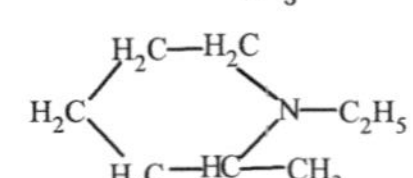	42.5	5.70	6.30 ± 0.1	0.90	n = 1 Σ 2.25/0.67
N-Ethyl-2,6-methyl-piperidine	$C_9H_{19}N$	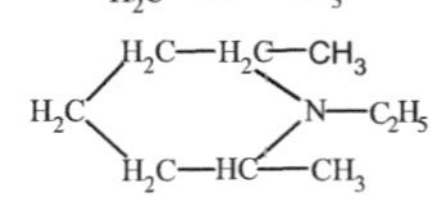	43.9	5.70	6.30 ± 0.1	2.30:3 = 1.15	n = 2 Σ 4.50/1.13

N-Ethyl-2,2,6,6-tetramethyl-piperidine	$C_{11}H_{23}N$	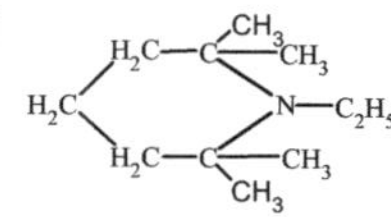	45.9	5.70	6.30 ± 0.1	4.90:4 = 1.22	n = 4 Σ 7.20 $DH_3C \rightarrow H–CH_2 = 0.90$
N-Propyl-2-methyl-piperidine	$C_9H_{19}N$	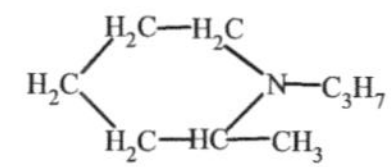	46.5	5.70	8.20 ± 0.1	1.60	n = 1 Σ 2.25/1.13
N-Propyl-2,6-methyl-piperidine	$C_{10}H_{21}N$	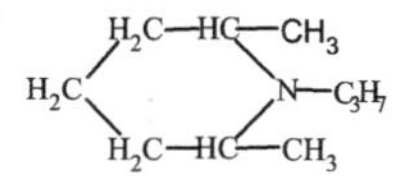	48.2	5.70	8.20 ± 0.1	3.30:2 = 1.65	n = 2 Σ 4.50/1.13
N-Propyl-2,2,6,6-tetramethyl-piperidine	$C_{12}H_{25}N$	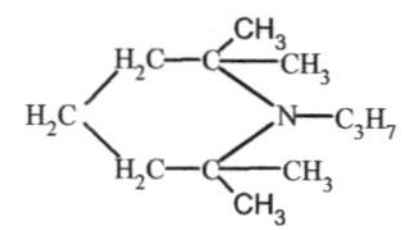	49.7	5.70	8.20 ± 0.1	4.80:4 = 1.20	n = 4 Σ 7.20 $DH_3C \rightarrow H–CH_2 = 0.90$
N-Butyl-2-methyl-piperidine	$C_{10}H_{21}N$	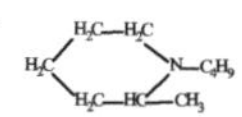	50.5	5.70	8.20 ± 0.1	1.40	n = 1 Σ 2.25/1.13
N-Butyl-2,6-methyl-piperidine	$C_{11}H_{23}N$	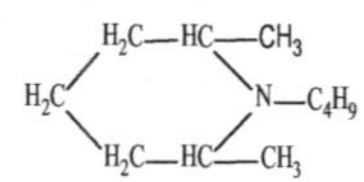	51.9	5.70	8.20 ± 0.1	3.00:2 = 1.50	n = 2 Σ 4.50/1.13
N-Butyl-l-2,2,6,6-tetramethylpiperidine	$C_{13}H_{27}N$	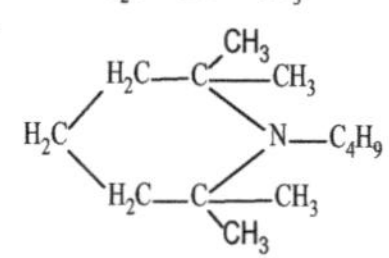	54.0	5.70	8.20 ± 0.1	5.10:4 = 1.28	n = 4 Σ 7.20 $DH_3C \rightarrow H–CH_2 = 0.90$

(43.0 ± .0.3 kJ mol^{-1}) at the standard conditions differ by 6.3 kJ mol^{-1}. It means that at an equal number of bond vacancies in the molecules of these compounds 1,4-dimethylpiperazine with reduced number of CH_2 groups forms the specific interaction N → CH_3–N with increased stability, leads to a significant shifting of the electron density in the molecule cycle, leading to the difference of the charges at the carbon atoms of methyl and methylene groups, and the nitrogen atom. As a result the specific interactions D–H_2C → CH_2 and the hydrogen bond for piperazine compared with those formed in liquid piperidines stabilize. It should be noted that the energy of the specific interaction N → CH_3–N of liquid *N*-methylpiperidine has less stability (4.10 kJ mol^{-1}) than liquid trimethylamine (4.25 kJ mol^{-1}). Each nitrogen atom in the molecule of 1,4-dimethylpiperazine is similar to Trimethylamine and is fringed by a methyl and two methylene groups. This location of the nitrogen atom is supposed to provide the stabilizing effect of the cycle by two methylene groups, which is comparable with the influence of the two methylene groups of trimethylamine.

The piperazine molecule and its derivatives with alkyl ligand and eight bond vacancies form the structure of liquid and crystal conditions with a stable network of the specific interactions (Fig. 5.3a) with four coordinating molecules. Each molecule of piperazine and 1,4-dimethylpiperazine forms the four hydrogen bonds D–N•••H–N and specific interactions D–N → CH_3–N, respectively, and its four CH_2 groups form four D–H_2C → CH_2 specific interactions. Using the correctness rule for the energies contributed by the CH_2 groups of cyclohexane to the vaporization enthalpy of piperazine, we can define the energy of the specific interaction D–N → CH_3–N in liquid condition with the help of Eq. 5.3, taking into account the equal number of this type of bond and interactions D–H_2C → CH_2,

$$\text{D-N}\rightarrow\text{CH}_3\text{-N} = (\Delta_{\text{vap}}H°(298\ \text{K})\text{pprz} - 4\text{D-H}_2\text{C}\rightarrow\text{CH}_2)/4 \qquad (5.3)$$

formed by four CH_2 groups.

The value of the energy of the specific interaction DN → CH_3–N obtained allows us to determine the energy of the hydrogen bond D–N•••H–N of liquid *N*-methylpiperazine with the help of Eq. 5.4

$$\text{D-N}\cdots\text{H-N} = (\Delta_{\text{vap}}H°(298\ \text{K})\ \text{pprz} - 4\text{D-H}_2\text{C}\rightarrow\text{CH}_2 - 2\text{D-N}\rightarrow\text{CH}_3\text{-N})/2 \quad (5.4)$$

The values of the energy of the hydrogen bond and the known value of the contributed energy of the CH_2 groups allows us to define the vaporization enthalpy of Piperazine

$$\Delta_{\text{vap}}H°(298\ \text{K}) = 4\text{D-N}\cdots\text{H-N} + 4\text{D-H}_2\text{C}\rightarrow\text{CH}_2)/4 \qquad (5.5)$$

The contributed energy of the specific interactions formed by the isostructural methyl groups of 2,5-dimethylpiperazine is calculated from the difference of vaporization enthalpies with piperazine and reduced by the energies contributed by saturated hydrogen atoms with the help of Eq. 5.2b.

a

b

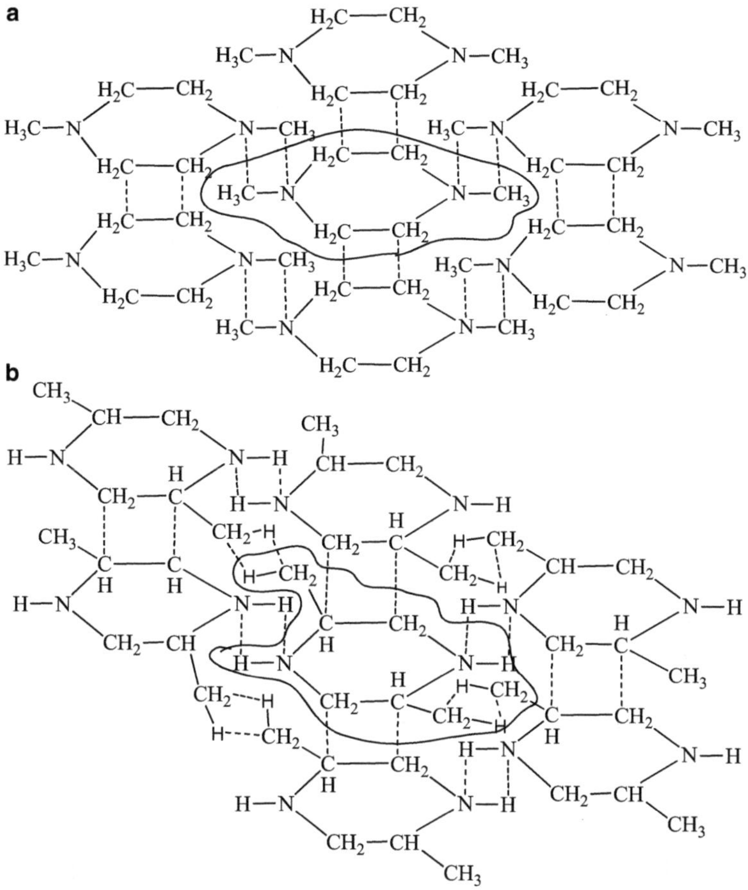

Fig. 5.3 Schematic picture of the liquid and solid structure of 1,4-Dimethyl-piperazine (**a**) and 2,5-Dimethylpiperazine (**b**) with the network of the specific interactions and hydrogen bonds

The established energies of the specific interaction D–N → CH_3–N and hydrogen bonds are described by the integral value of stabilization (Table 5.3)

D-N→CH_3-N 1,4-Dimethylpiperazine (5.25) < D-N•••H-N N-Methylpiperazine (6.35) < D-N•••H-N Piperazine (6.90 kJ mol^{-1})

For estimation of the energy of the hydrogen bond, formed in crystalline piperazine, one should use the the rule of the energy contribution noted above, setting the corresponding number of the methylene groups of cyclohexane (6.25 kJ mol^{-1}). The value of the energy of the hydrogen bond obtained points to the significant stabilization by the crystalline field (Table 5.3).

Table 5.3 Energies of the specific interactions (kJ mol^{-1}) of liquid methylpyrazines and methylpiperazine

Compounds	Formula	Structure	$\Delta_{vap}H°$ (298 K) [8]	D–N•••H–N	D–CH_2 → CH_2	D–N → CH_3–N	D*iso*CH_3/DH_3C → H–CH_2
1,4-Dimethyl-piperazine	$C_6H_{14}N_2$	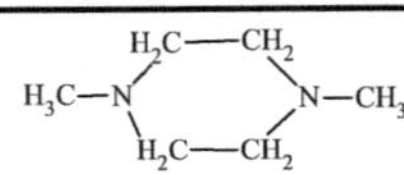	43.80.1 (298)	–	5.70	5.25	–
N-Methyl-piperazine	$C_5H_{12}N_2$	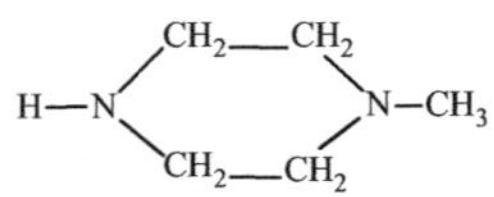	46.7 (289)	6.35	5.70	5.25	–
Piperazine	$C_4H_{10}N_2$	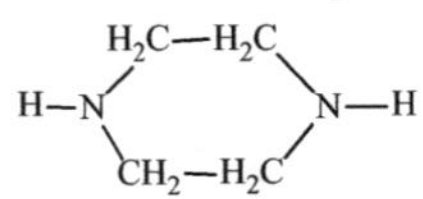	50.1 (298)	6.85	5.70	–	–
2,5-Dimethyl-piperazine	$C_6H_{14}N_2$	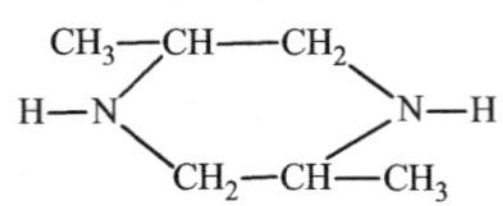	52.7[a]	6.90	5.70	–	Σ 2.30/2 = 1.15 DH_3C → H–CH_2 = 0.58
		Crystal	$\Delta_{sub}H°$(T) [3]				
Piperazine	$C_4H_{10}N_2$	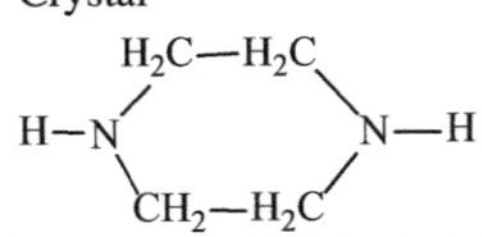	72.1 (298)	11.8	6.25	–	–

[a]Estimated on the base of additive contribution of an energies (298 K)

5.1.3 Piperidinonitriles

Vaporization enthalpies of piperidinonitriles are presented in the literature by a limited number of compounds (Table 5.4). Nevertheless, they are of interest in terms of the development of scientific ideas about specific interactions in liquid and crystal nitrogenous compounds. The structure of the cycle of piperidines in the piperidinoacetonitrile and piperidinepropionitrile molecules remains and the energies of the specific interactions of D–$H_2C \rightarrow CH_2$ and hydrogen bonds D–N•••H–N are invariable (Fig. 5.4).

The significant contribution to the enthalpy characteristics of vaporization of piperidinoacetonitrile and piperidinepropionitrile contributes to the -C≡N group, replaced by the hydrogen atom of the isostructural methyl and ethyl group, respectively. Thus, the contribution of the nitrile group to the vaporization enthalpy is the difference of vaporization enthalpies of 2-piperidinoacetonitrile and 2-methylpiperidine and 2-piperidinepropionitrile and 2-ethylpiperidine reduced by the energy contributed to the vaporization enthalpy by the substituted hydrogen atom. One should pay attention to the fact that contributions of the isostructural methyl and ethyl groups at its location C(2)H differ insignificantly. This is supported, for the example of the derivatives of piperidines, by

$$2,6\text{-Dimethylpiperidin}\ (1.60) < 2\text{-Methylpiperidin}\ \left(1.45\right) < N\text{-Ethyl} - 2,\ 6\text{-methylpiperidine}\left(1.15\ \text{kJ mol}^{-1}\right).$$

The presence at the isostructural ethyl group two carbon atoms lets us take the vaporization enthalpy of N-2-ethylpiperidine as equal to 43.0 kJ mol^{-1}.

The results of the calculations (Table 5.4) point to the fact that the energies of the specific interactions D≡N → C≡, realized in liquid 2-piperidinoacetonitrile and 2-piperidinepropionitrile, have equal value as caused by the small difference in the fragments of acetonitrile and propionitrile. The energies of the specific interactions D≡N → C≡ of liquid derivatives of piperidines have 2-Piperidinoacetonitrile (7.40) ≈ 2-Piperidinepropionitrile (7.40) < 2-Cyanopyridine (8.15 kJ mol^{-1}) a reduced stability compared with similar bonds of liquid pyridines.

5.2 Energies of Specific Interactions of Hexamerous Heterocyclic Compounds

5.2.1 Hexamerous Heterocyclic Compounds with Unsaturated Hydrocarbon

The first band with the wide oscillatory circuit of the IR-spectrum of C_5H_5N corresponds to the bending vibrations, and that points to the small changes of

Table 5.4 Energies of the specific interactions (kJ mol^{-1}) of liquid piperidinoacetonitrile and piperidinepropionitriles at 298 K

Compounds	Formula	Structure	$\Delta_{vap}H°(298\ K)$ [8]	D–$H_2C \rightarrow CH_2$	D–N•••H–N	$\Delta_{vap}H°(298\ K)$ ppn–DH	D≡N → C≡
2-Piperidino-acetonitrile	$C_7H_{12}N_2$	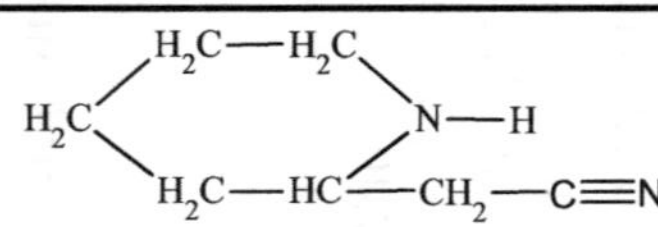	56.0 ± 0.5 (303–339 K)	5.70	5.60	41.35–0.60	7.60
2-Piperidino-propionitrile	$C_8H_{14}N_2$	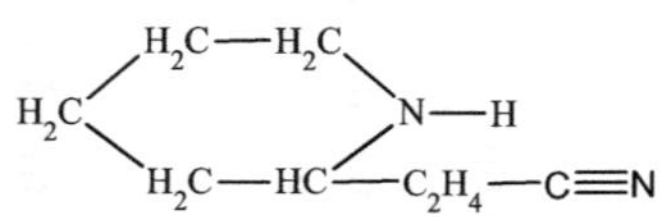	57.6 ± 0.3	5.70	5.60	43.0–0.60	7.60

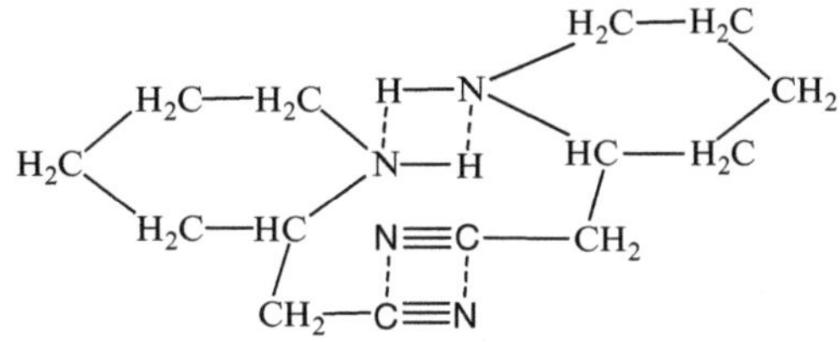

Fig. 5.4 Schematic picture of the liquid structure of 2-Piperidinoacetonitrile with the network of the specific interactions and hydrogen bonds

geometry of the nitrogen atom of aromatic type and imine nitrogen in the aliphatic case at the deleting of n-electrons [6, 7]. Shifting of the electron density at the nitrogen atom with more negative charge is accompanied by an increase in the vertical ionizing potential (Iv) at 0.5 eV.

The substituted pyridines are characterized by the shift of the three first bands, caused by distribution of the electron density of the molecular orbitals $a_1(n)$, $2b_1$, and $1a_2$. The location of the ethyl group at α-and β-locations leads to destabilization by overlinking of $1a_2$ –MO. As a result there is an inversion of values Iv(n) and Iv ($1a_2$) and an increase of $2b_1 - 1a_2$-splitting from 0.66 to1.0 eV [9]. Substituents of γ–type lead to the destabilization of $2b_1$-MO and reduction of its splitting [6, 7].

The most significant shifting of the electron density in the molecule of [10]

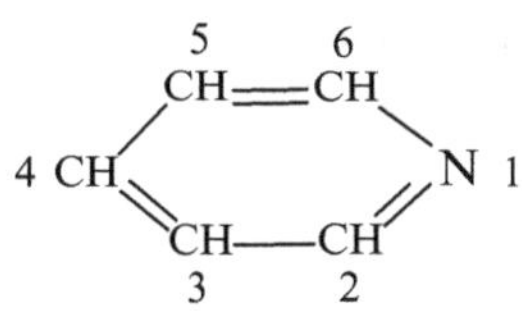

occurs from carbon atoms of the C(2)H and C(6)H groups to the nitrogen atom, providing it with positive charge and the nitrogen atom with a high negative charge. The resulting shift of the electron density from carbon atom C(4), C(3), and C(5) to the C(6) does not accompany this significant change in the charge of the carbon atom; nevertheless, it creates a negative charge at C(4) and a similar but lesser negative charge at the carbon atoms C(3) and C(5). Thus, in the pyridine molecule there has an alteration of the charge values at the carbon atoms takes place, causing the formation of two types of specific interactions with the molecules that are close in the environment, N → C(2)H or N → C(6), and the second H(4)C → C(3)H and H(5)C → C(6)H. All available six bond vacancies in the pyridine molecule should participate in the formation of the specific interactions under the condition that every molecule coordinates with two molecules. In this case each of the molecules forms three specific interactions, N → C(2)H, H(4)C → C(3)H, H(5) C → C(6)H, and N → C(6)H, H(3)C → C(2)H, H(4)C → C(5)H. The energies of these interactions correspond to the serial number N → C(6)H > H(3)C → C(2) H ≈ H(4)C → C(5)H. The formation of all types of interactions with the principle of interaction with maximum difference of the charges at the atoms is possible under the condition that a coordinating molecule that is close in the environment will be turned around its own axis at 60°, 90°, or 180°, so that the nitrogen atom interacts with atom C(2) or C(6) of the coordinating molecule. As a result the structure of the liquid condition and crystalline pyridine is the chain-network structure with the network of the specific interactions in which the molecules are connected by stable specific interactions, but the chains are crisscrossed by weak interactions, formed by hydrogen atoms of the CH groups (Fig. 5.5).

As was shown above (Table 3.1), the contribution of the CH group to the enthalpy characteristics

$$\text{Cyclobutane } (6.30) > \text{Cyclobutene } (6.15) > 1.3\text{-Cyclopentadiene}(5.94) >$$
$$\text{Cyclopentane}(5.84) > \text{Benzene } \left(5.63 \text{ kJ mol}^{-1}\right)$$

can be used for estimation of the energy contribution of CH groups or more correctly by the specific interactions to the enthalpy characteristics of liquid and crystalline nitrogenous cyclic compounds. The argument in favor of the correctness of this estimation is the fact that the wide oscillatory circuit of the first band, connected with deformation circuits and changes of geometry in the connection of "aromatic" nitrogen and immune nitrogen at aliphatic compounds remain close on removing the electron [7].

Presented in Fig. 5.5 is the schematic picture of the structure of liquid and crystalline pyridine with the network of the specific interactions illustrating the formation of the two D–N → CH–N and four specific interactions by the CH groups with the energy of the interaction of liquid benzene $D{=}HC \rightarrow CH = 5.63 \text{ kJ mol}^{-1}$that is formed.

The value of the energy of the second type is determined with the help of Eq. 5.6 from the difference of the enthalpy characteristics and the total contribution of the energy of HC groups

$$D{=}N{\rightarrow}CH\text{-}N = (\Delta_{vap}H°(298 \text{ K})pr - 4D{=}HC{\rightarrow}CH)/2 \tag{5.6}$$

The value of the energy contributed by the isostructural methyl group is defined as the difference of vaporization enthalpies of the corresponding compound and pyridine reduced by the energy of the substituted hydrogen atom (0.60 kJ mol^{-1})

$$D\mathit{iso}CH_3 = \left(\Delta_{vap}H^\circ(298 \text{ K})mpr - \left(\Delta_{vap}H^\circ(298 \text{ K})pr + DH\right)\right) \tag{5.7}$$

The energy introduced to the vaporization enthalpy of compounds of the pyridine derivatives by two or more isostructural methyl groups is determined with the help of Eq. 5.7a,

$$D\mathit{iso}CH_3 = \left(\Delta_{vap}H^\circ(298 \text{ K})mpr - \left(\Delta_{vap}H^\circ(298 \text{ K})pr - nDH\right)/n\right. \tag{5.7a}$$

where n is the number of isostructural methyl groups.

The stability of the specific interactions at the structure of liquid derivatives of pyridine is preserved and the isostructural functional methyl, ethyl, and propyl

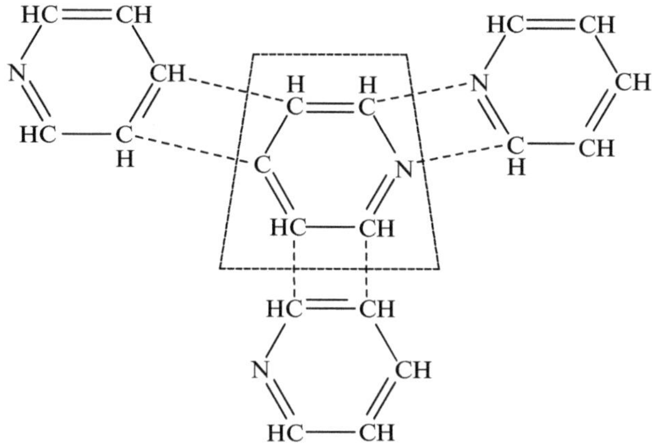

Fig. 5.5 Schematic picture of the liquid and crystalline structure of pyridine

groups form two specific interactions (Fig. 5.6). The results of the calculations (Table 5.5) illustrate the stabilization of the specific interaction of low stability formed by the isostructural methyl group at the replacement of the location at C(2) to C(3) and C(4) as the distance from the nitrogen atom,

$DH_3C{\rightarrow}H\text{-}CH_2$: 2-Methylpyridine (2.10) < 3-Methylpyridine (2.50) < 4-Methylpyridine (2.65 kJ mol^{-1}).

caused by an increase of the electron density at the carbon atom C(4) and its negative charge. As a consequence the donor property increases the essentially unshared $2s^2$ electron pair

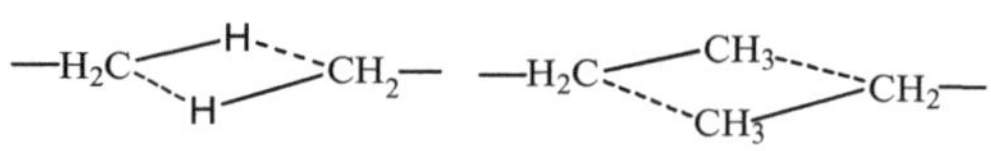

of the carbon atom, stabilizing the specific interaction $DH_3C \rightarrow H\text{–}CH_2$.

The similarity of the nature of the isostructural groups in the distribution of the electron density in the molecule does not allow us to clarify the role of each of them at the presence in the molecule with two or more. Hence, in the thermodynamic analysis it is possible to consider the average value of the energies of the specific interactions, formed by the isostructural methyl groups, located in the molecule by the different CH groups of the cycle. It draws attention to the wider range of the energy changes of the specific interactions of low stability, formed by the two isostructural groups.

$DH_3C{\rightarrow}H\text{-}CH_2$: 2,6–Dimethylpyridine (1.85) < 2,4–Dimethylpyridine (2.03 kJ mol^{-1}) < 2,5–Dimethylpyridine (2.13) ≈ 3,4–Dimethylpyridine (2.15) < 3,5 -Dimethylpyridine (2.45) < 2,3 –Dimethylpyridine (2.90 kJ mol^{-1})

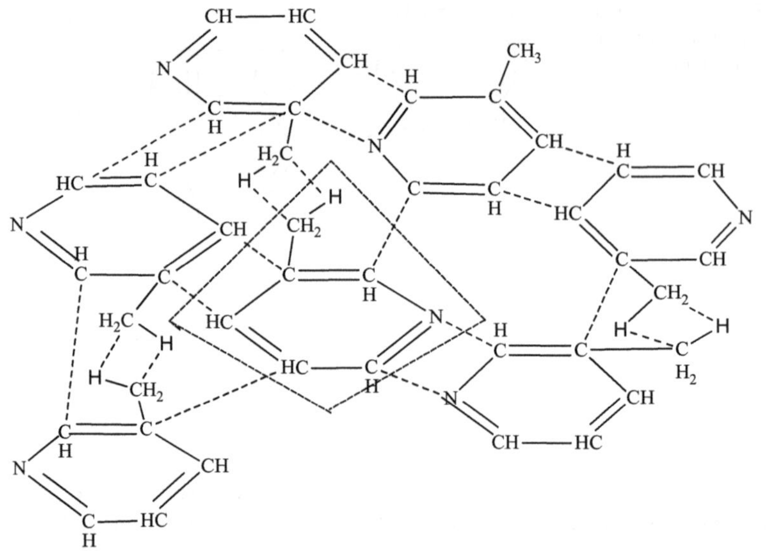

Fig. 5.6 Schematic picture of the liquid and crystalline structure of 3-methylpyridine with network of the specific interactions

The minimum influence on the distribution of the electron density in the pyridine molecular derivatives shows two isostructural methyl groups at the locations C(2) and C(6) with equally depleted electron density, partly shifted to the nitrogen atom. The maximum values of the energies contributed to the enthalpy characteristics by the isostructural methyl groups at the locations C(2)H, C(3)H (2.90 kJ mol^{-1}) are caused by the maximum difference in the charges of carbon atoms of the CH groups. For the same reason isostructural methyl groups located at the carbon atoms C(2)H or C(6)H form the less stable specific interaction $DH_3C \rightarrow H\text{-}CH_2$ (1.85 kJ mol^{-1}). Thus, the energies contributed by the isostructural methyl groups to the vaporization enthalpy of pyridine derivatives and the stability of the specific interactions reflects the difference in the charges of the carbon atoms of the CH groups, caused by the influence of the nitrogen atom of the cycle to the shifting of the electron density. It means that the carbon atoms of these isostructural groups have increased negative charges, which is why more stable bonds form with the hydrogen atoms of similar groups of molecules that are close in the environment.

The vaporization enthalpies of trimethylpyridines are established at different temperatures; nevertheless, the integral value of the energy contributed by three isostructural methyl groups to the vaporization enthalpy and the established values of the energies of low stability of the specific interactions $DH_3C \rightarrow H\text{–}CH_2$, located at the integral value of its stabilization, reflect not only the tendency of its change but also

Table 5.5 Energies of the hydrogen bonds and specific interactions (kJ mol^{-1}) of liquid pyridine and methylpyridines at 298 K

Compounds	Formula	Structure	$\Delta_{vap}H°(298\ K)$ [8]	D*iso*CH_3/D$H_3C \rightarrow H–CH_2$	D=HC → CH	D=N → CH–N
Pyridine	C_5H_5N	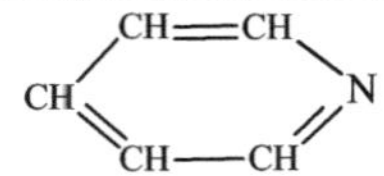	40.18 ± 0.16	–	5.63	8.80
2-Methyl-pyridine	C_6H_7N	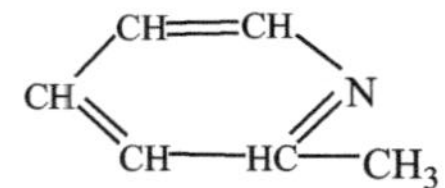	42.5 ± 0.1	4.20/2.10	5.63	8.80
3-Methyl-pyridine	C_6H_7N	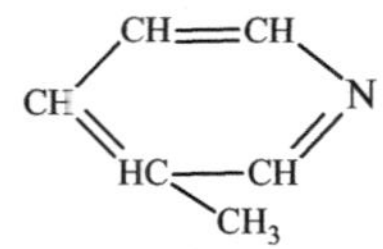	44.6 ± 0.1	5.02/2.50	5.63	8.80
4-Methyl-pyridine	C_6H_7N	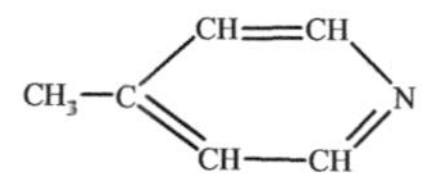	44.9 ± 0.1	5.32/2.65	5.63	8.80
2,3-Dimethyl-pyridine	C_7H_9N	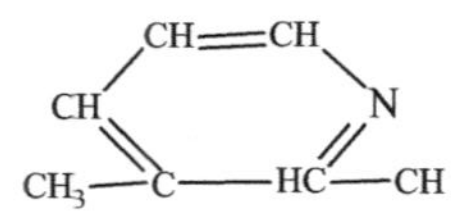	52.0 ± 0.1	5.74/2.90	5.63	8.80
2,4-Dimethyl-pyridine	C_7H_9N	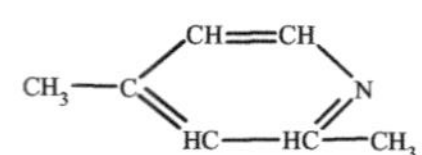	47.9	4.07/2.03	5.63	8.80
2,5-Dimethyl-pyridine[a]	C_7H_9N	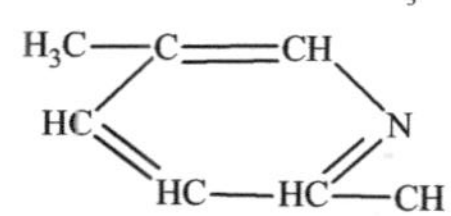	47.5	4.25/2.13	5.63	8.80

(continued)

Table 5.5 (continued)

Compounds	Formula	Structure	$\Delta_{vap}H°$(298 K) [8]	D*iso*CH_3/$DH_3C \rightarrow H–CH_2$	$D=HC \rightarrow CH$	$D=N \rightarrow CH–N$
2,6-Dimethyl-pyridine	C_7H_9N	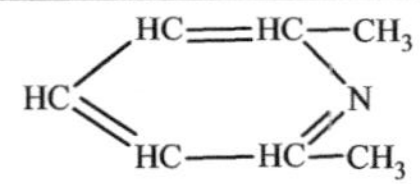	46.4	3.72/1.85	5.63	8.80
3,4-Dimethyl-pyridine	C_7H_9N	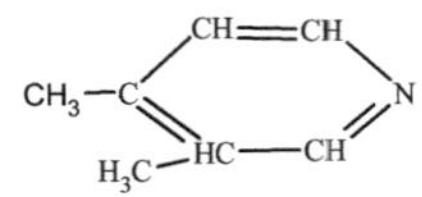	48.8	4.32/2.15	5.63	8.80
3,5-Dimethyl-pyridine[a]	C_7H_9N	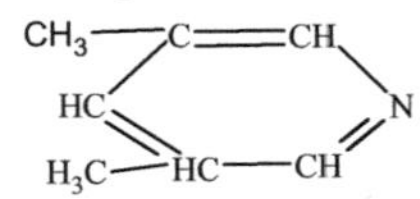	48.7	4.87/2.45	5.63	8.80
2,3,5-Trimethyl-pyridine	$C_8H_{11}N$	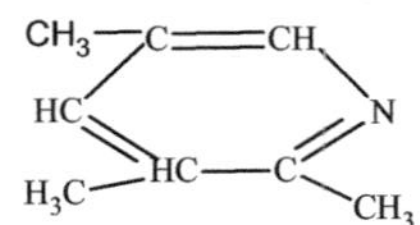	44.0[a] (359)	1.87/0.94	5.63	8.80
2,3,6-Trimethyl-pyridine	$C_8H_{11}N$	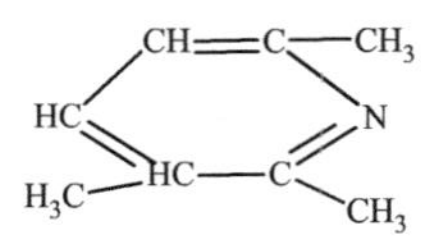	48.5 (328) 47.5 (343) 45.5 (368)	3.37/1.70 3.03/1.50 2.40/1.20	5.63	8.80
2,4,6-Trimethyl-pyridine	$C_8H_{11}N$	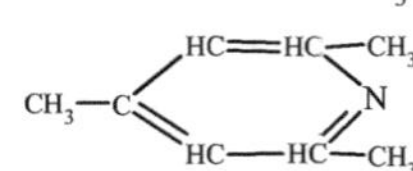	50.3 ± 0.2	4.00/2.00	5.63	8.80

[a]359 K

2, 3, 5-Trimethylpyridine (0.94.359 K) < 2, 3, 6-Trimethylpyridine (1.70.328 K) < 2, 4, 6-Trimethylpyridine (2.00 kJ mol^{-1}.298 K)

its influence on the distribution of the electron density in the trimethylpyridines molecules. From this natural tendency of the series of the specific interaction stabilization one can conclude that the increased influence of three isostructural methyl groups in the trimethylpyridine molecules is reached at the location of the two functional groups close to the nitrogen atom and the third one at the maximum removal. It should be mentioned that the carbon atoms of the isostructural methyl groups at the locations of C(2)H and C(6)H, shifting the electron density to the nitrogen atom, get a reduced negative charge. In its turn, the nitrogen atom gives from the $3p_z$ orbital part of the electron density to the carbon atom of the CH group, providing the latter with a specific interaction of increased stability. At the same time the distant carbon atom C(4)H from nitrogen with increased negative charge passes part of the electron density to the carbon atom of CH_3 group, stabilizing the specific interaction. Hence, more significant stabilization of the specific interactions by the isostructural methyl groups of the molecule 2,4,6-trimethylpyridine can be expressed at the transmission from positive charged carbon atom C(2) to the more negatively charged C(4)H.

Isostructural ethyl and propyl groups with two and three carbon atoms (Table. 5.6) form specific interactions with differing charges at the terminal carbon atoms of ethyl and propyl fragments.

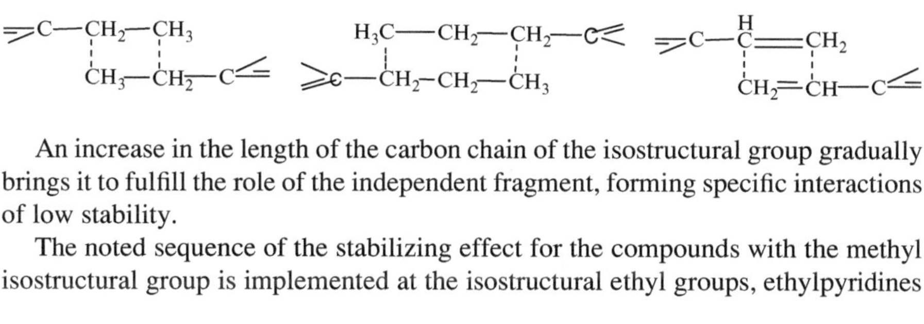

An increase in the length of the carbon chain of the isostructural group gradually brings it to fulfill the role of the independent fragment, forming specific interactions of low stability.

The noted sequence of the stabilizing effect for the compounds with the methyl isostructural group is implemented at the isostructural ethyl groups, ethylpyridines

D-CH_2-CH_3→CH_2-CH_3: 2-Ethylpyridine (2.07) < 3-Ethylpyridine (2.50) < 4-Ethylpyridine (2.95 kJ mol^{-1})

A small difference in the energies of the specific interactions formed by methyl and ethyl groups on the end members of this series relative to 0.60 and 0.30 kJ mol^{-1} points to the fact, that charges of the interacting carbon atom C(2) and C(4) of the cycle methylpyridines and ethylpyridines differ slightly.

The methyl and ethyl isostructural groups at the locations C(2)H and C(5)H of 5-ethyl-2-methylpyridine molecule make a joint contribution to the vaporization enthalpy 11.20 kJ mol^{-1}, significantly exceeding the similar contribution of the two isostructural methyl groups, located at the same carbon atoms of the cycle. Taking the contribution of the isostructural methyl group as equal to the contribution of

Table 5.6 Energies of the specific interactions (kJ mol^{-1}) of liquid ethylpyridine and propylpyridine at 298 K

Compounds	Formula	Structure	$\Delta_{vap}H°$(298 K) [8]	D*iso*C_2H_5/DCH_2–CH_3 → CH_2-CH_3	D=HC → CH	D=N → CH–N
2-Ethylpyridine	C_7H_9N	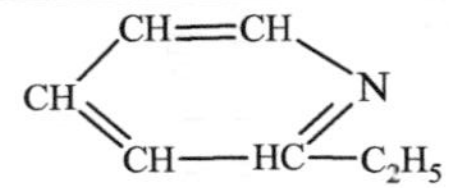	43.7 (338)	4.14/2.07	5.63	8.80
3-Ethylpyridine	C_7H_9N	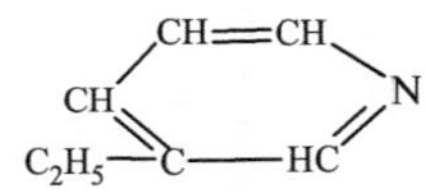	44.6 (349)	5.04/2.52	5.63	8.80
4-Ethylpyridine	C_7H_9N	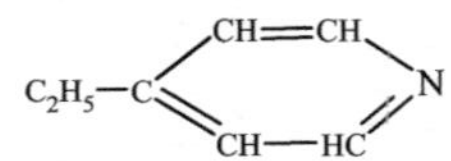	45.3 (348)	∑ 5.94/2.95	5.63	8.80
5-Ethyl-2-methylpyridine	$C_8H_{11}N$	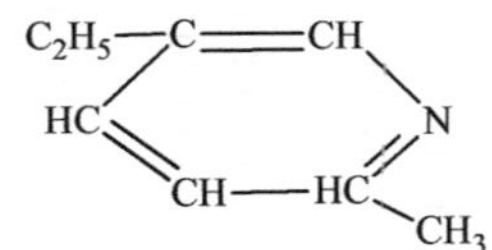	51.6 (264 K)	D*iso*C_2H_5 ∑ 7.00/3.50 D*iso*CH_3 4.20/2.10	5.63	8.80
				D*iso*C_3H_7/DC_2H_4–CH_3 → C_2H_4–CH_3		
2-Propyl-pyridine	$C_8H_{11}N$	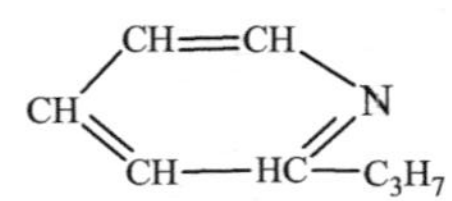	46.6 (359)	∑ 7.00/3.50	5.63	8.80
3-Propyl-pyridine	$C_8H_{11}N$	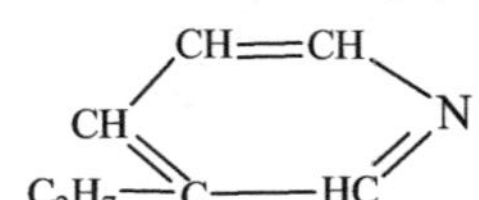	49.9 (365)	∑ 10.30/5.15	5.63	8.80
4-Propyl-pyridine	$C_8H_{11}N$	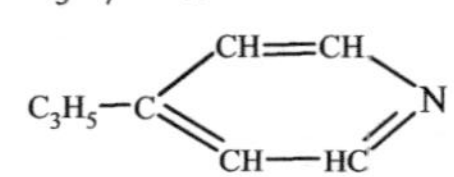	47.8 (368)	∑ 8.20/4.10	5.63	8.80
2-Methyl-5-vinylpyridine	C_8H_9N	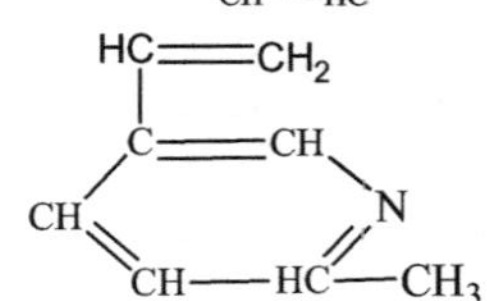	55.2 (357 K) 54.5	∑ 10.8/5.40	5.63	8.80

2-methylpyridine, we obtain the contribution of the isostructural ethyl group (7.00) that exceeds a similar contribution of the energy to the vaporization enthalpy of the same group of 3-ethylpyridine (4.14 kJ mol^{-1}). It follows that isostructural functional groups have a joint influence on the distribution of the electron density in the molecule. At the same time isostuctural methyl groups, 2-methylpyridine (4.20) and 3-methylpyridine (5.02), make a smaller energy contribution to the vaporization enthalpy compared with 2,3-methylpyridine (5.74) at 3.50 kJ mol^{-1}.

This conclusion comes from the analysis of the total energy contribution of isostructural methyl groups of the two compounds with a single functional group and compounds with two isostructural methyl groups. From this we can conclude that:

First, at two and three isostructural methyl groups in the cycle of the hexamerous hetero compound with nitrogen atom, its participation in the distribution of the electron density in the molecule is reduced,

Second, the presence in the hexamerous hetero molecule of the isostructural methyl and ethyl groups strengthens its participation in shifting the electron density and distribution of the charges at the carbon and nitrogen atoms.

The value of the energy contribution of isostructural propyl group to the vaporization enthalpy of propylpyridine is established at temperatures, differing from the standard one and without experimental error. Hence, one can suppose that these very conditions give rise to the exception to the natural character of the inadequate value of the energy of the specific interaction, formed by the isopropyl group

D-$(CH_2)_2$-CH_3→CH_2-CH_2-CH_3: 2–Propylpyridine (3.50) < 4–Propylpyridine (4.10) < 3–Propylpyridine (5.15 kJ mol^{-1})

of 4-propylpyridine. The character of the change of the energies of the specific interactions allows us to conclude that

$$\begin{aligned}
&\text{2-Methylpyridine (2.10)} \approx \text{2-Ethylpyridine (2.02)}\\
&\qquad < \text{2-Propylpyridine } \left(3.50\ \text{kJ mol}^{-1}\right)\\
&\text{3-Methylpyridine (2.50)} \approx \text{3-Ethylpyridine (2.52)}\\
&\qquad < \text{3-Propylpyridine } \left(5.15\ \text{kJ mol}^{-1}\right)\\
&\text{4-Methylpyridine (2.65)} \approx \text{4-Ethylpyridine (2.95)}\\
&\qquad < \text{4-Propylpyridine } \left(4.10\ \text{kJ mol}^{-1}\right)
\end{aligned}$$

the specific interactions formed by the isostructural methyl and ethyl groups have almost the same stability.

Isostructural propyl groups stabilize similar interactions at 1.0–2.50 kJ mol^{-1} even at the temperatures exceeding the standard one by 60–70°.

One can obtain the information to clarify the influence of unsaturated isostructural fragment to the energy of the specific interactions from thermodynamic analysis of 2-methyl-5-vinylpyridine. Molecules of this compound form, in liquid and crystalline conditions, the chains connected by specific interactions with formation of the network structure (Fig. 5.7).

The isostructural vinyl group forms two specific interactions D $CH{=}CH_2 \rightarrow CH_2{=}CH$ similar to the ethyl group. The total energy contributed by the specific interactions formed by vinyl and methyl isostructural groups is equal to the difference between the vaporization enthalpy of 2-methyl-5-vinylpyridine and the sum of the energies of the specific interactions D–HC → CH and D=N → C=(H)–N

$$D_{iso\text{-}CH=CH_2} + D_{isoCH_3} = (\Delta_{vap}H°(2987\ K)mvp - 2D{=}HC{\rightarrow}CH - 2\ D{=}N{\rightarrow}C{=}(H){-}N)/2. \quad (5.8)$$

Using the value of the energy contributed by the isostructural methyl group (4.20) or two specific interactions to the enthalpy characteristics, we obtained the energy contributed by the vinyl fragment and the specific interaction (Table 5.6). The established value of the energy of the specific interaction of the vinyl group greatly exceeds the energies of the interaction formed

2-Methylpyridine (2.10) ≈ 2-Ethylpyridine (2.07) < 2 –Propylpyridine (3.50) < 2-Methyl-5-vinylpyridine (5.40) < D=HC→CH (5.63 kJ mol^{-1})

by the isostructural methyl, ethyl, and propyl groups. The values of its energies approach the values of the energies of the intermolecular specific interaction, formed by the CH groups of contacting molecules. Considering the above, one can state that the influence of the isostructural group of the alkene series to the intermolecular interaction is close to the impact of the independent ligand.

5.2.2 *Crystalline Pyridines and Bipyridines*

The absence in the literature of the experimental value of sublimation enthalpy of pyridine complicates the conducting the thermodynamic analysis and establishing the energies of the specific interactions of its derivatives in the crystalline condition. Nevertheless, it can be estimated for 240 K, using the sublimation enthalpy of the methyl derivative of 3-methylpyridine (Table 5.8) and the equal contribution of the isostructural methyl group with similar compounds of the pentamerous cycle of tetrazoles to the enthalpy characteristics. Taking this contribution as equal to the average value (5.45 kJ mol^{-1}) contributed by the same isostructural methyl group for the derivatives of pentamerous hetero cycles of 2-methylimidazole (5.3 kJ mol^{-1}) and 5-methyltetrazole (5.6 kJ mol^{-1}), we obtain its value as equal to 57.4 kJ mol^{-1}. Note that this value corresponds to 240 K and is in excess of the value with the standard conditions. Analysis of the temperature dependence of enthalpies sublimation of the considered series points to the fact that around a

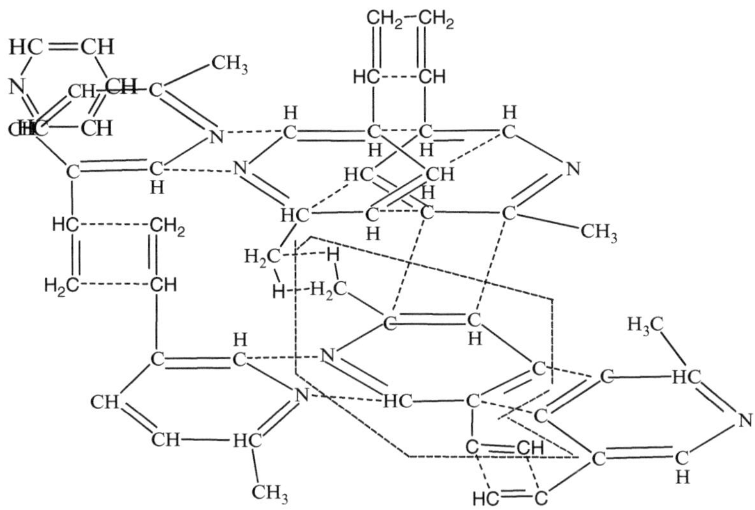

Fig. 5.7 Schematic picture of the liquid and crystalline structure of 2-methyl-5-vinylpyridine with the network of the specific interactions

temperature of 58^0 it changes by 2.0. Taking this value we obtain the value of sublimation enthalpy of pyridine as equal to 55.4 kJ mol^{-1} at the standard conditions. Note that using the value established above of the energy value contributed by the CH group of benzene to the vaporization enthalpy (5.63) and sublimation (7.43 kJ mol^{-1}) of benzene, we obtain the corresponding value of the total energy contributed by the specific interactions, formed by CH groups of pyridines molecules. The energy contributed by the isostructural methyl groups to the sublimation enthalpy of 3-methylpyridine and 4-methylpyridine are defined by the difference between its enthalpy characteristics with the sublimation enthalpy of Table 5.7. The values of the energy contributed by the specific interactions formed by the isostructural methyl groups do not depend on their location in the molecule.

Like benzene the crystalline and liquid structures of bipyridines are formed at the interaction with two molecules that are close in the environment with the formation of the network of the specific interactions (Fig. 5.8)

The significantly reduced value of the sublimation enthalpy (81.8) compared with the double value of the enthalpy characteristics of pyridine (110.8 kJ mol^{-1}) reflects the considerable changes in the charges of the carbon and nitrogen atoms and the relatively reduced values of the energies of the specific interactions.

Taking into account that one CH group of the pyridine cycle participates in the formation of the specific interaction =N → CH, the other four from the five groups participate in intermolecular interactions DHC → CH and DHC → H–C. Hence, the energy of the formed specific interaction D=N → CH–N bipyridines is defined by the equation

Table 5.7 Energies of the specific interactions (kJ mol^{-1}) of crystalline ethylpyridine and propylpyridine at 298 K

Compounds	Formula	Structure	$\Delta_{sub}H°(298\ K)$ [11]	T K	D=HC → CH	D=N → CH–N	D*iso*CH$_3$/DH$_3$C → H–CH$_2$
Pyridine	C_5H_5N	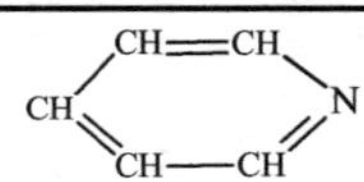	57.4. ± 2.0[a] 55.4 ± 2.0[b]		7.43	(13.8)[a] 12.8	–
3-Methyl-pyridine	C_6H_7N	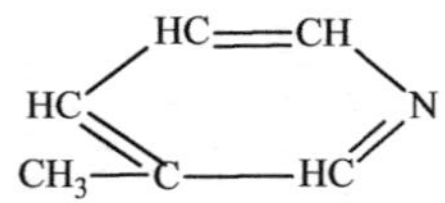	62.2	240	7.43	(13.8)[a] 12.8	4.8/2.4
4-Methyl-pyridine	C_6H_7N	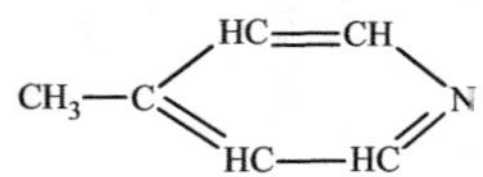	62.7	240	7.43	(13.8)[a] 12.8	5.30/2.65
2,2′-Bipyridine	$C_{10}H_8N_2$	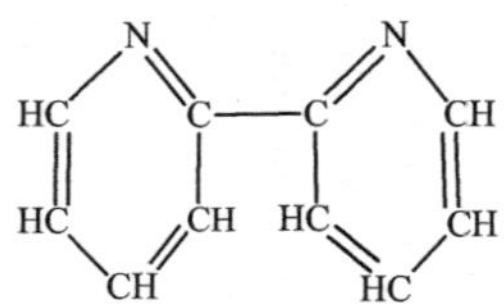	81.9 ± 0.3	298	7.43	5.6	–
2,4′-Bipyridine	$C_{10}H_8N_2$	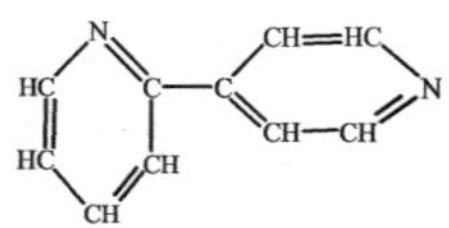	87.9 ± 1.7	298	7.43	7.1	–
4,4′-Bipyridine	$C_{10}H_8N_2$	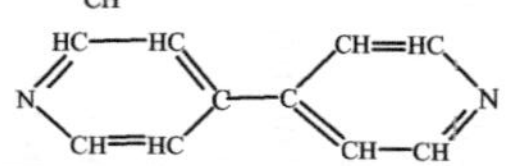	106.3 ± 2.8	298	7.43	11.6	–

[a]Estimated at 240 K
[b]At 298 K

$$D{=}N{\rightarrow}C{=}(H)\text{-}N = (\Delta_{sub}H°(298\ K)bp - 8D{=}HC{\rightarrow}CH) / 4. \qquad (5.9)$$

The values calculated are given in Table 5.8. Comparative analysis shows that the obtained energy values of the specific interaction 2D-N → CH–N

$$2,2'\text{-Bipyridine } (5.60) < 2,4'\text{-Bipyridine } (7.1) < 4,4' - \text{Bipyridine } \left(11.6\ \text{kJ mol}^{-1}\right)$$

reflect the significant stabilization of this type of interaction depending on the location of the bond C–C bipyridines at the second and fourth or at the fourth carbon atom of the cycles. This allows us to conclude that removal of this bond from the nitrogen atom causes an increase in negative charge of the nitrogen atom and a greater difference between the charge of this nitrogen atom and the charge of the carbon atom cycle interacting with it.

5.2.3 Aminopyridines and Methylaminopyridine

The nature of the interaction and formation of the structure of liquid and crystalline aminopyridines is also defined by the molecular structure, its constituent atoms, intermolecular interactions, isostructural groups, and free bond molecule vacancies. Perhaps one of the determining factors is the presence of double bonds between carbon atoms and nitrogen atoms, causing the planar structure of the molecule. Besides the six bond vacancies of the pyridine molecule considered and the two types of specific interactions, the amino group of aminopyridines adds four hydrogen bonds of low stability or in the presence of a methylamine group molecule two hydrogen bonds and two specific interactions N → CH_3–N of low stability (Fig. 5.9). The aminopyridine and methylaminopyridine molecules form the chains in liquid and crystalline conditions, criss-crossed into the network by weak specific interactions and hydrogen bonds.

Like other hexamerous nitrogenous cyclic compounds the formation in liquid and crystalline conditions of the specific interaction D=N → CH–N points to the participation of the four CH groups in the formation of four specific interactions of this series. Taking the general rule for pyridine and its derivatives as regards immutability of the energy of the specific interactions D=HC → CH or D=HC → CH, the total contribution of the amino groups to the enthalpy characteristics is defined by the difference of its vaporization enthalpies and pyridine or, with the help of Eqs. 5.10 and 5.10a, methylaminopyridine and aminopyridines, respectively. We have

$$2D\text{-}N{\cdots}H\text{-}N + 2\ D\text{-}N{\rightarrow}CH_3\text{-}N{=}\ (\Delta_{vap}H°(298\ K) - 4\ D{=}HC{\rightarrow}CH - 2\ D{=}N{\rightarrow}CH\text{-}N) \qquad (5.10)$$

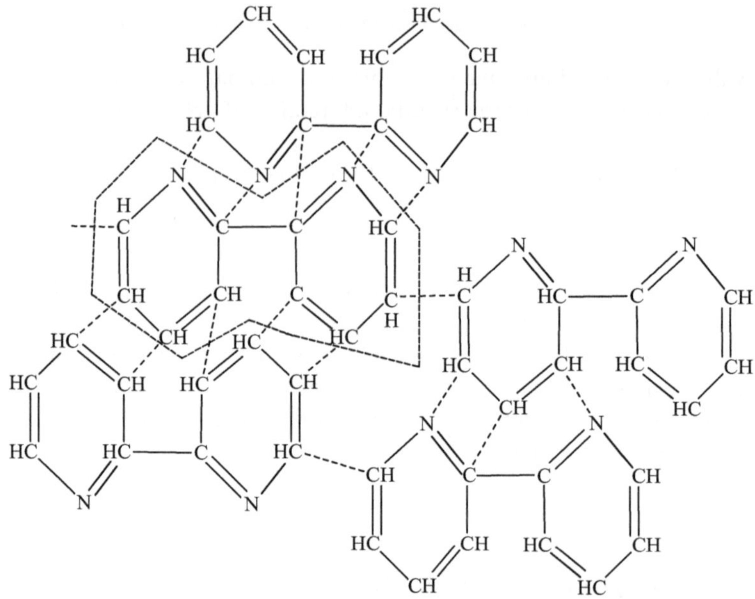

Fig. 5.8 Schematic picture of the crystalline structure of 2,2′-bipyridine

$$4\text{D-N}\bullet\bullet\bullet\text{H-N} = (\Delta_{vap}H°(298\ K) - 4\ \text{D=HC}\rightarrow\text{CH} - 2\ \text{D=N}\rightarrow\text{CH-N}) \quad (5.10a)$$

Because it is not possible to determine the values of the energies of the specific interactions D-N → CH_3–N and D–N•••H–N for methylaminopiridines, we have calculated (Table 5.9) the total contribution of two specific interactions and two hydrogen bonds (Table 5.9), from which it follows that in the case of 2-methylaminopiridine the total value of the energies of D–N•••H–N + D–N → CH_3–N (4.40 kJ mol^{-1}) is practically comparable with the energy of the same series of the hydrogen bond (3.86 kJ mol^{-1}) and specific interactions (4.25 kJ mol^{-1}) realized in liquid methylamines (Chap. 2). In other words the average value of the energy of the interaction D–N•••H-N=D–N → CH_3–N = 2.20 kJ mol^{-1} of liquid 2-methylaminopiridine is practically less than half the energy value of D–N → CH_3–N methylamine. The total energy values of the hydrogen bonds of 3.86 and specific interactions D–N → CH_3–N = 4.25 kJ mol^{-1} for 3-methylaminopiridine are more accurate.

Thus, the energies of the hydrogen bond (3.86 kJ mol^{-1}) and specific interaction (4.80 kJ mol^{-1}) reach the values corresponding to methylamine.

At the same time the energy values of the corresponding interactions of 2-methylamino-pyridine and 4-methylaminopyridine are more susceptible to introduced experimental errors; the original values for these are given without guidance in the literature sources. The results given in Table 5.8 of the calculations of the hydrogen bonds of the crystalline aminopyridines with the help of Eq. 5.11 are related to the number of sequences in its stabilization

Table 5.8 Energies of the hydrogen bonds and specific interactions (kJ mol^{-1}) of liquid methylaminopiperidines and crystalline aminopyridines (298 K)

Compounds	Formula	Structure	$\Delta_{vap}H°(298\ K)$ [8]	D=HC → CH	D=N → CH	D–N → CH_3–N/D–N•••H–N
Pyridine	C_5H_5N	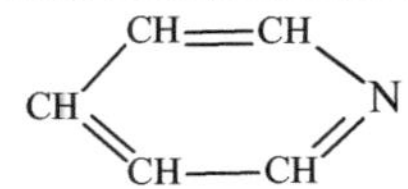	40.18 ± 0.16	5.63	8.83	–
2-Methylamino-pyridine	C_6H_8N	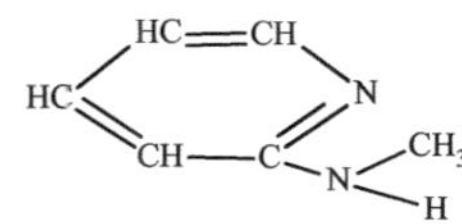	49.0 (316 K)	5.63	8.83	Σ 8.80, 2.20
3-Methylamino-pyridine	C_6H_8N	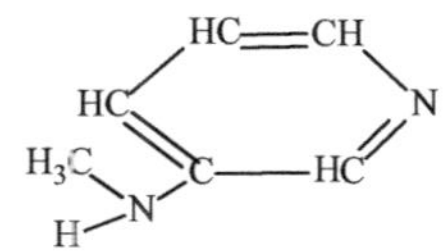	57.2 (326 K)	5.63	8.83	Σ 17.00, 3.86/4.25
4-Methylamino-pyridine	C_6H_8N	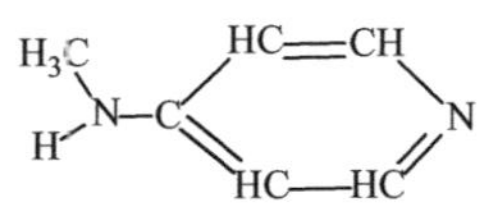	54.1 (328 K)	5.63	8.83	Σ 13.9 2.95/4.0 3.47
Crystal [11]						
				D=HC → CH	D=N → CH	D–N•••H–N
Pyridine	C_5H_5N	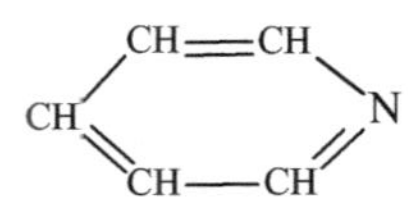	55.4 ± 2.0[a]	7.43	12.85	–
2-Aminopyridine	$C_5H_6N_2$	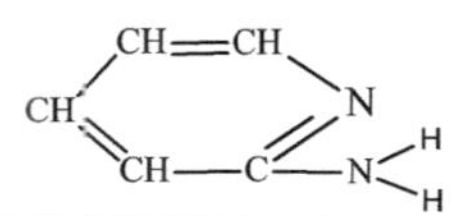	76.5 ± 0.4	7.43	12.85	Σ 21.1 5.42

(continued)

Table 5.8 (continued)

Compounds	Formula	Structure	$\Delta_{vap}H°(298\ K)$ [8]	D=HC → CH	D=N → CH	D–N → CH_3–N/D–N•••H–N
3-Aminopyridine	$C_5H_6N_2$	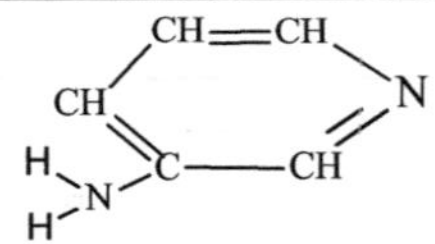	80.7 ± 0.3	7.43	12.85	Σ 25.3 6.32
4-Aminopyridine	$C_5H_6N_2$	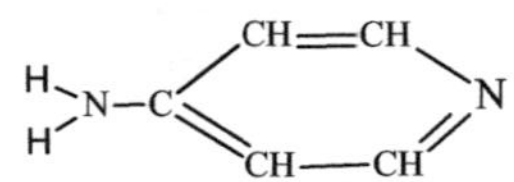	87.1 ± 0.4	7.43	12.85	Σ 31.7 7.92

[a]Estimated at 298 K

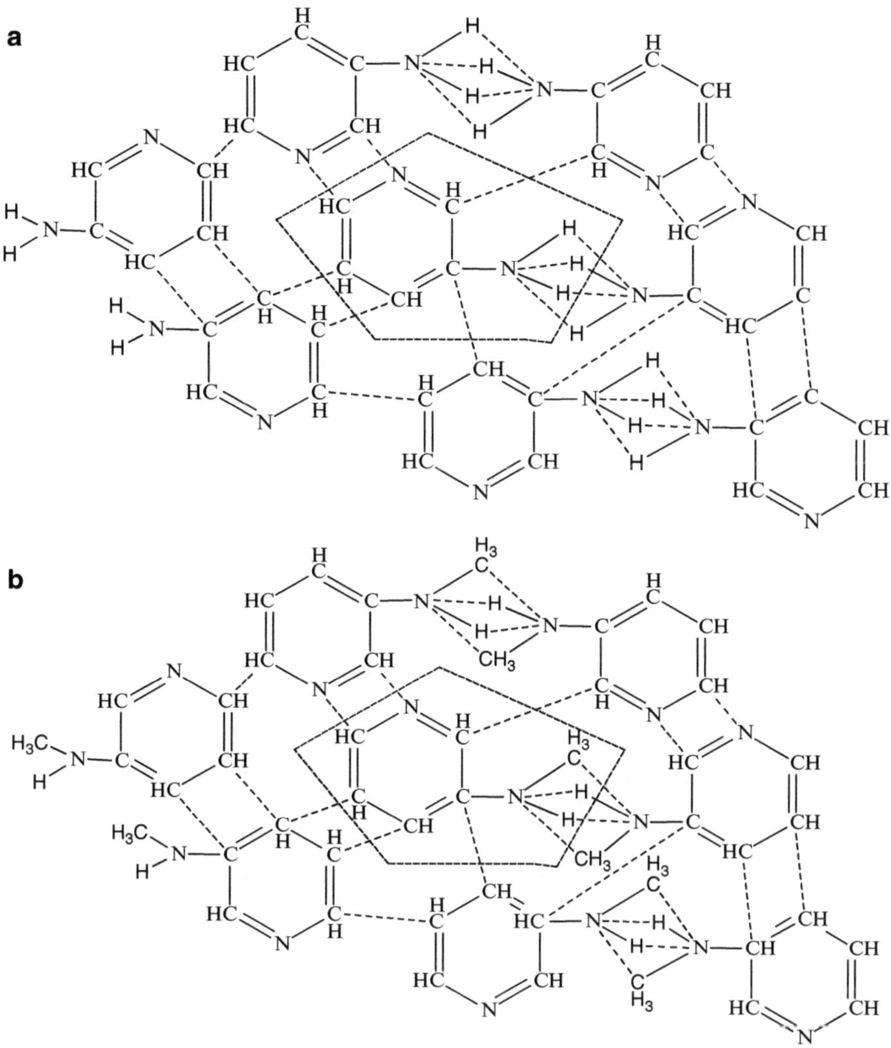

Fig. 5.9 Schematic picture of the liquid and crystalline structure of 3-aminopyridine (**a**) and 3-methylaminopyridine (**b**) with network of the specific interactions

D-N•••H-N: 2-Aminopyridine (5.42) < 3-Aminopyridine (6.32) < 4-Aminopyridine (7.92 kJ mol^{-1})

reflecting the significant influence of the removal of location of the amino group from the nitrogen atom of the cycle. This stabilizing effect is caused by the practical certainty of mixing of the electron density from the located carbon atom at the location (4) and the completion of the influence of the reverse dative bond. As a result this nitrogen atom has an increased electron density and the ability of its

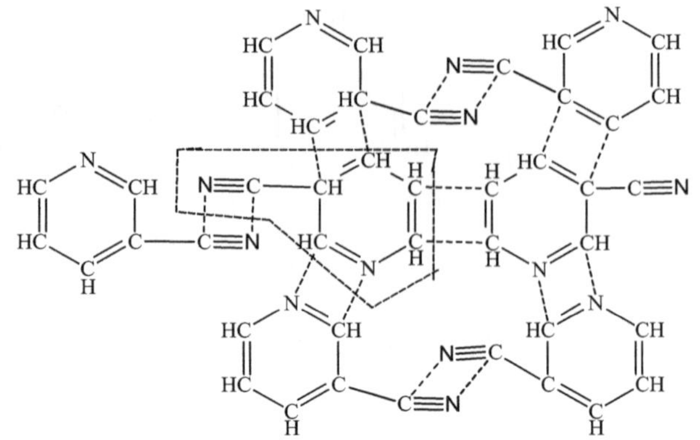

Fig. 5.10 Schematic picture of the crystalline structure of 3-cyanopyridine with the network of the specific interactions

transition to the nitrogen atom of the amine group, compared with the same carbon atoms, located at C(3) and C(2). It is the high electron density of the carbon atom at the location C(4) at hexamerous and at the location C(3) of pentamerous hetero cyclic compounds with one nitrogen atom in the cycle that makes it capable of forming the structure of liquid and crystalline conditions with a similar location of the nitrogen atoms at the interacting molecules that are close in the environment.

The cyanopiridine molecule with functional C≡N group is similar in the specific interactions D=HC → CH and D=N → CH–N, its number, and in the structure of the crystal, except that instead of four hydrogen bonds it forms two specific interactions ≡N → C≡ (Fig. 5.10).

The energy contribution of these interactions to the sublimation enthalpy is the difference between the enthalpies sublimation of cyanopyridine and pyridine reduced by the energies of the interaction formed by the saturated hydrogen atom of CH-group (1.0 kJ mol^{-1}). Consequently, the energy value of the the specific interaction for C≡N group, D≡N → C≡, is defined with thc help of Eq. 5.11:

$$D{\equiv}N{\rightarrow}C{\equiv}.= (\Delta_{sub}H°(298\ K) - 2D{=}N{\rightarrow}C{=}(H)\text{-}N - 4\ D{=}HC{\rightarrow}CH)/2 \quad (5.11)$$

The calculated value of the specific interaction energy (Table 5.9) illustrates the description by its integral value of stabilization from the location of the cyanide group at the carbon atoms C(2)H–C(3)H–C(4)H of molecular cycle

D≡N→C≡: 2-Cyanopyridine (7.65) < 3-Cyanopyridine (8.35) < 4-Cyanopyridine (8.95 kJ mol^{-1})

reflecting the sequence of the increasing difference in the charges at the carbon atoms. Thus, the conclusion that the influence of the reverse dative bond from the nitrogen atom of the cycle to the carbon atom at location C(4), increased by its electron density, donor properties, and the ability to form specific interactions of increased stability with the nitrogen atom of the coordinating molecule in the near environment, is confirmed by the energies of the specific interactions, D≡N → C≡, of the cyanopyridines formed.

Table 5.9 Energies of the hydrogen bonds and specific interactions (kJ mol^{-1}) of crystalline cyanopiridines

Compounds	Formula	Structure	$\Delta_{sub}H°(298\ K)$ [11]	T K	D=HC → CH	D=N → CH–N	D≡N → C≡
2-Cyano-pyridine	C_5H_6N	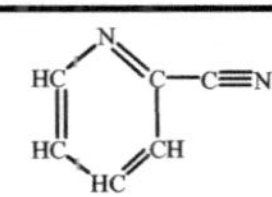	70.7 ± 1.7	298	7.43	12.85	7.65
3-Cyano-pyridine	C_5H_6N	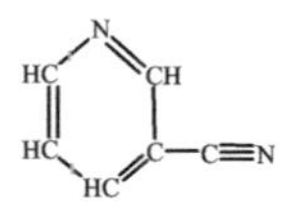	72.1 ± 1.8	298	7.43	12.85	8.35
4-Cyano-pyridine	C_5H_6N	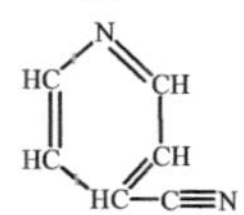	73.2 ± 0.6	298	7.43	12.85	8.95

5.3 Energies of Specific Interactions of Hexamerous Heterocyclic Compounds with Two Nitrogen Atoms

Hexamerous heterocyclic compounds with two nitrogen atoms differ by the location of the nitrogen atom in the heterocycle (Table 5.10), leading to the difference in vaporization enthalpies in the sequence pyridazine – pyrimidine – pyrazine. The reduction of the enthalpy characteristics in this series reflects the destabilization of the energies of specific interactions, realized in liquid azines. It follows that the location of the nitrogen atom in the heterocycle influences the shifting of the electron density in the cycle and to the atom's charges. In the pyridazine molecule with the nitrogen atoms located nearby the shifting of the electron density has taken place from carbon atoms of the CH group C(3)H and C(6)H, providing it with positive charges. As a result the carbon atoms C(4)H and C(5)H remain negatively charged, possibly with a small difference. From this we can conclude that the carbon atoms C(3)H and C(6)H of the molecules that are close in the environment interact with the nitrogen atoms of the contacting molecules with formation of the specific interactions N → CH–N. The remaining carbon atoms of the CH groups at the locations C(4)H and C(5)H interact similar to the benzene atoms with implementation of the principle of maximum difference between the charges in the carbon atoms of contacting molecules that are close in the environment with formation of the specific interaction HC → CH.

In the molecule of pyrimidine with the nitrogen atoms at locations N(1) and N (3) the biggest shifting of the electron density occurs from carbon atom C(2)H, fringed by these carbon atoms (Fig. 5.11). The influence of the nitrogen atoms on the fringe is so high that the carbon atom loses the electron density acquired from the hydrogen atom and part of its own electron density. As a result, the carbon atom at location C(2)H has a maximum value of positive charge (0.875).

Shifting the electron density from carbon atoms C(4) and C(6), which fringe the same nitrogen atoms, is equal in that it provides them with a positive charge, 0.898, and the nitrogen atoms a negative charge, 1.161. The high value of the charges in the nitrogen atoms is caused also by shifting of the electron density from carbon atom C(2)H and on the chain from C(5)H. Shifting of the electron density from carbon atom C(5)H is so significant that even shifting the electron density from the hydrogen atom provides it with an insignificant negative charge (1.007). It follows that the nitrogen atom N(1) forms the specific interaction N → CH–N with the carbon atom of the contacting molecules that are close in the environment at the location C(2)H, and nitrogen atom N(3) with the carbon atom located at C(4)H. The remaining atoms at the locations C(5)H with positive charge and C(6)H with minimum negative charge form the specific interaction HC → CH with a total difference between their charges of 0.109 with significantly reduced energy; then the specific interaction by the nitrogen atoms with atoms C(2)H and C(4)H with differences of 0.286 and 0.263 in the charges, respectively. In its turn, the energies of the specific interactions formed by the nitrogen atoms are different.

Table 5.10 Energies of the specific interactions (kJ mol^{-1}) of liquid and crystalline azines

Compounds	Formula	Structure	$\Delta_{vap}H°(298\ K)$ [8]	T K	D=HC → CH	D=N → CH–N	D*iso*CH_3/DH_3C → H–CH_2
Pyridazine	$C_4H_4N_2$	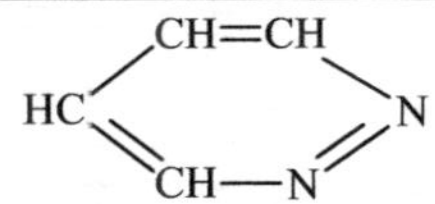	53.5 ± 0.4	298	5.63	10.55	–
Pyrimidine	$C_4H_4N_2$	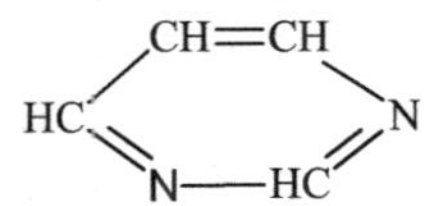	49.8 ± 0.3	298	5.63	9.65	–
Pyrazine	$C_4H_4N_2$	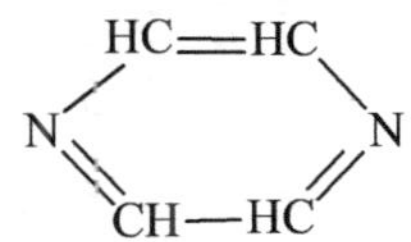	37.9 (44.1)[a]	352	5.63	8.20	–
2-Methyl-pyrazine	$C_5H_6N_2$	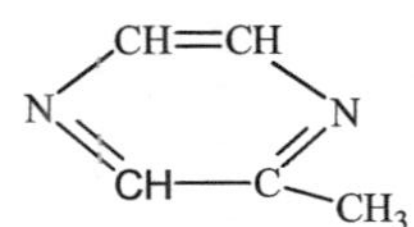	42.4 (46.8)[a]	340	5.63	8.20	n = 1 ∑ 3.30/1.65
2,5-Dimethyl-pyrazine	$C_6H_8N_2$	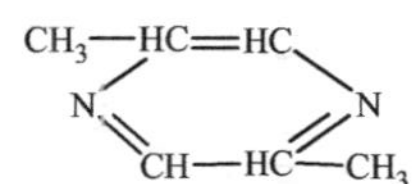	44.5 (357) (50.6)[a]	357	5.63	8.20	n = 2 ∑ 7.70 3.85/1.90
2.3.5-Trimethyl-pyrazine	$C_7H_{10}N_2$	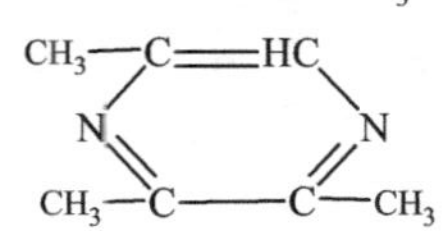	53.9 ± 1.6	298	5.63	8.20	n = 3 ∑ 11.8 3.93/1.95
Crystal [11]							
Pyrazine *cis*	$C_4H_4N_2$	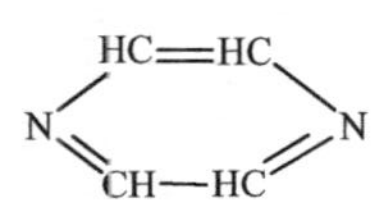	56.2 ± 0.5	303	7.43	10.30	–
Tetramethyl-pyrazine	$C_8H_{12}N_2$	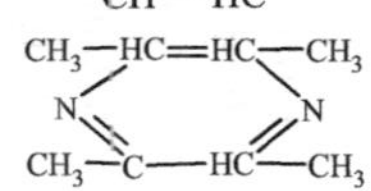	94.6 ± 4.0	298	7.43	10.30	∑ 38.4 = 9.60/4.80

[a]Estimated at 298 K

Fig. 5.11 Charges distribution on the atoms of molecule of Pyrimidine [10]

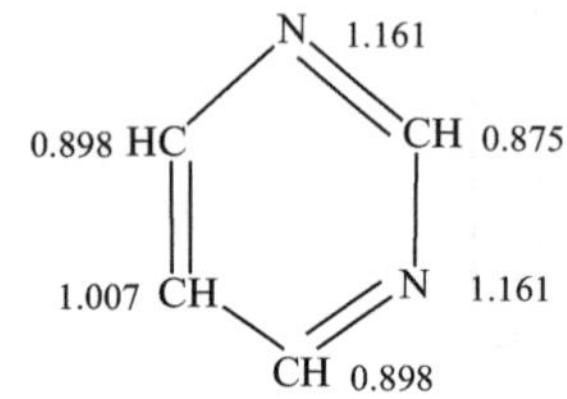

Considering the small differences in the charges of the nitrogen atoms, for the thermodynamic analysis the charges could be taken as equal.

In the pyrazine molecule with the greatest possible distance of the nitrogen atoms at hetero cycle, N(1) and N(4), the carbon atoms shift electron density equally and the charges, on the one hand, in nitrogen atoms, and, on the other, in carbon atoms should be equal. This leads to the conclusion that the energy of the specific interaction =N → C=(H)–N of liquid pyrazine should be less stable compared with the ones realized in liquid pyrimidine and pyridazine. This conclusion is proved by experimentally established enthalpy characteristics (Table 5.10).

The complex sequence of the change of the values of vaporization enthalpies of the compound series

$$\text{Pyridine } (40.18) < \text{ Pyridazine } (53.3) > \text{Pyrimidine } (49.8) \\ > \text{Pyrazine } \left(44.1 \text{ kJ mol}^{-1}\right)$$

of the hexamerous hetero cyclic compounds of nitrogen with the increase in the pyridazine molecule and then from the location of the nitrogen atom reflects, first, the maximum possible received by the nitrogen atom in the shifting of the electron density on the chain cycle with minimum loss of the latter by the carbon atoms at the C(3)H and C(5)H locations and especially at the C(4) location. Hence, the presence of the second nitrogen atom at the N(2) location is accompanied by the larger shifting of the electron density from carbon atom C(4) on the chain to the carbon atoms at the N(1) and N(2) locations, which result in the reduction of negative charges of the carbon atoms at the C(3) and C(6) locations up to a positive charge and a greater differentiation of the charges at the carbon atoms C(4) and C(5) is created. At the location of the nitrogen atoms at the N(1) and N(2), greater density shifts from two carbon atoms on the chain, than in the pyrazine molecule with the most remote location of the nitrogen atoms in N(1) and N(4) and with differing, but nearly equal charges of the carbon atoms, located in the cycle in trans-locations. Thus, as formed by the CH group pyridine molecules, the two series of specific interactions D=HC → CH (5.03) and D–HC → H–C (0.60) and nitrogen atom D=N → CH–N (8.83 kJ mol^{-1}) are less stable for all considered compounds series.

Pyridazine, pyrimidine, and pyrazine with six bond vacancies and two nitrogen atoms form four specific interactions N → CH–N and the remaining two CH groups form two specific interactions HC → CH and have low stability HC →

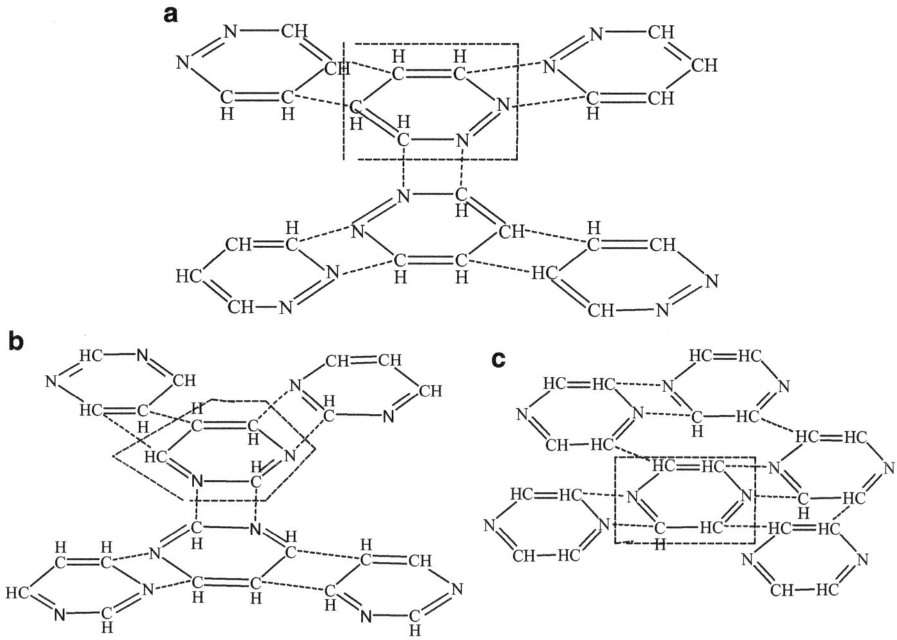

Fig. 5.12 Schematic picture of the liquid structure of pyridazine (**a**), pyrimidine (**b**), and pyrazine with the network of the specific interactions

H–C with values of the energies equal to the ones realized in liquid and crystalline conditions of pyridine relatively. The structure of the molecules of these compounds defines the formation of the structure of liquid and crystalline conditions with the network of the specific interactions (Fig. 5.12).

Realizing the two series of the specific interactions with the known energy of one of them allows us to determine the energy of the specific interaction D–N → CH–N with the help of Eq. 5.12,

$$D\text{–}N \rightarrow CH\text{-}N = (\Delta_{vap}H^{\circ}(298\ K) - 2D{=}HC \rightarrow CH)/4 \qquad (5.12)$$

The results of the thermodynamic calculations (Table 5.10) reflect the natural destabilization of the specific interactions D–N → CH–N, realized in the liquid condition, in the series of compounds

$$\text{Pyridazine } (10.55) > \text{Pyrimidine } (9.65) > \text{Pyrazine } \left(8.20\ \text{kJ mol}^{-1}\right)$$

This is caused by the reduction in the charges at the nitrogen atoms on the replacement of their location in the molecules from N(1) and N(2) by N(1) and N(3), or by N(1) and N(4), respectively.

The contribution of the energies of the isostructural methyl groups to the vaporization enthalpy of pyrazines and the specific interactions is set with the help of Eq. 5.13, which takes into account the energies of the specific interactions

$$DisoCH_3 = (\Delta_{vap}H°(298\ K)mpz - 4D\text{-}N{\to}CH_3\text{-}N - 2D{=}HC{\to}C\text{-}H - nDH)/n = (\Delta_{vap}H°(T)mpz - \Delta_{vap}H°(298\ K)pz - nDH)/n \qquad (5.13)$$

$4D–N \rightarrow CH_3$-N and 2D=HC → C–H and the contribution of the substituted hydrogen atoms in the (n) cycle, adequate to the difference of vaporization enthalpies, in particular, tetramethylpyrazine and pyrazine,

$$DisoCH_3 = (\Delta_{vap}H°(298\ K)mpz - \Delta_{vap}H°(298\ K)pz - nDH)/n \qquad (5.13a)$$

reduced by the contribution of the substituted hydrogen atoms.

The energies of low stability of the specific interactions $DH_3C \rightarrow H–CH_2$ (Table 5.11) are described by the integral value

$$\text{2-Methylpyrazine}(1.65) < \text{2,5-Dimethylpyrazine}(1.90) \approx \text{2,3,5-Trimethylpyrazine}\ (1.95\ kJ\ mol^{-1})$$

reflecting the small difference in their energies at the two and three methyl groups in the molecule. At the same time, the energy of the specific interaction of low stability of liquid 2.3.5-trimethylpyrazine (1.95) is stabilized by the crystal field in tetramethylpyrazine (4.80) at 2.85 kJ mol^{-1}.

5.4 Energies of Specific Interactions of Hexamerous Heterocyclic Compounds with Three Nitrogen Atoms

The most notable feature of the molecules with three nitrogen atoms of hexamerous cycle 1,3,5-triazine and 1.2.3-triazone is that they are characterized by the formation of one type of specific interaction independent of the location of the nitrogen atom in the cycle.

The 1,3,5-triazine molecule with the uniform location of the nitrogen atoms and CH groups in the cycle leads to equivalent shifting of the electron density from the hydrogen atoms to the carbon atoms and further to the nitrogen atoms. As a result, practically each of the atoms of nitrogen, carbon, and hydrogen have relatively equivalent charges and form specific interactions of similar stability.

A distinctive feature of 1,2,3-triazone molecule from 1.3.5-triazine is that the nitrogen atom at the location N(2) has a smaller surplus of the electron density and, consequently, the reduced negative charge and reduced donor ability in comparison with near located N(1) and N(3). At the same time the shifting of the electron

Table 5.11 Energies of the specific interactions (kJ mol^{-1}) of crystalline heterocyclic compounds with three nitrogen atoms at 298 K

Compounds	Formula	Structure	$\Delta_{sub}H°(298\ K)$ [11]	D=N → C=(H)–N	D–N•••H–N
1,3,5-Triazine	$C_3H_3N_3$	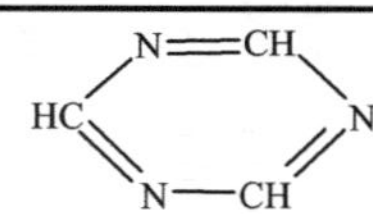	54.2 ± 0.2	9.03	–
1,2,3-Triazone	$C_3H_3N_3$	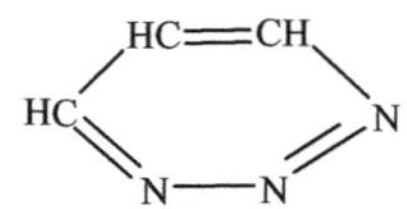	54.2 ± 0.3	9.03	–
2,4,6-Triamino-s-triazine (melamine)	$C_3H_6N_6$	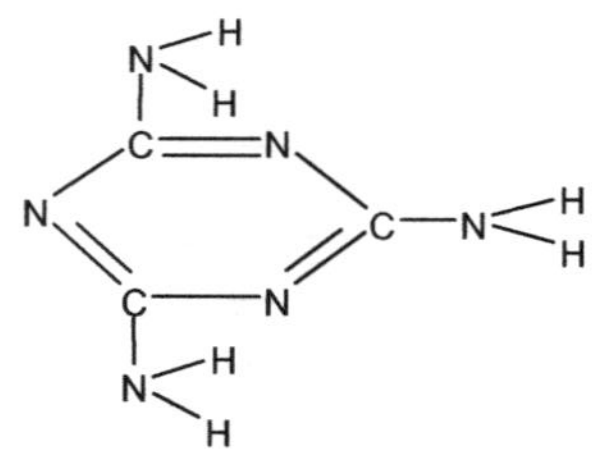	123.3 (432 K) 130.4[a]	9.03[a]	6.50[a]

[a]Elucidated

density of the carbon atom at the location C(5)H on the chain to the carbon atoms C(4) and C(6) provides to the last the shifting of the increased electron density to the nitrogen atoms N(1) and N(3), which part they transfer to the nitrogen atom at the location N(2). As a result, nitrogen atoms at the locations N(1) and N(3) keep the maximum negative charge and the carbon atoms at the locations C(4)H and C(6)H the maximum positive charge relatively. The nitrogen atom at the location N (2) and the carbon atom at the location C(5)H have reduced negative andpositive charge, respectively. This state of the charge at the nitrogen and carbon atoms provides the reduced stability of the specific interaction D=N → CH–N, formed by the nitrogen atom at the location N(2) and carbon atom at C(5)H. However, this position is often violated by the coordinating molecules with planar structure. Thus, nitrogen atoms with excess electron density at $3p_z$ – orbital and carbon atom with deficit of the electron density at the essentially unshared $2s^2$(c) electron pair provide the formation of stable specific interactions.

The 1,3,5-triazine and 1,2,3-triazone molecules with planar structure and six bond vacancies form the structure of liquid and crystal conditions with the network of the strong type of specific interactions with pentacoordinated carbon atom (Fig. 5.13). The energies of intermolecular interactions are established directly from the vaporization enthalpy divided by the number of the interactions formed (Table 5.11). The obtained values of the energies of the specific interactions of the considered crystalline

1.3.5-Triazine D=N→CH= (9.03) = 1,2,3-Triazone D=N→CH= (9.03 kJ mol^{-1})

compounds have equal values and are within the experimental error.

The results of the thermodynamic analysis of hexamerous heterocyclic compounds with the nitrogen atoms in the cycle allow us to draw the following conclusions:

At first, the specific interaction D=N → CH, formed by the two nitrogen atoms of the cycle are destabilized

$$\text{Pyridazine}\ (10.60) > \text{Pyrimidine}\ (9.65) > \text{Pyrazine}\ \left(8.20\ \text{kJ mol}^{-1}\right)\ (\text{liquid})$$

with replacement of location of the nitrogen atoms with N(2) to N(3) and N(5) at the permanent location of the second nitrogen atom at N(1).

Second, increase in the number of nitrogen atoms in the cycle is accompanied by the destabilization of the interaction

D=N→CH-N: Pyridine (12.85) > Pirazine (10.30) >1,3,5 –Triazine (9.03) = 1,2,3 –Triazone (9.03 kJ mol^{-1}) (crystal),

at the three nitrogen atoms in the cycle; but the energy of the given specific interaction does not depend on its location.

Third, implementation of the given regularities at the energies of the specific interactions D=N → CH–N is the argument for the correctness of the

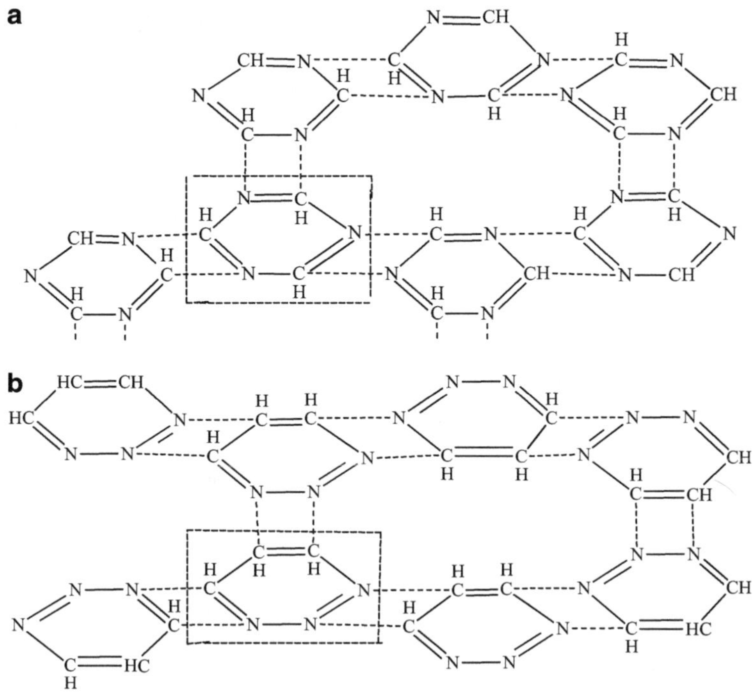

Fig. 5.13 Schematic picture of the liquid and crystalline structure of 1,3,5-triazine (**a**) and 1,2,3-triazone (**b**) with the network of the specific interactions

assumption of energy equality in the interaction D=HC → CH with the energy formed in liquid and crystalline benzene and D–H_2C → CH_2 cyclopentane or cyclohexane in dependence on the number of the carbon atoms in the heterocycle of aminopyridine.

The equality of the sublimation enthalpy of 1,3,5-triazine and 1,2,3-triazone reflects its stability from the location of the nitrogen atoms in the molecule. This allows the use of sublimation enthalpy of one these compounds, the average value of the energy contribution (26.0 kJ mol^{-1}) of three amine NH_2 fragments of aminopyridines (Table 5.8) reduced by the one contributed by three substituted hydrogen atoms (0.60 × 3 kJ mol^{-1}) to the enthalpy characteristics. The value of the sublimation enthalpy obtained (Table 5.8) at the standard conditions is higher than the one experimentally established at 432 K to 7.1 kJ mol^{-1}. The given high value of the enthalpy characteristic correctly describes its temperature dependence around 134 K. It is very attractive to assess the energies of the specific interactions for more complex cyclic compounds using the method of additive contributions.

5.5 Energies of Specific Interactions of Phenyl-pyridines

The totally adequate value of vaporization enthalpy of phenyl and pyridine (72.8 kJ mol^{-1}) exceeds the realistic value for 2-phenyl-pyridine at 4.1 kJ mol^{-1}. It means that in spite of the high stiffness of the benzene and pyridine ring, there appear changes in the atomic charges at the formation of phenylpyridines, leading to the reduction of the less stable specific interaction. If the specifics of these interactions remain unchangeable, then its energy is interrelated with the properties of the related fragments. In this case the molecule of pyridine benzene forms the structure of liquid condition by a similar series of the specific interaction D=HC → CH with equivalent value of the energy 5.52 kJ mol^{-1}. At the same time the molecule of 2-phenyl-pyridine forms the second and more stable specific interaction D=N → C= with energy value 8.83 kJ mol^{-1} (Table 5.8). It follows that the nitrogen atom shows significant acceptor properties and leads to the shifting of the electron density.

The reduced value of vaporization enthalpy of 2-phenylpyridine (Table 5.12) reflects the shifting of the electron density from the carbon atom of the benzene ring to the carbon atom of the pyridine ring, connected directly with the nitrogen atom, which, in its turn, imparts the obtained electron density to the nitrogen atom. Nevertheless, there appears a reduction in the difference of the positive and negative charges of the contacting atoms of carbon and nitrogen, which results in the reduction of stability of the specific interaction D=N → C= formed by these atoms with the molecule that is close in the environment. Since the shifting of the electron density from the carbon atom at the location C(2) to the nitrogen atom has already happened, the further shifting to the carbon atom at the location C(3) of pyridine in the 3-phenyl-pyridine molecule is accompanied by a reduced transition to the nitrogen atom and, consequently, the difference in charges of the same atoms C(2) and nitrogen are reduced to a greater extent. Hence, the value of the energy of the specific interaction D=N → C= should be lower. However, the errors in the experimentally obtained vaporization enthalpies of the considered compounds are large enough to trace the tendency of its change. Taking into account that at the functional groups, connected with the carbon atom at the location C(4), there is increased shifting of the electron density on the chain to the carbon atom C(2) and further to the nitrogen atom, which leads, to an even greater extent, to the difference in charges at the atoms of the given carbon and nitrogen. Of course, the shifting of the electron density from the nitrogen atom changes the energy of the specific interactions a little, formed by CH groups of the benzene ring; however, they are always expressed significantly to a small extent for the same reason as the rigidity of the benzene ring. In this regard the thermodynamic analysis of phenylpyridines should be implemented.

The molecule of 2-phenylpyridine with 12 bond vacancies forms the network structure of liquid and crystalline condition by the eight specific interactions D=HC → CH with energy 5.52 kJ mol^{-1}, two less stable D=HC → C at the expense of the substituted hydrogen atom and two more stable interactions of

Table 5.12 Energies of the specific interactions (kJ mol^{-1}) of liquid heterocyclic compounds with the nitrogen atoms at 298 K

Compounds	Formula	Structure	$\Delta_{vap}H°(298\ K)$ [8]	D=HC → CH	D=HC → C	D=N → C=
2-Phenylpyridine	$C_{11}H_9N$	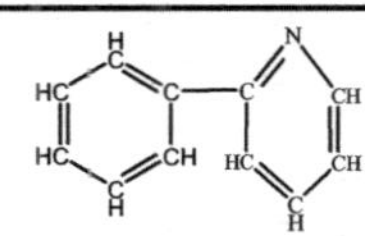	68.7 ± 4.6	5.63	5.03	6.8
3-Phenylpyridine	$C_{11}H_9N$	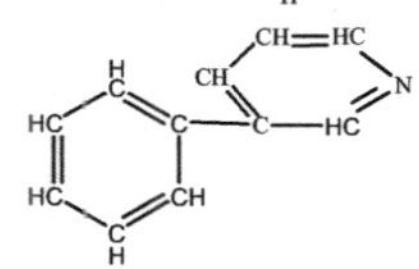	69.5 ± 4.5	5.63	5.03	7.2
4-Phenylpyridine	$C_{11}H_9N$	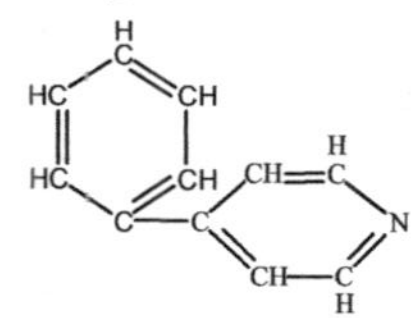	81.4 ± 1.6 Cryst [11]	7.40	6.4	5.4

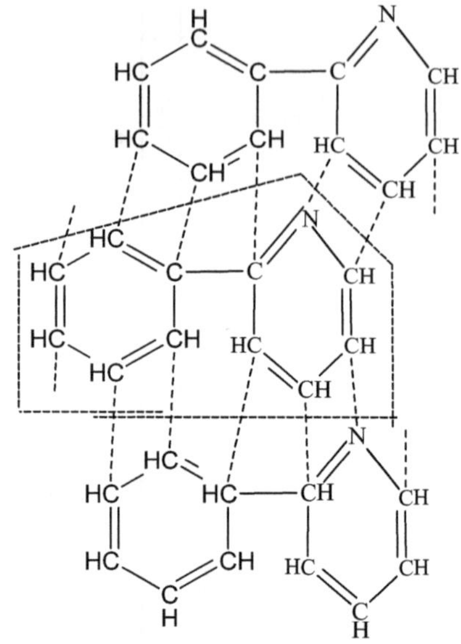

Fig. 5.14 Schematic picture of the liquid and crystalline structure of 2-phenylpyridine with the network of the specific interactions

the second series D=N → C= (Fig. 5.14a), the energy of which is calculated with the help of Eq. 5.14

$$D{=}N{\rightarrow}C{=} = (subH°(298\ K)php - DHC{\rightarrow}CH - DHC{\rightarrow}C)/n \qquad (5.14)$$

The calculations (Table 5.12) illustrate the reduction of stability of this series of the interaction of liquid 2-phenylpyridine at 2.0 kJ mol^{-1} compared with that realized in liquid pyridine. The energy values of this specific interaction of liquid 3-phenylpyridine and crystalline 4-phenylpyridine are calculated by the same equation. The energies of the specific interaction D=N → C= obtained point to the implementation of the natural character of its change,

$$D{=}N{\rightarrow}C{=}{:}\ \text{Pyridine}\ (8.85) > \text{2-Phenylpyridine}\ (6.8) \geq \text{3-Phenylpyridine}\ (7.2)$$

proposed on the basis of vaporization enthalpies of the compounds considered, the energies of this specific interaction of liquid pyridine, and peculiarities of shifting of the electron density. The low value of the energy of the same specific interaction of crystalline 4-phenylpyridine is symptomatic.

5.6 Energies of Specific Interactions of Quinolines and Their Derivative

5.6.1 *Quinoline and Benzoquinolines*

The compounds of the series of quinoline and quinoxaline with a rigid benzene ring and a fragment of the pyridine ring from the four components of a hexamerous heterocycle with one and two nitrogen atoms allow us to make relatively clear the role of the nitrogen atom from its location in the heterocycle and from the number of nitrogen atoms with location at the pyridine's cycle. Some difficulties are created by the examples in the sources of the literature indicating the enthalpy characteristics of vaporization and sublimation of these compounds without errors. It causes definite caution at its analysis, which is why we will use the research results of some authors, obtained by one method under comparable conditions. Molecules of quinoline and isoquinoline with one nitrogen atom, located at the first and second carbon atom of the benzene ring, respectively, with planar molecular structure and ten bond vacancies form the network structure of crystalline and liquid condition (Fig. 5.15) by specific interactions of the three series D=N → CH, D=HC → CH and D=HC → C with two of the first and four of the second and third series. The vaporization enthalpy of these compounds is 49.2 kJ mol^{-1} at 448 K and 51.6 kJ mol^{-1} at 454 K, differing by 2.4 kJ mol^{-1}. This value of the difference of vaporization enthalpies of these compounds at a similar series and the number of each series of the specific interactions implies that its energies should be different from the location of the carbon atoms in the cycle formed. The specific interactions 2D–N → CH are formed by the nitrogen atom with the carbon atom C(2) and interactions 2D=HC → CH by carbon atoms C(2) and C(3) and two others, C(5) and C(5), and combination C(8) and C(4) in the structure of quinolone, and by the nitrogen atom and carbon atom C(1) and binds 2D=HC → CH by carbon atoms C(3) and C(4) and two other bonds 2D=HC → CH by carbon atoms C(5) and C(5) in the structure of isoquinoline; some of them or all of them are at 0.2 kJ mol^{-1} more stable than in the structure of quinoline. However, it is practically unrealistic for these interactions to occur at three types and ten specific interactions, and we consider the average values of the energies of the specific interactions, obtained from its equivalent contribution with the energies of the specific interactions of crystalline and liquid benzene.

To implement the thermodynamic analysis of compounds of the quinoline series with the combined benzene and pyridine cycle it is necessary to estimate correctly the value of the sublimation enthalpy of the founder of these compounds. In this connection it is interesting to mention 2,2′-biquinoline (134.7 ± 1.3 kJ mol^{-1}), which molecule consists of two molecules of quinoline, with two substituted hydrogen atoms. Considering the energy contribution of the saturated hydrogen atom to the sublimation enthalpy 1.0 kJ mol^{-1}, we obtain the value of the enthalpy characteristics of quinoline to be equal to 68.7 ± 1.3 kJ mol^{-1}. Note that the

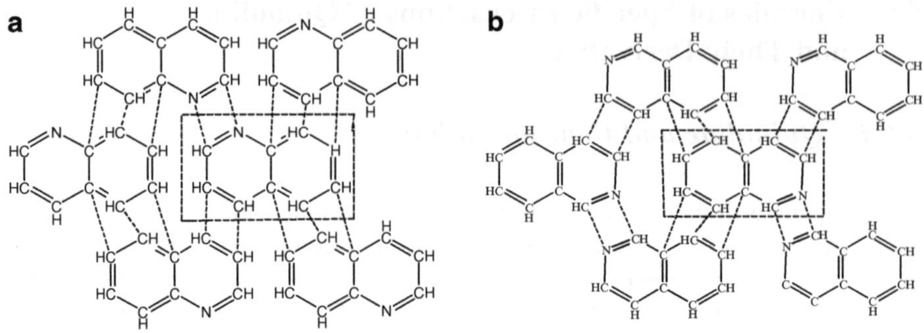

Fig. 5.15 Schematic picture of the liquid and crystalline structure of quinoline (**a**) and isoquinoline (**b**) with the network of the specific interactions

experimentally measured values of sublimation enthalpy of quinoxaline (69.4 ± 0.6, 66.9 ± 2.0 kJ mol^{-1}) with the benzene ring and two nitrogen atoms in hexamerous cycle at the standard conditions have higher accuracy. At the same time the most correct value of this characteristic is in good agreement with the estimated value of the sublimation enthalpy of quinoline using the additive contribution method.

The energy of the interaction DN → CH is determined by the equality

$$D{=}N{\rightarrow}CH = (\Delta_{sub}H°(298\ K)ql - nD{=}HC{\rightarrow}CH - mD{=}HC{\rightarrow}CH)/2 \qquad (5.15)$$

The energy of the specific interaction of D=HC → CH is determined by the sublimation enthalpy or benzene vaporization and the energy of the second type D=HC → C interaction has a reduced energy as regards the value of the contribution of the substituted hydrogen atom at 1.0 and 0.60 kJ mol^{-1} for the crystalline and liquid condition relatively. The molecule of 2,2′-biquinoline forms a similar grid structure of a crystal by the eight interactions of the same series. The results are given in Table 5.13.

The benzoquinoline compounds, differing by the location of the second benzene ring in the molecule of quinolone, allow us to trace its influence on the energy of the specific interaction D=N → CH. Independent of the location of the benzene ring at the carbon atoms C(3) and C(4), C(5) and C(6), or at the carbon atoms C(7) and C(8) with 14 bond vacancies in the molecules of 3,4-benzoquinoline, 5,6-benzoquinoline, and 7,8-benzoquinoline the total number of specific interactions and the number of interactions D=N → CH (two) is preserved. The planar structure of the molecules of benzoquinolines determines the network structure of a crystal of these compounds (Fig. 5.16a, b), for which the numbers of the realized specific interactions D=HC → CH and D=HC → C, respectively, 4 and 8 in 3,4-benzoquinoline, 6 and 6 in 5,6-benzoquinoline and 7,8-benzoquinoline are changed. The results of the calculations of the specific interaction energy D=N → CH in Eq. 5.15 are given in Table 5.13. The value of

Table 5.13 Energies of the specific interactions (kJ mol^{-1}) of crystalline quinoline and benzoquinolines

Compounds	Formula	Structure	$\Delta_{sub}H°(298\ K)$ [11]	T K	D=HC → CH	D=HC → C	D=N → CH
Quinoline	C_9H_7N	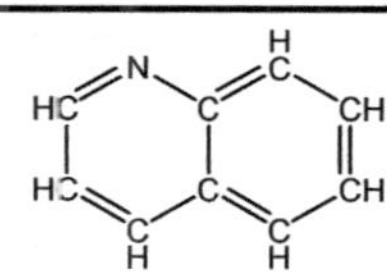	68.7 ± 1.3	298	7.40 × 4	6.40 × 4	6.9
Quinoline	C_9H_7N	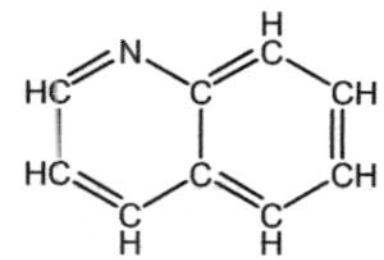	58.1[a]	286–309	5.63 × 4	5.03 × 4	7.75[a]
2,2′-Biquinoline	$C_{18}H_{12}N_2$	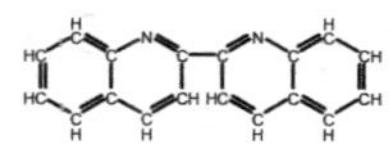	134.7 ± 1.3	298	7.40 × 8	6.40 × 8	6.9
3,4-Benzo quinoline	$C_{13}H_9N$	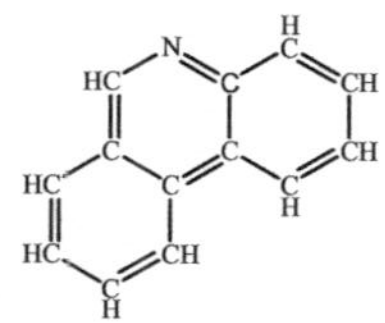	98.6	298	7.40 × 4	6.40 × 8	9.3
5,6-Benzo quinoline	$C_{13}H_9N$	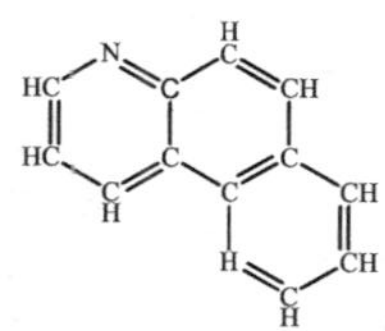	83.1 ± 3.6	308	7.40 × 6	6.40 × 6	≈2
7,8-Benzo quinoline	$C_{13}H_9N$	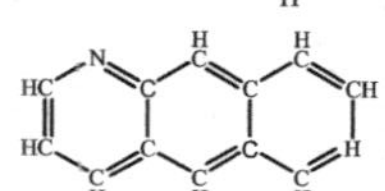	90.2 ± 2.0	298	7.40 × 6	6.40 × 6	3.6

[a]The vaporization enthalpy is given without in [9] without error

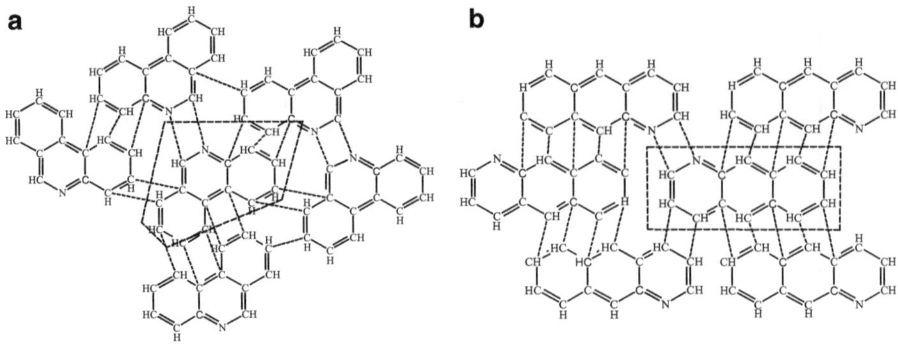

Fig. 5.16 Schematic picture of the crystalline structure of 3,4-benzoquinoline (**a**) 7,8-benzoquinoline (**b**) with the network of the specific interactions

the energy of the specific interaction D=N → CH of liquid quinoline obtained with the use of vaporization enthalpy, presented in the literature without error, significantly exceeds this value for its crystalline condition, which points to its unreliability. A more realistic value of these characteristics could be the value which does not exceeds 55 kJ mol^{-1}. From the data of the table it follows that the value of the energy of the specific interaction of crystalline quinoline and 2,2′-biquinoline occupy an intermediate position between the energies of the two other types of the specific interactions. We have

$$D{=}HC{\rightarrow}CH\ (7.40) > D{=}N{\rightarrow}CH\ (6.9) > D{=}HC{\rightarrow}C\ (6.40\ kJ\ mol^{-1}).$$

The location of the benzene ring at C(3) and C(4) in the molecule of 3,4-benzoquinoline causes the increases in the energy of the specific interaction with implementation of the integral value of stabilization.

$$D{=}N{\rightarrow}CH\ (9.3) > D{=}HC{\rightarrow}CH\ (7.40) > D{=}HC{\rightarrow}C\ (6.40\ kJ\ mol^{-1})$$

An increase of the energy of the specific interaction D=N → CH on 2.4 kJ mol^{-1} is caused by additional shifting of the electron density from the second benzene ring to the carbon atoms C(3) and C(4) of the heterocyclic fragment and further to the nitrogen atom. It results in the significant increase in the difference of the charges of nitrogen and carbon atoms at the location C(2), forming the specific interaction D=N → CH by the molecules that are close in the environment. The description of the energy of the specific interaction D=N → CH, obtained with high experimental errors, of the tendency in the series of compounds

$$3,4\text{-Benzoquinoline }(9.3) > 7.8\text{-Benzoquinoline }(3.6)$$
$$> 5,6\text{-Benzoquinoline }\left(\approx 2\ \text{kJ mol}^{-1}\right)$$

points to the fact that the bound rigid benzene ring with hexamerous heterocycle blocks the shifting of the electron density from the second benzene ring. On the other side, the shifted electron density is distributed in such a way that there appears a reduction of the charges at the nitrogen atom and carbon atom C(2), responsible for the formation of the considered specific interaction. The most significant destabilizing effect is shown at the location of the second benzene cycle at C(5) and C(6).

5.6.2 *Aminoquinolines*

The aminoquinoline molecule with hexamerous heterocycle should be considered in the capacity of a derivative of pyridine and dependent on the location of the amino group in the second, third, or fourth carbon atom there will be shown to a great extent the influence of the nitrogen atom of the cycle of quinoline at the shifting of the electron density, then at its location in the benzene cycle at the fifth, sixth, or at the eighth carbon atom C(8). The presence of nine atoms of carbon and nitrogen atoms in the aminoquinoline molecule keeps the specific interactions with the character energies bonds. Hence, the network structure of crystalline aminoquinoline repeats the structure of quinoline and gives the volume fragments, formed by the four hydrogen atoms of the amino group by hydrogen bonds (Fig. 5.17).

However, depending on the location of the amino group at the carbon atoms in the molecular cycle the differing number can take place in the types of the specific interactions at the stable total value. In the crystalline 3-aminoquinoline there are realized four specific interactions D=HC → CH, three D=HC → C, one D=C → C, two D=N → CH and four hydrogen bonds. The energy of the hydrogen bond is determined with the help of Eq. 5.15a,

$$\text{DN}\cdots\text{H-N} = (\Delta_{\text{sub}}H^{\circ}(298\ \text{K})\text{aql} - n\text{DHC}\rightarrow\text{CH} - m\text{DHC}\rightarrow\text{C} - q\text{DC}\rightarrow\text{C})/4 \quad (5.15a)$$

considering the energy contributions of all realized hydrogen bonds and specific interactions. From the results of the calculations given in Table 5.14 it follows that the energies of formed hydrogen bonds stabilize in dependence on the location of the amino group in the molecular cycle, i.e. there is an analogue with the benzene ring. We have

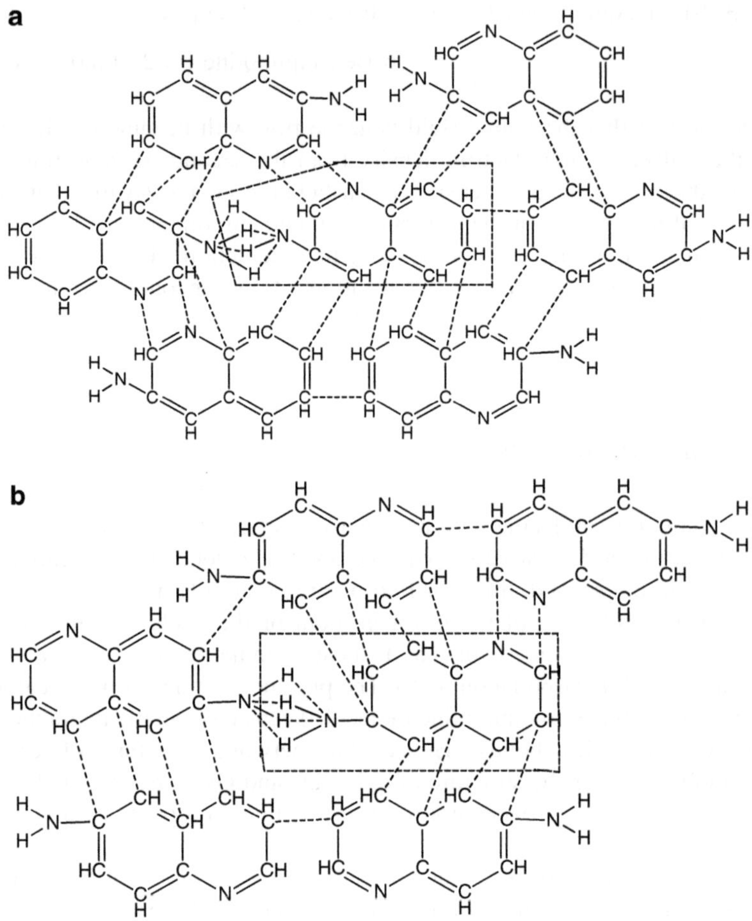

Fig. 5.17 Schematic picture of the crystalline structure of 3-aminoquinoline (**a**) and 6-aminoquinoline (**b**) with the network of the specific interactions and hydrogen bonds

$$\text{8-Aminoquinoline}\,(6.55) < \text{3-Aminoquinoline}\,(8.8) < \text{5-Aminoquinoline}\,(9.3) = \text{6-Aminoquinoline}\,(9.4) < \text{4-Aminoquinoline}\,\left(11.9\,\mathrm{kJ\,mol^{-1}}\right)$$

The formation of a less stable hydrogen bond at the location of the amino group in the benzene ring at C(8) with double carbon bond between the nitrogen atoms points to its significant influence as regards the shifting of the electron density. In contrast the lack of this double bond between the nitrogen atom and the carbon atom C(3) stabilizes the hydrogen bond. At the same time, the location of the amino group at C(5) or at C(6) of the benzene ring at the maximum distance from the nitrogen atom cycle leads to the shifting of the electron density to the nitrogen atom

Table 5.14 Energies of the hydrogen bonds and specific interactions (kJ mol^{-1}) of crystalline aminoquinolines at 298 K

Compounds	Formula	Structure	$\Delta_{sub}H°(298\ K)$ [11]	D=HC → CH	D=HC → C	D=HC → C	D=N → CH	D–N•••H–N
3-Aminoquinoline	$C_9H_8N_2$	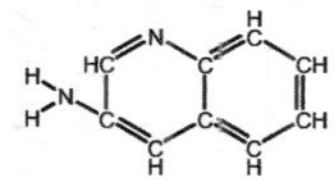	103.1 ± 0.9	7.40 × 4	6.40 × 3	5.40	6.9	8.8
4-Aminoquinoline	$C_9H_8N_2$	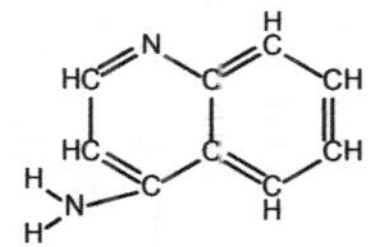	115.5 ± 0.8	7.40 × 4	6.40 × 3	5.40	6.9	11.9
5-Aminoquinoline	$C_9H_8N_2$	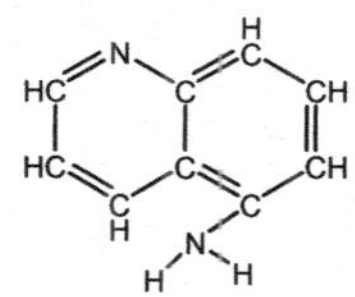	105.0 ± 0.7	7.40 × 4	6.40 × 3	5.40	6.9	9.3
6-Aminoquinoline	$C_9H_8N_2$	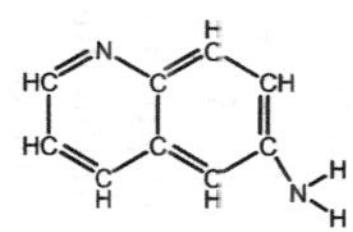	105.7 ± 1.0	7.40 × 3	6.40 × 4	5.40	6.9	9.4
8-Aminoquinoline	$C_9H_8N_2$	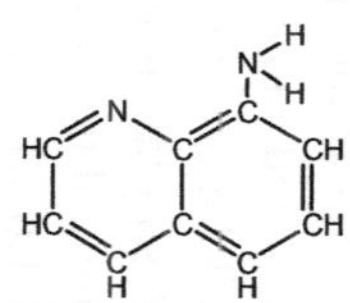	93.3 ± 0.5	7.40 × 3	6.40 × 4	5.40	6.9	6.55

of the amino group from the fringing carbon atoms. The same feature in the shifting of the electron density from the fringing carbon atoms at the location of the amino group at C(4) of the hexamerous hetero cycle is shown in maximum stabilization of the hydrogen bond. The fact that calculation of the energy of the hydrogen bonds, formed by amino group from the difference of sublimation enthalpy and for quinoline reduced by the contribution of the energy to the sublimation enthalpy of the substituted hydrogen atom

$$\text{D-N}\cdots\text{H-N} = (\Delta_{sub}H^{\circ}(298\ \text{K})\text{aql} - (\Delta_{vap}H^{\circ}(298\ \text{K})\text{ql} - \text{DH}))/4 \qquad (5.15\text{b})$$

also draws attention to the difference in the manifestation of the influence of pyridine's fragment and benzene ring on the stabilization of the energy of the specific interactions is noteworthy.

5.6.3 *Dimethylquinolines, Hydroxyquinolines, and Cyano Quinolines*

The uniqueness of the properties of the quinoline molecule is definitely manifest for the dimethylquinolines and methyl hydroxyquinolines compounds. It has been shown above that the contribution of the energy to the vaporization enthalpy of methylpyridine, obtained from the difference of the enthalpy characteristics of the corresponding methylpyridine and pyridine, is described by the integral value of its changes with the values changing according to

$$\text{2-Methylpyridine}\ (1.3) < \text{3-Methylpyridine}\ (4.8) < \text{4-Methylpyridine}\ \left(5.30\ \text{kJ mol}^{-1}\right)$$

within 1.3–5.3 kJ mol^{-1} for two coordinating molecules in the condensed state for two-coordinated molecules or at the one-coordinated isostructural methyl group 0.65–2.65 kJ mol^{-1} (Tables 5.2 and 5.8). The calculations of the contribution of the energy of similar methyl groups from the difference of sublimation enthalpy of dimethylquinolines and quinoline (Table 5.15) point to the high values, reaching 8.2 and 11.6 kJ mol^{-1} for one methyl group of the 2,6-dimethylquinoline and 2,7-dimethylquinoline molecules. Taking into account that the molecule with two isostructural methyl groups interacts with two similar groups of the molecules that are close in the environment in liquid and crystalline condition, then from each group we have 4.1 and 5.5 kJ mol^{-1} for the compounds considered. The given value of the energy corresponds to the contribution of the energy of the methylene group to the vaporization enthalpy with linear chain and cyclic hydrocarbons at a total number of carbon atoms chain of more than four. In other words the specific interaction of this group has high stability, significantly exceeding the energy of interaction of the isostructural methyl groups of methylpyridines. Each isostructural

Table 5.15 Energies of the contribution (kJ mol^{-1}) of methyl groups of dimethylquinolines at sublimation enthalpy

2,6-Dimethylquinoline		2,7-Dimethylquinoline	
ΣDCH_3		ΣDCH_3	
16.4		19.8	
C(2) $2DCH_3$	C(6) $2DCH_3$	C(2) $2DCH_3$	C(7) $2DCH_3$
8.2	8.2	8.2	11.6
C(2) DCH_3	C(6) DCH_3	C(2) DCH_3	C(7) DCH_3
4.1	4.1	4.1	5.8

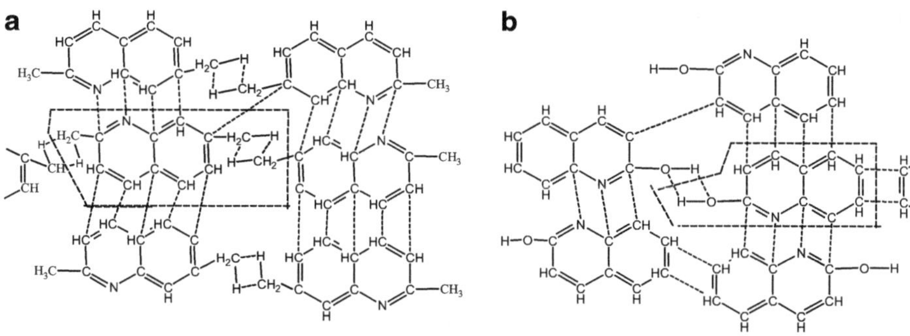

Fig. 5.18 Schematic picture of the crystalline structure of 2,6-dimethylquinoline (**a**) and 2-hydroxyquinoline (**b**) with the network of the hydrogen bonds and specific interactions

methyl group forms two specific interactions and the dimethylquinoline molecule forms the network structure of a crystal by the 14 specific interactions (Fig. 5.18), differing by the types and energies (Table 5.16).

Additionally to the considered specific interactions the hydroxyquinoline molecule forms by the hydroxyl group two hydrogen bonds with the molecules that are close in the environment in the crystal. The formed grid structure of crystalline hydroxyquinolines includes hydrogen bonds (Fig. 5.18b), the energies of which should change from the location at the carbon atoms in the cycle of the molecule whose energies should change from its location at the carbon atoms in the cycle of the molecule. Taking the energy contributions of the specific interactions, the energy of the hydrogen bond was determined with the help of Eq. 5.15c

$$D\text{-}O\cdots H\text{-}O = (\Delta_{sub}H^{\circ}(298\ K)hql - nD{=}HC\rightarrow CH - mD{=}HC\rightarrow C - qD{=}C\rightarrow C - 2D{=}N\rightarrow CH)/2 \quad (5.15c)$$

The molecule of 4-methyl-2-hydroxyquinoline contains a functional methyl group and additional to the noted interactions in its network structure there are two interactions, $DH_3C \rightarrow HCH_2$, and for determination of the energy value of the hydrogen bond in Eq. 5.15c one should add the energy contribution of these bonds,

D-O•••H-O = ($\Delta_{sub}H°(298\ K)hql$ –nD=HC→CH – mD=HC→C –qD=C→C - 2D=N→CH DH$_3$C→H-CH$_2$)/2

(5.15d)

The energy contribution of the functional methyl group to the sublimation enthalpy of 4-methyl-2-hydroxyquinoline is the difference of the sublimation enthalpy of 4-methyl-2-hydroxy quinoline and 2-hydroxyquinoline and is equal to 8.7 kJ mol^{-1}. The results of the calculation of the energy of the hydrogen bond for hydroxyquinoline presented in Table 5.16 are described by the integral value of stabilization from the location of the hydroxyl group in the molecular cycle

$$\text{8-Hydroxyquinoline } (10.6) < \text{2-Hydroxyquinoline } (25.8) = \text{4-Methyl-2-hydroxyquinoline } (25.8) < \text{4-Hydroxyquinoline } (33.7\ \text{kJ mol}^{-1})$$

caused by an increasing demand for shifting of the electron density from CH groups of the hexamerous cycle with their distance from the nitrogen atom, increasing the negative charge of the oxygen atom and the difference of the charges with the hydrogen atom of the hydroxyl group. On the basis of the thermodynamic analysis for hydroxyquinolines one should conclude that in the formation of the specific interactions between nitrogen atoms and carbon atoms of the cycle with the formation of bonds D=N → CH or D=N → C the value of the energy of the hydrogen bond and then the molecular structure are dominant.

The replacement of methyl and hydroxyl groups in the dimethylquinolines and hydroxyquinoline molecules by the cyano does not violate the grid structure of crystalline quinolines and cyanoquinolines with 12 bond vacancies that forms the character of the structure of the condensed condition with additionally two stable specific interactions D≡N → C≡ (Fig. 5.19).

The value of the energy contribution of the cyano group to the sublimation enthalpy for cyanoquinoline should be determined by the difference with enthalpy characteristics quinoline and energy of the specific interaction with help of Eq. 5.15c, taking the contributions of the energies of the other interactions. The values obtained of the energy D≡N → C≡ bonds (Table 5.16) allow us to state that replacement of location of the cyano group in the 2-cyanoquinoline molecule of C(2) by C(3) in the 3-cyanoquinoline molecule expresses the insignificance of this for the stabilization of this specific interaction and is within the experimental error.

Table 5.16 Energies of the hydrogen bonds and specific interactions (kJ mol^{-1}) of crystalline hydroxyquinolines at 298 K

Compounds	Formula	Structure	$\Delta_{sub}H°$ (298 K) [11]	D=HC → CH	D=HC → C	D=N → CH	D*iso*CH_3	D–O•••H–O
2,6-Dimethyl quinoline	$C_{11}H_{11}N$	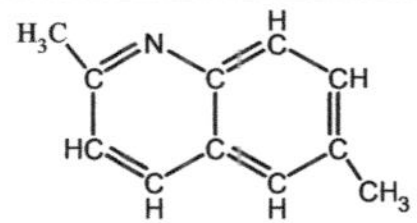	84.5 ± 1.5	7.40 × 3	6.40 × 5	6.9	7.9 × 2	6.40 × 4
2,7-Dimethyl quinoline	$C_{11}H_{11}N$	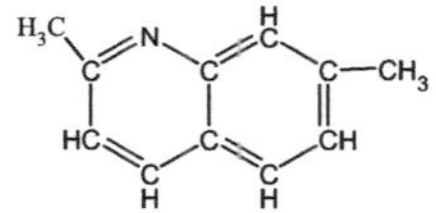	87.5 ± 1.5	7.40 × 4	6.40 × 4	6.9	4.1 × 2 5.8 × 2	–
2-Hydroxy quinoline	C_9H_7N	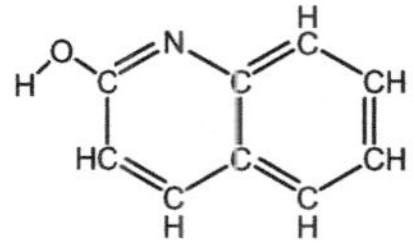	119.4 ± 0.6	7.40 × 4	6.40 × 4	D=N → C = 6.3 × 2	–	25.8
4-Hydroxy quinoline	C_9H_7N	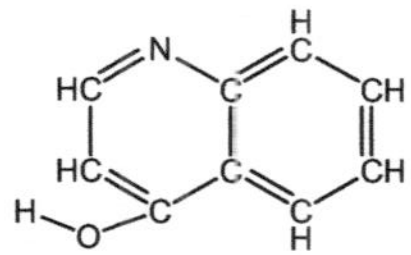	135.1 ± 1.1	7.40 × 4	6.40 × 4	D = N → C = 6.3 × 2	–	33.7
8-Hydroxy quinoline	C_9H_7N	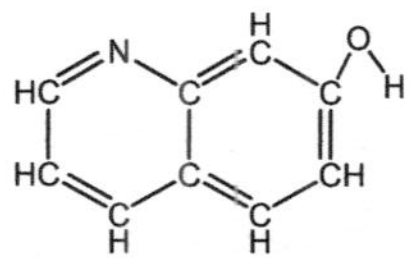	89.0 ± 1.4	7.40 × 4	6.40 × 4	D=N → C = 6.3x2	–	10.6
4-Methyl-2-hydroxy quinoline	$C_{10}H_9NO$	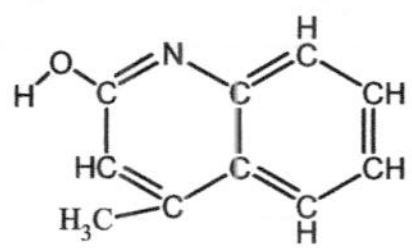	128.1 ± 1.6	7.40 × 4	6.40 × 4	D=N → C = 6.3x2	4.35 × 2	25.8

(continued)

Table 5.16 (continued)

Compounds	Formula	Structure	$\Delta_{sub}H^\circ$ (298 K) [11]	D=HC → CH	D=HC → C	D=N → CH	D*iso*CH_3	D–O•••H–O
2-Methyl-8-hydroxy quinoline	$C_{10}H_9NO$	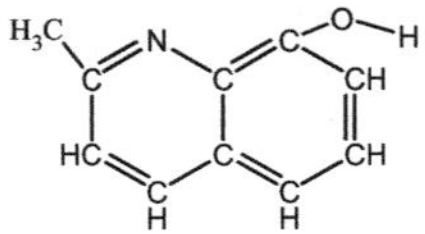	139.0 ± 1.0	7.40 × 4	6.40 × 4	D=N → C = 6.3x2	4.1 × 2	?
2-Cyano quinoline	$C_{10}H_6N_2$	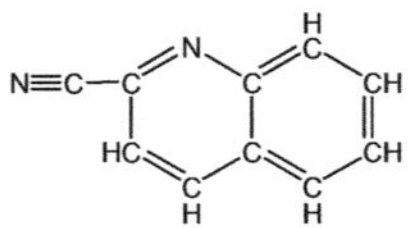	94.4 ± 0.7	7.40 × 4	6.40 × 4	6.9	–	D≡N → C≡13.3
3-Cyano quinoline	$C_{10}H_6N_2$	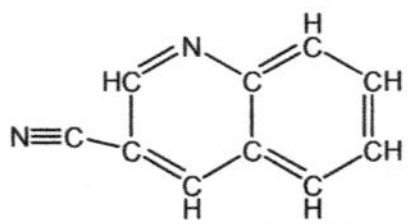	93.2 ± 0.8	7.40 × 4	6.40	6.9	–	13.6

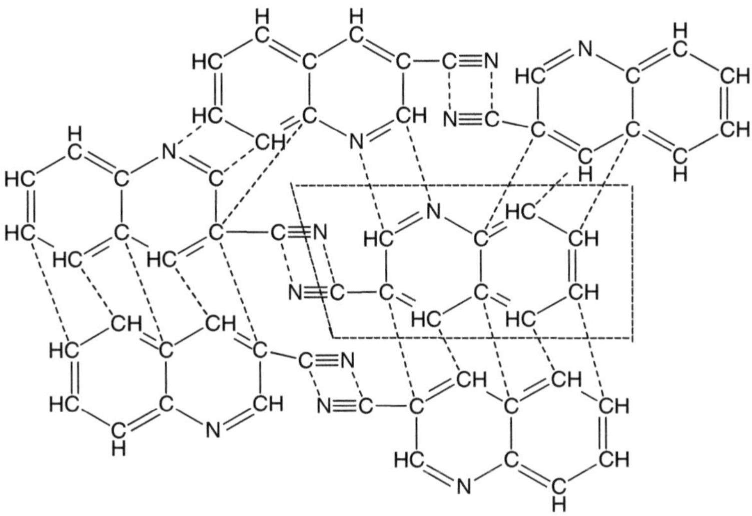

Fig. 5.19 Schematic picture of the crystalline structure of 3-cyanoquinoline with the network of the specific interactions

5.7 Energies of Specific Interactions of Quinoxaline and Their Derivative

5.7.1 Quinoxaline, Quinazoline, and Phthalazine

The compounds of the quinoxaline series allow us to follow the influence of the two nitrogen atoms in the compounds with a rigid cycle of the benzene ring and a pyridine's fragment to the stability of the specific interactions, formed by two nitrogen atoms cycle. The quinoxaline molecule with two nitrogen atoms forms the network structure of a crystal by the three series, ten specific interactions 4D=N → CH, 3D=HC → CH and 3D=HC → C (Fig. 5.20a). The energy of the specific interaction D=N → CH is determined by the difference of the total contribution of the specific interactions, formed by CH groups and carbon atoms of a cycle. We have

$$D{=}N{\rightarrow}CH = (\Delta_{sub}H^{\circ}(298\ K)qxl - nD{=}HC{\rightarrow}CH - mD{=}HC{\rightarrow}C - pD{=}C{\rightarrow}C)\ /q \quad (5.16)$$

Using the sublimation enthalpy of quinoxaline it is necessary to define the value of this characteristic of 2,2′-biquinoxaline, which is similar to 2,2′-biquinoline and equal to the sum of sublimation enthalpy of the two quinoxaline molecules reduced by the energy contribution of the two substituted hydrogen atoms (Table 5.17). The specific interactions in the network structure of a crystal of this compound of the

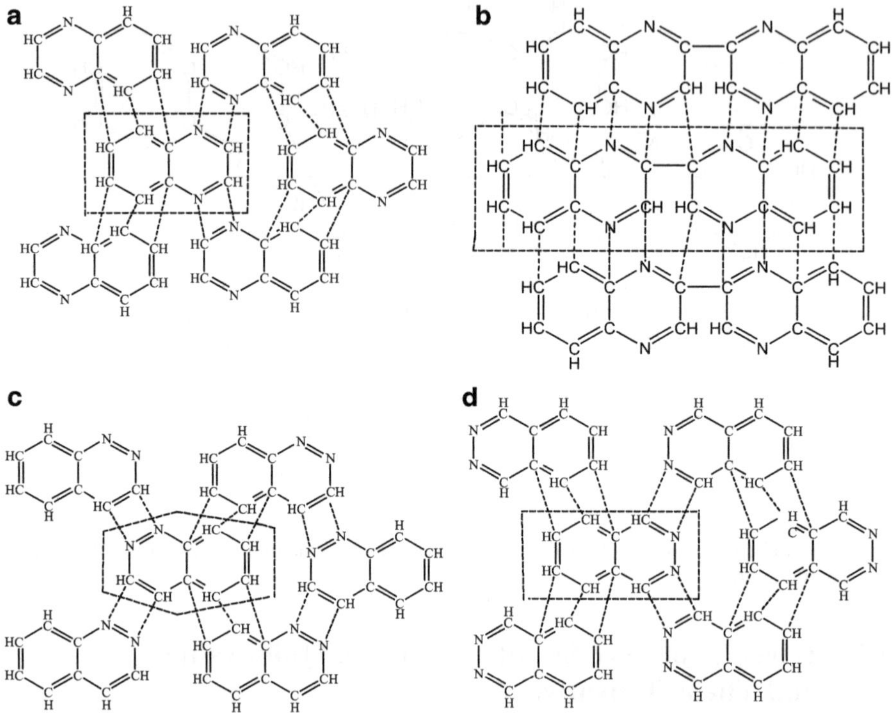

Fig. 5.20 Schematic picture of the crystalline structure of quinoxaline (**a**) 2,2′-biquinoxaline (**b**) quinazoline (**c**), and phthalazine (**d**) with the network of the specific interactions

four series (Fig. 5.20b) include 8D=N → CH. The specific interactions of this series include 2D=N → CH and 6D=N → C, of which the average of the values of the energies is defined with the help of an equation similar to Eq. 5.16. On the basis of the average value (7.1 kJ mol^{-1}) realistic values of each interaction series equal to 8.0 and 6.8 kJ mol^{-1}, respectively, were obtained (Table 5.17). The energy of the specific interaction D=N → CH of crystalline quinoxaline with reduced energy value (7.00 kJ mol^{-1}) is caused by the influence of the location of the carbon atom of the CH group.

The vaporization enthalpy of quinazoline and phthalazine increases and there is difference in values compared with quinoxaline, which under condition of stability of the number of bond vacancies, series, and number of the specific interactions points to the stabilization of the interaction D=N → CH (Fig. 5.20c, d). This implies that the location of the two nitrogen atoms in a cycle at the location N (1) and N(2) in the quinazoline molecule and N(2) and N(3) in the phthalazine molecule is accompanied by an equivalent shifting of the electron density to the nitrogen atoms, which is why fringing carbon atoms have equal or similar values of the charges and, at the same time, differ by the negative charge of the nitrogen atom rather than in the quinazoline molecule.

Table 5.17 Energies of the specific interactions (kJ mol^{-1}) of crystalline quinoxaline, quinazoline, and phthalazine at 298 K

Compounds	Formula	Structure	$\Delta_{sub}H°(298\ K)$ [11]	D=HC → CH	D=HC → C	D=C → C	D–N → CH
Quinoxaline	$C_8H_6N_2$	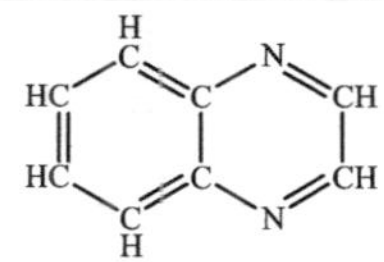	69.4 ± 0.6 66.9 ± 2.0	7.40 × 3	6.40 × 3	–	7.00
2,2′-Biquinoxaline	$C_{18}H_{12}N_2$	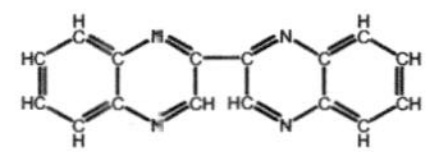	136.8[a]	7.40 × 5	6.40 × 6	5.40	Σ = 56.8 D=N → CH = 8.0 D=N → C = 6.8
Quinazoline	$C_8H_6N_2$	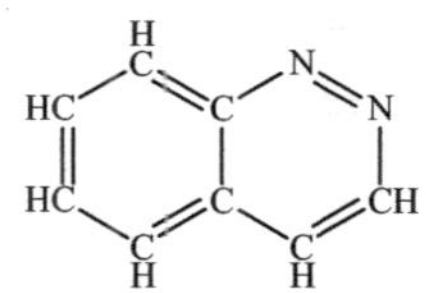	77.6 ± 0.5 76.61.4	7.40 × 2	6.40 × 4	–	9.30
Phthalazine	$C_8H_6N_2$	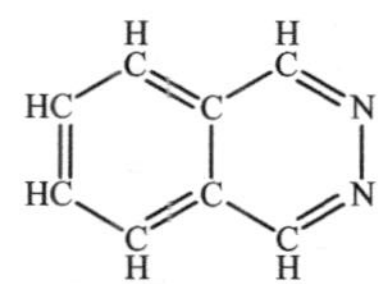	81.1	7.40 × 2	6.40 × 4	–	10.20

[a]Determined using the method of addictively contribution of enthalpies sublimation

Using Eq. 5.16 and considering the energy contributions of the two series of the specific interactions we calculated the energy of the interactions D=N → CH (Table 5.17), leading to the conclusion that there is a logical change in the stabilization of the specific interaction energy depending on the location of the two nitrogen atoms in a cycle: N(1) and N(4) < N(1) and N(2) < N(2) and N(3)

D=N→CH: Quinoxaline (7,00) < 2,2'-Biquinoxaline (8.0) < Quinazoline (9.30) < Phthalazine (10.20 kJ mol^{-1})

5.7.2 *Hydroxyquinoxalines*

In Sect. 5.6.3 it was shown that the formation of stable hydrogen bonds is the main factor in the coordination of molecules and in the formation of the specific interactions in the structure of crystalline hydroxyquinolines. This condition is the main one in the formation in crystalline 2-hydroxyquinoxaline with four specific interactions D=N → C instead of D=N → CH interactions (Fig. 5.21a). Two other series of the specific interactions are formed with more interactions with increased stability for 4D=HC → CH and 2D=HC → C. Thus, these series of the interactions and two hydrogen bonds form crystalline 2-hydroxyquinoxaline. In the crystalline 2,3-dihydroxyquinoxaline two additional hydrogen bonds are introduced, four D=N → C, 4D=HC → CH and 2D=HC → C. The increase in the number of hydroxyl groups in the 2,3-dihydroxyquinoxaline molecule inevitably leads to the reduction of its stability. The presence of the isostructural methyl group in the methyl-3-hydroxyquinoxaline molecule, saturating the hydrogen atom of the CH group, participating in the formation of the specific interaction, allows us to take it to the interaction D=N → CH. The energies of the hydrogen bond of 2-hydroxyquinoxaline and 2,3-dihydroxyquinoxaline can be obtained with the help of Eq. 5.15e

$$\text{D-O}\cdots\text{H-O} = (\Delta_{sub}H°(298\ K)hqxl - 4D{=}HC{\to}CH - 2D{=}HC{\to}C - 2D{=}N{\to}CH)/\ n \qquad (5.15e)$$

or from the difference of sublimation enthalpies of 2-hydroxyquinoxaline and quinoxaline

$$\text{D-O}\cdots\text{H-O} = (\Delta_{sub}H°(298\ K)hqxl - (\Delta_{sub}H°(298\ K)qxl)\ /n \qquad (5.16a)$$

For estimation of the energy contribution of the hydroxyl group to the sublimation enthalpy we draw attention to the fact that there is practically no difference in the contribution of the energy of the specific interactions (0.3 kJ mol^{-1}), formed by –C≡N by the group of 2-cyanoquinoline and 3-cyanoquinoline molecules. This is the basis for taking the contributions of the energy as equal to that of the hydrogen bonds to the sublimation enthalpy for 2-hydroxyquinoxaline and 3-hydroxyquinoxaline.

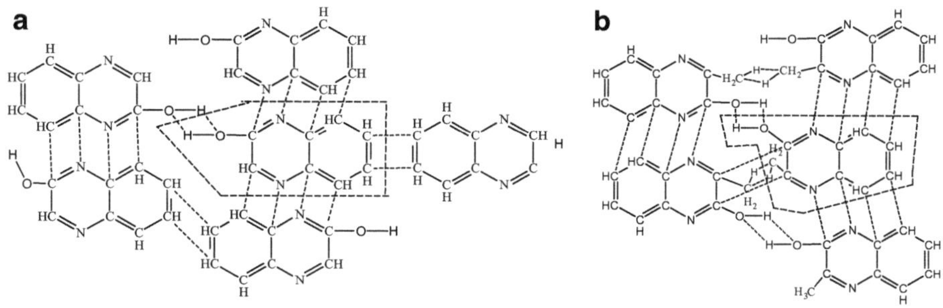

Fig. 5.21 Schematic picture of the crystalline structure of 2-hydroxyquinoxaline (**a**) and 2-methyl-3-hydroxyquinoxaline (**b**) with the network of the hydrogen bonds and specific interactions

Accepting this condition, we obtain from the difference in the sublimation enthalpy of 2-methyl-3-hydroxyquinoxaline and 3-hydroxyquinoxaline the fact that the contribution by the isostructural methyl group is equal to 5.6 kJ mol^{-1} and the energy of the specific interaction $DCH_3 \rightarrow H\text{–}CH_2$ formed by this group is equal to 2.8 kJ mol^{-1}. The results of the calculations given in Table 5.18 are described by the number of stabilizations of the energies of the hydrogen bond caused by

$$\text{2, 3-Dihydroxyquinoxaline}\,(22.2) < \text{2-Methyl-3-Hydroxyquinoxaline}\ \left(25.0\right) =$$
$$\text{2-Hydroxyquinoxaline}\,25.0\Big) \ \approx \text{4-Methyl-2-hydroxyquinoline}\ \left(25.8\ \text{kJ}\ \text{mol}^{-1}\right)$$

the reduction in the number of 2-methyl-3-hydroxyquinoxaline and 2-hydroxyquinoxaline 25.0 kJ mol^{-1} molecules formed by the hydrogen bonds from four till two. The energies of these bonds are within the experimental error in agreement with the value of the energy of the same bond, formed in crystalline 4-methyl-2-hydroxyquinoline.

5.8 Energies of Specific Interactions of Complicated Heterocyclic Compounds with Nitrogen Atoms

5.8.1 2-Phenylquinoline, Acridine, and Phenazine

The example of compounds of benzoquinolines was considered to illustrate the influence of the location of the benzene ring on the stabilization of the specific interaction D–N → CH (Sect. 5.6.1). The thermodynamic analysis of phenylquinoline, acridine, and phenazine will allow us to get information on the influence of the benzene ring, connected by =C–C= bond with quinoline and the influence of the two nitrogen atoms of the quinoxaline's fragment. Connected by

Table 5.18 Energies of the hydrogen bonds and specific interactions (kJ mol^{-1}) of crystalline hydroxyquinoxalines at 298 K

Compounds	Formula	Structure	$\Delta_{sub}H°$ (298 K) [11]	D=HC → CH	D=HC → C	D*iso*CH_3	D=N → C	D–O•••H–O
2-Hydroxyquinoxaline	$C_8H_6N_2O$	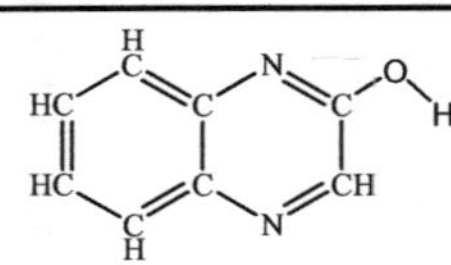	118.5 ± 3.1	7.40 × 4	6.40 × 2	–	6.25 × 4	25.0 × 2
2,3-Dihydroxyquinoxaline	$C_8H_6N_2O_2$	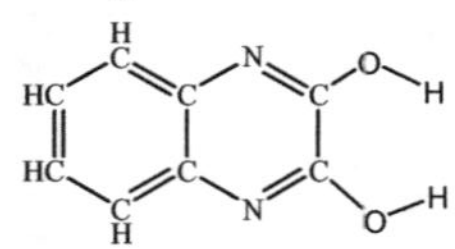	156.3 ± 5.5	7.40 × 4	6.40 × 2	–	6.25 × 4	22.2 × 4
2-Methyl-3-Hydroxy quinoxaline	$C_9H_8N_2O$	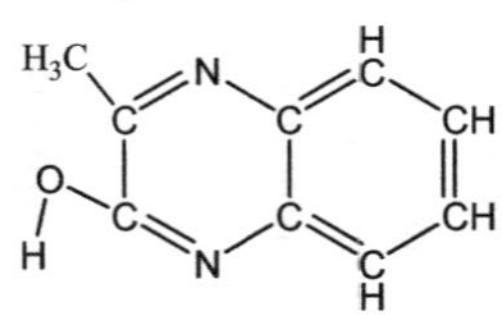	119.7 ± 2.8 123.0 ± 4.4 C	7.40 × 2	6.40 × 4	2.8 × 2	D=N → CH = 7.23 × 2 6.25 × 2	25.0 × 2

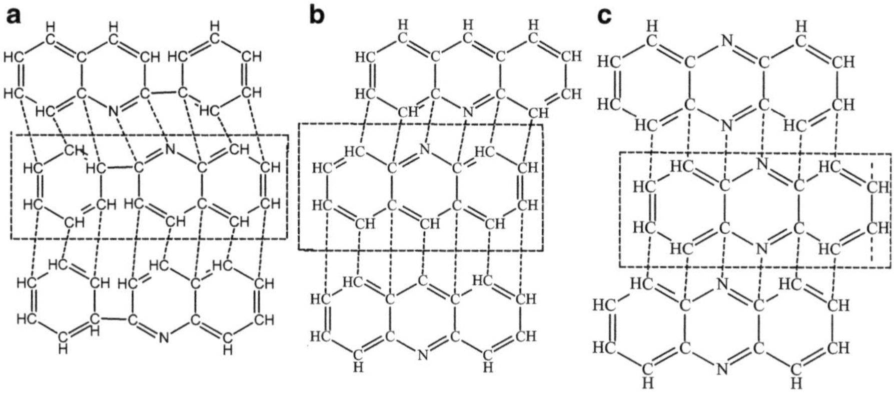

Fig. 5.22 Schematic picture of the crystalline structure of 2-phenylquinoline (**a**) Acridine (**b**) phenazine (**c**) with the network of the specific interactions

this bond the benzene ring has significantly less electron density under the influence of acceptor properties of the nitrogen atom of the molecular fragment of quinoline than the benzene ring of this fragment. In its turn, part of the electron density remains on the carbon atom of the pyridine's cycle, which could create a reduced difference in the charges of nitrogen and carbon atoms, forming the specific interaction of liquid and crystalline condition compared with the one realized in the quinoline molecule.

The 2-phenylquinoline molecule with rigid benzene rings located in one plane forms the network structure of a crystal by 16 specific interactions of four series, 11D=HC → CH, D=HC → C, 2D=C → C, and 2D–N → C (Fig. 5.22). The energy of the specific interaction is calculated with the help of the equation taking the energy contributions of all the series of the interactions to the sublimation enthalpy Eq. 5.16.

Acridine and phenazine molecules, with a similar number of bond vacancies and differing number of carbon and nitrogen atoms form the network structure of a crystal (Fig. 5.22b, c). The acridine molecule forms 7D=HC → CH, 3D=HC → C, 2D=C → C, and 2D=N → C interactions and phenazine molecule forms the 6D=HC → CH, 4D=HC → C, and 4D=N → C interactions , respectively. The results of the calculations of the energy values DN → C of the specific interactions are implemented with the help of Eq. 5.16b.

From the data presented in Table 5.19 it follows that energies of the specific interactions formed by the nitrogen atoms of the considered compounds in the crystalline condition, obtained with the values of experimentally measured values of sublimation enthalpy, are described by the inequality

D=N→C: 2-Phenylquinoline (3.4) < Acridine (8.85) > Phenazine (5.1 kJ mol^{-1})

Table 5.19 Energies of the specific interactions (kJ mol^{-1}) of crystalline 2-phenylquinoline, acridine, and phenazine

Compounds	Formula	Structure	$\Delta_{sub}H°(298\ K)$ [11]	D=HC → CH	D=HC → C	D=C → C	D=N → C
2-Phenylquinoline	$C_{15}H_{11}N$	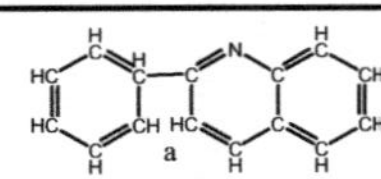	105.4 ± 0.9	7.40 × 11	6.40	5.40 × 2	3.4 × 2
Acridine	$C_{13}H_9N$	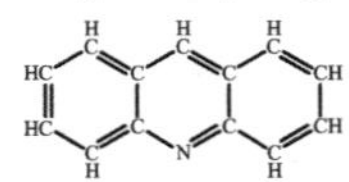	91.7 ± 0.4	7.40 × 7	6.40 × 3	5.40 × 2	8.85
Phenazine	$C_{12}H_8N_2$	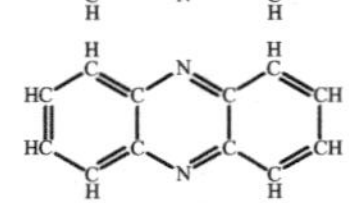	90.0 ± 1.5 90.4 ± 1.7	7.40 × 6	6.40 × 4	–	5.1 × 4

The low value of the energy of the specific interaction for 2-phenylquinoline proves the assumption above on the low difference between the charges of nitrogen and carbon atoms, forming this specific interaction, caused by insignificant shifting of the electron density from the phenol ring. At the same time the location of the nitrogen atom in the acridine molecule between two benzene cycles leads to the nitrogen atom with a similar shifting of the electron density with similar or equal values of the charges fringing the carbon atom and significantly increased differences with the negatively charged nitrogen atom. The value of the energy of the same series of specific interactions of phenazine approaching the average value points to the receiving by each of the nitrogen atoms of possibly equivalent, but, nevertheless reduced shifting of the electron density, leading also to the smaller difference between the charges of nitrogen and carbon atoms, forming the specific interaction D=N → C.

5.8.2 *Dibenzylideneazine*, p-*Azoxyanizole, and* N-*Phenylbenzophenoneimine*

The quite complex dibenzylideneazine molecule with two benzene cycles and two CH_2 groups should practically belong to the *sis-trans*-configuration, in which n-and π-orbital form bonding $\pi^+(a_u)$, n^- (b_u) and anti-bonding $\pi^-(b_g)$, $\pi^+(a_g)$ combinations. Splitting of the levels of the energies of π-and n-electrons is within 2–3 eV [12–15]. It could be the reason for the assumption on the weak influence of the benzene rings, comparable with manifestation of acyclic compounds, on the stabilization of the specific interactions.

The dibenzylideneazine molecule with double bonds forms the network structure of a crystal by three series of specific interactions, formed by 16 bond vacancies (Fig. 5.23a). The specific interactions formed by CH groups and carbon atoms of the benzene rings include the 8D=HC → CH and 4D=HC → C series and four interactions D=N → CH, formed by the fragment >C=N—N=C<. The energy of the specific interaction D=N → CH is calculated with the help of Eq. 5.15.

With 21 bond vacancies, the *N*-phenylbenzophenoneimine molecule (Fig. 5.23b) forms a similar number of specific interactions including 14 D=HC → C, 4D=HC → CH, one by the methylene group CH_2 with two saturated hydrogen atoms and two hydrogen bonds D–N•••H–N, formed by the imine group. In order to determine the energies of the hydrogen bond it is necessary to obtain the energy value of the specific interactions, formed by the methylene group. However, the mutual influence of benzene rings connected by the methylene group is expressed in such a way that the energy values of the specific interactions undergo changes. In this connection it is necessary to use the sublimation enthalpy of triphenylmethane minus the energy contribution of the two saturated hydrogen atoms of this group (2.0 kJ mol^{-1}). From the available sources of the literature [11] of the enthalpy

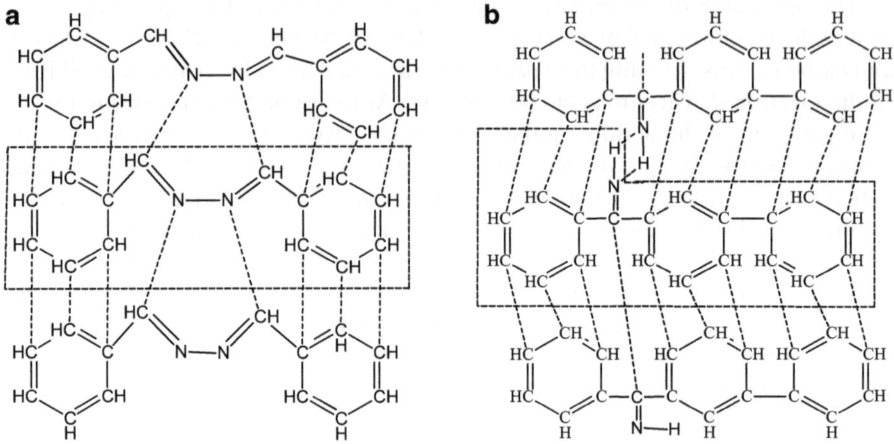

Fig. 5.23 Schematic picture of the crystalline structure of dibenzylideneazine (**a**) and *N*-phenylbenzophenoneimine (**b**) with the network of the specific interactions

characteristics the most reliable value of sublimation enthalpy is 109.1 kJ mol^{-1}. The energy value calculated with the help of Eq. 5.17

$$D\text{-}N\cdots H\text{-}N = \Delta_{sub}H°(298\ K)phbph - \Delta_{sub}H°(298\ K)tphe - 2DH \quad (5.17)$$

of the hydrogen bond is given in Table 5.20.

The value obtained for the energy of the specific interaction D=N → C=N of crystalline dibenzylideneazine (4.25 kJ mol^{-1}) at the three-coordinated nitrogen atom coincides with the energy of the specific interaction N → CH_3–N of liquid trimethylamine with hexamerous nitrogen atom, which should be connected with the influence of the methylene groups of the molecule, acting as an isostructural methyl group. A similar effect is shown for *N*-phenylbenzaldehyde imine (Table 3.12). On the other hand, the energy of the hydrogen bond of crystalline *N*-Phenylbenzophenoneimine is a little bit stabilized (6.3 kJ mol^{-1}) with reduction of the methylene groups up to one with the substituted hydrogen atoms at the imine group.

5.9 Energies of Specific Interactions of Acridone, Phenazine-*N*-oxide, and Quinoxaline-1,4-dioxides

The acridone molecule is a derivative of *N*-phenylbenzophenoneimine and additionally includes carbonyl oxygen atom with two free electron pairs, participating with the carbon atoms cycle in the formation of the specific interaction. The planar structure of the acridone molecule with 16 bond vacancies determines the formation

Table 5.20 Energies of the hydrogen bonds and specific interactions (kJ mol^{-1}) of crystalline dibenzylideneazine and *N*-phenylbenzophenoneimine

Compounds	Formula	Structure	$\Delta_{sub}H°$ (298 K) [11]	D=HC → CH	D=HC → C	D=C → C	D=N → CH	D–N•••H–N
Dibenzylideneazine	$C_{14}H_{12}N_2$	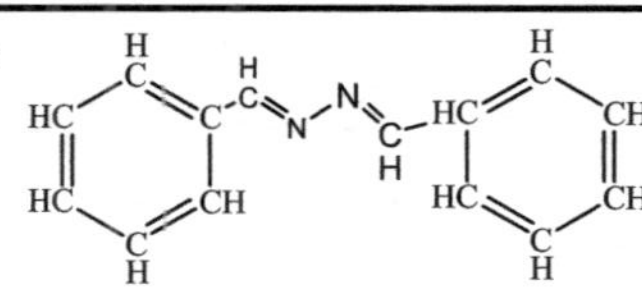	93.3 ± 2.1	7.40 × 8	6.40 × 4	–	4.25 × 6	–
N-Phenylbenzophenoneimine	$C_{19}H_{15}N_3$	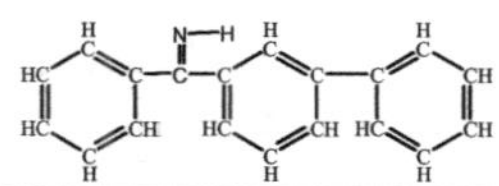	119.7 ± 1.8	107.1	–	–	–	6.3

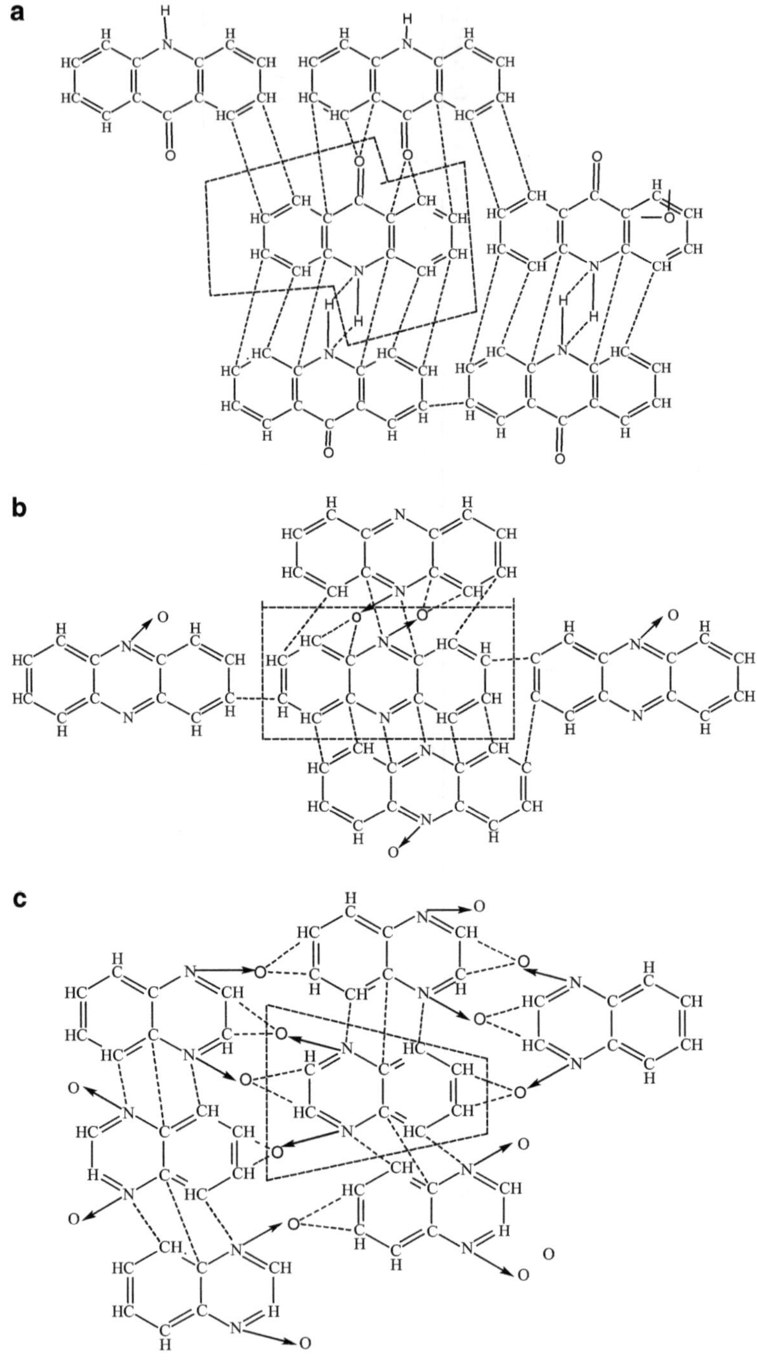

Fig. 5.24 Schematic Picture of the crystalline structure of acridone (**a**), phenazine-*N*-oxide (**b**), and quinoxaline-1,4-dioxide with the network of the specific interactions

of the network structure crystal (Fig. 5.24a) with six series of specific interactions 6D=HC → CH, 2D=HC → C, 2D=C → C, 2D–N•••H–N and (2D–O → CH, 2D–O → C). The energies of the first series of the specific interactions are known, the energy of the hydrogen bond was determined for phenylbenzophenoneimine, and this energy should be taken as equal to the one realized in crystalline acridone. We obtain the average value of the energies of the two series, formed by the oxygen atom, from the inequality

$$\text{D-O}\rightarrow\text{CH} = (\Delta_{sub}H°(298\ \text{K})\text{acr} - n\text{D=HC}\rightarrow\text{CH} - m\text{D=HC}\rightarrow\text{C} - q\text{D=C}\rightarrow\text{C} - 2\text{D-N}\bullet\bullet\bullet\text{H-N})\ /4 \quad (5.17a)$$

The phenazine-*N*-oxide and quinoxaline-1,4-dioxide molecules, unlike phenazine and quinoxaline, contain one and two N → O groups, acting as functional fragments, where the nitrogen atom shifts the electron density to the oxygen atom and provides it with negative charge, donor properties, and the ability to form specific interactions D–O → C. The phenazine-*N*-oxide molecule with 16 bond vacancies forms the planar structure of the crystal (Fig. 5.24b) by the four specific interactions 6D=HC → CH, 2D=HC → C, 4D=N → CH (2D=N → CH + 2 D=N → C), and 4D–O → CH (D–O → CH + D–O → C). The energies of the first two series are known and the energy of the third one was obtained, D=N → C, for phenazine, that is, the founder of phenazine-*N*-oxide. Thus, we obtain the energy of the specific interaction from the difference between sublimation enthalpy of phenazine-*N*-oxide and the total value of the energies of the last three series of the specific interactions, using an equation similar to Eq. 5.17a,

$$\text{D-O}\rightarrow\text{C} = (\Delta_{sub}H°(298\ \text{K})\text{pzo} - n\text{D=HC}\rightarrow\text{CH} - m\text{D=HC}\rightarrow\text{C} - q\text{D=C}\rightarrow\text{C} - 4\text{D=N}\rightarrow\text{CH})\ /2 \quad (5.17b)$$

The introduction into the phenazine-*N*-oxide molecule of the second oxide form (Fig. 5.24c) leads to the significant reorganization of the crystalline structure of quinoxaline-1,4-dioxide with preservation of the network form, with 14 specific interactions of three series 2D=C → C, D=N → CH and with unknown energy value for 8D–O → CH. We take the energy of the specific interaction D=N → CH (7.00 kJ mol^{-1}) as equal to the energy of the same series, realized in crystalline quinoxaline. We obtain the energy of the specific interaction D–O → CH with the help of the equation

$$\text{D-O}\rightarrow\text{CH} = (\Delta_{sub}H°(298\ \text{K})\text{qox} - 2\text{D=C}\rightarrow\text{C} - 4\text{D=N}\rightarrow\text{CH})\ /\ 8 \quad (5.17c)$$

The results of the calculations (Table 5.21) point to the high stability of the specific interactions, formed by the carbonyl oxygen atom of acridone. Considering the difference in the energies of the specific interactions with the saturated hydrogen atom of the CH groups of the cycle, we determined the energies of the specific interactions D–O → CH and D–O → C on the basis of the average calculated

Table 5.21 Energies of the hydrogen bonds and specific interactions (kJ mol^{-1}) of crystalline acridone, phenazine-*N*-oxide, and quinoxaline-1,4-dioxides at 298 K

Compounds	Formula	Structure	$\Delta_{sub}H°$ (298 K) [11]	D=HC → CH	D=HC → C	D=C → C	D–N•••H–N	D–O → CH
D=N → CH = 5.6; D=N → C = 4.6								
Acridone	$C_{13}H_9NO1$	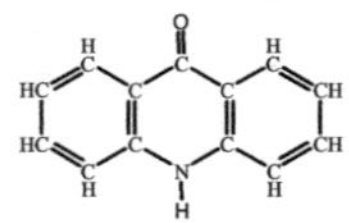	136.2 ± 0.5C	7.40 × 6	6.40 × 2	5.40 × 2	7.25 × 2	Midl 13.3 × 4 D–O → CH = 13.8/ D–O → C = 12.8
Phenazine-*N*-oxide	$C_{12}H_8N_2O$	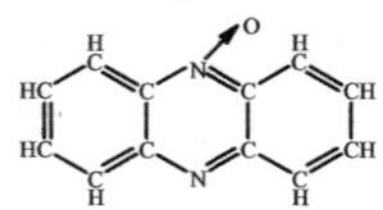	100 ± 1.3	7.40 × 6	6.40 × 2	–	D=N → CH = 5.1 × 4	5.8 × 4 6.3; 5.3
Quinoxaline-1,4-dioxide	$C_8H_6N_2O_2$	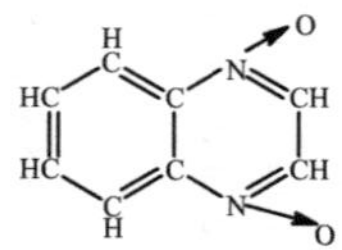	112.0 ± 1.9	–	–	5.40 × 2	7.00 × 4	9.15 × 8
2-Methyl quinoxal-ine-1,4-dioxide	$C_9H_8N_2O_2$ 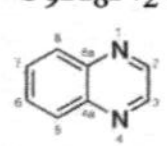	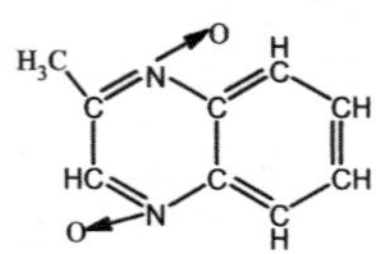	107.0 ± 6.2	–	D*iso*CH_3 = 5.6:2	5.40 × 2	7.00 × 4	7.83

value. The energies of the specific interactions of crystalline acridone are described by

D=C→C (5.40) < D=HC→C (6.40) < D-N•••H-N (7.25) < D=HC→CH (7.40) < D-O→C (12.8) < D-O→CH (13.8 kJ mol^{-1})

and the integral value of its stabilization with a high value of the energy interaction, formed by the oxygen atom of the carbonyl. It points to its high acceptor property, which leads to the increased shifting of the electron density from two rigid benzene rings and the high donor ability at the high difference of the charges with the carbon atom of the benzene rings.

The average value of the energies of the two series of interactions obtained, formed by the oxygen atom of phenazine-*N*-oxide, was used for obtaining the energies of the specific interactions of each series (Table 5.21). The energies of the specific interactions of crystalline phenazine-*N*-oxide illustrate the integral value of its change,

D=N→C (4.6) < D-O→C (5.3) < D=N→CH (5.6) < D-O→CH (6.3) ≤ D=HC→C(6.4) < D=HC→CH (7.4 kJ mol^{-1})

in which the energies of the formed specific interactions are due to the oxygen atom occupying an intermediate position and reflecting its low stability compared with

D-O→CH Phenazine-N-oxide < D-O→CH Acridone.

interactions formed by the carbonyl oxygen atom. It follows that the oxygen atom, connected with the nitrogen atom, gets a significantly reduced value of the electron density from the nitrogen atom and, in its turn, is smaller than the carbon atom.

The calculated value of the energy of the specific interaction D–O → C of crystalline quinoxaline-1,4-dioxide (9.15 kJ mol^{-1}) points to its significant increased stability compared with that realized in crystalline phenazine-*N*-oxide. One should conclude that the increase in the molecule of quinoxaline-1,4-dioxide of the number of oxide forms leads to the shifting of the electron density at the oxygen atoms and the difference in its charges with carbon atoms of the CH group of the cycle and, subsequently, the stabilization of this series of specific interactions.

The introduction into the quinoxaline-1,4-dioxide molecule of the isostructural methyl group at the C(2) location is accompanied by the reduction of the sublimation enthalpy of 2-methyl quinoxaline-1,4-dioxide at 5.0 kJ mol^{-1} on introduction into the molecule of the definite energy contribution to the enthalpy characteristics. This fact points to the significant difference of quinoxaline-1,4-dioxides from quinolines, which are characterized by a significant increase of sublimation enthalpy with the introduction of a methyl group to the same location in the molecule cycle. Taking the energy contribution of the isostructural methyl group in quinoxaline-1,4-dioxide as equal to the contribution of the CH_3 group in 3-hydroxyquinoxaline (5.6 kJ mol^{-1}), we determined the contribution of the energy of the eight specific interactions on Eq. 5.17b and the value of the energy of this series of interaction (Table 5.21).

References

1. Veillard A, Levy B, Danbll R, Gallais F (1967) Structural electroniqul of borazane BH_3-NH_3. Theor Chim Acta 8:312
2. Ehrenson S (1968) Theory of electron effect at formation of complex of trimethylamine with trimethylborium. Theor Chim Acta 10:193
3. Ehrenson S (1968) Effect environment at examination of a reactivity adding of trimethylamine to trimethylborium theory. Chem Acta 10:209
4. Moireau MG, Veillard A (1968) Quantum Mechanical calculations on barriers to internal rotation. II. The borazane molecular BH_3-NH_3. Theor Chim Acta (Bert) 11:344–357
5. Veillard A, Daudel R (1970) Nature et proprietes des liaison de Coordination. Colloques internationaux du centre National de la Recherche Scientifigue, Paris, p 22
6. Nefedov VI (1984) Infra red-spectroscopy of chemical compounds. Chemistry, Moscow, 255 p
7. Nefedov VI, Vovna VI (1989) Electronic structure of organic and elementorganic compounds. Nauka, Moscow, p 199
8. Chickos JS, Acree WE Jr (2002) Enthalpies of vaporization of organic and organometallic compounds, 1880–2002. J Phys Chem Rev Data 31:519–879
9. Karlson TA (1981) Photoelectron and OGE-spectroscopy. Mech. Eng. Russia, Leningrad, 431 p
10. Kalinin FL, Lobov VP, Zidkov VA (1971) References book on biochemistry. Nauka Dumka, Kiev, p 1012
11. Chickos JS, Acree WE Jr (2003) Enthalpies of sublimation of organic and organometallic compounds, 1910–2001. J Phys Chem Rev Data 32:537–698
12. Vovna VI (1974) Investigation by the method of FE-spectroscopy of electron structure of nitrogen and phosphorus contains compounds. Thesis, Cand. Phis. Mat. Science, Leningrad, p 154
13. Zverev VV, Vovna VI, Vilesov FI et al (1973) Photoelectron specters, electron and spatial structure uncyclic azines. Docl AN SSSR 213(5):1117–1120
14. Zverev VV, Vovna VI, Kitacv UyP, Vilesov FIJ (1975) Photoelectron specters and structure of azine, formaldehyde. Structur Chem 16(6):1032–1035
15. Kirste K, Poppek R, Rademacher R (1984) Photoelectronenspecten und Ronformatijns-verhalten von azinen. Chem Ber 117(3):1061–1068

Chapter 6
Specific Intermolecular Interactions of Methoxyamines, Methoxynitriles, and Amides

6.1 Energies of Specific Interactions of Oximes Methoxy and Ethoxy Compounds

6.1.1 *Methoxy and Ethoxy Amines*

Methoxyamines and ethoxyamines are the simplest hetero compounds with two hetero atoms of oxygen and nitrogen. The thermodynamic properties are known for a small number of these acyclic and cyclic compounds; however, one is interested in the diversity of the realized types of specific interactions in their condensed condition. Shifting of the electron density at the carbon atoms of the methyl groups leads to an insignificant negative charge [1–3], reducing the positive value as a result of the intermolecular reverse dative bond [4–8]

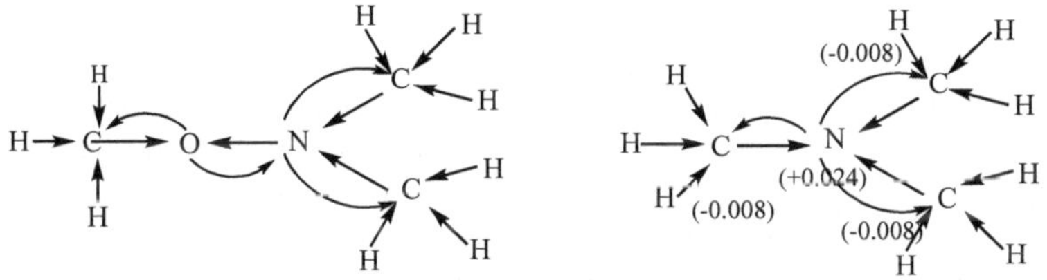

with transmission of the electron density from the $3p_z$-orbital to the essentially unshared $2s^2$ electron pair of the carbon atom [9–12]. The interaction of the nitrogen atom with the oxygen atom is accompanied by an additional change in the charges of carbon and nitrogen atoms, preserving for the last at least the same value of positive charge. Since the stability of the specific intermolecular interactions is correlated with the difference in the charges of bond vacancies, the energies of which are correlated with the vaporization enthalpies [13], it is of interest to study the correlation of the enthalpy characteristics of symmetric ether and alkylamines. This dependence allows us to obtain information on the relative

A.K. Baev, *Specific Intermolecular Interactions of Nitrogenated and Bioorganic Compounds*, DOI 10.1007/978-3-642-37472-2_6,

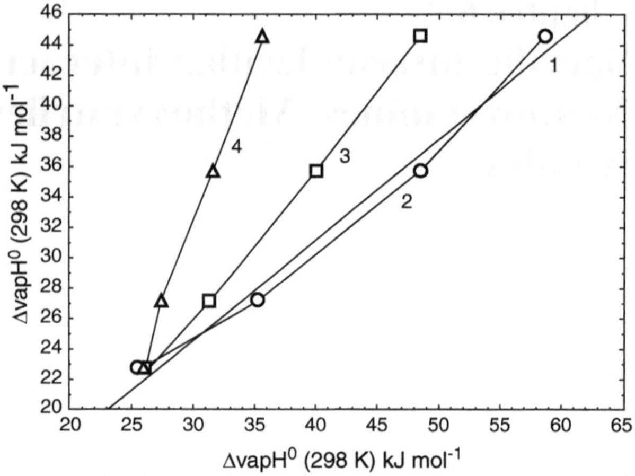

Fig. 6.1 Relation between vaporization enthalpy of symmetrical enters versus enthalpy versus enthalpy vaporization of alrylamines: (*1*) – dependency is described with using average quadratic error of enthalpy vaporization OR_2 versus NR_3, (*2*) – OR_2 versus NR_3, (*3*) – OR_2 versus NR_2H, (*4*) – OR_2 versus NRH_2

contributions of the energies, formed by amine and metoxy groups (Fig. 6.1). The special position in the figure of methylamines is caused by the endo effect of coordination of the carbon atoms of the methyl groups, which are directly connected with the nitrogen atom. The further increase of the number of the methylene groups in the alkyl chain is accompanied by stabilization of the specific interaction combined with the reduction of the influence of the reverse dative bond.

As a result, the dependence shown in Fig. 6.1 describes the enthalpy characteristics of compounds with ethyl – butyl fragments [14] as a straight line. The best correspondence of the linear dependence of the vaporization enthalpy of ethers and alkylamines is observed in compounds with one and two alkyl groups in the amine molecules. It follows that the energy contributions of the specific interactions formed by alkyl fragments of ethers and amines differ a little and for the corresponding pairs of compounds contribute equal energies to the enthalpy characteristics. From this we can conclude that the energies of the specific interactions, formed by ether fragments in liquid ethers and metoxy and ethoxy compounds on the one side and, on the other side, by an alkyl fragment in liquid alkylamines; corresponding metoxy and ethoxy compounds contribute practically equally, the values of which are within experimental error. Thus, we have reason to use the energies of the specific interactions $D–O \rightarrow CH_3–O = 5.63$ and $D–N \rightarrow CH_3–N = 4.25$ kJ mol^{-1} and furthermore of the fragments with a larger number of carbon atoms in the alkyl chain (C_2H_5, C_3H_7) for the determination of the energies of other types of specific interactions formed with hydrogen atom of ethers $D–O \rightarrow CH_3–N$, and $D–O \rightarrow C_2H_5–N$, $D–O \rightarrow C_3H_7–N$. Confirmation of this conclusion can highlight the high difference of inductive effect and field effect [15] of $ON(CH_3)_3$ (K), indicating the "pushing" of B3MO orbitals of the methyl groups. However, in compounds with phosphorus OPR_3 the low-lying vacant o*(PR)- and P3d-orbitals stabilize O2p-orbitals at 1.0–1.5 eV.

The *N*-methoxydimethyl amine molecules with a metoxy group and two methyl groups form the three differing types of specific interactions of the essentially unshared $2s^2$-electron pair of the carbon atom with the electron pair of the nitrogen atom and oxygen atom: 2D–O → CH_3–O, 2D–N → CH_3–N, and 2D–O → CH_3–N. . As a result the structure of the liquid is formed with a network of the specific interactions with a pentacoordinated carbon atom and four-coordinated nitrogen and oxygen atoms (Fig. 6.2a), in which the energy of the specific interaction D–O → CH_3–N with unknown energy is determined by the inequality of the vaporization enthalpy with the total energy contribution of all six specific interactions [16–18] shown in Eq. 6.1

$$\text{D-O}\rightarrow\text{CH}_3\text{-N} = (\Delta_{\text{vap}}H^0(298\text{ K})\text{mdm} - 2\text{D-N}\rightarrow\text{CH}_3\text{-N} - 2\text{D-O}\rightarrow\text{CH}_3\text{-O})/2 \quad (6.1)$$

The obtained value of the energy of the specific interaction DO → CH_3–N

$$\text{D-O}\rightarrow\text{CH}_3\text{-N} < \text{D-N}\rightarrow\text{CH}_3\text{-N} < \text{D-O}\rightarrow\text{CH}_3\text{-O}$$

illustrates its reduced stability, which is consistent with the conclusions given above (Table 6.1).

The presence of the metoxy group at location 2 and the hydrogen atom at location 1 for the 2-methoxyethylamine molecule leads to the formation of the hydrogen bond D–O•••H–N and the structure of the liquid and crystalline condition of this compound in the network including the given type of hydrogen bond and specific interactions D–N → CH_3–CH_2–N and D–O → CH_3–O (Fig. 6.2b). Taking into account that the ethyl group of ethylamine forms the specific interaction DN → CH_3–CH_2–N, with an energy of 6.00 kJ mol^{-1} (Chap. 2) and D–O → CH_3–O (5.63 kJ mol^{-1}), we obtain from Eq. 6.1a the energy of the hydrogen bond,

$$\text{D-O}\cdots\text{H-N} = (\Delta_{\text{vap}}H^0(298\text{ K})\text{meam} - 2\text{D-N}\rightarrow\text{CH}_3\text{-CH}_2\text{-N} - 2\text{D-O}\rightarrow\text{CH}_3\text{-O})/2 \quad (6.1a)$$

A molecule of 3-methoxypropylamine with metoxy and propyl groups at locations 3 and 2, respectively, also forms the hydrogen bond and two types of the specific interactions D–O → CH_3–O and D–N → CH_3–CH_2–CH_2–N. One of these interactions is formed with the same energy value of 5.63 kJ $mol^{-1}$and the second one with the energy of 8.1 kJ mol^{-1} (Chap. 2). That is why the energy of the hydrogen bond should be determined with the help of Eq. 6.1a. The results of the calculations (Table 6.1) point to the stabilization of the hydrogen bond, caused by weakening of the influence of the intermolecular hydrogen bond on the stabilization of the specific interaction with increasing number of methylene groups in the alkyl chain, which allows us to describe the stabilizing effect of these interactions D–O•••H–N by the integral value

$$\text{2-Methoxymethylamine} < \text{2-Methoxyethylamine } (7.8) < \text{3-Methoxypropylamine } \left(8.5\text{ kJ mol}^{-1}\right)$$

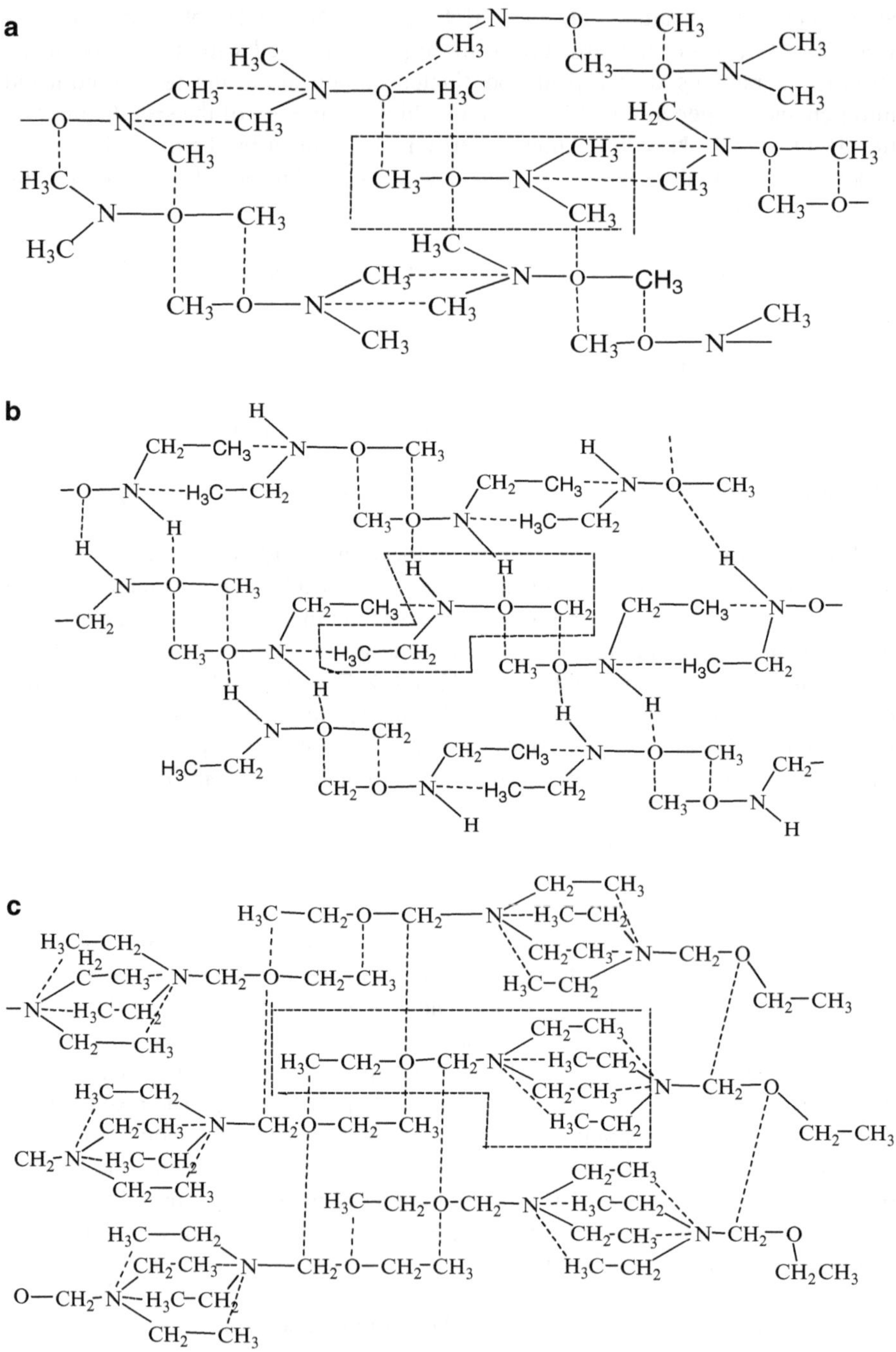

Fig. 6.2 Schematic picture of the liquid and crystalline structure of *N*-methoxydimethylamine (**a**), 2-methoxyethylamine (**b**) and *N*-(ethoxymethyl)diethylamine (**c**) with network of the specific interactions

Table 6.1 Energies of specific interactions (kJ mol^{-1}) of the liquid methozy- and ethoxyamines

Compounds	Formula	Structure	$\Delta_{vap}H°$ (T) [14]	T K	D–N → CH_3–(CH_2)n–N	D–O → CH_3–(CH_2)n–O	D–O → CH_3–(CH_2)n–N	D–O•••H–N
N-Methoxydimethyl amine	C_3H_9NO	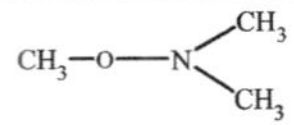	28.0	298	n = 0 4.25	n = 0 5.63	n = 0 4.15	–
2-Methoxymethylamine	C_2H_7NO	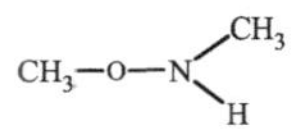	30.3[a]	298	n = 0 4.25	n = 0 5.63	–	5.3
2-Methoxyethylamine	C_3H_9NO	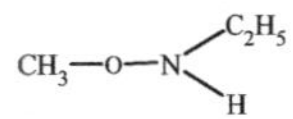	38.8	293	n = 1 6.00	n = 0 5.63	–	7.8
3-Methoxy propylamine	$C_4H_{11}NO$	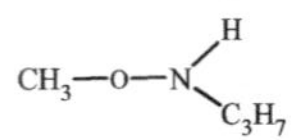	44.5	293	n = 2 8.1	n = 0 5.63	–	8.5
N-(Methoxymethyl) diethylamine	$C_6H_{15}NO$	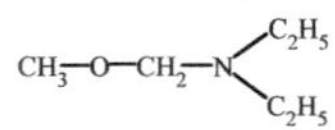	38.0 ± 0.1	302	n = 1 6.00 × 4	n = 0 5.63	D–O → CH_2–N 2.75	–
N-(Ethoxymethyl) diethylamine	$C_7H_{17}NO$	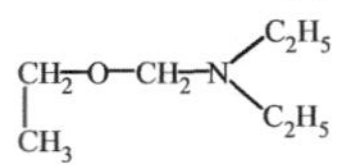	39.8	300	n = 1 6.00 × 4	6.50	D–O → CH_2–N 2.75	–
Morpholine	C_4H_9NO	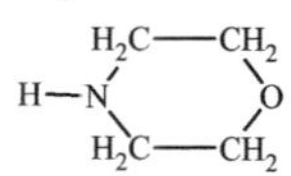	45.0 ± 0.4	298	–	DH_2C → CH_2 = 5.7	DO → CH_2–O = 5.75	5.30
N Methylmorpholine	$C_5H_{11}NO$	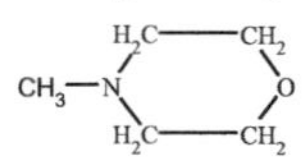	39.6 ± 0.2	298	4.25	5.7	5.75	D–N → CH_3–N = 2.6
N Ethylmorpholine	$C_6H_{13}NO$	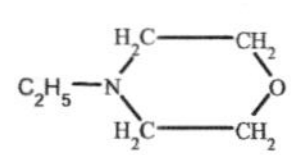	42.3 ± 0.3	298	DH_2C → CH_2 = 5.50	5.7	5.75	D–N → CH_3–CH_2–N = 4.0

[a]Elucidated

Taking the value of the energy of the hydrogen bond of liquid 2-methoxymethylamine as equal to the energy of the same bond in liquid ammonia with a four-coordinated nitrogen atom, equal to 5.3 kJ mol^{-1}, we estimated the vaporization enthalpy of this compound to be to 30.3 kJ mol^{-1}.

In a *N*-(methoxymethyl)diethylamine molecule the methylene group located between nitrogen and oxygen atoms of the fragment (methoxymethyl) expresses its bond vacancy and forms the specific interaction with the oxygen atom – D–O → CH_3–N. The appearance of an additional bond vacancy is accompanied by the transition of a nitrogen atom to the pentacoordinated condition. As a result the structure of the liquid condition with the network of the specific interactions includes an additional third type of interaction D–O → CH_3–N (Fig. 6.2c). Ethyl groups of the amine fragment form four equal specific interactions (6.00 kJ mol^{-1}), forming the tetragonal-bipyramidal structure. Thus, the structure of the liquid condition of *N*-(methoxymethyl)diethylamine and similarly the structure of *N*-(ethoxymethyl)diethylamine are more complex. The energy of the specific interaction, formed by the methylene group of the amine fragment should be determined by the equality of the vaporization enthalpy and total energy contribution of the specific interactions:

N-(Methoxymethyl)diethylamine

$$\text{D-O}\rightarrow\text{CH}_2\text{-N} = (\Delta_{vap}H^0(298\ \text{K})\text{mmde} - \text{D-O}\rightarrow\text{CH}_3\text{-O} - 2\text{D-N}\rightarrow\text{CH}_3\text{-CH}_2\text{-N})/4 \quad (6.1b)$$

The value of the energy of this type of interaction, formed by the methylene group, should be kept unchanged in the structure of liquid *N*-(ethoxymethyl) diethylamine. Hence, we have reason to use the similar equality in the transformed form, Eq. 6.1c to determine the energy of the specific interaction formed by the ethoxy group *N*-(ethoxymethyl)diethylamine

$$\text{D-O}\rightarrow\text{CH}_3\text{-CH}_2\text{-O} = (\Delta_{vap}H^0(298\ \text{K})\text{emde} - 4\text{D-N}\rightarrow\text{CH}_3\text{-CH}_2\text{-N} - 2\text{D-O}\rightarrow\text{CH}_2\text{- N})/2 \quad (6.1c)$$

The results of the calculations of the specific interactions energies, formed by the ethoxy group (6.5 kJ mol^{-1}), point to reduced values compared with the value of the energy of a similar type of interaction formed by ethyl ligand of diethyl ether (6.9 kJ mol^{-1}), which could be connected with the experimental error of the measurement of the vaporization enthalpy of *N*-(ethoxymethyl)diethylamine, given in the literature without indicating of error value. The energies of the specific interactions formed by metoxy and ethoxy groups are stabilized with the weakening of the influence of the reverse dative bond with an increasing number of carbon atoms in the alkyl ligand D–O → CH_3–O (5.63) < D–O → CH_3–CH_2–O (6.5 kJ mol^{-1}). The fact of the reduced energy deserves attention because of the specific interaction formed by the methylene group D–O → CH_2–N (2.75), located between electron acceptors of the nitrogen and oxygen atoms in *N*-(methoxymethyl)diethylamine and 3-methoxypropylamine molecules compared with the open methyl group, 4.25 kJ mol^{-1}, or hypothetically open CH_2 group. In this case it is necessary to account for the contribution of the hydrogen atom for the liquid condition 0.60 kJ mol^{-1}, and we obtain the difference in the contribution with the hypothetical methylene group, 3.60 kJ mol^{-1}. From this, one can draw the conclusion that on replacement of the electron density from the methylene group simultaneous with the acceptor electrons of nitrogen and oxygen there appears a reduction in the difference of the charges in the given elements and the carbon atom of the methylene group.

The morpholine molecule and its derivatives *N*-methylmorpholine and *N*-ethylmorpholine with two hetero atoms in the limit cycle lead to the increased stability of the specific interactions $D\text{–}H_2C \rightarrow CH_2$, formed by the CH_2 groups. At the small difference of the energy of the specific interaction $D\text{–}N \rightarrow CH_3\text{–}N = 4.25$ and the hydrogen bond D–N•••H–N = 5.30 kJ mol^{-1}, the difference of the vaporization enthalpies of morpholine and *N*-methylmorpholine at 5.4 kJ mol^{-1} (Table 6.1) points to the fact that the methyl group of the amine group forms the specific interaction with a significantly reduced value compared with the marked one for methylamine.

Molecules of these compounds with an equal number of bond vacancies form eight specific interactions, including six interactions of two types by oxygen atoms $4D\text{–}O \rightarrow CH_2\text{–}O$ and CH_2 groups, two $D\text{–}H_2C \rightarrow CH_2$ with molecules that are close in the environment and two hydrogen bonds by molecules of $D\text{–}N \rightarrow CH_3\text{–}N$ and $D\text{–}N \rightarrow CH_3\text{–}CH_2\text{–}N$, respectively, *N*-methylmorpholine and *N*-ethylmorpholine. The planar structure of molecules of these compounds forms the network structure of the liquid and crystal condition, of which chains are crisscrossed by stable specific interactions of $D\text{–}H_2C \rightarrow CH_2$ (Fig. 6.3). The value of the energy of the specific interaction, formed by the methylene groups in the compounds with a saturated cycle, is equal to the one realized in liquid hexane (5.50 kJ mol^{-1}) (Chap. 4). The energy of the specific interaction $D\text{–}O \rightarrow CH_2\text{–}O$ morpholine and its derivatives is determined by Eq. 6.1d

$$D\text{-}O{\rightarrow}CH_2\text{-}O = (\Delta_{vap}H^0(298\ K)mph - 2D\text{-}H_2C{\rightarrow}CH_3\text{-} - 2D\text{-}N{\bullet\bullet\bullet}H\text{-}N)/4 \quad (6.1d)$$

The obtained energy value (5.7 ± 0.4 kJ mol^{-1}) of liquid morpholine (Table 6.1) exceeds the energy of the same type of interaction of methyl ether (5.63 kJ mol^{-1}) at the accounted energy contribution of the saturated hydrogen atom (0.60 kJ mol^{-1}), which refers to the stabilizing influence of the hetero atom of nitrogen shifting the electron density, and increasing the difference in the charges of the oxygen atom and hydrogen atoms of the amino group. Using the value of the energy of this type of interaction, the energy contribution of the specific interactions, formed by the interaction of the methyl and ethyl group of *N*-methylmorpholine and *N*-ethylmorpholine is determined with the help of Eq. 6.1e

$$D\text{-}N{\rightarrow}R\text{-}N = (\Delta_{vap}H^0(298\ K)mph - 2D\text{-}H_2C{\rightarrow}CH_3 - 4D\text{-}O{\rightarrow}CH_2\text{-}O)/4 \quad (6.1e)$$

The results of the calculations of the energies of the specific interactions (Table 6.1) illustrate the reduced stability of the interactions $D\text{–}N \rightarrow CH_3\text{–}N$ (2.6) < $D\text{–}N \rightarrow CH_3\text{–}CH_2\text{–}N$ (4.0 kJ mol^{-1}).

Fig. 6.3 Schematic picture of the liquid structure of morpholine with network of the specific interactions

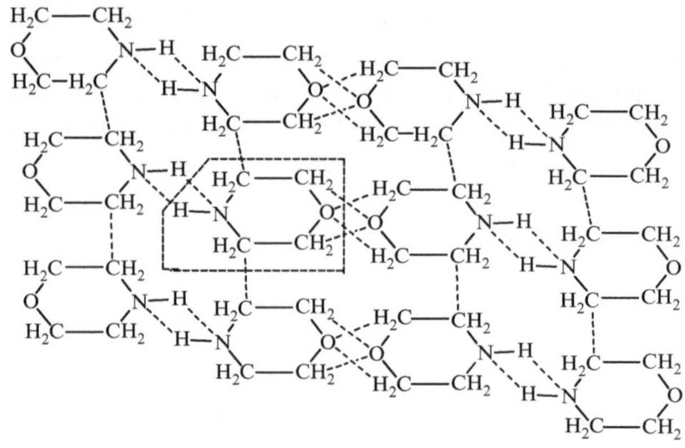

6.1.2 Methoxynitriles

The results of a thermodynamic analysis of metoxyamines and ethoxyamines point to the fact that the introduction to the molecule of alkylnitriles of methoxy and ethoxy fragments with the oxygen atom determines the energy of the specific interaction with the energy value equal to the one which is formed by the same fragment in symmetric simple ethers. With the characteristic high rigidity of the triple bond of fragment C≡N, alkylnitriles form the specific intermolecular interaction ≡N → C≡ with the energy of liquid acetonitrile (7.15 kJ mol^{-1}), reaching its maximum value at the liquid propionitrile (11.7 kJ mol^{-1}) in connection with the completion of the reverse dative bond with an increasing number of the methylene groups in the alkyl chain. This energy value remains practically constant at the further saturated and unsaturated nitriles (Chap. 2) and increase of the enthalpy characteristics is due to the energy contribution of the increasing number of methylene groups in the alkyl ligand. The presence of oxygen in the alkyl fragment of methoxyacetonitrile located between two carbon atoms defines it with the properties of a symmetric dimethyl ether, forming four similar specific interactions O → CH_3–O with molecules that are close contact in the environment (Fig. 6.4) with an energy equal to 5.63 kJ mol^{-1}.

Depending on the number of methylene groups in the alkyl chain and in the methoxy-, ethoxy-, or butoxy-group specific interactions with four similar chain lengths or pair interactions (Fig. 6.4b)are formed. Isostructural methyl groups of methoxymethylalkylnitriles form specific interactions of low stability. Thus, the compounds of this series form the structure of the liquid condition with the network of the specific interactions, the energies of which are determined by the difference of the vaporization enthalpy with the known total value of the energies of the second type of interaction with the help of Eqs. (6.2a, b),

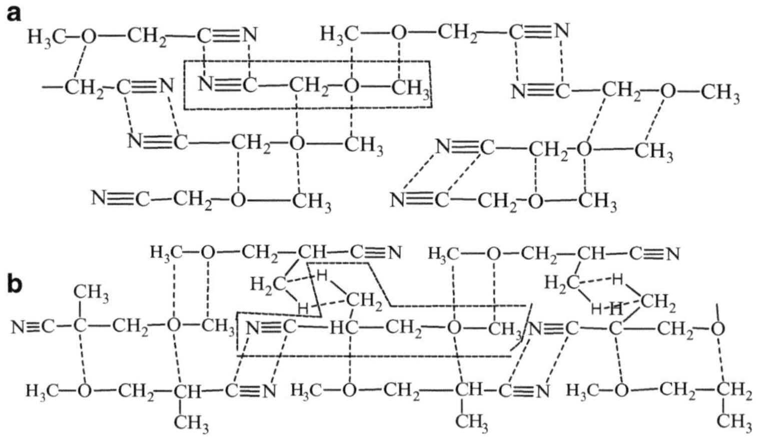

Fig. 6.4 Schematic picture of the liquid structure of methoxyacetonitrile (**a**) and α-methoxyisobutironitrile (**b**) with network of the specific interactions

$$D{\equiv}N{\rightarrow}C{\equiv} = (\Delta_{vap}H^0(298\ K)mac - 4D\text{-}O{\rightarrow}CH_3\text{-}O)/2 \qquad (6.2a)$$

$$D{\equiv}N{\rightarrow}C{\equiv} = (\Delta_{vap}H^0(298K) - 2D\text{-}O{\rightarrow}CH_3\text{-}(CH_2)n\text{-}O - 2D\text{-}O{\rightarrow}CH_3\text{-}(CH_2)q\text{-}O)/2 \qquad (6.2b)$$

The thermodynamic calculations of the energies of the specific interactions of methoxyacetonitriles (Table 6.2) allow us to draw the conclusion that saturated and unsaturated nytriles of metoxynitriles alike are characterized by the tendency to stabilize the specific intermolecular interaction $D{\equiv}C \leftarrow N{\equiv}$ with an increasing number of CH_2 groups of the alkyl chain of fragments reaching a maximum value

$$\begin{aligned}&\text{Methoxyacetonitrile}(9.6) < \text{Ethoxyacetonitrile}(10.7)\\&< \text{3-Methoxypropionenitrile}(11.4) \approx\\&\text{Propionitrile}\left(11.7\ \text{kJ mol}^{-1}\right)\end{aligned}$$

at the compound with the propyl ligand, with no significant difference from the energy value of the same type of liquid propionitrile. It suggests that the established value of the energy of the specific interaction $D{\equiv}N \rightarrow C{\equiv}$ of 3-Methoxypropionenitrile at 343 K has no significant difference from the standard value.

Using this value of the specific interaction $D{\equiv}C \leftarrow N{\equiv}$, we correlate the vaporization enthalpies of other compounds to the standard conditions (Table 6.2). Using the energy contribution of the isostructural methyl group to the vaporization enthalpy of isobutironitrile (2.20 kJ mol^{-1}), and the energies of the specific interactions, formed by methyl group of metoxy fragment and propyl ligand of the isobutyl fragment α-methoxyisobutironitrile, corrections were made to the

Table 6.2 Energies of specific interactions (kJ mol^{-1}) of the liquid methoxynitriles

Compounds	Formula	Structure	$\Delta_{vap}H°(T)$ [14]	T K	D–O → CH_3–(CH_2)n–O	D–O → CH_3–(CH_2)n–O	D≡N → C≡N
Methoxyacetonitrile	C_3H_5NO	H_3C—O—CH_2—C≡N	41.7 ± 0.6	298	n = 0; 5.63	n = 1; 5.63	9.6
Ethoxyacetonitrile	C_4H_7NO	CH_3—CH_2—O—CH_2—C≡N	46.5 ± 0.3	298	n = 1; 6.78	n = 1; 5.63	10.7
3-Methoxypropionenitrile	C_4H_7NO	CH_3—O—$(CH_2)_2$—C≡N	47.6	343	n = 0; 5.63	n = 2; 6.78	11.4[a]
2-Ethoxypropanenitrile	C_5H_9NO	CH_3—CH_2—O—(CH_2)—C≡N	46.7 50.0[b]	363	n = 1; 6.78	n = 2; 6.78	10.8[a]
2-butoxypropionitrile	$C_7H_{13}NO$	CH_3—$(OCH_2)_3$—$(CH_2)_2$—C≡N	46.2 54.2[b]	380	n = 3; 8.92	n = 2; 6.78	5.4[a] 11.4[b]
					D–O → CH_3–O		
α-Methoxyisobutironitrile	C_5H_9NO	H_3C—O—CH_2—CH(CH_3)—C≡N	37.4 ± 0.8 49.8[b]	298	5.63 D*iso*CH_3 2.2	n = 2; 6.78	3.2 11.4[a]
2-Methoxy-3, 3-dimethylbutanenitrile	$C_7H_{13}NO$	CH_3—O—CH_2—C(CH_3)₂—CH_2—C≡N	58.8 ± 1.1	298	5.63 D*iso*CH_3 3.45 × 2	n = 3 8.92	11.4[a]
2-Methoxy- 2-methylpentanenitrile	$C_7H_{13}NO$	CH_3—O—$(CH_2)_3$—CH(CH_3)—C≡N	48.5 ± 0.6 55.0[b]	298	5.63 D*iso*CH_3 2.2	n = 4 8.92 DCH_2 = 4.15	7.75

[a]It is not counted up the temperature dependency of enthalpy vaporization
[b]At Standard conditions

value of the vaporization enthalpy of this compound and 2-methoxy-2-methylpentanenitrile (Table 6.2). We draw attention to the high value of the energy, contributed by two isostructural methyl groups of metoxy-3,3-dimethylbutanenitrile, located at the third carbon atom, compared with its location at the second carbon atom in the pivalonitrile molecule.

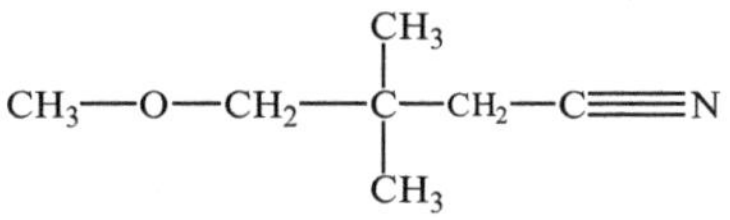

D*iso*CH_3 3.45 × 2 kJ mol^{-1}
Methoxy-3,3-dimethylbutanenitrile

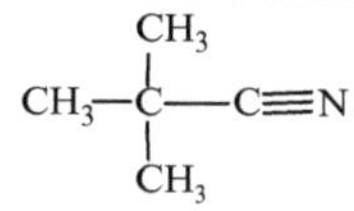

D*iso*CH_3 1.25 × 2 kJ mol^{-1}
Pivalonitrile

It follows from these data that removal of the isostructural methyl groups from nitrile to the third carbon atom of the alkyl chain is accompanied by a significant change in the shifting of the electron density in the molecule in the presence of the dimethyl fragment with high acceptor properties of the oxygen atom in the chain. In this connection there are doubts with regard to the correctness of such a sharp reduction in the values of the experimentally measured vaporization enthalpies for α-methoxyisobutironitrile and 2-methoxy-2-methylpentanenitrile with one isostructural methyl group located at the second carbon atom in the chain compared with the estimated value on the base of additive energy contributions of the specific interactions (Table 6.2).

6.2 Energies of Specific Interactions of Liquid Amides and Its Derivatives

6.2.1 Formamides

As was shown above the introduction of the hetero atom of oxygen to the alkylamines is accompanied by stabilization of the hydrogen bond, formed by the hydrogen bond of the amino group with the oxygen atom D–O•••H–N, caused by the increased negative charge of the latter. The energies of the hydrogen bond, formed by the oxygen atom 2-methoxyethylamine (7.60 kJ mol^{-1}) and carboxyl oxygen atom of aldehyde (7.85 kJ mol^{-1}), have comparable values [13].

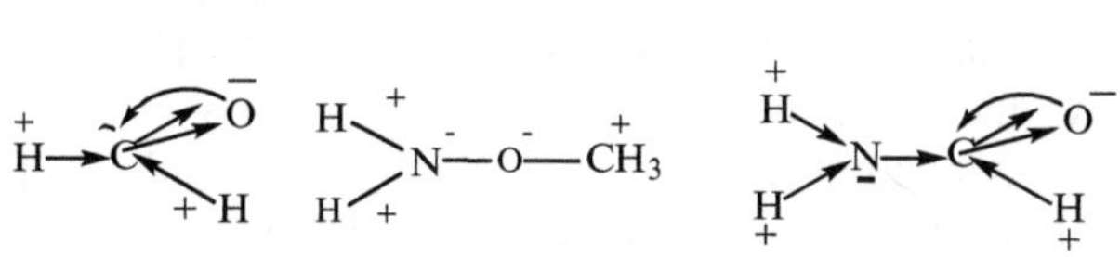

This implies that the amino group has a significant influence on the charge of the oxygen atom. Hence, the shifting of the electron density in the formamide molecule

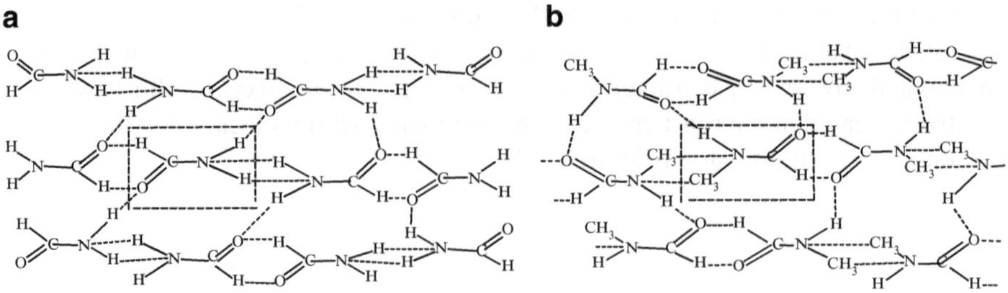

Fig. 6.5 Schematic picture of the liquid and crystalline structure of formamide (**a**) and *N*-methylformamide (**b**) with network of the specific interactions

leads to the change in the charge at the oxygen and hydrogen atoms, increasing its difference and stabilizing the forming interaction. The formamide molecule and its derivatives with six bond vacancies form six specific interactions. In the case of formamide the three hydrogen atoms, nitrogen atom, and oxygen atom with one and two electron pairs, respectively, form six hydrogen bonds of three types with the molecules that are close in the environment, forming the structure of the liquid condition with a network of hydrogen bonds D–N•••H–N, D=C=O•••H–N and D=C=O•••H–C=(Fig. 6.5a). The *N,N*-dimethylformamide molecule forms the structure of the liquid condition with a network of the specific interactions of the essentially unshared $2s^2$ electron pairs of the carbon atom of the methyl group D–N → CH_3–N, D=C=O → CH_3–N, and hydrogen bond D=C=O•••H–C=O. In liquid *N*-methylformamide and *N*-ethylformamide with one methyl and ethyl group, respectively, there forms the structure with the network of the specific interaction D–N → CH_3–N and hydrogen bonds D=C=O•••H–N and D=C=O•••H–C=O (Fig. 6.5b).

Taking into account the fact that hydrogen atoms of the amino group have higher positive chargesthan the hydrogen atoms of the carboxyl group, and considering the increased negative charge of the oxygen atom compared with the nitrogen atom, the further equality in the energies of the hydrogen bonds can be accepted:

$$\text{D-N}\cdots\text{H-N} < \text{D=C=O}\cdots\text{H-N} < \text{D=C=O}\cdots\text{H-C=O}.$$

This sequence in the change of bond stability occurs in the specific interactions formed for liquid *N,N*-dimethylformamide D–N → CH_3–N < D=C=O → H_3C–N and liquid *N*-methylformamide and *N*-ethylformamide D–N → CH_3–N (4.8) < D=C=O → H_3C–CH_2–N= (6.7 kJ mol^{-1}). The third type of hydrogen bonds in liquid formamide D=C=O•••H–C=O, D=C=O•••H–N, and D–N•••H–N is replaced by the specific interaction D–N → CH_3–N and D=C=O → H_3C–CH_2–N= in *N*-methylformamide and *N*-ethylformamide, respectively. The structure of liquid *N,N*-dimethylformamide includes two types of specific interactions, D–N → CH_3–N and D=C=O → H_3C–N and the most stable hydrogen bond D=C=O•••H–C=O.

It follows that the difference in the values of the vaporization enthalpies of the compounds considered depends on the energy of the realized types of specific interactions and hydrogen bonds (Table 6.3) introduced. Consequently, the replacement of the hydrogen atom in the amino group of formamide at the CH_3 group is accompanied by the replacement of the stable hydrogen bond D–N•••H–N by the specific interaction DN → CH_3–N with the energy of 4.8 kJ mol^{-1} and the formation of *N*-methylformamide. Hence, the difference in the vaporization enthalpies of these compounds, $\Delta_{vap}H°(298\ K)fa - \Delta_{vap}H°(298\ K)mfa$ (5.8 kJ mol^{-1}), corresponds to the difference in the energies of the two bonds of these types of interactions or DN → CH_3–N (4.8) < D–N•••H–N (7.7 kJ mol^{-1}). Replacement of the hydrogen atom in the amino group of *N*-methylformamide by the second methyl group occurs with the replacement of the stable hydrogen bond DC=O•••H–N by the specific interaction D=C=O → CH_3–N with reduced energy and formation of *N,N*-dimethylformamide with the smaller value of the vaporization enthalpy. This implies that the difference in the vaporization enthalpies of compounds $\Delta_{vap}H°(T)$ mfa – $\Delta_{vap}H°(T)$dmfa (7.5 kJ mol^{-1}) *N*-methylformamide and *N,N*-dimethylformamide corresponds to the difference in the stability of two specific interactions and two hydrogen bonds D=C=O•••H–N (9.0 kJ mol^{-1}). The energy of the hydrogen bond D=C=O•••H–N of liquid *N*-ethylformamide should be determined with the help of Eq. 6.3

$$D{=}C{=}O{\cdots}H\text{-}N = (\Delta_{vap}H^0(298\ K)ef - 2D\text{-}O{\rightarrow}CH_3\text{-}CH_2\text{-}O - 2D{=}C{=}O{\cdots}H\text{-}C{=}O)/2 \quad (6.3)$$

The results of the thermodynamic analysis (Table 6.3) allow us to draw the conclusion that the energies of the hydrogen bonds D=C=O•••H–C=O, D=C=O•••H–N and D–N•••H–N in liquid formamide and its derivatives are significantly stabilized under the mutual influence of the hetero atoms of nitrogen and oxygen compared with similar interactions of oxygen and nitrogenous compounds, respectively. Methyl and ethyl groups of the amide fragments form specific interactions with four-coordinated nitrogen atoms and oxygen atoms, the energies of which are close to the energies of the interactions of liquid methyl and ethyl ethers.

6.2.2 Acetamides

The acetamide molecule, directly connected with the nitrogen atom of the amino group and with the carbon atom of the carbonyl group, has a definite similarity to formamide and also forms six specific interactions by six bond vacancies. However, the presence of the acetate group and absence of aldehyde hydrogen atom in the acetamide molecule contributes greatly to the contrast in the formamide properties. In this connection we draw attention to the fact that movement of the hydrogen

Table 6.3 Energies of the hydrogen bonds and specific interactions (kJ mol^{-1}) of the liquid Formamides

Compounds	Formula	Structure	$\Delta_{vap}H°(T)$ [14]	T K	D–N → CH_3–N	D=C=O•••H–N	D=C=O••• H–C=O
Formamide	CH_3NO	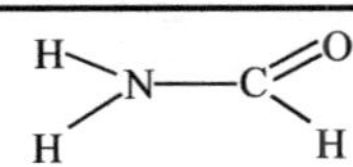	60.2	298	D–N–H•••N 7.7 × 2	9.0 × 2	13.4 ± 1.0
N-Methyl-formamide	C_2H_5NO	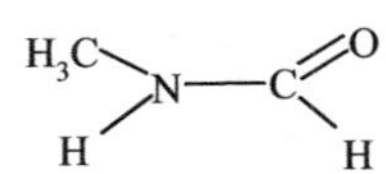	54.4 ± 1.3	298	n = 0;4.8 × 2	9.0 × 2	13.4 ± 1.0
N,*N*-Dimethyl-formamide	C_3H_7NO	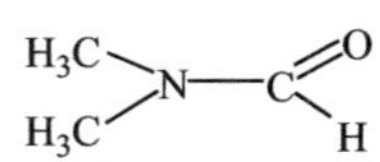	46.9	298	n = 0;4.8 × 4	D=C=O → CH_3–N 5.3 × 2	13.4 ± 1.0
N-Ethyl-formamide	C_3H_7NO	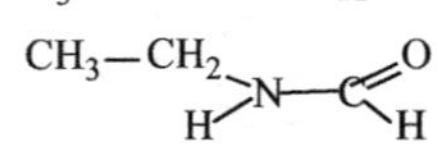	58.4	298	n = 1;6.8 × 2	9.5 × 2	13.4 ± 1.0

Table 6.4 Energies of the hydrogen bonds and specific interactions (kJ mol^{-1}) of the liquid compounds with the aldehyde group

Formaldehyde	Acetaldehyde	*O*-Methylhydroxylamine	Formamide	Acetamide
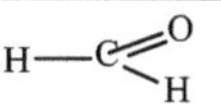	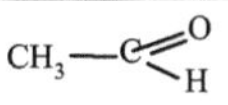	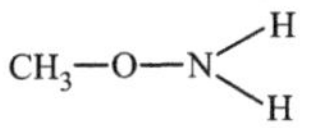	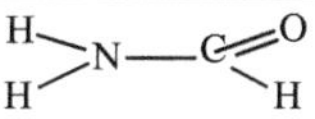	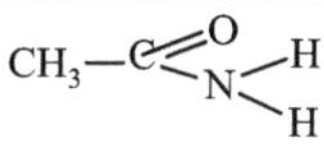
($\Delta_{vap}H°$ (T) = 24.3; 236 K	27.0; 307 K	36.3; 243 K	60.2; 298 K	64.1 K
DC=O•••H–C=O	DC=O•••H–C=O	DO•••H–N	DC=O•••H–C=O	DC=O•••H–N
6.70 kJ mol^{-1}	7.85	7.60	13.4	–

atom in formaldehyde to the methyl group in acetaldehyde is accompanied by the increase in the vaporization enthalpy of the latter at 2.7 kJ mol^{-1}, measured at the temperatures 298 and 236 K (Table 6.4). It follows that the methyl group with the essentially unshared $2s^2$ electron pairs of the carbon atom gives to the oxygen atom of the aldehyde group more electron density than the hydrogen atom of formaldehyde. In the result acetaldehyde, at the equal number of bond vacancies, forms the structure of the liquid condition with the network of specific interactions (Fig. 6.6) with more stable hydrogen bonds than formaldehyde.

It is interesting that the oxygen atom in the chain of methoxy compound with the amino group rather than the carbonyl atom of O–methylhydroxylamine forms a less stable hydrogen bond with the oxygen atom chain compared with the hydrogen atom of the aldehyde group acetaldehyde. An even more stable hydrogen bond is formed by the aldehyde group with the amide fragment in the formamide molecule (Table 6.4). It is obvious that as in acetaldehyde, the movement of the hydrogen atom of the aldehyde group in the formamide molecule to the methyl group should be accompanied by an increase of the vaporization enthalpy on preserving of the number of bond vacancies and the formed hydrogen bonds by stabilization of the latter. The available value of the vaporization enthalpy (60.9 kJ mol^{-1}) of this compound obtained at 353 K could be clarified at the standard conditions, accepting the difference in vaporization enthalpies of the related compound of *N,N*-dimethylacetamide at the standard conditions (50.2 kJ mol^{-1}) and at 386 K (45.1 kJ mol^{-1}), calculated at a temperature of about 55°. The obtained value of the enthalpy characteristics (Table 6.4) and analysis lead to the conclusion that, first, unlike in the amines the replacement of the hydrogen atom in the formaldehyde molecule by the methyl group, is accompanied by stabilization of the hydrogen bond and, second, the replacement of the hydrogen atom of the same group in the formamide molecule by the methyl group is accompanied by significant redistribution of the electron density and stabilization of the hydrogen bonds and acetate fragment.

The acetamide molecule with two hydrogen atoms of the amino group and acetate fragment forms four hydrogen bonds of two types DC=O•••H–N,

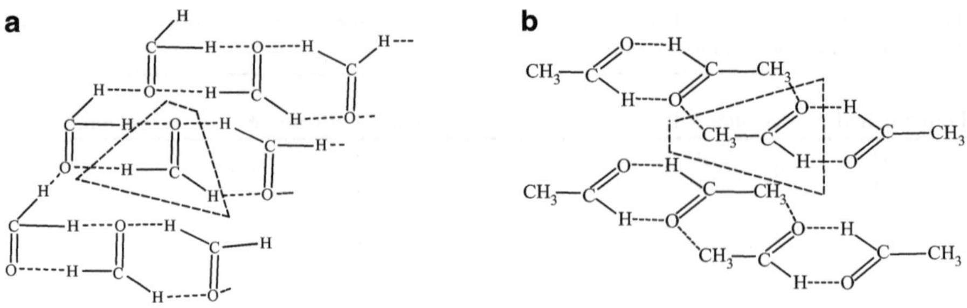

Fig. 6.6 Schematic picture of the liquid and crystalline structure of formaldehyde (**a**) and acetaldehyde (**b**) with network of the specific interactions

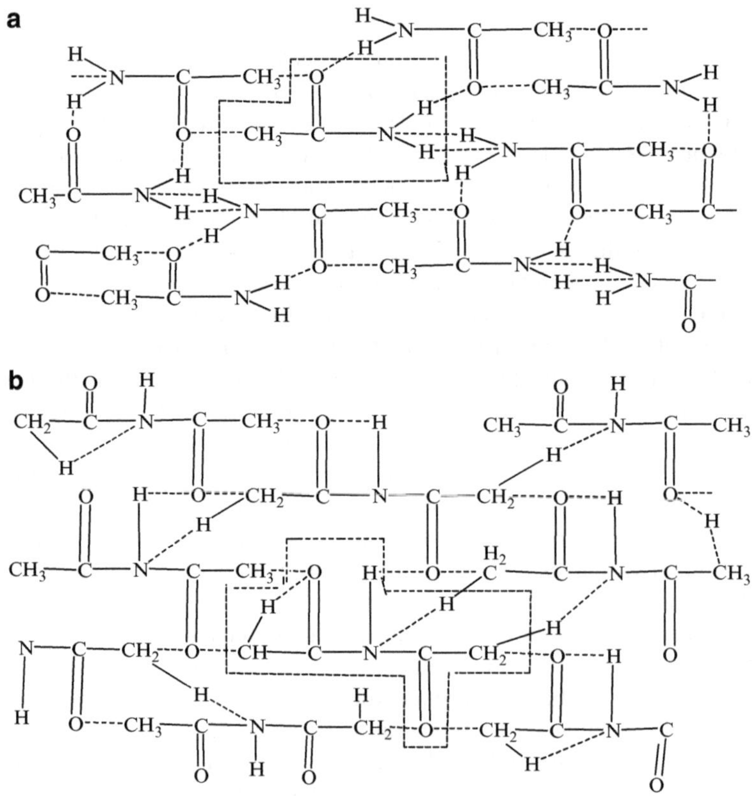

Fig. 6.7 Schematic picture of the liquid and crystalline structure of acetamide (**a**) and diacetamide (**b**) with network of the specific interactions

D–N•••H–N and two specific interactions D=C=O → CH_3–N, forming the structure of the liquid condition with the network of the specific interactions with a large display of equity of the hydrogen bonds (Fig. 6.7).

An approximately similar value of the energy contribution with the reduction of temperature from 343 to 298 K is introduced to the vaporization enthalpy of diacetamide. The extremely low value of the enthalpy characteristics (67.8 kJ mol^{-1}) of this compound with one hydrogen bond of high stability D=C=O•••H–N, two H-bonds of low stability D–N•••H–CH_2–C– and D=C=O•••H–CH_2–C, formed by hydrogen atoms of the methyl groups, and five specific interactions D=C=O → CH_3–C (Fig. 6.7b), illustrates the significant change in the shifting of the electron density and the charges at the atoms at two acetate groups and one amide group in the molecule. Consequently, for the amino group it is increased, but doubled at the two acetate fragments, providing the reduced negative charge to the nitrogen atom and increasing positive one for the hydrogen atoms, whereas on the other side, there is a reduced difference between the nitrogen and hydrogen atoms. As a result the specific interactions formed and the hydrogen bonds have less stability than those of liquid acetamide. The energy of the hydrogen bond D–N•••H–N with four-coordinated nitrogen atoms is accepted to be equal to the value of the energy of liquid ammonia (5.3 kJ mol^{-1}) and the energy of the specific interaction D–O → CH_2–N equal to the bond energy formed by liquid dimethyl ether 5.63 kJ mol^{-1}, and the energy of the hydrogen bond, formed by the aldehyde group, is determined with the help of Eq. 6.4:

$$D{=}C{=}O\cdots H\text{-}O{=}C = (\Delta_{vap}H^0(298K)dam - 4D\text{-}O{\to}CH_2\text{-}N - 2D\text{-}N\cdots H\text{-}N)/4 \qquad (6.4)$$

The energy values of the hydrogen bond are given in Table 6.4.

The presence of the amine groups at acetomides and alkylamines defines the similarity of their properties on alkylamines fragments with CH_3, C_2H_5, or R group and, on the other hand, the alkyl chain, connected with the carbonyl group. The analogue of the properties is in the manifestation of the reverse dative bond in alkyl fragments, reflected in the stability of the specific interaction with increase in the number of carbon atoms in the chains.

Trialkylamines are characterized by the weakening influence of the reverse dative bond on the energy of the specific interaction N → R–N with increasing number of carbon atoms at the transition from methyl to the methyl group and further to the propyl group (line 1) (Fig. 6.8). Further reduction of the energy contribution by the methylene group of butyl ligands reflects the completion of the influence of the reverse dative bond on the energies of the specific interactions. An increase of the enthalpy characteristics corresponds to the increasing energy contribution with an increasing number of methylene groups in the chain (2). As a result, the total dependence of the enthalpy characteristics on the number of carbon atoms is satisfactorily described by a straight line with values lying within the experimental error. A clear manifestation of the reverse dative bond appears at dialkylamines with similar changes in the vaporization enthalpy of the number of carbon atoms of the chain (line 3, Fig. 6.8). The presence of similar correlation in the chain of the amine fragment at the series of compounds of alkylacetamides

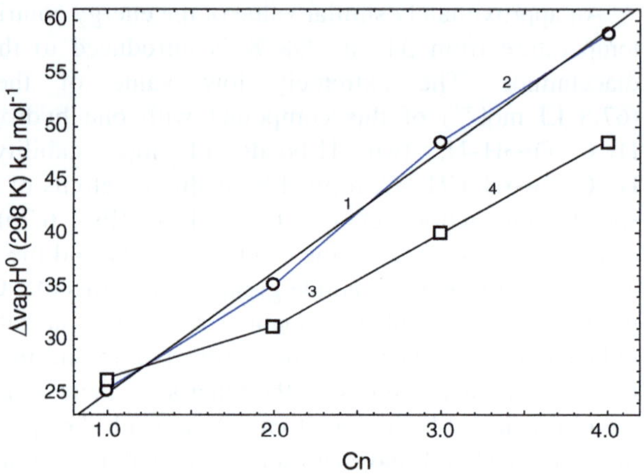

Fig. 6.8 Dependence of the vaporization enthalpy on the number of carbon atoms in the alkyl chain of trialkylamines and dialkylamines (*3*, *4*): trimethylamine – tributylamine (*1*, *2*), dimethylamine – dibutylamine (*3*, *4*)

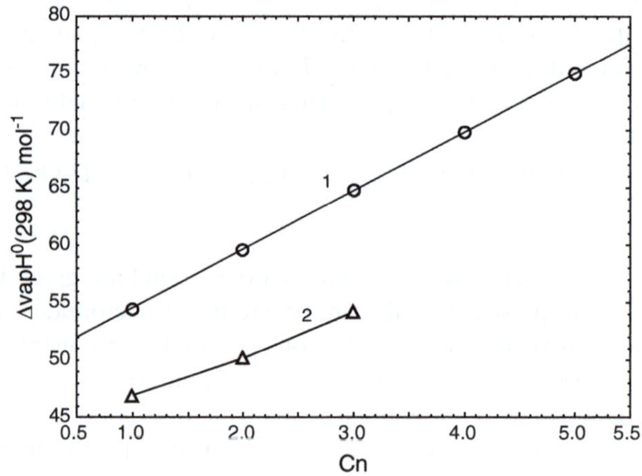

Fig. 6.9 Dependence of the vaporization enthalpy on the number of carbon atoms in the alkyl chain of *N*-methylacetamides: *N*-methylformamide – *N*-butylacetamide (*1*), *N*,*N*-dimethylformamide – *N*,*N*-dimethylacetamide – *N*,*N*-diethylacetamide (*2*)

reflects the natural change of the energy of the specific interaction, formed by methyl-butyl groups (Fig. 6.9). Ending of the influence of the reverse dative bond (line 4) at the propyl group is the experimental proof of reaching the maximum stability of the specific interaction and its constancy for all further compounds of the series.

Thus, the interrelation of the vaporization enthalpy of the liquid compound and energies of the specific interactions of the considered series is described by Eqs. 6.5a, 6.5b, 6.5c

$$\Delta_{vap}H^0(298K)am = 2DC{=}O\cdots H\text{-}N + 2DN\cdots H\text{-}N + 2DC{=}O\rightarrow CH_3\text{-}C \quad (6.5a)$$

Table 6.5 Energies of the hydrogen bonds and specific interactions (kJ mol^{-1}) of the liquid acetamides

Compounds	Formula	Structure	$\Delta_{vap}H°$ (T) [14]	T K	D–N → CH_3–(CH_2)n–N	D=C=O•••H–N	D=C=O → CH_3–C–
Formamide	CH_3NO	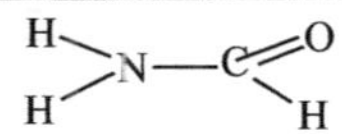	60.2	298	D–N•••H–N 7.7	9.0	D=C=O•••H–C=O 13.4 ± 1.0
Acetamide	C_2H_5NO	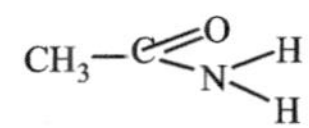	60.9 64.1[a]	353	D–N•••H–N 7.7	15.3	9.7[a]
N-Methyl-acetamide	C_3H_7NO	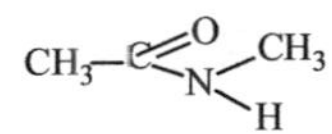	59.6 62.8[a]	348	n = 0; 7.1	15.3	9.0
N,N-Dimethyl-acetamide	C_4H_9NO	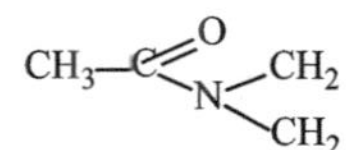	50.2 45.1	298 386	n = 0; 7.1	–	9.0
N-Ethyl-acetamide	C_4H_9NO	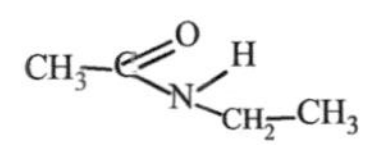	64.9 ± 0.2	298	n = 1; 8.45	15.3	9.0
N-Propyl-acetamide	$C_5H_{11}NO$	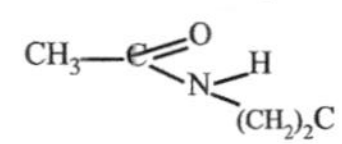	69.8 ± 0.2	298	n = 2; 10.6	15.3	9.0
N-Butyl-acetamide	$C_6H_{13}NO$	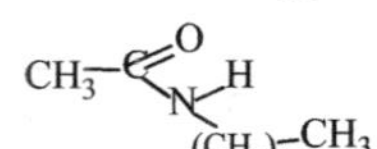	75.0 ± 0.3	298	n = 3; 10.6	15.3	9.0
N-Isopropyl-acetamide	$C_5H_{11}NO$	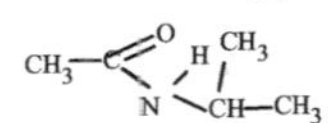	66.4 ± 0.3	298	8.45 D*iso*CH_3 = 1.5	15.3	9.0
N,N-Diethyl-acetamide	$C_6H_{13}NO$	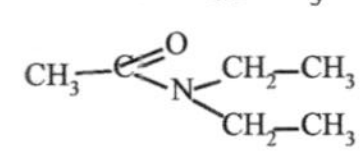	54.1	298	8.85; n = 1	D=C=O → C_2H_5–N = 9.15	9.0
Diacetamide	$C_4H_7NO_2$	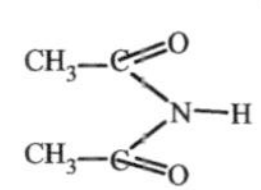	64.6 67.8[a]	358	D–N•••H–N = 5.3	D=C=O → CH_2–N = 5.63 × 4	8.6

[a]Elucidated at 298 K

$$\Delta_{vap}H^0(298K)mam = 2DC{=}O{\bullet\bullet\bullet}H\text{-}N + 2N{\to}CH_3\text{-}N + 2DC{=}O{\to}CH_3\text{-}C \quad (6.5b)$$

$$\Delta_{vap}H^0(298K)dmam = 2DC{=}OCH_3\text{-}N + 2N{\to}CH_3\text{-}N + 2DC{=}O{\to}CH_3\text{-}C \quad (6.5c)$$

The energy of the hydrogen bond DN•••H–N of liquid acetamide is accepted as being equal to the energy value of the same bond of liquid formamide (7.7 kJ mol^{-1}). Using Eqs. 6.5a, 6.5b, and 6.5c and analyzing the vaporization enthalpy of the compounds (Table 6.5), we draw the conclusion that the difference (1.3 kJ mol^{-1}) in the vaporization enthalpy of acetamide (64.1) and *N*-methylacetamide (62.8 kJ mol^{-1}) corresponds to the difference between the energy values of the two hydrogen bonds D–N•••H–N and two specific interactions D–N → CH_3–N. Thus, we obtain the energy of the specific interaction to be equal to the value 7.1 kJ mol^{-1}. A similar difference (13.9 kJ mol^{-1}) in the vaporization enthalpy with *N,N*-dimethylacetamide reflects the replacement of the four hydrogen bonds of two types of acetamide by the four specific interactions of two types D=C=O → CH_3–N and D–N → CH_3–N (7.1 kJ mol^{-1}). It follows that the energy of this hydrogen bond D=C=O•••H–N is equal to the value of 15.3 kJ mol^{-1}. Calculated with the use of Eq. 6.5, the values of the energies of the specific interactions are given in Table 6.5. They are calculated on the basis of the definition of the specific interactions, formed by ethyl and propyl ligands of the amino group and the energies of the specific interactions, formed by methyl groups of the amine fragment of *N*-ethylacetamide and *N*-propylacetamide with the help of Eq. 6.6, where R–C_2H_5, C_3H_7.

$$D{=}C{=}O{\to}R\text{-}N = (\Delta_{vap}H^0(298K)acam - 2D{=}C{=}O{\to}CH_3\text{-}C - 2D{=}C{=}O{\bullet\bullet\bullet}H\text{-}N)/2 \quad (6.6)$$

Since the energy value yielded by the propyl ligand reaches its maximum value and remains constant for all further compounds of this series the contribution of one methylene group of butyl fragment is taken into account in the calculation.

$$D{=}C{=}O{\to}R\text{-}N = (\Delta_{vap}H^0(298K)acam - 2D{=}C{=}O{\to}CH_3\text{-}C - 2D{=}C{=}O{\bullet\bullet\bullet}H\text{-}N - DCH_2)/2 \quad (6.7)$$

The results of the calculations of the marked specific interactions (Table 6.5) allow us to determine the contributions of the energies of two different types of specific interactions, formed by two ethyl groups of the amine fragment *N,N*-diethylacetamide

$$DC{=}O{\to}CH_3\text{-}CH_2\text{-}N = (\Delta_{vap}H^0(298K)acam - 2DN{\to}CH_3\text{-}CH_2\text{-}N - 2DC{=}O{\to}CH_3\text{-}C)/2 \quad (6.8)$$

The contribution of the isostructural methyl group of *N*-isopropylacetamide is determined by the difference between the vaporization enthalpy of *N*-ethylacetamide and the energies of the specific interactions with the help of Eq. 6.6.

The result of the calculated energies of the hydrogen bonds (Table 6.5) provides the proof for the theoretically predicted integral value of its stabilization at the compounds with amide groups and methoxy by the oxygen atom,

D-O•••H-N: O – Methylhydroxylamine (7.68 ; 243 K) < D=C=O•••H -C=O: Formamide (13.4 ; 298 K) < D=C=O•••H-N: Acetamide (15.3 kJ mol^{-1}; 298 K)

aldehyde group and carbonyl oxygen atom, reflecting the increasing ability of the nitrogen atom to shift electron density to the oxygen atom, leading to bigger differences in the charges at the atoms of oxygen and nitrogen, and the hydrogen atoms. As formed by methyl, ethyl, and propyl groups of the amine fragments of acetamides the specific interactions are most stable compared with those formed by the symmetric ethers

D-O→CH_3-O (5.63) < D=C=O→CH_3-N (7.1) < D-O→CH_3-CH_2-O (6.78) < D=C=O→CH_3-CH_2-N (8.45) < D-O→CH_3-CH_2-CH_2-O (8.52) < D=C=O→CH_3-CH_2-CH_2-N (10.6 kJ mol^{-1}),

which is the experimental proof of the significant mutual influence of the carbonyl carbon and nitrogen atoms on the stability of the specific interactions and hydrogen bonds.

6.2.3 Propionamides and Butyroamides

For the structures of the propionamide, butiroamide molecules of the acetamide series and those with a larger number of methylene groups in the alkyl chain in liquid and crystal condition we have a similar diversity in the hydrogen bond types and specific intermolecular interactions. If the hydrogen bonds formed by the hydrogen atoms of the amide groups with nitrogen and oxygen atoms of the carbonyl group do not undergo any change, then the elongation of the carbon chain of the amine fragment and alkyl ligand is accompanied by a change in the charge of the carbon atom at the end of the methyl group, and that inevitably leads to the stabilization of the specific interactions formed. In this connection it makes sense to analyze the dependence of enthalpies of sublimation and vaporization of alkylamides on the number of carbon atoms of the alkyl ligand (Fig. 6.10).

The available data in the literature on the vaporization enthalpies, obtained for a small number of alkylamides at temperatures different from standard conditions, allow us to speak about the end of the manifestation of the reverse dative bond at propylamide and of further increasing the energy contribution to the enthalpy characteristics with an increasing number of methylene groups in the molecules of the compounds of this series. The clearly defined transition of the dependence of sublimation enthalpy from propylamide to butylamide confirms the same type of course of dependence of the enthalpy characteristics of the vaporization processes of alkylamides and points to the end of the influence of the reverse dative bond on

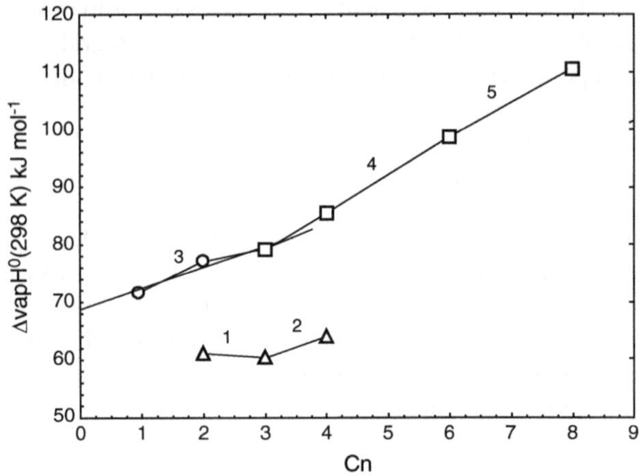

Fig. 6.10 Dependence of the vaporization enthalpy of acetamide – propionamide – butiroamide (*1*, *2*) and acetamide – octanamide (*3*–*5*) on the number of carbon atoms in the alkyl chain

the stabilization of the specific interaction at the compounds with three carbon atoms of the chain. Therefore there is a minimum in the dependence of the vaporization enthalpies on the number of carbon atoms of the chain, caused by the reduced values of the vaporization enthalpies of propylamide and butylamide, obtained at temperatures significantly increased from the standard conditions without any errors indicated. Thus estimation with the use of the related compound of *N*, *N*-dimethylacetamide with clear experimentally measured vaporization enthalpies at 298 and 388 K is not feasible.

The structure of the liquid and crystalline propionamides and butiroamides with the network of two types of the hydrogen bonds and two specific interactions (Fig. 6.11) of these compounds and *N*-methylpropionamide by two hydrogen bonds and four specific interactions of two types are transformed to the network with three types of the specific interactions in the structures of *N*,*N*-dimethylpropionamide, *N*,*N*-dimethylbutiroamide. Surely, such reorganizations lead to significant changes in the energies of the specific interactions. In this connection one should note that replacement of the hydrogen bond D–N•••H–N for propionamide to the specific interaction D–N → CH_3–N in *N*-methylpropionamide is connected with transformations in the amino group or in other words the replacement of the hydrogen atom of the amide group to the methyl group like those which occur at the replacement in tryamine of the hydrogen atom by the methyl group. In such transformations the energy of the hydrogen bond is similar to the energy of the specific interaction (Chap. 2), because the charges at the carbon atom of the same methyl group have a slightly reduced value [19–21]. Thus, we are forced to accept the energies of the hydrogen bond D–N•••H–N (7.7) and specific interaction D–N → CH_3–N (7.1 kJ mol^{-1}) obtained as correct. On the other hand the marked value of the energies of this specific interaction is permanent in the structures of the liquid compounds series *N*-methylpropionamide and *N*,*N*-dimethylpropionamide. It is obvious that the corresponding value of the energy of this specific interaction of

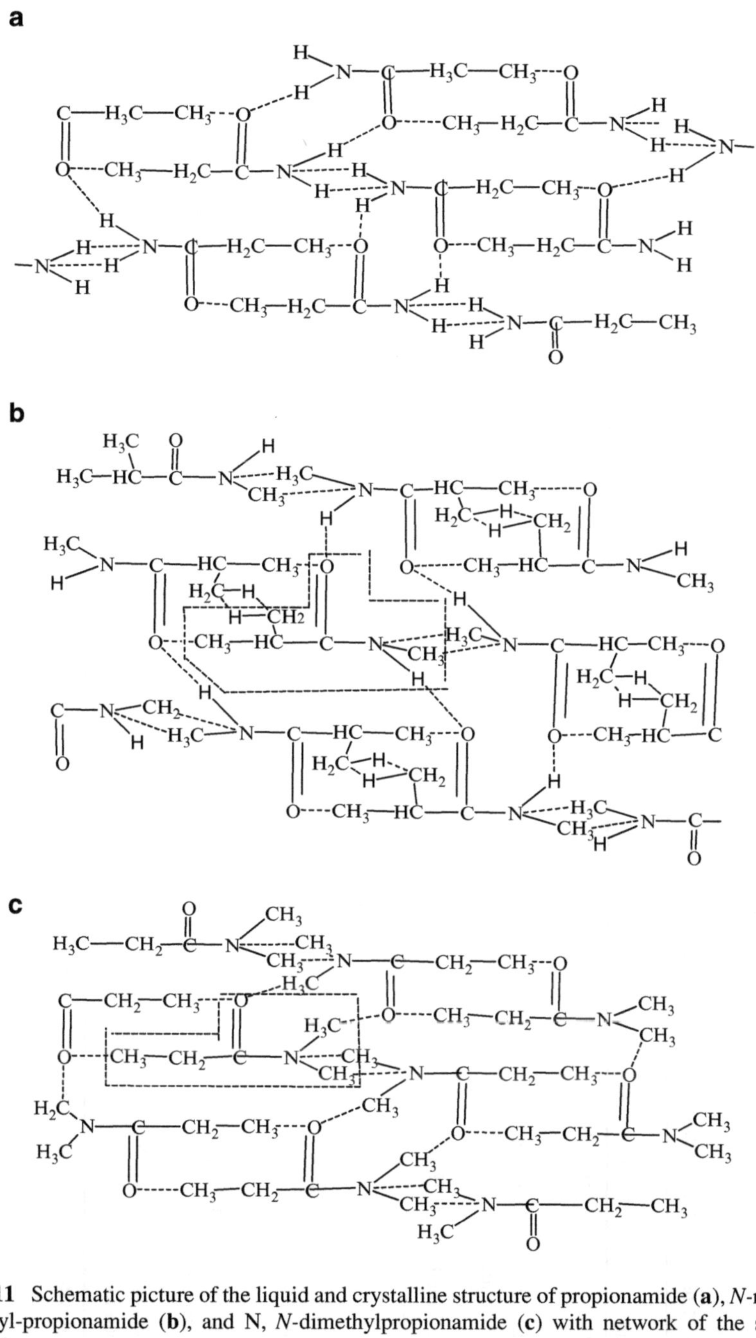

Fig. 6.11 Schematic picture of the liquid and crystalline structure of propionamide (**a**), *N*-methyl-2-methyl-propionamide (**b**), and N, *N*-dimethylpropionamide (**c**) with network of the specific interactions

Table 6.6 Energies of the hydrogen bonds and specific interactions (kJ mol^{-1}) of the liquid propionamides

Compounds	Formula	Structure	$\Delta_{vap}H°(T)$ [14]	T K	D–N → CH_3–N	D=C=O•••H–N	D=C=O → CH_3–$(CH_2)_n$–C=O
Propionamide	C_3H_7NO	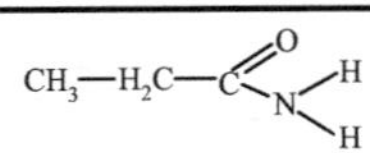	60.3 67.1[a]	358	DN•••H–N = 7.7	15.3	10.1
N-Methyl-propionamide	C_4H_9NO	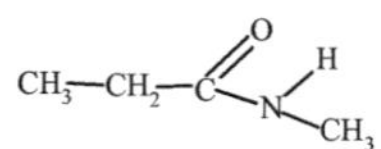	64.9 ± 0.3	298	7.1	15.3	10.1
N-Methyl-2-methyl-propionamide	$C_5H_{11}NO$	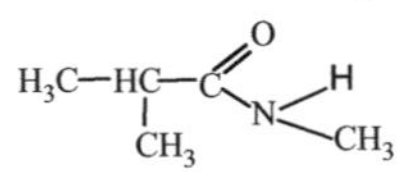	67.1 ± 0.2	298	7.1 D*iso*CH_3 × 2 = 1.7	15.3 D*iso* CH_3 = 2.2	10.25
						D=C=O → CH_3–N	
Dimethyl-*tert*-butylamide	$C_7H_{15}NO$	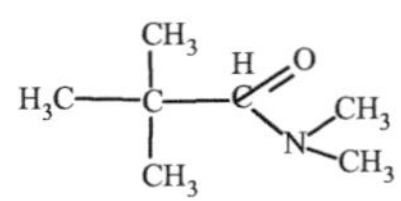	55.1 ± 0.1	298	7.1	8.4 2 D*iso* CH_3 = 3.45	10.1
N,N-Dimethyl-propionamide	$C_5H_{11}NO$	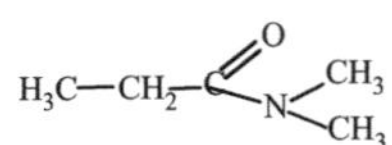	53.3 51.6[a]	341	7.1	8.4	10.1
N,N-Dimethyl-butiroamide	$C_6H_{13}NO$	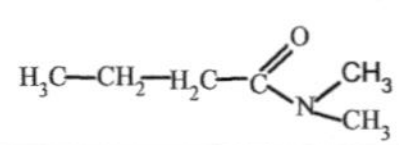	55.2	368	7.1	8.4 DCH_2 = 4.0	10.1

[a]Estimated without temperature dependency of enthalpy vaporization

crystalline *N*-methylacetamide, *N*-methylpropionamide, its series, and the series of compounds *N,N*-dimethylacetamide remain without change. A similar conclusion is obvious with regard to the hydrogen bond D=C=O•••H–N. However, the energies of the specific interactions, formed by the methylamine fragment with the nitrogen atom D–N → CH_3–N or with oxygen of the carbonyl group by oxygen D=C=O → CH_3–N should correspond to the inequality D–N → CH_3–N < D=C=O → CH_3–N.

The energy of the specific interactions D=C=O → CH_3–CH_2–CH_2–N of liquid *N*-methylpropionamide should be determined with the help of Eq. 6.5d

$$D{=}C{=}O{\to}CH_3\text{-}CH_2\text{-}CH_2\text{-}N = (\Delta_{vap}H^0(298\ K) - 2\ D\text{-}N{\to}CH_3\text{-}N - 2D{=}C{=}O{\cdots}H\text{-}N)/2 \quad (6.5d)$$

The energy value of this interaction was used for calculating the vaporization enthalpy of propionamide using Eq. 6.5a (Table 6.6). The energy contribution of the isostructural methyl group *N*-methyl-2-methyl-propionamide is determined by the difference with the enthalpy characteristics *N*-methylpropionamide and is equal to 2.2 kJ mol^{-1} and, consequently, the energy of the specific interaction has the value DH_3C → H–CH_2 = 1.1 kJ mol^{-1}. The energy of the specific interaction D=C=O → CH_3–CH_2–CH_2–N is determined with the help of Eq. 6.9.

$$D{=}C{=}O{\to}CH_3\text{-}CH_2\text{-}CH_2\text{-}N = (\Delta_{vap}H^0(298\ K) - 2\ D\text{-}N{\to}CH_3\text{-}N - 2D{=}C{=}O{\cdots}H\text{-}N - D\mathit{iso}CH_3)/2 \quad (6.9)$$

The energy contribution of the two isostructural methyl groups of dimethyl-tert-butylamide to the vaporization enthalpy should be accepted as equal to the similar contribution of the same two isostructural groups of 2-methoxy-3,3-dimethylbutanenitrile (3.45 kJ mol^{-1}) located at the third carbon atom, which is significantly lower than twice the value of contribution of one group of *N*-methyl-2-methyl-propionamide and better corresponds to the real value. Taking the energy of the specific interaction D=C=O → CH_3–CH_2–CH_2–N, we determined the value of the energy of the specific interaction D=C=O → CH_3–N

$$D{=}C{=}O{\to}CH_3\text{-}N = (\Delta_{vap}H^0(298\ K) - D{=}C{=}O{\to}CH_3\text{-}CH_2\text{-}CH_2\text{-}N - 2\ D\text{-}N{\to}CH_3\text{-}N)/2 \quad (6.10)$$

and with the help of Eq. 6.10 the values of the vaporization enthalpy *N,N*-dimethylpropionamide (Table 6.6).

The results of the thermodynamic analysis (Table 6.6) show the natural stabilization of the energies of the specific interactions, formed by the amide group and the oxygen atom of the carboxyl group D–N → CH_3–N)(7.1) < DC=O → CH_3–N (8.4 kJ mol^{-1}).

6.3 Energies of Specific Interactions of Crystalline Amides and Its Derivatives

6.3.1 Homological series of Acetamides, Propionamides, Amides, and Methyl Amides

Thermodynamic analysis of liquid formamides revealed the amazing property of these compounds to form aldehyde groups of highly stabilized hydrogen bonds, the energy values of which exceed the energy value of the hydrogen bond of methanol [13]. More than twofold excess energy of the hydrogen bond of acetamides and propionamides is provided by the mutual influence of the amino group and the oxygen atom of the carboxyl group. The high ability of the association in crystal leads us to expect the manifestation of a significant stabilizing effect by the crystalline field. This is indicated, in particular, by the difference in sublimation enthalpies and vaporization of formamide (11.5 kJ mol^{-1}) with average stabilizing effect for each of the six hydrogen bonds of three types 2.0 kJ mol^{-1} (Table 6.7). The data given above testify that intercorrection of the vaporization enthalpy with the number and energy of exploding specific interactions allow us to determine or estimate the value of the energy of the realized interactions in liquid and crystalline compounds. It is obvious that replacement of the hydrogen atom of the aldehyde formamide by acetate and the stable hydrogen bond DC=O•••H–C=O by the specific interaction D=C=O → CH_3–C reflect, on the one side, the contribution of this group to the sublimation enthalpy and, on the other side, the significant changes in the redistribution of the electron density in the molecule. As a result there appears a stabilization of the hydrogen bond D=C=O•••H–N and forming of a more stable specific interaction compared with the similar type, formed in crystalline diathyl ether (8.82 kJ mol^{-1}). It follows that the difference in sublimation enthalpies of formamide and acetamide reflects the stabilization of the hydrogen bonds D–N•••H–N and D=C=O•••H–N and is possibly a partial compensation for the reduced stability of the specific interaction compared with D=C=O•••H–C=O > D=C=O → CH_3–C. One should notice that replacement in the acetamide molecule of the hydrogen atom of the amino group by the methyl group is accompanied by reduction in the vaporization enthalpy of *N*-methylacetamide at 1.7 kJ mol^{-1}. At the same time the sublimation enthalpy of these compounds increases to 7.7–7.9 kJ mol^{-1} (Table 6.6). Such a difference in enthalpy characteristics is possible on the condition that the the *cis*- or *trans*-isomeric form of the *N*-methylacetamide molecule is formed in the liquid and crystalline conditions of these compounds. Even more striking differences in the sublimation enthalpies are observed in formamide and *N*-methylhydroxy (14.9 kJ mol^{-1}). It presupposes the replacement of the hydrogen atom of the aldehyde group, forming the most stable hydrogen bond, which principally cannot occur in the presence of the hydrogen atoms in the amino group, because there should be observed the principle of minimum energy of the system. Hence, the

Table 6.7 Energies of the hydrogen bonds and specific interactions (kJ mol^{-1}) of the crystalline formamide, acetamides, and propanamides

Compounds	Formula	Structure	$\Delta_{sub}H°(T)$ [22]	T K	D–N → CH_3–N	D=C=O•••H–N	D=C=O → CH_3–(CH_2)n–C
Formamide	CH_3NO	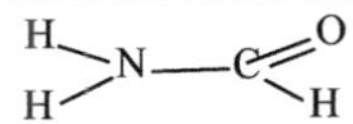	71.7	298	D–N•••H–N = 9.7	= 12.0	D=C=O•••H–C=O = 14.2
N-Methylhydroxy amine	CH_5NO	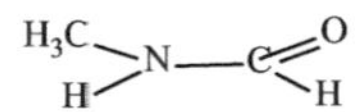	56.8 67.6	298	7.7	= 12.0	D=C=O•••H–C=O = 14.2
Acetamide	C_2H_5NO	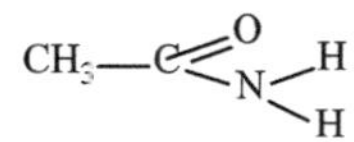	78.5 ± 0.3 78.7 ± 0.3	298	D–N•••H–N = 9.7	17.5	12.0 n = 0
N-Methylacetamide	C_3H_7NO	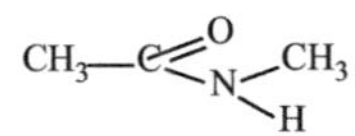	70.8 ± 2.0	298	7.7	17.6	10.45 n = 0
tert-Butylacetamide	$C_6H_{13}NO$	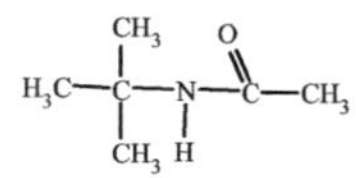	79.9 ± 0.4	298	7.7	17.6 ΣDisoCH_3 = 8.4	10.45 n = 0
N-Methylpropionamide	C_4H_9NO	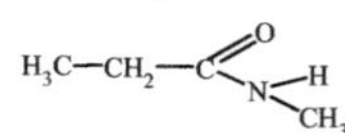	75.1	298	7.7	17.6	12.3 n = 1
Propanamide	C_3H_7NO	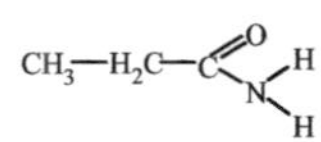	79.1	298	D–N•••H–N = 9.7	17.5	12.1 n = 1
2-Methylprop anamide	C_4H_9NO	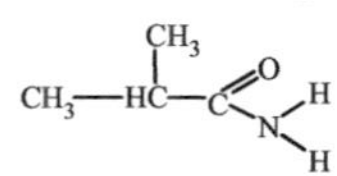	86.0 ± 0.2	298	D–N•••H–N = 9.7	17.5 DisoCH_3 = 6.9	12.1 n = 1
Diacetamide	$C_4H_7NO_2$	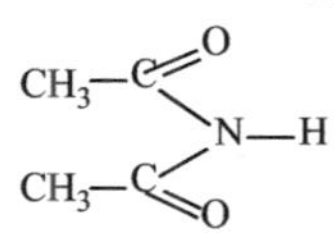	73.2 ± 0.8	298	D–N•••H–N = 6.34	6.0	9.2
					D–N•••H–N	D=C=S•••H–N	DC=S → CH_3–C

(continued)

Table 6.7 (continued)

Compounds	Formula	Structure	$\Delta_{sub}H°(T)$ [22]	T K	D–N → CH_3–N	D=C=O•••H–N	D=C=O → CH_3–(CH_2)n–C
Thioacetamide	C_2H_5NS	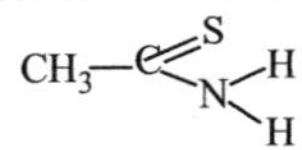	83.3 ± 0.3 82.8 ± 0.3		9.7	18.6	13.1

noted value of the sublimation enthalpy of *N*-methylhydroxy is not correct and, as a result, it is given in the literature without error indication. It is worth noting that the value of the difference of sublimation enthalpies (4.0 kJ mol^{-1}) of propanamide and *N*-methylpropionamide (Table 6.7) more correctly corresponds to the change of sublimation enthalpies of these compounds at the replacement of the stable hydrogen bond D–N•••H–N by the specific interaction N $\rightarrow$ CH_3–N with reduced stability. Taking this value as the change in the contribution of two hydrogen bonds for the two specific interactions in these compounds, the assignment of the value of this difference to the difference in the energies of these bonds D–N•••H–N (9.7) > N $\rightarrow$ CH_3–N (7.7 kJ mol^{-1}) is correct. The given structural-energetic parameters correspond to amide groups of crystalline formamides and amides, including those presented in Table 6.7. By analogue with liquid formamide, acetamide, and their derivatives and taking the difference in the sublimation enthalpies of crystalline formamide and acetamide, we take the energy of the hydrogen bond D=C=O•••H–N and specific interaction of D=C=O $\rightarrow$ CH_3–C of these compounds to be equal to 12.0 kJ mol^{-1}, respectively.

The results of the calculation of the interaction energies with this procedure, taking the total contribution of the energies of the specific interactions and hydrogen bonds in formamide

$$D{=}C{=}O\bullet\bullet\bullet H\text{-}O{=}C = (\Delta_{sub}H^0(298\ K) - 2D\text{-}N\bullet\bullet\bullet H\text{-}N - 2D{=}C{=}O\bullet\bullet\bullet H\text{-}N)/2 \quad (6.11)$$

acetamide and propanamide and with larger number of methylene groups than in CH_2–C

$$D{=}C{=}O\bullet\bullet\bullet H\text{-}N = (\Delta_{sub}H^0(298\ K) - 2D\text{-}N\bullet\bullet\bullet H\text{-}N - 2D{=}C{=}O\rightarrow CH_3\text{-}(CH_2)n\text{-}C)/2 \quad (6.12)$$

and *N*-methylacetamide, *N*-methylpropionamide

$$D{=}C{=}O\bullet\bullet\bullet H\text{-}N = (\Delta_{sub}H^0(298\ K) - 2D\text{-}N\rightarrow CH_3\text{-}N - 2D{=}C{=}O\rightarrow CH_3\text{-}(CH_3)n\text{-}C)/2 \quad (6.13)$$

where n is the number of methylene groups in alkyl fragment, given in Table 6.7.

The obtained energies of the hydrogen bonds and specific interactions allow us to determine the contribution of the energies of the two isostructural methyl groups *tert*-butylacetamide with the help of Eq. 6.7, taking into account the energies of the interactions of this compound. The results of the calculations illustrate the high contribution of the energies (8.4 kJ mol^{-1}) of the two isostructural methyl groups located at the first carbon atom *tert*-butylacetamide (Table 6.7). The value of the energy contribution of one isostructural methyl group is 4.2 kJ mol^{-1}. The high contribution to the vaporization enthalpy is given by the two isostructural methyl groups of methoxy-3,3-dimethylbutanenitrile, located at the second carbon atom of the alkyl fragment (Table 6.2). The energies of the specific interactions forming $H_3C \rightarrow H\text{–}CH_2\text{–}$ (2.1) and (1.7 kJ mol^{-1}) is, respectively, higher than the stability of the similar interactions, formed by isostructural methyl groups of other organic compounds.

As one can see from the data presented the energies of the hydrogen bonds in crystalline formamide,

D-N•••H-N (9.7) < D=C=O•••H-N (12.0) < D=C=O•••H-C=O (14.2 kJ mol^{-1})

acetamide

D-N•••H-N (9.7) < D-C=O→CH_3-C(10.45) < D-C=O•••H-C=O (17.5 kJ mol^{-1})

and propionamide are stabilized in the order of

D-N•••H-N (9.7) < D-C=O→CH_3-C (12.1) < D-C=O•••H-C=O (17.5 kJ mol^{-1})

with the character of the crystalline condition with increased energy of each type of the hydrogen bond rather than in liquid compounds of the considered series.

On ending the influence of the reverse dative bond at the third carbon atom of propilamide (Fig. 6.9) the maximum energy contribution to the enthalpy of vaporization and sublimation is reached. In this connection the difference of sublimation enthalpies of *N*-methylpropionamide (75.1) and *N*-methylacetamide (70.8 ± 2.0) 4.3 kJ mol^{-1} is not a major doubt. However, the similar difference of sublimation enthalpies of propanamide (79.1) and acetamide (78.7 ± 0.3) equal to 0.4 kJ mol^{-1} points to the understated value of the enthalpy characteristics of propanamide, presented in the literature with error indication. Hence, the obtained value of the energy contribution of the isostructural methyl group to the sublimation enthalpy of 2-methylpropanamide (86.0 ± 0.2 kJ mol^{-1}), equal to the difference in the enthalpy characteristics of propanamide (79.1 kJ mol^{-1}), has a significantly overstated value (6.9 kJ mol^{-1}).

As in the case of formamide the replacement of the oxygen atom of the carbonyl group in acetamide by the sulfur atom should result in a change of the energy of the hydrogen bond DC=S•••H–N and the specific interaction D=C=S → CH_3–C, formed by the thioacetamide molecule. The influence on the hydrogen bond D–N•••H–N will be minimal and the energy of this bond should be taken as equal to the one realized in crystalline acetamide. The contribution of the sulfur atom to the energies of the formed interactions is equal to the differences in sublimation enthalpies of thioacetamide and acetamide (4.3 kJ mol^{-1}). Based on the thermodynamic analysis of alkylamides, we can accept the stabilizing effect as equivalent to the estimation of the energy values of the hydrogen bond and the specific interactions (Table 6.7).

The energy of the hydrogen bond of crystalline diacetamide was determined by the proposed approach of a thermodynamic analysis of this compound in liquid condition using Fig. 6.6 as were the estimated energies of the hydrogen bond

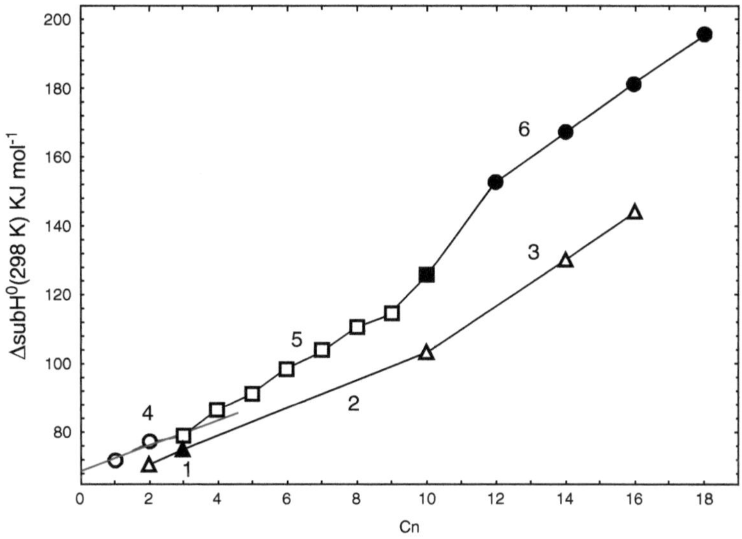

Fig. 6.12 Dependence of sublimation enthalpy on the number of carbon atoms in the alkyl chain of Methylalkylamides (*1–3*) and Alkylamides (*4–6*)

D–N•••H–N and the specific interaction D=C=O → CH_3–N of crystalline ammonia and the energies of interaction of crystalline methyl ether and Eq. 6.4.

The dependence of the sublimation enthalpies of alkylamides on the number of carbon atoms in the chain are complex (Fig. 6.12) and this reflects:

First, ending of the influence of the reverse dative bond (1, 4)
Second, stabilization of the energies of the specific interactions, formed by the alkyl chain, ending at the propyl ligand (2, 5)
Third, sharply increasing "splash" of stabilization and further equivalent stabilization by the crystalline field with the growth of the number of carbon atoms in the chain (3, 6)

The sharply differing energies of two types of hydrogen bonds and specific interactions, formed by the essentially unshared $2s^2$ electron pair of the carbon atom of the methyl group of the alkyl ligand with the oxygen atom of alkylamides and, on the other side, the hydrogen bond and two specific interactions in *N*-methylamides allow us to suppose the formation in the crystals of compounds of undecanamide – octadecanamide and *N*-methyldecanamide – *N*-methylhexadecanamide to have stable polymorphic forms. An indication of the existence of polymorphic forms is the marked "splash" at the enthalpy characteristics. The results of the calculations of the energies of the hydrogen bonds and specific interactions of alkylamides and methyl alkylamides with the help of Eqs. 6.12 and 6.13 are presented in Tables 6.8 and 6.9.

Strengthening of the stabilizing effect of the crystalline field is clearly seen on comparing the contributed energies of the methylene group of butanamide (5.1) and

Table 6.8 Energies of the hydrogen bonds and specific interactions (kJ mol^{-1}) of the crystalline Butanamide – Octadecanamide

Compounds	Formula	Structure	$\Delta_{sub}H°(T)$ [22]	T K	D–N•••H–N/ΣDCH$_2$	D=C=O•••H–N	D=C=O → CH$_3$–CH$_2$–C–
Butanamide	C_4H_9NO	$CH_3-(H_2C)_2-C(=O)-NH_2$	85.4 ± 1.7	298	9.7/5.2	17.6	12.8
Pentanamide	$C_5H_{11}NO$	$CH_3-(H_2C)_3-C(=O)-NH_2$	91.4	298	9.7/10.3	17.6	12.8
Hexanamide	$C_6H_{13}NO$	$CH_3-(H_2C)_4-C(=O)-NH_2$	98.7 ± 1.7 95.14	298 358	9.7/16.0	17.6	12.8
Heptanamide	$C_7H_{15}NO$	$CH_3-(H_2C)_5-C(=O)-NH_2$	99.6 103.8	345–363	9.7/18.4	17.6	12.8
Octanamide	$C_8H_{17}NO$	$CH_3-(H_2C)_6-C(=O)-NH_2$	110.5 ± 2.3		9.7/25.1	17.6	12.8
Nonanamide	$C_9H_{19}NO$	$CH_3-(H_2C)_7-C(=O)-NH_2$	114.6 ± 3.3 116.2	353–370	9.7/30.7	17.6	12.8
Decanamide	$C_{10}H_{21}NO$	$CH_3-(H_2C)_8-C(=O)-NH_2$	125.9 ± 1.3	361.5	9.7/50.5	17.6	12.8
Dodecanamide	$C_{12}H_{25}NO$	$CH_3-(H_2C)_{10}-C(=O)-NH_2$	152.7 ± 0.8	352	9.7/78.8	17.6	12.8
Tetradecanamide	$C_{14}H_{29}NO$	$CH_3-(H_2C)_{12}-C(=O)-NH_2$	167.4 ± 2.5	352	9.7/93.5	17.6	12.8
Hexadecanamide	$C_{16}H_{33}NO$	$CH_3-(H_2C)_{14}-C(=O)-NH_2$	181.6 ± 1.3	371	9.7/107.7	17.6	12.8
Octadecanamide	$C_{18}H_{37}NO$	$CH_3-(H_2C)_{16}-C(=O)-NH_2$	195.8 ± 4.2	373	9.7/121.9	17.6	12.8

Table 6.9 Energies of the hydrogen bonds and specific interactions (kJ mol^{-1}) of the crystalline *N*-Methylalkylamides

Compounds	Formula	Structure	$\Delta_{sub}H^{\circ}$ (T) [22]	T K	D–N → CH_3–N	D=C=O•••H–N = 17.6/ ΣDCH$_2$	D=C=O → CH_3–CH_2–C–
N-Methyl decanamide	$C_{19}H_{21}NO$	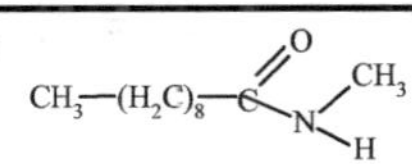	102.8 ± 0.8	314	7.7	17.6/27.7	12.3
N-Methyl dodecanamide	$C_{12}H_{25}NO$	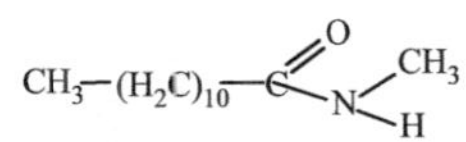	116.6 ± 0.8	330	7.7	17.6/41.5	12.3
N-Methyl tetradecanamide	$C_{14}H_{29}NO$	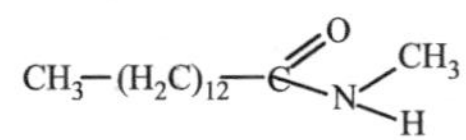	130.4 ± 0.8	340	7.7	17.6/55.3	12.3
N-Methyl hexadecanamide	$C_{16}H_{33}NO$	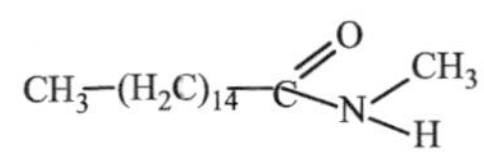	144.5 ± 50.8	350	7.7	17.6/69.4	12.3
cis-2-pentenoic acid amide	C_5H_9NO	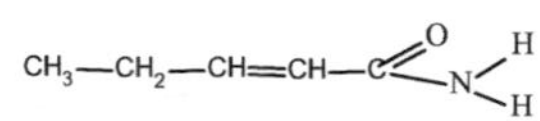	106.9	328	9.7 D*iso*-CH_3 3.45	17.6/ΣDCH$_2$ = 19	12.3

octadecanamide (7.1 kJ mol^{-1}). The same limits of change in the energies contributed by the methylene groups to the enthalpy characteristics are found for the methyl alkylamides. One can conclude that absolute and relative strengths of the hydrogen bonds and specific interactions at three differing types of interactions remain unchanged at the given series of compounds.

It should be noted that the available data in the literature on the sublimation enthalpy of *cis*-2-pentenoic acid amide (106.9) and *cis* 2-hexanoic acid amide (80 kJ mol^{-1}) without indication of experimental errors (Table 6.9) do not correspond to the dependence given in Fig. 6.11, which is used for estimation of its values.

6.3.2 Amides with Unsaturated Ligand

The results of the thermodynamic analysis considered in the previous chapters of liquid and crystalline alkylamides testify to the significant influence of the reverse dative bond of the alkyl chain of molecules of different series of these compounds on the stability of the specific interaction formed. In this connection we draw special attention to the inadequate change in sublimation enthalpies of compounds of the amides series with unsaturated carbon chain (Table 6.10). Considering the correctness of the energy values of the hydrogen bonds, the specific liquid and crystalline acetamides and propanamides, and the analogue of their structure with the structure of acrylamide (Fig. 6.11), it is possible to determine the energy of the specific interaction, formed by the unsaturated ligand $CH_2{=}CH{-}C$ of the acrylamide molecule with the help of Eq. 6.12. The obtained energies of the specific interaction $D{=}C{=}O \rightarrow CH_2{=}CH{-}C$ with double bond in alkene fragment at 330 K with the experimental error value shows the increased stability compared with the energy of the interaction, formed by the propyl ligand propanamide,

$$D{=}C{=}O{\rightarrow}CH_2{=}CH\text{-}C\ (13.6) > D{=}C{=}O{\rightarrow}CH_3\text{-}CH_2\text{-}C\ (12.1\ \text{kJ mol}^{-1})$$

The elongation of the carbon chain to the methyl and methylene groups and preserving the type of specific interaction and the number of formed hydrogen bonds by the crystalline δ-valerolactam molecule (Fig. 6.13) is not accompanied by the additional energy contribution by two redundant methylene groups to the sublimation enthalpy (Table 6.10). Oppositely, the energy of the formed specific interaction of all carbon chains $D{=}C{=}O \rightarrow CH_3{-}CH{=}CH{-}CH_2{-}C{-}$ reduces up to 10.0 kJ mol^{-1}. Based on this we can conclude that the double bond of the alkene ligand influences significantly the reduction in the shifting of the electron density and the charge of the terminal carbon atom, forming the specific interaction. Notice that the contribution of three excess methylene groups of the hexanamide molecule, after ending the influence of the reverse dative bond on the specific interaction, is equal to15.9 kJ mol^{-1}, which value could not be referred to the experimental error. A similar destabilizing effect of the double carbon bonds and

Table 6.10 Energies of the hydrogen bonds and specific interactions (kJ mol^{-1}) of the crystalline amides with unsaturated hydrocarbon ligand

Compounds	Formula	Structure	$\Delta_{sub}H°(T)$ [22]	T K	D–N•••H–N	D=C=O → CH_2=CH–C	D=C=O•••H–N
Acrylamide	C_3H_5NO	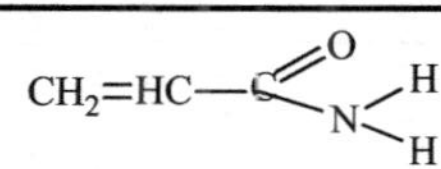	81.8 ± 0.8	330	9.7	13.6	17.6
δ-Valerolactam	C_5H_9NO	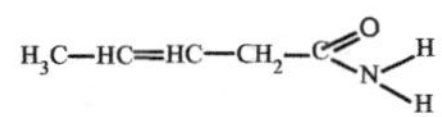	74.5	303	9.7	D=C=O → CH_3–CH=CH–CH_2–C 10.0	17.6

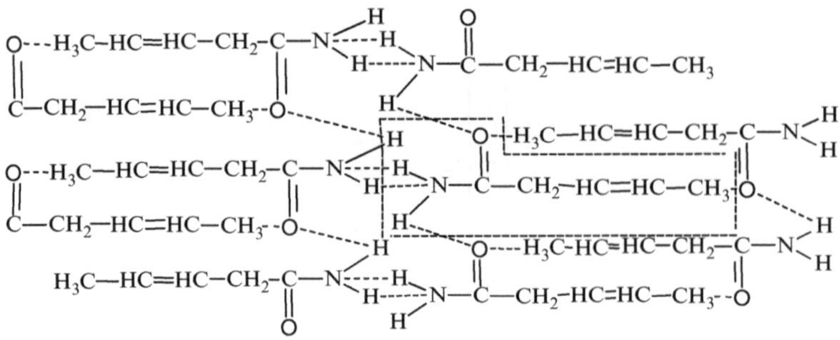

Fig. 6.13 Schematic picture of the crystalline structure of δ-valerolactam with network of the specific interactions

D≡N→C≡: N≡C—C≡N (5.9 kJ mol^{-1}) Cyanogen	>	D≡N→C≡: N≡C—N=C=C=N—C≡N (3.4 kJmol^{-1}) Dicyanoacetylene

the number of carbon atoms of the closed chain was seen at the compounds of the series cyanogen – dicyanoacetylene (Table 2.1).

6.3.3 Benzamides

The amino benzene molecule with two bond vacancies of the free electron pair of the carbonyl oxygen atom, two positive charged hydrogen atoms, the undivided electron pair of the nitrogen atom, and six CH-groups of the benzene ring with the essentially unshared $2s^2$ electron pair of each carbon atom forms 11 specific interactions with molecules that are close in the environment. As was shown above, the carbonyl group influences the stabilization of the specific interactions greatly compared with the amino group. This conclusion is in full compliance with the work [23] showing a strong influence of the same group in water environment compared with the amide group of urea. Thus the rigid structure of the benzene ring, as a rule, weakly influences the functional groups, and it is interesting to study the influence on the stabilization of the hydrogen bond, formed by hydrogen atoms of the amino groups. From the many bond vacancies of the benzolamide molecule, the amide fragment forms two hydrogen bonds DC=O•••H–N of increased stability with the oxygen atom of the carbonyl group (Fig. 6.14), two hydrogen bonds DN•••H–N with the nitrogen atom of the amino group and two specific interactions, formed by carbonyl oxygen with CH-group of the benzene ring D=C=O → CH. The remaining five CH groups of the benzene ring form specific interactions with the same rings of benzamide that are close in the environment, D=HC → CH, with

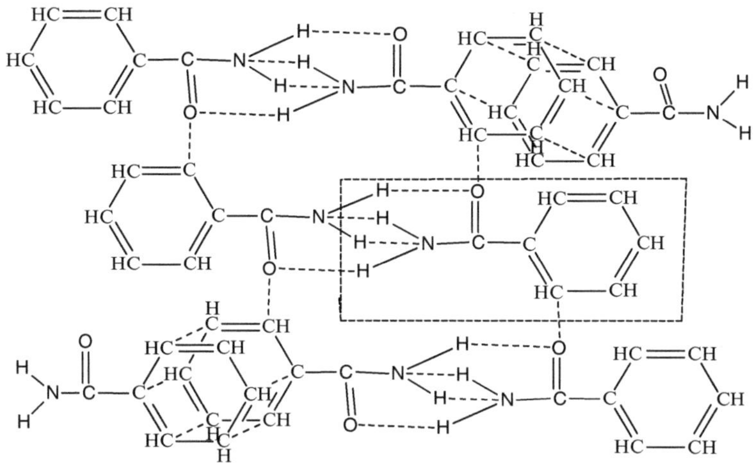

Fig. 6.14 Schematic picture of the crystalline structure of benzamide with network of the hydrogen bonds and specific interactions

an energy of 7.4 kJ mol^{-1}, obtained from the sublimation enthalpy of benzene minus the energy contribution to the sublimation enthalpy of the saturated hydrogen atom (1.0 kJ mol^{-1}). The value of the energy of the specific interaction D=C=O → CH should be taken as equal to the value of the energy of crystalline dimethyl ether (6.59 kJ mol^{-1}).

The hydrogen bonds and specific interactions form the structure of crystalline benzamide with the network of these types of interactions. The networks are criss-crossed by specific interactions, formed by CH-groups of the benzene rings, located in such a way that interacting carbon atoms have a maximum charge difference. Taking into account the fact that the energies of the hydrogen bonds D–N•••H–N in amides remain unchanged (9.7 kJ mol^{-1}), we have reason to use Eq. 6.14,

$$D{=}C{=}O\cdots H\text{-}N = (\Delta_{sub}H^0(298\ K) - 2D\text{-}N\cdots H\text{-}N - 2D{=}C{=}O{\rightarrow}CH - 5D{=}HC{\rightarrow}CH)/2 \quad (6.14)$$

We consider the number and the energies of all interactions for determining of the energy of the stable hydrogen bond DC=O•••H–N (Table 6.11).

This equation should be used for the determination of the energy of the hydrogen bond D=C=S•••H–N with a similar network of the specific interactions of crystalline thiobenzamide. The energies of the hydrogen bond presented in Table 6.10 have reduced values compared with those realized in crystalline alkylamides,

D=C=O•••H-N Benzamide (16.3) < Alkylamides (17.6) > D=C=S•••H-N Thiobenzamide (16.8 kJ mol^{-1})

This reflects the shifting of the reduced electron density from the CH group of

Table 6.11 Energies of the hydrogen bonds and specific interactions (kJ mol^{-1}) of the crystalline benzamides and its derivatives at T = 298 K

Compounds	Formula	Structure	$\Delta_{sub}H°$(298 K) [22]	D–N•••H–N	D–N → CH_3–N	D=C=O → CH	D=C=O•••H–N
Benzamide	C_7H_7NO	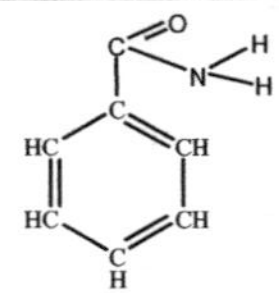	101.7 ± 1	9.7	–	6.59	16.3 17.1[a]
Thiobenzamide	C_7H_7NS	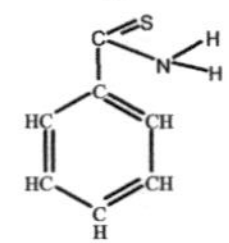	103.4 ± 2.2	9.7	–	6.59	16.8 17.6[a]
N-Methylbenzamide	C_8H_9NO	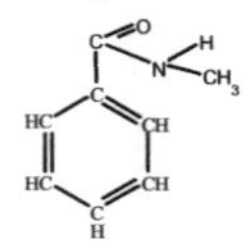	85.7 98.2[a]	318	7.7	6.59	16.3 17.1[a]
N,N-Dimethylbenzamide	C_9H_4NO	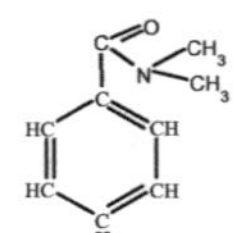	89.7 ± 0.3	–	7.7	6.59	D=C=O → CH_3–N 9.25
N,N-Dimethylthiobenzamide	$C_9H_{11}NS$	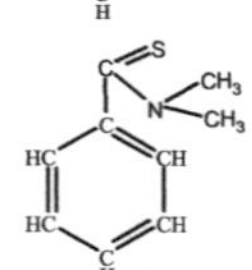	94.8 ± 2.0	–	7.7	8.4	9.5
N-Benzoyl-*N*-methylbenzamide	$C_{15}H_{13}NO_2$	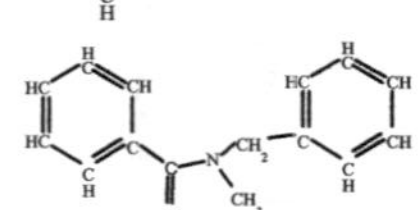	120.1 ± 0.4	–	7.7	5.8	–

[a]It is corresponded to T = 318 K

the benzene ring to the carbon atom of the carbonyl group and oxygen compared with the methyl group of acetate ligand acetamide and the reduced difference in the charges at the oxygen atom of the amide fragment, benzamide, and the hydrogen atom of the amino group, forming the most stable hydrogen bond, D=C=O•••H–N.

As was shown above the replacement of the hydrogen atom of the amino group by the methyl group is accompanied by reduction in the vaporization enthalpy of approximately 4–5 kJ mol^{-1}, which is caused by the change of the hydrogen bond for the less specific interaction D–N → CH_3–N. In this connection the sublimation enthalpy given in the literature of *N*-methylbenzamide (85.7 kJ mol^{-1}) without error indication is reduced at 10 kJ mol^{-1} compared with its real value. Hence, we estimated its value using the energy contribution of this type of interaction.

Similar to the case of the alkylamides the replacement of the second hydrogen atom in the amino group is accompanied by replacement of the more stable hydrogen bond by the specific interaction D=C=O → CH_3. We, therefore, determined the energy value of this type of interaction with the help of Eq. 6.13, using the energy value of two other types of interactions, realized in crystalline *N,N*-dimethylbenzamide and *N,N*-dimethylthiobenzamile. The results of the calculations (Table 6.11) show that the high stabilization of the specific interaction

$$\text{D-N}\rightarrow\text{CH}_3\text{-N}\ (7.7) < \text{D=C=O}\rightarrow\text{CH}_3\ (9.25) < \text{D=C=S}\rightarrow\text{CH}_3\ (9.5\ \text{kJ mol}^{-1})$$

of the amine fragment formed by methyl group, is almost twice the energy of the same type, formed by methylamine. Thus, the amide fragments of acyclic and cyclic amides possess high mutual influence of the carbonyl and amide groups on the shifting of the electron density, leading to a significant difference between the oxygen atoms and carbon atoms of the terminal methyl groups of acetate and propyl ligand and the hydrogen atom on one side and on the other side, hydrogen and nitrogen atoms of the amide fragment.

The crystalline structure of *N*-benzoyl-*N*-methylbenzamide for the network of the specific interactions (Fig. 6.15) is formed by eight specific interactions, D=C=O → CH, by electron pairs of the oxygen atoms with the significantly undivided $2s^2$ of electron pairs of the carbon atoms of CH groups of the benzene ring of the molecules that are close in the environment and two methyl groups D–N → CH_3–N of the amide fragment with the value of the energy of 7.7 kJ mol^{-1}. Networks are criss-crossed by stable specific interactions D=HC → CH, formed by the remaining free CH groups of each benzene ring with an energy of 7.40 kJ mol^{-1}. The energy of the specific interaction DC=O → CH is determined with the help of Eq. 6.15, considering the contributions of the energies of all interactions, realized

$$\text{D=C=O}\rightarrow\text{CH} = (\Delta_{\text{sub}}H^0(298\ \text{K}) - 2\text{D-N}\rightarrow\text{CH}_3\text{-N} - 8\text{D=HC}\rightarrow\text{CH})\,/\,2 \qquad (6.15)$$

in crystalline *N*-benzoyl-*N*-methylbenzamide. The value of the energy of this type of specific interaction (5.8 kJ mol^{-1}) is less stable compared with the energy of the same type of interaction of crystalline dimethyl ether (6.59 kJ mol^{-1}) and reflects

Fig. 6.15 Schematic picture of the crystalline structure of *N*-benzoyl-*N*-methylbenzamide with network of the hydrogen bonds and specific interactions

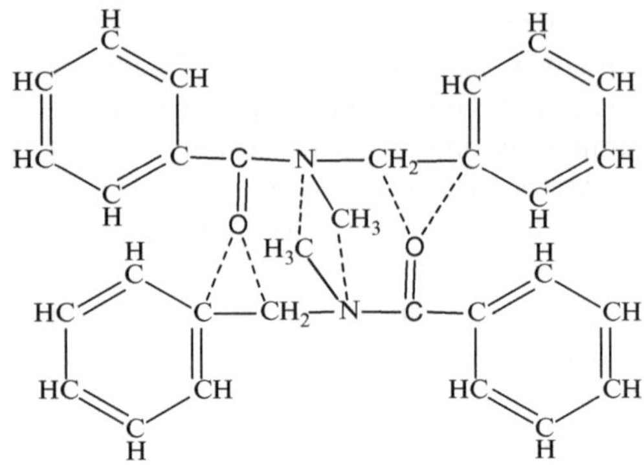

its nature accurately. This implies that it should be used for the correction of the obtained energy value of the hydrogen bond of crystalline *N*-benzoyl-*N*-methylbenzamide, instead of the energy value of the specific interaction of methyl ether.

6.3.4 Phenylacetamides

Compounds of phenol derivatives with the functional amide acetate group allow us to clarify the influence of the carbon atoms of the carbonyl and terminal methyl group and the nitrogen atom of the amide group, directly connected with benzene ring, to the hydrogen bond and specific interactions, formed by bond vacancies of the fragment. Lack of guidance in the errors of experimentally measured sublimation enthalpies of compounds of this series, obtained at different temperatures, differing from standard conditions, contribute definite difficulties to its discussion (Table 6.12). Nevertheless the thermodynamic analysis of the enthalpy characteristics allow us to create a common concept of the interactions pointed out. Taking into account the connection of the vaporization enthalpies of the compounds with the number and energy of bursting intermolecular interactions, we should draw attention to the reduced value of the sublimation enthalpy phenylacetamide at 8.1 kJ mol^{-1} compared with *N*-(2-methylphenyl)acetamide, of which the benzene ring together with the methyl group contributes less to the enthalpy characteristics (43.1 kJ mol^{-1}) than the acetamide fragment of the compound (44.6 kJ mol^{-1}). The phenylacetamide, *N*-(2-methylphenyl)acetamide, and *N*-(4-methylphenyl)acetamide molecules form the structure of a crystal with the network of an equal number of hydrogen bonds and specific interactions (Fig. 6.16a, b). The bond vacancies of molecules of these compounds form two hydrogen bonds of high stability D=C=O•••H–N, two specific interactions of the acetate group D=C=O → CH_3–C, two by the nitrogen atom with the CH group of the benzene ring D=N → CH and five interactions of the significantly undivided

Table 6.12 Energies of the hydrogen bonds and specific interactions (kJ mol^{-1}) of the crystalline phenylacetamides

Compounds	Formula	Structure	$\Delta_{sub}H°(298\ K)$ [22]	TK	$\Sigma D{=}HC \rightarrow CH/$ $D{-}N \rightarrow CH{-}$	D=C=O•••H–N	$D{=}C{=}O \rightarrow CH_3{-}C$
Acetanilide Phenylacetamide	C_8H_9NO	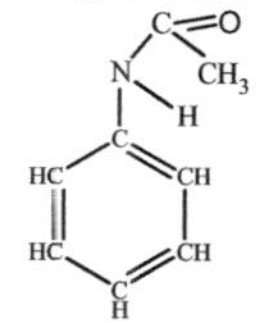	88.7 98.1[a]	326	36.3/D–N → CH = 5.8	17.1	9.20
N-(2-Methylphenyl) acetamide	$C_9H_{11}NO$	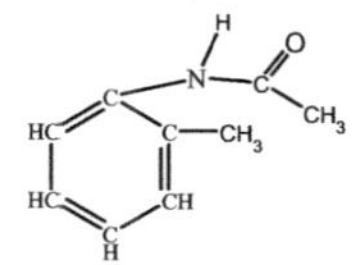	96.8	327.5	34.2/DN → CH = 5.8	17.1	9.20
N-(4-Methylphenyl) acetamide	$C_9H_{11}NO$	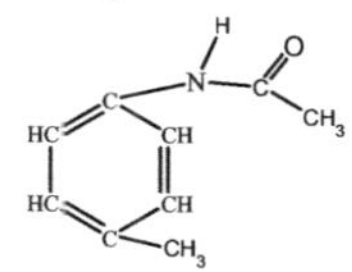	99.0	341	34.2/5.8	17.1	9.25
2-Phenylacetamide	C_8H_9NO	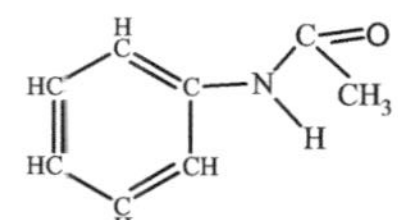	96.4	340	43.6/DN•••H–N = 9.7	7.8[a]	8.82
4-Aminoaceto phenone	C_8H_9NO	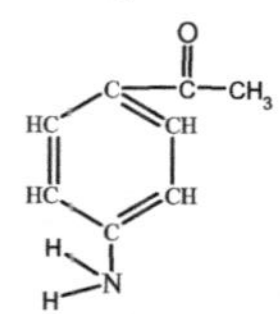	93.7 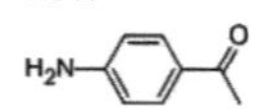	326	42.6/DN → CH	8.3[a]	8.82

(continued)

Table 6.12 (continued)

Compounds	Formula	Structure	$\Delta_{sub}H°(298\ K)$ [22]	TK	ΣD=HC → CH/ D–N → CH–	D=C=O•••H–N	D=C=O → CH_3–C
N-Phenyldiacetamide	$C_{10}H_{11}NO$	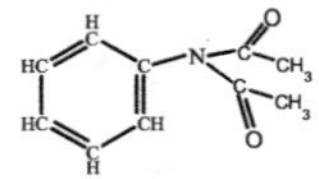	90 ± 0.8	298	21.8 D–N → CH = 5.0	D=C=O → CH 5.8	8.9

[a]Calculations do not account the temperature dependency of interactions energies

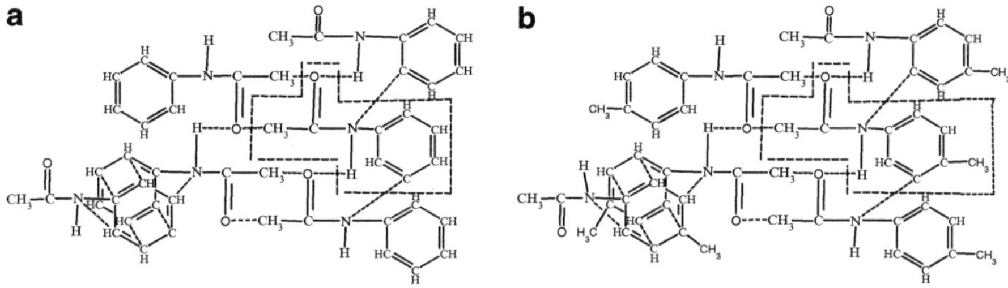

Fig. 6.16 Schematic picture of the crystalline structure of phenylacetamide (**a**) and *N*-(4-methylphenyl)acetamide (**b**) with network of the hydrogen bonds and specific interactions

$2s^2$ electron pair of the carbon atom of the remaining CH groups of the same benzene ring. We determine the value of the energy of the specific interaction D=HC → CH from the sublimation enthalpy of benzene reduced by the contribution of the substituted hydrogen atom (7.26 kJ mol^{-1}) and the benzene ring with the methyl group (toluene) reduced by the contribution of two substituted hydrogen atoms (2.0 kJ mol^{-1} and 6.85 kJ mol^{-1} , respectively) together with the isostructural group. Thus, five given interactions contribute to the sublimation enthalpy 36.3 and 34.25 kJ mol^{-1} of phenylacetamide and *N*-(2-methylphenyl) acetamide, *N*-(4-methylphenyl)acetamide and the energy of the specific interactions of these compounds D=C=O → CH_3–C should be determined with the help of Eq. 6.16

$$D{=}C{=}O{\rightarrow}CH_3\text{-}C = (\Delta_{sub}H^0(298\ K) - 2D{=}C{=}O{\cdots}H\text{-}N - 2D{=}N{\rightarrow}CH\text{-}N - 5D{=}HC{\rightarrow}CH)\ /\ 2 \tag{6.16}$$

Taking into account the above mentioned equivalence in the shift of the electron density of the methyl group and carbon atom of the benzene ring, connected with carbon atom of the carbonyl group, we can take the energy of the hydrogen bond as equal to the one formed in crystalline benzamide (17.1 kJ mol^{-1}). The energy of the specific interaction of low stability 2D N → CH–N should be higher than the values of the interactions of liquid amine with tetracoordinated nitrogen atom (5.33 kJ mol^{-1}) and lower than that realized in crystalline dimethyl ether and satisfactorily combined with the energy value, obtained for the similar interaction DC=O → CH (5.8 kJ mol^{-1}) of crystalline *N*-benzoyl-*N*-methylbenzamide. Calculated by Eq. 6.15, the energy value of the specific interaction, formed by the aceto group, is given in Table 6.12. The obtained values of the energies of the hydrogen bond and specific interaction D=C=O → CH_3–C were used for correct estimation of the values of sublimation enthalpy of phenylacetamide (Table 6.12).

We note that similarity of the types of the hydrogen bonds of crystalline *N*-(2-methylphenyl)acetamide and *N*-(4-methylphenyl)acetamide point to the fact that the difference in its sublimation enthalpies corresponds to the effect of stabilization of the contribution of the isostructural methyl group at the replacement of location from C(2) to C(4) and does not affect the changes in the energy of the hydrogen bond and specific interaction. We draw such a conclusion on the basis of

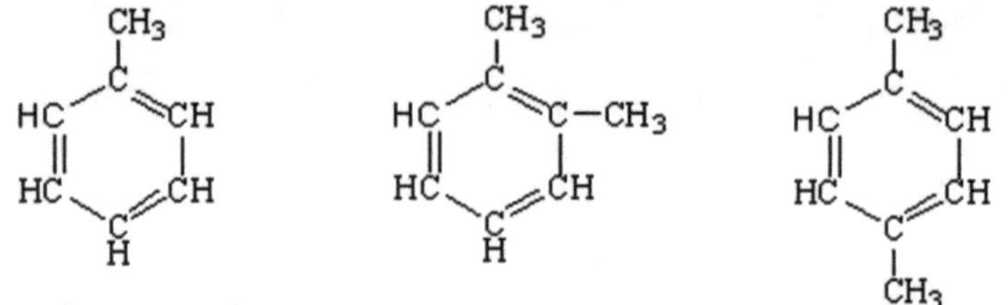

$\Delta_{sub}H^0(T)$ kJ mol^{-1} 43.1 60.1, 248 K; 56.6, 486 K 60.8, 286 K

the sublimation enthalpies of toluene, 1,2-dimethylbenzene, and 1,4-dimethylbenzene, by which the energy contributions of the isostructural methyl group increases by replacement of location C(2) to C(4) at 3.6 kJ mol^{-1}.

The 2-phenylacetamide molecule forms two types of hydrogen bonds of the amine group; one of them is formed inside the open amide group N•••H–N, and like the alkylamines, is least subjected to or practically not influenced by the benzene ring, because it is separated from the carbonyl methylene group with equal benzene ring by shifting of the electron density to the oxygen atom, and the second type is formed with the oxygen atom D=C=O•••H–N. Two other types of the specific interactions D=C=O → CH_3–C and D=HC → CH are formed by the acetate group being directly related to the benzene ring and CH groups of the same ring (Fig. 6.17a) unlike phenylacetamide, *N*-(2-methylphenyl)acetamide, and *N*-(4-methylphenyl)acetamide. On the grounds of the above the energy of the hydrogen bond D–N•••H–N, formed by the open amine group of crystalline structure with the network of the specific interactions 2-phenylacetamide (Fig. 6.17a), is taken as (9.7 kJ mol^{-1}) the one realized in crystalline benzamide. The energy contribution of six specific interactions formed by CH groups of the benzene ring reduced by the saturated hydrogen atom is equal to 43.6 or 7.26 kJ mol^{-1} per one bond.

Similar types of specific interactions form the network of crystalline 4-aminoacetophenone and contribute to the sublimation enthalpy (Fig. 6.17b). The energy of the hydrogen bond 2DC=O•••H–N of these compounds with the assumptions made should be determined with the help of Eq. 6.17

$$D{=}C{=}O{\cdots}H\text{-}N = (\Delta_{sub}H^0(298\ K) - D{=}C{=}O{\rightarrow}CH_3\text{-}C - 2D\text{-}N{\cdots}H\text{-}N - 6D{=}HC{\rightarrow}CH)\ /\ 2 \tag{6.17}$$

The results of the calculations (Table 6.12) do not take into account the temperature dependence of the energies of the hydrogen bonds and hence, specific interactions correspond to the minimum possible values.

The *N*-phenyldiacetamide molecule forms a crystal with the network of the specific interaction (Fig. 6.18), in which four D=C=O → CH_3–C and four D=C=O → CH interactions with an energy of 5.8 kJ mol^{-1} participate, and in crystalline *N*-benzoyl-*N*-methylbenzamide, three D=HC → CH contribute

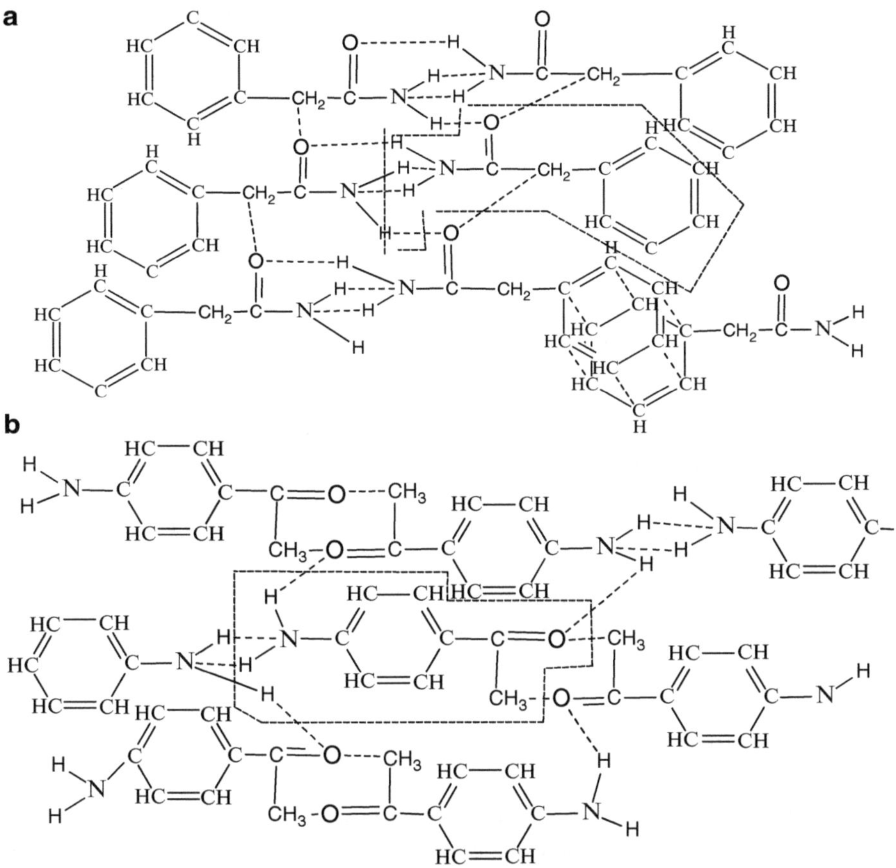

Fig. 6.17 Schematic picture of the crystalline structure of 2-phenylacetamide (**a**) and 4-aminoacetophenone (**b**) with network of the hydrogen bonds and specific interactions

21.8 kJ mol^{-1} to the sublimation enthalpy and one N → CH interaction. The energy value of this type of specific interaction in liquid methylamine (5.3 kJ mol^{-1}) with a four-coordinated nitrogen atom is close to the energy, formed by the carbonyl oxygen atom with a similar group of the benzene ring of the crystalline compound. The enthalpy characteristic of this compound is described by Eq. 6.18:

$$D{=}C{=}O{\to}CH_3\text{-}C = (\Delta_{sub}H^0(298\ K) - 4D{=}C{=}O{\to}CH - D{=}N{\to}C\ H - 3D{=}HC{\to}CH)\ /\ 4 \quad (6.18)$$

The calculations of the energies of the hydrogen bond and specific interactions are given in Table 6.12. This implies that at the location of the amide group inside the amino acetate fragment at the direct connection of the nitrogen atom with the benzene ring and the carbon atom of the carbonyl group a significant stabilization of

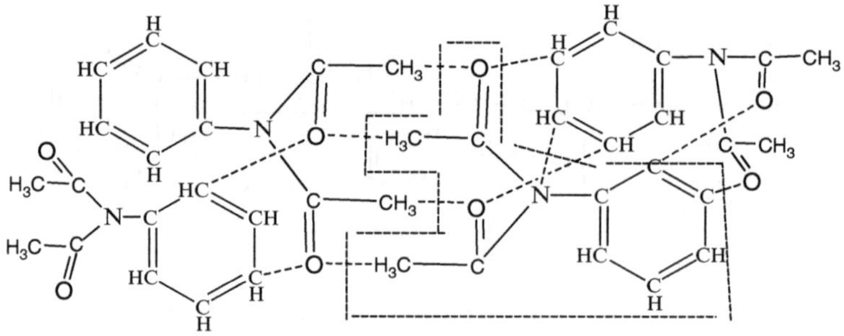

Fig. 6.18 Schematic picture of the crystalline structure of *N*-phenyldiacetamide with network of the hydrogen bonds and specific interactions

the hydrogen bond occurs, formed by the oxygen atom of the carbonyl group with the hydrogen atom of the amino group. When the methyl group of amino acetate fragment binds with the carbon atom of the benzene ring, the outer NH_2 group expresses the properties of the functional amino group and the energy of the hydrogen bond of the same type reduces to the energy value formed by this group in crystalline 4-aminoacetophenone. At the same sequence there appears an insignificant destabilization of the specific interaction, formed by the acetate group.

6.3.5 *Pyridinecarboxamides*

In pyridylacetamide and pyridinecarboxamides molecules, the nitrogen atom of the pyridine ring is a common factor that complicates the problem of determination of the hydrogen bond DC=O•••H–N, formed by the hydrogen atom of the amino group and the oxygen atom of the carbonyl group as the change of location of carboxamide fragment at C(2), C(3), or C(4) significantly influences the ability of the carbon atom of the CH groups of the ring to shift electron density to the nitrogen atom and further to the carbonyl oxygen atom. Surely, deletion at the location of the carboxamide fragment of nitrogen atom in the ring leads to the shifting of the electron density and as a result there is a relative increase of the oxygen atom charge and this leads to an increase of the differences in the charge of the nitrogen atoms relative to the hydrogen atoms of the amino group. The inadequate change of the sublimation enthalpy of pyridinecarboxamides derivatives

$$\text{2-Pyridinecarboxamide }(93.1) < \text{3-Pyridinecarboxamide }(121.2) > \text{4-Pyridinecarboxamide }\left(116.1\ \text{kJ mol}^{-1}\right)$$

points to the fact that maximum shifting of the electron density to the carbonyl oxygen atom is reached at the location of carboxamide fragment at C(3) in the cycle. This implies that preferential shifting of the electron density is carried out

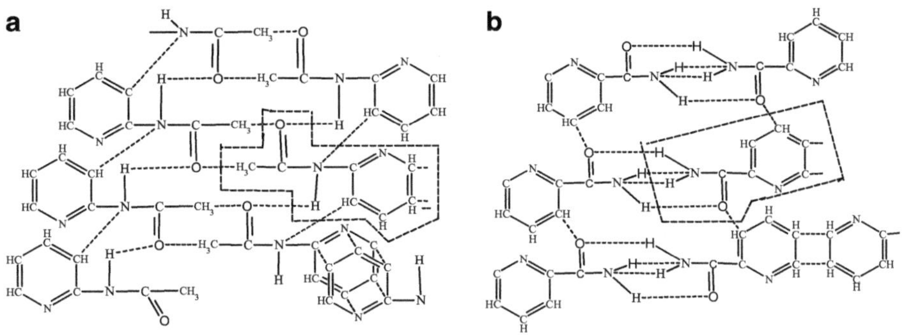

Fig. 6.19 Schematic picture of the crystalline structure of (2-pyridyl)acetamide (**a**) and 2-pyridinecarboxamide (**b**) with network of the hydrogen bonds and specific interactions

from C(4) and C(5). One can draw the conclusion that there is a natural stabilization of the hydrogen bond and marked specific interactions in the range of compounds: 2-pyridinecarboxamide < 4-pyridinecarboxamide < 3-pyridinecarboxamide.

The (2-pyridyl)acetamide molecule with a large number of bond vacancies forms five types of specific interactions, forming the crystal structure with the network of the hydrogen bonds and specific interactions, with the essentially unshared $2s^2$ electron pair of the carbon atom (Fig. 6.19a). The pyridine cycle with five CH groups leads to the formation of three specific interactions D=HC → CH (7.26 kJ mol^{-1}), two interactions D=N → CH with energies 12.8 kJ mol^{-1} and two DN → CH with an energy of 5.8 kJ mol^{-1} and likewise the one formed by crystalline *N*-benzoyl-*N*-methylbenzamide (Table 6.11), and the amide fragment forms the stable hydrogen bond D=C=O•••H–N and specific interaction D=C=O → CH_3–C. The energy of this interaction should be taken as 9.20 kJ mol^{-1}, equal to the value of the energy of the same type of interaction in crystalline phenylacetamides.

Thus, the sublimation enthalpy of (2-pyridyl)acetamide is described by the Eq. 6.19

$$D{=}C{=}O{\bullet\bullet\bullet}H\text{-}N = (\Delta_{sub}H^0(298\ K) - 2D{=}C{=}O{\rightarrow}CH_3\text{-}C - 2D{=}N{\rightarrow}C\ H - 3D{=}HC{\rightarrow}CH - 2D{=}N{\rightarrow}C({=})H) / 2 \tag{6.19}$$

The calculations in Eq. 6.19 show the integral values (Table 6.13) of stabilization formed in crystalline (2-pyridyl)acetamide with specific interactions and hydrogen bond

D=N→CH (5.8) < D=HC→CH (7.26) < D=C=O→CH_3-C (9.20) < D=N→C(=)H (12.8) < DC=O•••H-N (13.2 kJ mol^{-1})

The 2-pyridinecarboxamide molecules and its analogues form two types of hydrogen bonds, D–N•••H–N (9.7 kJ mol^{-1}), D=C=O•••H–N and specific

Table 6.13 Energies of the hydrogen bonds and specific interactions (kJ mol^{-1}) of the crystalline 2-pyridyl acetamides and pyridinecarboxamides

Compounds	Formula	Structure	$\Delta_{sub}H°$ (T) [22]	T K	$\Sigma D{=}N \rightarrow C$ (=)H) + D–N•••H–N	$D{=}C{=}O \rightarrow CH$	D=C=O•••H–N
(2-Pyridyl)acetamide	$C_7H_8N_2O$	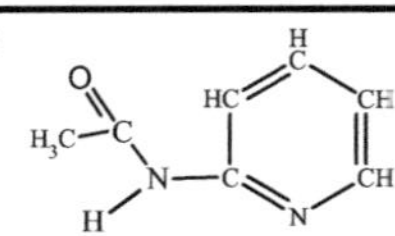	103.8	298	45.3/DHC → CH = 7.26 × 3	$D{=}C{=}O \rightarrow CH_3{-}C$ 9.20	13.2
2-Pyridinecarboxamide	$C_6H_6N_2O$	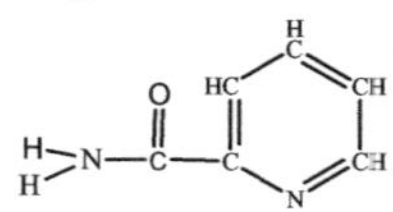	93.1 ± 3.3	298	45.3 + 21.8	5.8	7.4[a] 9.4
3-Pyridinecarboxamide	$C_6H_6N_2O$	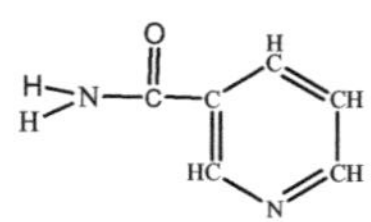	121.2 ± 3.3	298	45.3 + 21.8	5.8	21.3
4-Pyridinecarboxamide	$C_6H_6N_2O$	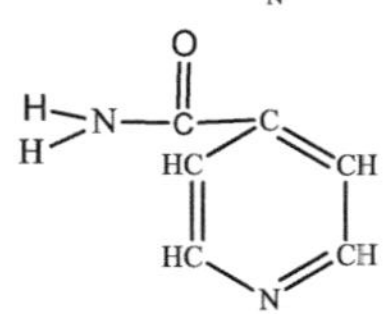	116.1 ± 3.3	298	45.3 + 21.8	5.8	18.8

[a]Reduced

interaction D=C=O → CH (5.8 kJ mol^{-1}) by the carboxamide fragment and two interactions D=N → C(=)H and three D=HC → CH (7.26 kJ mol^{-1}) by the pyridine ring, contributing to the sublimation enthalpy

$$(\Delta_{sub}H^0(298\ K) = 2D{=}C{=}O\cdots H\text{-}N + 2D\text{-}N\cdots H\text{-}N\ (9.7x2) + 2DC{=}O{\rightarrow}CH\ (5.8x2) + 3D{=}HC{\rightarrow}CH\ (7.26x3) + 2D{=}N{\rightarrow}CH\ (12.8x2) \quad (6.20)$$

Considering the definite convention of unchanged specific interactions and hydrogen bond at the change of location of carboxamide fragment, we should notice that formation of the specific interaction

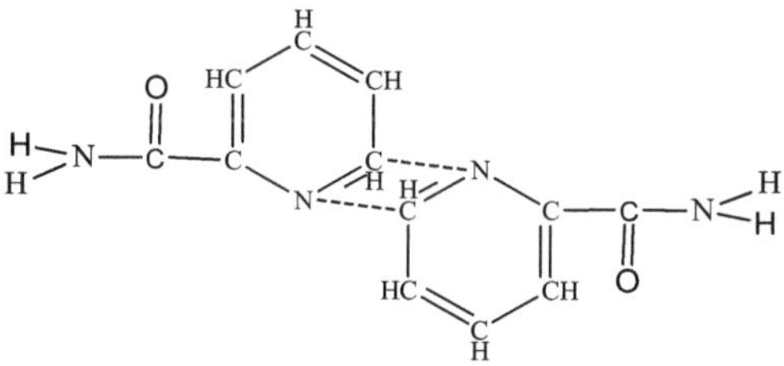

D=N → CH happens with maximum difference in the charges with the carbon atom in the cycle at the location C(6). Hence, the given type of interaction like the hydrogen bond D–N•••H–N is least influenced by the redistribution of the electron density in the pyridine cycle with the change of location of the carboxamide fragment. Nevertheless, the drastic reduction of the stability of these types of interactions by 1.0 kJ mol^{-1}approximates the energy value D=C=O•••H–N of the hydrogen bond for 2-pyridinecarboxamide, leading to the correct value (9.4 kJ mol^{-1}). The sharp change in sublimation enthalpies depending on the location of the carboxamide fragment of the considered compounds is accompanied by a similar change of character in one of the types of interactions, namely,the ones most exposed to the redistribution of the electron density and atomic charge, forming these interactions:

D=C=O•••H-N: 2-Pyridinecarboxamide (9.4) < 4-Pyridinecarboxamide (18.8) < 3-Pyridinecarboxamide (21.3 kJ mol^{-1}).

6.4 Energies of Specific Interactions of Alkylcyanates and Isocyanates

6.4.1 Alkylcyanates, 2-Cyanoacetamide

The alkylcyanate molecule with two oxygen atoms and two free electron pairs has an excess number of negatively charged bond vacancies compared with the two alkyl groups. Hence, the role of the two additional missing bond vacancies is implemented by positively charged hydrogen atoms of the methylene groups of acetate fragments and methyl or other alkyl groups. The hydrogen atoms of alkyl and alkene groups often act as functional groups in liquid and crystalline organic compounds, in which their contribution to the vaporization and sublimation enthalpy is not very high, being equal to 0.6 and 1.0 kJ mol^{-1}, respectively. As was shown above the formamide molecule forms a hydrogen bond of the aldehyde group (13.4 kJ mol^{-1}) almost three times higher than the energy value of the same bond formed by acetale and propimal. In this connection the compounds of these series are of great interest.

The simplest molecule of the first representative of these compounds of methylcyanoacetate includes the methylene and methyl groups, formed by the essentially unshared $2s^2$ electron pair of the carbon atom in the specific interaction with the oxygen atom of the chain or with carbonyl group D–O → CH_3, D=C=O → CH_2 and D–O•••H–CH_2, D=C=O•••H–CH, respectively. Four bonds of two types of the specific interaction, four hydrogen bonds, and two specific interactions D≡N → C≡ form the structure of the liquid condition with the network of the specific interactions (Fig. 6.20). It should be noticed that the analogue of the specific interactions, formed by methyl and ethyl groups with the oxygen atoms methylcyanoacetate, ethylcyanoacetate, methoxyacetonitrile, and ethoxyacetonitrile and with methyl and ethyl ether (Table 6.2), the equality of the energy values allows us to use these data for the determination of the energies of the hydrogen bonds, realized in liquid methylcyanoacetate and ethylcyanoacetate. However, the presence in the mesh structure of the liquid compounds of the three types of the specific interactions and two types of the hydrogen bonds make it difficult to determine the energies of the H-bonds of each type. In this connection we determined the value of the energy, using the contributions of each type of interaction to the vaporization enthalpy:

Methylcyanoacetate

$$\text{D-O}\cdots\text{H-CH}_2 = \text{D=C=O}\cdots\text{H-CH} = (\Delta_{vap}H^0(298\ \text{K}) - 2\text{D-O}\rightarrow\text{CH}_3 - 2\text{D=C=O}\rightarrow\text{CH}_2 - 2\ \text{D}\equiv\text{N}\rightarrow\text{C}\equiv)/4 \tag{6.21}$$

Ethylcyanoacetate

$$\text{D-O}\cdots\text{H-CH}_2\text{-CH}_2 = \text{D=C=O}\cdots\text{H-CH} = (\Delta_{vap}H^0(298\ \text{K}) - 2\text{D-O}\rightarrow\text{CH}_3\text{-CH}_2 - 2\text{D=C=O}\rightarrow\text{CH}_2 - 2\ \text{D}\equiv\text{N}\rightarrow\text{C}\equiv)/4 \tag{6.22}$$

The results of the calculations (Table 6.14) illustrate the high stability of the energy of the H-bond, formed by the hydrogen atom of the methylene, methyl, and terminal methyl group of ethyl fragment. Formed in liquid methylcyanoacetate, the specific interactions and hydrogen bonds are described by the integral values of its stability

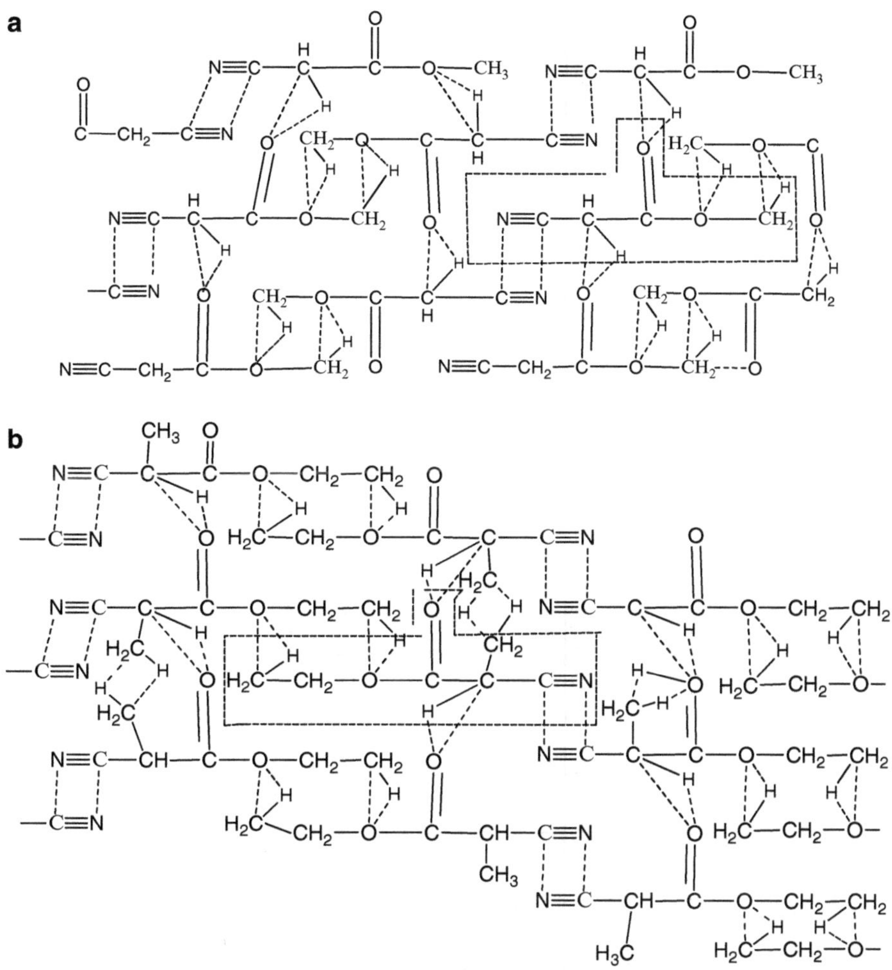

Fig. 6.20 Schematic picture of the liquid structure of methylcyanoacetate (**a**) and ethyl 2-cyanopropionate (**b**) with network of the specific interactions

D-O→CH_3-O (5.63) = D-O→CH_3-O (5.63) < D-O•••H-CH_2 = D=C=O•••H-CH (5.8) < D-O→CH_3-$(CH_2)_n$-O (6.78) < D≡N→C≡ (9.6kJ mol^{-1}).

We draw attention to the fact of the low value of the vaporization enthalpy of ethyl 2-cyanopropionate, differing at 8.1 kJ mol^{-1} from similar characteristics of the N≡C group. It follows that the change of location of the N≡C group, substituting the hydrogen atom at the terminal methyl group in the molecule of the last compound, to the location at the methylene group of propyl ligand in ethyl 2-cyanopropionate, is accompanied by a redistribution of the electron density in the molecule of this compound. As a consequence, there appears a reduction of the values of the positive and negative charge of carbon and nitrogen atoms of the N≡C

Table 6.14 Energies of specific interactions (kJ mol^{-1}) of the liquid cyanoacetates, ethyl 2-cyanopropionate, and 2-cyanoacetamide

Compounds	Formula	Structure	$\Delta_{vap}H°(T)$ [14]	T K	D≡N → C≡	D–O → CH_3–(CH_2) n–O	D=C=O•••H– CH_2=D=C=O•••H–CH
Methylcyanoacetate	$C_4H_5NO_2$	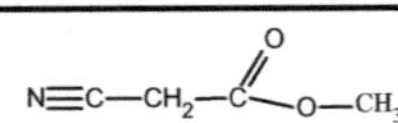	66.7 ± 0.9	298	9.6	n = 0 5.63/6.78	5.8
Ethylcyanoacetate	$C_5H_7NO_2$	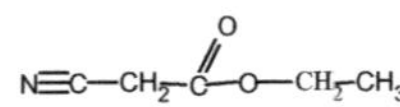	66.9	355	10.7	n = 1 6.78/6.78	5.1[a]
Ethyl 2-cyanopropionate	$C_6H_9NO_2$	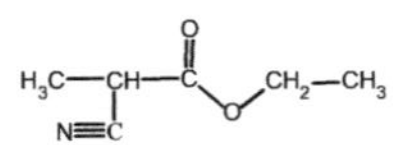	58.6 ± 0.3	298	9.6	n = 1 5.63/678 D*iso*CH_3 = 2.2	3.1
2-Cyanoacetamide	$C_3H_4N_2O$ solid	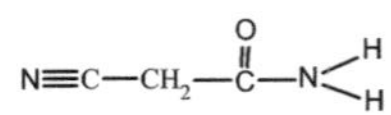	99.7[b] [22]	336	13.6	D=C=O → CH_2–C=O = 10.45	D=C=O•••H–N = 17.1 D–N•••H–N = 9.7

[a]Temperature dependence of the energy of specific interactions is not accounted
[b]Sublimation

group, respectively, the reduction of the difference in the value of its charges being important. At the same time, the replacement of the methyl group in methylcyanoacetate by the ethyl fragment in ethyl 2-cyanopropionate should be accompanied by an increase of the vaporization enthalpy of this compound additionally by 2.3 kJ mol^{-1}. Consequently, the total value of reduction of the vaporization enthalpy of ethyl 2-cyanopropionate in connection with the influence of the isostructural methyl group approaches 10.4 kJ mol^{-1}. Thus, we should draw the conclusion that the location of the N≡C group at the second carbon atom in propyl ligand is accompanied by the appearance of the isostructural methyl group, participating in a redistribution of the electron density. As a result, the specific interactions formed of the acetate group (Fig. 6.20b) are reduced by 2.3 kJ mol^{-1} up to the value of the contribution of the methylene group (5.63 kJ mol^{-1}) and the energy of the specific interaction D≡N → C≡ decreases to the value of 9.6 kJ mol^{-1}, formed in liquid methylcyanoacetate.

Besides the noted interactions of the ethyl 2-cyanopropionate molecule we have two specific interactions D–O → CH_3–CH_2 by the ethyl fragment, two hydrogen bonds D–O•••H–CH_2 and two D=C=O•••H–CH. In addition to the noted interactions of the isostructural methyl group we have two specific interactions DH_3C → H–CH_2 of low stability, the energies of which are equal to 2.2 kJ mol^{-1}, similar to metoxyalkylnitriles (Table 6.2). All the six types of specific interactions form the network structure for the liquid condition of ethyl 2-cyanopropionate (Fig. 6.19b). Considering the complexity of the energy estimation of each of the two types of H-bonds formed in liquid ethyl 2-cyanopropionate, we took them to be equal and, using the contribution of all the specific interactions, we determined their energies from the equation

$$\mathrm{DO{\cdots}H\text{-}CH_2\text{-}CH_2 = DC{=}O{\cdots}H\text{-}CH = (\Delta_{vap}H^0(298\ K) - 2DO{\rightarrow}CH_3\text{-}CH_2 - 2DC{=}O{\rightarrow}CH_2\text{–}C\text{–}2D{\equiv}N{\rightarrow}C{\equiv} - 2\,DH_3C{\rightarrow}H\text{-}CH_2)\,/\,4} \qquad (6.23)$$

The energy values of realized interactions of liquid ethyl 2-cyanopropionate (Table 6.14) are described by the number of its stability

D-O•••H-CH_2-CH_2 = D=C=O•••H-CH < D=C=O→CH_2–C (5.63) < D-O→CH_3-CH_2 (6.78) < D≡N→C≡ (9.6kJ mol^{-1}).

The 2-cyanoacetamide molecule with N≡C group, replacing the hydrogen atom of the methyl group in the acetate fragment (Fig. 6.21), forms the crystal structure with the network of stable hydrogen bonds (D=C=O•••H–N = 17.1 and D–N•••H–N = 8.7 kJ mol^{-1}), typical for acetamide compounds, specific interactions (D=C=O → CH_3–C), and for nitriles the D≡N → C≡ interaction (Fig. 6.21). The energies established for the hydrogen bonds correspond to the values of this series of compounds (Table 6.7).

From the difference of the energy contribution of three types of interactions with known energy values and enthalpy of sublimation for 2-cyanoacetamide

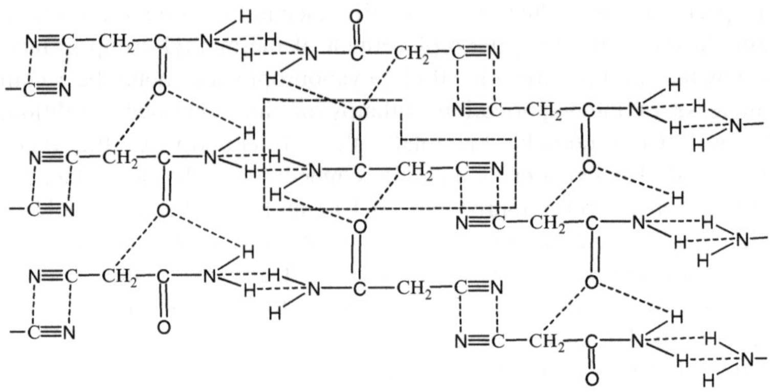

Fig. 6.21 Schematic picture of the crystalline structure of 2-cyanoacetamide with network of the specific interactions

$$D{\equiv}N{\rightarrow}C{\equiv} = (\Delta_{sub}H^0(298\ K) - 2D{=}C{=}O{\cdots}H\text{-}N - 2D\text{-}N{\cdots}H\text{-}N - 2D{=}C{=}O{\rightarrow}CH_2\text{-}C\text{-}) \quad (6.24)$$

we determined the energy value of the specific interactions D≡N → C≡, which at 4.0 kJ mol^{-1} exceeds the energy value of the same type of liquid methylcyanoacetate and ethoxyacetonitrile (10.7 kJ mol^{-1}) (Table 6.14) and this illustrates the significant stabilizing effect of the crystalline field. These values are in good agreement with the energies of the specific interactions and hydrogen bonds of the liquid and crystalline compounds with amide fragment, in which such sharp differences in the charges of the oxygen atoms of the carbonyl group of the acetate fragment are seen. As a sequence the energies of the interactions of crystalline 2-cyanoacetamide are described by integral values of

$D\text{-}N{\rightarrow}H\text{-}N$ (9.7) < $D{=}C{=}O{\rightarrow}CH_3\text{-}C{=}O$ (10.45) < $D{\equiv}N{\rightarrow}C{\equiv}$ (13.6) < $D{=}C{=}O{\cdots}H\text{-}N$ (17.1 kJ mol^{-1})

specific interactions, significantly differing from the energies of the interactions formed by oxygen-containing compounds [13].

6.4.2 *Isocyanates*

The transformation of the cyanates molecule in isocyanates is accompanied by the formation of the carbonyl group with its usual ability to form the specific interaction D=O → C= of the essentially unshared $2s^2$ electron pair of the carbon atom and the free electron pair of the oxygen atom. The remaining second free bond vacancy of the oxygen atom with the electron pair interacts with the hydrogen atom of the

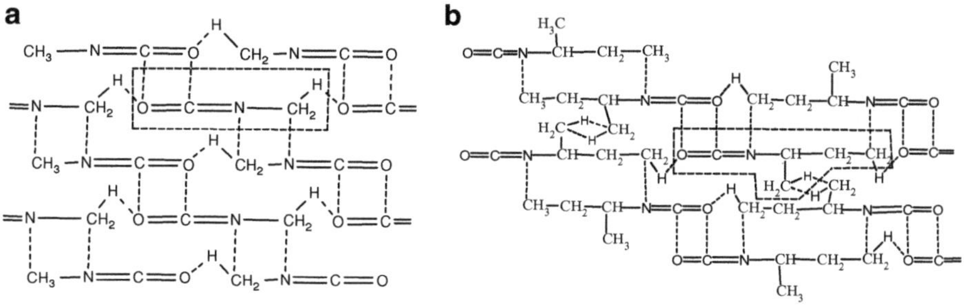

Fig. 6.22 Schematic picture of the liquid structure of methyl isocyanate (**a**) and isobutyl isocyanate (**b**) with network of the specific interactions

methyl group of the contacting molecule that is close in the environment in the case of methyl isocyanate or, in particular, with the similar terminal group of isobutyl isocyanate, having an increased positive charge. Hence, in liquid and crystalline conditions of isocyanates we have a network structure formed by specific interactions and the hydrogen bond D=C=O•••H–CH_2 (Fig. 6.22). The isobutyl isocyanate molecule with the isostructural methyl group contributes additional specific interaction of low stability, $DH_3C \rightarrow H–CH_2$, participating in stitching chains (Fig. 6.22b).

The energies of the specific interactions, formed by the methyl group of methyl isocyanate $D–N \rightarrow CH_3–N$ (4.25), or the ligand with a larger number of carbon atoms of the chain $D–N \rightarrow CH_3–CH_2–CH_2$ (8.1 kJ mol^{-1}), should be taken as equal to the value of the energy of the interaction, formed by the alkyl group of the corresponding alkylamine with the four-coordinated nitrogen atom. Ending of the influence of the intermolecular reverse dative bond to the specific interaction, formed by the alkyl chain, at the fourth carbon atom of the butyl group gives rise to the need to consider the energy contribution to the vaporization enthalpy of one methylene group of the butyl ligand or three methylene groups of heptyl isocyanate. The value of the contribution should be determined by the difference of the vaporization enthalpy of propylamine and butylamine or by the difference with the vaporization enthalpy of heptyl isocyanate. However, the vaporization enthalpy of this compound is defined at a temperature of 43 K difference from a standard one and 55 K difference from the conditions of the determination of the other isocyanates. The energy contribution of the hydrogen bond DC=O•••H–CH_2 (5.8 kJ mol^{-1}) to the vaporization enthalpy should be taken as equal to the value obtained for liquid methylcyanoacetate (Table 6.14). Such an assumption is correct because it was established for two compounds with different numbers of carbon atoms in the alkyl ligand. Taking the energy contributions of the specific interactions and hydrogen bonds to the vaporization enthalpy of methyl isocyanate and butyl isocyanate, the energy value of the interaction $D{=}O \rightarrow C{=}$ should be determined by the Eqs. 6.25a, 6.25b, and 6.25.

Methyl isocyanate

$$D{=}C{=}O{\rightarrow}C{=} = (\Delta_{vap}H^0(298\ K) - 2\ D{=}N{\rightarrow}CH_3\text{-}N - 2\ D{=}C{=}O{\cdots}H\text{-}CH_2) / 2 \quad (6.25a)$$

Butyl isocyanate

$$D{=}C{=}O{\rightarrow}C{=} = (\Delta_{vap}H^0(298\ K) - 2D{=}N{\rightarrow}CH_3\text{-}CH_2\text{-}CH_2\text{-}N - 2D{=}C{=}O{\cdots}H\text{-}CH_2 - DCH_2) / 2 \quad (6.25b)$$

Isobutyl isocyanate

$$D{=}C{=}O{\rightarrow}C{=} = (\Delta_{vap}H^0(298\ K) - 2D{=}N{\rightarrow}CH_3\text{-}CH_2\text{-}CH_2\text{-}N - 2\ D{=}C{=}O{\cdots}H\text{-}CH_2 - 2\ DH_3\text{-}C{\rightarrow}H\text{-}CH_2) / 2 \quad (6.25c)$$

We took the value of the energy contribution of the specific interaction of the isostructural methyl group to the vaporization enthalpy of isobutyl isocyanate as equal to the energy contribution of a similar group to the enthalpy characteristics of 2-methoxy-2-methylpentanenitrile (2.2 kJ mol^{-1}).

The results of the calculations (Table 6.15) illustrate the stabilization of the specific interaction $D{=}O \rightarrow C{=}$ with an increasing number of carbon atoms in the alkyl chain, which points to the shifting of the electron density from the nitrogen atom to the carbon atom and further to oxygen, which results in an increase in the difference in the charges of the carbon and oxygen atoms. Excluding the contribution of the energy of the isostructural methyl group from the vaporization enthalpy of isobutyl isocyanate, we determined the enthalpy characteristics of vaporization of unstudied propyl isocyanate and the value of the energy of the specific interaction $D{=}O \rightarrow C{=}$ (Table 6.15). Given in Fig. 6.23 is the dependence of the vaporization enthalpy on the number of carbon atoms in the alkyl fragment of isocyanate; it reflects the weakening of the influence of the intermolecular reverse dative bond on the specific interactions formed $D\text{–}N \rightarrow CH_3\text{–}(CH_2)n\text{–}N$ and $D{=}O \rightarrow C{=}$, reaching maximum stability at butyl isocyanate. The further increase of the enthalpy characteristics is determined by the energy contribution with an increasing number of methylene groups in the carbon chain of isocyanate.

The fragment $O{=}C{=}N\text{–}$ of the phenyl isocyanate molecule forms three types of specific interactions, one, $D{=}O \rightarrow C{=}$, which is common to all isocyanates and two other $D{=}C{=}O \rightarrow CH$ and $DC{=}N \rightarrow CH$ being formed with the CH groups of the benzene ring. The remaining unoccupied bond vacancies of four CH groups of the benzene ring form four specific interactions $D{=}HC \rightarrow CH$ with a total energy value 22.12 kJ mol^{-1}, calculated from the vaporization enthalpy of benzene reduced by the energy contribution of the saturated hydrogen atom (0.6 kJ mol^{-1}). Consequently, the structure of the liquid and crystalline phenyl isocyanate with the network of the specific interactions are formed by four types of interactions (Fig. 6.24), three of which contribute in total to the enthalpy characteristics

Table 6.15 Energies of specific interactions (kJ mol^{-1}) of the liquid isocyanates

Compounds	Formula	Structure	$\Delta_{vap}H°$ (298 K) [14]	T K	D=N → CH_3 ($C=H_2$)n–N	D=O•••H–CH_2	D=O → C=
Methyl isocyanate	C_2H_3NO	CH_3—N=C=O	29.9	265–308	4.25	5.8	4.9
Ethyl isocyanate	C_3H_5NO	CH_3—CH_2—N=C=O	35.7[a]	288	6.00	5.8	6.05
Propyl isocyanate	C_4H_7NO	CH_3—CH_2—CH_2—N=C=O	42.0[a]	288	n = 2; 8.1	5.8	7.1
Isobutyl isocyanate	C_5H_9NO	CH_3—CH_2—CH(CH_3)—N=C=O	44.2	288	n = 2; 8.1	5.8 D*iso*CH_3 = 2.2	7.1
Butyl isocyanate	C_5H_9NO	CH_3—$(CH_2)_2$—CH_2—N=C=O	46.8	288	8.1	5.8 DCH_2 = 3.4	7.8
Pentyl isocyanate	$C_6H_{11}NO$	CH_3—$(CH_2)_3$—CH_2—N=C=O	51.0[a]	288	8.1	5.8 ΣDCH_2 = 7.9	7.8
Hexyl isocyanate	$C_7H_{13}NO$	CH_3—$(CH_2)_4$—CH_2—N=C=O	55.9[a]	288	8.1	5.8 ΣDCH_2 = 12.8	7.8
Heptyl isocyanate	$C_8H_{15}NO_2$	CH_3—$(CH_2)_5$—CH_2—N=C=O	47.5 60.9[a]	341	n = 2; 7.9	5.8 ΣDCH_2 = 17.8	7.8
Phenyl isocyanate	C_7H_5NO	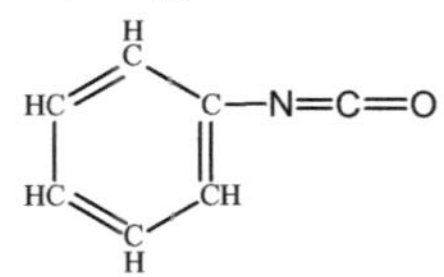	46.5 ± 0.3	298	4DCH = 5.53 × 4 = 22.12	DC=O → CH = 4.9	D=C=O → CH=DC=N → CH = 3.65

[a]Estimated at 288 K

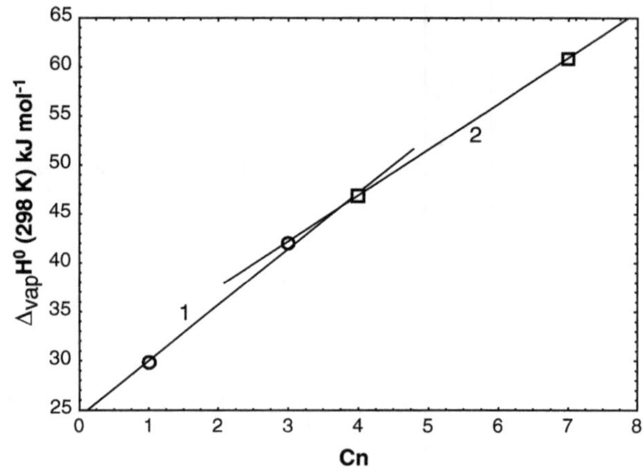

Fig. 6.23 Dependence of the vaporization enthalpy of *N*-methylacetamides versus on the number of carbon atoms in the alkyl chain of *N*-methylformamide – *N*-butylacetamide (*1*), *N,N*-dimethylformamide – *N,N*-dimethylacetamide – *N,N*-diethylacetamide (*2*)

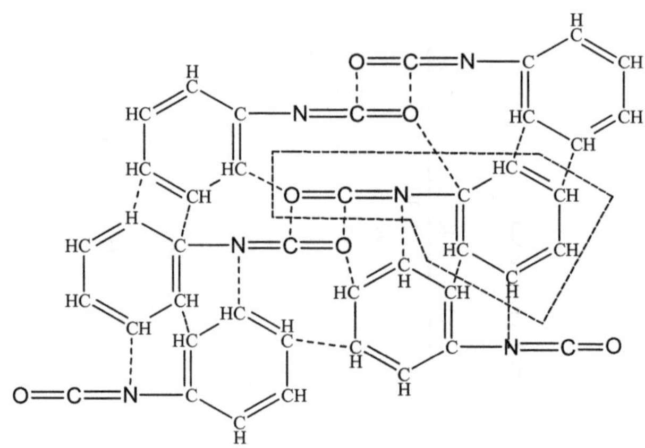

Fig. 6.24 Schematic picture of the liquid structure of phenyl isocyanate with network of the specific interactions

$$2D{=}C{=}O{\rightarrow}C{=} + D{=}C{=}O{\rightarrow}CH + D{=}C{=}N{\rightarrow}CH = (\Delta_{vap}H^0(298\ K) - 4\ D{=}HC{\rightarrow}CH) \tag{6.26}$$

The average value of the energies of these specific interactions is not high, being equal to 4.07 kJ mol^{-1}. We noted above that the low energy value of the specific interaction DC=O → C=, formed in liquid methyl isocyanate, reflects the low ability of the CH_3 group to provide electron density to the nitrogen atom and further to carbonyl oxygen. In its turn the rigid benzene ring also has a low ability to lead to shifting of the electron density. Therefore, it seems logical to take the energy value of the specific interaction D=C=O → C= as equal to the energy value of 4.9 kJ mol^{-1}. Taking into account the natural tendency to reduce the stability of the specific interactions in the order of D=C=O → C= ≥ D=C=O → CH ≈ D=C=N → CH, we have reason to take the equality of the specific interactions

D=C=O → CH=DC=N → CH and estimate the value of the energies (3.65 kJ mol^{-1}).

6.5 Energies of Specific Interactions of Amides and Imides with Two Carbonyl Groups

Acyclic amides with two carboxyl groups keep the ability of shifting of the electron density in the alkyl chain. Nevertheless, the reduction of the number of carbon atoms of the chain significantly influences the charges of the carbon and oxygen atoms of these groups, which determines the energies of the hydrogen bonds and specific interactions formed. Located between two carbonyl fragments, the methylene group of the malonamide molecule leads to the equivalent shifting of the electron density to their carbon atoms and further to the oxygen atoms. However, the obtained electron density is significantly reduced compared with the shifting of the electron density from the two methylene groups, connected directly with atoms of the carbonyl groups. Therefore, the electron density should obtain the excess needed from the amide groups. In this connection the amide groups form hydrogen bonds with small differences in the energy values, forming bonds by acyclic amides with one amide group. In this connection the energy value of this type of hydrogen bond D–N•••H–N should be taken as equal to 9.7 kJ mol^{-1} (Table 6.5). The participation rules of all 12 bond vacancies for the malonamide molecule forms the structure of the crystal with the network of the hydrogen bonds of two types D–N•••H–N and D=C=O•••H–N and specific interactions D=C=O → C=O (two) and D=C=O → CH_2–C (two) (Fig. 6.25a). Equal shifting of the electron density from the methylene group to the carbon atoms of the carbonyl groups and, consequently, the equal stability of the specific interactions formed allow us to take the condition DC=O → C=O ≈ DC=O → CH_2–C, corresponding to the imposition of equality of its average value. The energy value of the fragment CH_2–C=O formed by the acetate in the crystalline malonamide (Table 6.7) should be taken as equal to 9.2 kJ mol^{-1}, realized in the solid diacetamide. The unknown value of the energy of the second type of hydrogen bond is determined by the difference between sublimation enthalpies of malonamide and the energies contributed by the intermolecular interactions

Malonamide:

$$D{=}C{=}O{\cdots}H\text{-}N = (\Delta_{sub}H^0(298\ K) - 4D\text{-}N{\cdots}H\text{-}N - 2D{=}C{=}O{\rightarrow}C{=}O - 2D{=}C{=}O{\rightarrow}CH_2\text{-}C{=}O)\ /\ 4 \qquad (6.27)$$

The oxamide molecule with 12 bond vacancies forms a similar structure of crystal with the network of two types of the hydrogen bonds and one type of four specific interactions (Fig. 6.24b) DC=O → C=O, with reduced energy compared with the average value (9.2) and energy of the interaction, formed by the CH_2–C=O

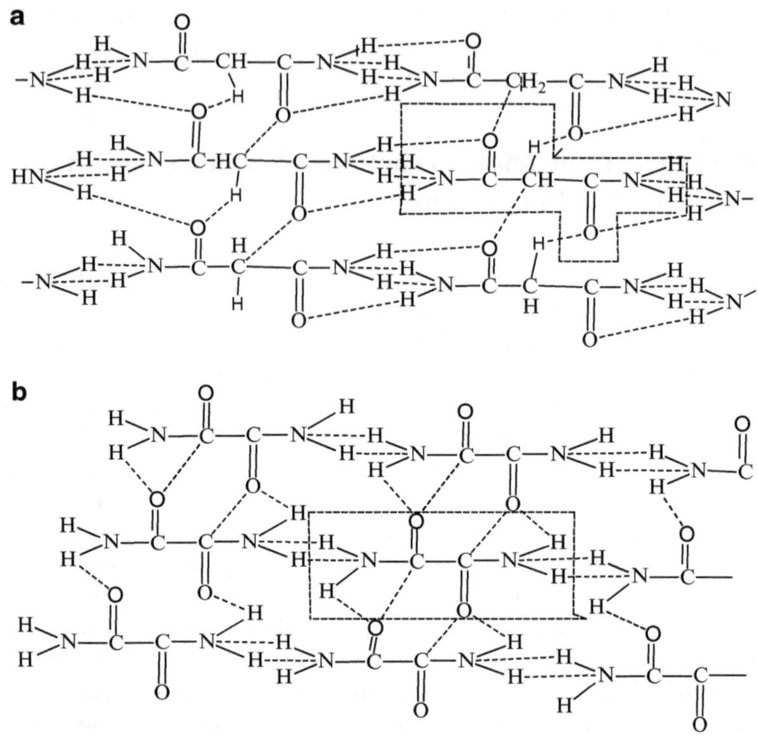

Fig. 6.25 Schematic picture of the liquid structure of malonamide (**a**) and Oxamide (**b**) with network of the specific interactions and hydrogen bonds

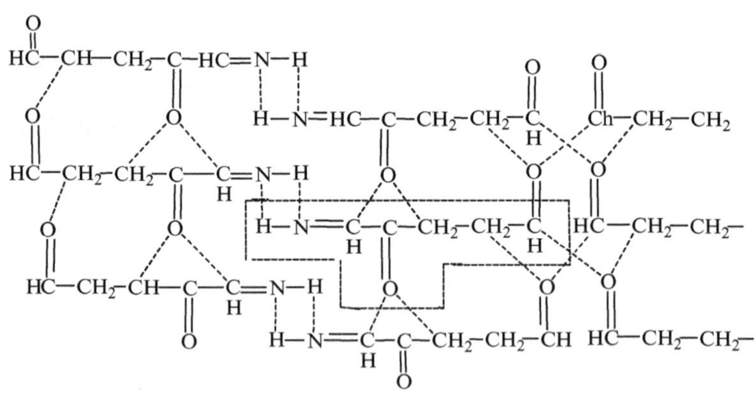

Fig. 6.26 Schematic picture of the liquid structure of glutarimide with network of the specific interactions and hydrogen bonds

group D=C=O → C=O < D=C=O → CH_2–C at 0.6 kJ mol^{-1}. Thus, the energy value of the specific interaction DC=O → C=O should be estimated as

8.5 kJ mol^{-1}. The given value is lower than the average value by 0.7 kJ mol^{-1} and the latter is lower at the same energy value of the interaction DC=O → CH_2–C=O. The energy of the hydrogen bond of increased stability should be determined with the help of the equation

Oxamide

$$D{=}C{=}O{\bullet\bullet\bullet}H\text{-}N = (\Delta_{sub}H^0(298\ K) - 4D\text{-}N{\bullet\bullet\bullet}H\text{-}N - 4DC{=}O{\rightarrow}C{=}O) / 4 \quad (6.28)$$

The glutarimide molecule with ten bond vacancies and imide group forms the structure of the crystal with the network of one type of the hydrogen bond D–N•••H–N with an energy of 9.7 kJ mol^{-1} and two types of specific interactions, D=C=O → CH_2–C=O and D=C=O → CH–CH=O (Fig. 6.26).

The energies of these two types of specific interactions are practically within the experimental error, which value should be determined by the Eq. 6.29

Glutarimide

$$D{=}C{=}O{\rightarrow}CH_2\text{-}C\text{-} = (\Delta_{sub}H^0(298\ K) - 2D\text{-}N{\bullet\bullet\bullet}H\text{-}N\ \text{-}) / 8 \quad (6.29)$$

The results of the calculations (Table 6.15) illustrate stabilization of the specific interactions

Oxamide Malonamide: Glutarimide

D=C=O→C=O (8.6) < (D=C=O→C=O +D=C=O→CH_2-C=O)/2 (9.2) < D=C=O→CH_2-C=O (9.35 kJ mol^{-1})

and the energy of the hydrogen bond DC=O•••H–N: Oxamide (11.0) < Malonamide (12.7 kJ mol^{-1})

6.6 Energies of Specific Interactions of Crystalline Cyclic Imides

Cyclic imides with four carbon atoms and two methylene groups in the cycle point to the fact that for the carbonyl groups (Table 6.17) the specific interactions formed by the oxygen atoms have high stability with a value of the energy which is little different from the energies of the corresponding specific interaction of acyclic imides. It has been proved that the high energies of the specific interactions, formed by the methylene groups of cyclic propane, butane, and pentaneare equal to 9.20, 8.28, and 7.20 kJ mol^{-1}, respectively (Chap. 3). For this reason we conclude that the energy of the hydrogen bond formed by the imide group also has a high stability and in the case of crystalline succinimide it is comparable with the energy value of the same type, formed by solid glutarimide. The structure of the crystal formed by the succinimide molecule with the network of the hydrogen bond D–N•••H–N

Table 6.16 Energies of the hydrogen bond and specific interactions (kJ mol^{-1}) of amides with tow carbonyl groups

Compounds	Formula	Structure	$\Delta_{sub}H°(T)$ [22]	T K	D=C=O → CH_2–C=	D–N•••H–N	D=C=O•••H–N
Malonamide	$C_3H_6N_2O_2$	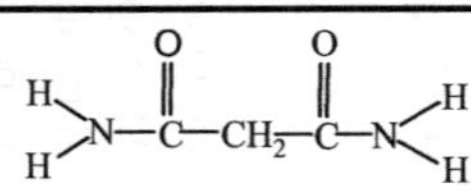	126.4 ± 0.5	298	9.2	9.7	12.7
Oxamide	$C_2H_4N_2O_2$	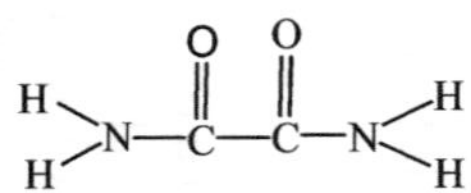	117.3 ± 1.2	298	8.6	9.7	11.0
Dithiooxamide	$C_2H_4N_2S_2$	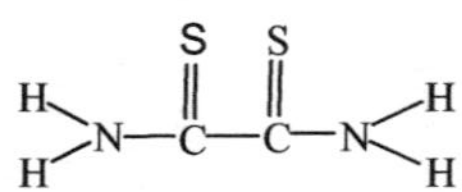	103.6	298	–	–	–
Glutarimide	$C_5H_7NO_2$	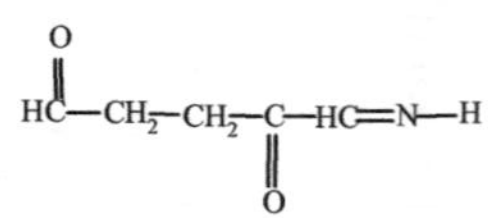	94.1 ± 1.6	298	DC=O → CH_2=DC=O → CH = 9.35	9.7	–

Table 6.17 Energies of specific interactions (kJ mol^{-1}) of cyclic crystalline imides

Compounds	Formula	Structure	$\Delta_{sub}H°(T)$ [22]	T K	D=C=O → CH_2–C=O	D–N•••H–N	DC=O → CH–
Succinimide	$C_4H_5NO_2$	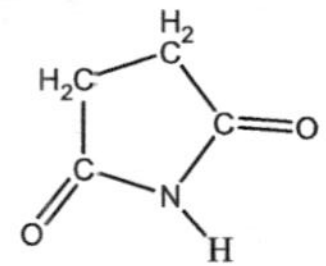	83.6 ± 1.5	298	8.03	9.7	–
N-Methylsuccinimide	$C_5H_7NO_2$	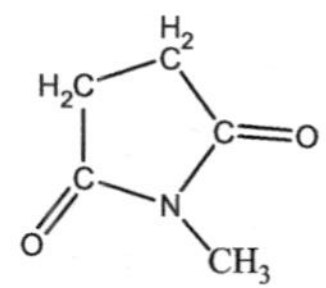	80.1 ± 0.3	298	8.03	7.95	–
N-Methylmaleimide	$C_5H_5NO_2$	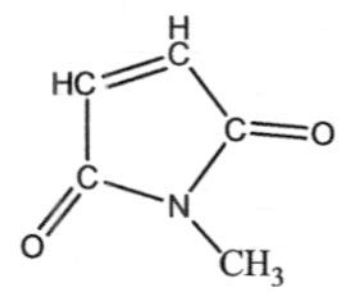	73.3 ± 0.5	298	–	7.95	7.2
N-Methylphthalimide	$C_9H_7NO_2$	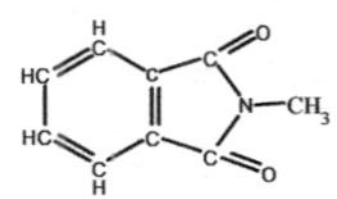	91.10.5	298	D=C=O → C= ≈ 2D=C → C = 5.2	D–N → CH_3–N = 7.7	7.2
2-Pyrrolidone	C_4H_7NO	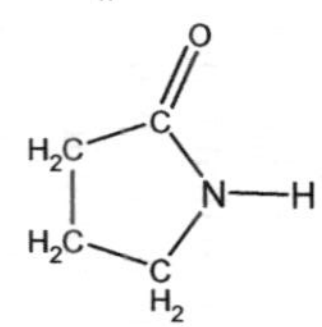	69.1 ± 0.5 Liquid [14]	298	DH_2C → C–H_2 = 7.2 D=C=O → C = 6.0	13.3	DC=O → CH_2 = 8.0

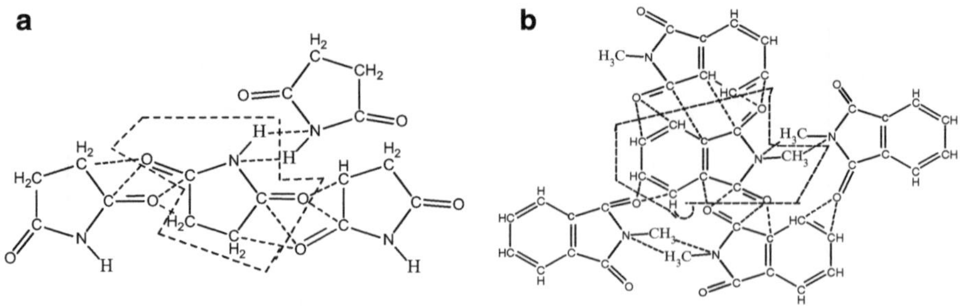

Fig. 6.27 Schematic picture of the crystalline structure of succinimide (**a**) and *N*-methylphthalimide (**b**) with network of the specific interactions

(9.7 kJ mol^{-1}) includes two types of specific interactions, D=C=O → CH_2–C= and D=C=O → C=O, different in the values of the energies (Fig. 6.27a). Thus, on the basis of the energy contribution of the interactions forming the sublimation enthalpy of succinimide we determine the average value of the energy of the two specific interactions (Table 6.17)

Succinimide

$$(\text{D=C=O}\rightarrow\text{CH}_2\text{-C =O}+\text{D=C=O}\rightarrow\text{C=O})/2 = (\Delta_{\text{sub}}H^0(298\ \text{K}) - 2\text{D-N}\cdots\text{H-N -})\ /\ 8 \quad (6.30)$$

Replacement of the hydrogen atom in the succinimide molecule by the methyl group is accompanied by reduction in the sublimation enthalpy *N*-methylsuccinimide at 3.5 kJ mol^{-1} or replacement of the hydrogen bond by the one formed by specific interaction D–N → CH_3–N (Fig. 6.27b), which is accompanied by the reduction of the enthalpy characteristics at 1.75 kJ mol^{-1}. It follows that the energy contribution of the specific interaction formed by the imide group of N–CH_3 is equal to 7.95 kJ mol^{-1}. Notice that by the method of RE–spectrometry [19–21] it was shown that replacement of the hydrogen atom by the methyl group in the amine group is accompanied by insignificant reduction of the electron density. However, the difference in the energies for the noted value points to the significant influence of the oxygen atoms on the shifting of the electron density, accompanied by the increase in the difference between the charges of the nitrogen and carbon atoms of the methyl group, increasing the energy value of the formed interaction from 5.3 kJ mol^{-1} in the imide group at the four-coordinated nitrogen atom up to 7.95 kJ mol^{-1} in the compound considered (Table 6.17).

Replacement of the methylene groups with a single bond in the *N*-methylsuccinimide molecule to the double –CH=CH– group in *N*-methylphthalimide is accompanied by the reduction of sublimation enthalpy at 6.8 kJ mol^{-1}. It follows that there is a significantly reduced transmission of the electron density from CH groups to the oxygen atoms and there is a reduced difference in the charges of the latter and the carbon atoms of the cycle. In this connection one should expect some

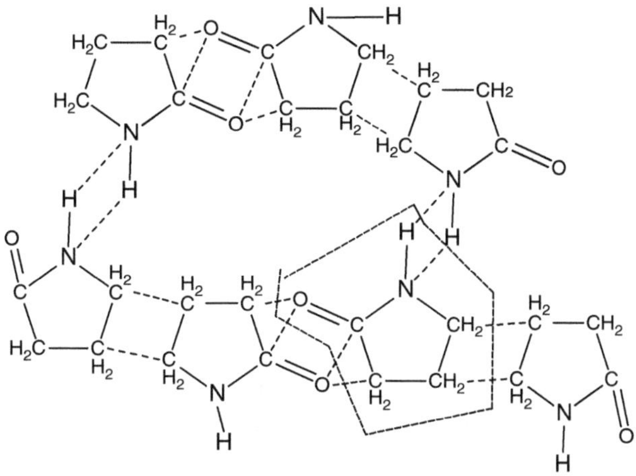

Fig. 6.28 Schematic picture of the crystalline and liquid structure of 2-pyrrolidone with network of the specific interactions

increase in the shifting of the electron density from the methyl group of imide fragment in the *N*-methylmaleimide molecule and one should suppose that the energy of the specific interaction D–N → CH_3–N practically does not change (7.95 kJ mol^{-1}). The obtained energy value of the specific interaction D=C=O → CH–C=O (Table 6.17) with the use of Eq. 6.30 points to the equality in the energies of the two types of the specific interactions, realized in crystalline *N*-methylsuccinimide and *N*-methylmaleimide

$$D{=}C{=}O{\rightarrow}CH_2\text{-}C{=}O\ (8.03) > D{=}C{=}O{\rightarrow}CH\text{-}C{=}O\ (7.2\ \text{kJ mol}^{-1}).$$

The *N*-methylphthalimide molecule forms the network structure of crystal (Fig. 6.26b), in which three types of specific interactions: D=C=O → CH= (six), DC=O → C= (four) and D=C → C= (two) and two stable hydrogen bonds D–N•••H–N participate. Such a large number of interactions of the oxygen atoms with carbon atoms and between carbon atoms at the two oxygen atoms of the carbonyl groups presupposes the maximum possible shifting of the electron density from the CH_3 imide group N → CH_3, leading to the highest stability of the specific interactions D–N → CH_3–N and the value of its energies (7.7 kJ mol^{-1}) of crystalline imides and amides. Taking the energy values of the specific interactions D=C=O → CH= as equal (7.2 kJ mol^{-1}) that realized in crystalline *N*-methylmaleimide, we can estimate the average value of the two types of interactions

$$(4D{=}C{=}O{\rightarrow}C{=}\ +2D{=}C{\rightarrow}C{=})/6 = (\Delta_{sub}H^0(298\ K)\ -\ 2D\text{-}N{\bullet\bullet\bullet}H\text{-}N\ -\ 6\ D{=}C{=}O{\rightarrow}CH{=})/6 \qquad (6.31)$$

with the inequality between them given by 4DC=O → C= > 2D=C → C= (Table 6.17). Liquid 2-pyrrolidone with a chain structure, criss-crossed by hydrogen bonds, formed by the amide (imide) group includes two hydrogen bonds

D–N•••H–N and three types of specific interactions with two bonds of each D=C=O → CH_2 > D–H_2C → C–H_2 > D–H_2C → C= (Fig. 6.28). Notice that at the eight specific interactions and single bonds in the cycle 2-pyrrolidone the vaporization enthalpy of this compound is less than the sublimation enthalpy characteristics of *N*-methylmaleimide at 4.2 kJ mol^{-1}, while the interactions are formed by CH groups. It means that the energies of the specific interactions D–H_2C → CH_2 in liquid 2-pyrrolidoneand crystalline D=HC → CH *N*-methylmaleimide are practically equal because of the significant influence of the oxygen atom of the carbonyl group to the shifting of the electron density from the methylene groups of the cycle. We estimate the value of the energy of the most stable specific interactions D=C=O → CH_2 to be equal to the energy of the same interaction in crystalline *N*-methylsuccinimide, which has a significant deficit in electron density, apparently compensated by the influence of the crystalline field. Taking into account the contribution of the substituted hydrogen atom to the vaporization enthalpy (0.6 kJ mol^{-1}), we estimated the energy of the specific interaction D=C=O → C– to be equal to 6.0 kJ mol^{-1} and the energy of the hydrogen bond to be equal to 13.3 kJ mol^{-1}, respectively.

6.7 Energies of Specific Interactions of Crystalline α-oxo benzene acetonitrile and Cyanophenones

The compounds A-oxo-benseneacetonitrile and cyanophenones with a rigid benzene ring, one functional carbonyl and one nitrile group show a definite similarity. Hence, there exists the possibility to follow the influence of acetate, propyl group and the locations of nitrile group in the molecules of these compounds on the stability of the specific interaction D≡N → C≡. The A-oxo-benseneacetonitrile molecule forms the network structure of a crystal (Fig. 6.29a), consisting of four specific interactions D=C=O → CH, four D=HC → CH with an energy of 7.27 kJ mol^{-1}, considering the calculated contribution of the substituted hydrogen atom of the benzene ring and two D≡N → C≡, formed by carbonyl oxygen, CH groups of the benzene ring, and the C≡N fragment. The value of the energy of the specific interaction DC=O → C–H is taken as equal to the energy value of the same type of the interaction, realized in crystalline *N*-methylmaleimide (7.2 kJ mol^{-1}) and *N*-methylphthalimide (Table 6.17). The energy of the specific interaction we determine from Eq. 6.32

$$D{\equiv}N{\rightarrow}C{\equiv} = (\Delta_{sub}H^0(298\ K)ban - 4D{=}HC{\rightarrow}CH - 4D{=}C{=}O{\rightarrow}CH\text{-})/2 \qquad (6.32)$$

The network structure of crystalline cyanoacetophenone (Fig. 6.29b) includes five specific interactions D=HC → CH, formed by CH groups of the benzene ring with an average energy value of 7.1 kJ mol^{-1}, with attention to the value of the contribution of two substituted hydrogen atoms, two specific interactions

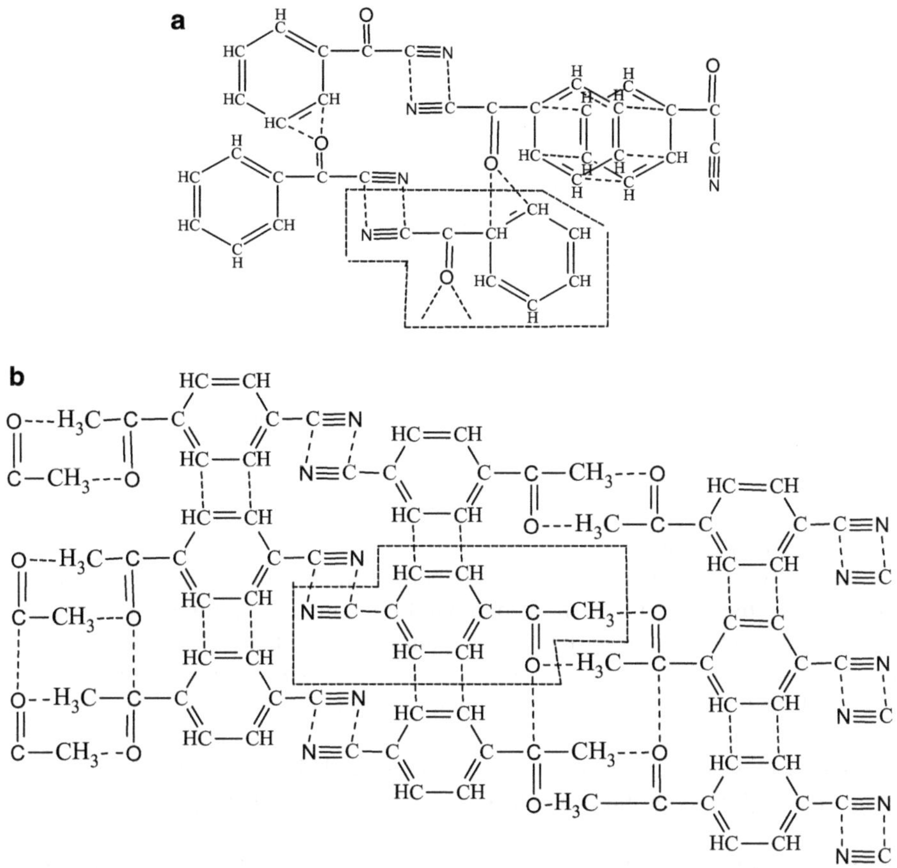

Fig. 6.29 Schematic picture of the crystalline structure of A-oxo-benseneacetonitrile (**a**) and cyanoacetophenone (**b**) with network of the specific interactions

D=C=O → C with the energy value of 5.8 kJ mol^{-1} obtained above or reduced by the energy contribution of the hydrogen atom (1.0 kJ mol^{-1}) to the specific interaction D=C=O → CH (6.2 kJ mol^{-1}), two specific interactions of D=C=O → CH_3 with the energy value equal to the interaction formed for crystalline dimethyl ether and an unknown value of the energy of two specific interactions D≡N → C≡.

The energy of the last specific interaction is determined by the equality

$$D{\equiv}N{\rightarrow}C{\equiv} = (\Delta_{sub}H^0(298\ K)caph - 5D{=}HC{\rightarrow}CH - 2D{=}C{=}O{\rightarrow}C - 2D{=}C{=}O{\rightarrow}CH_3\text{-}C)/2 \quad (6.33)$$

Similar specific interactions are realized in the network crystalline structure of β-cyanopropiophenone with propyl ligand replacing the interaction DC=O

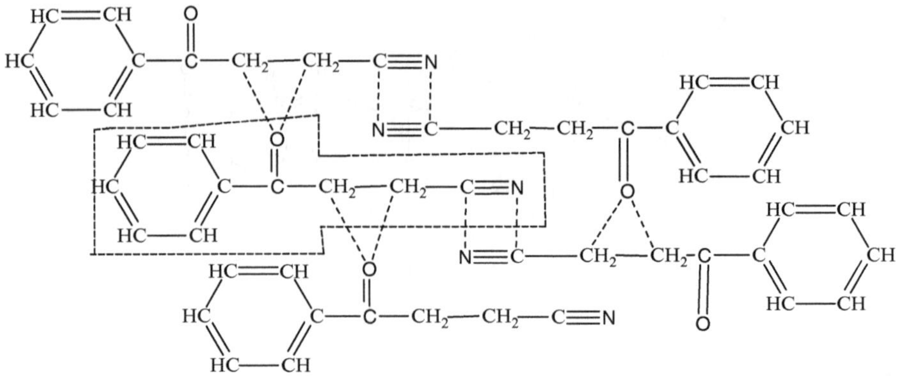

Fig. 6.30 Schematic picture of the crystalline structure of β-cyanopropiophenone with network of the specific interactions

→ CH_3–CH_2 in cyanoacetophenone to D=C=O → CH_3–CH_2–CH_2 with an energy of 11.61 kJ mol^{-1}.

The second form of β-cyanopropiophenone differs from the one considered by the fact that the molecule of this compound contains one fragment in which the propyl group is fringed by carbonyl and cyanide groups (Fig. 6.30). As a result, the propyl group participates in the formation of the same types of interactions, realized in crystalline β-cyanopropiophenone with the propyl and cyanide fragments located at C(1) and C(4), respectively. Thus, the energy of the unknown specific interaction D≡N → C≡ in crystalline cyanoacetophenone and the two different molecular structures of β-cyanopropiophenone are determined with the help of Eq. 6.33a

$$D{\equiv}N{\rightarrow}C{\equiv} = (\Delta_{sub}H^0(298\ K)cpph - 5DHC{\rightarrow}CH - 2DC{=}O{\rightarrow}C - 2DC{=}O{\rightarrow}CH_3\text{-}CH_2\text{-}C)/2 \qquad (6.33a)$$

The results of the calculations (Table 6.18) illustrate the natural stabilization of the energies of the specific interaction D≡N → C≡, realized in the crystalline acyclic and cyclic cyanides

D≡N→C≡: A-Oxo-benseneacetonitrile (10.5) < Cyanoacetophenone (13.55) ≈ 2-

Cyanoacetamide (13.6) < β-Cyanopropiophenone (15.5 kJ mol^{-1})

The values of the energy obtained for the considered specific interactions allow us to draw the following conclusions.

First, the energy of the specific interaction D≡N → C≡ has a minimum value on the condition of the location of the cyanide group at the carbonyl carbon atom, directly connected with the benzene cycle;

Table 6.18 Energies of specific interactions (kJ mol^{-1}) of cyanoacetophenones and β-cyanopropiophenones

Compounds	Formula	Structure	$\Delta_{sub}H°$ (298 K) [22]	T K	D=HC → C–H	D=C=O → C–H	D≡N → C≡
A-Oxo-benseneacetonitrile	C_8H_5NO	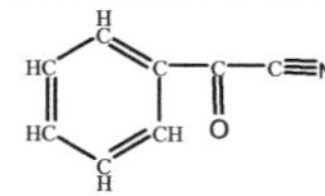	78.7 ± 4.2	298	7.27	7.2	10.5
Cyanoacetophenone	C_9H_7NO	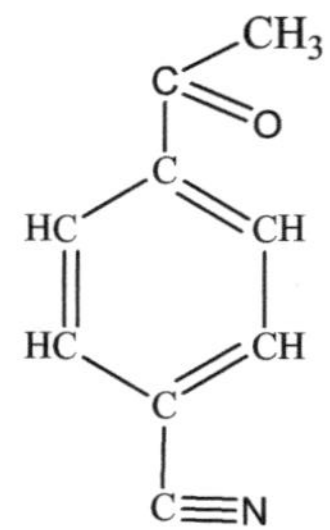	89.81.4	298	7.10	DC=O → C = 5.8/D=C=O→ CH_3–C = 6.59	13.55
β-Cyanopropiophenone	$C_{10}H_9NO$	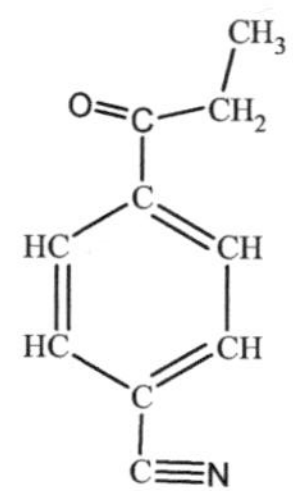	101.7 ± 4.2	325.5	7.10	D=C=O → C = 5.8/D=C=O → CH_3–CH_2–C = 11.61	15.5

(continued)

Table 6.18 (continued)

Compounds	Formula	Structure	$\Delta_{sub}H°$ (298 K) [22]	T K	D=HC → C–H	D=C=O → C–H	D≡N → C≡
β-Cyanopropiophenone	$C_{10}H_9NO$	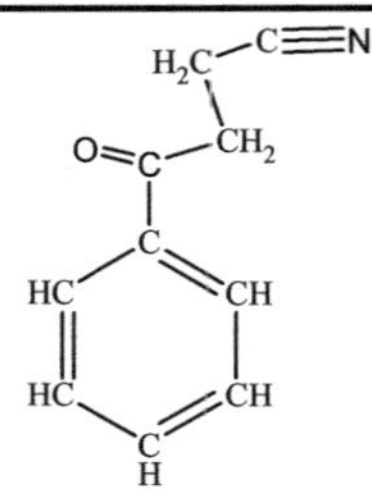	101.7 ± 4.2	325.5	7.27	D=C=O → C = 5.8/D=C=O → CH_2–CH_2–C = 11.61	15.3
4-Aminoacetophenone	C_8H_9NO 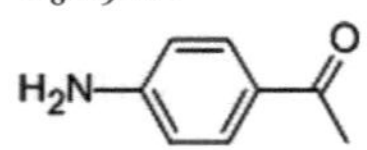	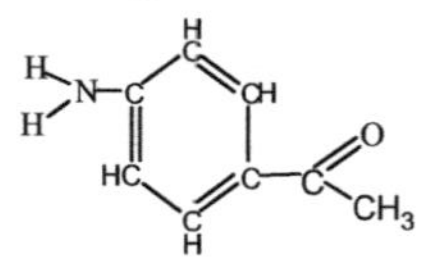	93.7	326	7.07 × 6	D=C=O•••H–N = 9.7 × 4	D–N → CH_3–C= = 6.3
N-Benzoyl-*N*-methylbenzamide	$C_{15}H_{13}NO_2$	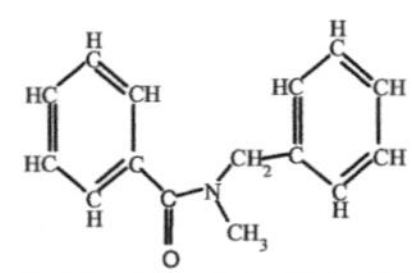	120.1 ± 0.4	298	8DHC → CH 7.13 × 8	2D–N → CH_3–N + 2D–N → CH_2–N = 7.7 × 2 + 6.7 × 2	2D=HC → C + 2D=C → C = 9.1 × 2 + 8.1 × 2

Second, the location of the methyl group with the carbon atom of the carbonyl fragment provides it with equal shifting of the electron density as the acetate group of the acyclic compound, which gives rise to the equality of the energy values of this specific interaction of cyanoacetophenone (13.55) ≈ 2-cyanoacetamide (13.6 kJ mol^{-1});

Third, the increase of the number of methylene groups in the propyl ligand is accompanied by a further increase in the shifting of the electron density to the cyanide group, leading to the further stabilization of the specific interaction of β-cyanopropiophenone (15.5 kJ mol^{-1});

Fourth, the completion of the influence of the intermolecular reverse dative bond at the third carbon atom of the propyl fragment causes a completion of the stabilization of the specific interaction D≡N → C≡. Hence, the further increase of the number of the methylene groups of the alkyl chain is accompanied practically by the adequate energy contribution by each of these groups to the vaporization enthalpy, leading to the use of the extrapolation method in order to estimate the enthalpy characteristics of unstudied compounds of this series.

The 4-aminoacetophenone molecule with 12 bond vacancies forms the network structure of the crystal by the two types of specific interactions and the hydrogen bond (Fig. 6.31a). The acetate and amine groups of the contacting molecules form four hydrogen bonds by carbonyl oxygen atoms and hydrogen atoms of the amine group D=C=O•••H–N with an energy of 9.7 kJ mol^{-1} and two specific interactions with participation of the methyl group of the acetate fragment D–N → CH_3–C=O. Six specific interactions form CH groups 4D=HC → CH and carbon atoms with the substituted hydrogen atom of the benzene ring 2D=C → C= have an average energy value of 7.07 kJ mol^{-1}, calculated from the sublimation enthalpy reduced by the contribution of two substituted hydrogen atoms. The unknown value of the energy of the specific interaction D–N → CH_3–C=O is determined with the help of Eq. 6.34, considering the contributions of all types of specific interactions in the crystalline 4-aminoacetophenone

$$\text{D-N}\rightarrow\text{CH}_3\text{-C=O} = (\Delta_{sub}H^0(298\ \text{K})\text{aaph} - (4\text{D=HC}\rightarrow\text{C-H} + 2\text{D=C}\rightarrow\text{C=}) - 4\text{D=C=O}\cdots\text{H-N})/2 \quad (6.34)$$

The energy value of this type formed by the nitrogen atom has a reduced stability compared with the formed oxygen atom

D-N→CH_3-C=O (6.3) 4-Aminoacetophenone < D=C=O→CH_3-C=O (10.45 kJ mol^{-1}) N-Methylacetamide

The *N*-benzoyl-*N*-methylbenzamide molecule with 16 bond vacancies and planar benzene rings forms the network structure of the crystal by the five types of specific interactions (Fig. 6.31b). The nitrogen atoms of the amide groups form two

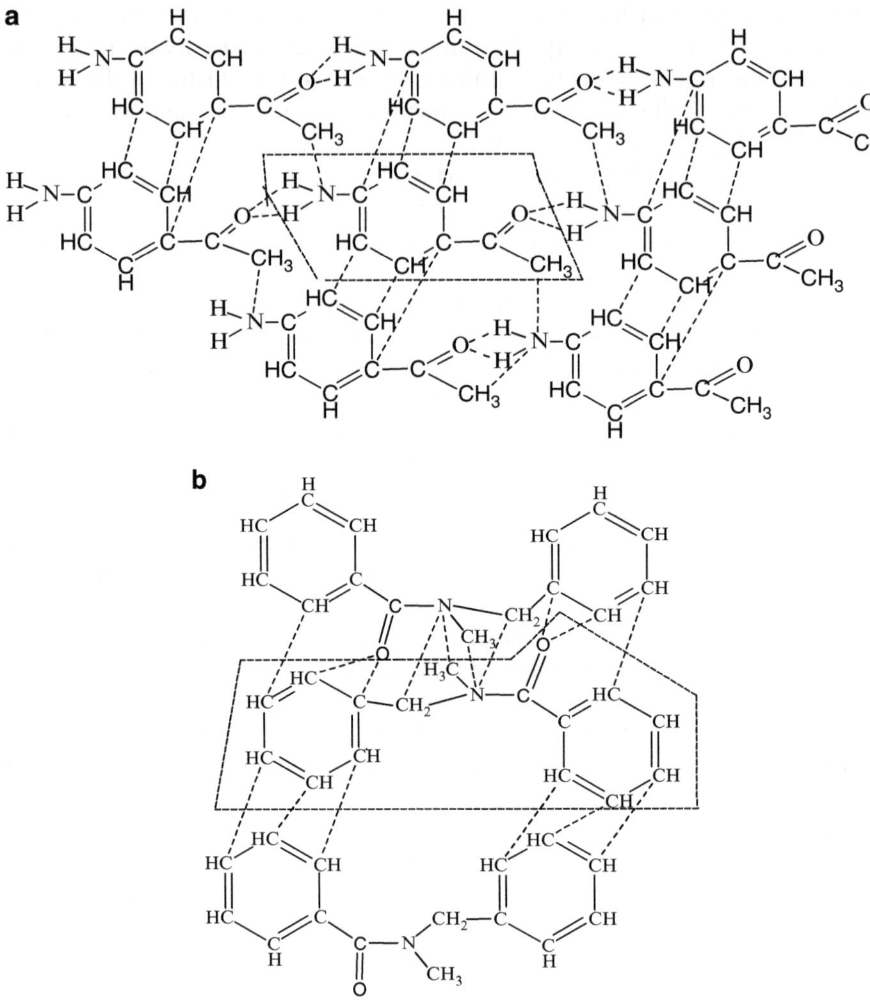

Fig. 6.31 Schematic picture of the liquid structure of 4-aminoacetophenone (**a**) and *N*-benzoyl-*N*-methylbenzamide (**b**) with network of the specific interactions

specific interactions with methyl groups D–N → CH_3–N and with methylene 2D–N → CH_2–N with energies 7.7 kJ mol^{-1} and reduced energy 6.7 kJ mol^{-1}, respectively, formed by the CH_2 group with a smaller number of hydrogen atoms. Four CH groups of each benzene ring participate in the formation of eight specific interactions 8D=HC → CH with an energy of 7.13 kJ mol^{-1}.

Amide fragments of the interacting molecules formed by the carbonyl oxygen atoms two specific interactions with CH groups 2D=HC → C= and two specific interactions with carbon atoms of the benzene rings 2D=C → C=, the energies of which could be obtained with the help of an equation considering the contributions of all specific interactions of crystalline *N*-benzoyl-*N*-methylbenzamide,

$$(2D{=}HC{\rightarrow}C + 2D{=}C{\rightarrow}C{=})/4 = (\Delta_{sub}H^0(298\ K)bmba - 8D{=}HC{\rightarrow}C\text{-}H - (2D\text{-}N{\rightarrow}CH_3\text{-}N + 2D\text{-}N{\rightarrow}CH_2\text{-}N)/4 \quad (6.35)$$

The obtained average value of the energies of two types of the specific interactions $(2D{=}HC \rightarrow C + 2D{=}C \rightarrow C{=})/4 = 8.6$ kJ mol^{-1} occupies an intermediate position, $D{=}HC \rightarrow C > 8.6 < D{=}C \rightarrow C{=}$, and differ from the natural values of the energies of these bonds at 0.5 kJ mol^{-1}.

References

1. Ehrenson S (1968) Theory of electron effect at formation of complex of trimethylamine with trimethylborium. Theor Chim Acta 10:193
2. Ehrenson S (1968) Effect environment at examination of a reactivity adding of trimethylamine to trimethylborium. Theor Chem Acta 10:209
3. Gurianova EN, Goldshtain IG, Romm IP (1973) Donor-acceptor bonds. Chemistry, Moscow, p 137
4. Baev AK (2003) Novel approaches to the structure of functional solvents and elementorganic compounds with abstract international conference on solution chemistry, Debrecen, p 79
5. Baev AK, Korolkov DV (2003) Phenomenon of dative bonds in elementorganic compounds with alkyl ligand. Abstract XVth FECHEM conference on organometallic chemistry. University of Zurich, Zurich, p 350
6. Baev AK (2004) Insolvence of model sp^3-hybridization of carbon atom and chemistry of non aqua solutions. IXth international conference on the problem of solvation and complex formation in Solutions, Plyos, pp 62–64
7. Baev AK (2004) Lack of co-ordination thermodynamic properties of elementorganic compounds with sp^3-hybridization of carbon atom and its insolence. Russian scientific symposium on thermo chemistry and calorimetry, Niznie Novgorod, pp 40–41
8. Baev AK (2009) Phenomenological significance of nonidentical sp^3 hybridization model and a nature of interaction in chemistry of liquid non-electrolites and chemical thermodynamics. Abstracts Xth international conference on physical and coordination Porphyrin's chemistry and their analogous (ICPC-10), Ivanovo, p 86
9. Veillard A, Levy B, Danbll R, Gallais F (1967) Structural electroniqul of borazane BH_3-NH_3. Theor Chim Acta 8:312
10. Moireau MG, Veillard A (1968) Quantum mechanical calculations on barriers to internal rotation. II. The borazane molecular BH_3-NH_3. Theor Chim Acta 11:344
11. Veillard A, Daudel R (1970) Nature et proprieties des liaison de Coordination. Colloques internationaux du centre National de la Recherche Scientifigue, Paris, p 22
12. Kudchadk SA, Rao CNR (1970) Extended Huckel calculations on N-V addition compounds. Theor Chim Acta 18:258
13. Baev AK (2012) Specific intermolecular interactions of organic compounds. Springer, Hiedelberg/Dordrecht/London/New York, 434p
14. Chickos JS, Acree WE (2002) Enthalpies of vaporization of organic and organometallic compounds, 1880–2002. J Phys Chem Rev 31(2):537

15. Nefedov VI, Vovna VI (1989) Electron structure of organic and elementorganic compound. Nauka, Moscow, 175p
16. Baev AK (1969) Problems of chemical nature of phase transformation. General and applied chemistry, vol 1. Vysheischaia Educating, Minsk, pp 197–206
17. Baev AK (1969) Phase condition and complex formation ability of halogenide metals. General and applied chemistry, vol 1. Vysheischaia Educating, Minsk, pp 207–218
18. Baev AK (1987) Chemistry of gas-heterogenic systems of elementorganic compounds. Science and Technology, Minsk, p 174
19. Karlson TA (1981) Photoelectron and OZE-spectroscopy. Mech. Eng. Russia. Leningrad, p 431
20. Nefedov VI (1984) Infra red-spectroscopy of chemical compounds, Chemistry. Moscow, p 255
21. Vovna VI, Vilesov FI (1974) Photoelectron specters and structure molecular orbital's methylamines. Opt Spectrosc 34(2):436–438
22. Chickos JS, Acree WE Jr (2003) Enthalpies of sublimation of organic and organometallic compounds, 1910–2001. J Phys Chem Rev Data 32:537–698
23. Ivanov EV, Abrosimov VK (2001) Contemporary state of investigations of structural and volumetric condition of urea and its aqueous solutions. Biological activity compounds in solutions. Structure, thermodynamics, reactionary ability. Science, Moscow, pp 110–183

Chapter 7
Specific Intermolecular Interactions of Nitrogen and Oxygen Containing Cyclic Hydrocarbons

7.1 Energies of Specific Interactions of Compounds with Pentamerous Heterocycle

7.1.1 Oxazole, Isoxazoles, and Thiazoles

The simplest oxazole molecules and its derivatives with two hetero atoms in the cycle exclude the possibility of understanding the specificity of the formed interactions with the use of ideas of sp^3-hybridization [1–7]. The increased ability of the oxygen atom to shift of the electron density predetermines the increased positive charge of the carbon atom, at location C(2), located between these atoms. Therefore, the interaction of the oxygen atom with carbon atoms at locations C(2) and C(5) leads to the four interactions of one type, D–O → CH, in the structure of liquid oxazole with the network of specific interactions with increased stability (Fig. 7.1). The second type of specific interaction is formed by the nitrogen atom of contacting molecules with double bond that are close in the environment, D=N → CH. The energy of the first type of specific interaction should be taken as equal to the value of the same type of 1,2-butadiene formed or the same as the energy of the double bond of cyclobutene (5.97 kJ mol^{-1}). The energy of the second type of interaction is determined by Eq. (7.1)

$$D{=}N{\rightarrow}C\text{-}H = (\Delta_{vap}H^0(298\ K)oz - 4D\text{-}O{\rightarrow}CH)/2 \tag{7.1}$$

considering the energy contributions of the realized types of specific interactions [9–11]. The results of the calculation are presented in Table 7.1.

The location of hydrogen and nitrogen atoms at locations O(1) and N(2) in the cycle of the isoxazole molecule contribute significantly to the shifting of the electron density from the CH group at locations C(5) and C(3), leading to the increase of the vaporization enthalpy of this compound at 4.0 kJ mol^{-1} compared with oxazole (Table 7.1). It means that charges of oxygen and carbon atoms C(5) and, on the other side, nitrogen and carbon atoms C(3) have an increased

A.K. Baev, *Specific Intermolecular Interactions of Nitrogenated and Bioorganic Compounds*, DOI 10.1007/978-3-642-37472-2_7,

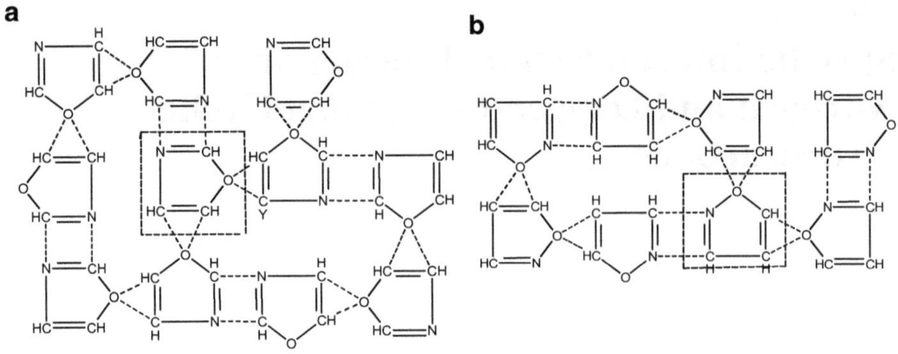

Fig. 7.1 Schematic picture of the liquid structure of oxazole (**a**) and isoxazole (**b**) with network of the specific interactions

difference in the negative and positive charges, respectively, and, as a result, the energies of the similar types of six specific interactions, forming the network structure of liquid isoxazole (Fig. 7.1b), increase on average by 0.65 kJ mol^{-1}.

The methyl derivatives of isoxazole form a similar structure of liquid condition with the network of stable specific interactions D–O → CH and D–O → C–H of the two types (Fig. 7.2a).

The difference of the vaporization enthalpies of 3-methylisoxazole and 5-methylisoxazole within the experimental error (0.1 kJ mol^{-1}) points to the equal energy contribution to these characteristics, made by the isostructural methyl group of these compounds. It leads to the conclusion that the value of the contributed energy of the isostructural methyl groups does not depend on its location at C(3) and C(5), which is possible on the condition of a small difference of the positive charges of these two carbon atoms. The value of the energy of the isostructural groups introduced into the enthalpy characteristics is determined by the difference of the vaporization enthalpies of 3-methylisoxazole and 5-methylisoxazole with isoxazole. The two specific interactions formed (Fig. 7.2a) have reduced stability, however, it is about twice higher than the stability of the similar interactions of the liquid compounds of other classes (Table 7.1). Replacement of the oxygen atom in the oxazole molecule by the sulfur atom preserves the network structure of liquid thyazole, formed by four specific interactions D–S → C–H and two interactions D=N → CH of the second type, causing the increase of the vaporization enthalpy at 7.0 ± 0.1 kJ mol^{-1} with the same number of interactions. Taking into account the energy contribution of the isostructural methyl group to the vaporization, which is equal to 3.2 kJ mol^{-1}, we obtain the correct value of enthalpy characteristics of thyazole from the difference in vaporization enthalpy of 4-methylthiazole (40.6 kJ mol^{-1}). Since the preferential stabilization of one type of specific interaction is always accompanied by low stabilization of the second type of interaction, we take the energy of the bond

Table 7.1 Energies of the specific interactions (kJ mol^{-1}) of the liquid oxazole, isoxazoles, and thyazoles

Compounds	Formula	Structure	$\Delta_{vap}H°(298\ K)$ [8]	T K	D–O → C–H	D=N → C–H	D*iso*H_2C → H–CH_2
Oxazole	C_3H_3NO	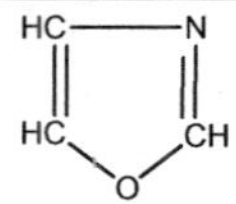	32.5 ± 0.1	298	5.97	4.30	–
Isoxazole	C_3H_3NO	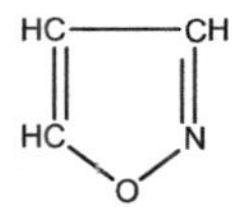	37.2 ± 0.2 36.5 ± 0.1	298 298	6.65	5.0	–
3-Methylisoxazole	C_4H_5NO	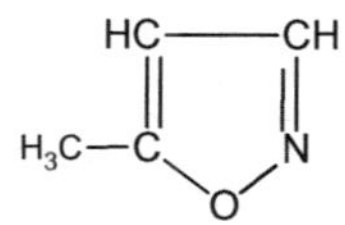	39.8 ± 0.2	298	6.65	5.0	1.6
5-Methylisoxazole	C_4H_5NO	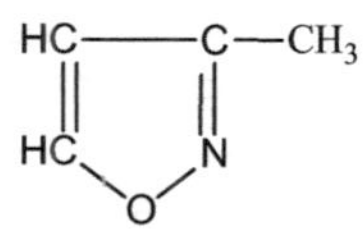	39.7 ± 0.2	298	6.65	5.0	1.55
Thyazole	C_3H_3NS	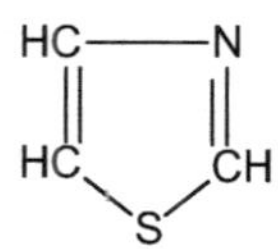	39.3 40.6[a]	348 298	DS → C–H 7.6	D=N → C–H 5.0	–
2-Methylthiazole	C_4H_5NS	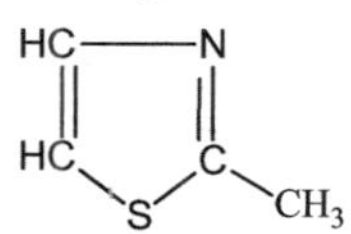	40.0 43.2[a]	357 298	7.6	5.0	1.6
4-Methylthiazole	C_4H_5NS	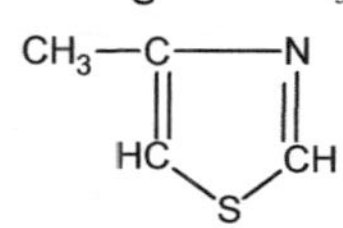	43.8 ± 0.2	298	7.6	5.0	1.6

[a]At T = 298 K

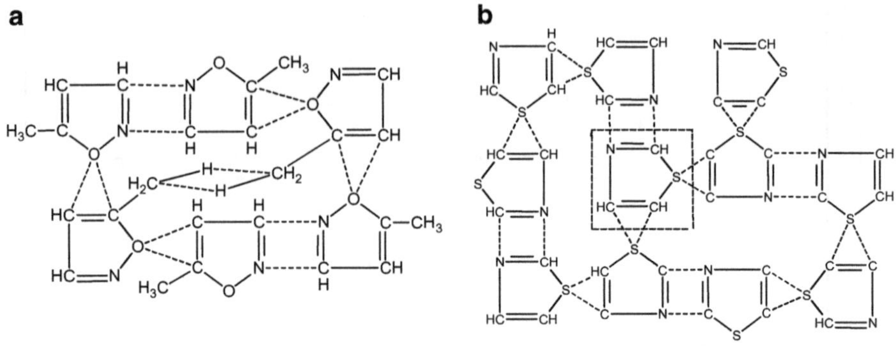

Fig. 7.2 Schematic picture of the liquid structure of 5-methylisoxazole (**a**) and thyazole (**b**) with network of the specific interactions

D=N → CH equal to 5.0 kJ mol^{-1} and obtain with the help of Eq. 7.1 the energy value of the DS → CH interaction (Table 7.1).

7.1.2 Oxazolines, Azetidine, and Aminoisoxazoles

Replacement of the two double bonds by the single ones in the cycle of oxazole molecule with preservation of the location of the oxygen atoms O(1) and nitrogen atom at location N(3) is accompanied by a significant influence on the shifting of the electron density from the CH_2 groups, with preservation of charge differences $C^+(5) > C^+(4)$ at the carbon atoms of these groups. Thus, the specific interactions formed by the oxygen atom, D–O → CH_2, should have some differences in their energies. The availability of the double bond at the nitrogen atom assumes the rigidity of this fragment and a small change of the charge at the nitrogen atom compared with the oxazole molecule. Corresponding with this one should suppose that the energies of the specific interactions D=N → CH in liquid oxazole and D=N → CH_2 in oxazolines should differ insignificantly. At the same time, the contribution of the one located at the second carbon atom of the methyl group to the vaporization enthalpy of oxazoline should correspond to the contribution of the isostructural functional group on its location at the carbon atom with an increased positive charge, which is C(2). In this connection the small difference is symptomatic of the vaporization enthalpies of 3-methylisoxazole and 5-methylisoxazole (Table 7.1) with the enthalpy characteristics of 2-methyl-2-oxazoline (Table 7.2), for which of its character is the formation of identical structures of liquid condition (Fig. 7.3) with six stable specific interactions of the two types as a full analogue of D=N → CH of the two interactions. On the basis of the general rule of a small difference of the energies of the specific interactions D–O → CH < D–O → CH_2, the practical equality of the vaporization enthalpies of methylisoxazole and 2-methyl-2-oxazoline points to the fact that forming the first type of interaction at the location of oxygen and nitrogen atoms at O(1) and N(2) provides them with the

Table 7.2 Energies of the specific interactions (kJ mol^{-1}) of the liquid and crystalline oxazolines

Compounds	Formula	Structure	$\Delta_{evap}H°$(298 K) [8]	T(K)	D–O → C–H_2	D=N → C–H	D*iso*H_3C → H–CH_2
2-Methyl-2-oxazoline	C_4H_7NO	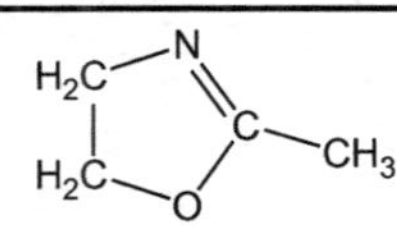	39.1 ± 0.3	298	6.5	5.0	1.6
2-Ethyl-2-oxazoline	C_5H_9NO	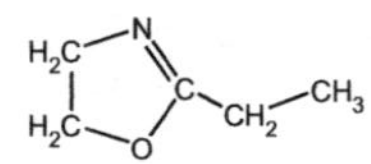	44.2 ± 0.4	298	6.5	5.0	4.05
			Crystal [12]				
3-Amino-5-methylisoxazole	$C_4H_6N_2O$	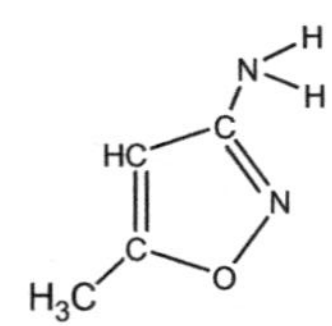	81.6 ± 2.5		7.2	5.5 D*iso*CH_3 = 3.2	D–N•••H–N = 9.7

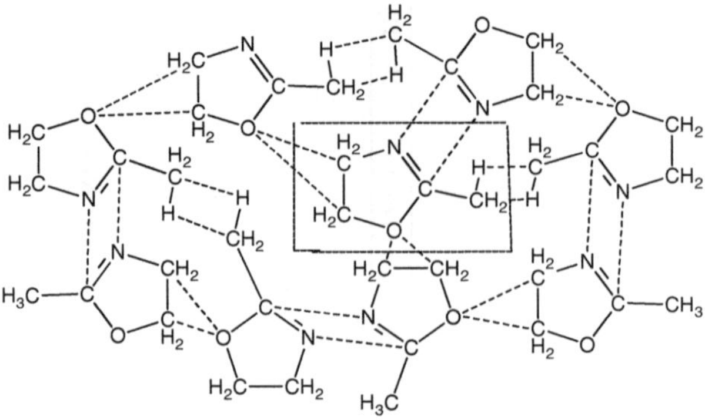

Fig. 7.3 Schematic picture of the liquid structure of 2-methyl-2-oxazoline with network of the specific interactions

equality of the interaction energies, D–O → CH ≈ D–O → CH_2. It is correct that the contribution of the energy of the methyl group (3.2 kJ mol^{-1}) and the interaction D=N → CH (5.0 kJ mol^{-1}) to the vaporization enthalpy of 2-methyl-2-oxazoline is equal to the contribution to the enthalpy characteristics of methylisoxazoles. The results of the calculations of the energy of the specific interaction D–O → CH_2 with the help of Eq. 7.2

$$\text{D-O}\rightarrow\text{CH}_2 = (\Delta_{vap}H^0(298\ \text{K})\text{mos - 2D=N}\rightarrow\text{CH - 2D}iso\text{H}_3\text{C}\rightarrow\text{H-CH}_2)/4 \quad (7.2)$$

are given in Table 7.2. The correctness of the assumptions is proved by the similarity of the calculated energy of the specific interaction D–O → CH_2 with the one realized in liquid isoxazole D–O → CH (6.65 kJ mol^{-1}), differing by the contribution (0.60) of the lacking second hydrogen atom. At the same time we notice that the increased stability compared with the energy of the same type of specific integration of the liquid dimethyl ether D–O → CH_3 = 5.63 kJ mol^{-1} reflects the mutual influence of oxygen and nitrogen atoms on the shifting of the electron density in the pentamerous cycle. The increased energy contribution to the vaporization enthalpy of 2-ethyl-2-oxazoline by the isostructural methyl group is in good agreement with the energy contribution by the functional ethyl group (8.1 kJ mol^{-1}), equal to the difference with the enthalpy characteristics 2-oxazoline (36.0 kJ mol^{-1}). The energy value of the contribution of the functional ethyl group on preservation of the energies of the specific interactions, formed by cycle components, directs attention to the stabilizing effect of the increasing number of the carbon atoms in this group (4.05 kJ mol^{-1}). The obtained energy values of the specific interaction D–O → CH in the compounds of derivatives of oxazole illustrate the good agreement with the noted characteristics

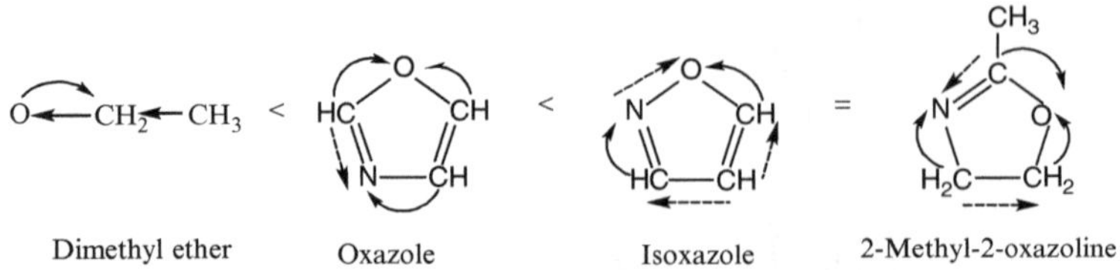

of shifting of the electron density in the molecules of these compounds.

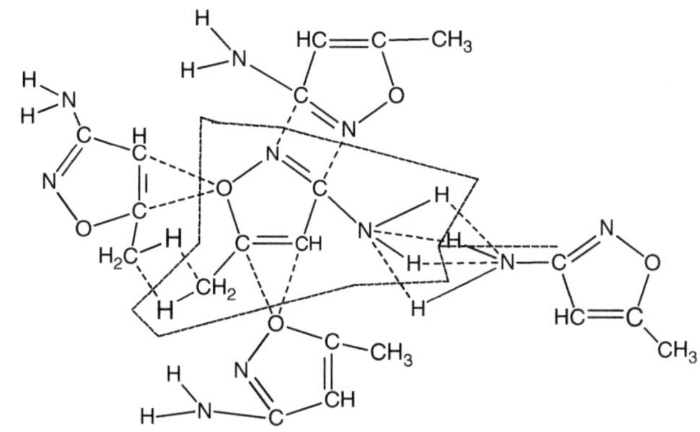

Fig. 7.4 Schematic picture of the liquid structure of 3-amino-5-methylisoxazole with network of the specific interactions

The 3-amino-5-methylisoxazol molecule with the donor properties of the two hydrogen atoms and three carbon atoms at the cycle has a lack of acceptors in the oxygen and nitrogen atoms. Hence, the nitrogen atom of the amine group acts as a pentacoordinated center and forms with hydrogen atoms four hydrogen bonds D–N•••H–N. Atoms of carbon, oxygen, and nitrogen of the cycle form with the molecule that is close in the environment four specific interactions D–O → CH and two of the second type, D=N → C=, forming the network structure of the crystal (Fig. 7.4a). This type of interaction of the nitrogen atom was formed by the carbon atom, directly connected with the oxygen atom and the derivatives of phenylacetamide; thus, the value of its energy should be taken as equal to the average value of 5.5 kJ mol^{-1} (Sect. 6.3.4). The rigid cycle with double bonds expresses for the amino group the acceptor properties and shifts the given electron density from the hydrogen atoms. As a result an increased difference in the charges of the hydrogen atoms is created and, as a sequence, the hydrogen bonds formed possess an increased stability. Hence, the energies of the hydrogen bonds are significantly more stable than the ones realized in crystalline ammonia (6.34 kJ mol^{-1}). As shown above, similar to the amides, the amino group in the compounds with the benzene cycle forms hydrogen bonds in the crystalline derivatives of benzene with energy of 9.7 kJ mol^{-1} (Sect. 6.3.3). The rest of the not yet estimated energy value of the specific interaction is determined by Eq. 7.3

$$\text{D-O}\rightarrow\text{CH}_2 = (\Delta_{\text{sub}}H^0(298\text{ K})\text{amioz} - 2\text{D=N}\rightarrow\text{CH} - 4\text{D-N}\bullet\bullet\bullet\text{H-N} - 2\text{D}\mathit{iso}\text{H}_3\text{C}\rightarrow\text{H-CH}_2)/4 \quad (7.3)$$

Finding a unique solution for such a complex system with three types of ten specific interactions is problematic, as well-founded assumptions are needed for a solution. The energy value of the specific interaction D–O → CH_2 (7.2 kJ mol^{-1}) obtained is quite correct and objectively reflects the nature of the interaction and the stabilizing effect of the combination of the amine functional group and nitrogen atoms and oxygen atoms of the cycle on the given specific interactions for the compounds

$$\text{2-Methyl-2-oxazoline } (6.50) = \text{2-Ethyl-2-oxazoline } (6.5) < \text{3-Amino-5-methylisoxazole } \left(7.2\ \text{kJ mol}^{-1}\right).$$

7.1.3 Energies of Specific Interactions of Crystalline Amides with Peptide Fragments

The 2-pyrrolidone molecule with carbonyl and imide groups located nearby should have significant mutually stabilizing influence on the formation of the specific interactions. Replacement of the hydrogen atom in the imide group by the functional methyl group in the *N*-methyl-2-pyrrolidone and 1-methyl-2-pyrrolidinone derivatives of this compound leads to the reduction in the vaporization enthalpy at 13.8 kJ mol^{-1} while saving the total number of the specific interactions in its liquid condition. It follows that hydrogen bonds are more stable, approximately at 6.9 kJ mol^{-1}, compared with the replaced specific interactions N → CH_3–N in liquid *N*-methyl-2-pyrrolidone and, second, the hydrogen atom of the imide group plays a special role in providing electron density to the nitrogen atom and further to the oxygen atom of the carbonyl group, the value of the positive charge of which ensures it has high acceptor properties and the capacity to form stable hydrogen bonds. It is connected with the fact that the reverse dative bond at such a remote $2p_z$-orbital practically has no known analog in the 3d-orbital of the nitrogen atom.

This assumption is accompanied by attempts to estimate the participation of the 3d-orbital in the NH_3 and BH_3 and their complex H_3B–NH_3 molecules [13–15]. Research results show that 3d-orbitals in the initial compounds and their complexes are insignificant because of the occupancy of the $2p_z$-orbital and, consequently, the 3d-orbital does not participate in the formation of the chemical bonds in the mentioned compounds [16–19]. Thus, the hydrogen atom loses its electron density forever and acquires acceptor properties in the formation of the hydrogen bond. A similar position is seen in the system's peptide fragment 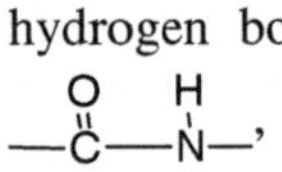, increasing by the presence of the acceptor carbon atom compared with the boron atom and in particular the acceptor's role of the oxygen atom with nitrogen of the peptide group. Significant changes occur in the occupancy of the $2p_z$-orbital. Because of the lack of data on the electron occupancy of atoms of the peptide fragment we consider the occupancy of atoms of its analogues $B(CH_3)_3$, $N(CH_3)_3$, and $(H_3C)_3B$-$N(CH_3)_3$, calculated by Erenson [20, 21] with the help of an improved [22] variant of Huckel's method.

We draw attention to the significant shifting of the electron density from the boron atom to the carbon acceptor atom of the methyl group, already having an increased electron density at the expense of shifting of the electron density from the hydrogen atom of the same groups. As a result the charge of the carbon atom reaches a significant negative value and the boron atom has a

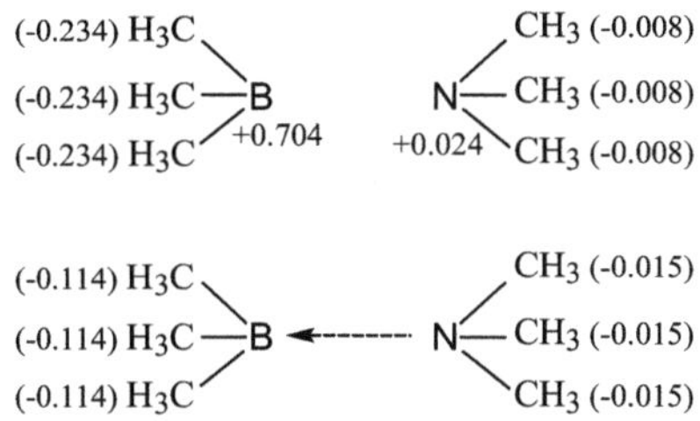

threefold excess of the positive charge of the methyl group. The boron atom in the molecule $B(CH_3)_3$ has a value near 0.3 electron, the source of which is mainly the carbon atom, forming the reverse dative bond, providing it to the $2p_z$-orbital of the boron atom. In its turn, the nitrogen atom having obtained the electron density from the carbon atom of the methyl group gives electron density to the essentially unshared $2s^2$-electron pair of the carbon atom and gets a positive charge (0.024) [20, 21]. It is important that we have a positive charge for the nitrogen atom, which, relatively could increase significantly in dependence on the interacting acceptor atom. Due to the interaction between trimethylamine and trimethylboron and formation of the $(H_3C)_3B–N(CH_3)_3$ complex, the positive charge at the nitrogen atom increases at +0.792 e and becomes higher than in the initial amine. In the bond formation in the complex

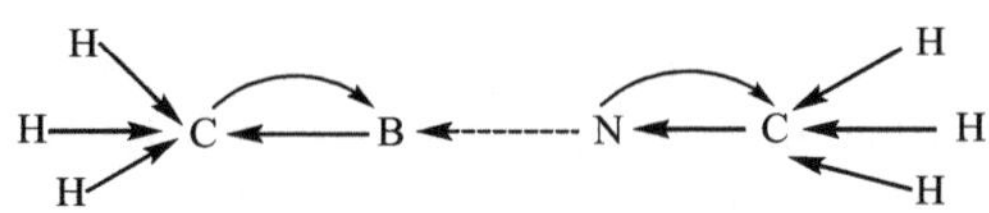

the $2p_z$-orbital of nitrogen participates. Accepting electron density, the boron atom reduces its positive charge at 0.411e. Regarding these data the carbon atom is a stronger acceptor compared with the boron atom, and considering the strong acceptor oxygen atom connected with the carbon atom, we are forced to recognize that close differences for the charges of boron and nitrogen in the $(H_3C)_3B–N(CH_3)_3$ complex are reached in the fragment 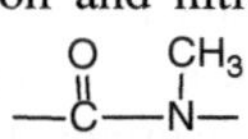 *N*-methyl-2-pyrrolidone and the peptide group of the 2-pyrrolidone molecule. In the latter it is strengthened additionally by carbonyl oxygen. If the charges of the carbon atoms of the carbonyl group and nitrogen are more aligned with the transfer of electron density with the $2p_z$-orbital of the latter, then the further transmission to the oxygen atom

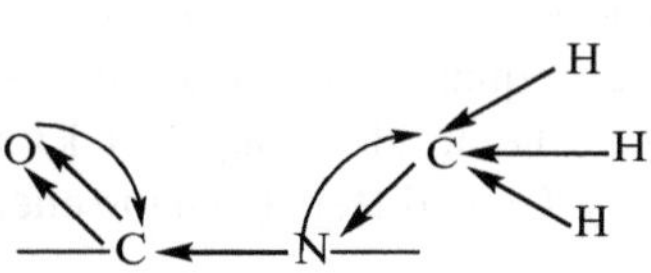

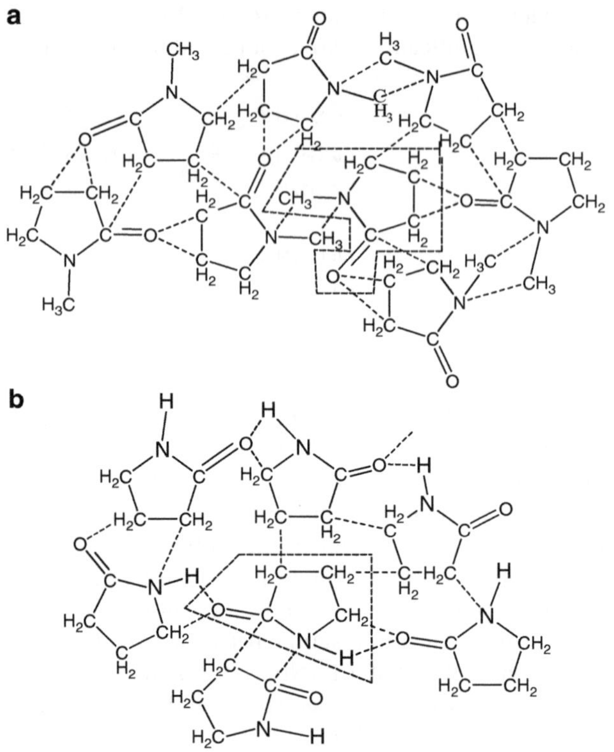

Fig. 7.5 Schematic picture of the liquid structure of 1-methyl-2-pyrrolidinone (**a**) and 2-pyrrolidone (**b**) with network of the specific interactions

leads to the increased difference in its negative charge with the positive charge of the hydrogen atom of the imide group and significantly less with the atom of the methyl group of the same peptide fragment. On the other side, there exists an analog in the difference of the charges of nitrogen and hydrogen atoms and the connected methyl group, as a result of which the energies of the hydrogen bonds and specific interaction D–N → CH_3–N formed are described by the integral values of stabilization D=C=O•••H–N > D–N•••H–N > D–N → CH_3–N with significant dominance of the first type.

Thus, the minimum of the energy system with peptide fragment is reached at the formation of a stable hydrogen bond by the oxygen atom of the carbonyl with the hydrogen atom of the imide group DC=O•••H–N. As a result, the network structure of liquid *N*-methyl-2-pyrrolidone is formed by four types of specific interactions 2D–N → CH_3–N, 2D=C=O → CH_2, D–H_2C → CH_2, and D–H_2C → C=O (Fig. 7.5a). The energy of the specific interaction D–H_2C → CH_2 (6.30 kJ mol^{-1}) is identical to the energy of the same type of interaction realized in liquid cyclobutane (Sect. 3.1). The energy of the second type of interaction, D–N → CH_3–N, should be taken as equal to the one realized in liquid amides with substituted hydrogen atom of the amide group (7.1 kJ mol^{-1}). The energy of the specific interaction D–H_2C → C=O differs from the energy value of the interaction

D–CH_2 → CH_2 by the contribution of the two substituted hydrogen atoms, equal to 1.2 kJ mol^{-1}. Thus, the energy of the specific interaction of D=C=O → CH_2 of liquid *N*-methyl-2-pyrrolidone should be obtained with the help of Eq. 7.4,

$$D{=}C{=}O{\rightarrow}CH_2 = (\Delta_{vap}H^0(298\ K)\ mprd - 2D\text{-}N{\rightarrow}CH_3\text{-}N - D\text{-}H_2C{\rightarrow}CH_2 + D\text{-}H_2C{\rightarrow}C{=})/4 \quad (7.4)$$

The structure of liquid 2-pyrrolidone is formed by two stable DC=O•••H–N hydrogen bonds (Fig. 7.5b), two more stable specific interactions D=C=O → CH_2, two interactions of D–H_2C → CH_2 with energy value, determined by the CH_2 groups of the cycle (6.5), with equal energies of similar interactions as formed in liquid cyclopentane by two single bonds of the nitrogen atom DN → C and methylene group D–CH_2 → C with the carbon atom of the cycle. The energy of the specific interaction D=C=O → CH_2 we take as equal to the one realized in liquid *N*-methyl-2-pyrrolidone (8.4 kJ mol^{-1}) and the interaction energy of D–N → C and D–CH_2 → C we accept to be equal to the contribution of the energy of methyl and methylene groups minus the energy contribution of three and two substituted hydrogen atoms, respectively: 1.80 and 1.20 kJ mol^{-1} (Fig. 7.5b).

The energy of the stable hydrogen bond is defined with the help of Eq. 7.5

$$D{=}C{=}O{\bullet\bullet\bullet}H\text{-}N = (\Delta_{vap}H^0(298\ K)pd - 2D\text{-}H_2C{\rightarrow}CH_2 - 2D{=}C{=}O{\rightarrow}CH_2 - D\text{-}N{\rightarrow}C - D\text{-}H_2C{\rightarrow}C)/2\ (7.5). \quad (7.5)$$

The results of the calculations of the energy of the specific interactions formed in liquid 1-methyl-2-pyrrolidinone and 2-pyrrolidone (Table 7.3)

1-Methyl-2-pyrrolidinone
D=O=C →C= (6.2) < D-H_2C→CH_2 (6.30) < D-N→CH_3-N (7.1) < D=C=O→CH_2 (8.4 kJ mol^{-1})

2-Pyrrolidone
D-N→C- (5.3) < D-N→C- (5.3) < D-H_2C→CH_2 (6.30) < D=C=O→CH_2 (8.4) < D=C=O•••H-N (14.3 kJ mol^{-1})

are described by the numbers, reflecting the stabilization of the interactions, formed with participation of the nitrogen atoms and carbonyl group for the compounds with heteroatoms in the presence of the carbonyl fragment and imide group. The significant stabilization is expressed in the energy of the specific interaction, formed by the methylene (8.4 kJ mol^{-1}) group and greater than one and a half times the energy of the interaction of liquid dimethyl ether (5.63 kJ mol^{-1}). The most significant stabilization is shown at the hydrogen bond in the presence of the peptide group of 2-pyrrolidone.

In Fig. 7.6 a network structure of crystalline 6-caprolactam, which illustrates the formation of the imide group of the two hydrogen bonds D=C=O•••H–N, five specific interactions, formed by free methylene groups of the heptyl ring D–H_2C → CH_2, two interactions DN → CH_2 and D=O=C → CH_2 is presented.

Table 7.3 Energies of the specific interactions and hydrogen bonds (kJ mol^{-1}) of the liquid and crystalline compounds with peptide fragment at 298 K

Compounds	Formula	Structure DN → CH_3–N	$\Delta_{vap}H°$ (298 K) [8]	D–N → CH_3–N	D–H_2C → CH_2/ DO=C → C=	D=C= O → CH_2	Peptide hydrogen bond DC=O•••H–N
1-Methyl-2-pyrrolidinone	C_5H_9NO	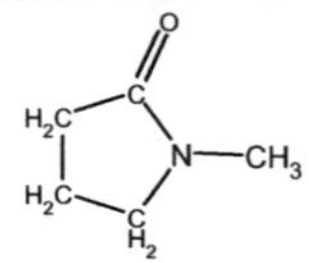	55.3[a]	7.1	6.30/5.2	8.4	–
N-Methyl-2-Pyrrolidone	C_5H_9NO	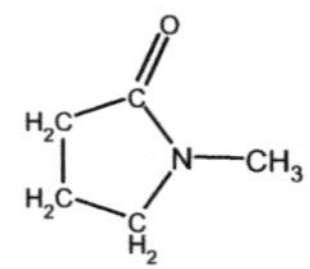	53.1	7.1	6.30	7.3	–
2-Pyrrolidone	C_4H_7NO	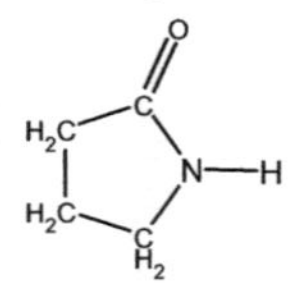	69.1 ± 0.5	DN → C = 5.3	6.5/DH_2C → C = 5.3	8.4	14.45
Crystal							
6-Caprolactam	$C_6H_{11}NO$	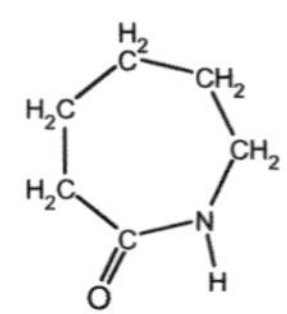	87.3 ± 0.2	6.7	6.30	9.35	16.5

[a]T is less standard

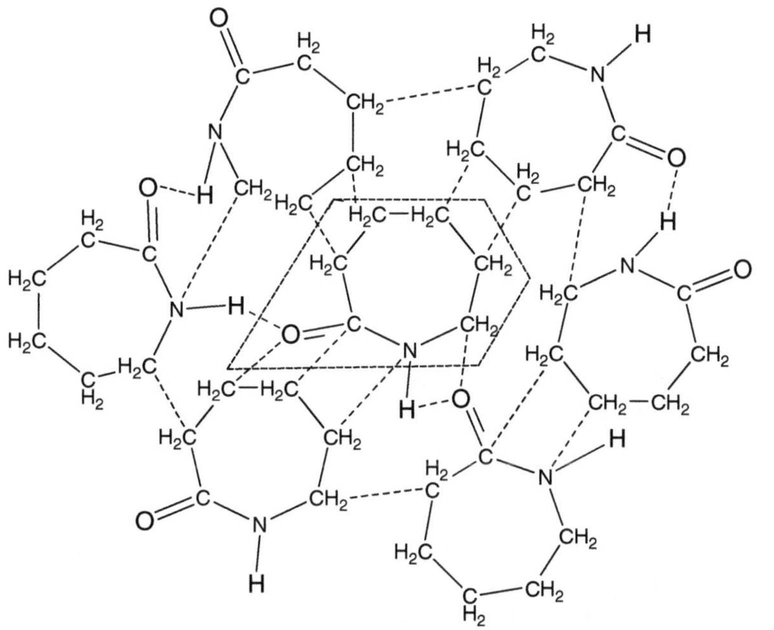

Fig. 7.6 Schematic picture of the crystalline structure of caprolactame with network of the specific interactions and hydrogen bonds

A great number of methylene groups in the cycle of the molecule 6-caprolactam are less influenced by the peptide group, which is why the energy of the specific interaction $DH_2C \rightarrow CH_2$ should be taken as equal to the energy of a similar type of interaction, realized in the crystalline cyclohexane – cyclododecane with an average value of 6.30 kJ mol^{-1}. In Sect. 6.5 the energy value of the specific interaction $D{=}C{=}O \rightarrow CH_2$ for methylene groups, located between two carbonyl fragments in the malonamide (9.2) < glutarimide (9.35 kJ mol^{-1}) molecules was determined. The energy of this type of interaction is small; it is stabilized at the two methylene groups in the molecule glutarimide, which implies completion of this process at the five similar groups in the ring chain. By taking the energy value of the specific interaction $D{=}C{=}O \rightarrow CH_2$ as equal to 9.35 kJ mol^{-1}, we contribute an insignificant error, which does not increase the error of the sublimation enthalpy of the compounds considered in the experiment. At the same time this value correlates closely with the energy of this type of interaction of liquid 1-methyl-2-pyrrolidinone (8.4 kJ mol^{-1}). The energy of the specific interaction $D\text{–}N \rightarrow CH_2$ cannot exceed the energy value of the specific interaction formed by the methyl group of amide fragment $D\text{–}N \rightarrow CH_3\text{–}N = 7.7$ kJ mol^{-1} of crystalline cyclic amides (Sect. 6.3.3). Therefore, from this value one should calculate the energy contributed by the substituted hydrogen atom (1.0 kJ mol^{-1}). Taking these assumptions, the energy of the hydrogen bond formed by the peptide group should be determined with the help of Eq. 7.6.

$$D{=}C{=}O{\bullet\bullet\bullet}H\text{-}N = (\Delta_{sub}H^0(298\ K) - 5D\text{-}H_2C{\rightarrow}CH_2 - D{=}C{=}O{\rightarrow}CH_2 - 2D\text{-}N{\rightarrow}CH_2)/2 \quad (7.6)$$

The results of the calculations of the energy values of the hydrogen bonds, formed by the peptide group in liquid 2-pyrrolidone (14.3) and crystalline 6-caprolactame (16.5 kJ mol^{-1}) point to its high stability, and the significant increase in the energy value of the hydrogen bond of liquid and (10.99 kJ mol^{-1}) crystalline water.

7.1.4 Energies of Specific Interactions of Crystalline Amides with Peptide Fragment and Carbonyl Group

The cyclic *N*-methyl-2-succinimide molecule with symmetric location of the oxygen atoms in the pentamerous cycle and with a similar influence on the shifting of the electron density from the methyl group and nitrogen atom of the amine fragment of NCH_3 and methylene groups provides it with a reduced positive charge compared with 2-pyrrolidone. At the same time the difference between the negative charge of the oxygen atoms and positive charge of the nitrogen, hydrogen, and carbon atoms of the methylene groups remain significantly lower than in the 6-caprolactam molecule. The *N*-methyl-2-succinimide molecule with ten bond vacancies (Fig. 7.7a) forms eight specific interactions $D{=}C{=}O \rightarrow CH_2$ and two interactions $N \rightarrow CH_3\text{–}N$ of reduced stability, formed by NCH_3 group in crystalline amides with the energy of 7.7 kJ mol^{-1}. The energy of the specific interaction of the second type should be determined by Eq. (7.7),

$$D{=}C{=}O{\rightarrow}CH_2 = (\Delta_{sub}H^0(298\ K) - 2D\text{-}N{\rightarrow}CH_3)/8 \quad (7.7)$$

The results of the calculations point to the inequality in the energies of this type of crystalline compounds *N*-methyl-2-succinimide (8.3) < 6-caprolactame (9.35 kJ mol^{-1}), reflecting the increased shifting of the electron density from the hydrogen atom of the imide group and a larger number of methylene groups in the cycle of the molecule 6-caprolactame. The succinimide molecule with ten bond vacancies forms the network structure of the crystal (Fig. 7.7b) with two hydrogen bonds and specific interactions formed by carbon atoms of the cycle $D\text{–}H_2C \rightarrow CH_2$, $2D\text{–}H_2C \rightarrow C$– and interaction with the nitrogen atom of the cycle and with the carbonyl oxygen atom with increased stability. In this connection it is important to clarify the influence of the second carbonyl group on the value of the energy of the peptide group formed. The energy of the specific interaction $D\text{–}H_2C \rightarrow CH_2$ should be taken as equal to the energy value (7.2 kJ mol^{-1}), realized in crystalline cyclopentane (Chap. 3). In the estimation of the energies of the specific interactions $D\text{–}H_2C \rightarrow C$ and $DC \rightarrow C$ the contributions of saturated hydrogen atoms (6.2 and 5.2 kJ mol^{-1}) were considered. The accepted correct values of the energies of the

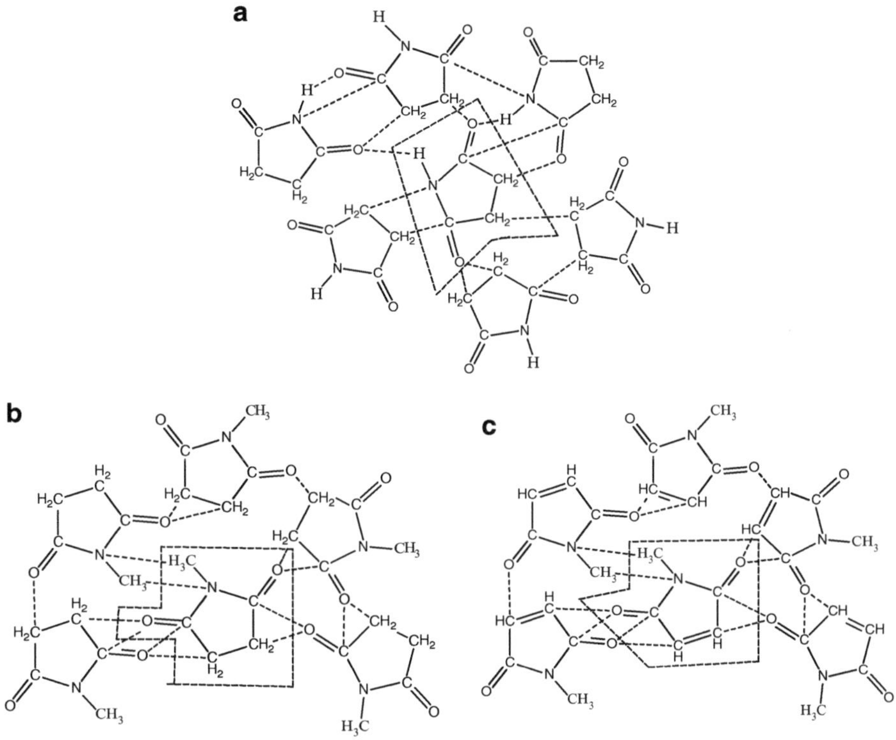

Fig. 7.7 Schematic picture of the crystalline structure of succinimide (**a**), *N*-methyl-2-succinimide (**b**), and *N*-methylmaleimide (**c**) with network of the specific interactions and hydrogen bonds

specific interactions allow us to determine the energy of the hydrogen bond of the peptide type using the equality of sublimation enthalpy with the energy contribution of the specific interaction and hydrogen bonds with the help of Eq. 7.8

$$D{=}C{=}O\cdots H\text{-}N = (\Delta_{sub}H^0(298\ K) - 4D{=}C{=}O\rightarrow CH_2 - D\text{-}H_2C\rightarrow CH_2 - D\text{-}H_2C\rightarrow C - DC\rightarrow C)/2 \quad (7.8)$$

The obtained energy value of the stable hydrogen bond (Table 7.4) of crystalline succinimide, formed by the peptide group DC=O•••H–N at the second symmetrically located carbonyl group in the molecule of this compound, points to the reduced energy value of the same bond of liquid 2-pyrrolidone only at

$$\text{2-Pyrrolidone}\ (14.45) < \text{Succinimide}\ \left(15.6\ \text{kJ mol}^{-1}\right) < \text{6-}\textit{Caprolactam}\left(16.5\ \text{kJ mol}^{-1}\right).$$

1.15 kJ mol^{-1} compared with crystalline succinimide; and at the same time it is less stable at 0.9 kJ mol^{-1} than crystalline 6-caprolactame. The difference of

Table 7.4 Energies of the specific interactions and hydrogen bonds (kJ mol^{-1}) of the crystalline compounds with pentamerous heterocyclic, peptide fragment, and carboxylic group (298 K)

Compounds	Formula	Structure D–N → CH_3–N	$\Delta_{sub}H^{\circ}$ (298 K) [12]	D–N → CH_3–N	D–H_2C → CH_2/ DH_2C → C	D=C=O → CH_2/ DC → C=	D=C= O•••H–N
N-Methylsuccinimide	$C_5H_7NO_2$	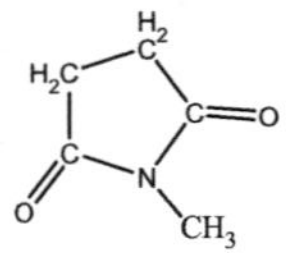	80.1 ± 0.3	7.7	–	8.3	–
Succinimide	$C_4H_5NO_2$	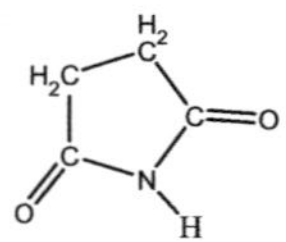	83.6 ± 1.5	–	7.2/6.2	8.3/5.2	15.6
N-Methylmaleimide	$C_5H_5NO_2$	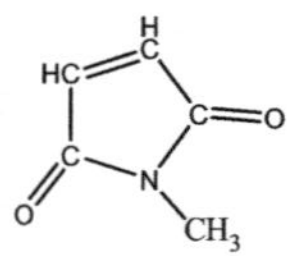	73.3 ± 0.5	7.7	D=C=O → C–(6.85)	D=C=O → CH = 7.85	–
N-Methylphthalimide	$C_9H_7NO_2$	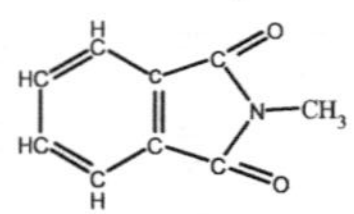	91.1 ± 0.5	7.7	D=HC → CH = 7.40	D=C=O → C = 5.8	–
2-Imadazoleidinone	$C_3H_6N_2O$	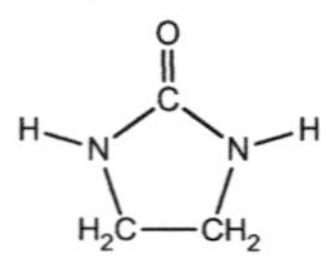	83.7	D–N•••H–N = 9.7 D–N → CH_2 =6.7	D–H_2C → CH_2 = 7.2/D–H_2C → C =	D=C=O → CH_2 = 8.3	13.8
2-Azetidinone	C_3H_5NO	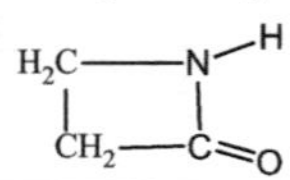	77.4 ± 0.3	D–N → CH_2 = 4.7	D–H_2C → CH_2 = 8.2	D=C=O → CH_2 = 8.3	17.5

stabilities of the formed hydrogen bonds of the peptide group of the succinimide molecules with two and five methylene groups of the molecule 6-caprolactame is natural and reflects the increased charges of oxygen and nitrogen atoms and the increased differences with hydrogen atoms of the amine group. The compound *N*-methylmaleimide with its molecule similar to the molecule structure of *N*-methylsuccinimide except for the presence of the double bond, with two methylene groups replacing the two CH groups, allow us to conclude that the significantly reduced value of the sublimation enthalpy at 6.8 kJ mol^{-1} reflects the reduced stability of the specific interactions formed. The interconnectivity of the sublimation enthalpy with the number and energy of the break bonds [9–11] and the presence of ten bond vacancies in the molecules of these compounds proves the correctness of this conclusion.

The network structure of crystal *N*-methylmaleimide is formed by two specific interactions D–N → CH_3–N, four D=C=O → CH=, and the same number of interactions D=C=O → C–. The first type of specific interaction, formed by the methyl group is not connected with the double bonds and can be estimated by the energy value formed by the same group in crystalline amides (7.7 kJ mol^{-1}). For the two types of specific interactions 4D=C=O → CH= + 4D=C=O → C– one can obtain the average value using the equation

$$\mathrm{D{=}C{=}O{\rightarrow}CH{=}} = \mathrm{D{=}C{=}O{\rightarrow}C{-}} = (\Delta_{sub}H^0(298\ K) - \mathrm{D{-}N{\rightarrow}CH_3{-}N})/8 \quad (7.9)$$

The obtained average value (7.35 kJ mol^{-1}) of these two types of specific interactions or more correctly the energy value of each from this type of interaction has reduced stability,

$$\mathrm{D{=}C{=}O{\rightarrow}C{-}}\ (6.85) < \mathrm{D{=}C{=}O{\rightarrow}CH}\ (7.85) < \mathrm{D{=}C{=}O{\rightarrow}CH_2{-}}\ (8.3\ kJ\ mol^{-1})$$

compared to crystalline *N*-methylsuccinimide D=C=O → CH_2–. Taking into account the established fact of the reduced stability of the specific interactions of compounds with the bond of the unsaturated ligand at different classes of organic compounds [6], we can state that the energies of the formed specific interactions of methylene and the CH group of saturated and unsaturated pentamerous hetero atomic cycles differ by approximately 0.5 kJ mol^{-1}.

The *N*-methylphthalimide molecule forms the network structure of the crystal (Fig. 7.8a), in which three types of specific interactions participate: D=C=O → C= (eight), two D–N → CH_3–N, and four D=HC → CH interactions, formed by the CH groups of the benzene ring with an energy of 7.40 kJ mol^{-1}. Such a large number of interactions between the oxygen atoms and carbon atoms and between carbon atoms at two oxygen atoms of the carbonyl groups presupposes a maximum possible shifting of the electron density from the CH_3 imide group N → CH_3, leading to the biggest stability of the specific integration D–N → CH_3–N with the energy value (7.7 kJ mol^{-1})of crystalline imides and amides. The obtained energy value of the specific interaction D=C=O → C= with the help of Eq. 7.10

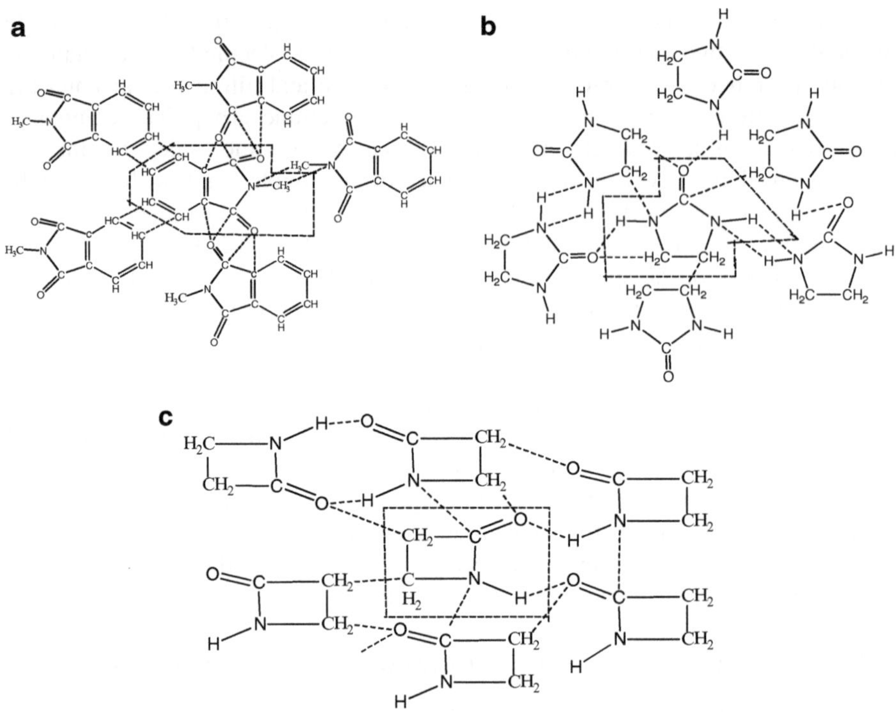

Fig. 7.8 Schematic picture of the crystalline structure of *N*-methylphthalimide (**a**), 2-imadazoleidinone (**b**), and 2-azetidinone (**c**) with network of the specific interactions and hydrogen bonds

$$D{=}C{=}O{\rightarrow}C{=} = (\Delta_{sub}H^0(298\ K) - 2D\text{-}N{\bullet\bullet\bullet}H\text{-}N - 4D{=}HC{\rightarrow}CH)/6 \quad (7.10)$$

is given in Table 7.4. The energies of the specific interactions of the carbon atom of the carbonyl fragment formed are described by the integral values of stabilization,

$DC{\rightarrow}C{=}$ (5.2) Succinimide < $DC{=}O{\rightarrow}C\text{-}$ (5.8) N-Methylphthalimide < $DC{=}O{\rightarrow}C\text{-}$ (6.85 kJ mol^{-1}) N-Methylmaleimide

caused by the donor properties of the carbon atom and its reduced negative charge in the *N*-methylphthalimide molecule under the influence of the rigid benzene ring of this compound.

The 2-imadazoleidinone molecule with the peptide group and the additional second imide group is able to present an increased shifting of the electron density to the oxygen atom. At the same time the succinimide molecule with the peptide group and the second carboxyl group has increased ability of the two oxygen atoms of electron acceptors with the reduced ability of one imide group to the shifting of the electron density. The overall similarity of the 2-imadazoleidinone and succinimide molecules is the equal number of CH_2 groups in the pentamerous hetero cycle.

In this connection it is of particular interest to obtain information about the energy of the hydrogen bond, formed by the peptide group on the condition, on one side, of the increased donor ability of the 2-imadazoleidinone molecule and, on the other side, of the ability of acceptor properties of the succinimide molecule in the formation of the hydrogen bond. The grid structure of crystal 2-imadazoleidinone (Fig. 7.8b) is formed by the molecule of this compound with nine bond vacancies. The principle of minimum energy of the system of the 2-imadazoleidinone molecule forms two stable D=C=O•••H–N and two hydrogen bonds of reduced stability 2DN•••H–N with energy of 9.7 kJ mol^{-1}, specific interactions D=C=O → CH_2 (two) with energy of 8.3 kJ mol^{-1}, and two types of specific interactions D–H_2C → CH_2 and D–H_2C → C with the above estimated energies of 7.2 and 6.2 kJ mol^{-1}, respectively. The energy value of the last type of specific interaction D–N → CH_2 should be taken as equal to 6.7 kJ mol^{-1}, the energy of interaction, reduced by the contribution of the substituted hydrogen atom in the methyl group (1.0 kJ mol^{-1}). The energy value of the hydrogen bond is obtained from the difference of the sublimation enthalpy for 2-imadazoleidinone and the total energy value and the number of all remaining

$$\text{D=C=O}\cdots\text{H-N} = (\Delta_{sub}H^0(298\text{ K}) - 2\text{D-N}\cdots\text{H-N} - 2\text{D=C=O}\rightarrow\text{CH}_2 - \text{D-H}_2\text{C}\rightarrow\text{CH}_2 - \text{D-CH}_2\rightarrow\text{C} - \text{D-N}\rightarrow\text{CH}_2)/2 \quad (7.11)$$

specific interactions, given in Table 7.4. The reduced energy value of this peptide hydrogen bond (13.8 kJ mol^{-1}) compared with the energy of the same type of hydrogen bond, realized in crystalline succinimide (15.6 kJ mol^{-1}), testifies to the predominant role of acceptors of the two oxygen atoms at one imide group in the molecule. It leads to the increased difference of charges of the oxygen atoms on one side and atoms of nitrogen and hydrogen, on the other side. On the basis of this one could conclude that the additional acceptors of electron density present in the molecules have a significant influence on the stability of the hydrogen bonds, formed by the peptide group.

The 2-azetidinone molecule with imide and two methylene groups possesses more significant donor properties in the presence of the oxygen atom of the peptide group compared with the succinimide molecule and less than the 2-imadazoleidinone molecule. The 2-azetidinone molecule with seven bond vacancies forms the network structure of the crystal with two stable hydrogen bonds, two specific interactions D=C=O → CH_2 (8.3 kJ mol^{-1}), two D–N → C and one D–H_2C → CH_2. The energy value of the interaction D–N → C determined by the energy contributed by the N → CH_3–N interaction, reduced by the contribution of three hydrogen atoms (3.0). The energy of the specific interaction D–H_2C → CH_2 we estimate as equal to the value of the energy (cyclic bond at cyclobutane) interaction, formed by the methylene group of cyclobytane (8.2 kJ mol^{-1}). The energy value of the hydrogen bond of crystalline 2-azetidinone calculated as the difference between the sublimation enthalpy and total energy value of the specific interactions (Table 7.4) gives rise to the integral value of stabilization of the

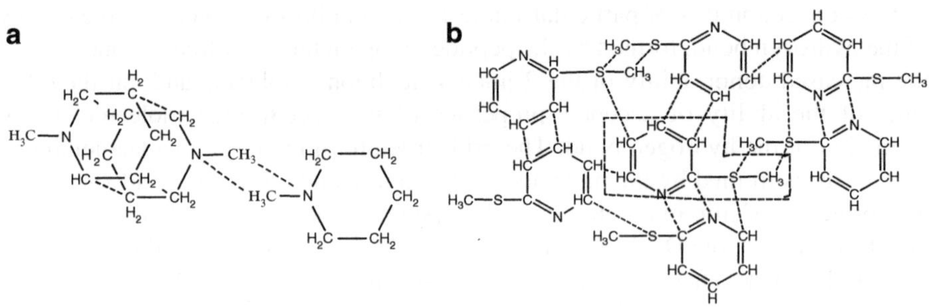

Fig. 7.9 Schematic picture of the liquid and crystalline structure of *N*-methylpiperidine (**a**) and 2,3,4,5-tetrahydro-(methylthio)pyridine (**b**) with network of the specific interactions

hydrogen bond, formed by the peptide group with the excess donors of electron density of the two imide groups by the 2-iadazoleidinone molecule, at the two oxygen atoms of the carbonyl group of strong acceptors

$$\text{2-Imadazoleidinone}\ (13.8) < \text{Succinimide}\ (15.6) < \text{2-Azetidinone}\ \left(17.5\ \text{kJ mol}^{-1}\right)$$

of electron density by the succinimide molecule with pentamerous cycles and with one peptide group and two methylene groups in the four-atomic cycle.

7.2 Energies of Specific Interactions of Compounds with Hexamerous Heterocycles

7.2.1 Alkylpiperidines and Methoxypyridines

The simplest alkyl derivative compounds with hexamerous hetero cycle of piperidine allow us to obtain the energy of the specific interaction, formed by the alkyl functional group. Compounds of this type form the network structure of liquid condition with specific interactions of the two types D–$H_2C \rightarrow CH_2$ and D–N $\rightarrow$ CH_3–(CH_2)n–N, differing by the energy contribution to the enthalpy characteristics of the vaporization process (Fig. 7.9a). The energy of the specific interaction, formed by the cycle methylene groups, is estimated to be equal to the energy value of the same type, formed by cyclopentane (5.70 kJ mol^{-1}).

The energy of the specific interaction, formed by the methyl, propyl group is determined by the difference between the vaporization enthalpy of the corresponding compound and the contribution of the five specific interactions.

$$D\text{-}N{\rightarrow}R\text{-}N = ((\Delta_{sub}H^0(298\ K) - 5\ D\text{-}H_2C{\rightarrow}CH_2)/2. \quad (7.12)$$

The energies obtained of the specific interactions (Table 7.5) correspond to the energies of liquid trimethylamine (4.25) and propylamine (8.1 kJ mol^{-1}), which points to the implementation by the methyl and propyl groups of the functions of the ligand. The high accuracy of the calorimetric determination of the vaporization enthalpy allows us to attribute the enthalpy characteristic of 2-methoxypyridine, given without any indication of error, to the incorrect one. Taking into account the low influence of the hydrocarbon ring of piperidine on the functional group, we have reason to attribute a similar ring of 2,3,4,5-tetrahydro-(methylthio)pyridine to the same type. It allows us to determine the energy contribution of the S–CH_3 group from the difference between the vaporization enthalpy of tetrahydro-(methylthio) pyridine and 4-(methylthio)pyridine and similar characteristics of pyridine. The energy contribution of the specific interaction D–S → CH_3 points to the significant influence on its stability expressed at the location of C(4). It follows that a strong acceptor such as the sulfur atom is able to shift the electron density from the nitrogen atom in the CH groups of the cycle at the maximum to form the specific interaction, increasing a similar interaction with the location of the sulfur atom at C(2) at 1.2 kJ mol^{-1}. Replacement of the S–CH_3 group from C(2) to the carbon atom at location C(4) is accompanied also by stabilization of the specific interaction D–N → CH (Table 7.5).

7.2.2 Energies of Specific Interactions of Liquid Hydroxypyridines

Piridine and its derivatives are the bases and in this connection it is of special interest to study the energies of the hydrogen bonds formed by hydroxypyridines. These compounds are characterized by significant increase in the sublimation enthalpy with the increase in the distance of the hydroxyl group in the cycle from the location of the nitrogen atom. It means that the acceptor properties of the nitrogen atom are not high and in the presence of the hydroxyl functional group its influence is distributed to the nearest carbon atoms of the chain. Thus, the location of the hydroxyl group at the fourth carbon atom C(4) is accompanied by significant shifting of the electron density from nearest CH groups of the chain to the oxygen atom and by the increase in its negative charge and the difference of the last with the charge of the hydrogen atom of the same hydroxyl group. Hence, there appears a stabilization of the hydrogen bond and an increase of the enthalpy characteristics at 17.2 kJ mol^{-1} compared with the sublimation enthalpy of 2-hydroxypyridine. Given the effect we saw with the example of the energy of the specific interaction DN → CH, formed by the nitrogen atom and CH groups of the cycle of the liquid compounds

Table 7.5 Energies of the specific interactions (kJ mol^{-1}) of the liquid piperidines and pyridines

Compounds	Formula	Structure	$\Delta_{vap}H°(298\ K)$ [8]	T K	D–N → CH_3–(CH_2)n–N	D–N → CH	D–S → CH_3
D–H_2C → CH_2 = 5.63							
N-Methylpiperidine	$C_6H_{13}NO$	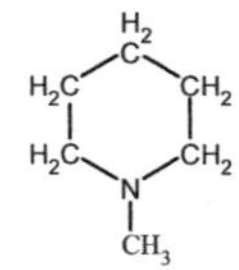	36.7 ± 0.1	298	n = 0 4.45	–	–
N-Propylpiperidine	$C_8H_{17}NO$	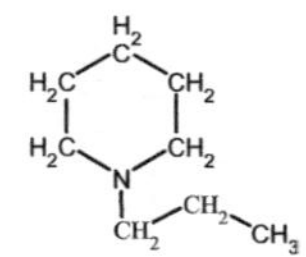	45.2 ± 0.4 44.3 ± 0.4	298	n = 2 8.6 8.15	–	–
Pyridine	C_5H_5N	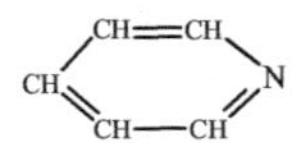	40.18 ± 0.16	298	–	8.80	–
2-Methoxypyridine	C_6H_7NO	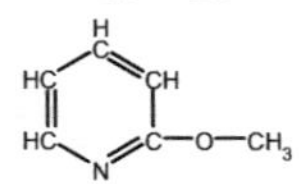	40.5?	319	–	–	–
2,3,4,5-Tetrahydro-(methylthio)pyridine	$C_6H_{11}NS$	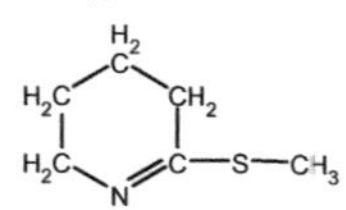	52.6	328	–	8.45	6.2
4-(Methylthio)pyridine	C_6H_7NS	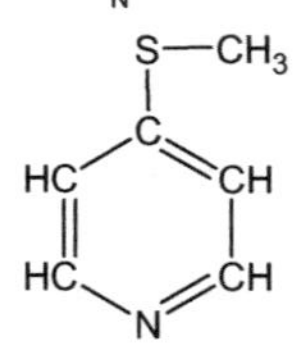	55.8	361	–	8.80	7.8

$$\text{Pyridine}\left(8.80\ \text{kJ}\,\text{mol}^{-1}\right) > 2,3,4,5\text{-Tetrahydro-}\left(\text{methylthio}\right)$$
$$\text{pyridine}\left(8.45\ \text{kJ}\,\text{mol}^{-1}\right) < 4\text{-}\left(\text{Methylthio}\right)\text{pyridine}\left(8.80\ \text{kJ}\,\text{mol}^{-1}\right)$$

at the simultaneous stabilization of the specific interaction D–S → CH_3, formed by the functional group $DSCH_3$ (6.2 and 7.8 kJ mol^{-1}, respectively). The results of these researches allow us to accept unchanged the energies of the specific interactions D=HC → CH and D–N → CH of crystalline pyridine and hydroxypyridines. In this connection it is reasonable to take the thermodynamic analysis for the determination of the energy of the hydrogen bonds and the pattern of its changes in the compounds of the hydroxypyridine series.

The hydroxypyridines molecules with nine bond vacancies form the crystalline structure with the network of seven specific interactions and two stable hydrogen bonds DO•••H–O (Fig. 7.10a). Two types of specific interactions D=HC → CH and D=N → CH of crystalline pyridine are preserved in the crystalline hydroxypyridines, the energies of which are equal to 7.40 and 12.9 kJ mol^{-1}, respectively (Chap. 5). Two additional types of specific interactions are formed by the CH groups of the cycle of pyridine with the oxygen atom of D–O → CH and, as a result, replacements of the hydrogen atom of the CH group of cycle D=CH → C, the energy of which should be taken as equal to the energy of the interaction D=HC → CH reduced by the energy contribution of the substituted hydrogen atom (1.0 kJ mol^{-1}) and equal to 6.4 kJ mol^{-1}. The energy of the specific interaction D–O → CH, formed by the oxygen atom of the functional hydroxyl group with the CH group of the cycle is not comparable with the energy of the specific interaction formed with the nitrogen atom of the cycle. Its absolute value always remains less and reflects a less stable interaction compared with the hydrogen bond, formed by the carboxyl group, and remains practically little different from the energy of the one formed by the corresponding symmetric ether [6]. However, this location undergoes significant rearrangements, caused by the influence of the hetero atoms of inter cyclic nitrogen and oxygen atoms in the presence of the double bonds between carbon atoms. In the case of liquid isoxazoles the energy of the specific interaction of D=CH → CH, formed by the CH group, exceeds the stability of the interaction of liquid dimethyl ether D–O → CH_3 at 1.0 kJ mol^{-1} on the condition that the contribution of the two saturated hydrogen atoms is 1.2 kJ mol^{-1}. Consequently, the contribution of the interaction D–O → CH_3 should be equal to 7.8 kJ mol^{-1}. Exactly this energycontribution to the sublimation enthalpy of hydroxypyridine should be contributed by the specific interaction D–O → CH, formed by the CH group with the oxygen atom of the hydroxyl group. Thus, Eq. 7.13 is correct on the condition of determination of the energy value of the hydrogen bond, formed in the crystalline hydroxypyridines

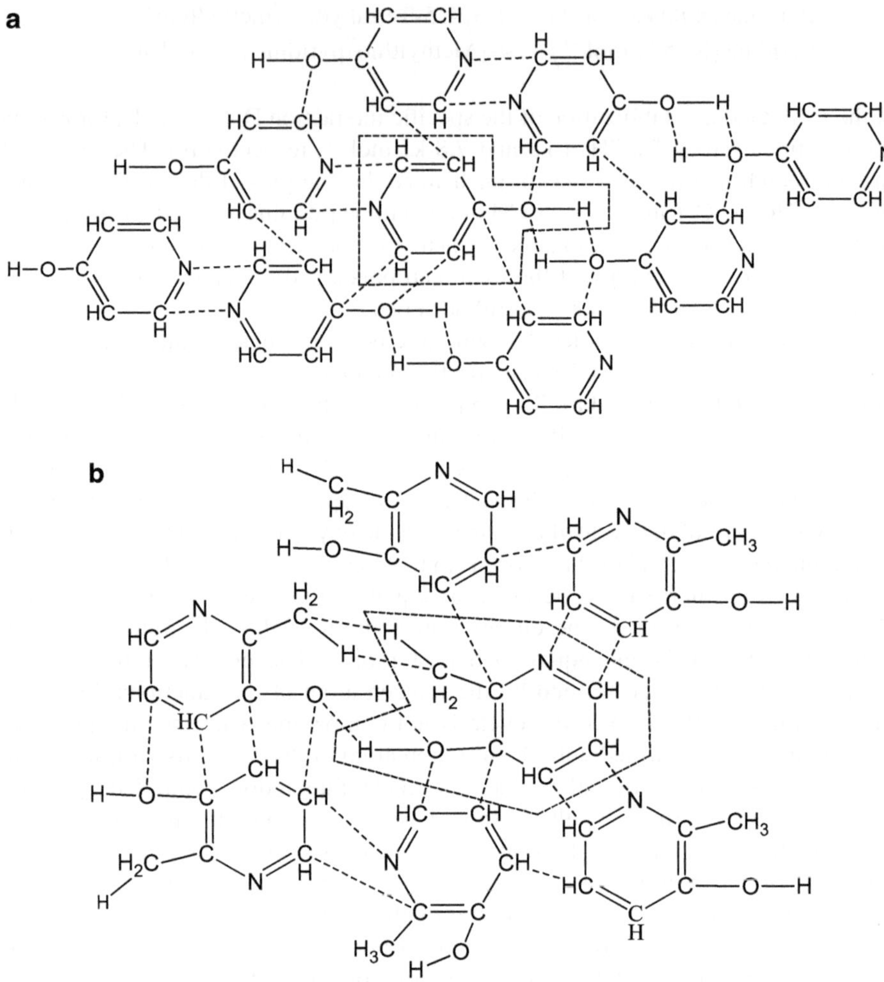

Fig. 7.10 Schematic picture of the crystalline structure of 4-hydroxypyridine (**a**) and 2-methyl-3-hydroxypyridine (**b**) with network of the specific interactions and hydrogen bonds

$$D\text{-}O\cdots H\text{-}O = (\Delta_{sub}H^{0}(298\ K) - 2D{=}CH{\rightarrow}CH - 2D{=}N{\rightarrow}CH - D{=}CH{\rightarrow}C - 2D\text{-}O{\rightarrow}CH)/2 \quad (7.13)$$

The results of the calculations of the energies of the hydrogen bonds (Table 7.6) prove the view expressed on the significant stabilization of the hydrogen bond for the compounds of the series

Table 7.6 Energies of the specific interactions and hydrogen bonds (kJ mol^{-1}) of the crystalline hydroxypyridines

Compounds	Formula	Structure	$\Delta_{sub}H°$ (298 K) [12]	D=HC → CH/ DHC → C	D=N → CH	D–O → CH	D–O•••H–O
2-Hydroxypyridine	C_5H_5NO	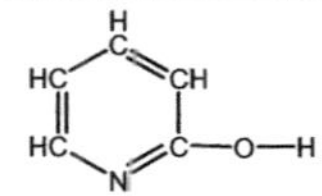	86.6 ± 1.3	7.40 × 2/6.6	12.9 × 2	7.8 × 2	12.5
3-Hydroxypyridine	C_5H_5NO	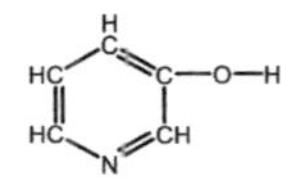	88.2 ± 1.3	7.40 × 2/6.6	12.9	7.8 × 2	13.3
4-Hydroxypyridine	C_5H_5NO	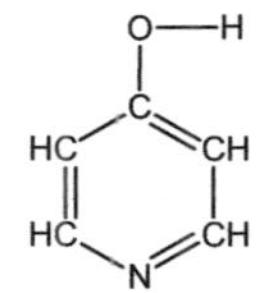	103.8 ± 1.7 118.6 ± 5.2	7.40 × 2/6.6	12.9	7.8 × 2	21.1
2-Methyl-3-hydroxypyridine	C_6H_7NO	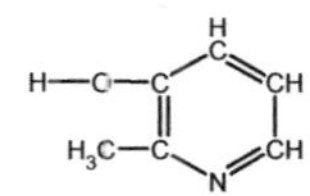	89.3 ± 1.3	7.40/6.40 × 2	D=N → C = 11.9 × 2	7.8 × 2	13.0
2-Methyl-4-hydroxypyridine	C_6H_7NO	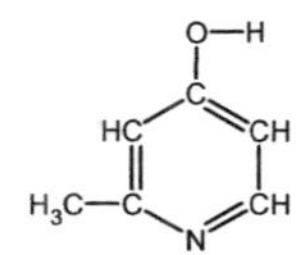	113.0 ± 1.3	7.40/6.40 × 2	D=N → C = 11.9 × 2	7.8 × 2	24.8
2-Methyl-5-hydroxypyridine	C_6H_7NO	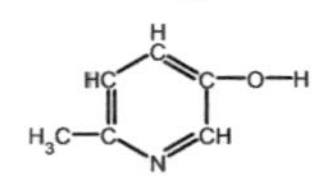	96.2 ± 2.1	7.40/6.40 × 2	D=N → C = 11.9 × 2	7.8 × 2	16.4

(continued)

Table 7.6 (continued)

Compounds	Formula	Structure	$\Delta_{sub}H°$ (298 K) [12]	D=HC → CH/ DHC → C	D=N → CH	D–O → CH	D–O•••H–O
2-Methyl-6-hydroxypyridine	C_6H_7NO	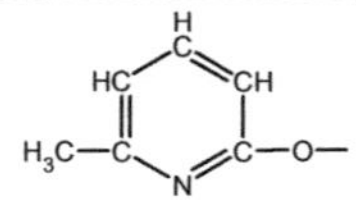	92.0 ± 1.3	7.40/6.40 × 2	D=N → C = 11.9 × 2	7.8 × 2	14.4
			$\Delta_{vap}H°$ (298 K) [8]	$DH_2C \rightarrow CH_2$	$D{=}N \rightarrow CH_2{-}N/D{-}O \rightarrow CH_2{-}O$	D–N•••H–N	
N-(Hydroxylethyl) piperazine	$C_6H_{14}NO$	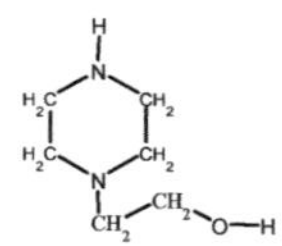	78.8 ± 0.7	5.70	5.75/6.35	6.85	9.05
1-Lactopiperidine	$C_8H_{15}NO_2$	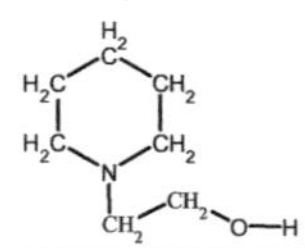	62.1	5.70	5.75/6.35	–	4.7

2-Hydroxypyridine < 3-Hydroxypyridine < 4-Hydroxypyridine

in connection with significant shifting of the electron density to the oxygen atom of the functional hydroxyl group with the distance of its location from the nitrogen atom of the cycle.

The methylhydroxypyridine molecules with nine bond vacancies form similar hydroxypyridines types of specific interactions and hydrogen bonds and grid structure of the crystal. The isostructural methyl group forms two interactions $DisoH_3C \rightarrow H\text{–}CH_2$ of low stability (Fig. 7.10b). The energy contribution of the isostructural methyl group to the sublimation enthalpy 2-methyl-3-hydroxypyridine, equal to the difference with sublimation enthalpy of 3-hydroxypyridine, has low value (1.1 ± 1.3 kJ mol^{-1}) and within the experimental error does not exceed 2.4 kJ mol^{-1}. The real value of this type of specific interaction is 0.6 kJ mol^{-1}. Taking the energy contributions of the isostructural methyl group as equal for all series of methylhydroxypyridines, we can determine the energies of the hydrogen bond of these compounds with the help of Eq. 7.13a.

$$D\text{-}O\cdots H\text{-}O = (\Delta_{sub}H^0(298\ K) - 3D{=}CH{\rightarrow}CH - 2D{=}N{\rightarrow}CH - D{=}CH{\rightarrow}C - D\text{-}O{\rightarrow}CH - DisoH_3C{\rightarrow}H\text{-}CH_2)/2 \quad (7.13a)$$

From the results of calculations given in Table 7.6 of the energies of the hydrogen bonds it follows that its extremely low value is at 2-methyl-3-hydroxypyridine with the location of the hydroxyl group at C(3) directly for the methyl group and nitrogen atom, which points to the special role of the isostructural methyl group in the distribution of electron density. The reduced energy value of the hydrogen bond (13.0 kJ mol^{-1}) compared with the energy of the same hydrogen bond, formed by the hydroxyl group at location C(5) or C(6) in the presence of the CH_3 group, points to the fact that this group prevents the shifting of the electron density in the cycle.

The *N*-(hydroxylethyl)piperazine molecule with hydroxyl group divided from thc piperazine ring of the ethylene fragment shows the special properties, forming each of the two ethylene groups, connected with the cycle nitrogen atom and, on the other side, with the oxygen atom the specific interaction $D\text{–}N \rightarrow CH_2\text{–}N$ and $D\text{–}O \rightarrow CH_2\text{–}O$, the energy values of which, in the presence of hetero atoms of the nitrogen of the cycle, exceed the stability of similar interactions formed by the methyl groups of methylamine and dimethyl ether 5.33 and 5.63 kJ mol^{-1}, respectively (Fig. 7.11). It follows, in particular, that the energy of the specific interaction $D\text{–}N \rightarrow CH_3\text{–}N = 6.35$ kJ mol^{-1} of liquid piperazine, replacing the fragment $\text{–}CH_2\text{–}CH_2\text{–}O\text{–}H$, contributes 44.8 kJ mol^{-1} to the vaporization enthalpy of *N*-(hydroxyethyl)piperazine (Sect. 5.1.2). Taking into account the contribution of the substituted hydrogen atom of the methyl group for $D\text{–}N \rightarrow CH_3\text{–}N$ for liquid condition (0.6 kJ mol^{-1}), we obtain the energy of the specific interaction $D\text{–}N \rightarrow CH_2\text{–}N$ equal to the value of 5.75 kJ mol^{-1}. Taking also the difference in the energies of the specific interactions D–O → CH_2–O > D–N → CH_2–N and the hetero cycle of the nitrogen atom to be the same value, we take the energy of the specific interaction with participation of the oxygen atom equal to 6.35 kJ mol^{-1}. Thus, the structure of liquid *N*-(hydroxylethyl)piperazine is the network formed by the specific interactions and hydrogen bonds (Fig. 7.11a), in which we have two hydrogen bonds of the two types DO•••H–O and D–N•••H–N (6.85 kJ mol^{-1}), four specific interactions D–H_2C → CH_2 (5.70 kJ mol^{-1}) and on two interactions D–N → CH_2–N, D–O → CH_2–O with the mentioned energy values. The unknown energy value of the hydrogen bond is determined from Eq. 7.13b

$$D\text{-}O\cdots H\text{-}O = (\Delta_{sub}H^0(298\ K)hpp - 2D\text{-}N\cdots H\text{-}N - 4D\text{-}H_2C{\rightarrow}CH_2 - 2D\text{-}O{\rightarrow}CH_2\text{-}O - D\text{-}N{\rightarrow}CH_2\text{-}N)/2 \quad (7.13b)$$

The energy value of the hydrogen bond of liquid *N*-(hydroxylethyl)piperazine (9.05 kJ mol^{-1}) obtained has the reduced value of 3.5 kJ mol^{-1} compared with crystalline 2-hydroxypyridine (12.5 kJ mol^{-1}), which reflects not only the stabilizing effect of the crystalline field, but to a certain extent the influence of the hetero atom cycle. The absence of the ethyl fragment between the piperazines cycle and the hydroxyl group of 2-methyl-3-hydroxypyridine (13.0) and 2-methyl-6-hydroxypyridine (14.4 kJ mol^{-1}) are additional arguments in favor of these views (Table 7.6).

The 1-lactopiperidine molecule forms the grid structure of liquid condition (Fig. 7.11b) with five specific interactions D–H_2C → CH_2, two interactions 2D–O → CH_2–O – and 2D–N → CH_2–N with values of the energies similar to those formed by *N*-(hydroxylethyl)piperazine (Table 7.6). The estimated value of the hydrogen bond energy of this compound at 361 K except for the temperature dependence of the specific interactions energies, points to the low stability of the hydrogen bonds at a given temperature.

7.2.3 *Energies of Specific Interactions of Uracil and Cytosines*

The most stable tautomeric form of uracil is dioxoform, which is a weak NH-acid, which points to the small difference between the two more stable specific interactions, including the hydrogen bond in its crystalline condition. The location of the nitrogen atoms in the cycle at positions (1) and (3) provides the carbon atoms C(6) and C(2) with reduced positive charges and the significant difference of charges with the nitrogen atoms to form the more stable four specific interactions D–N → CH (12.9) and D–N → C with established energy value above 12.9 and 11.9 kJ mol^{-1}, respectively, for crystalline compounds with the pyridine cycle. Two hydroxyl groups form four hydrogen bonds D–O•••H–O with unknown energy and two specific interactions DO → CH (7.8 kJ mol^{-1}), one D–O → C and one

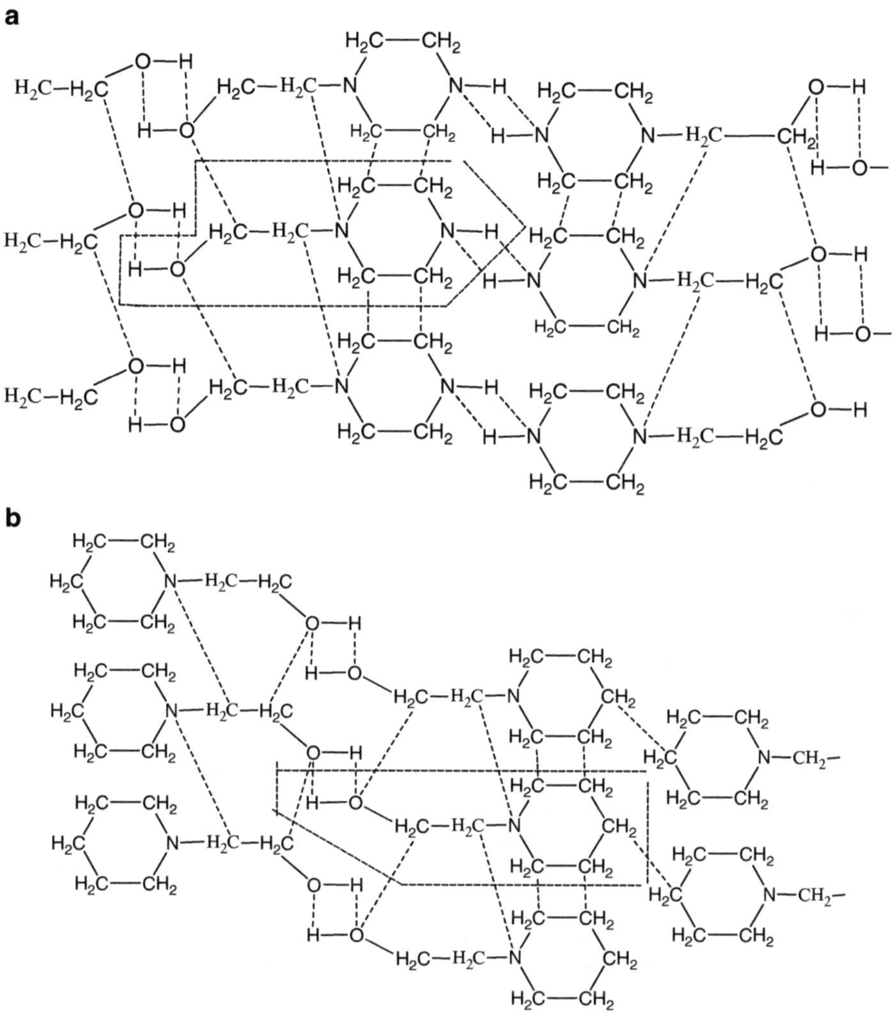

Fig. 7.11 Schematic picture of the liquid structure of *N*-(hydroxyethyl)piperazine (**a**) and 1-lactopiperidine (**b**) with network of the specific interactions and hydrogen bonds

D–C → C with energies 6.8 and 6.4 kJ mol^{-1}, respectively,. Thus, the network structure of crystalline uracil includes eight specific interactions and four hydrogen bonds (Fig. 7.12a). The energy of the unknown hydrogen bond should be determined by the equality of sublimation enthalpy and contributions of the energies of all types of interactions

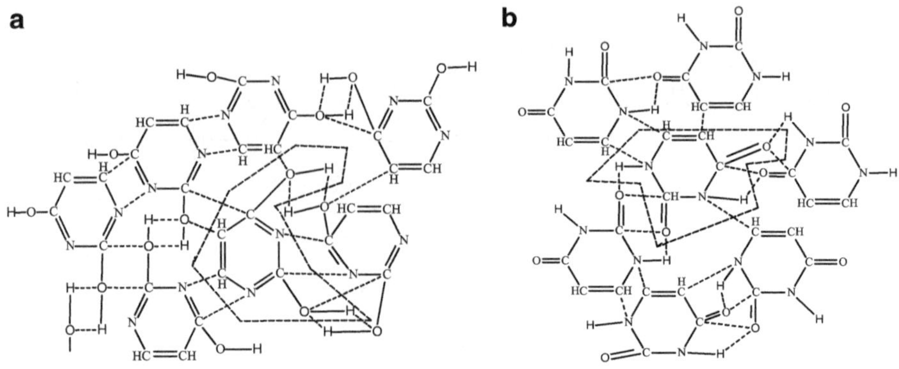

Fig. 7.12 Schematic picture of the crystalline structure of tautomeric forms uracil (**a**) and (**b**) with network of the specific interactions and hydrogen bonds

$$D\text{-}O\bullet\bullet\bullet H\text{-}O = (\Delta_{sub}H^0(298\ K)ur - (3D{=}N{\rightarrow}C + D{=}N{\rightarrow}CH)\ - (\ 2D\text{-}O{\rightarrow}CH + D\text{-}O{\rightarrow}C + DC{\rightarrow}C)/4 \qquad (7.14)$$

The energy value of the hydrogen bond (Table 7.7) obtained allows us to conclude that its increased energy value compared with the hydrogen bond of water (10.99 kJ mol^{-1}) at 2.4 kJ mol^{-1} complicates the destruction of the crystal structure of uracil and determines its low solubility.

The more stable tautomeric form of uracil in the dioxoform with two carbonyl and two imide = NH groups presents two peptide groups in one pirimidine cycle, significantly stabilizing hydrogen bonds and in this connection the sublimation enthalpies of the two tautomeric forms of uracil should have different values. The principal difference in the energies of the hydrogen bonds one can follow on the example of the enthalpy characteristics of this compound. The crystalline structure of the tautomeric form of uracil with two carboxyl groups has the network structure (Fig. 7.12b) formed by four hydrogen bonds D=C=O•••H–N and eight specific interactions of different types: 3D=C=O → C (6.8 × 3), D=C=O → CH (7.8), 3D=N → CH (12.9), D=HC → CH (7.40 kJ mol^{-1}). The calculated energy of the hydrogen bond with the use of an equation similar to Eq. 7.14 points to its increased value (Table 7.7), enough for the spontaneous transformation to the more stable tautomeric form.

The cytosine molecule with the hydroxyl and amino group forms three types of hydrogen bonds, significantly differing in stability. The less stable hydrogen bond is formed by amino groups D–N•••H–N with the above energy value of 9.7 kJ mol^{-1} for the crystalline compound (Chap. 6). Second, the more stable type of hydrogen bond D–O•••H–N, formed with the oxygen atom of the hydroxyl group significantly depends on the influence of the functional groups and the location of the cycle atoms. The third type of hydrogen bond D–O•••H–O forms hydroxyl groups of the molecules that are close in the environment. In the formation of the network

Table 7.7 Energies of the specific interactions and hydrogen bonds (kJ mol^{-1}) of the crystalline cytosines at 298 K

Compounds	Formula	Structure	$\Delta_{sub}H°$ (298 K) [12]	D=N → CH/ D=N → C	D=O → CH/ DC → C	D=O → C	D–O•••H–O
Uracil, 2,4-Dihydroxypirimidine	$C_4H_4N_2O_2$	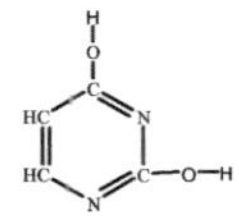	131 ± 5	12.9/11.9 × 3	7.8 × 2/6.4	6.8	13.4
Uracil, 2,4-Dihydroxypirimidine	$C_4H_4N_2O_2$	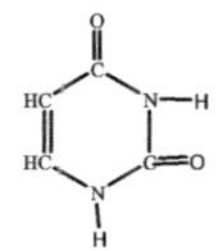	131 ± 5	12.9 × 3/7.40	D–O → CH 7.8 + DC → C 6.8 × 3	–	D=C=O•••H–N 14.2
Cytosine, 2-Hydrozy-4-aminopirimidine	$C_4H_5N_3O$	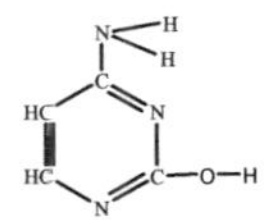	155.0 ± 3.0	12.9 × 2/11.9x2	/D–N•••H–N 9.7 × 2	DC → C 6.40 × 2	18.2 ± 3.0
1-Methylcytosine	$C_5H_7N_3O$	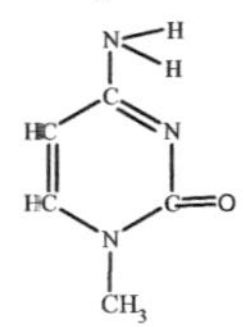	149.1 ± 0.9	/11.9 × 2 DHC → CH 7.4 + DHC → C 6.4	D–N•••H–N 9.7 × 2/D–N → CH_3–N 7.7 × 2	DC=O → CH 7.85 × 2	30.5 ±0.9
1-Methyl-*N*-Hydroxycytozine	$C_5H_7N_3O_2$	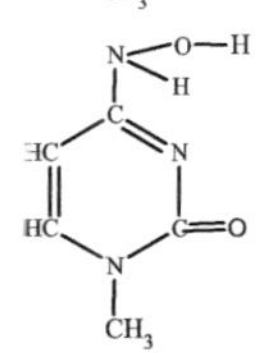	126.7 ± 1.5	11.9 × 2	D–N•••H–N 9.7 × 2/D–N → CH_3–N 7.7 × 2	D=C=O → CH 7.85 × 4 D=C=O → C 6.8	15.0 ± 1.5

(continued)

Table 7.7 (continued)

Compounds	Formula	Structure	$\Delta_{sub}H°$ (298 K) [12]	D=N → CH/ D=N → C	D=O → CH/ DC → C	D=O → C	D–O•••H–O
1-Methyl-4N-methoxycytosine	$C_6H_9N_3O_2$	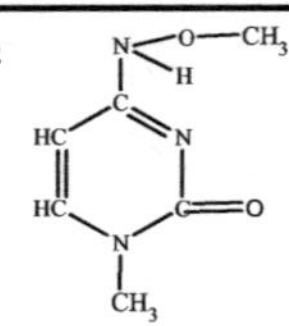	107.6 ± 0.3	/11.9 × 2 DOC 6.8 × 2	D–N•••H–N 9.7 × 2/D–N → CH_3–N 7.7 × 2	D=C=O → CH 7.85 × 4	D–O → CH_3 = 4.5

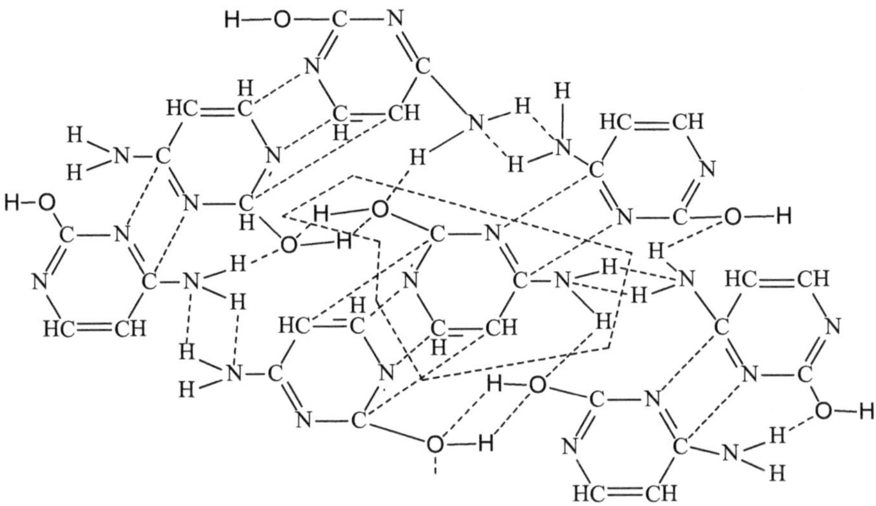

Fig. 7.13 Schematic picture of the. crystalline structure of cytosine with network of the specific interactions and hydrogen bonds

structure of crystal cytosine (Fig. 7.13) six specific interactions, formed by nitrogen atoms with CH and carbon atoms of the cycle with the energies 2D=N → CH (12.9), 2D=N → C (11.9), and 2DC → C (6.4 kJ mol^{-1}) participate.

Taking into account the presence of the two hydroxyl groups in the uracil molecule, one should note the influence of the increased number of coordinating OH groups on the stability of the formed hydrogen bonds compared with the reduced number of these groups in the cytosine molecule. However, in the presence of the two hydrogen atoms in the functional amino group and participation of the nitrogen atoms and CH groups of the cycle and 12 bond vacancies of the uracil and cytosine molecules the equality of the coordinating number is retained. In this connection the correctness of the use of the value of the energy of the hydrogen bond of crystalline uracil, formed by the hydroxyl group, in determining the energy of the hydrogen bond D–O•••H–N of crystalline cytosine is evident. Thus, taking the energy of the hydrogen bond D–O•••H–O, formed by the hydroxyl groups of uracil (13.4 kJ mol^{-1}) and cytosine, located within the error of experimentally measured values of enthalpies sublimation, we determine the energy value of the third type of hydrogen bond of crystalline cytosine with the help of Eq. 7.14a,

$$\text{D-O}\cdots\text{H-N} = (\Delta_{sub}H^0(298\ \text{K})\text{cts} - 2\text{D-O}\cdots\text{H-O} - 2\text{D-N}\cdots\text{H-N} - 2\text{D=N}\rightarrow\text{CH} - 2\text{D=N}\rightarrow\text{C} - 2\text{DC}\rightarrow\text{C})/2 \quad (7.14a)$$

The calculations of the energy of the hydrogen bond D–O•••H–N allow us to describe the energies of all realized specific interactions of crystalline cytosine by the general number of its stabilization,

DC→C (6.40) < D-N•••H-N (9.7) < D=N→C (11.9) < D=N→CH (12.9) < D-O•••H-O (13.4) < D-O•••H-N (18.2 kJ mol^{-1})

which shows that the tautomeric form of cytosine considered is more stable.

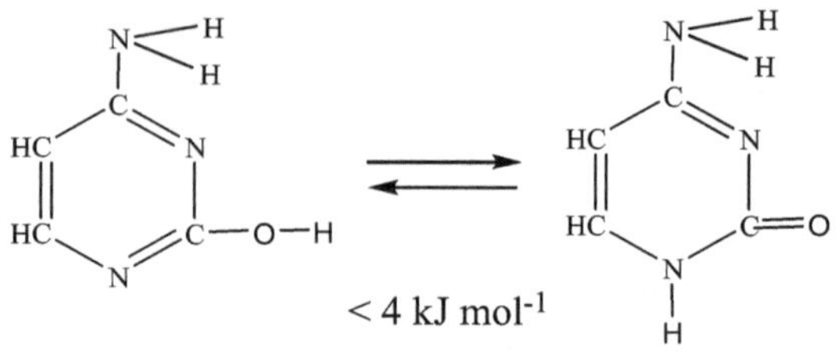

Such a high value of the energies of the hydrogen bond of crystalline cytosine compared with the energy of the hydrogen bond of water (10.99 kJ mol^{-1}) explains its low solubility in this compound.

The sublimation enthalpy of 1-methylcytosine is measured with a larger error, which contributes significant violations to the integral values of change of the hydrogen bonds of the homogeneous series of compounds. Nevertheless, it is reasonable to calculate the energies for clarifying the correctness of the measurements and the possibility of using similar research for estimation of the energies of the hydrogen bonds and specific interactions of high stability. The 1-methylcytosine molecule with 12 bond vacancies forms the network structure of the crystal (Fig. 7.14a), formed by two types of hydrogen bonds DC=O•••H–N and DN•••H–N with known energy (9.7 kJ mol^{-1}) and five types of eight specific interactions are also well-grounded energy values 2D–N → CH_3–N (7.7), 2D–C=O → CH (7.85), 2D–N → CH (11.9), D=HC → CH (7.4) and D=HC → C (6.4 kJ mol^{-1}). The energy of the more stable hydrogen bond with unknown value of D=C=O•••H–N is determined by the difference between the sublimation enthalpy and the total contribution of all noted specific interactions and the energy of the two hydrogen bonds D–N•••H–N using an equation similar to Eq. 7.14. The energy value of the hydrogen bond (Table 7.7) obtained significantly increases the value of the energy of the hydrogen bond of crystalline cytosine with respect to, in the first turn, the big experimental error in the enthalpy characteristics.

The 1-methyl-*N*-hydroxycytozine molecule with hydroxyl group in the amine fragment and carbonyl cycle, located between the nitrogen atoms, provides a highly negative charge to the atom of oxygen of carbonyl. Thirteen bond vacancies of the molecule of this compound form the structure crystal with a network of the two types of hydrogen bonds D–O•••H–O and D–N•••H–N with an energy of the latter of 9.7 kJ mol^{-1} and nine specific interactions of four types (Fig. 7.14b). The energies of the types of specific interactions 4D=C=O → CH, 2D=N → C, D–O → C and 2D–N → CH_3–N formed with known values 7.85, 11.9, 6.8, and 7.7 kJ mol^{-1}, respectively, are correct and are used in the calculations of the hydrogen bond energies of the number of compounds of analogs to the considered compounds. Using these data and Eq. 7.14b

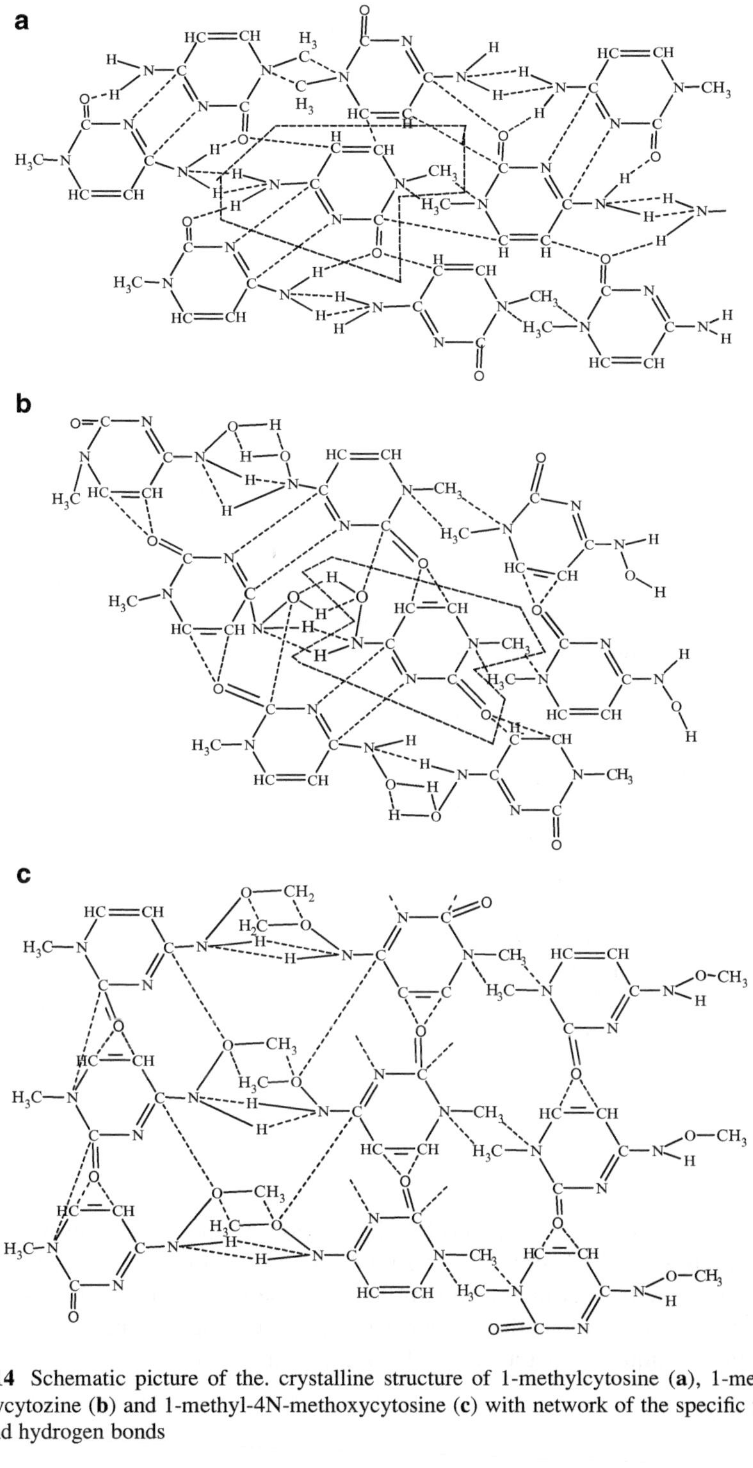

Fig. 7.14 Schematic picture of the. crystalline structure of 1-methylcytosine (**a**), 1-methyl-*N*-hydroxycytozine (**b**) and 1-methyl-4N-methoxycytosine (**c**) with network of the specific interactions and hydrogen bonds

$$D\text{-}O\cdots H\text{-}O = (\Delta_{sub}H^0(298\ K)mhxc - 2D\text{-}N\cdots H\text{-}N - 2D\text{-}N{\rightarrow}CH_3\text{-}N - 2D{=}N{\rightarrow}C - 4D{=}C{=}O{\rightarrow}CH - D\text{-}O{\rightarrow}C)/2 \quad (7.14b)$$

we calculated the energy of the hydrogen bond, formed by the hydroxyl group of the amine fragment of 1-methyl-*N*-hydroxycytozine (Table 7.7). The value obtained reflects the increased stability of the hydrogen bond, formed by the hydroxyl group

D-O•••H-O (13.4) Uracil < D-O•••H-O (13.4) Cytosine < D-O•••H-O (15.0) 1-Methyl-N-Hydroxycytozine < D-O•••H-N (18.2 kJ mol^{-1}) Cytosine

and reduced stability compared with the bond formed with the participation of the amino group of the cytosine molecule. One should notice that despite the high error in the value of the energy of the hydrogen bond, formed by the carbonyl oxygen atom with the amino group 1-methylcytosine, its increased stability can be assumed compared with the hydrogen bonds of the other compounds of this series considered.

The use of the sublimation enthalpy of 1-methyl-4N-methoxycytosine, measured at a temperature increasing the standard conditions by 27°, in the calculations certainly contributes definite errors to the values of the energies of the specific interactions, formed by the methoxy group. Evidently, those energies of the formed specific interactions of this fragment should depend on its location in the molecule. However, the data on thermodynamic properties allow us to make a thermodynamic analysis for 1-methyl-4N-methoxycytosine. This group and its location in the amine fragment do not violate the network structure of the crystal. It contributes only some details (Fig. 7.14c), consisting of the fact that two more stable specific interactions $D{=}N \rightarrow C$ with the energy of 11.9 kJ mol^{-1} are formed with similar cycles of the molecules of neighboring networks of the crystal and provide more stable criss-crossing of the last. The correct values of the energies of ten specific interactions and two hydrogen bonds (Table 7.7), realized in the crystalline 1-methyl-4N-methoxycytosine, allow us to determine the energy value formed by the methoxy group with the help of Eq. 7.14c

$$D\text{-}O{\rightarrow}CH_3 = (\Delta_{sub}H^0(298\ K)mmc - 2D\text{-}N\cdots H\text{-}N - 2D\text{-}N{\rightarrow}CH_3\text{-}N - 2D{=}N{\rightarrow}C - 4D{=}C{=}O{\rightarrow}CH - DO{\rightarrow}C)/2 \quad (7.14c)$$

The energy value of the specific interaction $D\text{–}O \rightarrow CH_3$ formed by the methoxy group (4.5 kJ mol^{-1}) at the total number of 14 specific interactions of crystalline 1-methyl-4N-methoxycytosine is obtained by the use of the sublimation enthalpy at 25° higher than the standard (5.63 kJ mol^{-1}) of the same type of specific interaction of liquid *N*-methoxydimethylamine (Table 6.1).

7.2.4 Energies of Specific Interactions of Compounds with Peptide's Group of 2-Piperidone and Its Derivatives

Compounds with sexatomic heterocycles unlike the pentatomic cycles contain additional methylene or CH groups with double bond, which increases the total shifting of the electron density from the donor atoms of carbon and hydrogen. The presence of one peptide group in the cycle of these two heterocyclic compounds provides for increased shifting of the total electron density from the carbonyl carbon atom. However, it does not mean that the hydrogen bond formed by the peptide group in the hexamerous heterocycle will have more stability than in the pentamerous cycle because the energy of this type of interaction is determined by the charge value at the oxygen atom and its difference from the charge value of the hydrogen atom. In this connection 2-piperidone is of interest, with one peptide group and five carbon atoms of the cycle.

The 2-piperidone molecule with nine bond vacancies forms the network structure of the liquid condition with two stable hydrogen bonds D=C=O•••H–N, two specific interactions of increased stability D=C=O → CH_2, four D–CH_2 → CH_2 and two single bonds D–N → CH_2 and D–H_2C → C with interaction of reduced stability (Fig. 7.15a). The energy of the specific interaction D=C=O•••H–N is taken as equal to the energy of the same type (8.3 kJ mol^{-1}), realized in liquid 1-methyl-2-pyrrolidinone. The energy value of the specific interaction D–CH_2 → CH_2 we take as equal to the energy of the same type (Chap. 3), formed in liquid cyclopentane (5.7 kJ mol^{-1}). The energies of the remaining two single bonds D–N → CH_2 (5.3 kJ mol^{-1}) and D–H_2C → C (5.3 kJ mol^{-1}) obtained from the energies of the interactions, formed by the methyl and methylene groups reduced by the contribution of three and two hydrogen atoms, respectively, are 1.8 and 1.2 kJ mol for the liquid compounds. The energy of the hydrogen bond is determined by the difference between the vaporization enthalpy and the total energy contribution of the remaining specific interactions with the help of Eq. 7.15

$$\text{D=C=O}\cdots\text{H-N} = (\Delta_{vap}H^0(298\ \text{K}) - 2\text{D=C=O}\rightarrow\text{CH}_2 - 4\text{D-H}_2\text{C}\rightarrow\text{CH}_2 - \text{D-N}\rightarrow\text{CH}_2)/2 \quad (7.15)$$

The obtained energy value of the hydrogen bond (Table 7.8) points to its reduced stability compared with the peptide group of liquid 2-pyrrolidone (14.45) > 2-piperidone (12.45 kJ mol^{-1}) formed. From this we can conclude that at the location of the four methylene groups "at cyclic bond with peptide fragment" we reach the maximum possible shifting of the electron density to the nitrogen and oxygen atoms in the cycle providing it with reduced difference in the charges. At the same time at the minimum shifting of the total electron density from the three methylene groups in the pentamerous cycle we have an increased difference between the charges of nitrogen and oxygen atoms and a difference in the charge of the hydrogen atom of the imide group. As a result the increased stability of the

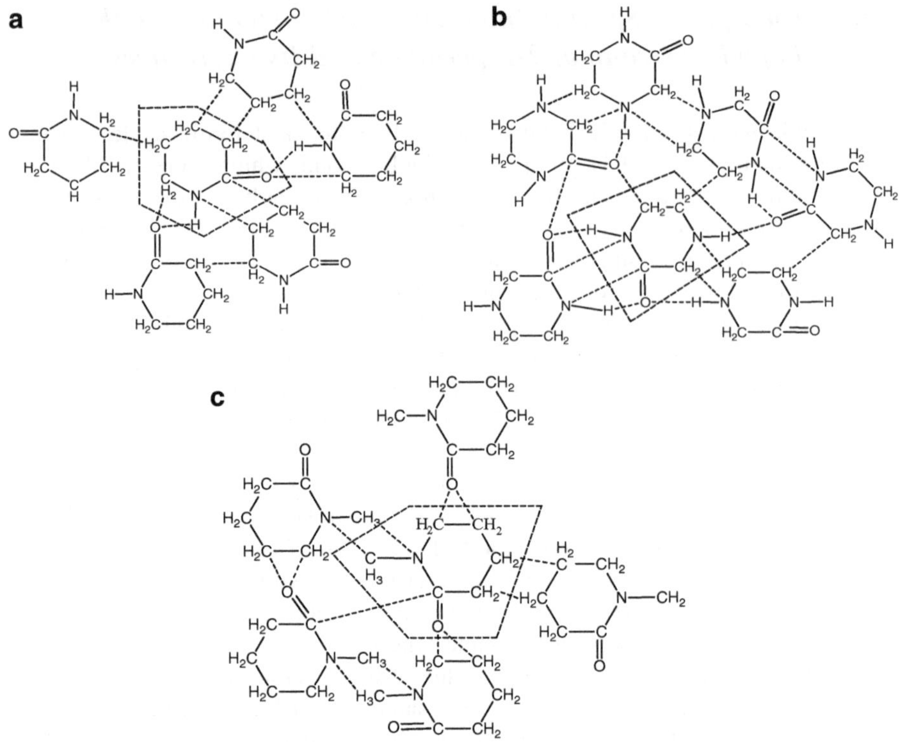

Fig. 7.15 Schematic picture of the liquid structure of 2-piperidone (a), tetrahydro-2-pyrinidone (b), and 1-methyl 2-piperidone (c) with peptide fragment and network of the specific interactions and hydrogen bonds

hydrogen bond, formed by the peptide group, is reached at the shifting of the electron density from the three methylene groups of the pentamerous hetero cycle.

The tetrahydro-2-pyrinidone molecule with two symmetrically located imide groups in the hexamerous hetero cycle and the carboxyl group undergoes the significant changes in the distribution of the electron density and creates certain depletion of charges at each nitrogen atom compared with the 2-piperidone molecule (Fig. 7.15b). However, the sublimation enthalpy, interrelated with the number and energies of the specific interactions of this compound, exceeds the same characteristics of 2-piperidone at 14.8 kJ mol^{-1} (Table 7.8). With a total number of bond vacancies of nine and ten at 2-piperidone and tetrahydro-2-pyrinidone molecules, respectively, for the last connection the formation of the more stable specific interactions is typical. Thus, the energies of these interactions are determined not so much by the charges of the nitrogen atoms, but by the charge of the oxygen atom of the carbonyl and charges of the hydrogen atoms of the imide N–H fragments. It follows that the tetrahydro-2-pyrinidone molecule forms an increased number, at least, for more stable hydrogen bonds, rather than the 2-piperidone

Table 7.8 Energies of the specific interactions and hydrogen bonds (kJ mol^{-1}) of the liquid 2-piperidone and its derivatives

Compounds	Formula	Structure	$\Delta_{vap}H°$ (298 K) [8]	T K	D–CH_2 → CH_2/DC → C	D=C=O → CH_2	D=C=O–•••H–N
2-Piperidone	C_5H_9NO	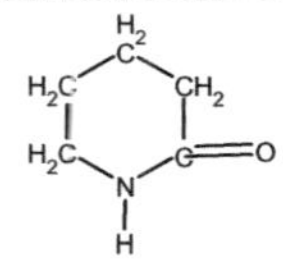	74.5	302	5.60/5.3	8.3	12.45
Tetrahydro-2-pyrinidone	$C_4H_8N_2O$	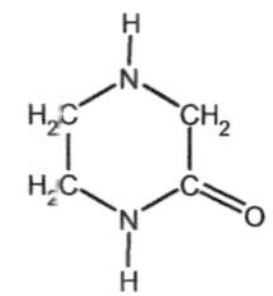	89.3	298	5.60 D–N → CH_2 = 5.7 × 2	8.3	13.3
1-Methyl 2-piperidone	$C_6H_{11}NO$	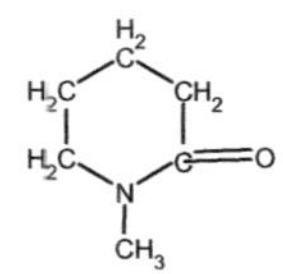	55.4	358	5.60/5.3	6.2	D–N → CH_3–N = 7.1
2-Piperidinethione	$C_6H_{11}NS$	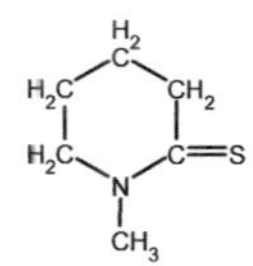	61.3	366	5.60/5.3	D=C=S → CH_2 = 7.7	D–N → CH_3–N = 7.1
1-Methyl-2(1H)-piridone	C_6H_7NO	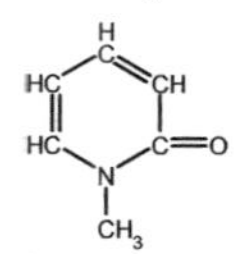	60.2	368	5.65/5.3	7.45	D–N → CH_3–N = 7.1

(continued)

Table 7.8 (continued)

Compounds	Formula	Structure	$\Delta_{vap}H^{\circ}$ (298 K) [8]	T K	D–CH_2 → CH_2/DC → C	D=C=O → CH_2	D=C=O–•••H–N
1-Nitrosopiperidone	$C_5H_{10}N_2O$	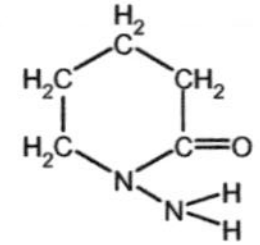	74.3	303	D–N → CH_2 5.7 × 2/DH_2C → C 5.1 × 2	D=C=O → CH_2 8.3 × 4	D–N•••H–N 4.9
		Crystal	[12]		D–N → CH_3–N	D=C=O → CH	–
1,3-Dimethylthymine	$C_7H_{10}N_2O_2$	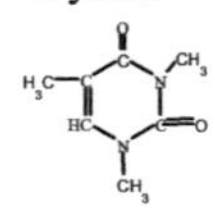	109.2 ± 2.1	338	7.7	9.53	D*iso*H_3C → H–CH_2 = 2.2
Phenyl-(2-Pyridyl) acetamide	$C_7H_8N_2O$	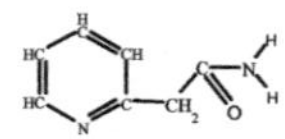	103.8	298	D–N → CH_2 12.8 × 2/ D–H_2C → CH_2 7.43 × 4	D=C=O → CH_2–O=C 9.45	(D=C=O•••H–N + DN–•••H–N)/ 2 = 13.6
Formanilide	C_7H_7NO	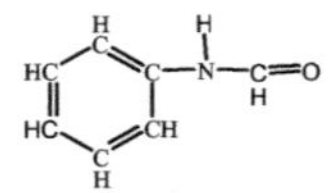	77.8	308	D–H_2C → CH_2 7.4 × 4	D=C=O → CH_2–O= C = 8.45	–

molecule. The tetrahydro-2-pyrinidone molecule with two hydrogen atoms of imide fragments and the oxygen atom of the carbonyl group with ten bond vacancies forms with the molecules that are close in the environment the network structure (Fig. 7.15b) of liquid condition with four stable hydrogen bonds and six specific interactions D–CH_2 → CH_2 (5.60), D=C=O → CH_2 (8.3), 2D–N → C (5.3) and D–N → CH_2 (5.7 kJ mol^{-1}), the energy values of which are equal to that realized in liquid *N*-methyl-piperazine (Table 5.4) and, practically, not different from the energies of the specific interactions of liquid piperidine (Table 5.1) and 2-piperidone (Table 7.8). The results of the calculations of the energy of the hydrogen bond with the help of Eq. 7.15a

$$D{=}C{=}O\cdots H\text{-}N = (\Delta_{vap}H^0(298\ K)tpr - D{=}C{=}O{\rightarrow}CH_2\ - D\text{-}H_2C{\rightarrow}CH_2 - 2D\text{-}N{\rightarrow}CH_2 - 2D\text{-}N{\rightarrow}C)/4 \tag{7.15a}$$

illustrate (Table 7.8) the increased stability of this type of interaction in liquid tetrahydro-2-pyrinidone (13.3) compared with liquid 2-piperidone (12.45 kJ mol^{-1}). The obtained data are an indication of the significant stabilization of the hydrogen bond, caused, at first, by the difference between charges of carbonyl oxygen atoms and the hydrogen atom of imide fragment and, second, by the special role of the nitrogen atom, which is able to transfer more electron density to the carbon atom of the carbonyl group and further to the oxygen atom, than it obtains from the hydrogen atom. As a consequence, such a high stabilization of the hydrogen bond is reached, the value of the energy of which exceeds by a half the energy value of the most stable specific interaction of liquid tetrahydro-2-pyrinidone.

The 1-methyl 2-piperidone, 2-piperidinethione, and 1-methyl-2(1H)-piridone compounds with the same molecular structure and the network structure of liquid condition formed (Fig. 7.15) allow us to follow the influence of the replacement of the oxygen atom in the carbonyl group by the sulfur atom and replacement of the methylene group to the double bond in the hexamerous cycle by the energy of the more durable specific interaction.

The 1-methyl 2-piperidone molecule forms in condensed condition (Fig. 7.15c) the network of nine specific interactions and includes four D=C=O → CH_2, two DN → CH_3–N (7.1 kJ mol^{-1}), two D–H_2C → CH_2 (5.6 kJ mol^{-1}) and one DC → C (5.3 kJ mol^{-1}) with corresponding energy values. The 2-piperidinethione molecule forms the same types and number of the specific interactions with above accepted values, except for the one formed by the carbonyl oxygen atom and sulfur atom, respectively. The value of the energy of the specific interactions, formed by the carbonyl oxygen and sulfur, we determine from Eq. 7.16,

$$D{=}C{=}O(S){\rightarrow}CH_2\ = (\Delta_{vap}H^0(298\ K) - 2D\text{-}H_2C{\rightarrow}CH_2 - 2D\text{-}N{\rightarrow}CH_3\text{-}\ N - DC{\rightarrow}C)/4 \tag{7.16}$$

The appearance of the inversion sign in the vaporization enthalpies of 1-methyl 2-piperidone and 1-methyl-2(1H)-piridone (Table 7.8) at the increased value at

4.8 kJ mol^{-1} for the compound with double carbon bonds in the cycle can serve as a concrete example of:

First, the significant influence of the total value of the electron density shifted on the stability of the specific interactions formed, to achievement of increased charge at the atom of the carbonyl oxygen and,

Second, a significant difference between the negative charge and the positive charge of the carbon atom of the methyl group of the peptide groups. Of course, such a condition on the charges at the oxygen atom and the methyl group is reached at a definite charge ratio, connected to the nitrogen atom and the carbon atom of the carbonyl at the shifting of the electron density from the CH groups with double bonds. The energies of the formed specific interactions by the CH groups of the chain 2D=HC → CH should be estimated from the vaporization enthalpy of 1,4-cyclohexadiene (Chap. 3) as equal to 5.65 kJ mol^{-1}. The energy of the specific interaction is determined with the help of the similar Eq.7.16a,

$$D{=}C{=}O{\rightarrow}CH = (\Delta_{vap}H^0(298\ K) - 2D{=}HC{\rightarrow}CH - 2D\text{-}N{\rightarrow}CH_3\text{-}\ N - DC{\rightarrow}C)/4 \tag{7.16a}$$

The results of the calculation (Table 7.8) of the energies of the specific interactions formed by the carbonyl oxygen atom of 1-methyl 2-piperidone and its analogs on the data of the vaporization enthalpies [8] are described by the inequality,

$D{=}C{=}O{\rightarrow}CH_2$ 1-Methyl 2-piperidone (6.2) < $D{=}C{=}O{\rightarrow}CH$ 1-Methyl-2(1H)-piridone (7.45) < $D{=}C{=}S{\rightarrow}CH_2$ 2-Piperidinethione (7.7 kJ mol^{-1}),

which reflects its stabilization in the hexamerous heterocycle at the replacement of the methylene groups by CH groups with double bonds and at the replacement of the carbonyl oxygen atom by the carbonyl atom of sulfur.

The 1,3-dimethylthymine molecule with two carboxyl and two imide groups with substituted hydrogen atoms to the methyl group has a definite deficit in the shifting of the electron density to the nitrogen atom from the methyl groups. Nevertheless, the deficit of the electron density provides relatively high capacity to the oxygen atom as regards the shifting of the electron density from the nitrogen atom to the carbon and further to the oxygen atom. The network structure of crystalline 1,3-dimethylthymine (Fig. 7.16a) is formed by eight specific interactions, including four D=C=O → CH types and four interactions of the second type DC=O → C with reduced energy. Taking into account the complexity of accounting for the differences in the energies of these two types of specific interactions, we restrict ourselves to the definition of its average value, which is possible, to estimate the energy of the specific interaction DN → CH_3–N, formed by the N–CH_3 group of crystalline imides, corresponding to an equal value of 7.7 kJ mol^{-1} (Table 6.7).

Thus, we determine the energy of the specific interaction DC=O → CH from the difference with the sublimation enthalpy for 1,3-dimethylthymine

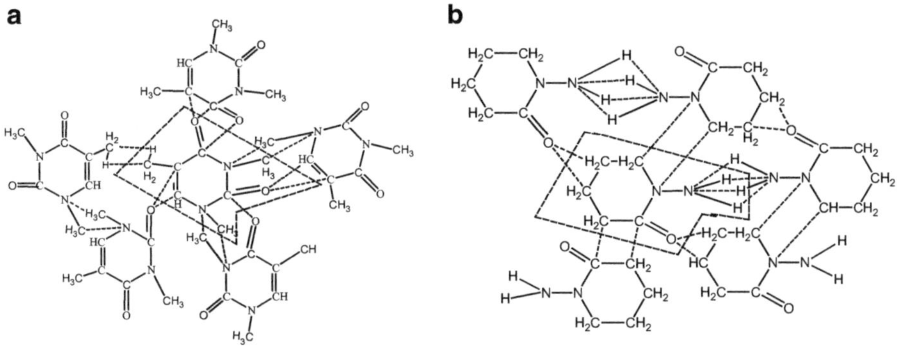

Fig. 7.16 Schematic picture of the crystalline structure of 1,3-dimethylthymine (**a**) and 1-nitrosopiperidone (**b**) with network of the specific interactions

$$D{=}C{=}O{\rightarrow}CH = (\Delta_{vap}H^0(298\ K) - 4D\text{-}N{\rightarrow}CH_3\text{-}N - 2D iso H_3C{\rightarrow}H\text{-}CH_2)/8 \quad (7.16b)$$

The value of the energy of the specific interaction D=C=O → CH of crystalline 1,3-dimethylthymine obtained (Table 7.8) exceeds the energy of the interaction D=C=O → CH_2 of liquid 2-piperidone at 1.2 kJ mol^{-1}, formed by the hexamerous cycle of the saturated hydrocarbon. This value is reduced, which should be explained by the significant deficit in the shifting of the electron density at two oxygen atoms of the carbonyl groups in the 1,3-dimethylthymine molecule and one hydrogen atom at the carbon atom in the hexamerous cycle.

Notice that the 1-nitrosopiperidone molecule with 12 bond vacancies is characterized practically by the same value of the vaporization enthalpy as 2-piperidone (74.5 kJ mol^{-1}) with nine bond vacancies. It means that, being connected with the nitrogen atom of the cycle, the amide group in the presence of the carbonyl fragment does not express the properties of the peptide group and does not form a stable hydrogen bond with the oxygen atom. In this connection we determine the energy of the hydrogen bond, formed by the amide group of 1-nitrosopiperidone, connected with the nitrogen atom of the cycle. The structure of liquid 1-nitrosopiperidone is the network of eight specific interactions and four hydrogen bonds D–N•••H–N (Fig. 7.16b). Being formed by the carbonyl's oxygen atom, the four stable specific interactions D=C=O → CH_2 have an increased stability (8.3 kJ mol^{-1}). The energies of the specific interactions of 2D–N → CH_2 (5.7) and D–H_2C → C (5.1 kJ mol^{-1}), formed by the nitrogen atom and CH_2 groups of the cycle were obtained from the energies of the methylene group of the cycle reduced by the contribution of one substituted hydrogen atom of the group with the energy value of 0.6 kJ mol^{-1} for the liquid condition. The results of the calculations (Table 7.8) point to a small difference in the hydrogen bond energy

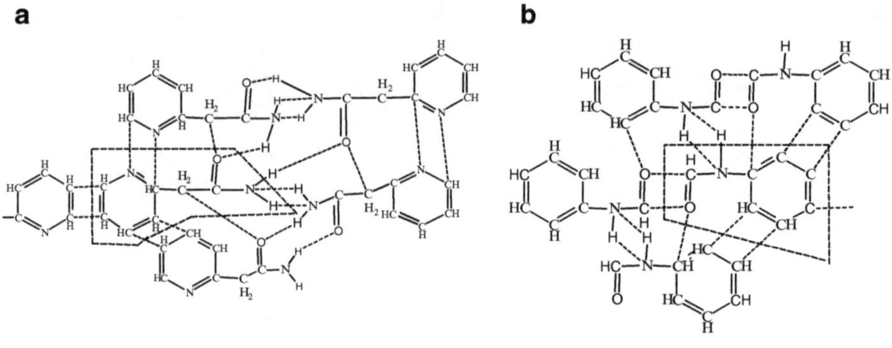

Fig. 7.17 Schematic picture of the crystalline structure of 2-pyridyl acetamide (**a**) and formanilide (**b**) with network of the specific interactions and hydrogen bonds

$$D\text{-}N\cdots H\text{-}N = (\Delta_{vap}H^0(298\ K) - 4D{=}C{=}O{\rightarrow}CH_2 - 2D\text{-}N{\rightarrow}CH_2 - 2D\text{-}H_2C{\rightarrow}C{=}O)/4 \quad (7.16)$$

compared with the one realized in liquid aniline (5.10 kJ mol^{-1}). We draw attention to the fact that estimated energies of the specific interactions, formed by the nitrogen atom of the hexamerous cycle are within the limits of the energy values of liquid dimethylamine (4.8) < D–N → CH_2 (5.1) < methylamine (5.33 kJ mol^{-1}).

It is fundamentally important to clarify the role of the carbonyl group of acetamide compounds in the stabilization of the hydrogen bond formed by amine and imide groups. In this connection we will use 2-pyridyl acetamide and formanilide and undertake a thermodynamic analysis of it.

The 2-pyridyl acetamide molecule with 12 bond vacancies form six specific interactions by hexamerous heterocycle 2D=N → CH and four 4D=CH → CH with energies 12.8 and 7.43 kJ mol^{-1} (Chap. 5, Table 5.8), respectively, and two specific interactions by the acetate fragment D=C=O → CH_2–C=O and four hydrogen bonds by the amide group of the two types with differing energies D–N•••H–N < D=C=O•••H–N (Fig. 7.17a). The energy of the specific interaction, formed by the acetate group with one saturated hydrogen atom, determined from the energy contribution of the given fragment (9.45 kJ mol^{-1}, Table 6.7) reduced by the contribution of the saturated hydrogen atom (1.0 kJ mol^{-1}) of crystalline 2-pyridyl acetamide. The total energy contribution of the two types of hydrogen bonds formulated by the amide group of 2-pyridyl acetamide should be determined by the difference of the sublimation enthalpy of this compound and total contribution of the energies of all specific interactions. Its total value can also determined by the difference of sublimation enthalpy of 2-pyridyl acetamide and pyridine (55.4 kJ mol^{-1}) reduced by the energy contribution of the substituted hydrogen atom of the last compound (49.4 kJ mol^{-1}). From the obtained value one should subtract the energy contributions of the specific interaction, formed by the acetate fragment

$$(D{=}C{=}O{\bullet\bullet\bullet}H\text{-}N + D\text{-}N{\bullet\bullet\bullet}H\text{-}N)/2 = (49.4 - 2D{=}C{=}O{\rightarrow}CH_2\text{-}O{=}C\ (9.45 \times 2))/4$$

The average energy value obtained of the two types of hydrogen bonds, formed by 2-pyridyl acetamide (7.6 kJ mol^{-1}) has an increased value compared with the value of the energy of crystalline ammonia. However, the energy is significantly lower than the energy formed by crystalline acetamide D–N•••H–N (9.7 kJ mol^{-1}) and has an even greater difference with a similar average value of the energy of the two types of hydrogen bonds (D=C=O•••H–N (17.5) + D–N•••H–N (9.7))/2, formed by the same crystalline acetamide (13.6 kJ mol^{-1}). It follows that the location of the methylene group of acetate fragment between pyridine's cyclic and carbonyl group greatly changes the properties of this group as a result of the changes in the distribution of electron density. In other words, the methylene group prevents the shifting of the electron density from pyridine's cyclic to the carbonyl oxygen atom.

The formanilide molecule with 11 bond vacancies forms the structure of the crystal with the network of nine specific interactions and two hydrogen bonds D–N•••H–N (Fig. 7.17b). Five specific interactions D=HC → CH are formed by the CH groups of the benzene ring with energies 7.40 kJ mol^{-1}. One of the CH groups of the benzene ring of the formanilide molecule that is close in the environment forms the specific interaction D=C=O → CH with the carbonyl oxygen atom. In its turn, the acetate fragment with two saturated hydrogen atoms of the molecule forms a similar interaction D=C=O → CH–O=C with increased energy value (8.45 kJ mol^{-1}), obtained from the value of the energy, formed by the same fragment of crystalline acetamide reduced by the contribution of the two substituted hydrogen atoms (2.0 kJ mol^{-1}). Taking into account that the energy of the specific interaction of CH groups with the oxygen atom is more stable than the ones formed by the same groups of the benzene ring, we nevertheless take them as equal to the value for D=C=O → CH=D=HC → CH = 7.40 kJ mol^{-1}.

$$D\text{-}N{\bullet\bullet\bullet}H\text{-}N = (\Delta_{vap}H^0(298\ K) - 5\ D{=}HC{\rightarrow}CH\ -\ 4D{=}C{=}O{\rightarrow}CH - 2D\text{-}N{\rightarrow}CH_2\ - 2D\text{-}H_2C{\rightarrow}C{=}O)/2 \tag{7.17}$$

Using the suggested approaches to the energy determination of the unknown type of interaction we obtained the energy of the hydrogen bond: 5.0 kJ mol^{-1}. The given value of the energy is a reduced one compared with the energy of hydrogen bond of crystalline ammonia, calculated for the tetracoordinated nitrogen atom. The value is determined by the measured value of the sublimation enthalpy of formanilide, given without any indication of error. The obtained data on the energies of the formed hydrogen bonds by imide with the closely located carbonyl functional group lead to the conclusion that in playing the role of the functional groups the stabilizing effect is practically not expressed.

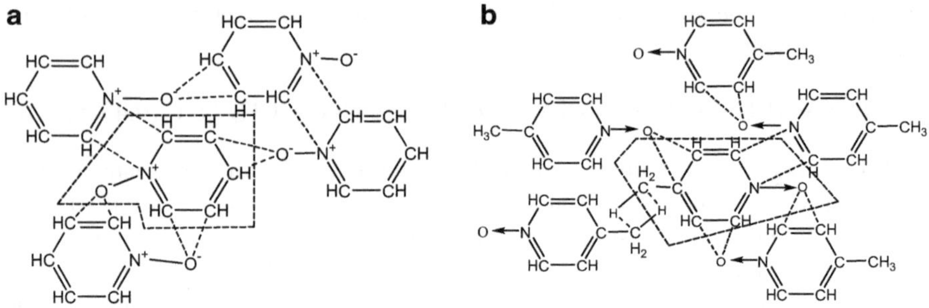

Fig. 7.18 Schematic picture of the crystalline structure of pyridine *N*-oxide (**a**) and methylpyridine *N*-oxide (**b**) with network of the specific interactions

7.2.5 Energies of Specific Interactions of Crystalline Pyridine N*-Oxides and Pyrazine-1,4-Dioxide*

The information on the energies of the specific interactions of liquid and crystalline hexamerous heterocycle pyridine and its derivatives obtained in Chap. 5 allow us to implement the thermodynamic analysis of more complicated compounds with positive and negative charges in a fragment's interconnected nitrogen and oxygen atom. The simplest compound from this series of compounds is pyridine *N*-oxide, of which the crystalline structure is formed at the coordination of four molecules near the central molecule (Fig. 7.18a). The increased negative charge of the oxygen atom in the pyridine *N*-oxide molecule provides high donor properties and high ability in the formation with the CH groups of the cycle of six stable specific interactions D–O → CH. The given interactions should possess higher stability compared with the realized D=HC → CH bonds of crystalline pyridine with an energy value of 7.40 kJ mol^{-1}. The second type of specific interactions, D=N → CH, is formed by contacting molecules with the formation of the interaction, realized in crystalline pyridine with the energy of 12.9 kJ mol^{-1}. The two types of specific interactions formed at eight specific interactions simplify and increase the correctness of the calculations of the interaction D–O → CH with an unknown energy value,

$$\text{D-O}\rightarrow\text{CH.} = (\Delta_{\text{vap}}H^0(298\text{ K}) - 2\text{D=N}\rightarrow\text{CH})/6 \qquad (7.18)$$

The energy value of this type of specific interaction D–O → CH (Table 7.9) is, at 2.5 kJ mol^{-1}, more stable compared with the interaction realized in crystalline pyridine (7.40 kJ mol^{-1}). It is caused by the influence of the high negative charge of the oxygen atom and practically by the preservation of the carbon atom charge of CH group, possibly remaining a stable one in the cycles of pyridine *N*-oxide and pyridine.

Table 7.9 Energies of the specific interactions (kJ mol^{-1}) of the crystalline cyanopyridine *N*-oxides and pyridinecarboxamide *N*-oxides at 298 K

Compounds	Formula	Structure	$\Delta_{sub}H^{\circ}$ (298 K) [12]	D≡N → C≡	D=N → CH	D=HC → CH	D–O → CH
Pyridine	C_5H_5N	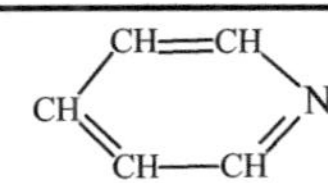	55.4. ± 2.0	–	12.90	7.40	–
Pyridine *N*-oxide	C_5H_5NO	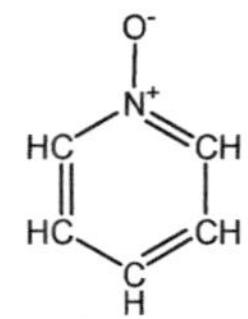	79.3 ± 1.0	298	12.90	–	8.9
2-Methylpyridine *N*-oxide 2-Picoline -*N*-oxide	C_6H_7NO	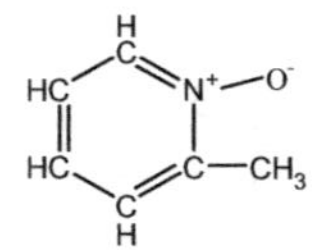	78.2 ± 2.2	298	12.9	isoDH_3C → H–CH_2 =0.55 × 2	8.55
3-Methylpyridine *N*-oxide, 3-Picoline-*N*-oxide	C_6H_7NO	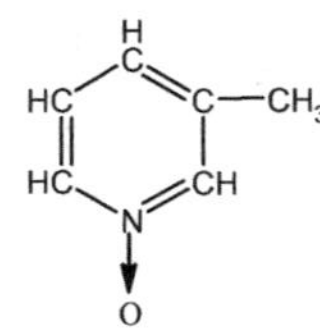	82.2 ± 2.4	298	12.9	D*iso*CH_3 = 0.55 × 2	9.1
4-Methylpyridine *N*-oxide	C_6H_7NO	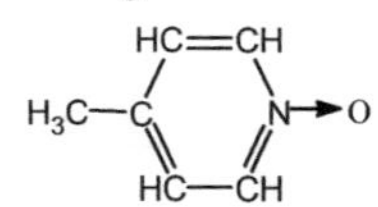	85.3 ± 2.6	298	12.9	D*iso*CH_3 = 0.55	10.0
Pyrazine1, 4-Dioxide	$C_4H_4N_2O_2$	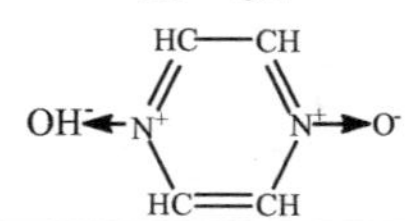	116.9 ± 0.8	298	12.9	–	11.25

The reduced value of the sublimation enthalpy of 2-methylpyridine *N*-oxide compared with the same characteristics of pyridine *N*-oxide at 1.1 kJ mol^{-1} points to the participation of the isostructural methyl group in the distribution of the electron density, which prevents its shifting.

In this case it is principally impossible to reliably determine the contribution of the isostructural methyl group to the sublimation enthalpy, because it participates in the formation of the two specific interactions of low stability. Nevertheless, estimating its contribution to the enthalpy characteristics from the difference of theproperties of these compounds, it is possible to determine the value of the energy contribution to the properties of this series of compounds. The conditions for the determination of the energy value of the second type of specific interaction D–O → CH have been created for these compounds. The schematic picture presented in Fig. 7.18b of the network structure of the crystal of 2-methylpyridine *N*-oxide with the same number and the same types of stable specific interactions and two interactions of low stability, $DH_3C \rightarrow H\text{–}CH_2$, allow us to describe the contributions of the specific interactions to the sublimation enthalpy by Eq. 7.17a,

$$\text{D-O}\rightarrow\text{CH.} = (\Delta_{vap}H^0(298\text{ K}) - 2\text{D=N}\rightarrow\text{CH} - 2\text{DH}_3\text{C}\rightarrow\text{H-CH}_2)/6 \qquad (7.17a)$$

Taking the contribution of the isostructural methyl group equal to the value of the reduction in the enthalpy characteristics (1.1 kJ mol^{-1}), we obtain the energy value of the specific interaction of low stability as equal to 0.55 kJ mol^{-1} and we define the energy value of the stable specific interaction, presented in Table 7.9. Taking the contribution of the energy of the isostructural methyl group to the sublimation enthalpy to be equivalent for all compounds of this series independent of the location of CH_3 group at the cycle carbon atom, we determined the energy of the specific interactions of the considered compounds with the help of Eq. 7.17a, the values of which are given in Table 7.9. The pyrazine1,4-dioxide molecule with two nitrogen atoms of the cycle and two oxygen atoms significantly increase the donor properties of the oxygen atoms at the reduced number of CH groups of the cycle rather than in the pyridine *N*-oxide molecule. The crystalline network structure of pyrazine 1,4-dioxide (Fig. 7.19) is formed by two stable specific interactions of DN → CH (12.9 kJ mol^{-1}) and eight interactions of reduced stability, the energies of which are determined by the equality, similar to Eq. 7.17. The calculation results presented in Table 7.9 point to the significant stabilization of the specific interactions, formed by the oxygen atom with carbon atoms of the CH cycle groups, and growing with increase in the nitrogen atoms and oxygen atoms of the cycle

D-O→CH: Pyridine N-oxide (8.9) > Pyrazine1,4-Dioxide (11.25 kJ mol^{-1})

and at the replacement of location of the isostructural methyl group

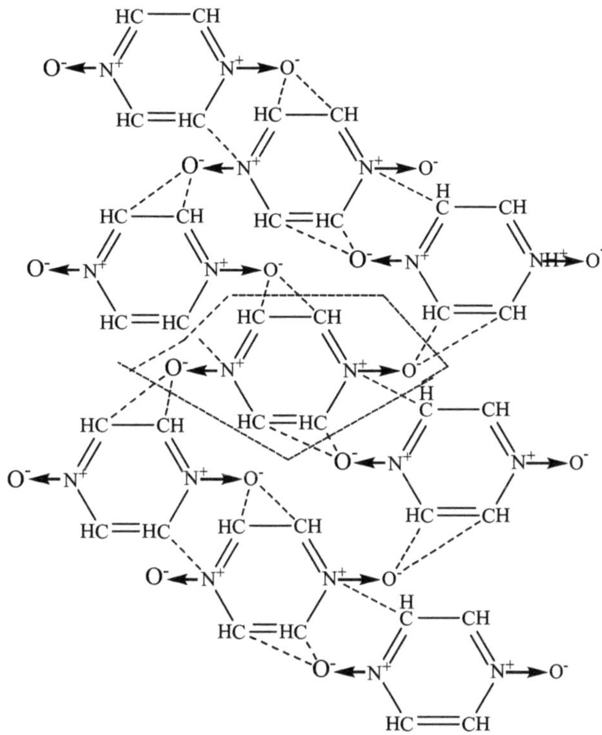

Fig. 7.19 Schematic picture of the crystalline structure of pyrazine 1,4-dioxide with network of the specific interactions

$$\begin{aligned}&\text{2-Methylpyridine}\,N\text{-oxide}(8.55) < \text{Pyridine}\,N\text{-oxide}(8.9)\\&< \text{3-Methylpyridine}\,N\text{-oxide}(9.1)\\&< \text{4-Methylpyridine}\,N\text{-oxide}\left(10.0\,\text{kJ}\,\text{mol}^{-1}\right).\end{aligned}$$

7.2.6 Energies of Specific Interactions of Crystalline Cyanopyridine N-Oxides and Pyridinecarboxamide N-Oxides

The thermodynamic analysis implemented in Chap. 5 of the crystalline pyridines, cyanopyridines, and pyridinecarboxamides allows us to move on to considerthe more complex compounds of the same series of cyanopyridine *N*-oxides and pyridinecarboxamide *N*-oxides. The task is somewhat simplified by the fact that we have to use the correctly obtained energy values formed by the N≡C group for each compound consisting of the derivatives of cyanopyridine, in particular, the energy of the specific interaction, formed by the CH groups of crystalline pyridine, cyanopyridine and correctly estimate the missing data. The 3-cyanopyridine *N*-oxide and 4-cyanopyridine *N*-oxide molecules form a similar network structure of the crystal (Fig. 7.20a), formed by four specific interactions D–O → CH, two

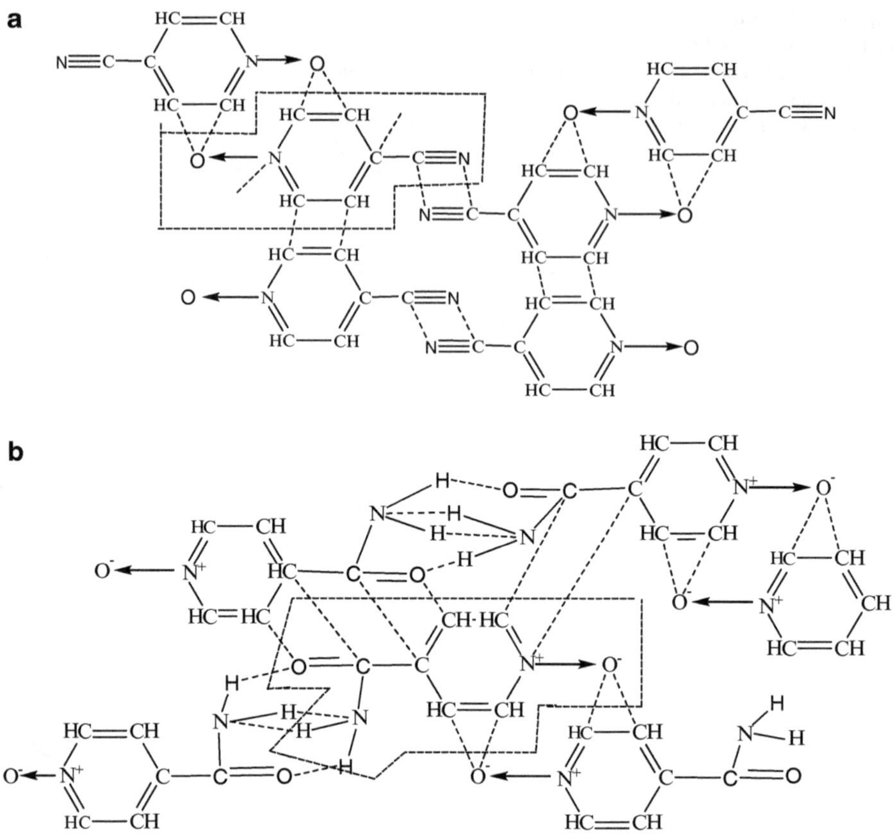

Fig. 7.20 Schematic picture of the crystalline structure of 4-cyanopyridine *N*-oxide (**a**) and 4-pyridinecarboxamide *N*-oxide (**b**) with network of the specific interactions and hydrogen bonds

CH groups of the cycle D=HC → CH = 7.40 kJ mol^{-1}, and two D≡N → C≡with energy of 8.85 and 9.40 kJ mol^{-1}, respectively, in crystalline 3-cyanopyridine *N*-oxide and 4-cyanopyridine *N*-oxide,

The fourth type of specific interaction formed in these crystalline compounds D=N → C should be determined by the energy of the specific interaction D=N → CH = 12.90 kJ mol^{-1} of crystalline cyanopyridine reduced by the contribution of the substituted hydrogen atom, 1.0 kJ mol^{-1}. From this we can conclude that the energy of the specific interaction D–O → CH, formed by the oxide form, should be determined using Eq. 7.19

$$\text{D-O}\rightarrow\text{CH} = (\Delta_{\text{sub}}H^0(298\text{ K}) - 2\text{D=N}\rightarrow\text{C} - 2\text{ D=HC}\rightarrow\text{CH} - 2\text{D}\equiv\text{N}\rightarrow\text{C}\equiv)/4 \quad (7.19)$$

The results of the calculations (Table 7.10) point to the increased stability of the

Table 7.10 Energies of the specific interactions (kJ mol^{-1}) of the crystalline cyanopyridine *N*-oxides and pyridinecarboxamide *N*-oxides at 298 K

Compounds	Formula	Structure	$\Delta_{sub}H°(298\ K)$ [12]	D≡N → C≡	D=HC → CH	D–O → CH
D≡N → C = 11.9						
3-Cyanopyridine *N*-oxide	$C_6H_4N_2O$	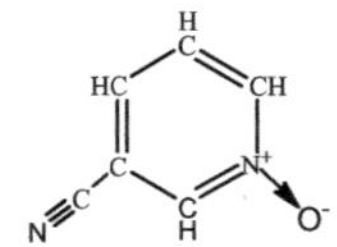	101.9 ± 2.0	8.85	7.40	11.4
4-Cyanopyridine *N*-oxide	$C_6H_4N_2O$	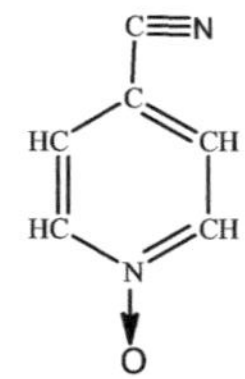	101.4 ± 4.3	9.40	7.40	11.0
3-Pyridinecarboxamide *N*-oxide	$C_6H_6N_2O_2$	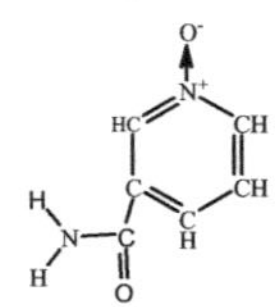	119.2 ± 2.3	DC=O → CH = 8.3	D=HC → C = DC → C=O = 6.40	D=C=0 •••H–N 9.2
4-Pyridinecarboxamide *N*-oxide	$C_6H_6N_2O_2$	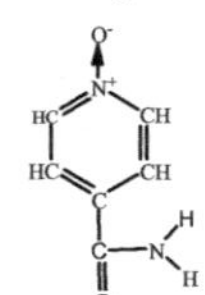	125.3 ± 1.3	8.3	6.40	D=C=0 •••H–N 13.0

specific interaction, formed by the nitrogen atom and the carbon atom of the cycle compared with the oxide form.

3-Cyanopyridine N-oxide D=N→C (11.9) > D-O→CH (11.4 kJ mol^{-1})

4-Cyanopyridine N-oxide D=N→C (11.9) > D-O→CH (11.0 kJ mol^{-1})

The 3-pyridinecarboxamide *N*-oxide and 4-pyridinecarboxamide *N*-oxide molecules have a more complicated structure with 14 free bond vacancies, forming the crystal structure with the network of the two types of stable hydrogen bonds and ten specific interactions of five types (Fig. 7.19b). The amine group forms the hydrogen bond of peptide type DC=O•••H–N and the second type D–N•••H–N with energy of 9.7 kJ mol^{-1}. At such large numbers the specific interactions have increased informational content and allow us to pick the correct choice of the energy properties but do not increase the error of its estimated values. In this case the energies of the specific interactions D=N → C, 2D=HC → CH are taken as equal to the ones realized in crystalline pyridines, cyanopyridines. The energy of the specific interaction DO → CH is taken as equal to one realized in crystalline 3-cyanopyridine *N*-oxide (11.40) and 4-cyanopyridine *N*-oxide (11.0 kJ mol^{-1}). The energy of the specific interaction, formed by carbonyl oxygen, takes a value equal to the energy value of the interaction, realized in crystalline succinimide (8.3 kJ mol^{-1}) with the peptide group.

The energies of the specific interactions 2D=HC → C and DC → C=O (6.4 kJ mol^{-1}) are taken as equal to the value of the energy of D=HC → CH; there is a smaller energy contribution on H substitution (1.0 kJ mol^{-1}). The energy of the hydrogen bond of the peptide type, D=C=O•••H–N, is determined by the difference between the sublimation enthalpy and the energy contribution of all remaining specific interactions and the second type of hydrogen bond

$$\text{D=C=O}\cdots\text{H-N} = (\Delta_{sub}H^0(298\ \text{K}) - 4\text{D-O}\rightarrow\text{CH} - \text{D=N}\rightarrow\text{C} - 2\text{D=HC}\rightarrow\text{C} - \text{DC}\rightarrow\text{C=O} - \text{D=C=O}\rightarrow\text{CH} - 2\text{D-N}\cdots\text{H-N})/2 \quad (7.20)$$

As is shown in Table 7.10 the energy of the hydrogen bond formed by the carbonyl's oxygen atom depends significantly on the location of the carboxamide fragment at the carbon atom of the cycle. The location of this group at the third carbon atom gives rise to an energy equal to the energy with the first type of hydrogen bond D–N•••H–N =9.7 kJ mol^{-1}. On the contrary, the movement of the hexamerous cycle from the location of the nitrogen atom to the center of the cycle at C(4) of the carbon atom leads to the energy of the considered type of hydrogen bond increasing (13.0 kJ mol^{-1}) significantly. It follows that the distance of the cycle from the nitrogen atom is connected to the presence of the increased electron density at the carbon atom at location C(4), which is transferred to the carboxamide fragment and leads to the increased difference in the charges of the nitrogen and hydrogen atoms and, on the other side, the carbonyl's oxygen atom. Consequently, this effect has nothing in common with the peptide bond of a general nature, reflecting the increase of the electron density of the carbon atom with removal from the acceptor atom.

7.2.7 *Energies of Specific Interactions of Crystalline Benzofurazan* N*-Oxides and Their Derivatives*

First of all, one should pay attention to the fact that the benzofurazan molecule with ten bond vacancies and sublimation enthalpy (64.4 kJ mol^{-1}) forms a specific interaction with the average energy value, which does not exceed 6.4 kJ mol^{-1}. At the same energy of the specific interaction, by the CH groups of the benzene ring formed in crystalline condition we have an energy of 7.40 kJ mol^{-1}. The pyridine molecule with one nitrogen atom in the hexamerous cycle (Table 7.9) forms specific interactions, D=HC → CH and D=N → CH, with energies of 7.40 and 12.9 kJ mol^{-1}. It follows that the carbon atoms of the benzene ring and the nitrogen and oxygen atoms of the second type of benzofurazan have definite differences in the formation of the specific interactions. At the same time, the oxygen atom located symmetrically between the nitrogen atoms with less expressed acceptor properties acquires a negative charge through the shifting of the electron density and even less difference in the charges is maintained between the nitrogen and carbon atoms of the benzene ring. Thus, the specific interaction D=N → C formed has less stability than the ones realized in crystalline benzofurazan. The oxygen atom with increased negative charge forms with the CH group of the benzene ring the specific interaction D–O → CH with increased stability. The increased rigidity of the benzene ring and other cyclic compounds with double bonds and the differing charges of the carbon atoms of the CH groups, forming the specific interactions D=HC → CH with

$$\text{Benzene}\,(7.40) > \text{Cyclodeca-1,2,6,7-tetraene}\,(7.30) \approx \text{Cyclododecahexadiene}\,(6.30) \approx \text{Cyclododecane}\,(6.37\,\text{kJ mol}^{-1})$$

the energies values, differing at the second digit, have an increased stability in crystalline benzofurazan and its derivatives.

In this connection the network structure of the crystal of benzofurazan (Fig. 7.21a), formed by three types of ten specific interactions, is described by the number of its stability D=N → C < D–O → CH < D=HC → CH. The energy of the last type of specific interaction we obtain from the sublimation enthalpy of benzene reduced by the energy contribution of the two substituted hydrogen atoms, equal to 7.1 kJ mol^{-1}. We estimate the energy of the specific interaction D=N → C on the basis of this value reduced by the contribution of the substituted hydrogen atom (1.0 kJ mol^{-1}) for crystalline condition (6.1 kJ mol^{-1}). We obtain the energy of the specific interaction D–O → CH from the difference of sublimation enthalpy

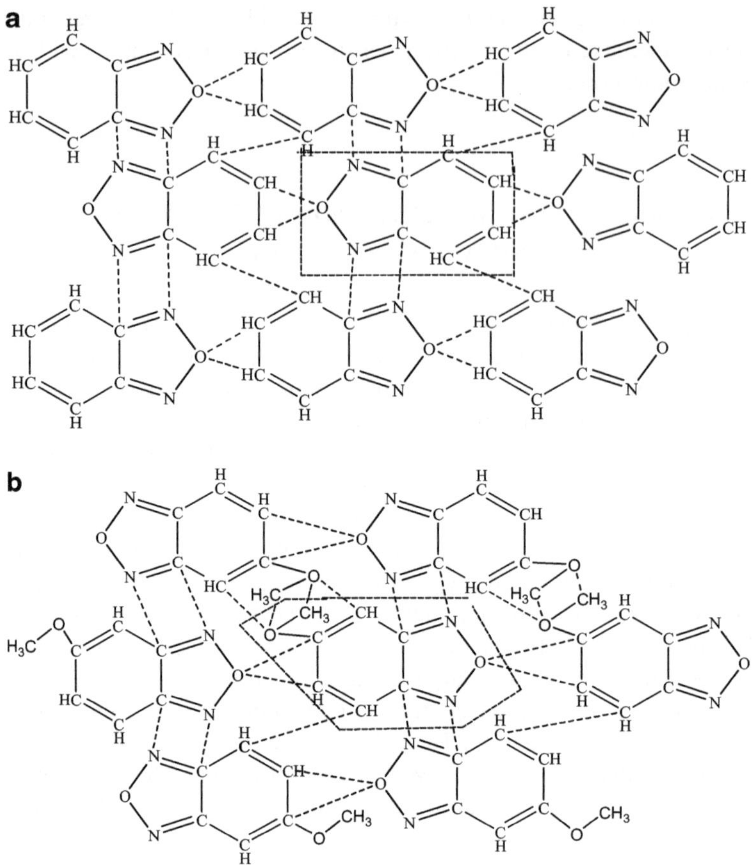

Fig. 7.21 Schematic picture of the crystalline structure of benzofurazan (**a**) and 5-methoxybenzofurazan (**b**) with network of the specific interactions and hydrogen bonds

of benzofurazan reduced by the contributions of the energies of four D=N → C and two D=HC → CH interactions, symmetrically located between the nitrogen atoms.

The 5-methoxybenzofurazan molecule forms 13 specific interactions of five types in the crystal: 3D–O → CH, 3D–O → C, 4D=N → C, 2D–O → CH_3–O and D=HC → CH. The value of the contribution of the methoxy group to the sublimation enthalpy of 5-methoxybenzofurazan should be determined by the difference of enthalpy characteristics with benzofurasan reduced by the energy contribution of the substituted hydrogen atom (1.0 kJ mol^{-1}).

The energy value (25.8 kJ mol^{-1}) obtained is contributed by two types of specific interactions D–O → CH_3–O and D–O → CH, formed by the methoxy group and the CH group of the benzene ring, participating in the formation of the structure of crystalline 5-methoxybenzofurazan with the network of specific interactions (Fig. 7.21b). A similar type of specific interaction forms the oxygen atom of the cycle. Fundamentally, these two types of specific interactions should differ in their energy values; however, because of the insignificant difference we take them as equal with the already accepted value of 6.45 kJ mol^{-1}.

The results of the calculations of the energies of the specific interactions (Table 7.11) of crystalline benzofurasane and 5-methoxybenzofurazane illustrate the natural numbers of its stabilization, respectively,

$$D{=}N{\rightarrow}CH\ (6.1) < D\text{-}O{\rightarrow}CH\ (6.45) < D{=}HC{\rightarrow}CH\ (7.1\ kJ\ mol^{-1})$$

$$D\text{-}O{\rightarrow}C\ (5.45) < D{=}N{\rightarrow}CH\ (6.1) < D\text{-}O{\rightarrow}CH\ (6.45) \approx D\text{-}O{\rightarrow}CH_3\text{-}O\ (6.45) < D{=}HC{\rightarrow}CH\ (7.1\ kJ\ mol^{-1})$$

The value of the energy of the specific interaction, formed by the methoxy group of D–O → CH_3–O (6.45 kJ mol^{-1}) is defined with the error of the measured sublimation enthalpy of 5-methoxybenzofurazan and is in good compliance with the value of the interaction energy of crystalline dimethyl ether (6.69 kJ mol^{-1}) [6]. It allows us to use this value in the estimation of the energies of the specific interactions of crystalline compounds with greater complexity.

The benzofurazan *N*-oxide molecule with planar structure forms the network structure of the crystal by three types of 12 specific interactions D=N → C, D=HC → CH and with unknown energy value DN → O → CH, formed by the oxide form of the fragment N → O (Fig. 7.22a). The energy of this type of specific interaction should be determined by the difference between the sublimation enthalpy and the total contribution of eight interactions of the two types

$$DN{\rightarrow}O{\rightarrow}CH. = (\Delta_{sub}H^0(298\ K)\ bfo - 4D{=}N{\rightarrow}CH - 4D\text{-}O{\rightarrow}CH)/4 \tag{7.21}$$

The obtained energy value (7.45 kJ mol^{-1}) reflects the stabilization of this type of specific interaction compared with the energies of other interactions in the crystalline compounds of benzofurazanes. This specific interaction is more stable than the one formed by the methoxy group, located at the carbon atom C(5), for which the increased shifting of the electron density to the oxygen atom of the methoxy group is characteristic, leading to the increased negative charge and increased difference in charges of the oxygen and carbon atoms of the methyl group.

The network structure of the crystalline 5-nethoxybenzofurazan-1-oxide is formed by 15 specific interactions and is more complicated (Fig. 7.22b). In its network structure four specific interactions 4D=N → CH, 4D–O → CH, 2D–O → C, 2D–O → CH_3–O and three interactions DN → 0 → CHparticipate, which are determined with the help of Eq. 7.22.

Table 7.11 Energies of the specific interactions (kJ mol^{-1}) of the crystalline benzofurazan and its derivative at 298 K

Compounds	Formula	Structure	$\Delta_{sub}H°$ (298 K) [12]	D=N → C	D=HC → CH	D–O → CH/ DO → C	D–O → CH_3
Benzofurazan	$C_6H_4N_2O$	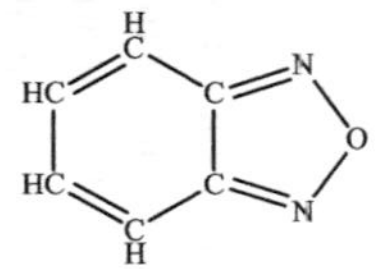	64.4 ± 1.6	6.1 × 4	7.1	6.45	–
5-Methoxybenz ofurazan	$C_7H_6N_2O_2$	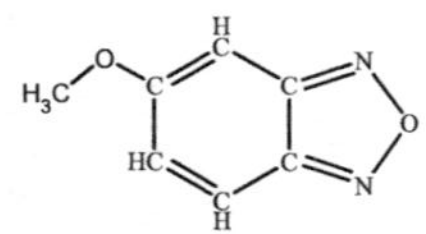	89.2 ± 0.7	6.1 × 4	7.1	6.45 × 3/5.45 × 3	6.45
Benzofurazan *N*-oxide	$C_6H_4N_2O_2$	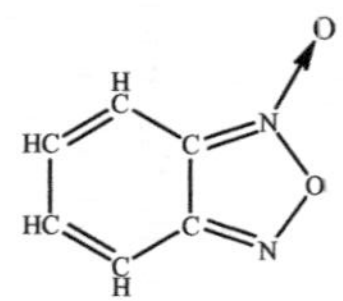	79.6 ± 1.7	6.1 × 4	–	6.45x4	DN → O 7.35
5-Methoxybenz ofurazan-1-oxide	$C_7H_6N_2O_3$	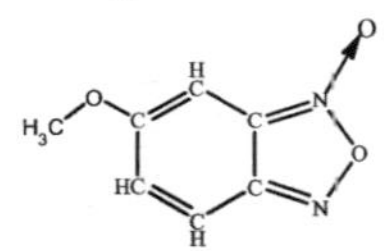	96.0 ± 1.6	6.1 × 4	–	6.45 × 4/5.45 × 2	6.45/4.1
5-Methylbenzo furazan-1-oxide	$C_7H_6N_2O_2$	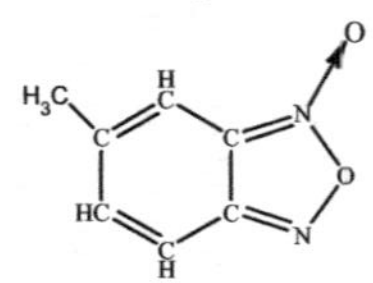	92.2 ± 1.2	6.1 × 4	–	6.45 × 4	10.2
Dipropyldiazene *N*-oxide	$C_6H_{14}N_2O$	Liquid 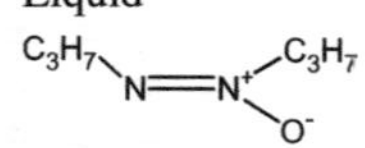	[8] 51.7 ± 0.1	–	–	–	5.9 × 2

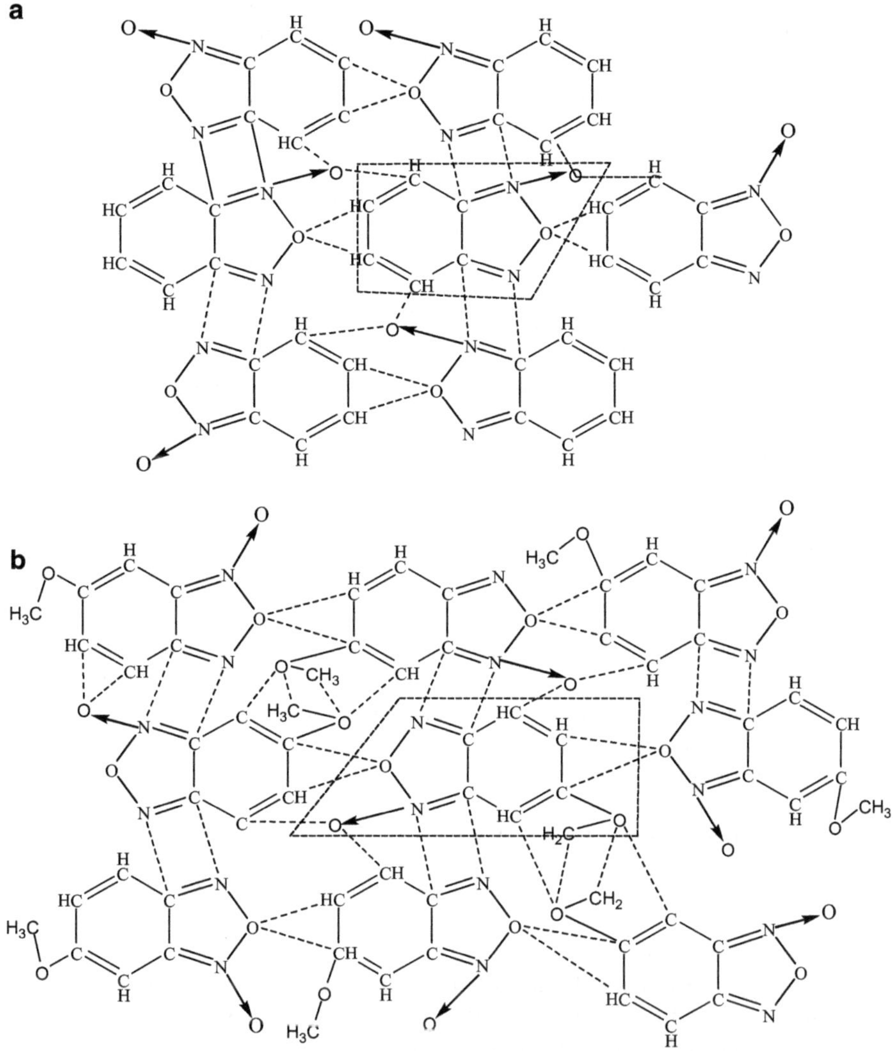

Fig. 7.22 Schematic picture of the crystalline structure of benzofurazan (**a**) and 5-methoxybenzofurazan (**b**) with network of the specific interactions and hydrogen bonds

$$D_{N\rightarrow O\rightarrow CH} = (\Delta_{sub}H^0(298\ K)mbfo - 4D{=}N{\rightarrow}CH - 4D\text{-}O{\rightarrow}CH - 2D\text{-}O{\rightarrow}C - 2D\text{-}O{\rightarrow}CH_3\text{-}O)/3 \quad (7.22)$$

The obtained energy value of the specific interaction DN → O → CH(4.1) is significantly reduced compared with the energy value realized for the same type of interaction of crystalline benzofurazan *N*-oxide (7.35 kJ mol^{-1}), because the location of the methoxy group at the carbon atom C(5) prevents the shifting of the electron density, at least, from the carbon atoms C(4), C(6), and C(7). It is reflected significantly on the charge of the oxygen atom at N(2). In this it is indicative that 5-methylbenzofurazan-1-oxide, the carbon atom C(5) is located at the isostructural methyl group, participating in the distribution of electron density in the molecule. The benzofurazan *N*-oxide and 5-methylbenzofurazan-1-oxide molecules form three types of interactions and in each of them four stable specific interactions DN → O → CH, 4D=N → CH and 4D–O → CH, form an identical crystalline structure. In the case of 5-methylbenzofurazan-1-oxide a weak specific interaction of the isostructural methyl group is formed, the energy contribution of which to the sublimation enthalpy of this compound could be taken as equal to the energy contribution of the same group as the sublimation enthalpy of 2-methylpyridine *N*-oxide (Table 7.9).

Using Eq. 7.21a, added by the energy contribution of the isostructural methyl group

$$DN \to O \to CH. = (\Delta_{sub}H^0(298\ K)mbfo - 4D{=}N{\to}CH - 4D\text{-}O{\to}CH - DisoCH_3)/4 \quad (7.21a)$$

we obtained the energy of the specific interaction DN → O → CH Methylbenzofurazan-1-oxide (10.2 kJ mol^{-1}), which value exceeds the energy value of this type of interaction of crystalline benzofurazan *N*-oxide (7.35 kJ mol^{-1}). A similar stabilization of the energy of the hydrogen bond is for hydroxypyridines, changing in the number of compounds

$$\text{2-Hydroxypyridine}\ (12.3) < \text{4-Hydroxypyridine}\ (21.1) < \text{2-Methyl-4-hydroxypyridine}\ \left(24.8\ \text{kJ mol}^{-1}\right)$$

respectively (Table 7.6).

The energy of the contribution of the oxide's group (NO) to the vaporization enthalpy of dipropyldiazene *N*-oxide and the energy of the specific interaction DO → N formed by this group should be determined by the difference of the vaporization enthalpies of dipropyldiazene *N*-oxide and dipropyldiazene

$$D\text{-}O{\to}N = (\Delta_{vap}H^0(298\ K)dpao - \Delta_{vap}H^0(298\ K)dpa)/2 \quad (7.21b)$$

The values of the energy contribution of the DN → O group to the vaporization enthalpy and this type of specific interaction, D–O → N, of liquid dipropyldiazene *N*-oxide are equal to 11.8 and 5.9 kJ mol^{-1}, respectively.

References

1. Baev AK, Korolkov DV (2003) Phenomenon of dative bonds in elementorganic compounds with alkyl ligand. Abstract XVth FECHEM conference on organometallic chemistry. University of Zurich, p 350
2. Baev AK (2003) Novel Approaches to the structure of functional solvents and elementorganic compounds with abstract international conference on solution chemistry, Debrecen. p 79
3. Baev AK (2004) Insolvence of model sp^3-hybridization of carbon atom and chemistry of non aqua solutions. IX international conference. The problem of solvation and complex formation in solutions, Plyos, pp 62–64
4. Baev AK (2004) Lack of co-ordination thermodynamic properties of elementorganic compounds with sp^3-hybridization of carbon atom and its insolence. Russian scientific symposium on thermo chemistry and calorimetry, Niznie Novgorod, pp 40–41
5. Baev AK (2009) Phenomenological significance of nonidentical sp3-hybridization model and a nature of interaction in chemistry of liquid non-electrolites and chemical thermodynamics. Abstracts. In: X international conference on physical and coordination porphyrin's chemistry and their analogous (ICPC-10), Ivanovo, p 86
6. Baev AK (2012) Specific intermolecular interactions of organic compounds. Springer, Hiedelberg/Dordrecht/London/New York, 434p
7. Schustorovich EM (1978) Chemical bonds in coordination compounds. Knowledge, Moscow, p 434
8. Chickos JS, Acree WE (2002) Enthalpies of vaporization of organic and organometallic compounds, 1880–2002. J Phys Chem Rev 31(2):537
9. Baev AK (1969) Problems of chemical nature of phase transformation. In: General and applied chemistry, vol 1. Vysheischaia Educating, Minsk, pp 197–206
10. Baev AK (1969) Phase condition and complex formation ability of halogenide metals. In: General and applied chemistry, vol 1. Vysheischaia Educating, Minsk, pp 207–218
11. Baev AK (1987) Chemistry of gas-heterogenic systems of elementorganic compounds, Science and technology. Vysheischaia Educating, Minsk, 174p
12. Chickos JS, Acree WE (2003) Enthalpies of sublimation of organic and organometallic compounds, 1910–2001. J Phys Chem Rev 32(2):519
13. Armstrong DR, Perkins PG (1969) Calculation of the electronic structures and the gas-phase heats of formation of BH_3, NH_3, and BH_3,CO. Inorganic physical theoretical. J Chem Soc 7(7):1044–1048
14. Kahn O (1969) Semi-empirical determination of electron transfer in complexes by a quantum chemical method. J Chim Phys 66:869
15. Gurianova EN, Goldshtain IG, Romm IP (1973) Donor-acceptor bonds. Chemistry, Moscow, p 397
16. Veillard A, Levy B, Danbll R, Gallais F (1967) Structural electroniqul of borazane BH_3-NH_3. Theor Chim Acta 8:312
17. Moireau MG, Veillard A (1968) Quantum mechanical calculations on barriers to internal rotation. II. The borazane molecular BH_3-NH_3. Theor Chim Acta 11:344
18. Veillard A, Daudel R (1970) Nature et proprieties des liaison de Coordination. Colloques internationaux du centre National de la Recherche Scientifigue, Paris, p 22
19. Kudchadk SA, Rao CNR (1970) Extended Huckel calculations on N-V addition compounds. Theory Chim Acta 18:258
20. Ehrenson S (1968) Theory of electron effect at formation of complex of trimethylamine with trimethylborium. Theor Chim Acta 10:193
21. Ehrenson S (1968) Effect environment at examination of a reactivity adding of trimethylamine to trimethylborium. Theory Chem Acta 10:209
22. Boer FP, Newton MD (1966) Molecular orbital's for boron hydrides parameterized from SEF – model calculations. J Am Chem Soc 88:2361–2366

Chapter 8
Specific Intermolecular Interactions of Amino Alcohols and Their Derivatives

8.1 The Energies of Hydrogen Bonds and Specific Interactions of Liquid Hydroxylamines

According to modern ideas about the configuration *N,N*-hydroxylamine and its derivatives $n_N(5a')$-$n\sigma$ $(2a'')$-orbital are orthogonal [1], which reflects the character of n_N- and $n\sigma$-electrons of vibration band contours. The replacement of the hydrogen atom by the methyl group pushes n-orbitals up to the values of the vertical ionization potentials 1.45 and 1.83 eV, leading to convergence of OH-groups up to 0.47 eV [1–3]. Such significant changes in the electron configuration of the alkyl derivatives of hydroxylamine should be expressed at the energies of the hydrogen bonds and the formed specific interactions and are supposed to be at higher energies of D–N•••H–N connections in H_2N–OH in comparison with the realized specific interactions of D–N → CH_3–N. As we mentioned at Chap. X [25], that an insignificant reduction of the electron density at the nitrogen atom when replacing a hydrogen atom with the ethyl group was found using X-ray electron spectroscopy [4, 5].

On the basis of the values of the energies of the vertical ionization potentials and binding energy C(1s) in the work one can draw a [3] conclusion on the total acceptor nature of the OH group and the weak donor NH_2 group, which should affect the energies of the formed hydrogen bonds. Taking the electron configuration of *N,N*-hydroxylamine H_2N–OH, the structure of its liquid and crystalline condition should be presented as a network of three types of hydrogen bonds (Fig. 8.1a), three types 2D–O•••H–O, 2D–O•••H–N, and 2D–N•••H–N, reflecting the participation in the interactions of all bond vacancies. The increased acceptor properties of the oxygen atom, the positive charge of the hydrogen atom of the OH group, and the reduced electron density at the nitrogen atom when the hydrogen atom in NH_2 is replaced by the CH_3 group determines the sequence in the energies of the hydrogen bonds of D–O•••H–O > D–O•••H–N > D–N•••H–N.

The molecule of *N*-methylhydroxylamine with one methyl group (Fig. 8.1b) forms two types of the hydrogen bonds 2D–O•••H–O, 2D–O•••H–N, and the specific interaction of reduced stability 2D–N → CH_3–N. In liquid and crystalline

A.K. Baev, *Specific Intermolecular Interactions of Nitrogenated and Bioorganic Compounds*, DOI 10.1007/978-3-642-37472-2_8,

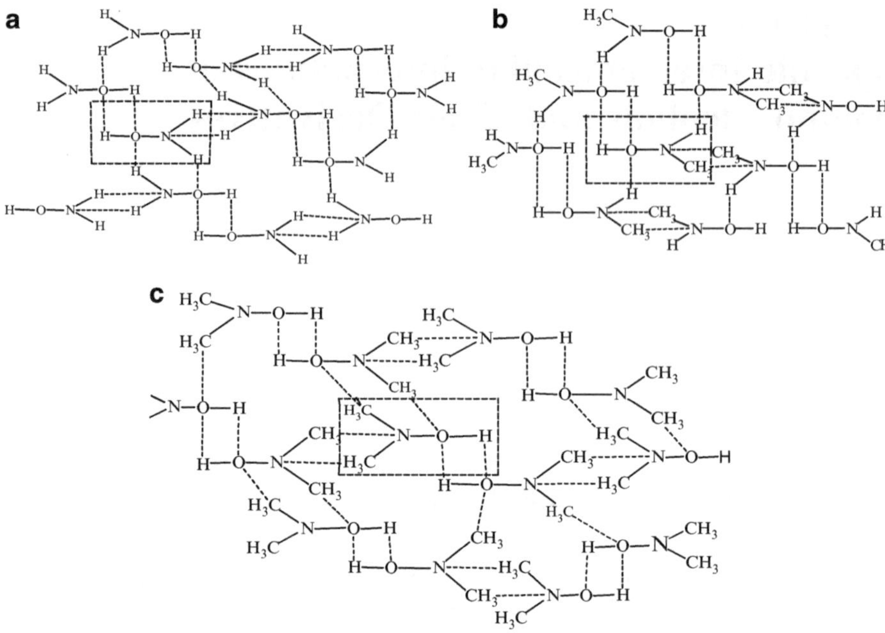

Fig. 8.1 Schematic picture of the liquid and crystalline structure of aminohydroxyl (**a**), *N*-methylhydroxylamine (**b**), and *N,N*-dimethyhydroxylamine (**c**) with the network of the specific interactions

N,N-dimethylhydroxylamine (Fig. 8.1c) there forms a hydrogen bond by the hydroxyl group and two types of the specific interactions by the two methyl groups 2D–O•••H–O, 2D–O → CH_3–N, and 2D–N → CH_3–N. The stability of these interactions is also described by the natural number of 2D–O•••H–O > 2D–O → CH_3–N > 2D–N → CH_3–N all because of the reduced ability of the carbon atom of the CH_3 group to transfer electron density in comparison with the hydrogen atom of NH_2 group and increased donor ability of the oxygen atom at the formation of the hydrogen bond and the specific interaction, compared with the nitrogen atom. It follows that the most stable energies of interactions are formed at *N,N*-hydroxyl-amine and the difference of the vaporization enthalpies of this compound with *N*-methylhydroxylamine is caused only by the different contribution of the energies of the substituted hydrogen bond D–N•••H–N to the reduced stability of the specific interaction of D–N → CH_3–N. The replacement of the hydrogen atom in the amino group to the methyl group is accompanied by the change of the hydrogen bond D–O•••H–N to the specific interaction D–O → CH_3–N, crosslinking chains, formed by the molecules of hydroxylamines, to the network of the specific interactions. Thus, the difference of enthalpies of vaporization or sublimation for *N*-methylhydroxylamine and *N,N*-dimethylhydroxylamine is determined by the different energy contribution of the hydrogen bond D–O•••H–N and the corresponding specific interaction D–O → CH_3–N, respectively. It means that the of 4.0 kJ mol^{-1} difference in vaporization enthalpies of *N*-methylhydroxylamine and *N,N*-dimethylhydroxylamine (Table 8.1) is equal to the difference in the

Table 8.1 The energies of the hydrogen bonds and the specific interactions (kJ mol^{-1}) of the liquid hydroxylamines

Compounds	Formula	The structure	$\Delta_{vap}H°$(298 K) [6]	T K	D–N → CH_3–N	D–O•••H–N	D–O•••H–O
N,*N*-Dimethylhydroxylamine	C_2H_7NO	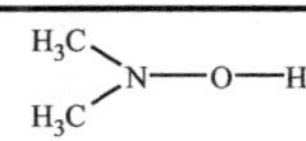	45.7	305	4.80	D–O → CH_3–N 5.63	12.45
N-Methyl Hydroxylamine	CH_5NO	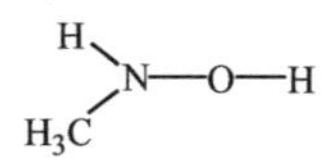	49.7	308	4.80	7.60	12.45
N,*N*-Hydroxylamine	NH_3O	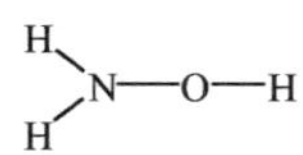	(52.1)	308	D–N•••H–N 5.80	7.60	(12.45)

energies of 2D–O•••H–N – 2D–O → CH_3–N or, in other words, the energy of the hydrogen bond D–O•••H–N is more stable at 2.0 kJ mol^{-1} than the specific interaction D–O → CH_3–N. Thus, the amino alcohol forms three types of bonds with differing energies in the liquid and crystal conditions. It should be noted that the hydrogen atom in the amino group has a reduced positive charge compared with the similar hydrogen atom in the hydroxyl group and the negative charge of the oxygen atom significantly exceeds the charge of the nitrogen atom in the amino group (Chap. 2). In this connection the nitrogen atom of the same amino group is able to form a hydrogen bond with reduced stability, D–N–H•••N < D–N–H•••O. At the same time methylamines form less stable specific interactions with the methyl group than dimethyl ether. We have

D-N-H•••N (3.86) Ammonium (c.n..=6) < D-N-H•••N (4.23) Ammonium (c.n.= 4) ≈ D-N-CH_3←N (4.25) Trimethylamine < D-N-CH_3←N (4.8) Dimethylamine < D-N-CH_3←N (5.33) Methylamine < D-O-CH_3←O (5.63 kJ mol^{-1}) Dimethyl ether.

On the basis of the principle of energy minimum at the interaction of particles, we should take into account that exactly in the mentioned sequence of stability of the specific interactions and the hydrogen bonds there should be formed the bonds. However, one must be aware that in compounds with an amino group and an O–H group, a nitrogen atom (Fig. 8.1c) exceeds its negative charge and, consequently, the positive charge of the hydrogen atom. In turn the hydrogen bond D–N•••H–N formed is stabilized because of the reduction in the coordination number of the nitrogen atom, of which the energy in liquid *N,N*-Hydroxylamine

D-N-H•••N (3.86) Ammonium (c. n.= 6) < D-N-H•••N (5.80 kJ mol^{-1}) Ammonium (c.n.= 4)

approaches the energy of the same hydrogen bond of liquid ammonia with the tetracoordinaated nitrogen atom. Similarly, there are changes in the energy of the specific interaction, formed by the methyl group of trimethylamine and amine fragment with a reduction in the number of these groups and, consequently, a the reduction of the coordinating number,

D-N→CH_3-N (4.25) Trimethylamine < D-N→CH_3-N (4.8) Dimethylamine < D-N→CH_3-N (5.33) Methylamine < D-O→CH_3-O (5.63 kJ mol^{-1}) Diethyl ether

This is a correct condition for accepting the energy of the specific interaction D–N → CH_3–N of liquid *N,N*-dimethylhydroxylamine and *N*-methylhydroxylamine to be equal to 4.8 kJ mol^{-1}. At one methyl group in the fragment NCH_3 the energy of the specific interaction D–N → CH_3–N stabilizes and reaches the value of 5.33 kJ mol^{-1}, which is only 0.33 kJ mol^{-1} lower than the energy value of the specific interaction of liquid dimethyl ether D–O → CH_3–O. Taking into account the increase in the negative charge of the nitrogen atom at the formation of the specific interaction H–O → CH_3–N by the oxygen atom, we should take the energy value of this type of the specific interaction in liquid *N,N*-dimethylhydroxylamin as equal to 5.63 kJ mol^{-1}. Thus, the replacement of this type of the specific interaction by the hydrogen bond D–O•••H–N of liquid *N*-methylhydroxylamine is accompanied by the increase in the vaporization enthalpy at 4.0 kJ mol^{-1}, and we have reason to accept its energy to be equal to 7.60 kJ mol^{-1}. Taking the realized types of H-bonds and the specific interactions, the energy of the hydrogen bond D–O•••H–O should be determined by the difference of the vaporization enthalpies and the total contribution of the two realized types of the specific interactions

N,*N*-dimethylhydroxylamine

$$\text{D-O}\cdots\text{H-O} = (\Delta_{vap}H^{\circ}(298\ \text{K})\ \text{dmha-2D-N}\rightarrow CH_3\text{-N - 2D-O}\rightarrow CH_3\text{-N})/2 \qquad (8.1)$$

N-Methylhydroxylamine.

$$\text{D-O}\cdots\text{H-O} = (\Delta_{vap}H^{\circ}(298\ \text{K})\text{mha -2 D-N}\rightarrow CH_3\text{-N - 2D-O}\cdots\text{H-N})/2 \qquad (8.1a)$$

Using the estimated values of the energies of the specific interactions and H-bonds (Table 8.1), we define the energies of the hydrogen D–O•••H–O bonds of *N*-methylhydroxylamine (12.45) and *N*,*N*-dimethylhydroxylamine (12.45 kJ mol^{-1}), slightly differently from the standard conditions. The invariability of the energies of the hydrogen bonds D–O•••H–O of hydroxylamine is explained by the two bonds observed at the methylation effect of the amino group apart from the n-orbital up to 45 and 1.83 eV and, on the other side, the OH group pulls together up to 0.45 eV [1, 2]. Since the temperature dependence of the vaporization enthalpy, which is not accounted for, does not contribute significant errors to the calculations of the energies of the specific interactions and H-bond, the assessment made of the vaporization enthalpy of *N*,*N*-hydroxylamino, using the values of the energies of the hydrogen bonds of liquid *N*-methylhydroxylamine, is correct (Table 8.1). The results of the calculations of the energies formed in liquid hydroxylamines are described by the integral values of their stabilization,

N,*N*-Dimethylhydroxylamine:

$$\text{D-N}\rightarrow CH_3\text{-N (4.80)} < \text{D-O}\rightarrow CH_3\text{-N (5.63)} < \text{D-O}\cdots\text{H-O (12.45 kJ mol}^{-1})$$

N-Methylhydroxylamine:

$$\text{D-N}\rightarrow CH_3\text{-N (4.80)} < \text{D-O}\cdots\text{H-N (7.60)} < \text{D-O}\cdots\text{H-O (12.45 kJ mol}^{-1})$$

N,*N*-Hydroxylamine:

$$\text{D-N}\cdots\text{H-N (5.80) D-O}\cdots\text{H-N (7.60)} < \text{D-O}\cdots\text{H-O (12.45 kJ mol}^{-1})$$

Similar integral values of stabilization of the energies of the specific interactions and the hydrogen bonds can be calculated in the crystalline condition of these compounds. We draw attention to the good compliance of the energies of the specific interaction of *N*-methylhydroxylamine

D-N→CH_3-N (4.80) N-Methylhydroxylamine < D-N•••H-N (5.80) N,N-Hydroxylamine < D-O•••H-N (7.60 kJ mol^{-1})

and the hydrogen bond of *N,N*-hydroxylamine with established position by the method of RE-spectroscopy [4, 5], where the replacement of the hydrogen atom by the methyl group is accompanied by insignificant reduction in the electron density at the nitrogen atom.

The obtained values of the energies of the hydrogen bond of liquid *N,N*-hydroxylamine with the oxygen atom of the alcohol hydroxyl group have a reduced stability compared with formamide with the oxygen atom of the carbonyl.

N,N-Hydroxylamine D-N•••H-N (5.80) < Formamide D-N•••H-N (7.7 kJ mol^{-1});

N,N-Hydroxylamine D-O•••H-N (7.60) < Formamide D-C=O•••H-N (9.0 kJ mol^{-1});

N,N-Hydroxylamine D-O•••H-O (12.45) < Formamide D=C=O•••H-O=C (13.4± 1.0 kJ mol^{-1});

N-Methylhydroxylamine D-O•••H-N (7.60) < N-Methylformamide D=C=O•••H-N (9.0 kJ mol^{-1});

N-Methylhydroxylamine D-O•••H-O (12.45) < N-Methylformamide D=C=O•••H-O=C (13.4± 1.0 kJ mol^{-1});

At the stable energy of the specific interaction, formed by the methyl group, on replacing the hydrogen atom of the amide group (4.80 kJ mol^{-1}) the energies of the hydrogen bonds preserve the mentioned regularity at *N*-methylhydroxylamine and *N*-methylformamide.

8.2 The Energies of Hydrogen Bonds and Specific Interactions of Amino Alcohols

8.2.1 Liquid Aminoethanols and Aminopropanol

Besides the effects of the spreading of the n-orbital as mentioned earlier [1, 2] one should draw attention to the fact that in allilamines there is expressed difficulty of n-π-conjugation by the CH_2 group, causing the reduction of ΔI_{1-2} from the value of 3.5 eV in vinylamine to 0.6 eV in the case of diamines with the value of splitting of n-orbitals ($\Delta I(n)$) at the isolated nitrogen atom being caused by competing interactions "through the space" and "through the bonds" [7, 8]. Manifestations similar to those of the methylene groups can be expected in aminoethanols with the isolated atoms of nitrogen and oxygen. Shifting of the electron density in the molecule of 2-aminoethanol has complex features. At first, it shifts from the hydrogen atoms of the amino group to the nitrogen atom and partly from

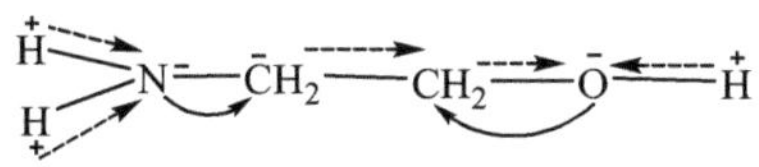

the carbon atom of the methylene group in contact with it. As an example, we note that at the $2p_x$, $2p_y$, and $3p_z$ orbitals of the nitrogen atom of ammonia there are 3.516, 1.210, and 1.861e [9], respectively and at the boron atom of the BH_3 molecule [10] 3.129, 0.990, 0.029e, respectively. The part of the electron density obtained and located at the p_z orbital of the nitrogen atom is transferred to the essentially unshared $2s^2$ electron pair of the carbon atom; the charges of atoms of nitrogen, carbon, and hydrogen in the molecule of $N(CH_3)_3$ becoming equal to 0.024, −0.113, and 0.033e, respectively, [11]. Second, under the influence of the high acceptor ability of the oxygen atom the electron density shifts from the hydrogen atom of the hydroxyl group and contacting carbon atom and partly from the carbon atom of the second methylene group. A definite part of the electron density of oxygen is also transferred to the essentially unshared and depleted $2s^2$ electron pair of the contacting carbon atom, reducing the positive charge with preservation of the high negative charge at the oxygen atom. Under the influence of the oxygen atom increased difference of the charges are created at the hydrogen and nitrogen atoms, leading to the stabilization of the hydrogen bond D–N•••H–N formed or at the hydrogen atom substituted by the methyl group with increased stability of the specific interaction D–N → CH_3–N. Thus, the hydrogen bonds formed by an isolated atom of nitrogen and hydrogen or a saturated CH_3 group have a small specific interaction; it is influenced by the oxygen atom and this makes it keep its stability. Charges of the carbon atoms of the methylene groups have different values with preservation of increased positive direction at the carbon atom, in contact with the oxygen atom. Located between two acceptors, the chain of two methylene groups with the influence of the reverse dative bond from the oxygen atom is able to form compounds like dyamines (Chap. 2), and we have the additional bond with the nitrogen atom, increasing its coordination up to the pentacoordinated condition. Being formed by the fragment $-CH_2-CH_2-$, the specific interaction D–N → CH_2–CH_2–N should have reduced stability. Thus, the molecules of aminoethanols including the substituted hydrogen atom of the amino group by the methyl and ethyl group with seven bond vacancies form the chain-network structure of the liquid and crystalline condition (Fig. 8.2a, b), fundamentally differs little from the structure of the condensed condition of hydroxylamines (Fig. 8.1).

The molecule of 2-aminoethanol forms spirit O–H groups and two stable hydrogen bonds D–O•••H–O and two hydrogen bonds of reduced stability with the hydrogen atom of the amino group D–O•••H–N and the remaining hydrogen atom of the same amino group forms two hydrogen bonds of the third type, D–N•••H–N. The 2-(methylamino)ethanol molecule with the substituted hydrogen atom of the amino group by the methyl group forms the same two types of hydrogen bonds and the specific interaction and instead of the hydrogen bond D–N•••H–N, formed by the amino group, forms the specific interaction D–N → CH_3–N of

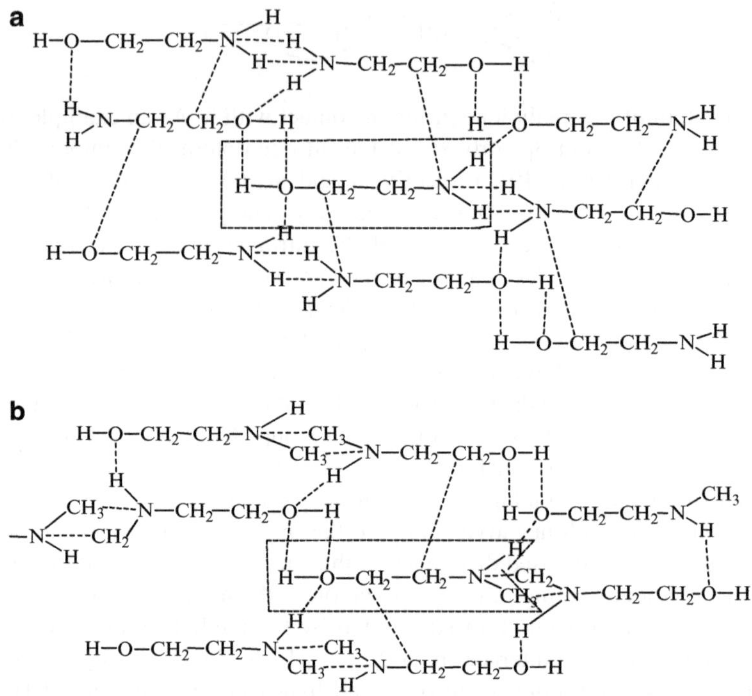

Fig. 8.2 Schematic picture of the liquid structure of 2-aminoethanol (**a**) and 2-(methylamino) ethanol (**b**) with the network of the specific interactions

reduced stability with an energy of 4.80 kJ mol^{-1}. Consequently, reduction of the vaporization enthalpy of 2-aminoethanol from 61.7 to 4.7 kJ mol^{-1} up to the value of the vaporization enthalpy of 2-(methylamino)ethanol, 57.0 kJ mol^{-1}, is caused by the replacement of the stable D–N•••H–N (7.15 kJ mol^{-1}) hydrogen bond by the less stable specific interaction. The energy of the hydrogen bond, formed by the spirit hydroxyl group, should be determined with the help of equality (8.2), including the energy contributions of the hydrogen bonds and the specific interactions (Table 8.2.)

2-Aminoethanol

$$D\text{-}O\cdots H\text{-}O = (\Delta_{vap}H^\circ(298\ K)amet - 2D\text{-}N\cdots H\text{-}N - 2D\ \text{-}N{\rightarrow}CH_2\text{-}CH_2\text{-}N - 2D\text{-}O\cdots H\text{-}N)/2 \quad (8.2)$$

2-(Methylamino)ethanol

$$D\text{-}O\cdots H\text{-}O = (\Delta_{vap}H^\circ(298\ K)maet - 2D\text{-}N{\rightarrow}CH_3\text{-}N - 2\ DN{\rightarrow}CH_2\text{-}CH_2\text{-}N - 2D\text{-}O\cdots H\text{-}N)/2 \quad (8.2a)$$

The replacement in 2-(methylamino)ethanol of the second hydrogen atom of the amino group by methyl is accompanied by a sharp (12.3 kJ mol^{-1}) reduction of the

Table 8.2 The energies of the hydrogen bonds and the specific interactions (kJ mol^{-1}) of the liquid aminoethanoles and 1-amino-2-propanol, D–O → CH_2–CH_2–N = 3.6 kJ mol^{-1}

Compounds	Formula	The structure	$\Delta_{vap}H°$(298 K) [6]	T K	D–N•••H–N	D–O•••H–N	D–O•••H–O
2-Aminoethanol	C_2H_7NO	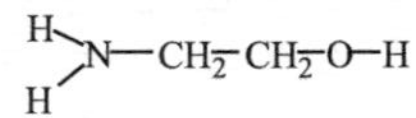	61.7	325	7.15	7.60	14.35
2-(Methylamino)ethanol	C_4H_9NO	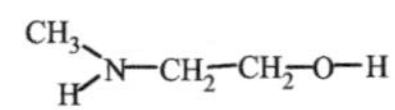	57.0	298	D–N → CH_3–N 4.80	7.60	14.35
2-Dimethylaminoethanol	$C_4H_{11}NO$	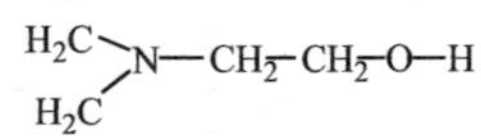	42.7	338	D–N → CH_3–N 4.80 × 2	D–O → CH_3–N 5.63 × 2	9.1
N,N-Diethylethanolamine	$C_6H_{15}NO$	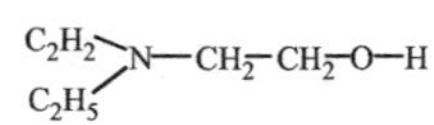	48.5 ± 0.2	340	D–N → CH_3–N 4.80 × 2	D–O → C_2H_5–N 6.72 × 2	9.8
1-Amino-2-pro-panol	C_3H_9NO	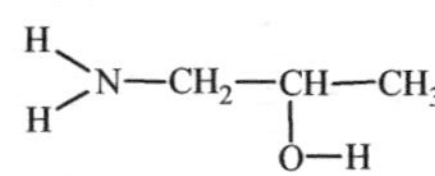	51.8	338	7.15 × 2	7.60 × 2	8.3
N,N-bis(2-Hydroxyethyl) ethylenediamine	$C_6H_{16}N_2O_2$	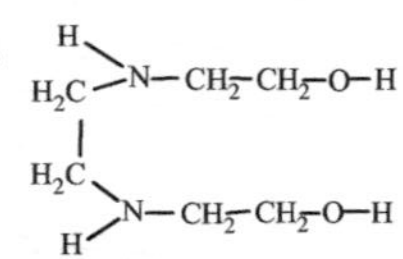	106.4 ± 6.4	298–335	7.15 × 2 D–N → CH_2–N 4.20 × 4	D–O → C_2H_5–N 6.72 × 4	8.3

vaporization enthalpy of 2-dimethylaminoethanol (Table 8.2), which is connected to the replacement of the hydrogen bond 2D–O•••H–N with the energy of 7.60 kJ mol^{-1} by the specific interaction D–O → CH_3–N (5.63 kJ mol^{-1}) and the significant reduction of the energy of the most stable hydrogen bond, D–O•••H–O.

The latter is caused by the reduced ability of the essentially unshared $2s^2$ electron pair of the carbon atom, replacing the hydrogen atom of the amino group. As a result, the charge of the nitrogen atom is reduced and the contacting carbon atom of the methylene group is to pass more electron density, which, in its turn, is reflected in the reduction of the oxygen atom of the alcohol group. The energy of the hydrogen bond of 2-dimethylaminoethanol can be determined with the help of Eq. 8.2b

$$\text{D-O}\cdots\text{H-O} = (\Delta_{vap}H°(298\ \text{K})\text{dmet} - 2\text{D-N}\rightarrow\text{CH}_3\text{-CH}_2\text{-N} - 2\text{D-N}\rightarrow\text{CH}_2\text{-CH}_2\text{-N} - 2\text{D-O}\rightarrow\text{CH}_2\text{-CH}_2\text{-N})/2 \quad (8.2b)$$

This equation is valid for calculation of the energy of the hydrogen bond of *N,N*-diethylethanolamine, the energies of the specific interaction D–N → CH_3–CH_2–N, whose value is equal to 6.0 kJ mol^{-1} (Chap. 2, Table 2.1) corresponding to the tetracoordinated nitrogen atom. The energy of the specific interaction D–O → CH_2–CH_2–N, formed by the ethyl fragment, we take in analogy with the specific interaction equal to the energy contribution of ethyl ligand of the diethyl ether. In the calculations of the energies of the hydrogen bonds of aminoethanoles the values of the experimentally established vaporization enthalpies found in the literature without indication of errors, at temperatures not corresponding to the standard conditions, and having some reduced values were used. In this connection the correct value of the energy of the hydrogen bond is the energy value obtained for this type bond of 2-(methylamino)ethanol and the energy value of 2-aminoethanol is relatively reduced as it is obtained without using the temperature dependence of the vaporization enthalpy. For the same reasons the energy of the hydrogen bond of 2-dimethylaminoethanol is also reduced and the most correct energy with the reduced value is the same as that of the bond in *N,N*-diethylethanolamine.

The molecule of compounds similar to 1-amino-2-propanol forms the structure of a liquid, where the methyl group functions as the isostructural group. The contribution of this group to the vaporization enthalpy of this compound should be estimated on the basis of *N*-methyl-2-methyl-propionamide, equal to the value of 2.2 kJ mol^{-1}. The structure of the liquid condition of this compound with the network of the hydrogen bonds is similar to the structure of 2-aminoethanol with inherent types of interactions and additionally specific interactions of low stability, D–H_3C → H–CH_2, with an energy of 1.1 kJ mol^{-1} (Fig. 8.3a). The calculated energy of the hydrogen bond with the use of Eq. 8.2 reduced by the contribution of the isostructural methyl group has a reduced value compared with the one obtained for 2-dimethylaminoethanol and *N,N*-diethylethanolamine, which is possibly

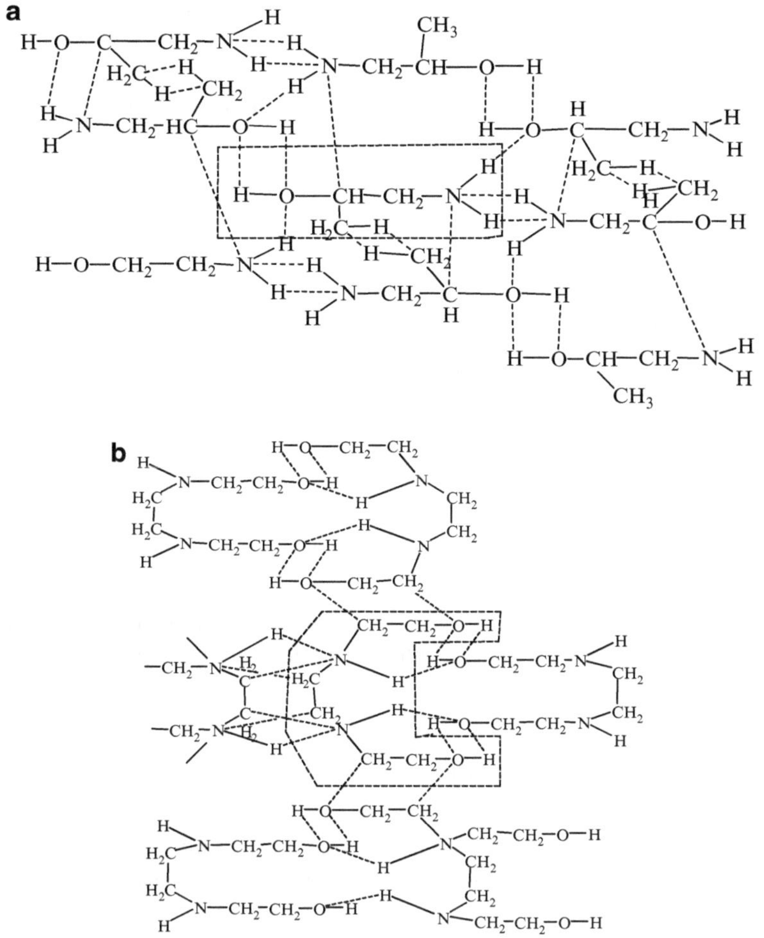

Fig. 8.3 Schematic picture of the liquid structure of 1-amino-2-propanol (**a**) and *N,N-bis* (2-hydroxyethyl)ethylenediamine (**b**) with the network of the specific interactions

connected with the increased errors in the experimentally measured vaporization enthalpy of this compound.

The molecule of *N,N-bis*(2-hydroxyethyl)ethylenediamine (Fig. 8.3a) with 16 bond vacancies, four of which are expressed by nitrogen atoms, in pentacoordinated condition, forms four specific interactions 4D–N → CH_2–N by the fragment D–N–CH_2–CH_2–N– by the coordinating molecule (Fig. 8.3b). The energies of these bonds are equal to the energy of the specific interaction N → CH_2–N (4.80 kJ mol^{-1}) reduced by the energy contribution of the substituted hydrogen atom of 0.60 kJ mol^{-1} for the liquid structure of the compounds. The hydrogen atoms of the amine group form two hydrogen bonds with the oxygen atoms of spirit hydroxyl groups 2D–O•••H–N with, based on the above, an energy value of 7.60 kJ mol^{-1}, and the remaining two hydrogen atoms of the amino groups form two hydrogen bonds, 2DN•••H–N. In addition, chain fragments –CH_2–CH_2– form four specific interactions 4D–O → CH_2–CH_2–O with an energy of 6.72 kJ mol^{-1} of liquid ethyl ether. In its turn the two hydroxyl groups of the molecule form four hydrogen bonds with contacting molecules that are close in the environment: 4D–O•••H–O. Within the frame of the considered procedure for the energy calculation of this type of hydrogen bond for *N,N-bis*(2-hydroxyethyl) ethylenediamine, we use Eq. 8.3

$$D\text{-}O\cdots H\text{-}O = (\Delta_{vap}H^{\circ}(298\ K)heea - 4D\text{-}N{\to}CH_2\text{-}N - 2D\text{-}N\cdots H\text{-}N - 4D\text{-}O\cdots H\text{-}N - 2D\text{-}O{\to}CH_2\text{-}CH_2\text{-}O)/4 \tag{8.3}$$

The calculated energy value of this hydrogen bond (8.3 kJ mol^{-1}) reflects its reduced stability compared with one ethylene group in

$$\text{2-(Methylamino)ethanol}\ (14.35) > N,N\text{-}Diethylethanolamin(9.8) > N,N\text{-}bis(\text{2-hydroxyethyl})\text{ethylenediamine}\ (8.3\ \text{kJ mol}^{-1})$$

the molecule of 2-aminoethanol and 2-(methylamino)ethanol and at the same time its value coincides with the energy value of liquid 1-amino-2-propanol.

8.2.2 Liquid Cyan Alcohols

In the molecule of cyan alcohols with isolated atoms of nitrogen and oxygen, like dyamines [2], the competition of interactions "through the space" and "through the bonds" [7, 8] at the values of splitting of n-orbitals ($\Delta I(n)$) can be expressed. At the same time the role of σ-(CC)-n-interaction in acyclic compounds is less significant compared with cyclic $N(CH_2–CH_2)_3N$. The value of the vertical ionizing potential I_v at the bonding combination n* has the reduced value of 2.04 eV compared with $I_v(n^-)$ [2]. It is caused by the fact that the significantly increased acceptor ability of the oxygen atom compared with the nitrogen atom shifts the increased electron density from the ethyl fragment. Thus, the methylene group in contact with the carbon atom of the –C≡N group gives less electron density to the carbon atom and further to the nitrogen atom.

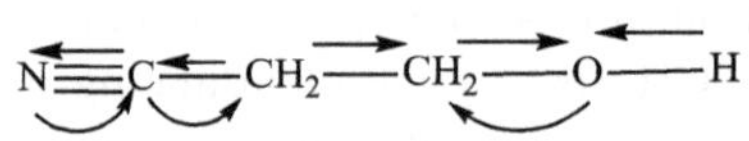

At the same time, the reduced ability of transmission of part of the electron density from the p_z-orbital of the nitrogen atom to the carbon atom causes an increased difference in the charges of the carbon atoms of the methylene groups, in contact with atoms of oxygen and nitrogen. In the result the opposing shifts of the electron densities in the formation of the intermolecular reverse dative bond create

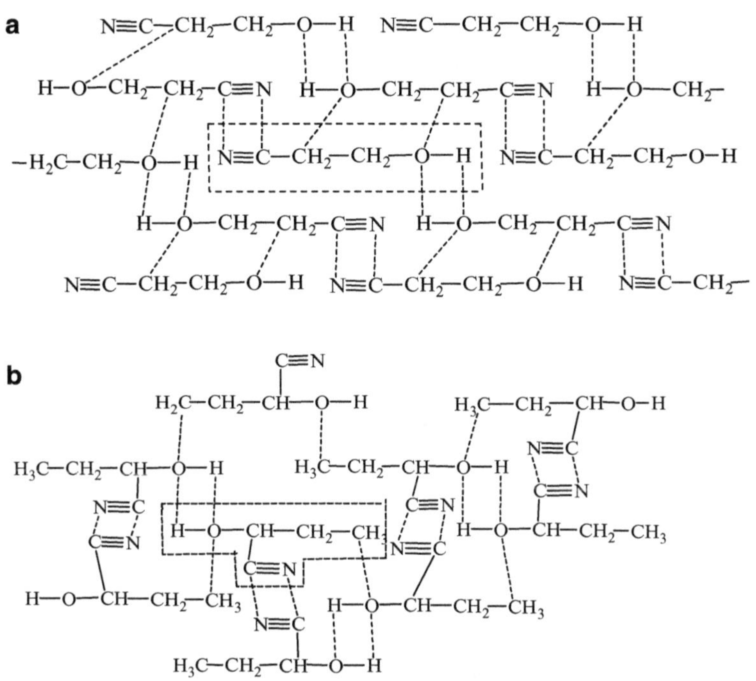

Fig. 8.4 Schematic picture of the liquid structure of 2-cyanoethanol (**a**) and 2-hydroxybutironitrile (**b**) with the network of the specific interactions

less difference in the charges of the carbon atoms of the fragment $–CH_2–CH_2–$ compared with the open chain and in the cycle $N(CH_2–CH_2)_3N$. This effect takes place at the aminoethanols (Table 8.2).

The molecule of 2-cyanoethanol with six bond vacancies (Fig. 8.4a) forms the network structure of the liquid condition with two hydrogen bonds DO•••H–O and the specific interactions of the two types 2D≡N → C≡ and 2D–O → $CH_2–CH_2–O$. The energy of the specific interaction N≡C formed is accepted to be equal to the value of the energy interaction, formed by 3-methoxypropionenitrile (11.4 kJ mol^{-1}) with two methylene groups of the fragment, with isolated atoms of nitrogen and oxygen. The energy of the specific interaction D–O → $CH_2–CH_2–O$ formed by the isolated fragment was based on it being equal to the value of 3.6 kJ mol^{-1}. The energy value of the hydrogen bond D–O•••H–O is determined by Eq. 8.4 as

$$\text{D-O}\cdots\text{H-O} = (\Delta_{vap}H°(298\ \text{K})\text{cel} - 2\text{D}{\equiv}\text{N}{\rightarrow}\text{C}{\equiv} - \text{D-O}{\rightarrow}\text{CH}_2\text{-CH}_2\text{-O})/4 \quad (8.4)$$

equal to 11.4 kJ mol^{-1}.

The molecule of 2-hydroxybutironitrile with functional N≡C– group and six bond vacancies forms the grid structure of the liquid condition by similar types of specific interactions and the hydrogen bond (Fig. 8.4b) with the 2-cyanoethanol

Table 8.3 The energies of the hydrogen bonds and the specific interactions (kJ mol^{-1}) of the liquid cyanoethanols, D–O → CH_2–CH_2–N = 3.6 kJ mol^{-1}

Compounds	Formula	The structure	$\Delta_{vap}H°$(298 K) [6]	T K	D≡N → C≡	D–O → CH_2–CH_2–N	D–O•••H–O
2-Cyanoethanol	C_3H_3NO	N≡C–CH_2–CH_2–O–H	53.4	346	11.4	3.6	11.4
2-Hydroxybutironitrile	C_4H_7NO	H_3C–CH_2–CH(O–H)–C≡N	57.9	329	11.4	6.15	11.4
2-Azidoethanol	$C_2H_5N_3O$	$N^{\ominus}$=$N^{\oplus}$=N–CH_2–CH_2–O–H	33.9	–	3.3[a]	3.6	11.4

[a]It is not correct

molecule. Hence, the difference in the values of the vaporization enthalpies of 2-cyanoethanol and 2-hydroxybutironitrile (Table 8.3) is caused by the introduction of an additional contribution of the energy of a third methylene group of the butyl isolated fragment. Using the analog of methoxypropionenitrile and α-methoxyisobutironitrile, the energy values of the hydrogen bonds and the specific interaction D≡N → C≡, we obtain the energy contribution of the isolated propyl fragment of the 2-hydroxybutironitrile molecule (Table 8.3).

Increase in the energy contribution with increasing numbers of the methylene group in the isolated fragment D–O → CH_2–CH_2– (3.6) < D–O → CH_2–CH_2–CH_2– (6.15 kJ mol^{-1}) points to the reduction in the transmission of the electron density from the p_z-orbital of the nitrogen and oxygen atoms to the contacting carbon atoms of the molecule at the formation of the intermolecular reverse dative bond by each of these atoms. As presented in the literature [6], the value of the vaporization enthalpy of 2-azidoethanol, without indication of errors and temperature, does not give the correct value of the energy contribution of the azides group for the vaporization enthalpy.

8.2.3 *Crystalline Amino Alcohols*

The 2-methyl-2-amino-1-propanol molecule with methyl and amine functional groups does not contain an isolated fragment and forms the network structure of the crystal by the two types of hydrogen bonds 4D–N•••H–N and D–O•••H–O and two types of the specific interactions of high stability 2D–O → CH_3–CH_2–CH_2–O and low stability DH_3C → H–CH_2 of the specific interactions (Fig. 8.5a). The hydrogen bonds of the amine group and propyl alcohol fragment formed in crystalline amides and dipropyl ether have a high stability of 9.7 (Chap. 6) and 11.61 kJ mol^{-1}, respectively. The value of the energy contribution of the isostructural methyl group to the sublimation enthalpy of 2-methyl-2-amino-1-propanol we take as equal to the contributing 2-methyl-1-propanol (4.55 kJ mol^{-1}) or at one specific interaction of this group, 2.27 kJ mol^{-1}. The calculated energy value of the hydrogen bond, formed by the spirit hydroxyl group with the help of Eq. 8.5

$$\text{D-O}\cdots\text{H-O} = (\Delta_{vap}H°(298\text{ K})\text{map} - 4\text{D-N}\cdots\text{H-N} - 2\text{D-O}\to\text{CH}_3\text{-CH}_2\text{-CH}_2\text{-O} - 2\,\text{DH}_3\text{C}\to\text{H-CH}_2)/2 \quad (8.5)$$

is given in Table 8.4.

For the three-coordinated nitrogen atoms the structure of its molecule is adopted as belonging to the planar structure [2]; thus, one should represent the structure of crystalline diethanolamine with the network of the specific interactions, formed by six bond vacancies of the molecule of this compound (Fig. 8.5b). The presence in the molecule of diethanolamine of the isolated fragment from two methylene groups leads to the formation of the specific interaction D–O → CH_2–CH_2– with

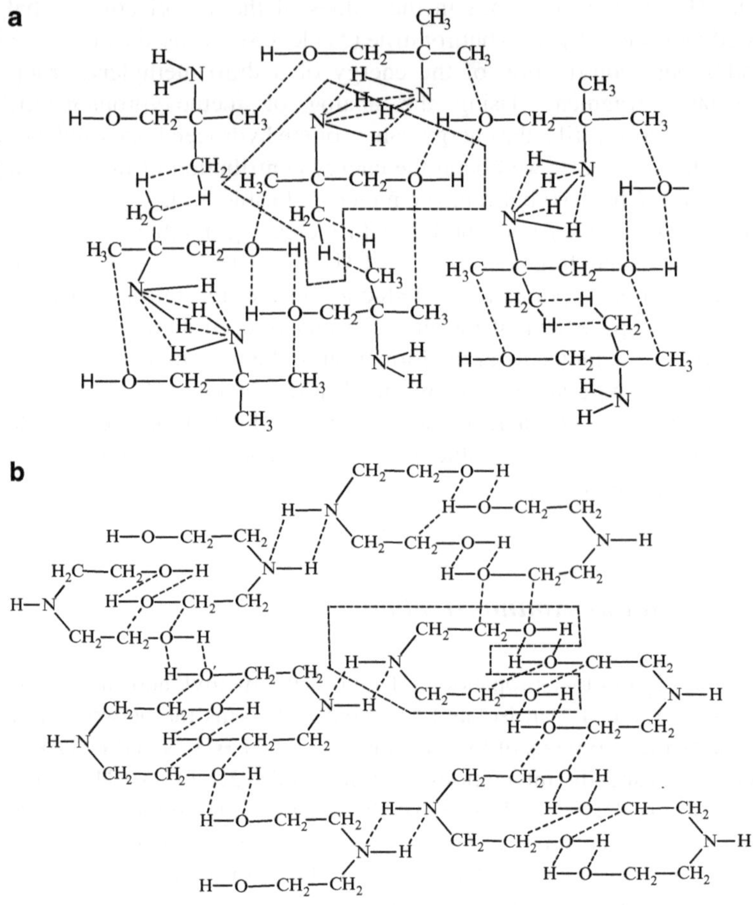

Fig. 8.5 Schematic picture of the liquid structure of 2-methyl-2-amino-1-propanol (**a**) and diethanolamine (**b**) with the network of the specific interactions

an energy of 3.6 kJ mol^{-1} for liquid condition and of 4.7 kJ mol^{-1} for the crystal condition, considering the energy of the specific interaction of 8.82 kJ mol^{-1}, formed by the ethyl ligand. Taking into account the rules of the equality of the total energy values of the specific interactions and sublimation enthalpy [13, 14],

$$D\text{-}O\cdots H\text{-}O = (\Delta_{vap}H°(298\ K)deap - 2D\text{-}N\cdots H\text{-}N - 2D\text{-}O\rightarrow CH_3\text{-}CH_2\text{-})/2 \quad (8.5)$$

we calculated the energies of the hydrogen bond for diethanolamine (Table 8.4).

The molecule of *N*-methylhydroxylamine forms the structure of the crystal similar to the liquid condition with the network of specific interactions and hydrogen bonds (Fig. 8.1). The difference of the vaporization enthalpy and sublimation enthalpy of this compound at 6.9 kJ mol^{-1} points to the fact that even at

Table 8.4 The energies of the hydrogen bonds and the specific interactions (kJ mol^{-1}) of the crystalline amino alcohols

Compounds	Formula	The structure	$\Delta_{sub}H°(298\ K)$ [12]	T K	D–N•••H–N	D–O → CH_3–(CH_2)n–O	D–O•••H–O
D-O→CH_2-CH_2-N = 4.7 kJ mol^{-1}							
2-Methyl-2-amino-1-propanol	$C_4H_{11}NO$	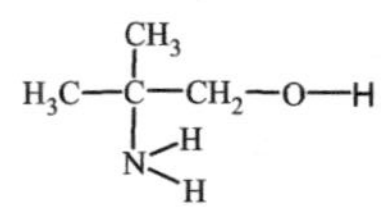	114.5 ± 5.2	339	9.7 × 4 D*iso*CH_3 =4.55	n = 2 11.61 × 2	24.0
Diethanolamine	$C_4H_{11}NO_2$	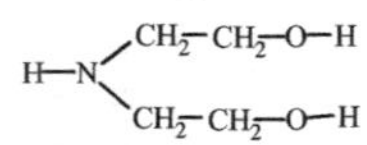	105.9 ± 2.0	298	9.7 × 2	n = 1 4.7 × 4	16.9
N-Methyl hydroxylamine	CH_5NO	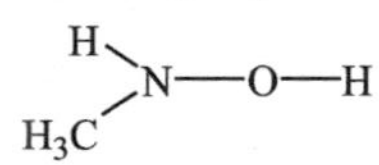	56.6	288	D–N → CH_3–N 6.0 × 2	D–O•••H–N 82 × 2	13.5

proportional stabilization of the hydrogen bonds D–O•••H–O, D–O•••H–N, and the specific interaction D–N → CH_3–N, the energy of the most stable hydrogen bond cannot be set at an energy of less than 12.45 kJ mol^{-1}. Using the above limit for the energy values of liquid *N*-methylhydroxylamine, we accept the energies of the hydrogen bond DO•••H–N and the specific interaction to be equal to the values as realized in its crystalline condition, 6.6 and 8.80 kJ mol^{-1} , respectively. The energies of the hydrogen bonds presented in Table 8.4 are described by the integral values of its stabilization,

D-O•••H-O: 2-Methyl-2-amino-1-propanol (24.0) > Diethanolamine (16.9) > N-Methylhydroxylamine (13.5 kJ mo^{-1})

reflected, at first, by the influence of the endo-effect of the coordinating methylene groups, interacting with the nitrogen atom at diethanolamine, and, second, by the absence of the alkyl groups in the molecule of *N*-methylhydroxylamine, leading to the shifting of the electron density to the atoms of oxygen and nitrogen, and the difference of the charges with the hydrogen atoms of the hydroxyl group and the amine fragment.

8.2.4 Crystalline Amino Phenols

For a thermodynamic analysis of aminophenols with the present position of the hydrogen bonds and the specific interactions one should draw attention to the photoelectron spectra of these compounds. On the basis of the photoelectron spectra with π-donor deputies OR_2 [15–17] and NR_2 [15, 16, 18, 19] the mesomeric shift B3MO significantly exceeds the mixing of the a_2(e1g)-level. The values of I_v and Ebo (1s) led the author [2] to the conclusion of the total acceptor nature of the OH-group and weak q donor ability of the NH_2-fragment. In the case of phenol there is insignificant predominance of the $\pi_1(a_2u)$-n_o-interaction under $\pi_3(e_{1g})$- n_o and for $C_6H_5NH_2$ the first of the interactions mentioned is approximately by 0.1 eV less effective compared with the second interaction. It follows that the amine group is insignificantly expressed at the energy of the hydrogen bond in crystalline aminophenols. It leads to the conclusions that, first, the total contribution of the four hydrogen bonds D–N•••H–N formed by the amine group is obtained from the difference between the vaporization enthalpy or sublimation enthalpy of liquid and crystalline aminophenol and phenol, respectively, from which

$$4\text{D-N}\cdots\text{H-N} = (\Delta_{sub}H°(298\ K)aph - (\Delta_{sub}H°(298\ K)phl - DH) \qquad (8.6)$$

we have determined the energy value of the hydrogen bond. It is necessary to take into account the energy of the substituted hydrogen atom (0.60 or 1.0 kJ mol^{-1}) in the molecule of phenol at the formation of the corresponding liquid and crystalline aminophenol, and, second, the total energy contribution of the hydrogen bond of liquid and crystalline phenol is equal to the difference

$$2D\text{-}O\cdots H\text{-}O = \Delta_{sub}H°(298\ K)\ phl - (\Delta_{sub}H°(298\ K)bz\text{–}DH) \qquad (8.6a)$$

between the enthalpy characteristics of this compound and benzene reduced by the contribution of the substituted hydrogen atom. From these calculations (Table 8.5), it is possible to specify the values of the energies of all other specific interactions, formed in aminophenol crystals.

The phenol molecule with ten bond vacancies and a planar structure forms the grid structure for the crystal (Fig. 8.6a) by the two hydrogen bonds D–O•••H–O, five specific interactions 5D=HC → CH with energies of interactions, formed by the CH groups of the benzene rings (7.40 kJ mol^{-1}) and by the energies of the two interactions 2D–O → C of the carbon atoms of the benzene rings with the oxygen atom of the hydroxyl group of molecules that are close in the environment. The energy of the last type of the specific interaction is calculated with the help of Eq. 8.6b

$$D\text{-}O\rightarrow C = (\Delta_{sub}H°(298\ K)phl - 2D\text{-}O\cdots H\text{-}O - 5D{=}HC\rightarrow CH)/2 \qquad (8.6b)$$

This specific interaction with reduced stability (3.3 kJ mol^{-1}) points to the reduced acceptor and donor ability of the carbon atom at location C(1) and to the reduction in the difference between its charge and the charge of the oxygen atom.

The molecule of aminophenol with 13 bond vacancies with planar structure of the crystal forms the network structure with the appearance of a volume fragment, formed by hydrogen bonds of the amine group with five-coordinated carbon atoms (Fig. 8.6b, c). The carbon atoms of the benzene rings form four specific interactions D=HC → CH and two D–O → C similar types of crystalline phenol and one DC → C by the carbon atoms of the benzene ring, connected with the amine group. The energies of these interactions are described by the Eq. 8.6c

$$\Delta_{sub}H°(298\ K)aph = 2D\text{-}O\cdots H\text{-}O + 4\ D\text{-}N\cdots H\text{-}N + 4D{=}HC\rightarrow CH + 2D\text{-}O\rightarrow C + DC\rightarrow C \qquad (8.6c)$$

The calculations of the energy contributions of two types of specific interactions 2DO → C and DC → C (13.0 kJ mol^{-1}) and its average value (4.13 kJ mol^{-1}) is within the error of the experimental sublimation enthalpy with the energy interaction D–O → C of phenol (3.3 kJ mol^{-1}). This implies that replacement of the specific interaction in crystalline aminophenols D=HC → CH at D=HC → C and at DC → C is accompanied by more significant destabilization than at 1.0 kJ mol^{-1} of one substituted hydrogen atom in the benzene cycle, i.e. approximately up to 3.3 kJ mol^{-1} at the replacement of two hydrogen atoms. At the same time the specific interaction DO → C is destabilized compared with the one formed in crystalline dimethyl ether (6.59 kJ mol^{-1}) up to 4.8 kJ mol^{-1} in crystalline aminophenols. It follows that in liquid and crystalline aminophenols there appear

Table 8.5 The energies of the hydrogen bonds and the specific interactions (kJ mol^{-1}) of the crystalline aminophenols, T = 298 K

Compounds	Formula	The structure	$\Delta_{sub}H°(298\ K)$ [12]	$\Delta_{sub}H°(298\ K)$ –DH	D–O•••H–O	D–N•••H– N	2D–O → C + DC → C
Phenol	C_6H_6O	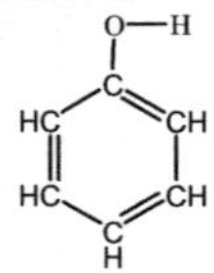	69.7 ± 0.9	43.6	13.05	–	D–O → C = 3.3
2-Aminophenol	C_6H_7NO	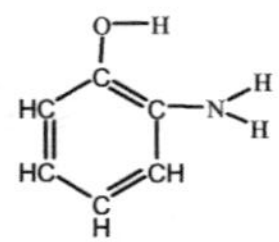	103.9 ± 0.9	68.7	13.05	8.81	13.0:3 = 4.13
3-Aminophenol	C_6H_7NO	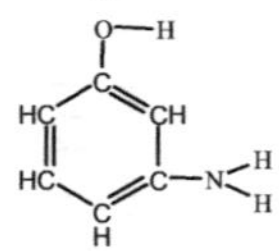	104.7 ± 1.2	68.7	13.05	9.00	13.0:3 = 4.13
4-Aminophenol	C_6H_7NO	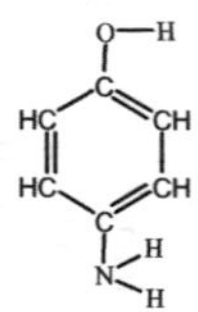	109.1 ± 1.4	68.7	13.05	10.10	13.0:3 = 4.13

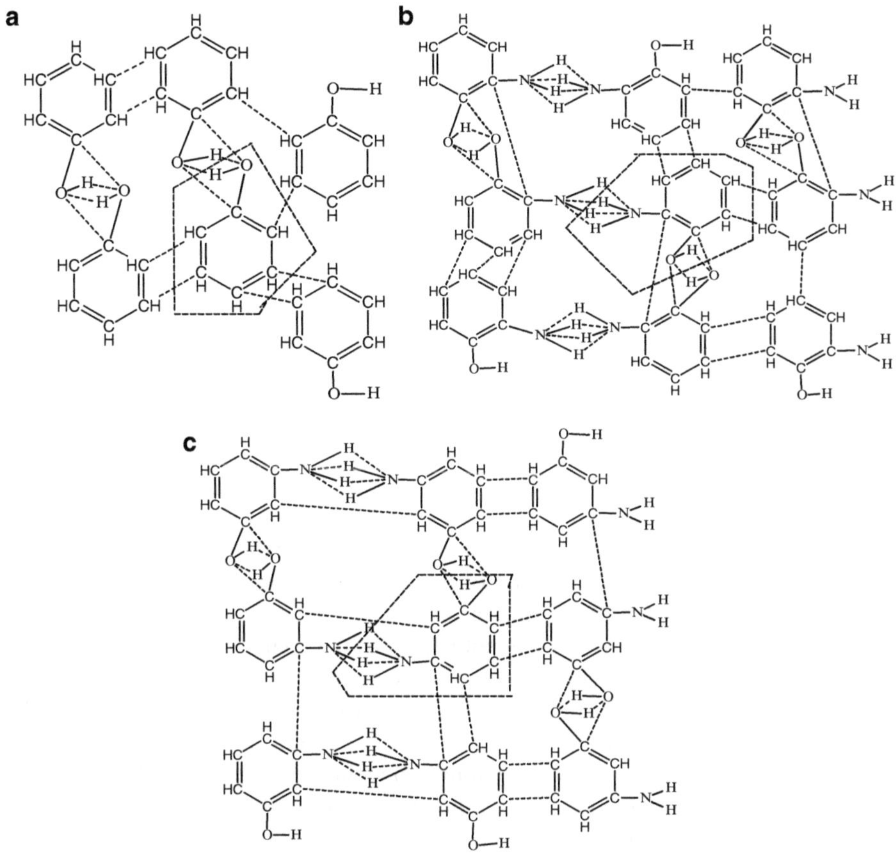

Fig. 8.6 Schematic picture of the crystalline structure of 2-methyl-2-amino-1-propanol (**a**) and diethanolamine (**b**) with the network of the specific interactions

significant changes in the charges of the oxygen, nitrogen, and contacting carbon atoms in the number of the compounds

Phenol < 2-Aminophenol < 3-Aminophenol < 4-Aminophenol.

8.3 The Energies of Specific Interactions of Liquid Oximes

8.3.1 Homological Series of Acetaldehyde Oximes

A simple compound of oximes is formaldehyde $H_2C{=}N{-}O{-}H$, for which the strong mixing of π(CN)- and n_o-oprbitals are characteristic [20]. Splitting of levels of the energies of two π-orbitals reaches 3.7 eV at energies close to those of the

Fig. 8.7 Dependence of the vaporization enthalpy versus the number of the carbon atom in the alkyl chine of acetaldehyde oxime's series (*1*, *2*) and acetone oxime's series (*3*, *4*)

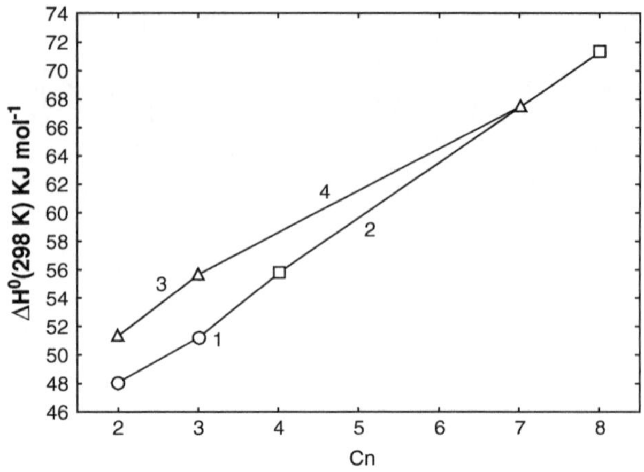

fragmentary orbitals from 12.4 to 12.6 eV. Thus, the energies of the hydrogen bonds of this compound should have high stability at one methylene group CH_2, giving one hydrogen atom for the formation of the hydrogen bond with the oxygen atom and the hydrogen atom of the alcohol group.

A limited number of vaporization enthalpies of the compounds of this series are described by two intersecting lines of the showing the dependence of the vaporization enthalpy on the number of the carbon atoms of the alkyl chain (Fig. 8.7). The intersection of lines (1) and (2) at the vaporization enthalpy of butyraldehyde oxime corresponds to the influence of the intermolecular reverse dative bond on the reduction of the energy of the specific interaction, formed by the terminal methyl group of the butyl ligand.

The achievement of maximum stability for this specific interaction for the compound with three carbon atoms of the alkyl chain points to the fact that further increase of the vaporization enthalpy of the compounds butyraldehyde oxime – octanal dehyde oxime is related to the energy contribution of the increasing number of methylene groups of the alkyl chain. A similar course of dependence $\Delta_{vap}H°$ (T) = f(Cn) is observed at the compounds of acetone oxime series (3, 4), which is characterized by the intersection of two correlation lines to the vaporization enthalpy of 2-butanone oxime, reflecting the influence of the reverse dative bond on the forming energies of the specific interaction at three carbon atoms in a long chain and a further increase in the enthalpy characteristics is caused by the introduction of energy to the enthalpy characteristics by increasing the number of methylene groups in the same long alkyl chain.

Using the dependence $\Delta_{vap}H°(T) = f(Cn)$, we determined the vaporization enthalpy of propylaldehyde oxime (Fig. 8.7), pentylaldehyde oxime – heptylaldehyde oxime by the method of interpolation and that of formaldehyde oxime and nonanyldehyde oxime (Table 8.6) by the method of extrapolation.

The molecule of the compound formaldehyde oxime with six specific interactions forms the network structure of liquid condition by the two stable

Table 8.6 The energies of the hydrogen bonds and the specific interactions (kJ mol^{-1}) of the liquid acetaldehyde oxime's series

Compounds	Formula	The structure	$\Delta_{vap}H°$ (298 K) [6]	T K	D–N → CH_3 (CH) n=N	D–O–•••H–CH	D–O•••H–O
Formaldehyde oxime	CH_3NO	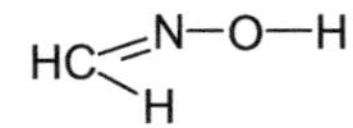	44.1[a]	303	4.30	5.50	12.45
Acetaldehyde oxime	C_2H_5NO	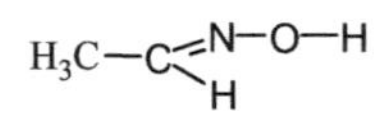	48.0	303	5.55	6.00	12.45
Propylaldehyde oxime	C_3H_7NO	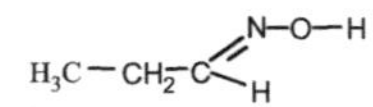	51.2 52.0[a]	326 308	7.55	6.00	12.45
Butyraldehyde oxime	C_4H_9NO	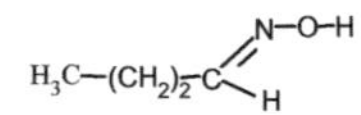	55.8	328	DCH_2 = 3.8	6.00	12.45
Pentanalaldehyde oxime	$C_5H_{11}NO$	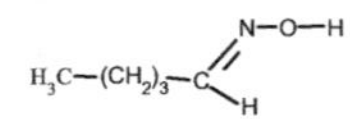	59.7[a]	328	2DCH_2 =7.7	6.00	12.45
Hexanalaldehyde oxime	$C_6H_{13}NO$	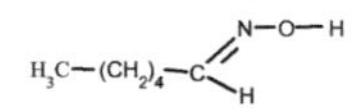	63.6[a]	328	3DCH_2 =11.6	6.00	12.45
Heptanalaldehyde oxime	$C_7H_{15}NO$	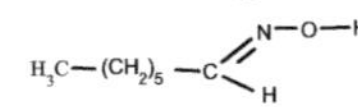	67.5[a]	328	4DCH_2 =15.5	6.00	12.45
Caprilaldehyde oxime, Octanal dehyde oxime	$C_8H_{17}NO$	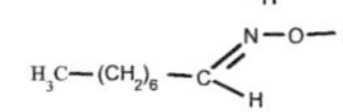	71.3	328	5DCH_2 =19.3	6.00	12.45
Nonanaldehyde oxime	$C_9H_{19}NO$	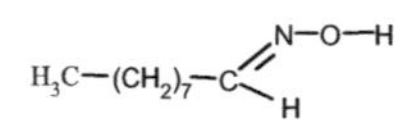	75.3[a]	328	6DCH_2 = 23.3	6.00	12.45
			Crystal		D–N•••H–N		
2-Hydroxybenzaldoxime	$C_7H_7NO_2$	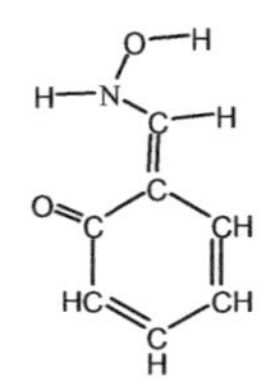	105.2 ± 10	298	9.7	6.00	8.9

[a]It is used method interpolation

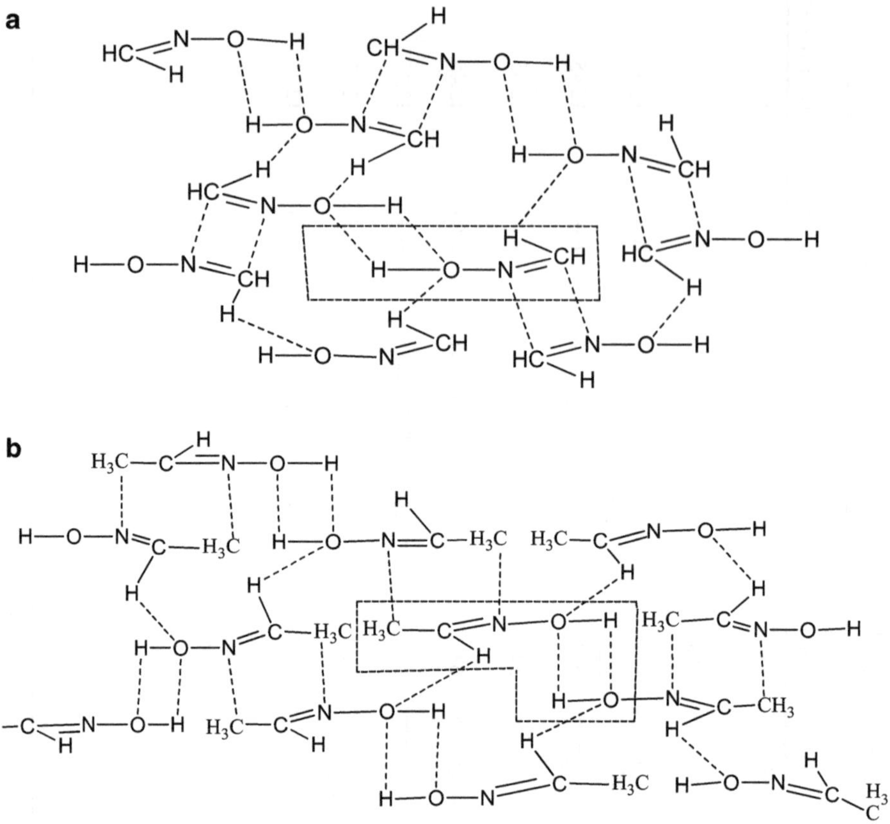

Fig. 8.8 Schematic picture of the liquid structure of formaldehyde oxime (**a**) and acetaldehyde oxime (**b**) with the network of the specific interactions

hydrogen bonds D–O•••H–O, two of reduced stability D–O•••H–CH and two specific interactions D–N → CH–N (Fig. 8.8a). In the case of the compounds of the acetaldehyde oximes series the specific interactions are formed with a larger number of carbon atoms of the alkyl chain D–N → CH_3–C=N and D–N → CH_3–CH_2–CH_2–C=N, of acetaldehyde oxime and butyraldehyde oxime, respectively (Fig. 8.8b).

The hydrogen atom of the H–CH group of the formaldehyde oxime molecule and compounds of the acetaldehyde oximes series with hydrogen atom substituted by the N–OH group acts as the missing (the lacking) vacancy as in formaldehyde and acetaldehyde molecules. The series of compounds forms the hydrogen bond with the oxygen atom of the hydroxyl group. In the literature a similar type of specific interaction is discussed. [21] It is noted that the existence of rather weak complexes with the hydrogen bonds of type CH•••O is possible. The formation of such bonds was assumed on the basis of a redistribution of the electron density and as a result different spectral effects [22] occur. Analyzing the experimentally observed differences of the spectral lines, describing the extension of the bonds of a different nature in gaseous and liquid methanol, the authors of [23] came to the conclusion of the possibility of the existence in condensed condition of this compound by one such bond for each molecule. The authors [24] made an analysis of the quantum- chemical calculations of the monomer of methyl spirit by different methods, including Hartree-Fock (HF) approximations, with the use of the electron correlation on the basis of the method of MP2, MP4, interconnected clusters (CC), and configuration interactions (CI), and the functional density method (B3LYP, B3PW9I, MPWIPW9I).A comparison of the results of the calculation of the monomer structure, including the lengths of the intermolecular bonds, valence, and torsion angles, the electron density at the atoms, and spectral characteristics leads one to the conclusion, at first, that there is disparity of protons in the methyl group of the methanol molecule and, second, that the proton, located in *trans*-position toward the hydroxyl hydrogen atom, has differing characteristics from the two other hydrogen atoms of this methyl group. The estimated value of the energy barrier of inner rotation of the CH_3 group is close to 4 kJ mol^{-1}. It should be mentioned that this value of the energy is close to the energy of the specific interaction of D–O → CH_3–O (5.63 kJ mol^{-1}), formed by the essentially unshared $2s^2$ of the electron pair of the carbon atom of the methyl group of liquid methanol [25].

This problem was considered [25] in Chaps. 1 and 2 and the contribution by each hydrogen atom to the energy of this specific interaction was determined by the contribution of the electron density by each hydrogen atom, changing the charge of the carbon atom and its donor properties. Of course, the reduction of the hydrogen atoms in the methyl group is accompanied by replacement of the less connected hydrogen atom and redistribution of the electron density at the atoms, which inevitably leads to the increase of positive charge in the remaining hydrogen atoms. As a result, the energy contribution of the methylene group of cyclohexane has a significantly reduced value compared with the introduced CH group as regards the sublimation enthalpy of benzene (Table 3.2).

The participation by the oxygen atom of the formaldehyde and acetaldehyde series and the formaldehyde oxime and acetaldehyde oxime series leads to a certain similarity in the energies of the hydrogen bonds, formed by the hydrogen atom of the =CH group of the aldehyde fragment

$$\text{Formaldehyde } (5.5) < \text{Acetaldehyde } (6.1) \approx \text{Propanal } (6.1) \approx \text{Butanal } (6.10)$$
$$\approx \text{Pentanal } (6.20) \approx \text{Hexanal } (5.80 \text{ kJ mol}^{-1})$$

with reduced value at the basis of the compounds of this series with further change with error in the experimentally obtained enthalpy characteristics (6.00 kJ mol^{-1}) [25]. The possible factor, expressed at the energies of this type of hydrogen bonds of the formaldehyde and acetaldehyde series, can be the double bond =C=O in aldehyde compounds, absent in the formaldehyde and acetaldehyde oxime series. At the same time coherence of the alcohol hydroxyl group with the nitrogen atom in the last compound creates some difference with the aldehyde class of the compounds and contributes a definite similarity with hydroxylamines in the formation of the hydrogen bond of another type, D–O•••H–O, and in the formation of the specific interaction with CH and R–CH of the alkyl chain.

We ought to pay attention to alkilamines, for which at the three-coordinated condition of nitrogen the energy of the formed specific interaction D–N → CH_3–N has a value of 4.80 kJ mol^{-1} or at one substituted hydrogen atom of the methyl group D–N → CH_2–N it should be equal to the value reduced by the contribution of the substituted hydrogen atom (0.60 kJ mol^{-1}).

Taking the similarity of the formaldehyde and acetaldehyde oxime series in the formation of the hydrogen bond D–O•••H–C(H)=N = (5.50 kJ mol^{-1}) and, on the other side, the analog with alkilamines in the formation of the hydrogen bond D=O•••H–O (12.45 kJ mol^{-1}), and taking into account the contributions of two types of hydrogen bonds with known energy values to the enthalpy characteristics, we calculated the energy of the specific interaction D–N → CH_2=N of liquid formaldehyde oxime with the help of Eq. 8.7

$$\text{D=N}\rightarrow\text{CH}_2\text{=N} = (\Delta_{\text{vap}}H^\circ(298\ \text{K})\text{fdho} - 2\text{D-O}\cdots\text{H-O} - 2\text{D-O}\cdots\text{H-CH=N})\,/\,2 \tag{8.7}$$

The obtained energy value of this type of specific interaction allows us to conclude on the correctness of the assumption and the possibility of fulfillment of the energies of the specific interactions D–N → CH_3–CH–N and D–N → CH_3–CH_2–C(–H)–N, realized in liquid acetaldehyde oxime – propylaldehyde oxime. The results of the calculations of the energy of the specific interaction (Table 8.6) reflect its stabilization with an increasing number of carbon atoms in the chain

$$\text{Formaldehyde oxime (4.3)} < \text{Acetaldehyde oxime (5.55)} < \text{Propylaldehyde oxime } \left(7.55\ \text{kJ mol}^{-1}\right)$$

with the weakening effect of the intermolecular reverse dative bond and completion

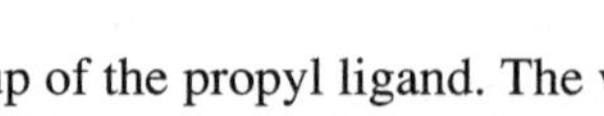

at the terminal methyl group of the propyl ligand. The values of the energies of the specific interactions, formed by methyl, ethyl, and propyl groups D–N → CH_3–N, D–N → CH_3–CH_2–N, and D–N → CH_3–CH_2–CH_2–N, have energy values of 4.8, 6.0, and 8.0 kJ mol^{-1}, respectively (Table 2.1). A further increase in the vaporization enthalpy of all subsequent compounds is connected to the contribution of the energy by the increasing number of methylene groups. The calculations of the contributions of the energies of the specific interactions with substituted hydrogen

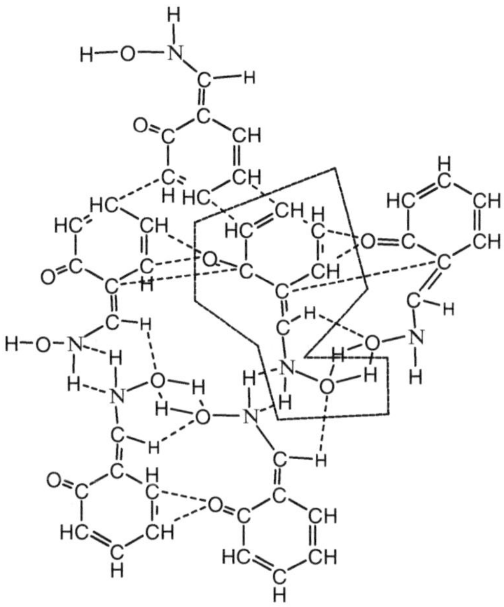

Fig. 8.9 Schematic picture of the crystalline structure of 2-Hydroxybenzaldoxime with the network of the specific interactions

atoms in the methylene group points to the reduction of stability in these interactions.

The molecule of 2-hydroxybenzaldoxime with 14 bond vacancies forms the network structure of the crystal by three types of the specific interactions and three types of the hydrogen bonds (Fig. 8.9). The experimentally obtained value of the sublimation enthalpy of this compound 105.2 ± 10 has an increased error level and, thus, it is of definite interest for finding a basis for the more correct value of the enthalpy characteristics on these energies of the hydrogen bonds and various types of the specific interactions, formed by the essentially unshared $2s^2$ electron pair of the carbon atom.

The four specific interactions 4D−C=O → CH are formed by the carbonyl oxygen atom of the benzene ring with CH groups of the same ring of molecules that are close in the environment with the energy of 7.85 kJ mol^{-1} (Chap. 7, Table 7.7). The second type of the specific interaction is formed by the CH groups 2D=HC → CH of the benzene rings with the energy of 7.10 kJ mol^{-1} reduced by the energy contribution (2.0 kJ mol^{-1}) of the two substituted hydrogen atoms of the cycle. The third type of the specific interaction 2DC → C is formed by carbon atoms of the cycle with substituted hydrogen atoms with the energy of 5.2 kJ mol^{-1}. The energy of the hydrogen bond, formed by the amino group 2D–N•••H–N of crystalline 2-hydroxybenzaldoxime is equal to 9.7 kJ mol^{-1}. The third type of the hydrogen bond is formed by the aldehyde hydrogen atom with the oxygen atom of the hydroxyl spirit group of 2D–O•••H–C= with the energy value of 6.0 kJ mol^{-1}.

The energy of the hydrogen bond, formed by the alcohol hydroxyl group, we obtain with the help of Eq. 8.7a

$$DO\cdots H\text{-}O = (\Delta_{vap}H^{\circ}(298\ K)hbox - 2D\text{-}N\cdots H\text{-}N - 2D\text{-}O\cdots H\text{-}C{=} - 4D\text{-}C{=}O{\rightarrow}CH - 2D{=}HC{\rightarrow}CH - 2DC{\rightarrow}C) / 2 \quad (8.7a)$$

The results of the calculation point to the significantly reduced energy value of this type of hydrogen bond D–O•••H–O = 8.9 kJ mol^{-1} compared with the energy of the hydrogen bond of the same type of crystalline 2-aminophenol (13.05 kJ mol^{-1}). On this ground we can conclude that the energy of this type of the hydrogen bond should be higher than the calculated one, approximately at 5 kJ mol^{-1}.

8.3.2 Homological Series of 2-Propenal Oxime and Octanone's Oximes

As presented in Fig. 8.7 the functional dependence $\Delta_{vap}H^{\circ}(T) = f(Cn)$ on the carbon atoms in the long alkyl chain allows us to use the method of interpolation and extrapolation for determination of the vaporization enthalpies of unstudied compounds, which is of interest for a thermodynamic analysis and for obtaining the energy values of the hydrogen bond and the specific interactions of this series of compounds (Table 8.7).

The molecule of acetone oxime, 2-butanone oxime and this series of the compounds with double > C=N– bond and located in the plane long alkyl chain, methyl group, and oxime's fragment with six bond vacancies (Fig. 8.10a, b) form the network structure of liquid alkyl oximes by the two hydrogen bonds 2D–O•••H–O, four specific interactions of two types 4D–N → CH_3–C=N and with long alkyl chain 2D–O → CH_3–(CH_2)n–C=N, the energy value of which reaches its maximum with three carbon atoms of the chain 2-butanone oxime because of the completion of the influence of the intermolecular reverse dative bond of the terminal methyl group. The replacement of the nitrogen atom by the oxygen atom in this type of specific interaction results in its stabilization. The interaction energy reaches the value of 6.00 and 8.00 kJ mol^{-1} for the two substituted hydrogen atoms in the methylene group, respectively, acetone oxime, and compounds with three or more of the carbon atoms in the long chain. Using the energies of these specific interactions and Eqs. 8.8a and 8.8b

Acetone oxime

$$D\text{-}O\cdots H\text{-}O = (\Delta_{vap}H^{\circ}(298\ K)ao - 2D\text{-}N{\rightarrow}CH_3\text{-}C{=}N - 2D\text{-}O{\rightarrow}CH_3\text{-}C{=}N) /2 \quad (8.8a)$$

2-Butanone oxime

Table 8.7 The energies of the hydrogen bonds and the specific interactions (kJ mol^{-1}) of the liquid and crystalline acetone oxime's series

Compounds	Formula	The structure	$\Delta_{vap}H°(298\ K)$ [6]	T K	D=N → CH_3–C=N	D–O → CH_3 (CH_2)n–C=N	D–O•••H–O
Acetone oxime	C_3H_7NO	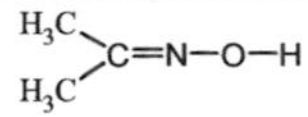	51.4	345	5.55 × 2	n = 0 6.00	14.15
2-Butanone oxime	C_4H_9NO	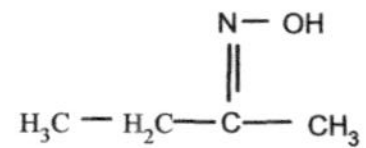	55.7	330	5.55 × 2	n = 1 8.00 × 2	14.3
2-Pentanone oxime	$C_5H_{11}NO$	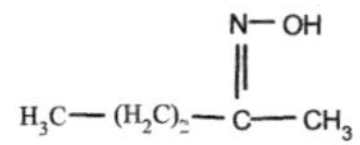	58.7[a]		5.55 × 2 DCH_2 = 3.0	n = 1 8.00 × 2	14.3
2-Hexanone oxime	$C_6H_{13}NO$	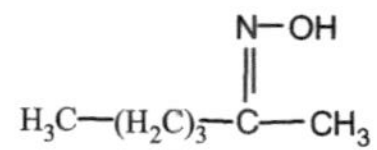	61.7[a]	308	5.55 × 2 $2DCH_2$ = 6.0	n = 1 8.00 × 2	14.3
2-Heptanone oxime	$C_7H_{15}NO$	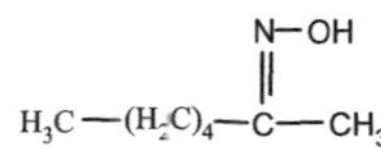	64.7[a]	308	5.55 × 2 $3DCH_2$ = 9.0	n = 1 8.00 × 2	14.3
2-Octanone oxime	$C_8H_{17}NO$	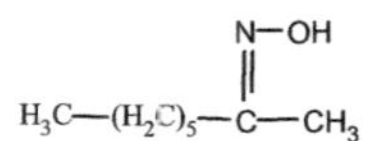	67.5	308	5.55 × 2 $4DCH_2$ = 11.8	n = 1 8.00 × 2	14.3
2-Nonanone oxime	$C_9H_{19}NO$	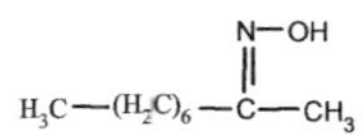	70.6[a]	308	5.55 × 2 $5DCH_2$ = 14.9	n = 1 8.00 × 2	14.3
3-Pentanone oxime	$C_5H_{11}NO$	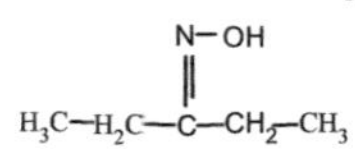	55.8	333	7.55 × 2	n = 1 8.00 × 2	12.35[b]
3-Octanone oxime	$C_8H_{17}NO$	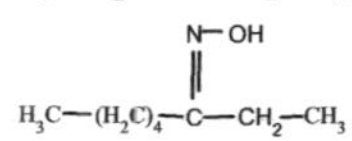	67.2	308	7.55 × 2 $2DCH_2$ = 6.0	n = 1 8.00 × 2	15.05
4-Octanone oxime	$C_8H_{17}NO$	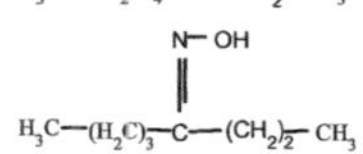	68.8	308	$2DCH_2$ = 6.0	n = 1 8.00 × 2	15.85
			Crystal				
Acetone oxime	C_3H_7NO	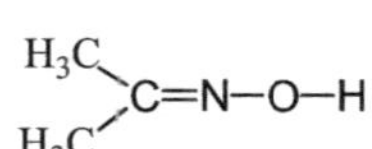	59.6	323	6.4	6.9	16.5

[a]Estimated by interpolation method
[b]T = 333 K

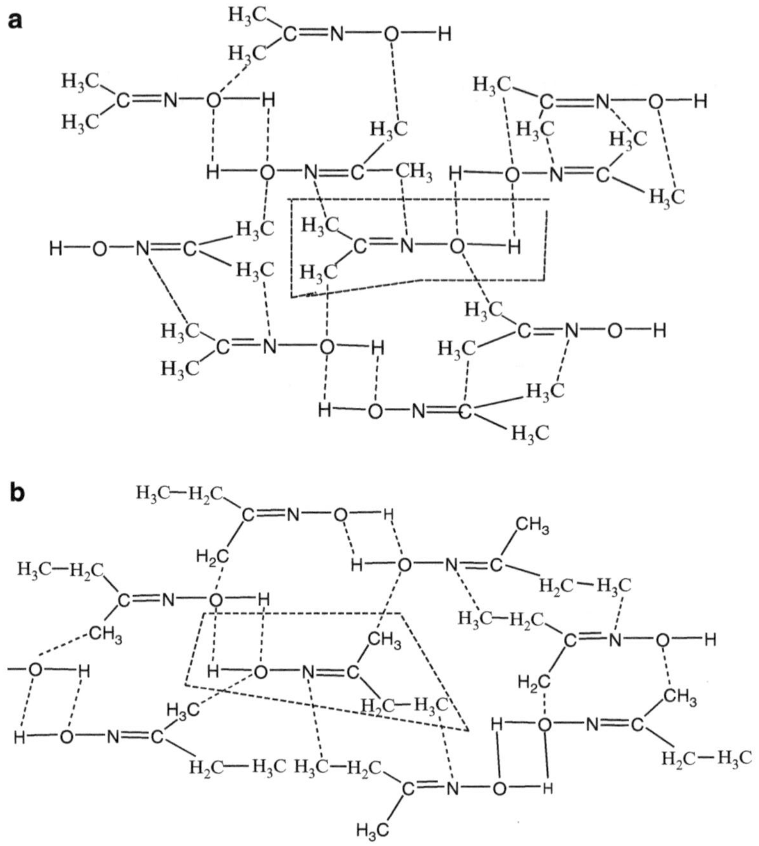

Fig. 8.10 Schematic picture of the liquid structure of acetone oxime (**a**) and 2-Butanone oxime (**b**) with the network of the specific interactions

$$D\text{-}O\cdots H\text{-}O = (\Delta_{vap}H°(298\ K)bo - 2D\text{-}N\rightarrow CH_3\text{-}C{=}N - 2D\text{-}O\rightarrow CH_3\text{-}CH_2\text{-}C{=}N)/2 \quad (8.8b)$$

We made necessary calculations of the energies of the hydrogen bonds of these compounds. From the given in Table 8.7 data it follows that independent of the length of the alkyl chain and formed specific interactions, the energy of the hydrogen bonds remains practically stable for all series of the compounds. The energy of the hydrogen bonds of the compounds with fragment N–O–H are described by the integral value of its stabilization

$$\textit{Hydroxylamins}\ (12.45) = \text{Acetaldehyde oxime's series}\ (12.45) < \text{Acetone oxime's series}\ \left(14.3\ \text{kJ mol}^{-1}\right)$$

Equal values of the energies of the hydrogen bonds of the first two series of the compounds reflects a similar influence of the methyl group of hydroxylamines and the hydrogen bond of the methylene group of the formaldehyde oxime and acetaldehyde oxime series to the redistribution of the electron density in the molecules and charges at atoms, formed the corresponding specific interactions and the hydrogen bonds.

The stabilization of the hydrogen bonds at 1.9 kJ mol^{-1} at compounds of the acetone oxime series is caused by replacement of the hydrogen atom in the acetaldehyde oxime series to the CH_3–C= group, changing the charge of the nitrogen atom and increasing of the charge of the oxygen atom of the hydroxyl group.

The energies of the hydrogen bond 3-pentanone oxime and derivatives of octanone oxime are determined with the help of Eq. 8.8c

$$\text{D-O}\cdots\text{H-O} = (\Delta_{vap}H°(298\ K) - 2D{=}N{\to}CH_3\text{-}CH_2\text{-}C{=}N - 2D\text{-}O{\to}CH_3\text{-}CH_2\text{-}C{=}N - \Sigma DCH_2)/2 \quad (8.8c)$$

Data presented in Table 8.7 point to the stabilization of the hydrogen bond

$$\text{2-Octanone oxime } (14.3) < \text{3-Octanone oxime } (15.05) < \text{4-Octanone oxime } \left(15.85\ \text{kJ mol}^{-1}\right)$$

with the distance of N–OH fragment to the fourth carbon atom of the alkyl chain, leading to the completion of the influence of the intramolecular reverse dative bond to the methyl group, forming the specific interaction.

Sublimations enthalpies of crystalline compounds of the acetone oxime's series are presented in the literature as the origin of these compounds (Table 8.7). Thus, for the estimation of the energy values of the specific interactions D=N → CH_3–C=N, D–O → CH_3–C=N, and the hydrogen bond of this condition in the crystal there exists only the method of equal contributions under the influence of the crystalline field to the realized interactions. The results of the calculations with the use of the energies of the specific interactions and the hydrogen bond of liquid acetone oxime are given in Table 8.7.

8.3.3 Cyclohexanone Oxime

Cyclohexanone oxime is the only representative of the cyclic oximes, for which there is reliable thermodynamic characteristic of the vaporization process, leading to thermodynamic analysis and calculation of the energy of the hydrogen bond. The molecule of this compound with ten bond vacancies and planar structure forms the network structure of the crystal (Fig. 8.11) by the two hydrogen bonds 2D–O•••H–O, specific interactions of two types 3D–H_2C → CH_2, 2D–H_2C → C, formed by CH_2 groups of cyclohexane with the energies of 6.25 and 5.0 kJ mol^{-1},

Fig. 8.11 Schematic picture of the liquid structure of cyclohexanone oxime with the network of the specific interactions

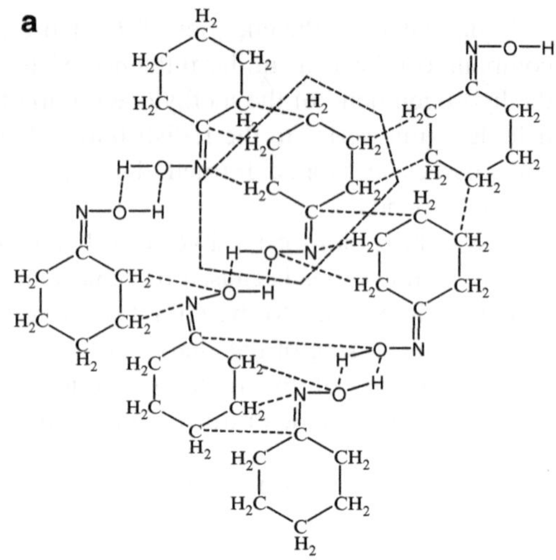

respectively, reduced by the contribution of the substituted hydrogen atoms, two interactions 2D=N → CH_2 (5.5 kJ mol^{-1}) and the specific interaction D–O → CH_2 with the energy of specific interactions of crystalline methyl ether. The mentioned types of the specific interactions with its energies determine the value of sublimation enthalpy of cyclohexanone oxime

$$\Delta_{sub}H°(298\ K)cho = 2D\text{-}O\cdots H\text{-}O + 3DH_2C\rightarrow CH_2 + 2DH_2C\rightarrow C + D\text{-}O\rightarrow CH_2 + 2DN\rightarrow CH_2 \quad (8.9)$$

The energy value of the hydrogen bond of crystalline cyclohexanone oxime obtained determines its solubility in water and in other solvents and the specificity of its interaction in the solution (Table 8.8).

8.4 The Energies of Specific Interactions of Liquid Amino-2-Propanone, Amino-1-Phenylethanone, and Amino-1-Phenyl-1-Propanone

Aminooxocompounds, which include α- and β-aminoketones, allow us to clarify the influence of the amino group on the change of the energy of the specific interaction, formed by the carboxyl oxygen atom. The 1-(dimethylamino)-2-propanone molecule with symmetrical methyl groups of the amine fragment and also symmetrically located at a certain angle by the methyl groups of the ketone fragment forms the structure of liquid condition of four specific interactions DN → CH_3–N with the fifth coordinated nitrogen atom and four more stable interactions D=C=O → CH_3–C=, formed by the carbonyl oxygen atom, respectively (Fig. 8.12a). Taking into account that formed specific interactions by

Table 8.8 The energies of the hydrogen bonds and the specific interactions (kJ mol^{-1}) of the crystalline cyclohexanone oxime

Compounds	Formula	The structure	$\Delta_{sub}H°(298\ K)$ [12]	T K	D–O → CH_2/D=N → CH_2	D–H_2C → CH_2/D–H_2C → C	D–O•••H–O
Cyclohexanone oxime	$C_6H_{11}NO$	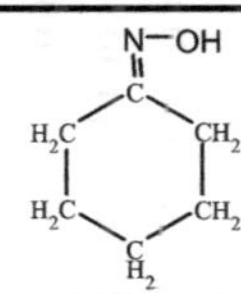	79.9 ± 0.7	317	6.40/5.5 × 2	6.25 × 3 5.3	16.6

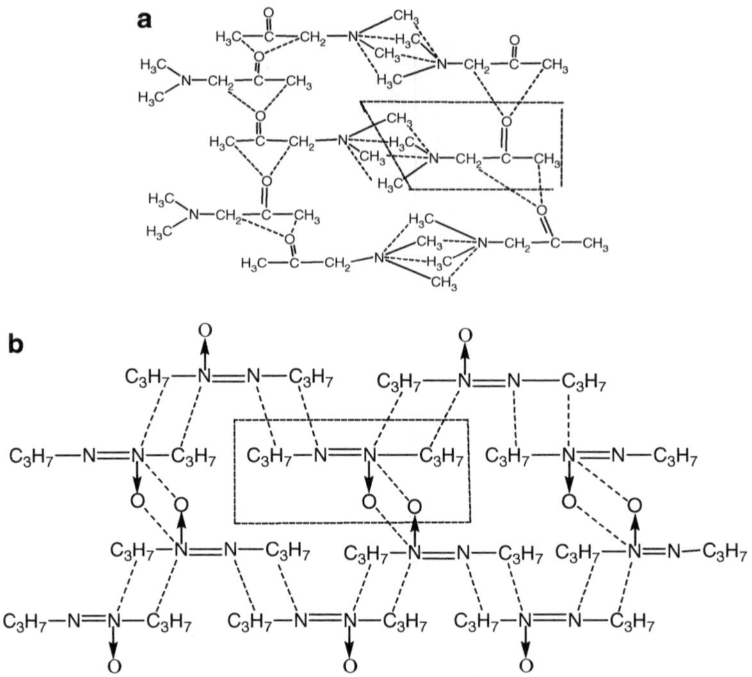

Fig. 8.12 Schematic picture of the liquid structure of 1-(dimethylamino)-2-propanone (a) and dipropyldiazene *N*-oxide (b) with the network of the specific interactions

dimethylamino D–N → CH_3–N have energies equal to 4.80 kJ mol^{-1} (Table 2.2), we can possibly calculate the energy specific interaction 1-(dimethylamino)-2-propanone, formed by carbonyl oxygen atom also with the methyl group with the help of Eq. 8.10

$$D{=}C{=}O{\rightarrow}CH_3\text{-}C{=} \; = \; (\Delta_{sub}H°(298\ K)dmap - 4DN{\rightarrow}CH_3\text{-}N)\,/\,4 \qquad (8.10)$$

This equation is also used for calculation of the energy of the same type of the specific interaction 1-(diethylamine)-2-propanone, taking the energy value of the specific interaction D–N → CH_3–CH_2–N, formed by the ethyl group diethyl fragment, equal to the energy of the same type of diethyl amine (Table 2.1). Presented in Table 8.9 results of the calculations of the energies of the specific interaction D=C=O → CH_3–C= with the error of experimental calculations points to the stability of the energy values of this interaction at compounds

$$\text{1-(Dimethylamino)-2-propanone}\ (6.10) = \text{1-(Diethylamine)-2-propanone}\ (6.10)$$
$$< \text{2-Propanone (Aceton)}\ \left(7.7\ \text{kJ mol}^{-1}\right)$$

which is significantly less stable compared with the energy of the same type, formed by liquid 2-Propanone and reliably experimentally measured value of the vaporization enthalpy (30.84 kJ mol^{-1}).

Table 8.9 The energies of the hydrogen bonds and the specific interactions (kJ mol^{-1}) of the liquid cyclohexanone oxime amino-2-propanone, amino-1-phenylethanone, and amino-1-phenyl-1-propanone at T = 298 K

Compounds	Formula	The structure	$\Delta_{vap}H°$(298 K) [6]	D–N → CH_3–N	D–N → CH_3–CH_2–N	D=C=O → CH_3–C=
1-(Dimethylamino)-2-propanone	$C_5H_{11}NO$	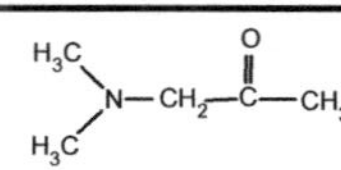	43.6 ± 0.3	4.80	–	6.10
1-(Diethylamine)-2-propanone	$C_7H_{15}NO$	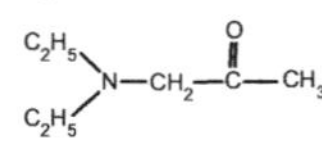	47.7 ± 0.3	–	5.85	6.10
				D–N → CH_3–CH_2–CH_2–N		DN–O → N–O
Dipropyldiazene *N*-oxide	$C_6H_{14}N_2O$	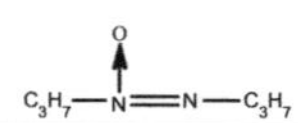	51.7 ± 0.1	8.1		9.65

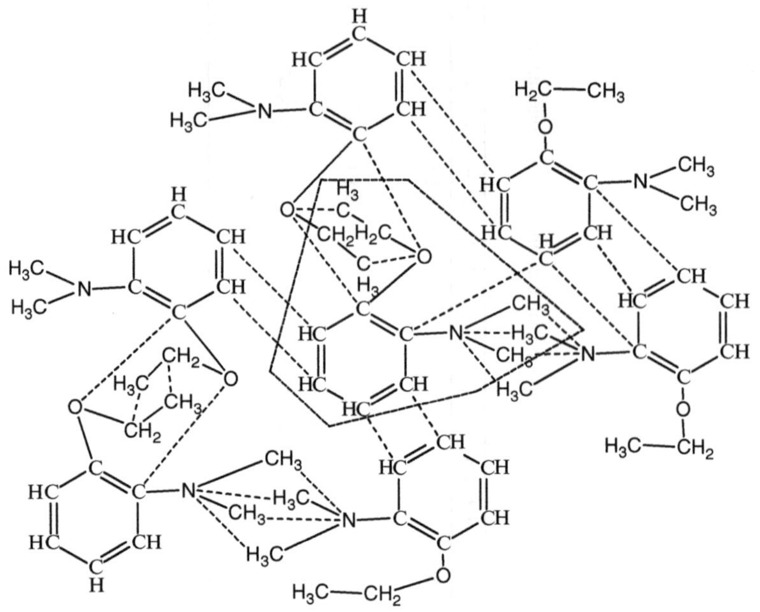

Fig. 8.13 Schematic picture of the liquid structure of 2-(dimethylamino)-1-phenylethanone with the network of the specific interactions

On the base of these data we have the possibility to conclude that the presence of the amine group at 1-(dimethylamino)-2-propanone and 1-(diethylamine)-2-propanone causes the significant redistribution of the electron density in its molecules. It is accompanied by the part shifting the electron density from the carbon atoms of methyl and methylene group to the nitrogen atom and transmission of the electron density from p_z-orbital of the nitrogen atom to the same carbon atoms, increasing its negative charge. Thus, the reduction of the difference in the negative charge of the oxygen atoms and its fringing of the carbon atoms of methyl and methylene groups occurs. As a result, the formed specific interactions D=C=O → CH_3–C= of liquid 1-(dimethylamino)-2-propanone and 1-(diethylamine)-2-propanone are less stable at 1.6 kJ mol^{-1}, than liquid 2-propanone.

The molecule of dipropyldiazene *N*-oxide with two propyl groups forms in liquid condition four specific interactions D–N → CH_3–CH_2–CH_2–N (Fig. 8.12b) with the energies of 8.1 kJ mol^{-1} (Chap. 2, Table 2.1) and two specific interactions $D_{N\rightarrow O\dashrightarrow N\rightarrow O}$, whose energies are calculated with Eq. 8.10a, dividing,

$$D_{N\rightarrow O\dashrightarrow N\rightarrow O} = (\Delta_{sub}H°(298\ K)dpdz - 4D\text{-}N\rightarrow CH_3\text{-}CH_2\text{-}CH_2\text{-}N) / 2 \qquad (8.10a)$$

the difference between the sublimation enthalpy and total value of four specific interactions by the number of interactions. The obtained energy value of this type of specific interaction (Table 8.9) points to its increased stability.

The compounds 2-(dimethylamino)-1-phenylethanone and 2-(diethylamino)-1-phenylethanone and the derivatives with 13 bond vacancies have a similarity in the structure of the molecules and form the network structure of the liquid condition (Fig. 8.13). In this connection it allows us to reveal the influence of the amine groups on the energies of the specific interactions, formed by alkyl ligands with the oxygen atom with the nature of simple ethers. The character of 2-(dimethylamino)-1-phenylethanone, 2-(dimethylamino)-1-phenyl-1-propanone, and 2-(dimethylamino)-2-methyl-1-phenylpropanone is due to the formation of four specific interactions D–N → CH_3–N with an energy of 4.80 kJ mol^{-1}. The compounds 2-(diethylamino)-1-phenylethanone and 2-(diethylamino)-1-phenylethanone form four specific interactions by the ethyl ligand of the amine group D–N → CH_3–CH_2–N with energy of 5.85 kJ mol^{-1} (Table 8.10). The location of the fragment of molecules of the compounds mentioned at the same carbon atoms of the benzene ring is the condition for the invariability of the energies of four specific interactions, formed by the CH groups D=HC → CH with an energy of 5.65 kJ mol^{-1}, one specific interaction D=HC → C with substituted hydrogen atom (0.60 kJ mol^{-1}) to the vaporization enthalpy and two interactions DO → C = 4.15 kJ mol^{-1}. The results of the calculations of the energies of the specific interactions of oxygen with the ethyl and propyl ligands D–O → CH_3–(CH_2)n–O

2-(Dimethylamino)-1-phenylethanone

$$\text{D-O}\rightarrow\text{CH}_3\text{-CH}_2\text{-O} = (\Delta_{vap}H°(298\ \text{K})\text{dmpe} - 4\text{D-N}\rightarrow\text{CH}_3\text{-N} - 4\text{D=HC}\rightarrow\text{CH} - \text{D=HC}\rightarrow\text{C} - 2\text{DO}\rightarrow\text{C})/2 \quad (8.10b)$$

2-(Dimethylamino)-1-phenyl-1-propanone

$$\text{D-O}\rightarrow\text{CH}_3\text{-CH}_2\text{-CH}_2\text{-O} = (\Delta_{vap}H°(298\ \text{K})\text{dmpp} - 4\text{D-N}\rightarrow\text{CH}_3\ \text{-N} - 4\text{D=HC}\rightarrow\text{CH} - \text{D=HC}\rightarrow\text{C} - 2\text{DO}\rightarrow\text{C})/2 \quad (8.10c)$$

2-(Dimethylamino)-2-methyl-1-phenylpropanone

$$\text{D-O}\rightarrow\text{CH}_3\text{-CH}_2\text{-CH}_2\text{-O} = (\Delta_{vap}H°(298\ \text{K})\text{dmmp} - 4\text{D-N}\rightarrow\text{CH}_3\ \text{-N} - 4\text{D=HC}\rightarrow\text{CH} - \text{D=HC}\rightarrow\text{C} - 2\text{D-O}\rightarrow\text{C} - \text{DisoCH}_3)/2 \quad (8.10d)$$

2-(Diethylamino)-1-phenylethanone

$$\text{D-O}\rightarrow\text{CH}_3\text{-CH}_2\text{-O} = (\Delta_{vap}H°(298\ \text{K})\text{depe} - 4\text{D-N}\rightarrow\text{CH}_3\text{-CH}_2\text{-N} - 4\text{D=HC}\rightarrow\text{CH} - \text{D=HC}\rightarrow\text{C} - 2\text{DO}\rightarrow\text{C})/2 \quad (8.10e)$$

Table 8.10 The energies of the hydrogen bonds and the specific interactions (kJ mol^{-1}) of the liquid 2-(dimethylamino)-1-phenylethanone and its derivatives

Compounds	Formula	The structure	$\Delta_{vap}H°$ (298 K) [6]	D–N → CH_3–N/ DC → C	D=HC → CH/ DHC → C	D–O → CH_3–CH_2–O	D–O → CH_3–CH_2–CH_2–O
2-(Dimethylamino)-1-phenylethanone	$C_{10}H_{15}NO$	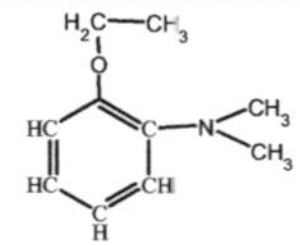	69.7 ± 0.5	4.80 × 4/4.15	5.65 × 4/5.05	7.3 × 2	–
2-(Dimethylamino)-1-phenyl-1-propanone	$C_{11}H_{17}NO$	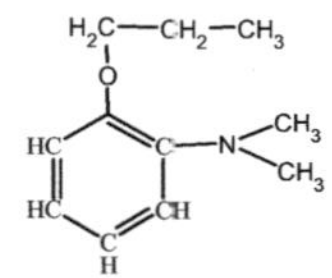	64.8 ± 1.2	4.80 × 4/4.15	5.65 × 4/5.05	–	4.8 ± 1.2 × 2
2-(Dimethylamino)-2-methyl-1-phenylpropanone	$C_{12}H_{19}NO$	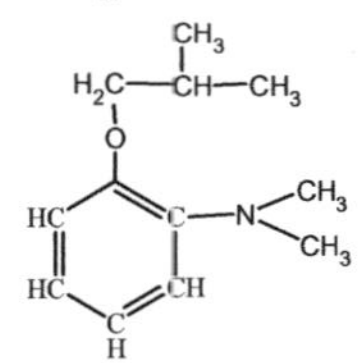	66.7 ± 0.4	4.80 × 4/4.15	5.65 × 4/5.05	D*iso* = 2.50	4.8 ± 1.2 × 2
				DN → CH_3–CH_2–N/ DC → C		–	–
2-(Diethylamino)-1-phenylethanone	$C_{12}H_{19}NO$	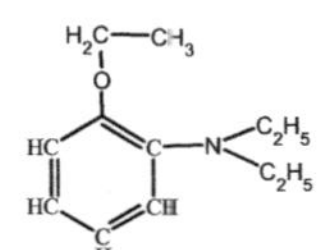	71.6 ± 0.9	5.85 × 4/4.15	5.65 × 4/5.0.5	6.3 × 2	–
2-(Diethylamino)-1-phenyl-1-propanone	$C_{13}H_{21}NO$	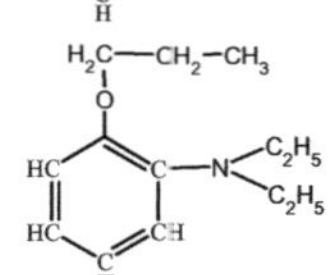	71.6 ± 1.0	5.85 × 4/4.15	5.65 × 4/5.0.5	–	6.3 × 2

2-(Diethylamino)-1-phenyl-1-propanone

D-O→CH_3-CH_2-CH_2-O = ($\Delta_{vap}H°$(298 K)depp - 4D-N→CH_3-CH_2-N - 4D=HC→CH – D=HC→C - 2DO→C) /2

(8.10f)

point to the increased stabilization of the specific interaction, formed by the ethyl ligand 2-(dimethylamino)-1-phenylethanone (7.3 kJ mol^{-1}), rather than with propyl 2-(dimethylamino)-1-phenyl-1-propanone and 2-(dimethylamino)-2-methyl-1-phenylpropanone (4.8 ± 1.2 kJ mol^{-1}). The compounds 2-(diethylamino)-1-phenylethanone and 2-(diethylamino)-1-phenyl-1-propanone form specific interactions with the ethyl and propyl ligand D–O → CH_3–CH_2–O=D–O → CH_3–CH_2–CH_2–O = (6.3 kJ mol^{-1}) < D–O → CH_3–CH_2–O (6.78 kJ mol^{-1}) and the diethyl ether of similar stability and is reduced compared with the liquid diethyl ether.

References

1. Rademacher P, Freckmann BJ (1980) Photoelektronenspektren und konformationsverhalten von hydroxylamin und methylhydroxylaminen. J Electron Spectrosc 19(3):251–259
2. Rademacher P, Freckmann B (1978) Photoelectron spectra and conformation of tetrahedron-1,2-oxasines and isozasines. Tetrahedron Lett 9:841–842
3. Nefedov VI, Vovna VI (1989) Electronic structure of organic and elementorganic compounds. Nauka, Moscow, p 199
4. Karlson TA (1981) Photoelectron and OZE- spectroscopy. Mech. Eng. Russia, Leningrad, p 431
5. Nefedov VI (1984) Infra red – spectroscopy of chemical compounds. Chemistry, Moscow
6. Chickos JS, Acree WE Jr (2002) Enthalpies of vaporization of organic and organometallic compounds, 1880–2002. J Phys Chem Rev Data 31:519–879
7. Gleiter R (1979) Photoelectron spectra and bonding in small ring hydrocarbons. Top Curr Chem 86:197–285
8. Honegger E, Yang ZZ, Hellbronner E et al (1985) Lone-pair ionization energies of diazabicycloalkanes. J Electron Spectrosc 36:297–304
9. Armstrong DR, Perkins PG (1969) Calculation $\sigma - \pi$ and $\pi - \sigma^*$ – transfer at vinyl borium. J Chem Soc 7:1044
10. Kato H, Yamaguche K et al (1965) Electron structure some galogenides and alkyl borium and aluminum. Bull Chem Soc Japan 36:2144
11. Ehrenson S (1968) Theory of electron effect at formation of complex of trimethylamine with trimethylborium. Theor Chim Acta 10:193
12. Chickos JS, Acree WE Jr (2003) Enthalpies of sublimation of organic and organometallic compounds, 1910–2001. J Phys Chem Rev Data 32:537–698
13. Baev AK (1969) To problems of chemical nature of phase transformation. In: General and applied chemistry. Vysheischaia Educating, Minsk, pp 197–206
14. Baev AK (1969) Phase condition and complex formation ability of halogenide metals. In: General and applied chemistry, vol 1. Vysheischaia Educating, Minsk, pp 207–218
15. Vovna VI, Vilesov FI, Vovna VI, Vilesov FI (1974) Photoelectron specters and structure molecular orbital's methylamines. Opt Spectrosc 34(2):436–438

16. Rao CNR, Basu PK, Hefre MS (1979) Systematic organic UV. Photoelectron spectroscopy. Appl Spectrosc Rev 15(1):1–193
17. Palmer MH, Moyes W, Spiers M, Ridyard JNF (1979) The electronic structure of substituted: benzenes: ab initio calculationsand photoelectron spectra for phenol, the methyl- and fluoro-derivatieves, and the dihydroxybenztnes. J Mol Struct 52(2):293–307
18. Debies TP, Rabalais JW (1972/1973) Photoelectron spectra of substituted benzenes: II. Seven valence electron substituents. J Electron Spectrosc 1(4):355–370
19. Palmer MH, Moyes W, Spiers M, Ridyard JNF (1979) The electronic structure of substituted: benztnes: a study of aniline, the toluidines, phenilentdiamines and fluoroaniline by photoelectrone spectroscopy and ab unitio calculations. J Mol Struct 53(2):235–249
20. Frost DC, Lau WN, McDowell CA, Westwood NPC (1982) A study by He I photoelectron spectroscopy of monomericnitrosomethane, the *cis* and *trans* dimmers and formaldoxime. J Phys Chem 86(18):3577–3581
21. Legon AC, Roberts BP, Wallwork FL (1990) Do methyl groups form hydrogen bonds? An answer from the rotational spectrum of ethane – hydrogen cyanide. Chem Phy Lett 191:97
22. Hobza P, Hovlas Z (2000) Blue-shifting hydrogen bodes. Chem Rev 100:4253
23. Yukhnevich GV, Tarakanova EG (1998) Hydrogen bond CH...O in liquid methanol. J Mol Struct 447:25
24. Ivanov EV, Puchovskie Yu P, Abrosimov VK (2004) About formation of weak hydrogen bonds CH•••O in liquid methanol. Abstract. IX international conference. The problem of solvation and complex formation in solutions, Plyos, p 345
25. Baev AK (2012) Specific intermolecular interactions of organic compounds. Springer, Heidelberg, p 434

Chapter 9
Specific Intermolecular Interactions and Energies of Amino Acids and Esters

9.1 Energies of Hydrogen Bonds and Specific Interactions of Amino Acids

9.1.1 Series of Glycine

The thermodynamic properties of important proteins amino acids have been studied a little in recent decades and the enthalpy and entropy values of the vaporization processes are still known for only a few (Table 9.1). These properties, along with the melting temperature carry important information because the process of phase transformation, with its chemical nature [1–3], is interconnected with the number of severe bonds and their energies. Therefore, even the melting temperature reflects the depth of the transformation of the crystal structure and its difference from the structure of the liquid condition [1–3]. On the basis of these ideas we have reasons to discuss the increase of the melting temperature of alanine at 2° in comparison with the same property of glycine amino acetic acid connected with stabilization of the same type of specific interaction. This stabilization is caused by the contribution to the structure of alanine of the methyl group, implementing the functional properties of the isostructural group, participating in the redistribution of the electron density in the molecule. Further replacement of the methyl group by the fragment $CH_3(CH)CH$– in the molecule of L-valine and $CH_3(CH)CH–CH_2$– in the molecule of D-(L)-leucine is accompanied by stabilization of the specific interaction, formed by these fragments with the oxygen atom of the hydroxyl group of the glycine fragment D–O → $CH_3(CH_3)CH$– < D–O → $CH_3(CH_3)CH–CH_2$–. Such a sharp reduction of the melting temperature of isoleucine points to the implementation of theethyl ligand with regard the functional properties of the isostructural group. Reduced melting temperatures of diacids L-ashartic and L-glutamic acids reflect the reduction of the energies of the hydrogen bonds with an increasing number of carboxyl groups and the significant change in the difference in the charges at the oxygen and hydrogen atomsof the hydroxyl groups. The hydrogen bonds introduced in a new type at

A.K. Baev, *Specific Intermolecular Interactions of Nitrogenated and Bioorganic Compounds*, DOI 10.1007/978-3-642-37472-2_9,

Table 9.1 Melting point and sublimation enthalpy of amino acids of the derivative glycine

Compounds	Formula	Structure	Tm. K [4]	$\Delta_{sub}H°(298 K)$ [5]	Tb. K
Glicine, Aminoacetic acid	$C_2H_5NO_2$	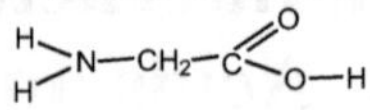	565	138.1 ± 4.6	298
L-(D)-Alanine	$C_3H_7NO_2$	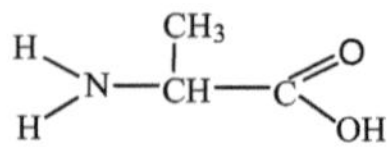	570	144.8 ± 4.2	298
L-Valine	$C_5H_{11}NO_2$	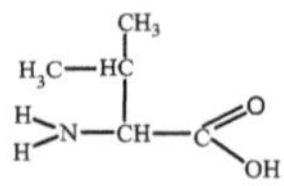	588	162.6 ± 8	455
L-Leucine	$C_6H_{13}NO_2$	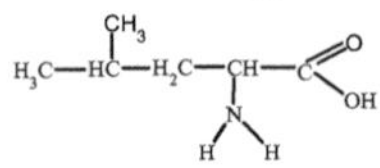	610	150.6 ± 0.8	485
L-Isoleucine	$C_6H_{13}NO_2$	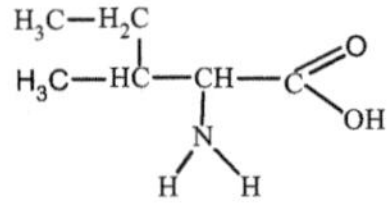	557	–	–
L-Aspartic acid	$C_4H_7NO_4$	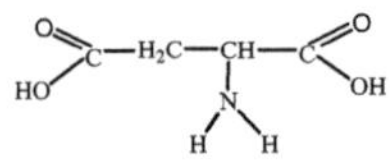	543	U96 ± 4.2	–
L-Glutamic acid	$C_5H_9NO_4$	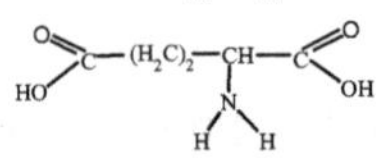	516	U121 ± 3.4	403
Asparagin	$C_4H_8N_2O_3$	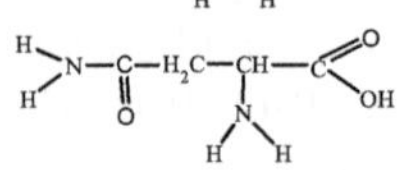	509	–	–
L-Glutamin	$C_5H_{10}NO_3$	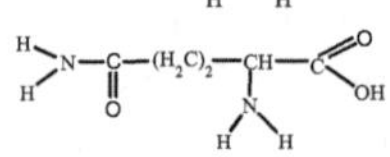	458	–	–
L-Opintin	$C_5H_{12}N_2O_2$	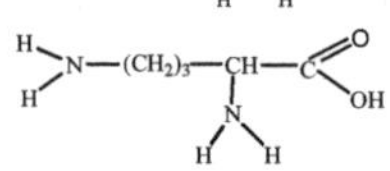	–	–	–
L-Lysine	$C_6H_{14}N_2O_2$	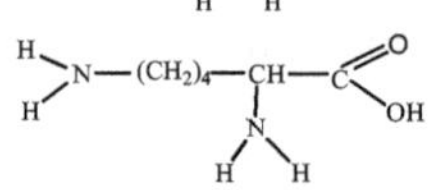	–	*88 ± 8*	442
L-Arginime	$C_6H_{14}N_4O_2$	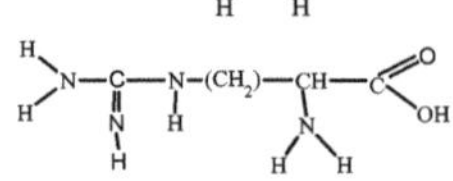	511	U134 ± 8	491
DL-Serine	$C_3H_7NO_3$	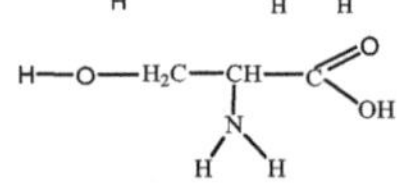	–	U83.7 ± 4	404
DL-Threonine	$C_4H_9NO_3$	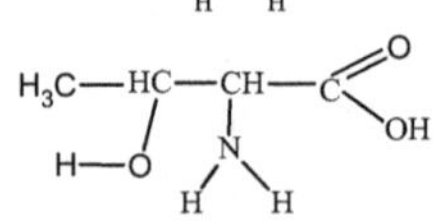	526	U96 ± 8	664
L-Cysteine	$C_3H_7NSO_2$	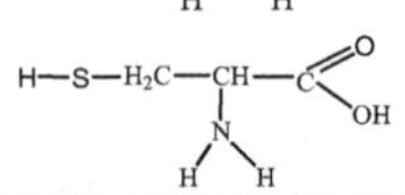	451	U96.2 ± 4.2	660

(continued)

Table 9.1 (continued)

Compounds	Formula	Structure	Tm. K [4]	$\Delta_{sub}H°$(298 K) [5]	Tb. K
L-Methionine	$C_5H_{11}NSO_2$	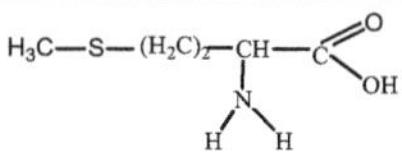	556	125 ± 0.8	478
L-Histidine	$C_6H_{10}N_3O_2$	–	550	141 ± 8	442
L-Tyrosine	$C_9H_{11}NO_3$	–	617	101 ± 3	462

asparagine and glutamine with reduced energy values, formed by the amide group D–O•••H–N > D–N•••H–N, and methylene groups, fringed by the amide groups, are responsible for the reduction of the melting temperature. In turn, the increasing number of methylene groups, fringed by the same amide groups, causes the reduction of the melting temperatures of the mentioned diacids in the sequence asparagine – glutamine. Similar changes in the energies of the specific interactions formed of crystalline and liquid DL-threonine, L-cysteine, L-arginine, and other compounds reflect the melting temperature of important protein amino acids. The insolubility of α-amino acids in hydrocarbons and ethers points to the high stability of the hydrogen bonds formed and the water solubility of most acids in this range indicates the comparability of the energies of the hydrogen bonds with a similar bond with water molecules (10.99 kJ mol^{-1}) [6].

Many amino acids are degraded upon heating with the release of water and formation of cyclic substituted amides. The low volatility and the susceptibility of amino acids to degradation are important difficulties for correct measurement of the vapor pressure of amino acids and thermodynamic properties of the vaporization processes, which explains the high errors of the enthalpy characteristics of this class of compounds (Table 9.1).

The high error of the sublimation enthalpy of amino acetic acid (glycine) and increasing its value for the value of the enthalpy characteristics of aminopropionic acid (α-alanine) points to the unreliability of the experimentally measured value (Table 9.2). The sublimation enthalpies of this series of compounds are well described by the dependence $\Delta_{sub}H°(298\ K) = f(Cn)$ on the number of carbon atoms in the molecule by two crossing lines (Fig. 9.1).

On the first line is aminoformic acid – 4-aminobutanoic acid, which is stabilized by the specific interactions, formed by carbon atom of the terminal group of the alkyl chain by the weakening influence of the intermolecular interaction of the reverse dative bond. Completion of this influence and achievement of a maximum hydrogen bond, realized at the four carbon atoms in the alkyl chain of 4-aminobutanoic acid, points to the fact that all other compounds of this series are described by a similar line, reflecting the increasing contribution of the energy on increasing the number of methylene groups in the alkyl chain of amino acids. The location of the sublimation enthalpy of 5-aminopentanoic acid (dotted line 3) is lower than the line 2; this is caused by the high error of the measured sublimation enthalpy of this compound. The functional dependency $\Delta_{sub}H°(298\ K) = f(Cn)$ of the sublimation enthalpy

Table 9.2 Energies of the hydrogen bonds and specific interactions (kJ mol^{-1}) of crystalline saturated amino acids at 298 K

Compounds	Formula	Structure	$\Delta_{sub}H°$ (298 K) [5]	D–N•••H–N/ ΣDCH_2	D=C=O → C=O/ DO → CH_2–$(CH_2)n$–	D=C=O•••H–N	D–O•••H–O
Aminoformic acid	CH_3NO_2	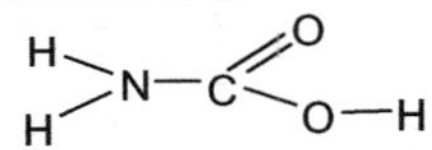	121.3[a]	9.7	5.7 × 2	17.5	27.75
Glicine, aminoacetic acid	$C_2H_5NO_2$	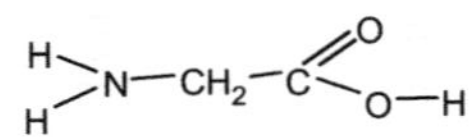	138.1 ± 4.6 127.5[a]	9.7	5.7 × 2 n = 0 3.6 × 2	17.5	27.25
β-Alanine, aminopropionic acid	$C_3H_7NO_2$	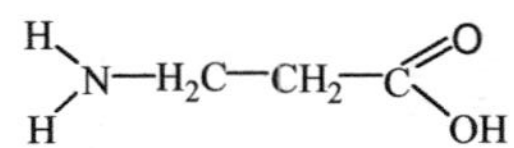	134.2 ± 2	9.7	5.7 × 2 n = 1 4.8 × 2	17.5	29.0 ± 2
4-Aminobutanoic acid	$C_4H_9NO_2$	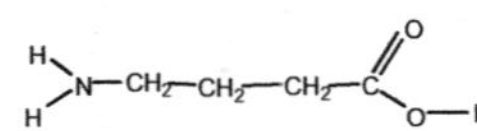	140.2	9.7	5.7 × 2 n = 2 6.4 × 2	17.5	30.2 ± 2
5-Aminopentanoic acid	$C_5H_{11}NO_2$	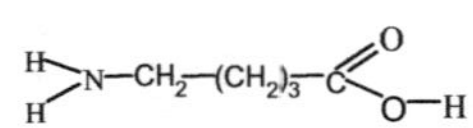	144 ± 3 147.7[a]	9.7/7.7	5.7 × 2 n = 2 6.4 × 2	17.5	30.2 ± 2
6-Aminohexanoic acid	$C_6H_{13}NO_2$	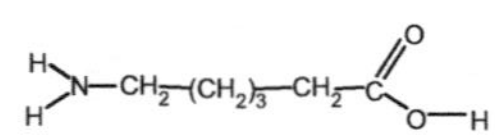	155 ± 3	9.7	5.7 × 2 n = 2 6.4 × 2	17.5	30.2 ± 2
7-Aminoheptanoic acid	$C_7H_{15}NO_2$	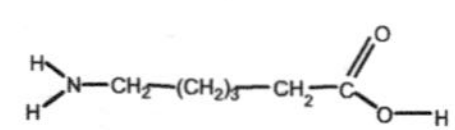	162.7[a]	9.7	5.7 × 2 n = 2 6.4 × 2	17.5	30.2 ± 2
8-Aminooctanoic acid	$C_8H_{17}NO_2$	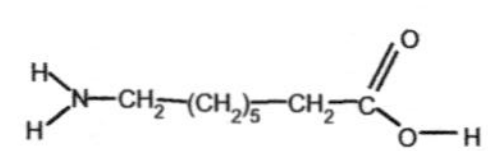	170.4	9.7	5.7 × 2 n = 2 6.4 × 2	17.5	30.2 ± 2
9-Aminononanoic acid	$C_9H_{19}NO_2$	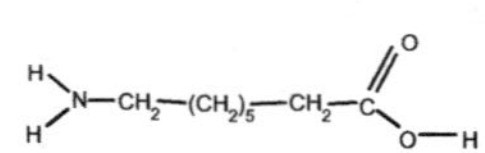	177.7[a]	9.7	5.7 × 2 n = 2 6.4 × 2	17.5	30.2 ± 2
10-Aminodecanoic acid	$C_{10}H_{21}NO_2$	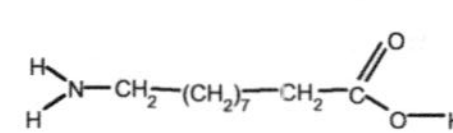	185.2[a]	9.7	5.7 × 2 n = 2 6.4 × 2	17.5	30.2 ± 2

[a]Elucidated by interpolations and extrapolations methods

Fig. 9.1 Dependence of the sublimation enthalpy of amino acids versus the number of the carbon atoms at alkyl chine

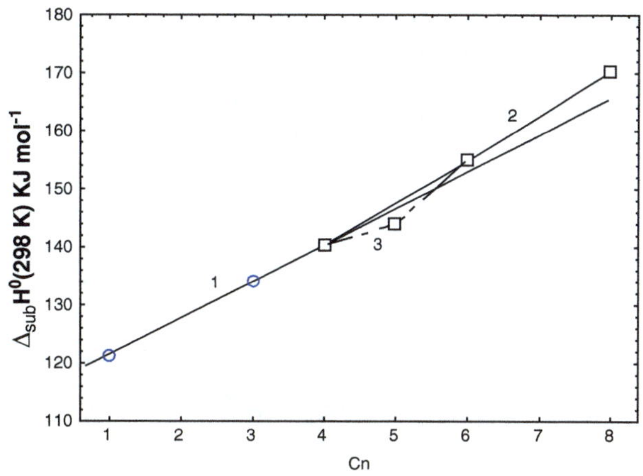

from the number of carbon atoms of the chain allows us to use the method of extrapolation and interpolation for the assessment of the enthalpy characteristics of aminoformic and aminoacetic acids (glycine) and, with the help of the second line of the same dependency to estimate the sublimation enthalpies of 7-aminoheptanoic, 9-aminononanoic, and 10-aminodecanoic acid (Table 9.2).

The molecule of aminoformic acid with three bond vacancies of the hydroxyl group, by the two bond vacancies of carbonyl oxygen atom, one vacancy of the carbon atom of the same group and with the maximum possible four bond vacancies of the amide group due to the lack of atom donors is able to express eight bond vacancies with the four-coordinated nitrogen atom and one vacancy if the oxygen atom =C=O is to participate in the formation of the hydrogen bond with the hydrogen atom of the amide group. The remaining unoccupied second unshared electron pair of the oxygen atom of the carbonyl group is forced to participate in the formation of the cooperative interaction. Therefore the aminoformic acid molecule forms two hydrogen bonds with the acid hydroxyl group D–O•••H–O, two hydrogen bonds of the second type of carboxyl oxygen atom with the hydrogen atom of the group of contacting molecules close in the environment, two specific interactions D–O → C=O and two hydrogen bonds D–N•••H–N, formed by the amide group (Fig. 9.2a).

The planar structure of the aminoformic acid molecule causes the formation of the network structure of crystal with three types of hydrogen bonds and one type of specific interaction, formed by the essentially unshared $2s^2$ electron pair of the carbon atom of the carbonyl group. The energy of this type of interaction is taken as equal to the energy contribution of the methylene group to the sublimation enthalpy minus the energy contribution of the hydrogen atoms replaced (2.0 kJ mol^{-1}). The energies of the hydrogen bonds D=C=O•••H–N and D–N•••H–N we take to be equal to the energies, formed of the same types of interactions of the amide group and oxygen atom of the carbonyl fragment, respectively, 17.5 and 9.7 kJ mol^{-1}. The energy value of the hydrogen bond D–O•••H–O is determined by the equality (9.1),

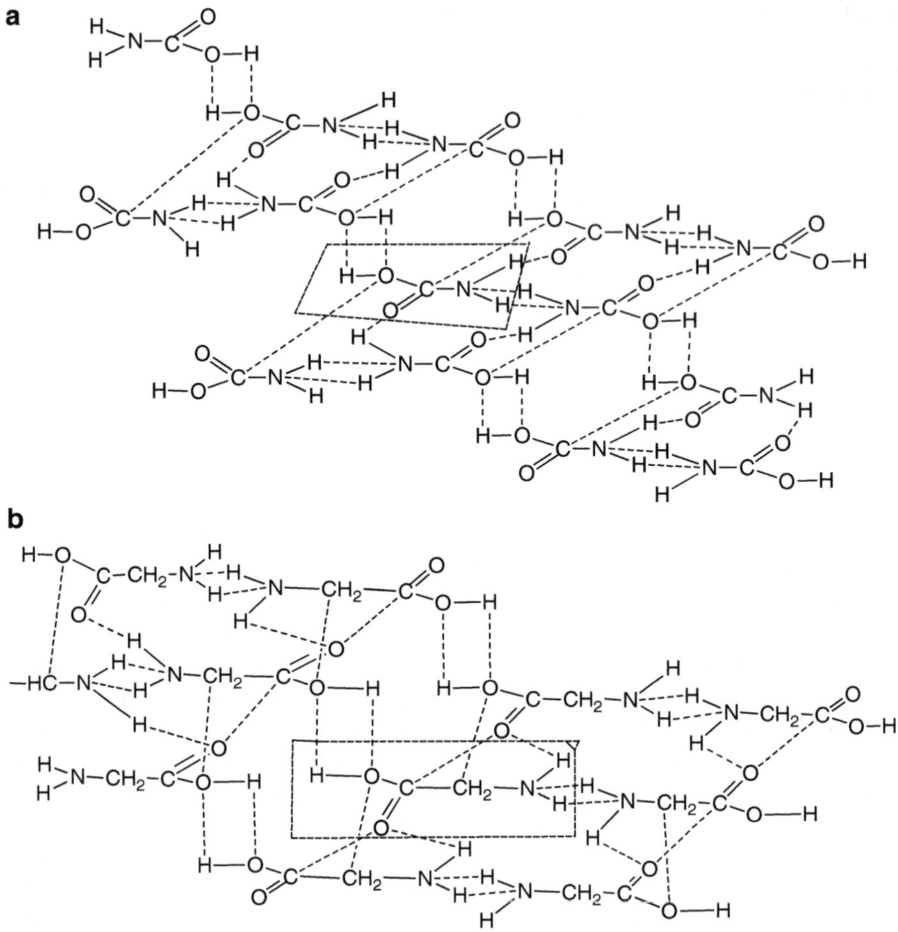

Fig. 9.2 Schematic picture of the crystalline structure of aminoformic acid (a) and amino acetic acid's (glicine) series (b) with the network of the hydrogen bonds and specific interactions

considering the energy contributions of all types of interactions to the enthalpy characteristics

$$D\text{-}O\cdots H\text{-}O = (\Delta_{sub}bH^{0}(298\ K)fa - 2D{=}C{=}{=}O\cdots H\text{-}N - 2D\text{-}N\cdots H\text{-}N - 2D\text{-}O{\rightarrow}C{=}O)/2 \quad (9.1)$$

The results of the calculations (Table 9.2) point to the high stability of this type of hydrogen bond, which is possible with the lack of donors of electrons in the molecule and strong acceptors of electrons by the oxygen atom of hydroxyl and carbonyl groups. The high deficit of electrons in the molecule ensures the increased shifting of the electron density to the oxygen atoms, which results in the increased difference of the charges of the interacting atoms.

The the amino acetic acid (glycine) molecule contains the additional methylene group CH_2, equalizing the number of donors and acceptors of electrons, significantly changing the redistribution of the electron density. In Sect. 8.2 it was shown that in the result of the redistribution of the electron density and its shifting from the nitrogen atom of the amide fragment to the carbon atom $-CH_2$—CH_2- group and further to the oxygen atom of the carboxyl group (Chap. 8) conditions are created where the $-CH_2$—CH_2- group located between two fragments of acceptors of electrons of the nitrogen atom of the amide group and oxygen atom forms a specific interaction D–O → CH_2–CH_2–N = 3.6 kJ mol^{-1} that is approximately two times less stable than that of crystalline diethyl ether (6.59 kJ mol^{-1}). Thus, this destabilizing effect of two fragments of acceptors of electrons to the specific interaction located between them, at the alkyl group, is also expressed by amino acids and participates in the general rule. In the case of crystalline amino alcohol this group forms the same type of specific interactions with the energy of D–O → CH_2–CH_2–N = 4.7 kJ mol^{-1}. The energy of this specific interaction is stabilized in the number

D-O→CH_2-N < D-O→CH_2-CH_2-N < D-O→CH_2-CH_2--CH_2-N

by the weakening of the influence of the reverse dative bond from the nitrogen atom with the increasing number of carbon atoms of the alkyl chain, ending at the amino acids at 4-aminobutanoic acid (Fig. 9.1). The energies estimated of the specific interactions, formed by the alkyl group with the location between two fragments of electron acceptors of the amino alcohols, amino acids, and the analogs of the formed interactions of ethers and amines, are described by the integral values of its stabilization

D-O→CH_2-N (3.6) < D-O→CH_2-CH_2-N (4.7) < D-O→CH_2-CH_2-CH_2-N (6.4 kJ mol^{-1})

Taking into account the identity of the crystal structures of glycine and amino acetic acids of glycine series (Fig. 9.2), formed by three types of hydrogen bonds with its energies of D–O•••H–O, D=C=O•••H–N (17.6), D–N•••H–N (9.7 kJ mol^{-1}) and specific interactions of two types D–O → C=O (5.7 kJ mol^{-1}) and D–O → (CH_2) n–N and using Eq. 9.2

$$\text{D-O}\cdots\text{H-O} = (\Delta_{sub}H^0(298\ \text{K})\text{saa} - 2\text{D=C=O}\cdots\text{H-N} - 2\text{D-N}\cdots\text{H-N} - 2\text{DO}\rightarrow\text{C=O} - 2\ \text{DO}\rightarrow(\text{CH}_2)\text{n-N}) / 2 \quad (9.2)$$

we calculated the energies of the hydrogen bonds DO•••H–O (Table 9.2).

From the data presented in the table it follows that the energies of the hydrogen bonds formed

D-O•••H-O: Aminoformic acid (27.75) < Glicine,Aminoacetic acid (27.25) < *β*-Alanine, Aminopropionic acid (29.0) < 4-Aminobutanoic acid (30.2 кJ mol^{-1})

by the hydroxyl group of saturated amino acetic acids of the glycine series are stabilized with the increasing number of carbon atoms of the alkyl group, located between two acceptor fragments. The stabilizing effect is completed with the termination of the influence of the intermolecular interaction of the reverse dative bond at the third carbon atom of the chain $–CH_2–CH_2—CH_2–$ and further increase of the enthalpy characteristics is due to the contribution of the increasing number of methylene groups. The contribution of the energy of the enthalpy characteristics of each CH_2-group is equal to 7.7 kJ mol^{-1}.

9.1.2 *Amino Acids with Fragment of Molecule Glycine*

In previous chapters we have proved that the reduced enthalpy characteristics of the vaporization processes of amines with the hydrogen atom in the amine group substituted by a methyl one is caused by the reduced energy value of the specific interaction of this group in comparison with the energy value of the hydrogen bond formed by the substituted hydrogen atom D–N•••H–N (9.7) > D–N → CH_3–N (7.7 kJ mol^{-1}). This is consistent with results of the research by RE-spectroscopy, pointing to a slight reduction of the electron density at the nitrogen atom of methylamine [7, 8]. In this connection it is necessary to introduce corrections to the sublimation enthalpy of the derivatives of glycine and amino acids on the basis of the available energies of the hydrogen bonds and specific interactions and to compensate for errors in the calculation of the energies of interactions of a number of compounds.

We draw attention to the fact of the increased value of sublimation enthalpy of sarcosine (*N*-methylglicyne) with the substituted hydrogen atom (Table 9.3) [5], experimentally obtained by the calorimetric method, compared with the value of the enthalpy characteristics of glycine. These compounds with the network structure of the crystal, similar types of hydrogen bonds and specific interactions (Fig. 9.3a), except for the hydrogen bond D–N•••H–N substituted by the interaction D–N → CH_3–N, allow us to clarify the sublimation enthalpy of sarcosine. Considering the contributions of all realized interactions in the crystalline condition (Table 9.3), described by Eq. 9.2, we obtained the correct value of this characteristic. The energy value of the specific interaction of the methylene group, located between the amide and carboxyl fragment, is accepted to be equal to 3.6 kJ mol^{-1}.

The L-(D)-α-alanine molecule differs from glycine by the presence of the isostructural methyl group with its location between two fragments with strong acceptor properties. The value of the energy contribution to the sublimation enthalpy is low and can be estimated to be equal to the energy contribution of the same group, *cis*-2-pentenoic acid amide 3.45 kJ mol^{-1} (Table 6.9). At the same time the value of the energy contribution of the isostructural methyl group of this compound, equal to the difference between the sublimation enthalpy $\Delta_{sub}H°$(298 K) L-(D)-α-A- $\Delta_{sub}H°$ (298 K)gl and the contribution of substituted hydrogen atom in the molecule of glycine, is equal to 7.7 kJ mol^{-1}. The two specific interactions $2DH_3C$ → H–CH_2 (3.85 kJ mol^{-1}) formed by this group have a reduced value compared with the energy value of the specific interaction of crystalline dimethyl ether, 6.59 kJ mol^{-1}.

Table 9.3 Energies of the hydrogen bonds, specific intermolecular interactions, and the corrected sublimation enthalpy (kJ mol^{-1}) of crystalline amino acids of derivative glycine at 298 K

Compounds	Formula	Structure	$\Delta_{sub}H°$ (298 K) [5]	D–N•••H–N/ DN → CH_3–N	D=C=O → C=O/ DO → CH_2–(CH_2)n–	D=C=O•••H–N	D–O•••H–O
Sarcosine (*N*-methylglicyne)	$C_3H_7NO_2$	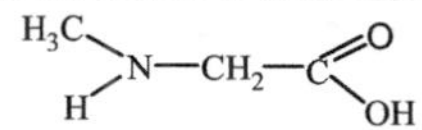	146 ± 1 123.5[a]	7.7	5.7 × 2 n = 0 3.6 × 2	17.5	27.25
L-(D)- α-Alanine, aminopropionic acid	$C_3H_7NO_2$	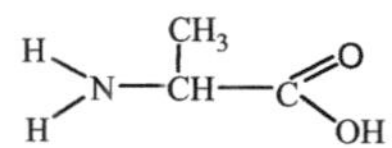	144.8 ± 4.2 135.2[a]	9.7 D*iso* 3.45 × 2	5.7 × 2 n = 0 3.6 × 2	17.5	27.25
L-Glutamic acid	$C_5H_9NO_2$	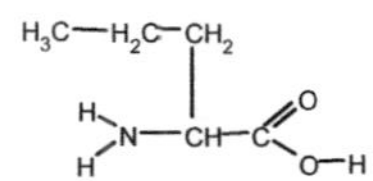	121.8 ± 4.4 133.9[a]	9.7 D*iso* CH_2–CH_2–CH_3 = 6.4	5.7 × 2 n = 0 3.6 × 2	17.5	27.25
2-Aminohexanoic acid	$C_6H_{13}NO_2$	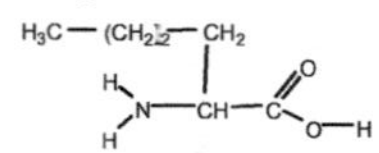	114.5 ± 0.4 135.7[a]	9.7 D*iso* CH_2–$(CH_2)_2$–CH_3 = 8.2	5.7 × 2 n = 0 3.6 × 2	17.5	27.25
N-Acetylglycine	$C_4H_7NO_3$	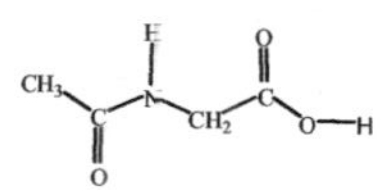	127.0 ± 1.0[b] 134.7[a]	D=C=O → CH_3–C=O = 10.45 × 2	5.7 × 3 n = 0 3.6 × 2	17.5	27.25
Oxalic acid, monoxamide	$C_2H_3NO_3$	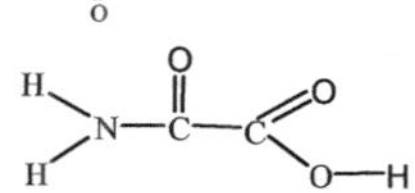	108.9 ± 2.1	9.7	5.7 × 5	12.2	18.8

[a]It is used the energies of hydrogen bonds and specific interactions at 298 K
[b]Experimental value at 389 K

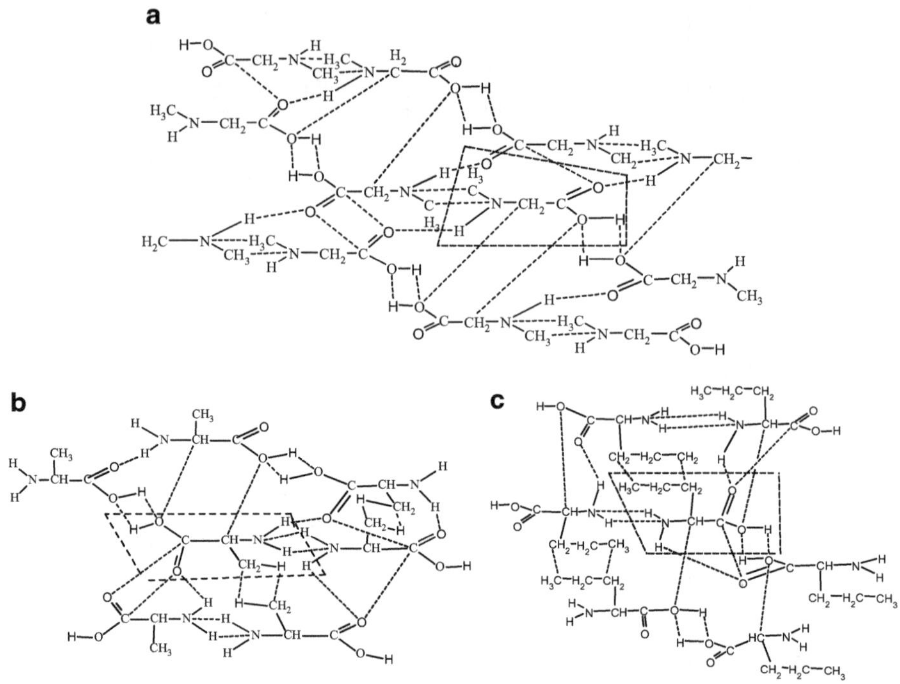

Fig. 9.3 Schematic picture of the crystalline structure of sarcosine(*N*-methylglicyne) (**a**), L-(D)-α-alanine (**b**) and L-glutamic acid (**c**) with the network of the hydrogen bonds and specific interactions

The molecule of L-(D)-α-alanine with ten bond vacancies, stable hydrogen bonds and specific interactions, and two bond vacancies of the isostructural methyl group, formed by the specific interactions of low stability, forms the grid structure of the crystal with energy contributions of all types of interactions, described by Eq. 9.3

$$\Delta_{sub}H^0(298\ K)L\text{-}(d)\text{-}\alpha\text{-}A = (2D\text{-}O\bullet\bullet\bullet H\text{-}O - 2D{=}C{=}O\bullet\bullet\bullet H\text{-}N - 2D\text{-}N\bullet\bullet\bullet H\text{-}N - 2DO{\rightarrow}C{=}O - 2D\text{-}O{\rightarrow}(CH_2)n\text{-}N) - Diso\ CH_3) / 2 \quad (9.3)$$

The results of the calculations of the enthalpy characteristic, energies of the hydrogen bonds, and specific interactions are given in Table 9.3.

The molecules of L-glutamic acid and 2-aminohexanoic acid of one series of compounds with propyl and butyl ligand, respectively, connected with the methylene group in the location between the amide and carboxyl fragment, act as analogs of L-(D)- α-alanine in the manifestation of the properties by the isostructural methyl, propyl, and butyl ligands. The increase of the carbon chain of these ligands comes with isostructural properties, and, consequently, a reduced energy contribution to the sublimation enthalpy compared with similar chain ligands. Therefore, the sublimation enthalpies of these amino acids cannot have more reduced values than the sublimation enthalpy of glycine, (aminoacetic acid 127.5 kJ mol^{-1}). The molecule of L-(D)-α-alanine with butyl isostructural ligand with four carbon atoms cannot have a more reduced value of the enthalpy characteristic than L-glutamic acid, especially on 7.3 kJ mol^{-1} (Table 9.3). The location of the isostructural ligands of these amino acids at the methylene group, located between the fragments with strong acceptor properties of the electron density, impose on them the same properties of the reduced energy contribution to the enthalpy characteristics of the vaporization process. In this connection we have reason to take its contributions as equal to the energy contribution of the ligands with three and four carbon atoms of the chain, 6.4 and 8.2 kJ mol^{-1}, respectively, forming the specific interactions D–CH_2–(CH_2) n–CH_3 → CH_2–(CH_2)n–CH_3 (Fig. 9.3c). Stabilization of the specific interactions by these alkyl groups leads to regularities; nevertheless, the energies remain practically twice less stable compared with the energies of the specific interaction of dipropyl ether, 11.61 kJ mol^{-1}.

The molecule of *N*-acetylglycine with 12 bond vacancies forms 11 intermolecular interactions due to the deficit of electron donor atoms. The grid structure of the crystal formed (Fig. 9.4a) includes two types of hydrogen bonds 2D–O•••H–O, D=C=O•••H–N with the energies of 27.25, 17.5 kJ mol^{-1}, and three types of specific interactions 3D=C=O → C=O (5.7), 2D–N → CH_2–O (3.6) and D=C=O → CH_3–C=O (10.45 kJ mol^{-1}) with indicated energy values (Table 6.7).

The energy contributions of the hydrogen bonds and specific interactions to the sublimation enthalpy of *N*-acetylglycine is described by Eq. 9.4

$$\Delta_{sub}H^0(298\ K)ag = (2D\text{-}O\cdots H\text{-}O - 2D{=}C{=}O\cdots H\text{-}N - 2D{=}C{=}O{\rightarrow}CH_3\text{-}C{=}O - 3D\text{-}O{\rightarrow}C{=}O - 2D\text{-}O{\rightarrow}CH_2\text{-}N)/2 \tag{9.4}$$

The calculated value of the sublimation enthalpy of *N*-acetylglycine 134.7 kJ mol^{-1} is in good compliance with the experimentally obtained value at 389 K considering its temperature dependence.

The oxalic acid (monoxamide) molecule with two carbon atoms, two carbonyl oxygen atoms, and one oxygen atom of the hydroxyl group experiences a great shortage of electron donor atoms. It allows us to expect the reduced difference in the charges of the oxygen atoms and electron donor atoms, which inevitably leads to the more significant reduction in the energies of the most stable hydrogen bonds. The molecule of this compound with 12 bond vacancies could at maximum form 11 specific interactions; the last bond vacancy participates in the formation of a cooperative interaction of low stability. The network structure of crystalline oxalic acid (Fig. 9.4b) is formed by two types of hydrogen bonds of high stability 2D–O•••H–O, 2D=C=O•••H–N, one hydrogen bond D–N•••H–N of average stability and five specific interactions of two types 3D=C=O → C=O and 2D–O → C=O with a small difference in the energy values of reduced stability. In the absence of necessary data we take the energy of the three last types of interactions to be equal to the values of the energies of the same types of interactions of crystalline glycine and estimate the values of the energies of the most stable

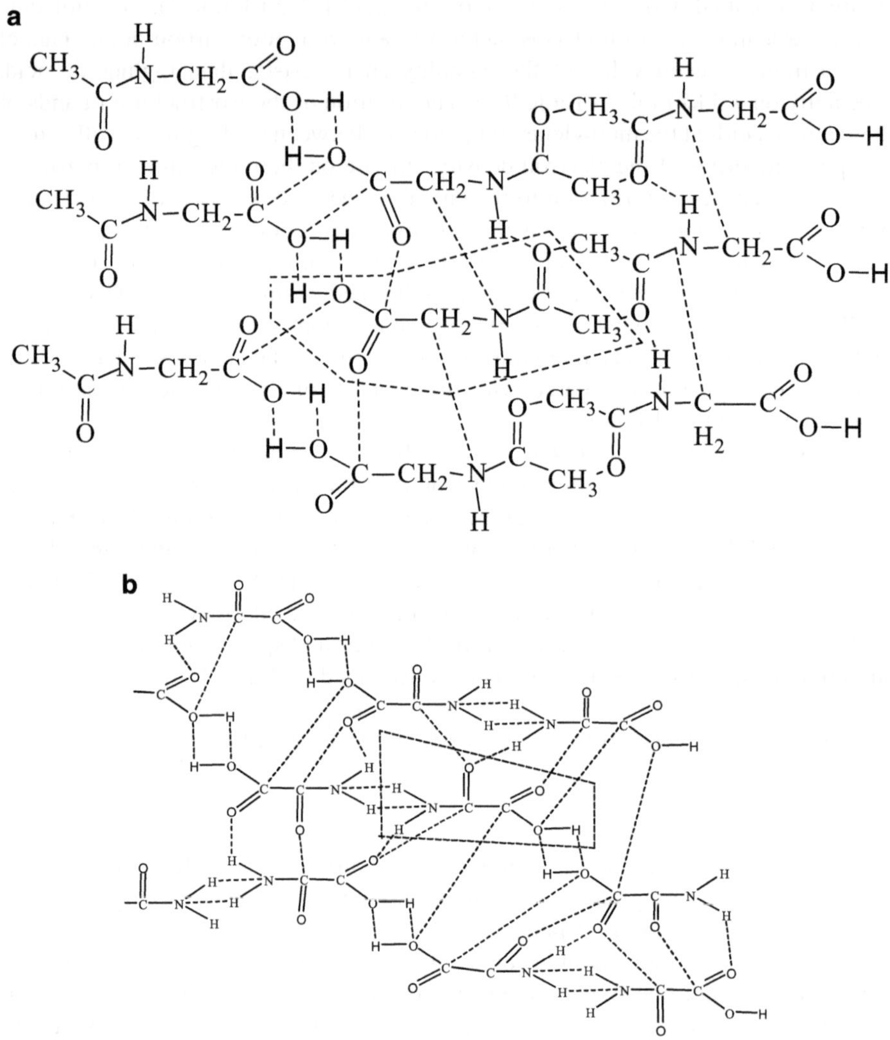

Fig. 9.4 Schematic picture of the crystalline structure of *N*-acetylglycine (**a**) and oxalic acid, monoxamide (**b**) with the network of the hydrogen bonds and specific interactions

hydrogen bonds, using a correlative ratio of the values of the energies with the sublimation enthalpy (Table 9.3).

From the considered material and consistent with experiment and correlations of sublimation enthalpy with the number of carbon atoms of the alkyl chain of compounds of the series of aminoacetic acids and derivatives of glycine with isostructural alkyl groups we obtain the system of consistent energy properties, which acts as reliable material for the energy characteristics of unstudied important protein amino acids.

9.1.3 *Energies of Specific Interactions and Sublimation Enthalpies of Significant Amino Acids of Protein*

At the core of the molecules and structures of the most important amino acids of proteins there is the glycine fragment with substituted hydrogen atom with values of the energies of the hydrogen bonds and specific interactions, formed by bond vacancies contributing energy to the enthalpy characteristic of each derivative, 125.6 kJ mol^{-1}. The compounds L-valine, L-leucine, and L-isoleucine with the alkyl isostructural fragment are analogs of compounds L-(D)-alanine and 2-aminohexanoic acid with corresponding isostructural propyl and butyl ligands. Considering the contribution of the isostructural methyl group (3.45 kJ mol^{-1}) and energy of the CH_3–CH of the ligand D*iso*C_2H_4 = 4.7 kJ mol^{-1}, we obtain the sublimation enthalpy of L-valine equal to 135.7 kJ mol^{-1}. The energy contributions of the isostructural methyl group and propyl ligand (6.4 kJ mol^{-1}) and, on the other side, two DCH_3–CH ligands of the molecules L-leucine and L-isoleucine, respectively, lead to the small difference in the values of the sublimation enthalpy of these compounds (Table 9.4). One should mention that calculations of the sublimation enthalpy of *N*-acetylglycine (Table 9.3) pointed to the equal contribution of energy from the acetate group CH_3–C=O of acetamide to the enthalpy characteristic of this compound. Certainly, that 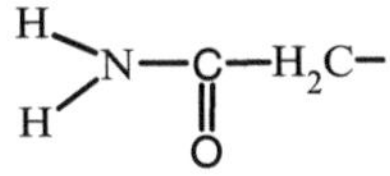 fragment, fringed by the amide groups to asparagin, contributes full energy to the enthalpy characteristic of the hydrogen bonds formed and specific interactions by its bond vacancies. As given in Table 9.4, the schematic picture of the network structure of crystal illustrates the formation of two types of hydrogen bonds 2D–N•••H–N, 2D=C=O•••H–N, specific interactions 2D=C=O → C=O with the energies of 17.5, 9.7, and 5.7 kJ mol^{-1}, respectively. The second type of specific interaction is formed by the methylene group, located between amide fragments, which points to the reduced stability of the specific 2D–N → CH_2–O (3.6 kJ mol^{-1}) interaction. A similar network structure of crystals with the same types and energies of the hydrogen bonds and specific interaction DC=O → C=O is formed by the fragment

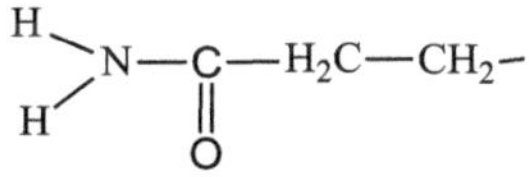

with two methylene groups of L-glutamin, forming the interaction 2D–N → CH_2–CH_2–O with an energy of 4.7 kJ mol^{-1}.

The calculated total contribution of the energies of the interactions formed by this fragment is equal to 75.2 kJ mol^{-1}. The obtained sublimation enthalpies of L-asparagin and L-glutamin are given in Table 9.4.

The L-opnitine and L-lysine molecules contain an amide fragment with three and four methylene groups instead of the substituted hydrogen atom in the methylene

Table 9.4 Energies of the specific interactions and sublimation enthalpy (kJ mol^{-1}) of the most important protein amino acids

Compounds	Structure	DC=O → C=O	D–O → CH_2–$(CH_2)n$–	D–N•••H–N	D=C=O•••H–N	D–O•••H–O	D*iso*CH_3–$(CH_2)_n$	$\Delta_{sub}H°$ (298 K)
Glycine, aminoacetic acid	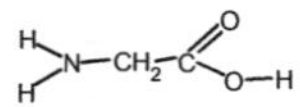	5.7 × 2	n = 0 3.6 × 2	9.7	17.5	27.25	–	127.5[a]
L-Valine	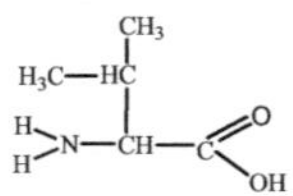	5.7 × 2	n = 0 3.6 × 2	9.7	17.5	27.25	D*iso*CH_3 = 3.45 D*iso*C_2H_4 = 4.7	135.7
L-Leucine	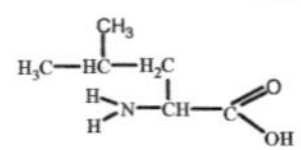	5.7 × 2	n = 0 3.6 × 2	9.7	17.5	27.25	D*iso*CH_3 = 3.45 D*iso*C_3H_6 =6.4	137.8
L-Isoleucine	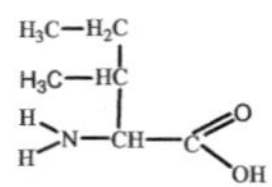	5.7 × 2	n = 0 3.6 × 2	9.7	17.5	27.25	D*iso*C_2H_4 = 4.7 × 2	136.4
L-Asparagin	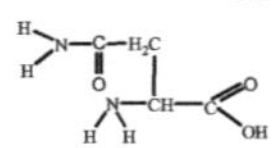	5.7 × 4	n = 0 3.6 × 2	9.7 × 2	17.5 × 2	27.25	DN → CH_2–N = 3.6 × 2	200.5
L-Glutamin	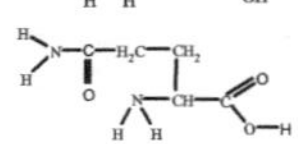	5.7 × 4	n = 0 3.6 × 2	9.7 × 2	17.5 × 2	27.25	DN → CH_2–CH_2– N = 4.7 × 2	201.7
L-Opnitin	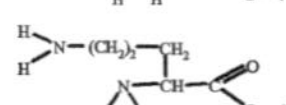	5.7 × 2	n = 0 3.6 × 2	9.7 × 6	17.5	27.25	DN → CH_2–$(CH_2)_2$– N = 6.4 × 2	178.1
L-Lysine	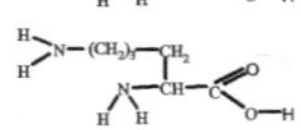	5.7 × 2	n = 0 3.6 × 2	9.7 × 6	17.5	27.25	DN → CH_2–$(CH_2)_3$– N = 8.2 × 2	182.1
L-Arginine	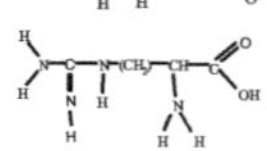	5.7 × 2 + 5.0 × 2	n = 0 3.6 × 2	9.7 × 2 + 9.7 × 8	17.5	27.25	DN → CH_2–CH_2– N = 4.7 × 2	223.5
L-Aspartic acid	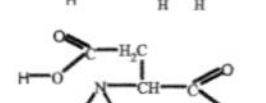	5.7 × 4	n = 0 3.6 × 3	9.7 × 2	17.5 × 3	27.25 × 4	2DO → C=O = 5.7 × 2	214.2
L-Glutamic acid	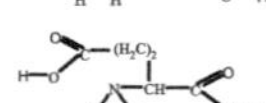	5.7 × 4	n = 1 4.7 × 3	9.7 × 2	17.5 × 3	27.25 × 4	2DO → C=O = 5.7 × 2	220.0

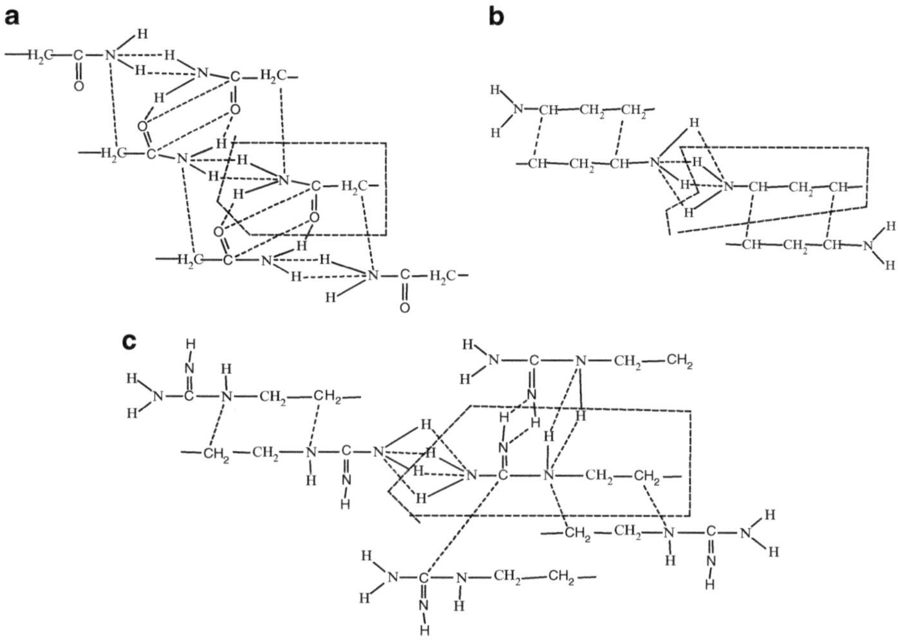

Fig. 9.5 Schematic picture of the crystalline structure of fragment of N-Acetylglycine (a) and Oxalic acid, Monoxamide (b) and L-Arginime (c) with the network of hydrogen bonds and specific interactions

group and forms four hydrogen bonds 4D–N•••H–N (Fig. 9.5b) and two specific interactions 2D–CH_2–CH_2–CH_2 → CH_2–CH_2–CH_2 and 2D–CH_2–$(CH_2)_2$–CH_2 → CH_2–CH_2–CH_2 with an energy of 6.4 and 8.2 kJ mol^{-1}, respectively. The nature of the stable intermolecular interactions formed adequately complements the hydrogen bonds and specific interactions, formed by the glycine molecule, the contributions of which allow us to obtain the correct value of the sublimation enthalpy of these compounds (Table 9.4).

The L-arginime molecule could be presented as a glycine molecule with the hydrogen atom substituted by the fragment with amine and two imide groups (Fig. 9.5c) from three- and pentacoordinated nitrogen atoms in the latter, forming the network structure of a crystal with eight hydrogen bonds D–N•••H–N and two types of specific interactions 2DN=C → C=N, 2D–N → CH_2–CH_2–N with the energies of 9.7, 5.0, and 4.7 kJ mol^{-1}. Considering the energy contribution of all hydrogen bonds and specific interactions, forming crystalline L-arginime, we obtained the value of sublimation enthalpy of this compound to be equal to 223.5 kJ mol^{-1} (Table 9.4).

The L-aspartic acid molecule with two carboxyl groups and an amine group located practically in the central part of the molecule

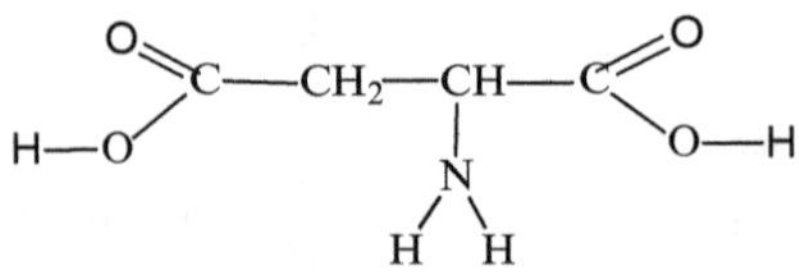

points to the equal shifting of the electron density from the methylene groups to the strong acceptors of electrons, the carboxyl oxygen atoms. Located in the central part of the molecule, the amine group with less acceptor ability shifts the electron density from the nitrogen atom as well, including that partly acquired from the hydrogen atom, reducing the charge possibly up to a positive value. Therefore the charges of the carbon atoms of the methylene groups are determined and regulated by the nitrogen atom of the amine group. The L-aspartic acid molecule with 16 bond vacancies forms three types of hydrogen bonds with four bonds 4D–O•••H–O, two 2D–N•••H–N, and three bonds 3D=C=O•••H–N with the energies of 27.25, 17.5, and 9.7 kJ mol^{-1} respectively. Four types of specific interactions are formed by the essentially unshared $2s^2$ electron pair of the carbon atom 2D=C=O → C=O (5.7), 2D–O → C=O (5.7) and 3D–O → CH_2 (3.6 kJ mol^{-1}) with corresponding energy values (Fig. 9.6).

The existing deficit in the electron density in the molecule with two carbonyl oxygen atoms and two oxygen atoms of acid –O–H groups is accompanied by a reduction of the energy of the formed hydrogen bond 4D–O•••H–O. In this connection the energy value obtained for this type of hydrogen bond has an increased error.

The L-glutamic acid molecule differs from L-aspartic acid by the increase at one methylene group of the fragment

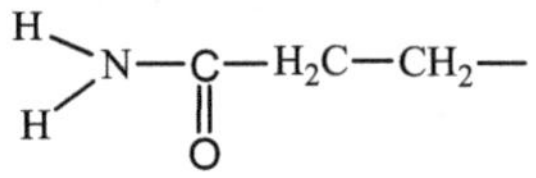

and thus its 16 bond vacancies form the network structure of the crystal by similar types and by the same number of hydrogen bonds and specific interactions with similar energy values except the interaction $3DCH_2$–CH_2– with an energy of 4.7 kJ mol^{-1}. The results of the calculations of the sublimation enthalpies of the considered compounds are given in Table 9.4.

9.1.4 Phenylglycines

It was shown above that the fragment of the molecule glycine (aminoacetic acid) with substituted hydrogen atom of the methylene group contributes identically to the enthalpy characteristic of the vaporization process of derivatives of this acid with isostructural alkyl ligands (Table 9.4).

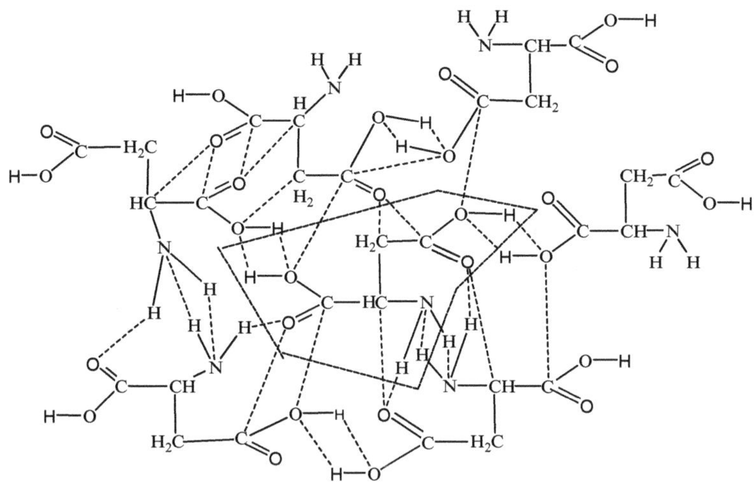

Fig. 9.6 Schematic picture of the crystalline structure of L-aspartic acid with the network of the hydrogen bonds and specific interactions

The presence of a similar fragment in the molecules with the rigid benzene cycle and substituted hydrogen atom of the amine group in *N*-phenylglycine and of the methylene group D-phenylglycine leads to a significant change in the sublimation enthalpy of these compounds (Table 9.5), reflecting the inadequacy of its crystal structures.

The molecule of *N*-phenylglycine with 14 bond vacancies forms the network structure of the crystal of this compound (Fig. 9.7a), including two most stable hydrogen bonds 2D–O•••H–O, two hydrogen bonds 2D–N–•••H–N with an energy of 9.7 kJ mol^{-1}, two specific interactions 2D=HC → CH–, formed by the carbon atom of the CH group of the benzene ring with an energy of 7.40 kJ mol^{-1}, one D=O=C → C=O with an energy of 5.7 kJ mol^{-1}. Located between acceptor electrons of the nitrogen and oxygen atoms of the carboxyl group the methylene group forms specific interaction 2D–N → CH_2–N with the reduced energy of 3.6 kJ mol^{-1}, lying within the error with the energy value of the same type of interaction D–O → CH_2–O. The energies of the two remaining types of specific interactions, formed by carbon atoms of the CH groups of the benzene ring with the oxygen atoms D–O → CH= and DC=O → CH= of the alcohol and carbonyl group, are significantly dependent on the fragment of this group and the location of the nitrogen atom in the cycle.

The energies of these types of specific interactions in crystalline hydroxypyridines and 1-methylcytosine have the values of 7.80 and 7.85 kJ mol^{-1}, respectively, which reflect the influence of the oxygen atom on the stabilization of interaction D=HC → CH (7.40 kJ mol^{-1}) at the replacement of the CH group by the oxygen

Table 9.5 Energies of the hydrogen bonds and specific interactions (kJ mol^{-1}) of crystalline phenylglyciness

Compounds	Formula	Structure	$\Delta_{sub}H°$ (298 K) [5]	DO=C → C=O/ D–N → CH_2–N	D=HC → CH–/ DC=O → CH	D–N•••H–N/ DO → CH	D–O•••H–O
N-Phenylglycine	$C_8H_9NO_2$	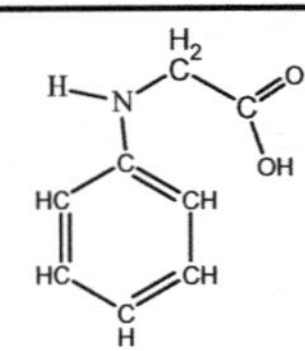	128.0 ± 2.0	5.7 × 2/3.8 × 2	7.43/7.8	9.7/7.8	20.8
				DO=C → C=O/ DO=C → CH–C=	D=HC → CH–/ D=C=O•••H–N	–	–
D-Phenylglycine	$C_8H_9NO_2$	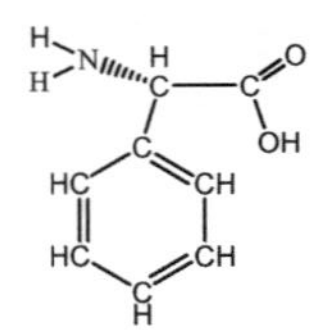	165.0 ± 1.0	5.7 × 2/3.3 × 2	7.43/17.5	9.7/7.8	26.7 D=C → C=(3.3)

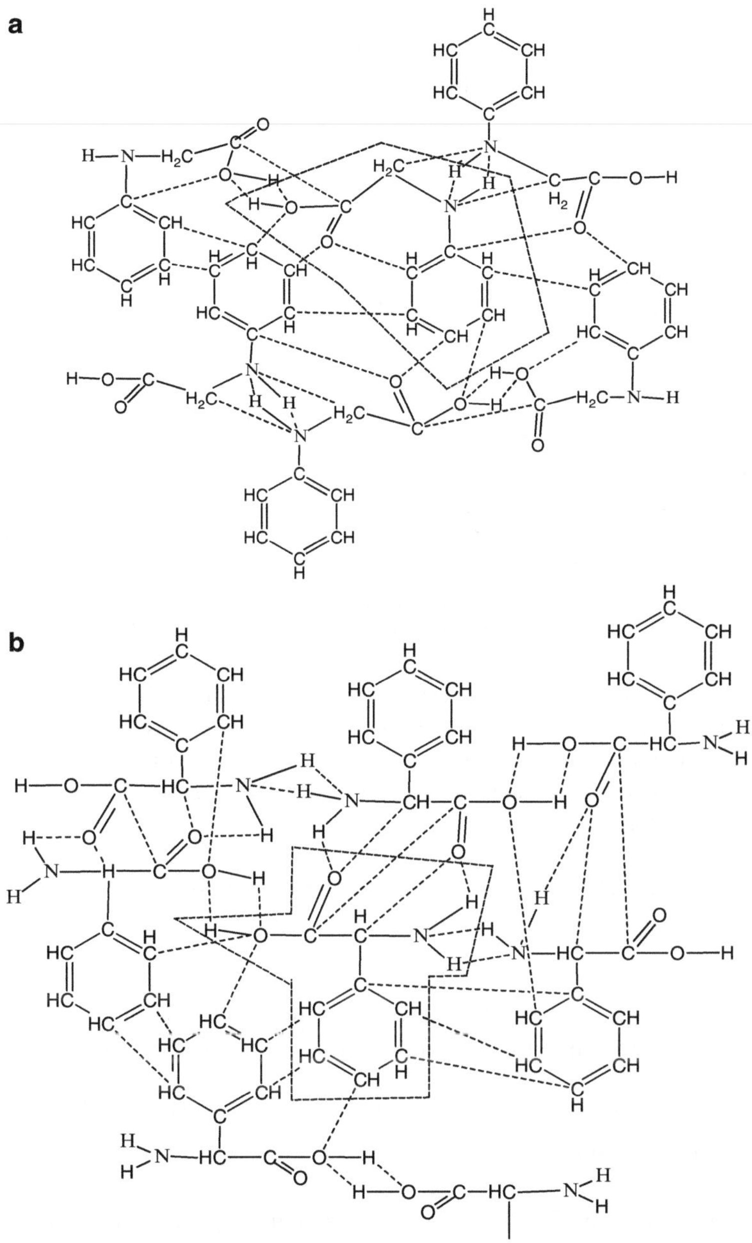

Fig. 9.7 Schematic picture of the crystalline structure of *N*-phenylglycine (**a**) and D-phenylglycine (**b**) with the network of the hydrogen bonds and specific interactions

atom or C=O group. Considering the values of these energies of two types of specific interactions, we calculated the energy of the hydrogen bond of *N*-phenylglycine with the help of Eq. 9.5

$$2D\text{-}O\cdots H\text{-}O = (\Delta_{sub}H^0(298\ K)nfg - 2D\text{-}O{\rightarrow}CH - 2D\text{-}N\cdots H\text{-}N - 2D\text{-}N{\rightarrow}CH_2\text{-}N - 2D{=}HC{\rightarrow}CH - 3DC{=}O{\rightarrow}CH\text{-} - DC{=}O{\rightarrow}C{=}O)/2 \quad (9.5)$$

The energy value of the hydrogen bond D–O•••H–O (Table 9.5) obtained points to its reduced stability compared with that realized in crystalline glycine (aminoacetic acid).

The molecule of D-phenylglycine with 16 bond vacancies forms the network structure of the crystal (Fig. 9.7b) by three types of hydrogen bonds 2D–O•••H–O, 2D=C=O•••H–N (17.5), 2D–N•••H–N (9.7 kJ mol^{-1}), two stable types of specific interactions 2D–O → CH and 4D=HC → CH, formed by the oxygen atom of the hydroxyl group and the carbon atom of the benzene ring, with the energy values of 7.8 and 7.4 kJ mol^{-1}, respectively. In addition to the interactions mentioned in the crystal of this compound there are three types of specific interactions of reduced stability of the carbon atom of the benzene ring with the substituted hydrogen atom D=C → C= (3.3), carbonyl carbon atoms DO=C → C=O (5.7) and two 2DO=C → CH–C= interactions of the methylene group with the substituted hydrogen atom (3.3 kJ mol^{-1}). The energy of the hydrogen bond 2D–O•••H–O of crystalline D-phenylglycine is determined by the difference between the sublimation enthalpy and total energy values of two types of hydrogen bonds and five types of specific interactions, formed by the essentially unshared 2s^2 electron pairs of the carbon atom,

$$2D\text{-}O\cdots H\text{-}O = (\Delta_{sub}H^0(298\ K)dfg - 2D{=}C{=}O\cdots H\text{-}N - 2D\text{-}O{\rightarrow}CH - 2D\text{-}N\cdots H\text{-}N - 4D{=}HC{\rightarrow}CH - 3DC{=}O{\rightarrow}CH\text{-}C{=} - DO{=}C{\rightarrow}C{=}O - D{=}C{\rightarrow}C{=})/2 \quad (9.5a)$$

The energy values of the hydrogen bond of *N*-phenylglycine and D-phenylglycine obtained (Table 9.5) point to its natural location in a number of compounds

$$N\text{-Phenylglicine } (20.8) < \text{D-Phenylglicine } (26.7) \approx \text{Glicine, Aminoacetic acid } \left(27.25\ \text{kJ mol}^{-1}\right)$$

caused, on the one side, by the difference of glycine fragments in the molecules of *N*-phenylglycine and D-phenylglycine and, on the other side, by the similarity of the fragments in the molecules of D-phenylglycine and glycine, which, practically, leads to adequate energy contribution to the sublimation enthalpy of D-phenylglycine. It means that the benzene ring with the substituted hydrogen atom does not make any significant changes in the shifting of the electron density and charges of the atoms of the glycine fragment.

9.2 Energies of Hydrogen Bonds and Specific Interactions of Cyanic Acids

Acids with the cyanide group with known vaporization enthalpies are limited to three compounds (Table 9.6). The cyanic acid molecule with five bond vacancies is the most simple from the given series of compounds. On its bond vacancy as the free electron pair of the oxygen atom, it remains unoccupied at the formation of the structure of the liquid condition of this compound. The molecular linear structure leads to the formation of the chain structure of liquid conditions with the formation of two hydrogen bonds D–O•••H–O and two specific interactions D≡N → C≡. The chains are connected with each other by unstable cooperative interactions (Fig. 9.8).

The cyanic acid molecule acts as an analog of cyanogens with a similar structure, N≡C–C≡N, and an energy value of the specific interaction D≡N → C≡ equal to 5.9 kJ mol^{-1}. The four interactions formed of two types in the liquid condition allow us to obtain the energy of the hydrogen bond from the difference between the vaporization enthalpy and the total value of the energies of two specific interactions

$$\text{D-O}\bullet\bullet\bullet\text{H-O} = (\Delta_{vap}H^0(298\ \text{K}) - 2\text{D}{\equiv}\text{N}{\rightarrow}\text{C}{\equiv})\,/\,2 \qquad (9.6)$$

The presence in the cyanoformic acid molecule of the methylene group with the location between electron acceptors by the nitrogen atom of group N≡C– and oxygen atoms of the carboxyl group points to the fact that the methylene group CH_2 forms the specific interaction D=C=O → CH_2 with an energy of 3.6 kJ mol^{-1}. The molecule of this compound with nine bond vacancies and prevailing number of electron acceptors makes only eight of it form two stable hydrogen bonds of the hydroxyl group and six specific interactions including two of them are formed by the cyanide group 2D≡N → C≡, two interactions 2DC=O → CH_2 by the carbonyl oxygen atom (3.6 kJ mol^{-1}), and two 2DO → C=O with the oxygen atom of the hydroxyl group of low stability (3.3 kJ mol^{-1}) (Fig. 9.8b). Considering the energy value of the specific interaction D≡N → C≡, realized in liquid cyanic acid, as the most correct one we obtained the energy of the hydrogen bond of liquid cyanoformic acid with the help of Eq. 9.6a

$$\text{D-O}\bullet\bullet\bullet\text{H-O} = (\Delta_{vap}H^0(298\ \text{K}) - 2\text{D}{\equiv}\text{N}{\rightarrow}\text{C}{\equiv} - 2\text{D}{=}\text{C}{=}\text{O}{\rightarrow}\text{CH}_2 - 2\text{D-O}{\rightarrow}\text{C}{=}\text{O})\,/\,2 \qquad (9.6a)$$

is equal 6.9 kJ mol^{-1}.

The 2-cyanoacrilic acid molecule with the cyanide group at the second carbon atom points to the fact that the fragment CH_3–CH= acts as the functional group and implements the role of the isostructural group with double bond. This group forms with the carbonyl oxygen atom with two specific interactions D=C=O → CH_3 (3.6) and D=C=O → CH= (1.1) with the total value of the energy contribution to the vaporization enthalpy equal to the contribution of two methylene groups, located between atoms that are electron acceptors. Such an interaction leads to the

Table 9.6 Energies of the hydrogen bonds and specific interactions (kJ mol^{-1}) of liquid cyanic acid

Compounds	Formula	Structure	$\Delta_{vap}H°(298\ K)$ [5]	T K	D≡N → C≡	D–O•••H–O
DC=O → CH_2 = 3.6 × 2, DC=O → C=O = 3.3 × 2; 2DC=O → CH_3 = 3.6 × 2 + 2DC=O → CH = 1.1 × 2						
Cyanic acid	CHNO	N≡C—O—H	30.7	253	5.9	9.3
Cyanoformic acid	$C_3H_3NO_2$	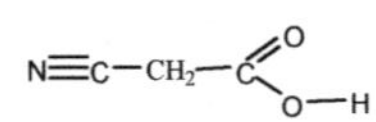	39.3	288	5.9	6.9
2-Cyanoacrilic acid	$C_5H_5NO_2$	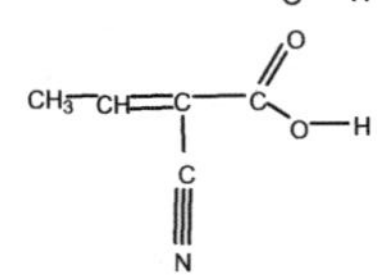	57.8 2DO → C–C = 3.0 × 2	270	8.7	9.4

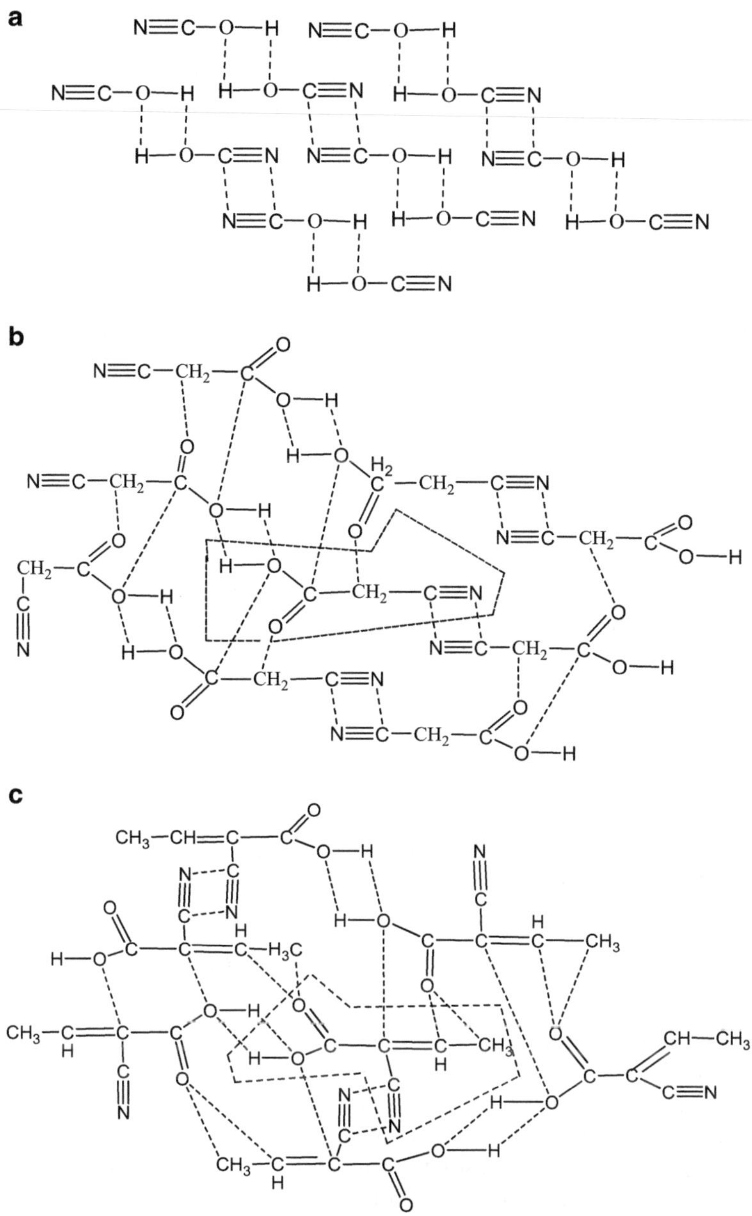

Fig. 9.8 Schematic picture of the liquid structure of cyanic acid (**a**), cyanoformic acid (**b**), and 2-cyanoacrilic acid (**c**) with the network of the hydrogen bonds and specific interactions

2-cyanoacrilic acid molecule with ten bond vacancies to form the structure of liquid condition with the network of five types of specific interactions (Fig. 9.8c). The location of the cyanide group at the second carbon atom of the 2-cyanoacrilic acid molecule not containing a hydrogen atom significantly reduces its ability to shift the electron density to the nitrogen atom and leads to the similarity with the hydrogen atom of the hydrogencyanide molecule. It allows us to take the energy of the specific interaction D≡N → C≡ 2-cyanoacrilic acid to be equal to the energy value of the same type of interaction, realized in liquid hydrogencyanide (8.9 kJ mol^{-1}). The energy of the formed specific interaction D–O → C–C by the oxygen atom of the hydroxyl group 2-cyanoacrilic acid with the carbon atom, connected with cyanide group (Fig. 9.8c), has a close energy value (3.0 kJ mol^{-1}) with specific interaction of the D–O → C=O cyanoformic acid. The energy of the hydrogen bond of 2-cyanoacrilic acid is calculated with the help of Eq. 9.6b

$$\text{D-O}\cdots\text{H-O} = (\Delta_{vap}H^0(298\ \text{K}) - 2\text{D}{\equiv}\text{N}{\rightarrow}\text{C}{\equiv} - 2\text{D}{=}\text{C}{=}\text{O}{\rightarrow}\text{CH}_3 - 2\text{D}{=}\text{C}{=}\text{O}{\rightarrow}\text{CH} - 2\text{DO}{\rightarrow}\text{C})/2 \quad (9.6b)$$

The results of the calculations of the energies of the hydrogen bond point to stabilization

$$\text{Cyanoformic acid } (6.3) < \text{2-Cyanoacrilic acid } (9.4)\ \text{Cyanic acid } \left(9.3\ \text{kJ mol}^{-1}\right)$$

at the replacement of one methylene group in the cyanoformic acid molecule, located between the electron acceptor nitrogen and oxygen atoms of the carboxyl group, by the fragment CH_3–CH= with two carbon atoms and a double bond with the functional properties of the isostructural group in the 2-cyanoacrilic acid molecule, participating in the redistribution of the electron density.

9.3 Energies of Hydrogen Bonds and Specific Interactions of Aminobenzoic Acids and Heterocyclic Acids

9.3.1 Benzoic Acids and Aminobenzoic Acids

To obtain a more correct idea of the energies of the hydrogen bonds and the specificity of interaction on a thermodynamic analysis of aminobenzoic acids we should draw attention to benzoic acid and methylbenzoic acids. In this connection it is necessary to notice that in the case of mono-substituted benzene the splitting value at e_{1g} ($\pi_{2,\,3}$)-orbital and absolute value of the vertical ionizing potential Iv (π) [9] lead to the possibility of the spectroscopic inductive and mesomeric effects of the substituents. Nevertheless, aiming at simplifying the analysis of the photo electron specters led to the necessity to consider inductive shifts at the b_1 (e_{1g})- and a_2 (e_{1g})-levels as equal, with others being neglected [9]. Such assumptions could affect certain weak interactions, and for objective reasons could not be taken to include lack of untraditional ideas on the specificity interactions with the participation of the pentacoordinated carbon atom, hydrogen atoms of the CH_3-, CH_2-, and CH- groups of aromatic compounds. In particular, at present there is lack of information of experimental groundwork and denial of participation in the formation of the hydrogen bond of low stability or interaction of the carbon atom of the CH group of the benzene ring with carbonyl carbon atom of liquid and crystalline benzoic acid.

In one of the first works to measure the balance liquid – vapor of bensoic acid [10] the dimerization in vapors of this compound was pointed out. Further study of this balance proved the partial dimerization in the vapors of bensoic acid [11–16]. Research results of the sublimation process with differing grades of separation efficiency of bensoic acid, by different methods and with errors of the experiment measured by thermal conductivity manometer [11], torsion effusion [12], effusion [17–20], transpiration [21], and calorimetry [10, 11, 21, 22], led to a quite wide range of enthalpy characteristics, (88–92). Five of this compound, widened to 94 kJ mol^{-1} by the research of the author [23]. Study of the process of sublimation of benzene acid of sublimation-calorimetric systems incorporating the Calve microcalorimeter [24] allowed us to obtain the sublimation enthalpies in the interval of 88.9 $\pm$ 0.3–89.3 $\pm$ 0.3 kJ mol^{-1}, being in good compliance with the ones established in the latter work [25]. The results of the research [24] established the presence in the vapors of monomer and dimer molecules forms of benzoic acid in the considered temperature interval and determined the dissociation enthalpy of the dimeric molecules (68 $\pm$ 5 kJ mol^{-1}) with the energy of breaking of one hydrogen bond, 34 $\pm$ 5 kJ mol^{-1}. The incomparably low value of the enthalpy characteristic of dissociation of the dimer molecule at high error, exceeding approximately one-and-a half times the error in the established sublimation enthalpy of bensoic acid, is the result of the low partial component of the dimeric form in vapors. The value of the dissociation enthalpy of the dimer, significantly lower than the experimentally obtained sublimation enthalpy of the total process of sublimation of monomer and dimer forms, reflects the low level of polymerization in the vapors. Low partial vapor pressure of the latter points to the fact that its partial pressure is reduced with the increase in temperature, the general pressure, and with increasing monomer pressure [26]. Thus, a semilog dependence, reflecting the partial vapor pressure of the dimeric forms, should have the form of a bent curve with the maximum value in the field of low temperatures. It follows that calculation of the sublimation enthalpy of the dimer on the base of semilog dependence is not possible. That is exactly the reason for the difficulty in defining the thermodynamic characteristics of the sublimation process [19] of monomer and dimer forms of bensoic acid and the reason for its absence in the literature. The development of dissociation of the dimer molecules with the increase in temperature causes the reduction of the slope on the semilog curve, reflecting the total sublimation process, which leads to the reduced value of the enthalpy characteristic [26]. Therefore, the presence in the vapors of the dimer forms of the bensoic acid molecule at low temperatures with low partial pressure does not lead to an increase of enthalpy error based on the presence in the vapors of only monomer molecules of this compound [26]. Thus, all calculations in the thermodynamic analysis of benzoic acid and its derivatives should be carried out taking into account only molecules of the monomer forms, the accuracy of which is actually determined by the errors of the experimentally obtained sublimation enthalpy, presented in [24, 25].

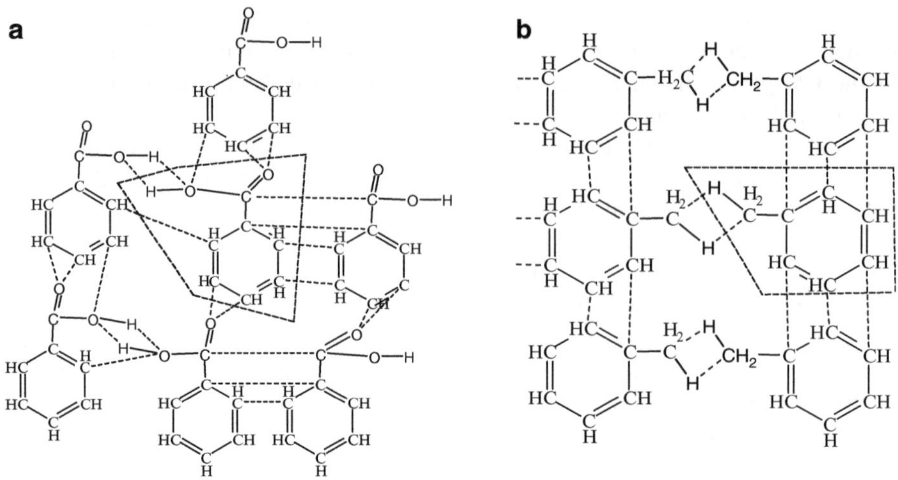

Fig. 9.9 Schematic picture of the crystalline structure of benzoic acid (**a**) and toluene (**b**) with the network of the specific interactions

The total energy contribution of the carboxyl group to the sublimation enthalpy of bensoic acid could be obtained from the difference between its characteristic and the benzene sublimation enthalpy reduced by the energy contribution of the substituted hydrogen atom (1.0 kJ mol^{-1}). However, in this approach the obtained energy value (45.6 kJ mol^{-1}) takes into account the replacement of stable specific interactions of the CH groups of the benzene ring by the interactions formed by the oxygen atoms of the carboxyl groups with CH groups of the molecules that are close in the environment (Fig. 9.9a). The correct energy value of the hydrogen bond is reached in calculations, taking into account the contributions of all specific interactions, formed by bond vacancies of the benzoic acid molecule. The 12 bond vacancies of the bensoic acid molecule form the network structure of crystalline and liquid condition (Fig. 9.9a) by the two stable hydrogen bonds 2D–O•••H–O with unknown energies, three specific interactions 3D=HC → CH, one D–O → CH, DO=C → C=O with energies of 7.40, 7.80, and 5.7 kJ mol^{-1}, respectively (Sect. 9.1.4), and the specific interactions D=C → C= and DC=O → CH. The energy of the specific interaction D=C → C=, formed by the carbon atom of the benzene ring, is determined by the value of benzene sublimation enthalpy reduced by the energy contribution of the substituted hydrogen atom and the energies of five specific interactions D=HC → CH, equal to

$$\mathrm{D{=}C{\rightarrow}C{=}} - (\Delta_{sub}H^0(298\ \mathrm{K})\mathrm{b} - 5\mathrm{D{=}HC{\rightarrow}CH} - \mathrm{DH})$$

The contributed energy of the latter type of specific interaction, DC=O → CH, formed by the carbonyl oxygen atom with significantly reduced negative charge compared with the oxygen atom of the hydroxyl group is essentially of two types

Table 9.7 Vaporization enthalpies of benzene, its derivative and energies (kJ mol^{-1}) of contribution of the methyl groups at T = 298 K

Compounds	Formula	$\Delta_{vap}H°$(298 K) [27]	$\Sigma D iso\, CH_3$	$D iso H_3C \rightarrow H–CH_2$
Benzene	C_6H_6	33.8	–	–
Toluene	C_7H_8	38.0	4.8	2.4
1,2-Dimethylbenzene	C_8H_{10}	43.4 ± 0.1	11.0	2.75
1,3-Dimethylbenzene	C_8H_{10}	42.7 ± 0.1	10.1	2.5
1,4-Dimethylbenzene	C_8H_{10}	42.3 ± 0.1	9.7	2.4
Tetramethylbenzene	$C_{10}H_{14}$	52.0 ± 0.2[a]	10.0	1.25

[a]Crystalline

of interactions DC → C with the energy of 3.6 kJ mol^{-1}, realized in crystalline phenols, and the formation of the hydrogen bond of low stability DC=O•••H–C with the energy value of 1.0 kJ mol^{-1}; we have a total value of 4.6 kJ mol^{-1}. As a result the energy of the stable hydrogen bond we determine from Eq. 9.7.

$$D\text{-}O\cdots H\text{-}O = (\Delta_{sub}H^0(298\ K) - 3D{=}H\ C{\rightarrow}CH - D\text{-}O{\rightarrow}CH - DO{=}C{\rightarrow}C{=}O - 4DC{=}O{\rightarrow}CH\ DC{\rightarrow}C - D{=}O{\rightarrow}CH{=})/2 \quad (9.7)$$

The obtained energy value of the stable hydrogen bond is equal to 14.3 kJ mol^{-1}.

Before turning to the consideration of methylbenzoic acids one should turn to benzene, toluene, and the other derivatives with methyl groups. In the case of sublimation enthalpies of these compounds we draw attention to the fact of the reduced enthalpy characteristic of toluene (43.1, T = 298 K) with the methyl group compared with the same characteristic of benzene (44.3 kJ mol^{-1}) obtained at standard conditions [5]. The sublimation enthalpies of 1,2-dimethylbenzene (60.1, T = 248 K) and 1,4-dimethylbenzene (60.8 kJ mol^{-1}, T = 286 K) were obtained at the conditions differing from the standard conditions and are also given in the literature without the errors of experimentally obtained values [5]. Thus we analyze the most correct values of the vaporization enthalpy of these compounds (Table 9.7). The difference in the values of the vaporization enthalpies of toluene and benzene reduced by those from the contributed values of the substituted hydrogen atom (0.6 kJ mol^{-1}) to the methyl group is equal to 4.8 kJ mol^{-1}. This value corresponds to two $DH_3C \rightarrow H–CH_2$ interactions, formed by the isostructural methyl group of the fragment $=C–CH_3$ of the benzene ring of toluene, equal to 2.4 kJ mol^{-1} [6]. A similar type of specific interaction of liquid and crystalline ethyl is equal to 3.78 and 5.12 kJ mol^{-1}. It follows that the energy of the specific interaction $DH_3C \rightarrow H–CH_2$ of liquid ethyl is approximately more stable by half than liquid toluene. Thus, we have reasons to conclude that the isostructural methyl group, participating in the distribution of the electron density, diminishes the negative charge of its carbon atom, leading to the increased negative charge of the bound carbon atom of the benzene ring. The appearance of the second isostructural group at location C(2) of the 1,2-dimethylbenzene molecule is accompanied by more than twice the increase of the energy contribution of these groups, indicating its mutual influence on the slight increase of negative charges of the carbon atoms of the cycle C(1) and C(2), as a result of which the negative charges of the carbon atoms of the isostructural methyl groups increase and the specific interactions are stabilized

$DH_3C \rightarrow H\text{-}CH_2$ of liquid 1,2-dimethylbenzene.

The remote location of the second isostructural methyl group in the benzene ring at location C(3) and further at C(4) is accompanied by the declining tendency of the reduced energy contribution at these locations of the methyl groups, pointing to the reduction of the charges at the carbon atoms of these groups and carbon atoms C(3) and C(4) and carbon atoms of the cycle fringing it, respectively. Therefore, despite the remoteness of the relative location of the methyl groups in the benzene ring they form specific interactions with similar energy values. The sequence of changes of the values of the energies of the specific interaction $DH_3C \rightarrow H\text{–}CH_2$ (Table 9.7), formed by the isostructural methyl group, is described by the number of

$$\begin{aligned} &1,2\text{-Dimethylbenzene } (2.75) > 1,3\text{-Dimethylbenzene } \left(2.5\right) \\ &> 1,4\text{-Dimethylbenzene } \left(2.4\right) = \text{Toluene } \left(2.4 \text{ kJ mol}^{-1}\right) \end{aligned}$$

The latter points to the fact that at the maximum remoteness of the isostructural methyl groups in the benzene cycle in the 1,4-dimethylbenzene molecule they express the influence on the distribution of the electron density only on the fringing CH groups, i.e., at the location C(1) the influence is expressed by the the influence at C(2) and C(6), and the location at C(4) is accompanied by the influence at C(3) and C(5), respectively. The increase in the number of isostructural methyl groups up to four in the molecule of tetramethylbenzene is accompanied by significant reduction in the difference of the charges at the carbon atoms of these groups. As a result there is a significant destabilization of the specific interaction $DH_3C \rightarrow H\text{–}CH_2$; the energy in crystalline condition of this compound (1.25 kJ mol^{-1}) takes a lower value compared with the liquid condition of toluene (2.4 kJ mol^{-1}).

In the 2-methylbenzoic acid molecules the presence of the carboxyl group with strong acceptor properties of electrons of the oxygen atoms and its donor properties at the formation of the specific interactions significantly change the properties of the carbon atom of the isostructural methyl group, transforming them to the acceptor ones at the interaction with the carbonyl oxygen atom. The methylbenzoic acid molecules have 14 bond vacancies, which number does not depend on the location of the isostructural methyl group at the carbon atoms of the benzene cycle (Fig. 9.10). These acid molecules form two stable hydrogen bonds 2D–O•••H–O by the oxygen atom of the hydroxyl group and two hydrogen bonds of low stability by the hydrogen atom of the isostructural methyl group $2D{=}C{=}O{\cdots}H\text{–}CH_2$ with the carbonyl oxygen atom. The oxygen atom of the hydroxyl group forms two specific interactions, 2D–O → CH=, with the carbon atom of the CH group of contacting benzene cycles and carbonyl oxygen atom forming the specific interaction D=C=O → CH_3 with the carbon atom of the isostructural methyl group

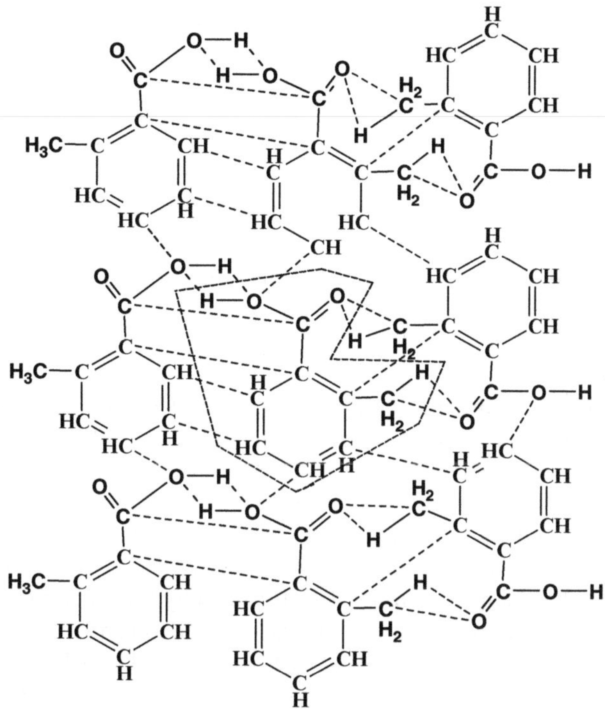

Fig. 9.10 Schematic picture of the crystalline structure of 2-methylbenzoic acid with the network of the hydrogen bonds specific interactions

of low stability. The remaining six specific interactions are formed by carbon atoms of the carbonyl group D=O=C → C=O, two interactions 2D=C → C= are formed by carbon atoms of the benzene cycle with substituted hydrogen atoms and three interactions, 3D=HC → CH, are formed by CH groups of the benzene rings.

The energies of the specific interactions D=HC → CH, DO → CH, D=C → C= and D=O=C → C=O we take to be equal to the values of the energies of the same types, realized in crystalline benzoic acid 7.4, 7.8, 6.4, and 5.7 kJ mol^{-1}. The value of the energy of low stability D=C=O•••H–C_2, formed by the hydrogen atom of the isostructural methyl group with the carbonyl oxygen atom, could be taken as equal to 1.0 kJ mol^{-1}, with the substituted hydrogen atom of the CH group of the benzene cycle of the crystalline compound. Taking into account that the specific interactions D=C=O → H–C= and D=C=O → CH_3 formed by the carbonyl oxygen atom are little affected by the change of stability at the replacement of the location of the isostructural methyl group in benzene cycle and the influence of the carboxyl group on the stability of the hydrogen bond at the location of CH_3 groups at C(2) and the energy of the hydrogen bond being dominant, we take its energy as equal to the energy of the same bond of crystalline benzoic acid. Such an assumption is justified by the analysis of the mutual influence of the isostructural methyl groups 1,2-dimethylbenzene (Table 9.7). Thus, the total energy contribution D=C=O → CH_3, introducing the isostructural methyl group, to the sublimation enthalpy of 2-methylbenzoic acid should be obtained with the help of Eq. 9.8,

$$DC{=}O{\rightarrow}CH_3 = (\Delta_{sub}H^0(298\ K)mba - 2D\text{-}O{\bullet\bullet\bullet}H\text{-}O - 3D{=}HC{\rightarrow}CH - 2DO{\rightarrow}CH{=} - DC{=}O{\rightarrow}C{=}O - 2D{=}C{\rightarrow}C{=}) / 2 \tag{9.8}$$

The results of the calculation of the energy contribution of the isostructural methyl group (Table 9.8) point to its low value. Using the obtained energy value of the interaction and the modified Eq. 9.9a, we calculated the energy

$$D\text{-}O{\bullet\bullet\bullet}H\text{-}O = (\Delta_{sub}H^0(298\ K)mba - 3D{=}HC{\rightarrow}CH - 2D\text{-}O{\rightarrow}CH{=} - DC{=}O{\rightarrow}C{=}O - 2D{=}C{\rightarrow}C{=} - DisoCH_3) / 2 \tag{9.9a}$$

of the hydrogen bonds of 3-methylbenzoic acid and 4-methylbenzoic acid (Table 9.7), reflecting the tendency of its stabilization at the removal from the location of the isostructural methyl group of the carboxyl group.

Replacement of the isostructural methyl group by the amino group is accompanied by significant increase in the sublimation enthalpy of aminobenzoic acids and in the case of 2-aminobenzoic acid this characteristic exceeds the enthalpy characteristic of benzoic acid, including the substituted hydrogen atom, at 13.0 kJ mol^{-1}. This value of the energy contribution to the sublimation enthalpy practically corresponds to the energy contribution of four hydrogen bonds DN•••H–N, formed by the amino group, which, by its value (3.25 kJ mol^{-1}), has significantly reduced energy compared with liquid ammonia (3.86 kJ mol^{-1}). It follows that the location of carboxyl and anime groups, on the one side, with strong electron acceptors of carbonyl and hydroxyl oxygen atoms and, on the other side, acceptor nitrogen atom at locations C(1) and C(2), lead to the fact that the nitrogen atom acts as the donor of electrons to the marked oxygen atoms. Electron density from the hydrogen atom provided to the nitrogen atom does not give it a negative charge due to the shifting of significantly increased value to the carbon atoms and further to the oxygen atoms.

The aminobenzoic acid molecules regardless of the location of the amino group at the carbon atoms of the benzene cycle have 16 bond vacancies, forming the structure of the crystal with a network of two types of hydrogen bonds and five types of specific interactions (Fig. 9.11), including hydrogen bonds 2D–O•••H–O (14.3) and 4D–N•••H–N, and the specific interactions 2D–O → C= (6.8), 2D=HC → CH (7.4), 4DC=O → CH (4.6), DO=C → C=O (5.7), D=C → C= (6.4 kJ mol^{-1}).

The energy of the hydrogen bond, formed by the amino group, should be obtained with the help of Eq. 9.9b

$$D\text{-}N{\bullet\bullet\bullet}H\text{-}N = (\Delta_{sub}H^0(298\ K) - 2D\text{-}O{\bullet\bullet\bullet}H\text{-}O - 2DO{\rightarrow}C{=} - 2D{=}HC{\rightarrow}CH - 4DC{=}O{\rightarrow}CH - DO{=}C{\rightarrow}C{=}O - D{=}C{\rightarrow}C{=})/ 4 \tag{9.9b}$$

The results of the calculations implemented (Table 9.9) were obtained with the use of the assumption of unchanged energy value of the hydrogen bond, formed by the hydroxyl group, due to the lack of possibility of conducting a more correct

Table 9.8 Energies of the hydrogen bonds and specific interactions (kJ mol^{-1}) of benzoic acid and methylbenzoic acids at 298 K

Compounds	Formula	Structure	$\Delta_{sub}H°(298\ K)$ [5]	D*iso*CH_3	DO → CH	D–O•••H–O
DO=C → C=O = 5.7; D=C → C= = 6.4; D=HC → CH = 7.4						
Benzoic acid	$C_7H_6O_2$	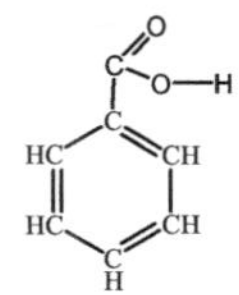	89.0 ± 0.5	DC=O → CH = 4.6 × 4	7.8	14.3
2-Methylbenzoic acid	$C_8H_8O_2$	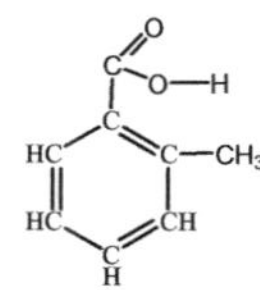	95.8 ± 0.3	1.4	7.8	14.3
3-Methylbenzoic acid	$C_8H_8O_2$	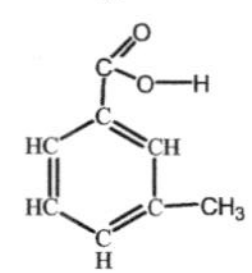	97.0 ± 0.8	1.4	7.8	14.9
4-Methylbenzoic acid	$C_8H_8O_2$	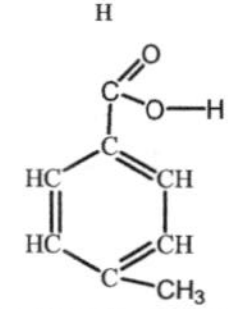	98.1 ± 0.9	1.4	7.8	15.4

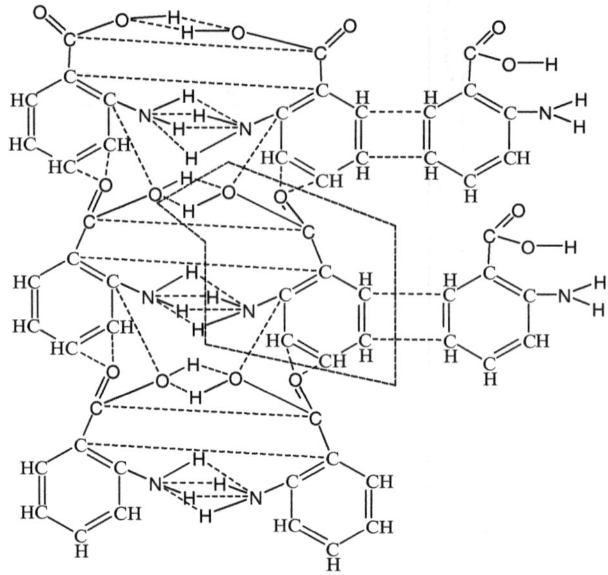

Fig. 9.11 Schematic picture of the crystalline structure of 2-aminobenzoic acid with the network of the specific interactions

assessment. Thus, the values of the energies of the hydrogen bonds D–O•••H–O of crystalline 3-aminobenzoic acid and 4-aminobenzoic acid are reduced and the values of the energies of the bonds D–N•••H–N are increased. Nevertheless the values of the energies of the hydrogen bonds, formed by the amino group, correctly illustrate the stabilization from its location at C(2), C(3), and C(4). A similar character of its changes is observed at the energies of the hydrogen bonds, formed by the hydroxyl group.

Note that the values of the energies obtained of the hydrogen bonds are within the experimental error of measurement of the sublimation enthalpies of the compounds considered.

A higher value of the energies of two types of hydrogen bonds at 3-aminobenzoic acid is caused, first, by the influence of the oxygen atoms of the carboxyl group on the shifting of the electron density from the carbon atom of the CH group at location C(2) and significant reduction of its influence on the shifting of the electron density from the nitrogen atom at its location at C(3) and, second, the nitrogen atom of the amino group gets the possibility to shift the electron density from the carbon atom at locations C(4) and C(5). The additional role in the increase of the hydrogen bonds at the location of carboxyl and amine groups at C(1) and C(3), respectively, is caused by the increase of positive charge of the carbon atoms at locations C(1) and C(2) and by the difference in the charges of the latter due to the shifting of the electron density from them, which, in turn, form in the crystal the specific interactions with CH groups at C(4) and C(5). However, its contribution to the enthalpy characteristic in the conducted calculations is introduced to the energy value of the hydrogen bond. This effect is absent at the location of the amino group at C(4) in the molecule of 4-aminobenzoic acid, which is one of the reasons for the

Table 9.9 Energies of the hydrogen bonds and specific interactions (kJ mol^{-1}) of crystalline aminobenzoic acids at 298 K

Compounds	Formula	Structure	$\Delta_{sub}H°(298\ K)$[5]	DC=O → CH	D–O•••H–O	D–N•••H–N
D=HC → CH = 7.4; DO → C= = 6.8; D=C → C= = 6.4; DO=C → C=O = 5.7						
2-Aminobenzoic acid	$C_7H_7NO_2$	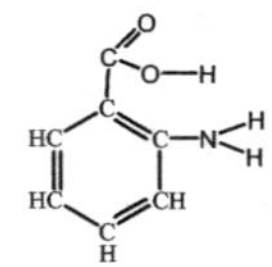	101.0 ± 1.4	4.6	14.3 ± 1.4	3.25 3.3
3-Aminobenzoic acid	$C_7H_7NO_2$	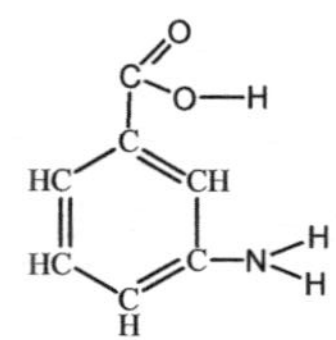	123.9 ± 2.1	4.6	14.3 ± 2.1	9.0
4-Aminobenzoic acid	$C_7H_7NO_2$	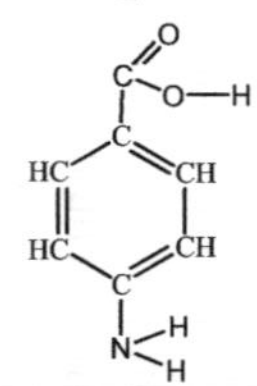	114.2 ± 2.2	4.6	14.3 ± 2.2	6.6

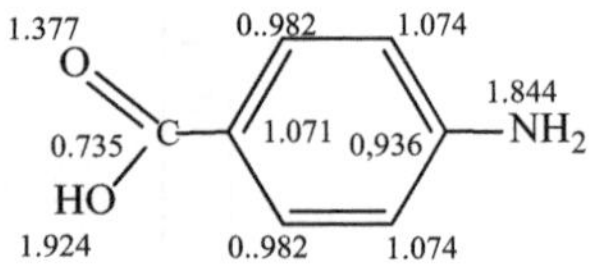

Fig. 9.12 Distribution of the charge on atoms of the molecule 4-aminobensoic acid [28]

reduced value of the energy of the hydrogen bond D–N•••H–N and reduced values of the energies of the hydrogen bond D–O•••H–O, approximately at 1–1.5 kJ mol^{-1}, at the high difference of the charges of the nitrogen atoms and the contacting carbon atom of the benzene cycle (Fig. 9.12). The second reason for the reduced energies of the hydrogen bonds at the location of the carboxyl and amine groups at atoms C(1) and C(4) at the maximum distance from each other is the lack of difference in the positive charges at the carbon atoms C(2) and C(6) and the quite high difference compared with the negative charges of the carbon atoms C(3) and C(5), forming the specific interactions D=HC → CH between each other in crystalline condition. These interactions also contribute to the sublimation enthalpy of 4-aminobenzoic acid, but also takes the energy value of the hydrogen bond into account in the calculations and in this case, the energy value of the hydrogen bond –DN•••H–N. One should pay attention to the fact that the difference of the charges in the molecule of 4-aminobenzoic acid between carbon atoms at locations C(2) and C(3), C(6), and C(5) amounts to 0.082e, practically equal to the value of the charges difference (0.080e) at oxygen atoms (1.924e) of the hydroxyl group and nitrogen atom (1.844e) of the amino group at such high value of negative charge of the nitrogen atom. Therefore, the high stability of the hydrogen bond D–N•••H–N, formed by the molecule of 4-aminobenzoic acid in crystalline condition, considering the increase in its real value, is caused by the high difference in the charges of the hydrogen atoms of the same amino group, not mentioned in the literature [28].

Taking account the sublimation enthalpy of 3-methylbenzoic acid (97.0 kJ mol^{-1}) and 2-amino-3-methylbenzoic acid (107.3 kJ mol^{-1}), we obtain the energy contribution of the amino group reduced by the contribution of the substituted hydrogen atom to the amine group equal to 11.3 kJ mol^{-1}. It follows that the value of the energy of the hydrogen bond D–N•••H–N is equal to 2.8 kJ mol^{-1}. By similar calculations of the energy contribution of the amino group from the sublimation enthalpy of 2-aminobenzoic acid and benzoic acid (Tables 9.8 and 9.9), we obtained the energy value of the same type of hydrogen bond to be equal to 3.25 kJ mol^{-1}. In more accurate calculations (Table 9.9) the energy value was obtained as 3.30 kJ mol^{-1}. So, the obtained energy values of DN•••H–N, formed by the amino group of the molecule 2-amino-3-methylbenzoic acid (2.80 kJ mol^{-1}) and 2-aminobenzoic acid (3.30 kJ mol^{-1}), point to the participation of the isostructural methyl group of the 2-amino-3-methylbenzoic acid molecule in the distribution of the electron density. The energy contribution of this group to the sublimation enthalpy of this compound could be obtained from the difference between the enthalpy characteristics of 2-amino-3-methylbenzoic acid and 2-aminobenzoic acid (Table 9.10), the energy value of which could be accepted as identical to the contribution of the isostructural methyl groups to the sublimation enthalpies of aminomethylbenzoic acids (Table 9.10).

Table 9.10 Energies of the hydrogen bonds and specific interactions (kJ mol^{-1}) of crystalline aminomethylbenzoic acids at 298 K

Compounds	Formula	Structure	$\Delta_{sub}H°(298\ K)$ [5]	D=HC → CH	D=C → C=/D*iso*CH_3	D–O•••H–O	D–N•••H–N
DO → C= = 6.8; DO=C → C=O = 5.7							
2-Amino-3-methylbenzoic acid	$C_8H_9NO_2$	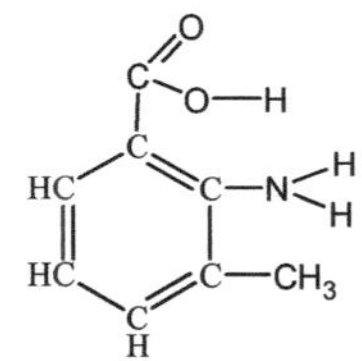	107.3 ± 1.8	7.4	6.4/5.3	14.3	3.3[a] 4.65[b]
2-Amino-5-methylbenzoic acid	$C_8H_9NO_2$	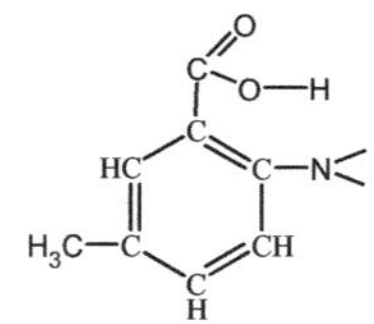	110.6 ± 1.9	7.23	6.8/5.3	14.3	5.5
2-Amino-6-methylbenzoic acid	$C_8H_9NO_2$	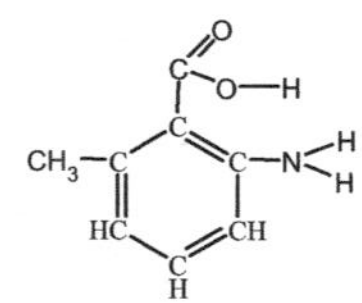	116.1 ± 2.0	7.23	6.8/5.3	14.3	6.85
3-Amino-2-methylbenzoic acid	$C_8H_9NO_2$	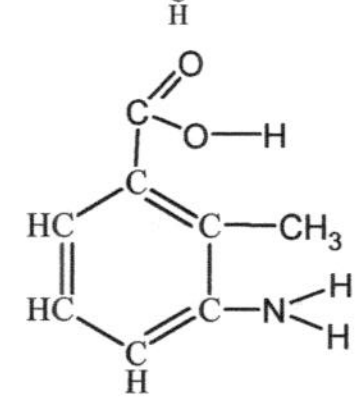	127.8 ± 2.8	7.23	6.8/5.3	14.3	9.7

(continued)

Table 9.10 (continued)

Compounds	Formula	Structure	$\Delta_{sub}H°(298\ K)$ [5]	D=HC → CH	D=C → C=/D*iso*CH_3	D–O•••H–O	D–N•••H–N
3-Amino-4-methylbenzoic acid	$C_8H_9NO_2$	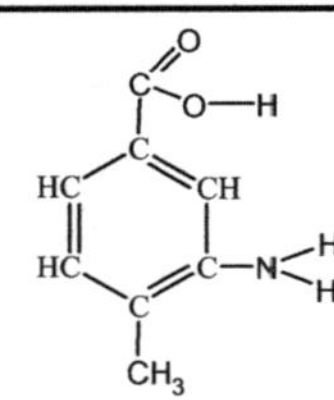	119.4 ± 0.7	7.23	6.8/5.3	14.3	7.7
4-Amino-3-methylbenzoic acid	$C_8H_9NO_2$	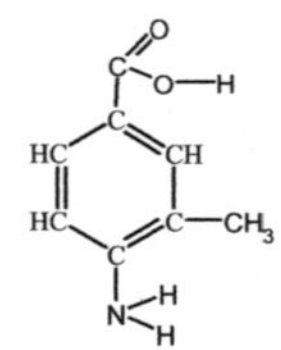	122.0 ± 2.6	7.23	6.8/5.3	14.3	8.5

[a]Calculation
[b]Calculation with correction

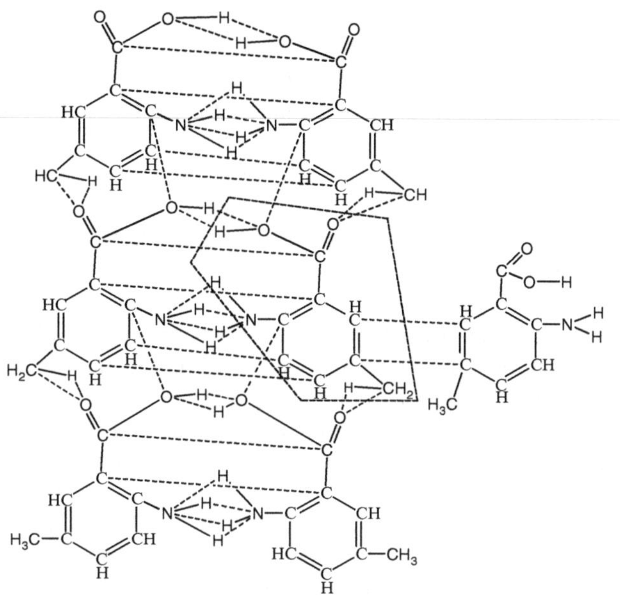

Fig. 9.13 Schematic picture of the crystalline structure of 2-amino-5-methylbenzoic acid with the network of the hydrogen bonds and specific interactions

The aminomethylbenzoic acid molecules with 18 bond vacancies form the grid structure of the crystal by 18 specific interactions (Fig. 9.13), including stable hydrogen bonds 2D–O•••H–O (14.3), 4D–N•••H–N and specific interactions 4D=HC → CH (7.4), 2D–O → C= (6.8), DO=C → C=O (5.7) and D=C → C= (6.4 kJ mol^{-1}) and introducing the contribution of the isostructural methyl group D*iso*CH_3 by four specific interactions 2DC=O → H–CH_2, 2DC=O → CH_3 of low stability.

The energy of the hydrogen bond, formed by the amino group, is calculated with the help of Eq. 9.9c

$$\text{D-N}\cdots\text{H-N} = (\Delta_{sub}H^0(298\ \text{K})\text{amba} - 2\text{D-O}\cdots\text{H-O} - 2\text{D-O}\rightarrow\text{C=} - 4\text{D=HC}\rightarrow\text{CH} - \text{DO=C}\rightarrow\text{C=O} - \text{D=C}\rightarrow\text{C=} - \text{DisoCH}_3)/4 \quad (9.9c)$$

The results of the calculations given in Table 9.10 are described by the integral values of stabilization of the hydrogen bonds D–N•••H–N, formed by the amino group

$$\begin{aligned} &\text{2-Aminobenzoic acid } (3.30) \le \text{2-Amino-3-methylbenzoic acid } (4.65) \\ &< \text{2-Amino-5-methylbenzoic acid } (5.5) < \text{4-Aminobenzoic acid } (6.6) \\ &\approx \text{2-Amino-6-methylbenzoic acid} (6.85) < \text{3-Amino-4-methylbenzoic acid } (7.7) \\ &< \text{4-Amino-3-methylbenzoic acid } (8.5) < \text{3-Aminobenzoic acid } (9.0) \\ &< \text{3-Amino-2-methylbenzoic acid } \left(9.7\ \text{kJ mol}^{-1}\right) \end{aligned}$$

A similar natural course of dependence of the energies of the hydrogen bond D–O•••H–O, formed by hydroxyl group, changing within 1.5 kJ mol^{-1}, with less clear manifestation of the structural changes, is implemented for the compounds of his series. This series subdivides structural changes in the molecules with the shifting of the electron density and energies of the hydrogen bonds, formed by the amino group with the location of the latter at C(2) close to the location of the carboxyl group at C(1) in the 2-aminobenzoic acid molecule and its derivatives and the second group of compounds with the location of the amino group at C(3) and C(4).

The difference in the energies of the hydrogen bonds at 2-aminobenzoic acid (3.30) $\approx$ 2-amino-3-methylbenzoic acid (4.65 kJ mol^{-1}) points to the influence of the isostructural methyl group with its location at C(3) close to the amine group at C(2). Removal of the methyl group from the amine one with its location at C(5) in the molecule of 2-amino-5-methylbenzoic acid and further at C(6) in the molecule of 2-amino-6-methylbenzoic acid is accompanied by the increase of the charge to the corresponding carbon atoms of the cycle, which leads to the additional stabilization of the intermolecular interactions and its contributions to the sublimation enthalpy of these compounds. The special role in the 2-amino-6-methylbenzoic acid molecule is played by the mutual influence of carboxyl and methyl groups in the increase of the positive charge of the carbon atom C(6) of the benzene cycle. The special role of the isostructural methyl group in the distribution of the electron density is expressed by its location at C(2) between the carboxyl C(1) and amine groups C(3) at the carbon atoms of the cycle. Such a precise location of the isostructural methyl group is accompanied by the formation of a possibly higher negative charge at the nitrogen atom of the 3-amino-2-methylbenzoic acid molecule from the shifting of the electron density from hydrogen atoms of the amino group and from the carbon atom of the cycle at locations C(4) and possible C(5). Lack of the isostructural methyl group in the 3-aminobenzoic acid molecule leads to the greatest possible shifting of the electron density and increased differences in the charges at the donor atoms and acceptor of electrons in the molecule, the highest values of sublimation enthalpy being calculated on the basis of the value of the energy of the hydrogen bond D–N•••H–N (9.0 kJ mol^{-1}). The replacement of the location of the isostructural methyl group from C(2) at location C(4) is accompanied by a reduction of the sublimation enthalpy and the energy of the hydrogen bond for 3-amino-4-methylbenzoic acid, which reflects the weakening influence of the oxygen atoms of the carboxyl group on the reduction of the negative charge of the nitrogen atom and positive charge of the hydrogen atom of the amine group (Table 9.10) at the replacement of NH_2 group from C(3) to C(4) with the removal from the carboxyl group. The replacement of the location of the amine group from C(2) at the isostructural methyl group at C(3) in the molecule of 2-amino-3-methylbenzoic acid at location C(4) of the molecule 4-amino-3-methylbenzoic acid is accompanied by the increased value of the sublimation enthalpy and energy of the hydrogen bond (Table 9.10), which points to the strengthening of role of the amino groups in the shifting of the electron density, leading to an increase of negative charge of the nitrogen atom and the difference in the charges with the hydrogen atoms of this group.

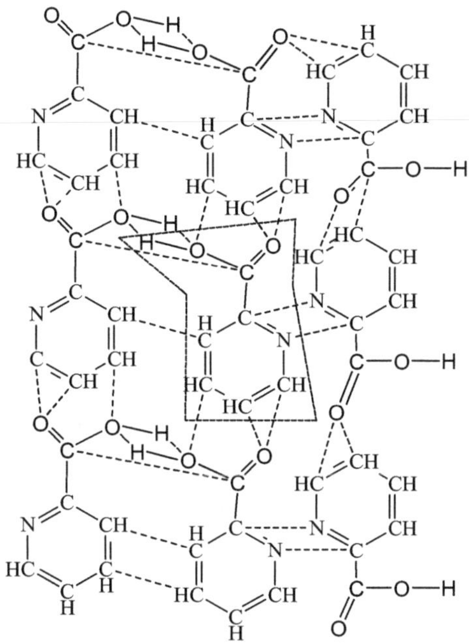

Fig. 9.14 Schematic picture of the crystalline structure of 2-pyridinecarboxylic acid with the network of the hydrogen bonds and specific interactions

9.3.2 *Pyridinecarboxylic Acids, Pyridine-3-Carboxylic acid N-Oxide, and Pyrazinecarboxylic Acid*

The experimentally obtained sublimation enthalpies of pyridinecarboxylic acids presented in the literature allow conducting a reliable thermodynamic analysis and obtaining of information on hydrogen bonds and specific interactions, realized in its crystalline condition. The molecule of this series has 12 bond vacancies despite the location of the carboxyl group at the carbon atom of the hetero cycle. The schematic picture given in Fig. 9.14 of the structure of crystalline 2-pyridinecarboxylic acid illustrates the formation of two hydrogen bonds, three types of specific interactions, formed by the carboxyl group 2D–O → CH, 4DC=O → CH, DO=C → C=O with the energies of 7.8 × 2, 4.6 × 4, 5.7 kJ mol^{-1} and ones formed by the hetero cycle of pyridine D=HC → CH=, 2D=N → CH, with the energy values of 7.4, and 12.85 kJ mol^{-1}, respectively.

The energy of the hydrogen bond is calculated with the help of Eq. 9.10,

$$D\text{-}O\cdots H\text{-}O = (\Delta_{sub}H^0(298\ K)pca - 2D{=}N{\rightarrow}C{=} - D{=}HC{\rightarrow}CH - DO{=}C{\rightarrow}C{=}O - 2DO{\rightarrow}CH - 4D{=}C{=}O{\rightarrow}CH) / 2 \qquad (9.10)$$

The obtained values of the energies presented in Table 9.11 of the hydrogen bonds allow us to conclude that its low stability at the location of the carboxyl fragment at carbon atom C(2) close to the nitrogen atom is caused by the significant

Table 9.11 Energies of the hydrogen bonds and specific interactions (kJ mol^{-1}) of crystalline pyridinecarboxylic acids and pyridinecarboxylic acids *N*-oxide at 298 K

Compounds	Formula	Structure	$\Delta_{sub}H°$(298 K) [5]	D=HC → CH	D=N → CH	D–O•••H–O	D=N–O → N
DO=C → C=O = 5.7, D=C=O → CH = 4.6, DO → CH = 7.8 kJ mol^{-1}							
2-Pyridinecarboxylic acid	$C_6H_5NO_2$	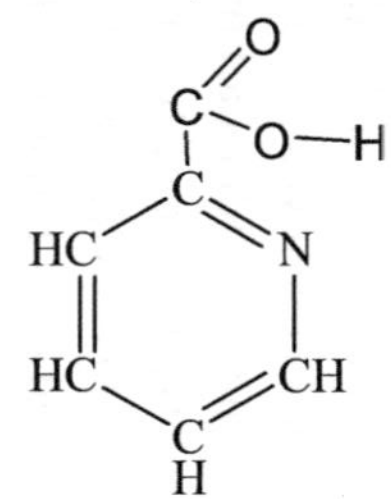	92.7 ± 0.5	7.4	12.85	10.3	–
3-Pyridinecarboxylic acid	$C_6H_5NO_2$	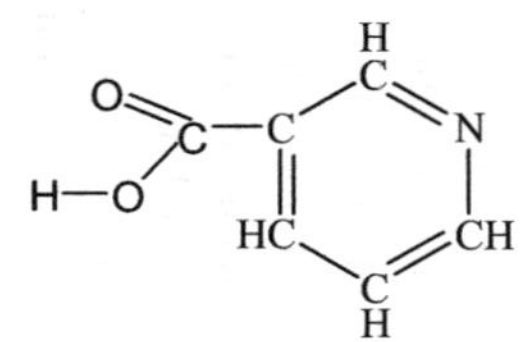	123.4 ± 1.2	7.4	12.85	25.6	–
4-Pyridinecarboxylic acid	$C_6H_5NO_2$	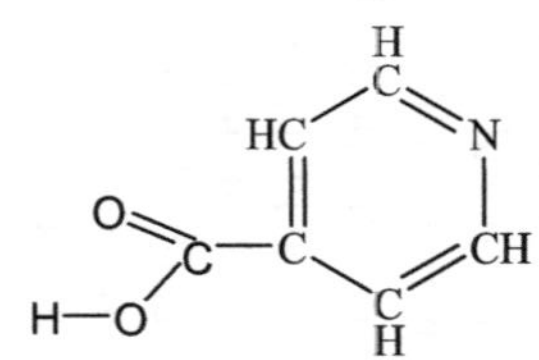	111.3 ± 0.6	7.4	12.85	19.5	–
Pyridine-2-carboxylic acid *N*-oxide	$C_6H_5NO_3$	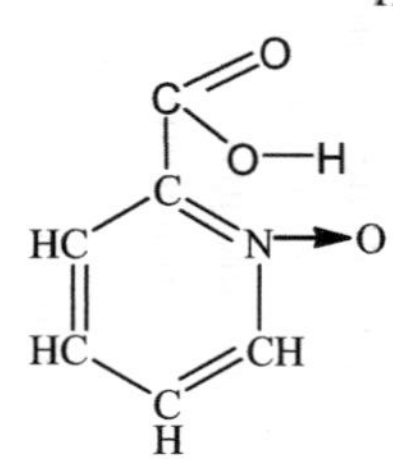	94.4 ± 4	7.43	12.85	10.3	10.0

Pyridine-3-carboxylic acid *N*-oxide	$C_6H_5NO_3$	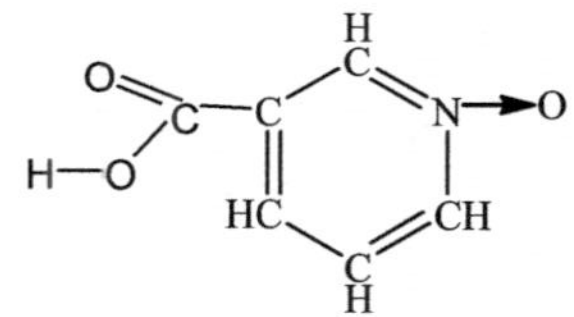	152.3 ± 1.9	7.43	12.85	25.6	23.7
Pyridine-4-carboxylic acid *N*-oxide	$C_6H_5NO_3$	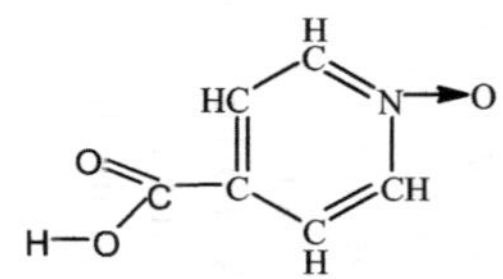	136.1 ± 1.2	7.43	12.85	19.5	21.7
DO=C → C=O = 5.7, 2D=C → C=(6.4), DO → C = 6.8 kJ mol^{-1}							
2-Methyl-5-pyrazine carboxylic acid	$C_6H_6N_2O_2$	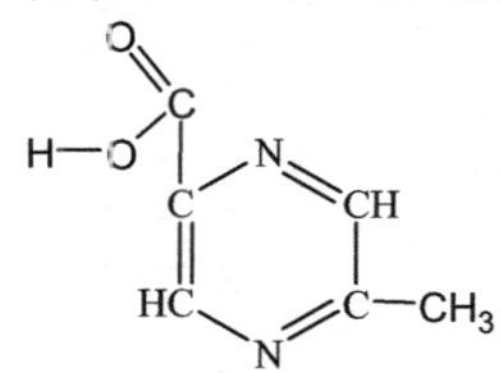	100.9 ± 1.3	–	10.30	18.0	–

competing influence of the acceptor of electrons atoms, on the one side, oxygen atoms of this fragment, and, on the other side, the nitrogen atom of the cycle. The greater stability of the energy of the hydrogen bond of 2-amino-3-methylbenzoic acid (14.3 kJ mol^{-1}) compared with the energy of the same bond 2-pyridinecarboxylic acid (19.3 kJ mol^{-1}) points to the fact that the location of the nitrogen atom in the cycle leads to its having an increased ability to shift the electron density from the fringing carbon atoms of the cycle rather than the nitrogen atom of the amino group, located at C(2) of the benzene ring in the 2-amino-3-methylbenzoic acid molecule. Removal of the carboxyl group from the nitrogen atom of the cycle of pyridine with location at C(3) sharply increases the ability of the oxygen atom of the hydroxyl group of the carboxyl fragment to shift the electron density and increase the negative charge and difference of the charges from the hydrogen atom of the same hydroxyl group. As a result, the stability of the hydrogen bond formed in crystalline 3-pyridinecarboxylic acid dramatically increases up to the value of 25.6 kJ mol^{-1}. Of course, the location of the carboxyl fragment at C(3) is expressed at the increase of positive charge of this carbon atom, leading to the formation of the specific interaction D=HC → CH of increased stability of 7.4 kJ mol^{-1}, not considered in the implemented calculation of the energy value of the hydrogen bond, reflecting its increased value compared with the value calculated. In other words, the energy value of the hydrogen bond 25.6 kJ mol^{-1} includes the value of the difference of the accepted energy of the specific interaction D=HC → CH, realized in the cycle of crystalline pyridine and actually realized in crystalline 3-pyridinecarboxylic acid. This stabilizing effect should be significantly expressed at the charges of the carbon atoms, fringing the nitrogen atom of the cycle of 4-pyridinecarboxylic acid with location of the carboxyl fragment at C(4) and the energy of the hydrogen bond, formed by a molecule of this compound practically corresponding to its real value. Thus, we could conclude that in fact the high value of calculated energy of the hydrogen bond of 3-pyridinecarboxylic acid includes the additional stabilizing effect due to the increase of positive charge in the carbon atom C(2) at the shifting of the electron density from oxygen atoms, forming the specific interactions with the stability increased more than in crystalline pyridine. A similar carbon atom of the CH group at location C(2) of the β-4(5)-imidazolylacrilic acid molecule, located between two nitrogen atom acceptors of the electron density, has a positive charge, while carbon atoms at locations C(4) and C(5) possess

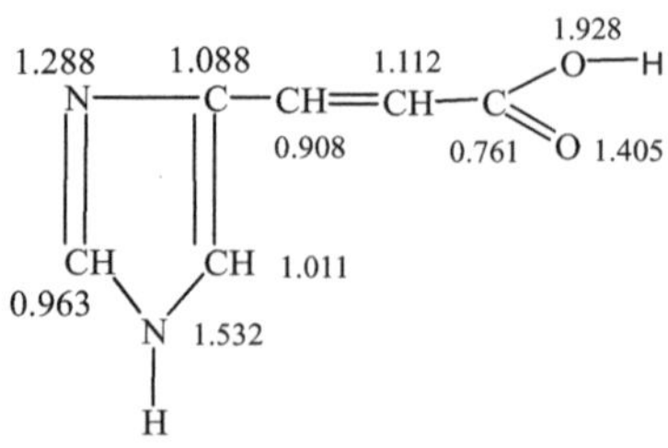

negative charges, staying at nitrogen atoms with the charges of 1.288e and 1.532e. It follows that the presence of the positive charge at the carbon atom C(2) points to

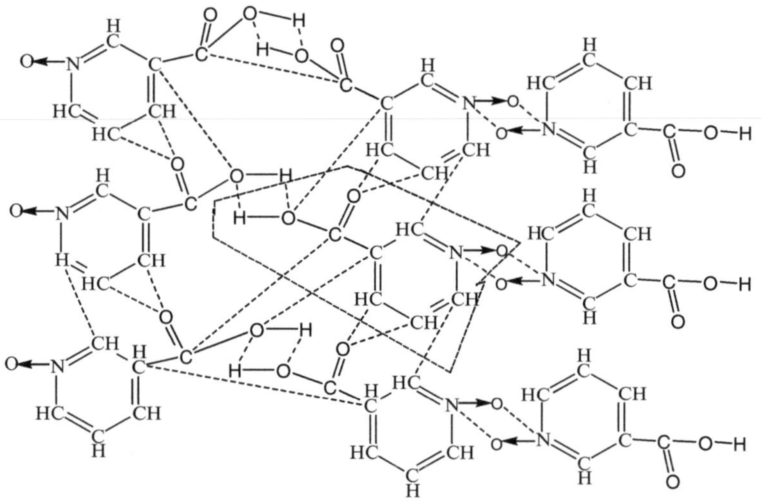

Fig. 9.15 Schematic picture of the crystalline structure of pyridine-3-carboxylic acid *N*-oxide with the network of the hydrogen bonds and specific interactions

the significant difference of its charge and the positive charge of the connected hydrogen atom. So, interacting CH groups at the C(2) and C(4) locations in liquid and even more in crystalline condition of the compound form significantly more stable specific D=HC → CH interactions. Such a high charge of the oxygen atom (1.928e) of the hydroxyl group presupposes the high positive charge of the hydrogen atom of the same group and the high value of the energy of the hydrogen bond formed.

The molecules of pyridinecarboxylic acid *N*-oxides with 13 bond vacancies forms the crystal structure with the network of two hydrogen bonds and 11 specific interactions of four types (Fig. 9.15) with known energies DO=C → C=O (5.7), D=C=O → CH (4.6), DO → C= (6.8 kJ mol^{-1}) and one specific interaction 2D=N–O → N with unknown energy. The energy value of this type of specific interaction should be obtained with the use of Eq. 9.11,

$$D{=}N\text{-}O{\to}N = (\Delta_{sub}H^0(298\ K)pcao - 2D\text{-}O{\cdots}H\text{-}O - 2D{=}HC{\to}CH - DO{=}C{\to}C{=}O - 2DO{\to}C{=} - 4D{=}C{=}O{\to}CH)/2 \quad (9.11)$$

The values of the energies of the specific interaction D=N–O → N, formed by the oxygen atom of the N–O group,

$$\text{Pyridine-2-carboxylic acid N-oxide}\,(10.0) < \text{Pyridine-4-carboxylic acid N-oxide}\,(21.7) < \text{Pyridine-3-carboxylic acid N-oxide}\,(23.7\ \text{kJ mol}^{-1})$$

are described by the integral value of the change of the stability from the location of the carboxyl fragment at the carbon atoms of the pyridine cycle with the nitrogen

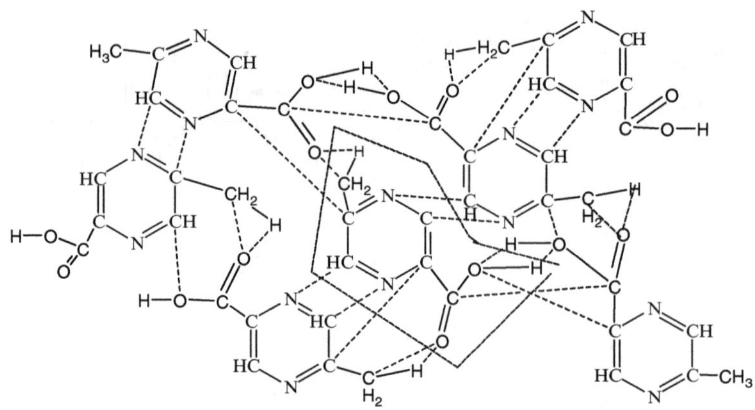

Fig. 9.16 Schematic picture of the crystalline structure of 2-Methyl-5-pyrazine carboxylic acid with the network of the hydrogen bonds and specific interactions

atom in the role of donor of the electron density, with a similar character of the change of

$$\text{2-Pyridinecarboxylic acid}\ (10.3) < \text{Pyridine-4-carboxylic acid N-oxide}\ (19.5) \\ < \text{Pyridine-3-carboxylic acid N-oxide}\ \left(25.6\ \text{kJ mol}^{-1}\right)$$

energies of the hydrogen bonds with the nitrogen atom with acceptor properties.

The increase in the number of acceptor atoms of electrons in the 2-methyl-5-pyrazine molecule from one to two nitrogen atoms leads to the necessity for the isostructural methyl group to form, by analogy with methylbenzoic acids, in addition to a stable hydrogen bond D–O•••H–O a second type of hydrogen bond of low stability with participation of the carbonyl oxygen atom and hydrogen atom D=C=O•••H–CH_2 (1.0) and specific interaction D=C=O → CH_3 by the same methyl group (Table 9.8), with contribution of the isostructural methyl group to the sublimation enthalpy of the compound considered. Therefore, the 2-methyl-5-pyrazine carboxylic acid molecule contains 14 bond vacancies, forming the crystal structure with the network of 14 specific interactions (Fig. 9.16). In addition to the mentioned interactions in the crystalline structure of 2-methyl-5-pyrazine carboxylic acid specific interactions 2D=C → C= (6.4), DO=C → C=O (5.7), four specific interactions 4D=N → CH, formed by nitrogen atoms and carbon atoms of the pyrazine cycle with the energies of 10.30 kJ mol^{-1} (Table 5.10), and the specific interaction of the oxygen atom of the hydroxyl group formed with the carbon atom with substituted hydrogen atom D–O → C (6.8 kJ mol^{-1}) for the CH group of the hetero nuclear cycle of pyrazine, respectively, are formed. The energy of the hydrogen bond is calculated with the help of Eq. 9.12

D-O•••H-O = ($\Delta_{sub}H^0$(298 K)mpz - 4D=N→CH – D-O→C= - 2D=C→C= - DO=C→C=O - DisoCH$_3$) / 2

(9.12)

The energy contribution of the isostructural methyl group in liquid 2-methylpyrazine is equal to 3.3 and the average contribution value of the same CH_3 group in crystalline tetramethylpyrazine is equal to 4.8 kJ mol^{-1}, which value can be accepted to be equal to the contribution of the isostructural methyl group of crystalline 2-methyl-5-pyrazine. The energy value of the hydrogen bond (Table 9.11) of crystalline 2-methyl-5-pyrazine carboxylic acid obtained (18.0 kJ mol^{-1}) exceeds the energy value for the same type of interaction of crystalline 2-methylbenzoic acid (14.3) and 4-methylbenzoic acid (15.4 kJ mol^{-1}) and has a reduced value compared with that realized in crystalline 4-pyridinecarboxylic acid (19.5 kJ mol^{-1}). It testifies to the stabilizing effect of two symmetrically located nitrogen atoms of the hetero nuclear cycle which is expressed to a lesser extent than one nitrogen atom located at the carboxyl group at C(4) cycle of pyridine.

9.3.3 *Pyrrolecarboxylic Acids*

The pyrrole molecule forms specific interactions D=HC → CH in liquid and crystalline conditions with the energies of 7.70 and 13.85 kJ mol^{-1}, significantly increasing the energy values of the same type of benzene interaction and hetero cyclic compounds with the hydrocarbon cycle as the limit. It follows that the distribution of the electron density in the pentamerous hetero cycle of pyrrole, caused by the influence of the nitrogen atom, leads to the increased differences in the charges of the carbon atom cycle, significantly different from the charges in hetero cycle of pyridine. The pyrrole-2-carboxylic acid molecule with 12 free bond vacancies forms the crystalline structure (Fig. 9.17) with the network of four differing types of specific interactions including that formed by the carbon atom of the CH groups of the pentamerous hetero cycle with the oxygen atom of the hydroxyl 2DO → C= (6.8) and carbonyl 4D=C=O → CH (4.6) group, the CH groups of cycle D=HC → CH (13.85), the carbon atom of the carboxyl group DO=C → C=O (5.7 kJ mol^{-1}) and hydrogen bond D–N•••H–N (9.7 kJ mol^{-1}) with known energies and with unknown value of the energy of the hydrogen bond D–O•••H–O. The energy value is calculated with the help of Eq. 9.13,

D-O•••H-O = ($\Delta_{sub}H^0$(298 K)pca - 2D-N•••H-N – D=HC→CH - 4D=C=O→CH - DC=O→C=O - 2DO→C) / 2

(9.13)

and is given in Table 9.12.

The energy of the hydrogen bond of pyrrole-2-carboxylic acid (28.2) significantly exceeds the energy value of the same bond for 2-pyridinecarboxylic acid

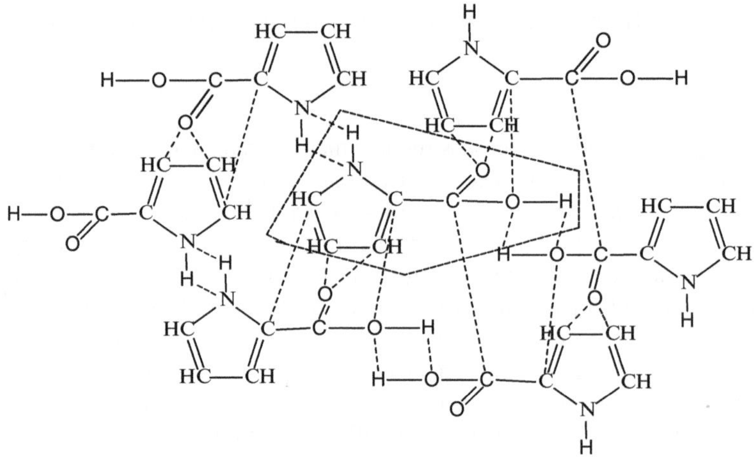

Fig. 9.17 Schematic picture of the crystalline structure of pyrrole-2-carboxylic acid with the network of the hydrogen bonds and specific interactions

(12.2) with the location of the carboxyl group at C(2), which is possibly caused by the influence of the hydrogen atom of the fragment of N–H cycle, transferring the electron density and leading to high negative charge of the nitrogen atom, respectively, of the oxygen atom of the hydroxyl.

9.4 Energies of Hydrogen Bonds and Specific Interactions of Amino Esters

9.4.1 Liquid Methyl Carbamates and Ammonium Carbamate

Complex ethers with amino groups are a small class of nitrogenous compounds with known values of the vaporization enthalpies, leading to the implementation of thermodynamic analysis in order to clarify the stabilizing effect of the mutual influence of the carbonyl oxygen atom and the amino group at their location in the molecule that is close to each other. This influence of the combination of two groups with acceptor atoms of nitrogen of electron density and carbonyl oxygen can be considered as a stabilizing effect of the peptide bond.

The methyl carbamate molecule with ten bond vacancies, including six vacancies of the nitrogen atom and two oxygen atoms, lacks the atom donors of the electron density at the formation of the specific interactions. Therefore the hydrogen atom of the methyl group acts in the capacity of atom donor, leading to the formation of a hydrogen bond of low stability with energy of 0.60 kJ mol^{-1} with the electron pair of the carbonyl oxygen atom, D=C=O•••H–CH_2. This type of hydrogen bond in combination with four other interactions forms the network structure of its liquid

Table 9.12 Energies of the hydrogen bonds and specific interactions (kJ mol^{-1}) of crystalline pyrrolecarboxylic acid

Compounds	Formula	Structure	$\Delta_{sub}H°(298\ K)$ [5]	T K	D=HC → CH	D–N•••H–N	D–O•••H–O
D=C=O → CH (4.6), DC=O → C=O (5.7), DO → C (6.8 kJ mol^{-1})							
Pyrrole	C_4H_5N	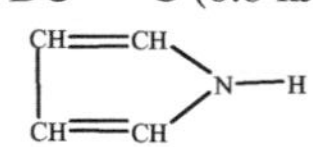	67.3	298	13.85	6.34	–
Pyrrole-2-carboxylic acid	$C_5H_5NO_2$	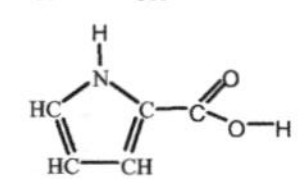	126.8	352	13.85	9.7	27.8

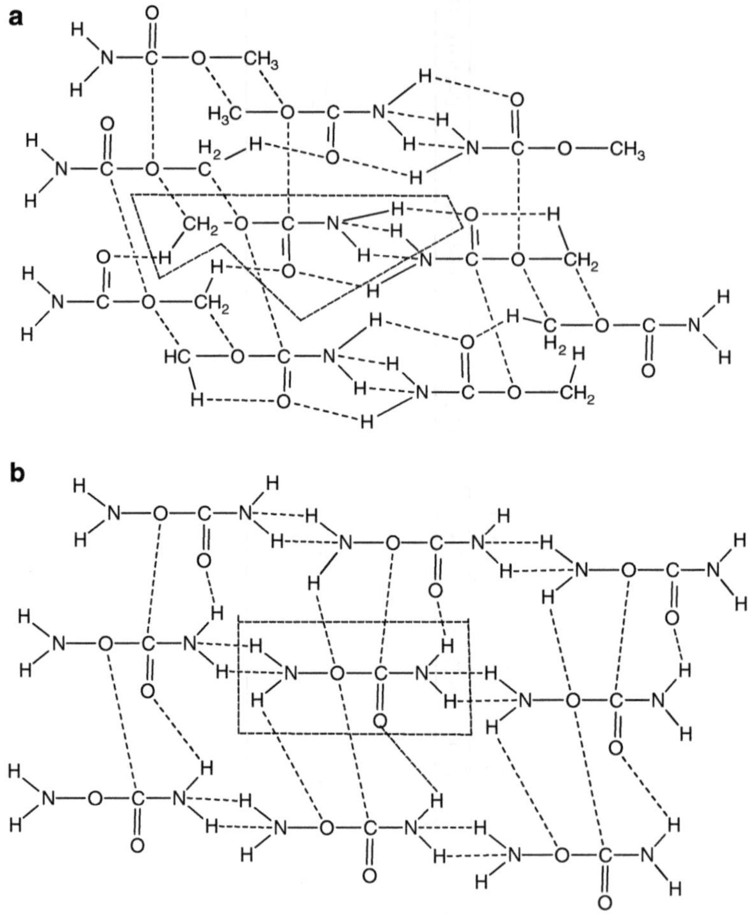

Fig. 9.18 Schematic picture of the liquid structure of methyl carbamate (**a**) and ammonium carbamate (**b**) with the network of the hydrogen bonds and specific interactions

condition (Fig. 9.18a). The carbon atom of the methyl group forms the specific interaction of the essentially unshared $2s^2$ electron pair with the oxygen atom D–O → CH_3–O with the energy of the same type of the specific interaction of liquid dimethyl ether (5.63 kJ mol^{-1}). The second electron pair of the same oxygen atom forms the specific interaction D–O → C=O with the energy of 4.0 kJ mol^{-1}. The nitrogen atom in pentacoordinated condition forms the hydrogen bond D–N•••H–N with the hydrogen atom of similar amino groups of the molecule in contact in the environment with an energy of 4.63 kJ mol^{-1} (Chap. 2). It should be mentioned that the increased energy value of this type of hydrogen bond in comparison with the energy value of the specific interaction D–N → CH_3–N (4.25 kJ mol^{-1}), formed by the methyl group of trimethylamine, is proved by the method of RE-spectroscopy [7, 8]. The energy of the third type of hydrogen bond, formed by the second hydrogen atom of the same amino group with the remaining electron pair of the carbonyl oxygen atom D=C=O•••H–N is determined with the help of Eq. 9.14, taking into account the energies of all hydrogen bonds and specific interactions to the vaporization enthalpy

$$D{=}C{=}O\cdots H\text{-}N = (\Delta_{vap}H^0(298\ K)ac - 2D\text{-}N\cdots H\text{-}N - 2D\text{-}O{\to}CH_3\text{-}O - 2D\text{-}O{\to}C{=}O - 2D{=}C{=}O\cdots H\text{-}CH_2)/2 \quad (9.14)$$

Replacing the energy contribution of the specific interaction D–O → CH_3–O by the vaporization enthalpy to the energy contribution of interactions D–O → CH_3–CH_2–O and D–O → CH_3–CH_2–CH_2–O, realized in liquid diethyl (6.78) and dipropyl ether (8.92 kJ mol^{-1}), this equation should be used for calculation of the energy of the hydrogen bond D=C=O•••H–N of liquid ethyl carbamate and propyl carbamate. The results presented in Table 9.13 of the calculations allow us to state that there occurs significant stabilization of the hydrogen bond D=C=O•••H–N in a number of compounds.

$$\text{Methyl carbamate } (7.8) < \text{Ethyl carbamate } (12.3)$$
$$< \text{Propyl carbamate } \left(13.1\ \text{kJ mol}^{-1}\right)$$

In the first representatives of compounds, overlapping errors of the measured values of the vaporization enthalpies obtained at the differing temperatures in the interval of 340–351 K, clearly reflect the joint influence of the nitrogen atom, carbonyl oxygen, and the oxygen connecting to the alkyl fragment on the stabilizing effect. The nature of this effect is to shift the electron density to the carbonyl atom of oxygen, leading to the increased negative charge of this oxygen atom and increased difference in its charge and the charge of the nitrogen atom. At the same time, the sharp reduction in the stabilizing effect on the hydrogen bond considered for propyl carbamate is caused by the completion of the influence of the reverse dative bond at the propyl fragment with three carbon atoms. Therefore for all further compounds of this series of compounds the energy of this D=C=O•••H–N type of hydrogen bond has a value of 13.1 kJ mol^{-1}, which in combination with other realized hydrogen bonds, specific interactions, and the energy contribution of the methylene groups allows us to estimate the value of its vaporization enthalpy. Note that the energies of the hydrogen bonds of methyl carbamate (7.8) and 2-aminotropone (2,4,6-cycloheptatriene-1-one, 2-amine, 8.4 kJ mol^{-1}) have a light difference, lying within the error of measurement of the enthalpy characteristics.

The molecule of ammonium carbamate with ten bond vacancies and two amino groups forms the network structure of liquid condition by ten intermolecular interactions, including three types of hydrogen, 4D–N•••H–N (4.63), 2D–O•••H–N, and 2D=C=O•••H–N bonds, and one specific interaction 2D–O → C=O (4.0 kJ mol^{-1}) (Fig. 9.18b). With the known values of the energies of the same type of hydrogen bond and specific interaction we determined the average energy value of (D–O•••H–N + D=C=O•••H–N)/2 from Eq. 9.15

Table 9.13 Energies of the hydrogen bonds and specific interactions (kJ mol^{-1}) of crystalline alkyl carbamates

Compounds	Formula	Structure	$\Delta_{vap}H°$ (298 K) [27]	T K	D–O → CH_3 (CH_2)n–O	D–N•••H–N	D–O → C=O/ D=C=O•••H–CH_2	D=C=O•••H–N
Methyl carbamate	$C_2H_5NO_2$	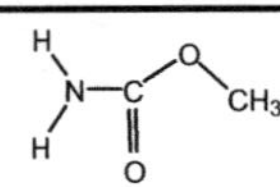	45.7	348	n = 0 5.63	4.63	4.0/5.0	7.8
Ethyl carbamate	$C_3H_7NO_2$	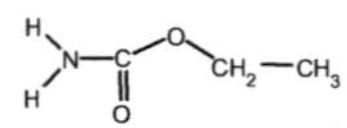	56.6	351	n = 1 6.78	4.63	4.0/5.0	12.3
Propyl carbamate	$C_4H_9NO_2$	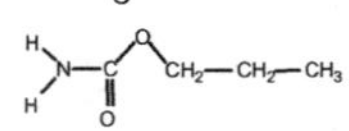	61.6	340	n = 2 8.93	4.63	4.0/5.0	13.1
					D=C=O•••H–N ≈ DO•••H–N = 6.9			
Ammonium carbamate	$CH_4N_2O_2$	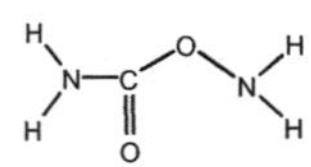	54.1	262	–	4.63	4.0/–	6.9

$$D\text{-}O\bullet\bullet\bullet H\text{-}N \approx D{=}C{=}O\bullet\bullet\bullet H\text{-}N{=} (\Delta_{vap}H^0(298\ K)amc - 4D\text{-}N\bullet\bullet\bullet H\text{-}N - 2D\text{-}O{\rightarrow}C{=})/4 \quad (9.15)$$

to be equal to 6.9 kJ mol^{-1}. The obtained value of the average energy of two types of hydrogen bonds (Table 9.13) differs little from the energy value of the peptide hydrogen bond D=C=O•••H–N of liquid methyl carbamate (7.8 kJ mol^{-1}). At this average value the increased stability is characterized by the hydrogen bond D–O•••H–O > D=C=O•••H–N.

9.4.2 *Energies of Specific Interactions of Liquid Amino Ethanoates and Amino Propanoates*

The main characteristic feature of amino esters with functional fragment 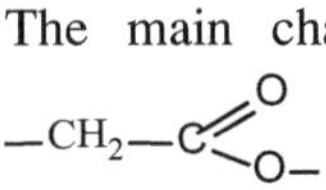 is the location of the methylene CH_2 group between acceptor electrons of the nitrogen atom of the amino group and, on the other side, by the carbonyl oxygen atom, forming ether at methyl 2-(*N*,*N*-dimethylamino)ethanoate and ethyl 2-(*N*,*N*-dimethylamino)ethanoate (Table 9.14). Such a location of the methylene group gives to it a reduction of the energy contribution to the enthalpy characteristic of vaporization, caused by the distribution of the electron density in the molecules, reducing the difference in the charges at the oxygen and carbon atoms of CH_2 group. It was shown above that the energy of the formed specific interaction D=C=O → CH_2–N is equal to 3.6 kJ mol^{-1}. The molecules of methyl 2-(*N*,*N*-dimethylamino)ethanoate and ethyl 2-(*N*,*N*-dimethylamino)ethanoate with ten bond vacancies form a similar structure of liquid condition with the network of five types of specific interactions (Fig. 9.19). The specific interaction formed by the methyl group of the ether fragment D–O → CH_3–O of the methyl 2-(*N*,*N*-dimethylamino)ethanoate molecule has an energy equal to the energy of liquid dimethyl ether (5.63 kJ mol^{-1}). The specific interaction D–O → C=O formed by the oxygen atom of the same hydroxyl group with the carbon atom of C=O group is equal to 3.3 kJ mol^{-1}. The contribution energy of the specific interaction, formed by the methylene group DC=O → CH_2–N, located in the chain between acceptor atoms of nitrogen and oxygen of the carboxyl group to the vaporization enthalpy is equal to 3.6 kJ mol^{-1} (Sect. 9.1.2). The energy of the formed specific interaction D–N → CH_3–N= by nitrogen atom with two methyl groups in pentacoordinated condition is equal to 4.8 (Chap. 2). The distinctive feature in the formation of the specific interactions of liquid ethyl 2-(*N*,*N*-dimethylamino)ethanoate is to replace the interaction D–O → CH_3–O by the interaction formed by the ethyl group D–O → CH_3–CH_2–O with the energy (6.78 kJ mol^{-1}), equal to the value of a similar specific interaction of liquid dimethyl ether. In turn, the methyl 2-(*N*,*N*-dimethylamino) propanoate molecule differs from the ones considered by the location of two methylene groups between acceptor atoms of nitrogen and oxygen of the carboxyl

Table 9.14 Energies of the specific interactions (kJ mol^{-1}) of liquid amino ethanoates and propanoates at 298 K

Compounds	Formula	Structure	$\Delta_{vap}H°$ (298 K) [27]	D=C=O → CH_2–N	D–N → CH_3 (CH_2)n–N	D–O → CH_3–(CH_2)n–O	D=C=O → CH_3–(CH_2)n–N
D–O → C = O = 3.3							
Methyl 2-(*N*,*N*-dimethylamino) ethanoate	$C_5H_{11}NO_2$	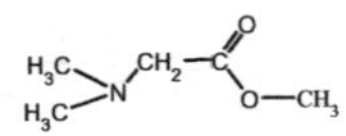	43.9 ± 0.4	3.6	n = 0 4.8	n = 0 5.63	n = 0 4.65
Ethyl 2-(*N*,*N*-dimethylamino) ethanoate	$C_6H_{13}NO_2$	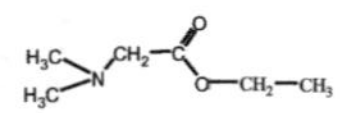	47.6 ± 0.8	3.6	n = 0 4.8	n = 1 6.78	n = 0 5.15
Methyl 2-(*N*,*N*-dimethylamino) propanoate	$C_6H_{13}NO_2$	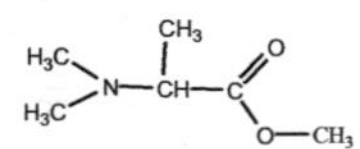	46.1 ± 1.1	3.2	n = 0 4.8	n = 0 5.63	n = 0 4.65 DH_3C → H–CH_2 = 1.3 × 2
Methyl 2-(*N*,*N*-dimethylamino)-2-methylpropanoate	$C_7H_{15}NO_2$	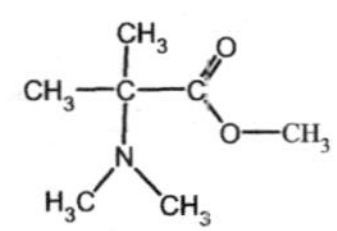	49.2 ± 1.0	3.0	n = 0 4.8	n = 0 5.63	n = 0 4.65 DH_3C → H–CH_2 = 1.6 × 4
Ethyl 2-(*N*,*N*-dimethylamino)-2-methylpropionate	$C_8H_{17}NO_2$	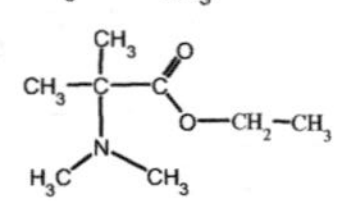	55.6 ± 0.4	3.0	n = 0 4.8	n = 1 6.78	n = 0 5.15 DH_3C → H–CH_2 =2.1 × 4
Ethyl 2-(*N*,*N*-Dimethylamino)-2-methylpropanoate	$C_8H_{16}NO_2$	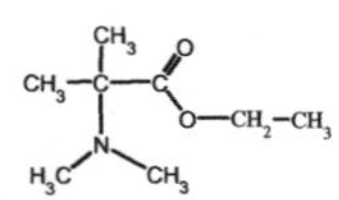	51.6 ± 0.3	3.6	n = 0 4.8	n = 1 6.78	n = 0 5.15 DH_3C → H–CH_2 =1.6 × 4
Propyl 2-(*N*,*N*-dimethylamino)-2-methylpropanoate	$C_9H_{19}NO_2$	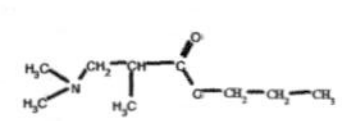	54.0 ± 0.5	3.6	n = 0 4.8	n = 2 8.92	n = 0 4.7 DH_3C → H–CH_2 =1.1 × 4

Ethyl 2-(*N,N*-diethylamino)-2-propanoate	$C_9H_{19}NO_2$	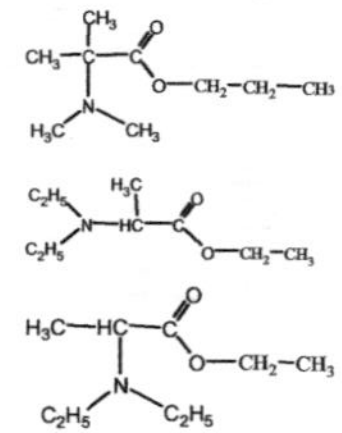	54.9 ± 0.6	3.6	n = 1 6.0	n = 1 6.78	n = 1 6.2 $DH_3C \rightarrow H–CH_2 = 1.6$
Ethyl 2-(*N,N*-diethylamino)-2-butanoate	$C_{10}H_{21}NO_2$	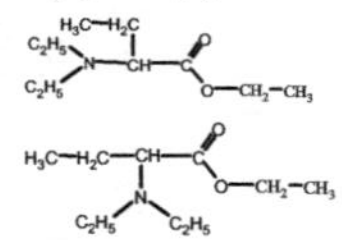	57.3 ± 0.2		n = 1 6.0	n = 1 6.78	n = 1 6.2 $DCH_2–CH_3 \rightarrow CH_2–CH_3 = 3.9$

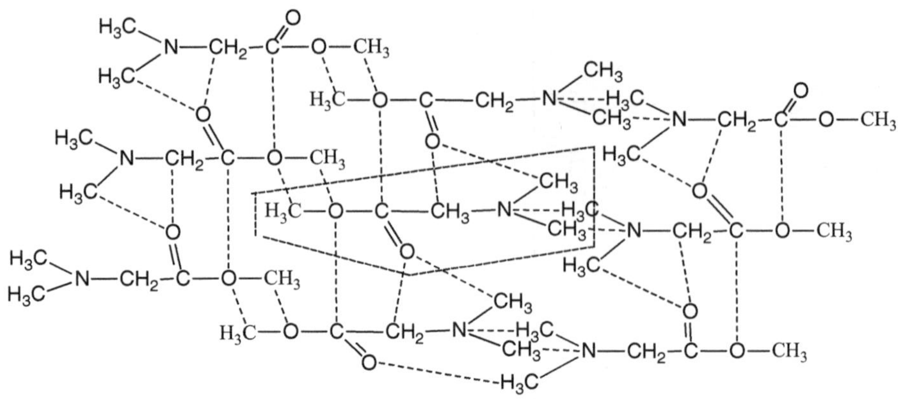

Fig. 9.19 Schematic picture of the liquid structure of methyl 2-(*N,N*-dimethylamino)ethanoate with the network of the specific interactions

group, forming the specific interaction DC=O → CH_2–CH_2–N with an energy of 4.7 kJ mol^{-1}. The energy of the fifth type of specific interaction DC=O → CH_3–N with unknown value of the energy of liquid methyl 2-(*N,N*-dimethylamino) ethanoate is calculated with the help of Eq. 9.16, and is the amount of the energy contributions of all specific interactions to the vaporization enthalpy of the compounds.

$$D{=}C{=}O{\rightarrow}CH_3\text{-}N = (\Delta_{vap}H^0(298\ K)mdmae - 2D\text{-}N{\rightarrow}CH_3\text{-}N - 2D\text{-}O{\rightarrow}CH_3\text{-}O - 2D{=}C{=}O{\rightarrow}CH_2 - 2D\text{-}O{\rightarrow}C{=}O)/2 \quad (9.16)$$

The distinctive feature in the formation of the specific interactions of liquid ethyl 2-(*N,N*-dimethylamino)ethanoate is in the replacement of the interaction D–O → CH_3–O to the interaction formed by the ethyl group D–O → CH_3–CH_2–O with the energy of 6.78 kJ mol^{-1}, equal to the value of the energy of a similar specific interaction of liquid diethyl ether, and the energy of interaction D=C=O → CH_3–N is calculated with the help of Eq. 9.17.

Ethyl 2-(*N,N*-dimethylamino)ethanoate

$$D{=}C{=}O{\rightarrow}CH_3\text{-}N = (\Delta_{vap}H^0(298\ K)edmae - 2D\text{-}N{\rightarrow}CH_3\text{-}N - 2D\text{-}O{\rightarrow}CH_3\text{-}CH_2\text{-}O - 2D{=}C{=}O{\rightarrow}CH_2 - 2D\text{-}O{\rightarrow}\ C{=}O)/2 \quad (9.17)$$

The results of the calculation of the energy values of the specific interaction D=C=O → CH_3–N (Table 9.14) reflects the increase in its stability with the replacement of the methyl group in methyl 2-(*N,N*-dimethylamino)-2-methylpropanoate (4.65) by the ethyl one in the ethyl 2-(*N,N*-dimethylamino)ethanoate (5.15 kJ mol^{-1}) molecule. This effect is caused by the increased shifting of the electron density from the ethyl group to the oxygen atoms of the carboxyl group, ensured the difference of the charges at the atoms of carbonyl oxygen and the carbon atom of the methyl group of the amine fragment. The inequality in the energies of the specific interaction, and the hydrogen bond DC=O → CH_3–N (4.65) < D=C=O•••H–N (7.8 kJ mol^{-1}) is the natural factor, because the replacement of the hydrogen atom in amine by the methyl group is accompanied by an insignificant reduction in the electron density [7, 8]. Participation of the oxygen carbonyl atom in the formation of these bonds contributes an additional effect to the stabilization of this type of hydrogen bond. The methyl 2-(*N,N*-dimethylamino)propanoate molecule with the location of the amino group at the second carbon atom of the propyl ligand has a structure similar to the structure of methyl 2-(*N,N*-dimethylamino)ethanoate, in which the atom of the methylene group is substituted by the methyl group, acting as the isostructural group. Therefore, molecules of these two compounds have an equal number of bond vacancies, forming ten specific interactions of five identical types, forming the network structure of the liquid condition (Fig. 9.20a). Thus, the contribution of energy of the isostructural methyl group is obtained from the difference of the vaporization enthalpy of methyl 2-(*N,N*-dimethylamino)propanoate and methyl 2-(*N,N*-dimethylamino)ethanoate reduced by the energy contribution of the substituted hydrogen atom of the methylene groups. With the energy contribution (0.60) of one substituted hydrogen atom to the vaporization enthalpy the reduced value of the energy contribution of the methylene group, of 3.6 kJ mol^{-1} instead of 5.0 kJ mol^{-1}, allows us to estimate the energy contributed by the hydrogen atom of this group as equal to 0.4 kJ mol^{-1}. Thus, the value of the energy contribution of the isostructural methyl group of methyl 2-(*N,N*-dimethylamino)propanoate is equal to 2.6 kJ mol^{-1} and the energy of the specific interaction of low stability DH_3–C → H–CH_2, formed by this group, is equal to 1.3 kJ mol^{-1}.

The methyl 2-(*N,N*-dimethylamino)-2-methylpropanoate molecule with the location of the fragment 2-(*N,N*-dimethylamino)- at the second carbon atom of the propyl ligand (Table 9.14) changes the properties of the terminal methyl group, rearranging it to the functional isostructural methyl group.

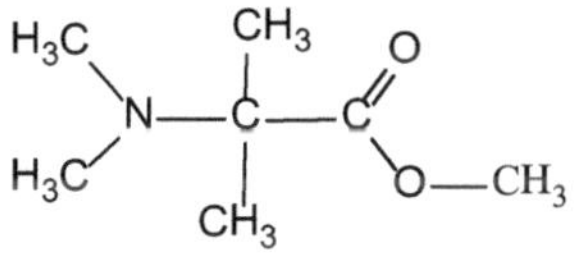

In this connection, the molecule of methyl 2-(*N,N*-dimethylamino)-2-methylpropanoate is a derivative of the methyl 2-(*N,N*-dimethylamino)propanoate molecule, in which two hydrogen atoms of the methylene group are substituted by two isostructural methyl groups. The energy contribution of the methyl groups is equal to the difference in the vaporization enthalpy of these compounds reduced by the contribution of two substituted hydrogen atoms to the vaporization enthalpy of methyl 2-(*N,N*-dimethylamino)propanoate (5.6 kJ mol^{-1}). Therefore, the methyl 2-(*N,N*-dimethylamino)-2-methylpropanoate molecule forms the same specific

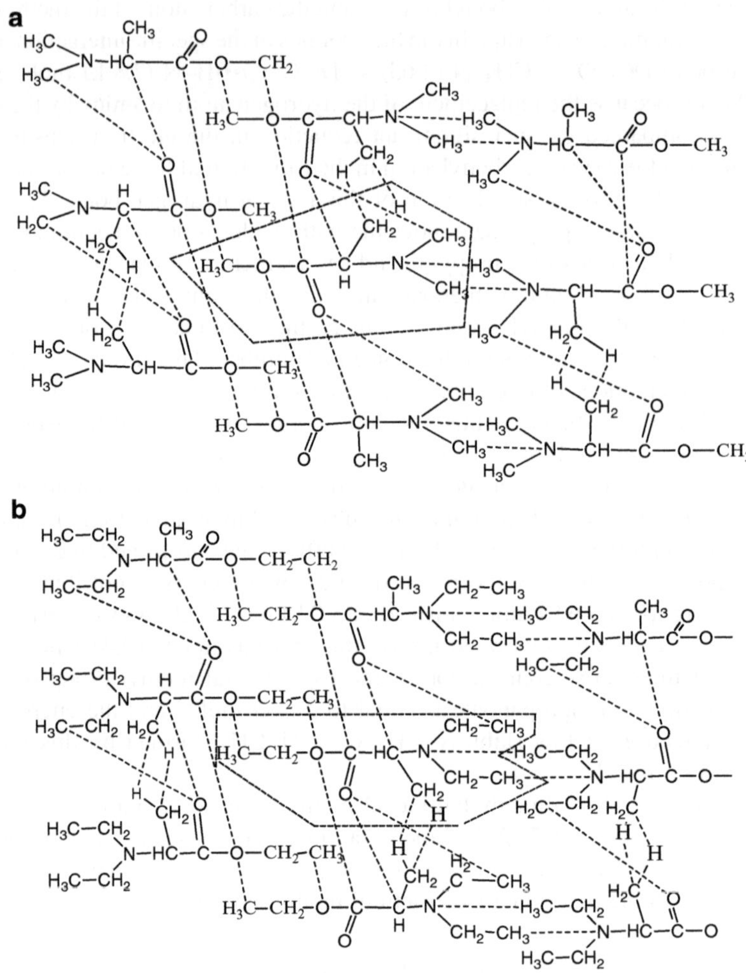

Fig. 9.20 Schematic picture of the liquid structure of methyl 2-(*N*,*N*-dimethylamino)propanoate (**a**) and Ethyl 2-(*N*,*N*-diethylamino)-2-propanoate (**b**) with the network of the specific interactions

interactions with the same energy values and additionally four specific interactions of low stability DH_3–C → H–CH_2 with an energy of 1.4 kJ mol^{-1}.

This approach with the use of the analog of the molecule structures and structures of liquid condition with the network of the specific interaction of amino ethanoates and propanoates, the energy contribution of the isostructural methyl groups to the vaporization enthalpy, and the energies of the specific interactions of low stability could be used for ethyl 2-(*N,N*-dimethylamino)-2-methylpropionate, ethyl 2-(*N,N*-dimethylamino)-2-methylpropanoate, and propyl 2-(*N,N*-dimethylamino)-2-methylpropanoate with replacement of the specific interaction D–O → CH_3–O at D–O → CH_3–CH_2–O or D–O → CH_3–CH_2–CH_2–O, formed by the methyl group to ethyl and propyl in the compounds considered or to use Eqs. 9.18 and 9.19 ethyl 2-(*N,N*-dimethylamino)-2-methylpropionate, ethyl 2-(*N,N*-Dimethylamino)-2-methylpropanoate:

$$DH_3\text{-}C{\rightarrow}H\text{-}CH_2 = (\Delta_{vap}H^0(298\ K) - 2D\text{-}N{\rightarrow}CH_3\text{-}N - 2D\text{-}O{\rightarrow}CH_3\text{-}CH_2\text{-}O - D{=}C{=}O{\rightarrow}CH_3\text{-}N - 2D{=}C{=}O{\rightarrow}CH_2 - 2D\text{-}O{\rightarrow}C{=}O + 2DH)/4 \quad (9.18)$$

propyl 2-(*N,N*-dimethylamino)-2-methylpropanoate

$$DH_3\text{-}C{\rightarrow}H\text{-}CH_2 = (\Delta_{vap}H^0(298\ K) - 2D\text{-}N{\rightarrow}CH_3\text{-}N - 2D\text{-}O{\rightarrow}CH_3\text{-}CH_2\text{-}CH_2\text{-}O - D{=}C{=}O{\rightarrow}CH_3\text{-}N - 2DC{=}O{\rightarrow}CH_2 - 2D\text{-}O{\rightarrow}C{=}O + 2DH)/4 \quad (9.19)$$

As given in Table 9.14 the results of the calculations point to the following:

First, to the low values of the energies of the specific interactions, formed by the isostructural methyl groups, proving the analogy in the molecule structures and liquid condition, the energies of the specific interactions of amino ethanoates and propanoates and,

Second, to the tendency of stabilization of the specific interactions, formed by the isostructural methyl groups with an increase up to two in

$$\begin{aligned}&\text{Methyl 2-(N, N-dimethylamino)propanoate} \quad (1.3)\\ &< \text{Methyl 2-(N, N-dimethylamino)-2-methylpropanoate } (1.6)\\ &< \text{Ethyl 2-(N, N-dimethylamino)-2-methylpropionate } (2.1\ \text{kJ mol}^{-1})\end{aligned}$$

the methyl 2-(*N,N*-dimethylamino)-2-methylpropanoate molecule and replacement of the methyl group by ethyl one in ethyl 2-(*N,N*-dimethylamino)-2-methylpropionate.

The schematic picture presented in Fig. 9.20b of the structure of liquid ethyl 2-(*N,N*-diethylamino)-2-propanoate illustrates its analogy with the structure of methyl 2-(*N,N*-dimethylamino)propanoate and ethyl 2-(*N,N*-dimethylamino)ethanoate. It allows us to use the value of the energy contribution of the specific interaction $DH_3\text{–}C \rightarrow H\text{–}CH_2$ (1.6 kJ mol^{-1}) at the calculation of the energy interaction $DC{=}O \rightarrow CH_3\text{–}CH_2\text{–}N$

$$DC{=}O{\rightarrow}CH_3\text{-}CH_2\text{-}N = (\Delta_{vap}H^0(298\ K) - 2D\text{-}N{\rightarrow}CH_3\text{-}CH_2\text{-}N - 2D\text{-}O{\rightarrow}CH_3\text{-}CH_2\text{-}O - - 2D{=}C{=}O{\rightarrow}CH_2 - 2D\text{-}O{\rightarrow}C{=}O - 2DH_3\text{-}C{\rightarrow}H\text{-}CH_2 + DH)/2 \quad (9.20)$$

of liquid ethyl 2-(*N,N*-diethylamino)-2-propanoate with the help of Eq. 9.20, given in Table 9.14.

The presence of such obvious contradictions in the data on the structural-thermodynamic characteristics of vaporization of amino ethanoates and propanoates

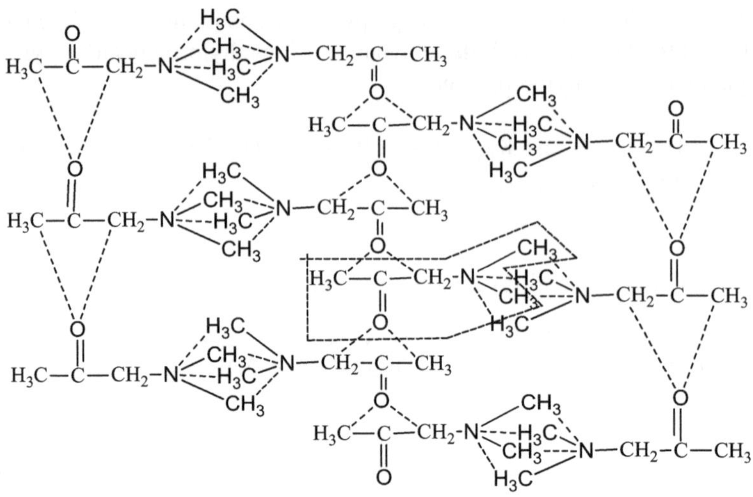

Fig. 9.21 Schematic picture of the liquid structure of 1-(Dimethylamino)-2-propanone with the network of the specific interactions

suggests the use of more correct approaches in terms of uniqueness of the interpretation. In this connection we draw attention to ethyl 2-(*N,N*-diethylamino)-2-butanoate, in which the (*N,N*-diethylamino) molecule-fragment is connected with the second carbon atom of the butyl ligand leading to the remaining ethyl part of the properties of the isostructural group, forming the specific interaction $DCH_2–CH_3 \rightarrow CH_2–CH_3$. The energy of this type should be calculated with the help of Eq. 9.21, considering the contributions of all types of specific interactions of liquid

$$D\text{-}CH_2\text{-}CH_3 \rightarrow CH_2\text{-}CH_3 = (\Delta_{vap}H^0(298\ K) - 2D\text{-}N \rightarrow CH_3\text{-}CH_2\text{-}N - 2D\text{-}O \rightarrow CH_3\text{-}CH_2\text{-}O - 2D{=}C{=}O \rightarrow CH_3\text{-}CH_2\text{-}N - 2DC{=}O \rightarrow CH_2 - 2D\text{-}O \rightarrow C{=}O + DH) \quad (9.21)$$

ethyl 2-(*N,N*-diethylamino)-2-propanoate. The obtained value of the energy contribution of the isostructural ethyl group and the specific interaction $DCH_2–CH_3 \rightarrow CH_2–CH_3$ formed with the energy of 3.9 kJ mol^{-1} is in good compliance with the energy value of the same type of interaction (3.78 kJ mol^{-1}), formed by the molecule of ethane [6].

9.4.3 Energies of Specific Interactions of Liquid 1-(Dialkylamino)-2-Propanones

The location of the methylene group between electron acceptor atoms of nitrogen and carbonyl oxygen in the 1-(dimethylamino)-2-propanone molecule leads to it having a reduced ability in the formation of the specific interaction

Table 9.15 Energies of the specific interactions (kJ mol^{-1}) of liquid 1-(dialkylamino)-2-propanones and 2-(dimethylamino)-1-phenylalkanone at 298 K

Compounds	Formula	Structure	$\Delta_{vap}H°(298\ K)$ [27]	D–N → CH_3 (CH_2)n–N	D=C=O → CH_3–C=O
1-(Dimethylamino)-2-propanone	$C_5H_{11}NO$	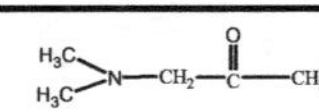	43.6 ± 0.3	n = 0 4.8	6.2 ± 0.3
1-(Diethylamino)-2-propanone	$C_7H_{15}NO$	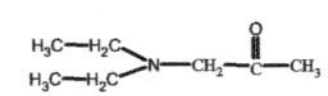	47.7 ± 0.3	n = 1 6.0	5.9 ± 0.3

D=C=O → CH_2–C=O (Fig. 9.21). Therefore, the specific interactions of ketone CH_2 group with substituted hydrogen atom and methyl group of the molecule of 1-(dimethylamino)-2-propanone have significantly different energy contributions to the vaporization enthalpy. Being formed by the diamine group the second type of specific intermolecular interaction D–N → CH_3–N with pentacoordinated nitrogen atom and an energy value of 4.8 kJ mol^{-1} allow us to determine the total energy value, contributed by the ketone fragment and the average energy value of the methylene and methyl group of the specific interaction D=C=O → CH_3–C=O, using the Eq. 9.22

$$\text{D=C=O}\rightarrow\text{CH}_3\text{-C=O} = (\Delta_{vap}H^0(298\text{ K})\text{dmap} - 4\text{D-N}\rightarrow\text{CH}_3\text{-N})/4 \quad (9.22)$$

Having replaced the specific interaction 4D–N → CH_3–N by 4D–N → CH_3–CH_2–N, formed by the diethyl fragment of the 1-(diethylamino)-2-propanone molecule in the equation, we calculated the energy value of the same type of interaction of liquid 1-(diethylamino)-2-propanone. The results of the energy calculations of D=C=O → CH_3–C=O given in Table 9.15 point to its reduced stability (6.2 kJ mol^{-1}) compared with that realized in liquid acetone (7.70 kJ mol^{-1}).

References

1. Baev AK (1969) Problems of chemical nature of phase transformation. In: General and applied chemistry, vol 1. Vysheischaia Educating, Minsk, pp 197–206
2. Baev AK (1969) Phase condition and complex formation ability of halogenide metals. In: General and applied chemistry, vol 1. Vysheischaia Educating, Minsk, pp 207–218
3. Baev AK (1972) Complex formation ability halogenides of second – six groups periodical system. In: General and applied chemistry, vol 5. Vysheischaia Educating, Minsk, pp 35–51
4. Nesmeyanov AN, Nesmeyanov NA (1974) Beginning of organic chemistry, vol 2. Chemistry, Moscow, p 744
5. Chickos JS, Acree WE Jr (2003) Enthalpies of sublimation of organic and organometallic compounds, 1910–2001. J Phys Chem Rev Data 32:537–698
6. Baev AK (2012) Specific intermolecular interactions of organic compounds. Springer, Hiedelberg/Dordrecht/London/New York, 434p
7. Karlson TA (1981) Photoelectron and OGE- spectroscopy. Leningrad, Mech. Eng. Russia. Leningrad. 15:431 p
8. Nefedov VI (1981) Infra red – spectroscopy of chemical compounds. Chemistry, Moscow, p 255
9. Nefedov VI, Vovna VI (1989) Electronic structure of organic and elementorganic compounds. Nauka, Moscow, p 199
10. Armitage JW, Gray P (1962) Temperature coefficient vaporization heats and curvature lines of vapor pressure formic acid, acetates acid and another compounds formed dissociated vapor. Trans Faraday Soc 58:1746
11. de Kruif CG, Sonk HA (1973) The determination of enthalpies of sublimation by mean of thermal conductivity manometer. J Chem Ing Technol 45:455
12. Stull DR, Westrum EF Jr, Sinke GC (1969) The chemical thermodynamic of organic compounds. Wiley, New York

13. de Kruif CG (1980) Enthalpies of sublimation and vapour pressures of 11 polycyclic hydrocarbons. J Chem Thermodyn 12:243
14. de Kruif CG, Kuipers T, van Miltenburg JC, Schaake RCF, Stevens G (1981) The vapour pressure of solid and liquid naphthalene. J Chem Thermodyn 13:1081
15. de Kruif CG, Schaake RCF, van Miltenburg JC, van der Klauw K, Blok JQ (1982) Thermodynamic properties of the normal alkanoic acids III. Enthalpies of vaporization and vapour pressures of 13 normal alkanoic acids. J Chem Thermodyn 14:791–798
16. Galis-van Ginkel CHD, Galis GHM, Timmermans CWM, de Kruif CG, Oonk HA (1978) Enthalpies of sublimation and dimerization in the vapour phase of formic, acetic, propanoic and butanoic acids. J Chem Thermodyn 10:1083
17. Malaspina L, Gigli R, Bardi G (1973) Microcalorimetric determination of the enthalpy of sublimation of benzoic acid and anthracene. J Chem Phys 59:387
18. Colomina LM, Laynez JL, Peres-Ossorio R, Turrion C (1972) Enthalpies of combustion and formation of six methyl esters of benzene carboxylic acids. J Chem Thermodyn Data 4:499–506
19. van Ginkel CHD, de Kruif CG, de Waal FEB (1975) The need for temperature control in effusion experiments (and application to heat of sublimation determination). J Phys E Sci Instrum 8:490
20. Colomina LM, Jimi P, Turrion C (1982) Thermodynamic properties of benzoic acid derivatives. X. Enthalpies of combustion and formation of 2.3-, 2.4-, 2.6-, 3.4- and 3.5-dimetoxybenzoic acid. Academic press. Inc. 1982.
21. Maraweiz E (1972) Vaporization enthalpy n-alkanes from C_{10} till C_{20}. J Chem Thermodyn Data 4:139–144
22. Cox JD (1964) The calorimetry of combustions and related reactions: organic compounds. Pure Appl Chem 8:143–156
23. Cox JD (1982) Notation for states and processes, significance of the word standard in chemical thermodynamics, and remarks on commonly tabulated forms of thermodynamic functions. Pure Appl Chem 54:1239–1250
24. Murata S, Sakiym M, Seki S (1982) Enthalpy of sublimation of benzoic acid and dimerization in the vapor phase in the temperature range from 320–370 K. J Chem Thermodyn Data 14:723–731
25. Monte MJS, Hillesheim DM (2001) Thermodynamic study of the sublimation of six aminomethyl benzoic acids. J Chem Thermodyn 38:745–754
26. Baev AK (1987) Chemistry of gas-heterogenic systems of elementorganic compounds. Science and Technology, Minsk, p 174
27. Chickos JS, Acree WE Jr (2002) Enthalpies of vaporization of organic and organometallic compounds, 1880–2002. J Phys Chem Rev Data 31:519–879
28. Kalinin FL, Lobov VP, Zidkov VA (1971) Reference book on biology. Naukiva Dumka, Kiev, p 1012

Chapter 10
Specific Intermolecular Interactions and Energies of Compounds with Peptidic Structure

10.1 Energies of Hydrogen Bonds and Specific Interactions of Simple Peptides

10.1.1 Compounds with Peptide Fragment of N-*Methyl Carbamic Acid Ethyl Ester and Aminogroup of* N-*Acetyl*-N-*butylacetamide and δ-Valerolactam*

Derivatives of amino acids are used as model compounds, reflecting the nature of polypeptides and proteins [1–3]. Therefore the analog of structures of these acids, simple peptides, dipeptides, and polypeptides, allow us to conduct a correct thermodynamic analysis of each series of compounds, using the enthalpy characteristics of the processes of vaporization and sublimation to obtain the energies of the hydrogen bonds and the specific interactions, formed in its liquid and crystalline conditions.

In previous chapters we have considered the peptide group by the example of the number of compounds and revealed the nature of the formation by carboxyl oxygen and the hydrogen atom of the amine group of the stable hydrogen bond D=C=O•••H–N. Being essentially a combination of the carboxyl group and the presence of a hydrogen atom connected with a nitrogen atom, we can consider the fragment

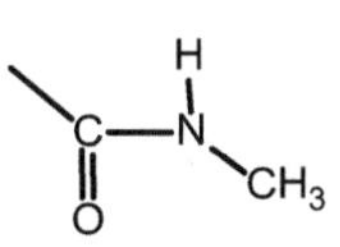

We have *N*-methyl carbamic acid ethyl ester as a peptide group, forming a hydrogen bond of increased stability. The molecule of *N*-methyl carbamic acid ethyl ester molecule with eight bond vacancies and two oxygen atoms and a nitrogen atom as electron acceptors lacks bond vacancies at the formation of the specific interactions.

A.K. Baev, *Specific Intermolecular Interactions of Nitrogenated and Bioorganic Compounds*, DOI 10.1007/978-3-642-37472-2_10,

Therefore oxygen atoms of the hydroxyl and carbonyl groups form by one hydrogen bond with the hydrogen atom of the terminal methyl group of the ethyl and the methyl group of the amine fragment D–O•••H–CH_2, D=C=O•••H–CH_2–N with low stability and energy of 0.60 kJ mol^{-1} for liquid conditions. The remaining eight bond vacancies form four types of specific interactions, three of which are formed by the essentially unshared $2s^2$(c) electron pair of the carbon atom of the terminal methyl group of the ethyl fragment D–O → CH_3–CH_2–O, the methyl group of the amine fragment D–N → CH_3–N, and carbon atoms of the carbonyl groups D=O=C → C=O (Fig. 10.1a). The energies of these interactions are equal to the energies of the specific interactions of liquid diethyl ether (6.78 kJ mol^{-1}) and dimethylamine with a four-coordinated nitrogen atoms (4.2 kJ mol^{-1}) reduced by the energy contribution of the third hydrogen atom, forming a hydrogen bond of low stability, and the third type of interaction of 3.3 kJ mol^{-1} (Chap. 9). The fourth type of interaction with the peptide bond is formed by the carbonyl atom of oxygen and hydrogen atom of the amide group with unknown energy value, which is calculated with the help of the Eq. 10.1a,

DC=O•••H-N = ($\Delta_{vap}H^0$(298 K)mca - 2DO→CH_3-CH_2-O - 2DN→CH_2-H-N - 2DC=O•••H-CH_2- - 2DO•••H-CH_2-N - 2DO=C→C=O) / 2 (10.1a)

The results of the calculations (Table 10.1) point to the increased stability of the hydrogen bond compared with other types of specific interactions of liquid *N*-methyl carbamic acid ethyl ester.

The *N*-acetyl-*N*-butylacetamide molecule with two oxygen atoms of the carbonyl groups and a nitrogen atom as electron acceptors simultaneously, acting as donors to provide electron pairs at the formation of the specific interactions, has a deficit in acceptor atoms at the formation of stable interactions. These functions are implemented by the hydrogen atoms of the methyl groups of the amine fragment, forming four hydrogen bonds with low stability, D–O•••H–CH_2–N, with energy of 0.60 kJ mol^{-1}. The butyl fragment forms a specific interaction with an energy of 8.1 kJ mol^{-1} of liquid dipropylamine and an additional contribution of energy of the methylene group (4.6 kJ mol^{-1}) or 10.4 kJ mol^{-1}. The energy of the specific interaction, formed by the carbonyl oxygen atom and the methyl group at the pentacoordinated carbon atom, is calculated with the help Eq. 10.1b and is presented in Table 10.1.

DC=O→CH_2-H-N = ($\Delta_{vap}H^0$(298 K)aba - 2DN→CH_3-$(CH_2)_2$-CH_2-N - 4DC=O•••H-CH_2-N - 2DO=C→C=O) / 4 (10.1b)

The value of the energy of the interaction DC=O → CH_2(H)–N is comparable with the value of the energy of the specific interaction formed by the propyl ligand of liquid dipropilamine and diethyl ether (6.78 kJ mol^{-1}) and its reduced value

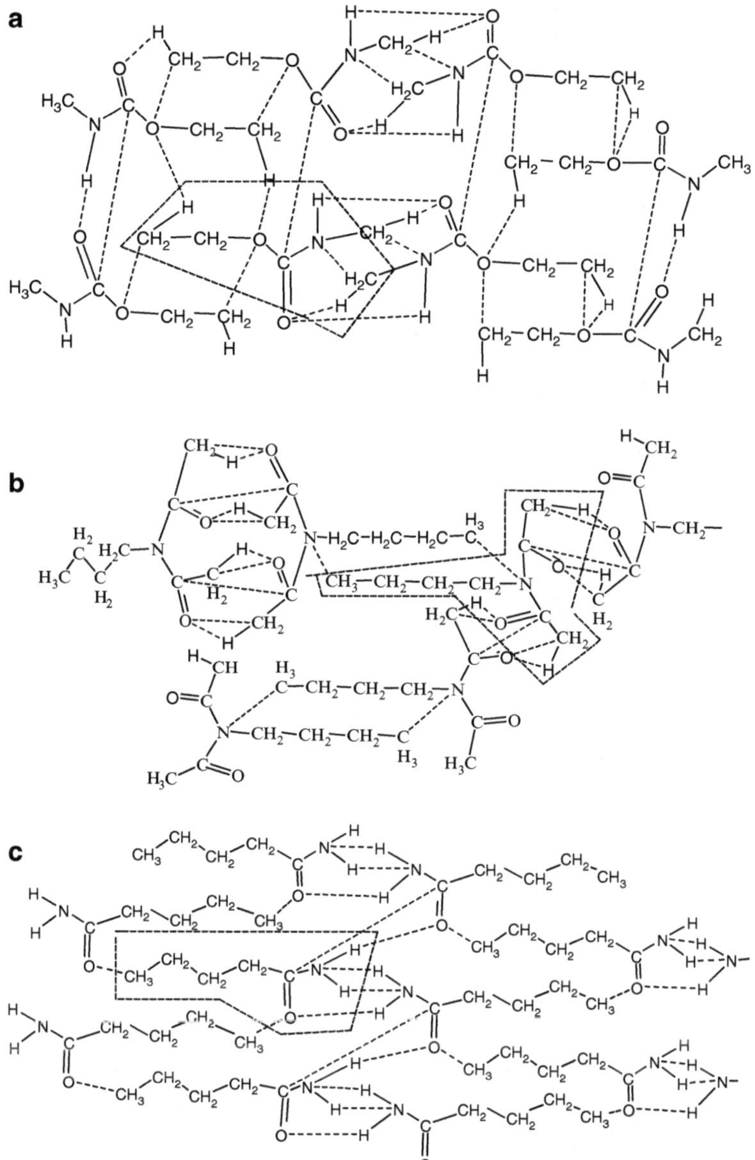

Fig. 10.1 Schematic picture of the crystalline structure of *N*-methyl carbamic acid, ethyl ester (**a**) and δ-valerolactam (**b**) with the network of the hydrogen bonds and the specific interactions

compared with the energy of the hydrogen bond D=C=O•••H–N of liquid *N*-methyl carbamic acid, ethyl ester, is consistent with the results of research of RE-spectroscopy, pointing to the reduction of the electron density at the nitrogen atom when hydrogen is replaced by the methyl group [6, 7]. The δ-valerolactam

Table 10.1 Energies of hydrogen bonds and specific interactions (kJ mol^{-1}) of *N*-acetyl-*N*-butylacetamide, δ-valerolactam

Compounds	Formula	Structure	$\Delta_{evap}H°$ (298K) [4, 5]	T K	DO → CH_3–CH_2–O	DN → CH_2(H)–N	DC=O•••H–N
		Liquid [4]		DC=O•••H–CH_2=DO•••H–CH_2–O = 0.60; DO=C → C=O = 3.6			
N-methyl carbamic acid, ethyl ester	$C_4H_9NO_2$	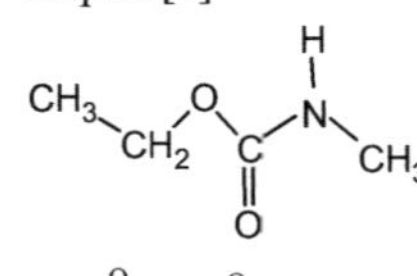	51.2 Liquid	314	6.78	4.2	9.1
N-acetyl-*N*-butylacetamide	$C_8H_{15}NO$	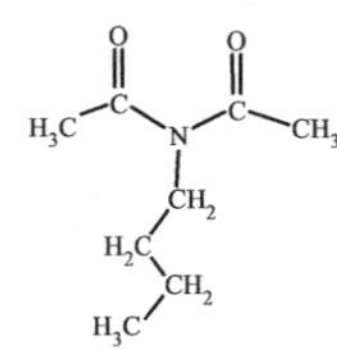	64.4 ± 0.4 Liquid	298	4.8 × 2 8.1 × 2 + 2	DC=OCH_2–N = 8.6	–
		Crystal [5]				DN•••H–N	
δ-valerolactam	C_5H_9NO	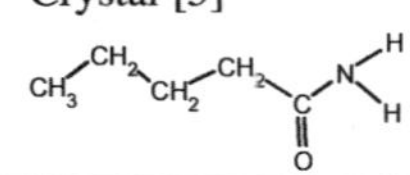	74.5	303	7.2	9.7	17.5

molecule with seven bond vacancies forms the crystal with the network for amides with two types of hydrogen bonds, D=C=O•••H–N and D–N•••H–N, and the specific interactions DO=C → C=O with known energies of 17.5, 9.7, and 5.7 kJ mol^{-1}, respectively, and the interaction D=C=O → CH_3–$(CH_2)_2$–CH_2–, formed by butyl ligand with the carbonyl oxygen atom. The energy of this type of interaction could be calculated with the help of Eq.10.1c:

$$DC{=}O{\rightarrow}CH_3\text{-}(CH_2)_2\text{-}CH_2\text{-} = (\Delta_{vap}H^0(298\ K)vl - 2DC{=}O{\cdots}H\text{-}N - 2DN{\cdots}H\text{-}N - DO{=}C{\rightarrow}C{=}O)/2 \quad (10.1c)$$

One should mention that such low energy values of the specific interactions formed by the alkyl group and carbonyl oxygen are characteristic for other compounds. Nevertheless they are characterized by stabilization with an increasing number of carbon atoms in the chain.

10.1.2 *Energies of Specific Interactions of Crystalline* N-*Acetylglycineamide,* N-*Acetyl-L-alanineamide, and* N-*Acetyl-L-valinamide*

Peptides have special significance in the development of scientific ideas, and therefore interest in them requires the determination of correct sublimation and vaporization enthalpies for a thermodynamic analysis and reliable spectral data on the structural characteristics of the condensed condition and behavior in the vapor phase. In this connection the process of developing thermodynamic research of peptides illustrates the transition from a cautious attitude toward the study of this abundance of molecular forms of compounds with complex composition [8–16] to obtaining a low value of the error of the correct thermodynamic properties of the simplest peptides [10, 12].

For the compounds of *N*-acethyl-glycineamides series it must be noted that at the small difference in the alkyl group of molecules of the simple peptides *N*-acetyl-L-alanineamide and *N*-acetyl-L-valinamide enthalpies, sublimations of these compounds are different at15.0 kJ mol^{-1} (Table 10.2). According to the data of [12] the *N*-acethyl-glycineamide molecule forms in the crystal the structure with a three-dimensional network of hydrogen bonds for the two-dimensional network structure of crystal *N*-acetyl-L-valinamide. The *N*-acethyl-glycineamide (NAGA) molecule with 12 bond vacancies, six of which are expressed by two oxygen atoms and two nitrogen atoms, lacks acceptor atoms taking the obtained excess electron density at the formation of the specific interactions and hydrogen bonds in crystal and liquid condition. In this connection the hydrogen atom of the methyl group, lying in the *trans*-position relative to the carbonyl oxygen atom [16], participates in the formation of the hydrogen bond with low stability D=C=O•••H–CH_2–N with the energy value in crystal equal to 1.0 kJ mol^{-1}. First a bulk crystal structure of *N*-acethyl-glycineamide is formed with three types of stable hydrogen bonds

Table 10.2 Energies of hydrogen bonds and specific interactions (kJ mol^{-1}) of crystalline – *N*-acethyl-glycineamide, *N*-acetyl-L-alanineamide, and *N*-acety-L-valinamide

Compounds	Formula	Structure	$\Delta_{sub}H°(T)$ [4]	D–N•••H–N,	D–N–CH_3 (CH_2) n ← N	D=C=O•••H–N	D=C=O•••H–N–CH_2–
DO=C → C=O = 5.7, D–N → CH_2–O = 5.6, D=C=O•••H–CH_2–N = 1.0							
N-acethyl-glycineamide, NAGA	$C_4H_8N_2O_2$	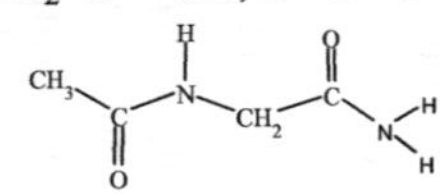	126.3 ± 2.3	9.7	n = 0 3.6	17.5	20.1
							D=C=O•••H–N–CH_2–*iso*
N-acetyl-L-alanineamide, L-NAAA	$C_5H_{10}N_2O_2$	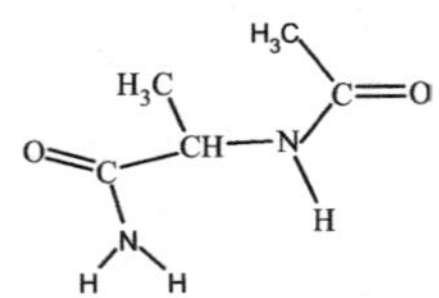	118.1 ± 1.6	9.7	n = 0 3.6	17.5	1.6.0
N-acety-L-valinamide, L-NAVA	$C_7H_{14}N_2O_2$	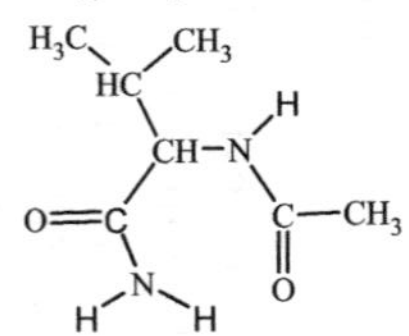	133.1 ± 2.2	9.7	n = 0 3.6	17.5	16.0
N-acetyl-L-leucineamide, L-NALA	$C_8H_{16}N_2O_2$	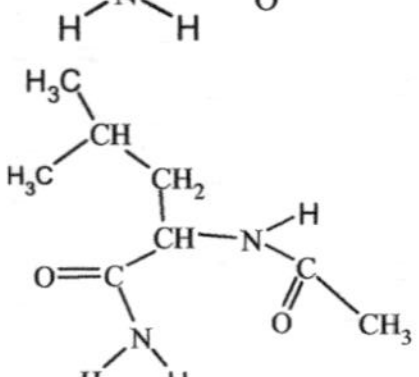	119.8 ± 1.5		9.1	17.5	1.6.0

N-acetyl-D-leucineamide D-NALA	$C_8H_{16}N_2O_2$	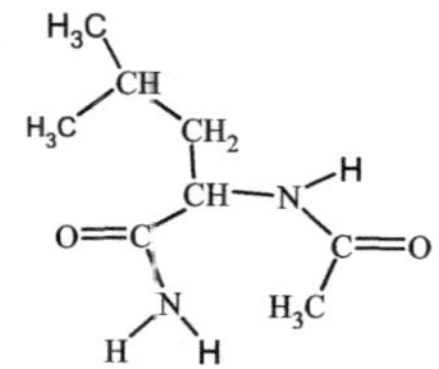	120.4 ± 0.4	9.7	9.5	17.5	1.6.0
Acetyl-L-isoleucineamide, L-NAIA	$C_8H_{16}N_2O_2$	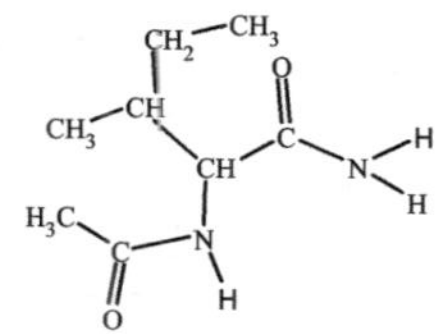	147.4 ± 0.3	9.7	14.3	17.5	20.1

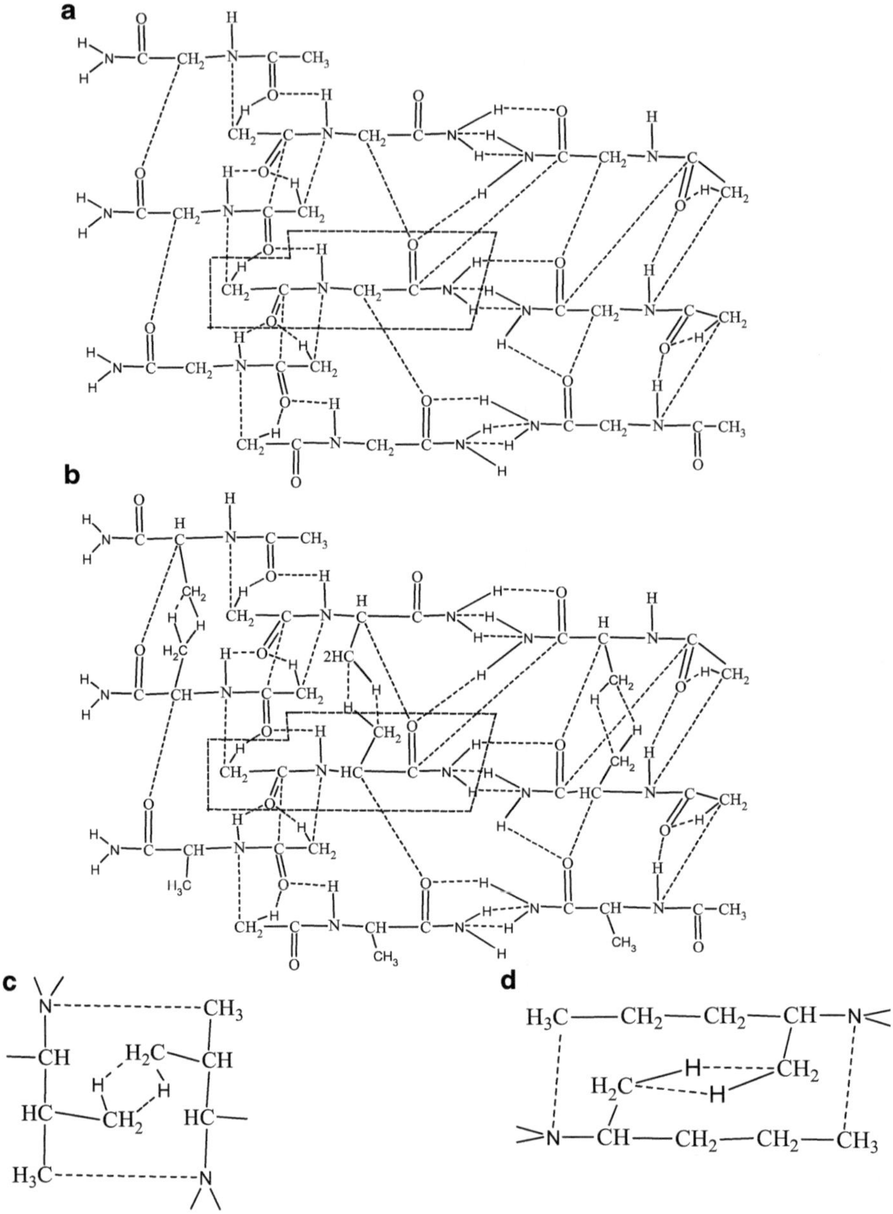

Fig. 10.2 Schematic picture of the crystalline structure of *N*-acethyl-glycineamide (**a**), *N*-acetyl-L-alanineamide (**b**), fragments of *N*-acety-L-valinamide (**c**), and acetyl-L-isoleucineamide (**d**) of the three-dimensional network of the hydrogen bonds and the specific interactions

D–N•••H–N, D=C=O•••H–N, and D–C=O•••H–N–CH_2–, ,and one with low stability, =DC=O•••H–CH_2–N (Fig. 10.2a). Second, three types of specific interactions are formed by the essentially unshared $2s^2$ electron pair of the carbon atom of the methylene group with the oxygen atom of the carbonyl group D=C=O → CH_2–N, the methyl group with the nitrogen atom of the amide group D–N → CH_2–H and a third specific interaction DO=C → C=O of the carbon atom of the carbonyl groups with energy value of 5.7 kJ mol^{-1}. It was shown earlier (Chap. 9), that the methylene group located between fragments of electron acceptors gives to the sublimation enthalpy a reduced energy contribution (3.6 kJ mol^{-1}) compared with the contribution of the similar group of the alkyl chain, therefore the energy of the specific interaction D=C=O → CH_2–N, formed by this group, takes an equal value. The energy of the specific interaction D–N → C(H)H_2–C with tetracoordinated nitrogen atom (5.6 kJ mol^{-1}) we take equal to the energy of crystalline dimethyl ether minus the energy of the interaction formed by the hydrogen atom with carbonyl oxygen of low stability (1.0 kJ mol^{-1}). The mentioned three types of hydrogen bonds formed differ by the values of the energies. The first two types, D–N•••H–N and D=C=O•••H–N, are formed by pentacoordinated nitrogen with the energies of 9.7 and 17.5 kJ mol^{-1} and its energies should have reduced values compared with the D=C=O•••H–N–CH_2– amide group formed with tetracoordinated nitrogen atom, which energy value should be calculated with the help of Eq. 10.2.

$$\text{D=C=O}\cdots\text{H-N-CH}_2\text{-} = (\Delta_{vap}H^0(298\text{ K})\text{naga} - 2\text{D=C=O}\cdots\text{H-N} - 2\text{D-N}\cdots\text{H-N} - 2\text{D=C=O}\rightarrow\text{CH}_2\text{-N} - 2\text{DO=C}\rightarrow\text{C=O} - 2\text{D-N}\rightarrow\text{CH}_2\text{-H} - 2\text{D=C=O}\cdots\text{H-CH}_2\text{-N}) / 2 \tag{10.2}$$

The energy value of the third type of hydrogen bond obtained (Table 10.3) points to its increased stability, which allowed the authors of crystal research of *N*-acethyl-glycineamide by the method of X-ray diffraction to verify the presence of the three-dimensional network with six hydrogen bonds [17].

We draw attention to the reduced value of the sublimation enthalpy of *N*-acetyl-L-alanineamide (L-NAAA) at 8.2 kJ mol¯1 compared with the same characteristic of *N*-acethyl glycineamidc (Table 10.3). It is a hint that the following holds:

The isostructural methyl group at the methylene group, located between two fragments with strong acceptor properties in direct connection with the nitrogen atom, first, provides the negative charge to the carbon atom and increased positive charge to the carbon atom of the same methylene group and, second, influences the shifting of the electron density in the chain and the reduction in the charges of the carbonyl oxygen atom and the nitrogen atom. As a result there appears a reduction in the difference of the negative and positive charges of the oxygen atoms and the hydrogen atom of the amide group, providing the reduction of stability of the peptide hydrogen bond D=C=O•••H–N–CH_2–*iso*. It follows that at the analogs of these types of specific interactions and hydrogen bonds in the *N*-acetyl-l-alanineamide and *N*-acethyl-glycineamide molecules (Fig. 10.2a, b), the main influence of the isostructural methyl group to the change in the shifting of the electron density in the molecule is expressed by the reduction of stability of the

Table 10.3 Energies of hydrogen bonds and specific interactions (kJ mol^{-1}) of crystalline of (2-pyridyl)acetamide at 298 K

Compounds	Formula	Structure	$\Delta_{sub}H°(T)$ [4]	D=HC → CH	D=N → CH	D–N•••H–N	D=C=O •••H–N
Pyridine	C_5H_5N	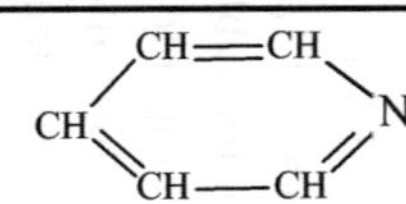	55.4 ± 2.0	7.4	12.85	–	–
DO=C → C=O = 5.7, DC=O → C = 3.3							
(2-pyridyl)acetamide	$C_7H_8N_2O$	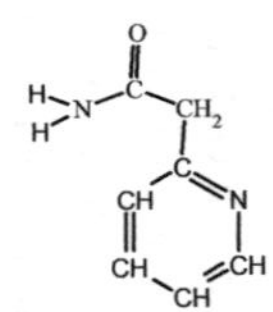	103.8	7.4	12.85	9.7	10.9

peptide hydrogen bond D=C=O•••H–N–CH_2–*iso*, formed by the amide group –N–H. Consequently, the value of the difference of the sublimation enthalpies *N*-acethyl-glycineamide and *N*-acetyl-L-alanineamide does not contain energy contributions of the isostructural methyl group and it only reflects destabilization of the peptide hydrogen bond D=C=O•••H–N–CH_2–*iso*. Consequently, the peptide group with the isostructural methyl group at the methylene group

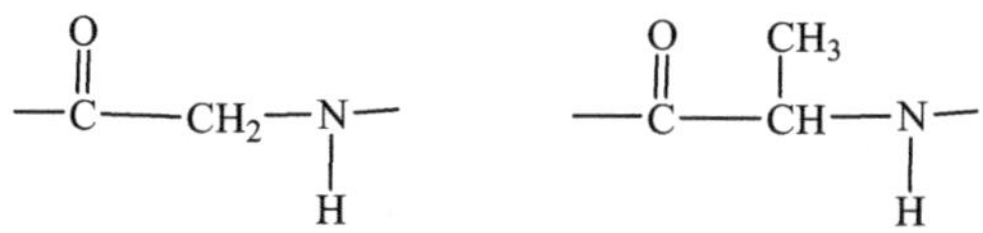

significantly changes the energy values of the peptide hydrogen bond formed. Thus, without any calculations, we obtain the energy value of this type D=C=O•••H–N–CH_2–*iso* of the peptide hydrogen bond as equal to 16.0 kJ mol^{-1}. Therefore, *N*-acethyl-glycineamide and *N*-acetyl-L-alanineamide amide molecules and the amino group form three types of hydrogen bonds D–N•••H–N and D=C=O•••H–N; each of the two types has the energy values of 9.7 and 17.5 kJ mol^{-1} and give rise to two types of peptide hydrogen bonds D=C=O•••H–N–CH_2–*iso* and D=C=O•••H–N–CH_2– with differing energy values, ranging in the integral values of stabilization.

D-N•••H-N (9.7) < D=C=O•••H-N-CH_2-*iso* (16.0) < D=C=O•••H-N (17.5) < D=C=O•••H-N-CH_2- (20.1 kJ mol^{-1})

Comparing the enthalpies sublimation of *N*-acetyl-L-alanineamide, *N*-acetyl-L-leucineamide, and *N*-acetyl-D-leucineamide (Table 10.2), we have to accept a small difference in the values of this thermodynamic characteristic of 1.7 and 0.6 kJ mol^{-1}, which points to the low contribution of the isostructural methyl group to the sublimation enthalpy. The sharp increase of energy contribution to the considered parameter of the isostructural group with two carbon atoms is caused by the change of specificity

of interaction of the methyl group, forming specific interactions of low stability, different from the ethyl and, more so, the propyl group, forming a different type of specific interaction with increased stability. Therefore, the accepted value of contribution of energy of the isostructural methyl group of 1.5 kJ mol^{-1}, which is within the experimental error, has a slightly reduced value compared with the value of the energy contribution of the ethyl (2.7 kJ mol^{-1}) and propyl groups (3.3 kJ mol^{-1}), obtained including the substituted hydrogen atom of the methylene group (1.0 kJ mol^{-1}). This value reflects the general integral value of the contributed energy to the sublimation enthalpy, in which the group is located between strong acceptors of the electron density by the carbonyl oxygen and nitrogen atoms.

When you consider the number and sequence of the location of the chain groups in the acethyl-glycineamide molecule and the derivatives of *N*-acetyl-L-alanineamide and *N*-acetyl-L-valinamide molecules, the difference of these compounds lies in the lengths and the branching of alkyl groups, located between two fragments that are strong acceptors of the electron density. The analogous crystal structure of these compounds with the network of six stable and two hydrogen bonds of low stability, three specific interactions with mentioned energies points to the following:

First to the fact that the location of oxygen of the carbonyl atom is in close proximity to the amino group of the *N*-acetyl-L-valinamide molecule.

Second, the increasing number of carbon atoms up to the isopropyl ligand provides it with practically the function of an independent ligand and to a lesser extent this depends on atoms with strong acceptor properties of electron density. This implies that this ligand, forming specific interactions (Fig. 10.2c), contributes energy to the sublimation enthalpy equal to the difference in the enthalpy characteristics of *N*-acetyl-L-valinamide and *N*-acetyl-L-alanineamide, i.e., equal to 15.0 kJ mol^{-1}. In other words, the total energy contributed by the specific interaction D–N → CH_3–CH_2–CH_2–N and the reduced stability of DH_3C → H–CH_2 is 7.5 kJ mol^{-1}. Note that the location of the propyl group between two fragments with the carbonyl oxygen atom and the nitrogen atom leads to a contribution to the sublimation enthalpy of compounds equal to 6.4 kJ mol^{-1} (Sect. 9.1). At the same time, the contribution of the propyl ligand to the vaporization enthalpy of dipropilamine is 8.1 kJ mol^{-1} and the isostructural methyl group is 2.3 kJ mol^{-1} (Chap. 2). Based on this, one can arrive at a conclusion of the influence of the location of the amino group, the amide group –N–H, the carbonyl groups in *cis*-, *trans*-positions, and the alkyl group on the distribution of the electron density in the molecule and to the value of the energy contribution of the alkyl fragments to the vaporization enthalpy of the derivatives of *N*-acethyl-glycineamides. This is confirmed by the values of the sublimation enthalpies of *N*-acetyl-L-alanineamide, L-NAAA (118.1), *N*-acetyl-L-leucineamide, L-NALA (119.8), and *N*-acetyl-D-leucineamide, D-NALA (120.4 kJ mol^{-1}). The increase in the number of chain carbon atoms of the alkyl fragment with the methylene group and with the isostructural methyl group at the second or third carbon atom of these compounds up to the propyl or butyl ligand CH_3–CH(CH_3)–CH_2–CH_2–, respectively, is accompanied by an increase in the energy contribution of 1.7 and 2.3 kJ mol^{-1}. Note that the total contribution of this ligand is equal to 9.0 and 9.5 kJ mol^{-1}, respectively, in crystals of other compounds. The energies of the hydrogen bonds and the specific interactions of the crystals of these compounds have identical values with crystalline *N*-acetyl-L-alanineamide (Table 10.2).

The *N*-acetyl-L-isoleucineamide molecule forms a space-lattice structure by the same types of hydrogen bonds and specific interactions similar to that of *N*-acetyl-L-alanineamide. The structure formed by the alkyl fragment (Fig. 10.2d) adds to the general structure of the crystal of *N*-acetyl-L-isoleucineamide. The increased sublimation enthalpy of *N*-acetyl-L-isoleucineamide, exceeding the value of the enthalpy characteristic of *N*-acethyl-glycineamide at 14.3 kJ mol^{-1}, allows us to consider

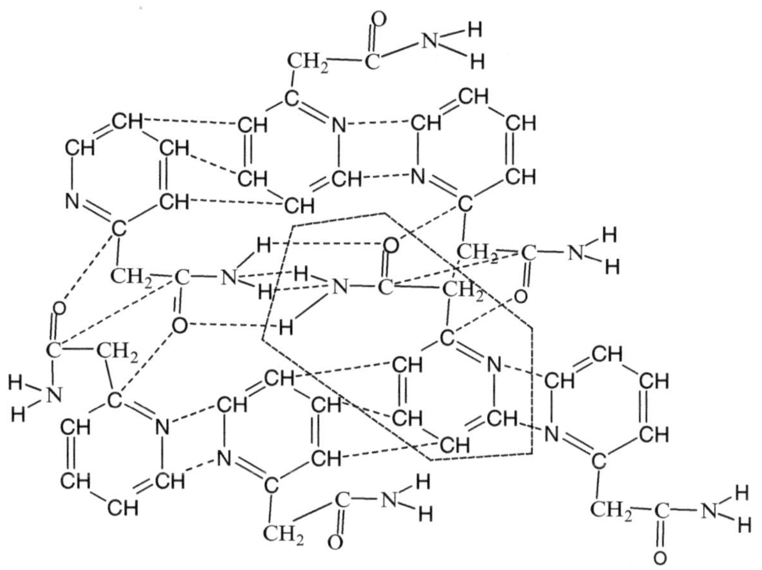

Fig. 10.3 Schematic picture of the crystalline structure of (2-pyridyl)acetamide with network of the hydrogen bonds and the specific interactions

this value as the total contribution of the propyl ligand (8.1 kJ mol^{-1}), the fourth methylene group (4.8 kJ mol^{-1}), and the isostructural methyl group (1.5 kJ mol^{-1}). The remaining energy value is the sum of the energy contributions of the hydrogen bonds and the specific interactions, corresponding to the energies formed by *N*-acetyl-L-isoleucineamide and *N*-acethyl-glycineamide molecules, given in Table 10.2.

10.1.3 Energies of Hydrogen Bonds and Specific Interactions of (2-Pyridyl)acetamide

The (2-pyridyl)acetamide molecule with the peptide group and increased acceptor ability experiences certain difficulties in the shifting of the electron density from the rigid ring of the hexamerous hetero cycle to the oxygen atom of the carboxyl group and the nitrogen atom of the amino group, and therefore has reduced negative charges and reduced differences in the charges of its atoms and hydrogen, forming peptide hydrogen bonds. As a result they form these bonds with reduced stability. The (2-pyridyl)acetamide molecule with 13 bond vacancies (Fig. 10.3) forms a hydrogen bond, D–N•••H–N, with an energy of 9.7 kJ mol^{-1} and a peptide hydrogen bond D=C=O•••H–N and the specific interaction DO=C → C=O and DC=O → C= with known energies of 5.7 and 3.3 kJ mol^{-1} and D=HC → CH, D=N → CH interactions formed by the CH groups and the nitrogen atom of the hexamerous hetero cycle with energy values of 7.4 and 12.85 kJ mol^{-1} (Table 10.3). The value of the energy contribution of the methylene group is estimated to be equal to 2.4 kJ mol^{-1} on the basis of the energy of the specific interaction of CH_3 group of 2-methylpyridine reduced by the contribution of the substituted hydrogen atom (1.0 kJ mol^{-1}) introduced for the crystalline condition. The energy of the peptide hydrogen bond is calculated with the help of the Eq. 10.3.

$$DC{=}O\bullet\bullet\bullet H\text{-}N = (\Delta_{vap}H^0(298\ K)paa - 2DN\bullet\bullet\bullet H\text{-}N - 3DHC{\rightarrow}CH - 2D{=}N{\rightarrow}CH - 2DC{=}O{\rightarrow}C{=} - DO{=}C{\rightarrow}C{=}O - 2DCH_2)/2 \tag{10.3}$$

The calculated energy value of this type of hydrogen bond illustrates its reduced stability compared with the one realized in crystalline acethyl-glycineamides, which is caused by the rigidity of the hexamerous hetero cycle.

10.2 Correlation of Temperature's Fusion with Energetically Parameters of Small Peptides

The measured vapor pressures of the peptides' sublimation process with low molecular mass by the torsion-effusion method for *N*-acethyl-glycineamide, *N*-acetyl-L-alanineamide, and *N*-acetyl-L-leucineamide, and sublimation enthalpies obtained on the basis of these results as well as measurements of enthalpies and fusion temperatures by differential scanning calorimetry and calculated entropies of this process are very important and necessary for various parts of chemistry, biochemistry, and biophysics [17–24]. The work of the authors [21] increased information on enthalpies and entropies of fusion of the simplest peptides and some substituted dipeptides, which were clarified by the use of cleaner starting compounds [23, 24]. With a view to estimate quantitatively the enthalpies and the entropies of fusion and sublimation obtained by analysis and from the results of the X-ray diffraction method [25] the authors [23] used the possible correlation between the fusion temperatures of thermodynamic properties of compounds and the number of identified hydrogen bonds. However, the results of the research of the X-ray diffraction method [25] were not correct for all compounds; in particular, for L-NAPA and L-NALA molecules. The L-NALA molecule forms six hydrogen bonds of three types, two of which have a difference in the energy values of 1.5 kJ mol^{-1} (Table 10.2), complicating decoding of the spectral data. This negative fact has an important positive direction, proving the correctness of the results of the thermodynamic analysis in the determination of the energy values of the hydrogen bonds and the specific interactions of crystalline small peptides, given in Table 10.2.

Having solved the problem of the chemical nature of phase changes and having based this phenomenon in the works of [26–28], we consider the temperature of polymorphic transformation and fusion, interconnected with the number of sever ties, its energies, and depths of the processes and entropy characteristic. In turn, the last characteristic is interconnected with the energy gaps of the specific intermolecular interactions or ionic bonds and its number in the crystal at the corresponding phase transformation. A polymorphic transformation proceeds as a result of the energy reduction of interaction with an increase of temperature and reduction of coordination number at the coordinating center and the number of ligands or molecules in molecular crystals. The two and almost three times reduced value of the energies of the specific interactions, formed by the essentially unshared $2s^2$ electron pair of the carbon atom, compared with the energies of the hydrogen bonds of three types, formed by *N*-acethyl-glycineamide and its derivatives, points to the prevailing influence of the hydrogen bonds of the mentioned types on the value of the fusion temperature. Therefore the low value of the fusion temperatures of *N*-acetyl-D-leucineamide and *N*-acetyl-L-leucineamide reflect a preferential gap of the specific interactions with reduced stability and total low energies contributed by possibly a very small number of less stable peptide hydrogen bonds D=C=O•••H–N–CH_2–*iso* (16.0), formed by the amide group –N–H with reduced stability compared to D=C=O•••H–N (17.5 kJ mol^{-1}) formed by the amino group of NH_2. Besides, the energies of the specific interactions formed by the isopropyl group of molecules of these compounds contribute significantly to the reduced energy values compared with other compounds with a similar isostructural group and practically do not exceed the energy value, introduced by the methylene group, located between the carbonyl oxygen atom and the amide group. The increased values of temperature (412.2 K), enthalpy (26.71 kJ mol^{-1}), and entropy (64.88 $JK^{-1}mol^{-1}$) of fusion of *N*-acethyl-glycineamide (Table 10.4) are caused by the gap of the same specific interactions, the stability of which is subsequently increased with the reduction of its number [26–28] and with the decrease in the coordinating molecules. A further increase of all parameters considered in the fusion of *N*-acetyl-L-prolinamide (L-NAPA) and the increased value of the fusion temperature with two types of four stable hydrogen bonds D=C=O•••H–N, formed by the carbonyl oxygen atom with location of the amide group in the hexamerous hetero cycle, reflects the partial gap of the specific interactions, formed with the participation of the CH groups of the hetero cycle and the methyl group with the carbonyl oxygen atom (Table 10.4). These interactions provide the exact depths of the vaporization process of the crystalline structure at the melting of the *N*-acetyl-L-prolinamide compound. Obviously, the depth of the vaporization process of the crystalline structure points to the low value of the energy of the four specific interactions D=C=O → CH_2, formed by the oxygen atom of the carbonyl group and the carbon atom of the CH_3 group of the hexamerous hetero cycle (Fig. 10.4a), the gap for which significantly increases the value of the entropy of melting and contributes to the enthalpy characteristic. However, the increased melting temperature of 1,4-dimethyl-2,5-piperazine dione(cyclo-glycilglycine) at the reduced values of enthalpy and entropy points to the increased stability of the same specific interactions, formed by the carbonyl oxygen atom and the methyl groups, located in *trans*-position at the hexamerous cycle of piperazine and to the reduction of the degree of reorganization of crystal structure at melting.

Table 10.4 Temperatures, enthalpies, and entropies of fusion of small peptides [23]

Compounds	Formula	Structure	T_{fus}, K	$\Delta_{fus}H^\circ$ (T) kJ mol^{-1}	$\Delta_{fus}S^\circ$ (T) J K^{-1}mol^{-1}
N-acetyl-D-leucineamide D-NALA	$C_8H_{16}N_2O_2$	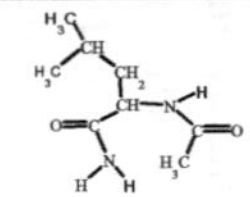	401.4 ± 0.3	16.2 ± 0.3	40.3 ± 0.7
N-acetyl-L-leucineamide, L-NALA	$C_8H_{16}N_2O_2$	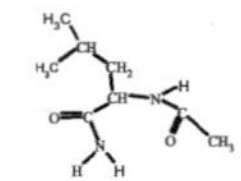	404.4 ± 0.2	16.55 ± 0.14	40.92 ± 0.35
N-acethyl-glycineamide, NAGA	$C_4H_8N_2O_2$	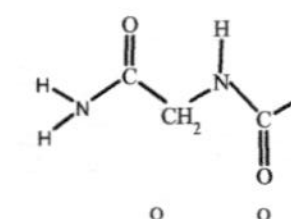	412.2 ± 0.2	26.71 ± 0.12	64.88 ±0.29
N-acetyl-L-prolinamide L-NAPA	$C_7H_{12}N_2O_2$	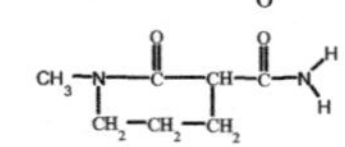	417.5 ± 0.2	29.34 ± 9.24	70.28 ± 0.58
1,4-dimethyl-2,5-piperazine dione(cyclo-glycilglycine), c-SARSAR	$C_6H_{10}N_2O_2$	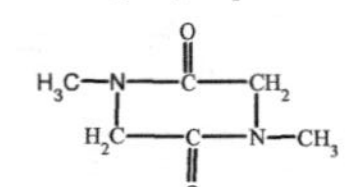	418.2 ± 0.35	22.04 ± 0.35	52.70 ± 0.85
N-acetyl-L-alanineamide, L-NAAA	$C_5H_{10}N_2O_2$	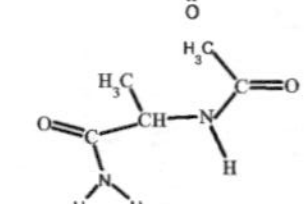	436.4 ± 0.2	23.59 ± 0.13	54.06 ± 0.30
N-acety-L-valinamide, L-NAVA	$C_7H_{14}N_2O_2$	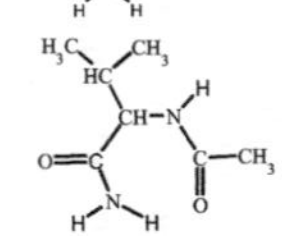	509.0 ± 0.3 509.0	39.10 ± 0.23	76.82 ± 0.45
N-acetyl-L-isoleucineamide, L-NAIA	$C_8H_{16}N_2O_2$	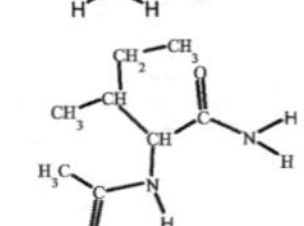	529.6 ± 0.2 529.6	41.81 ± 0.14	78.93 ± 0.26

2,5-piperazine dione(cyclo-glycilglycine), c-GG	$C_4H_6N_2O_2$	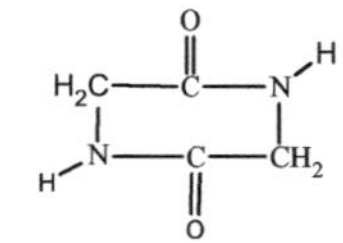	Dec 560	–	–
3-methyl-2,5-piperazine dione(cyclo-glycilglycine), c-GA	$C_5H_8N_2O_2$	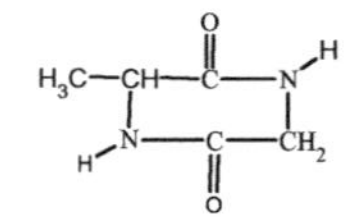	543.9 ± 0.2	30.62 ± 0.22	56.30 ± 0.40
3,6-dimethyl-2,5-piperazine dione(cyclo-glycilglycine), c-AA	$C_6H_{10}N_2O_2$	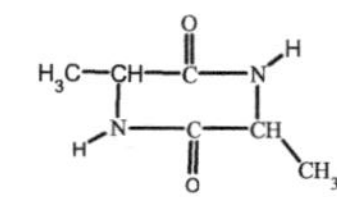	556.1 ± 0.3	30.64 ± 0.25	55.10 ± 0.45

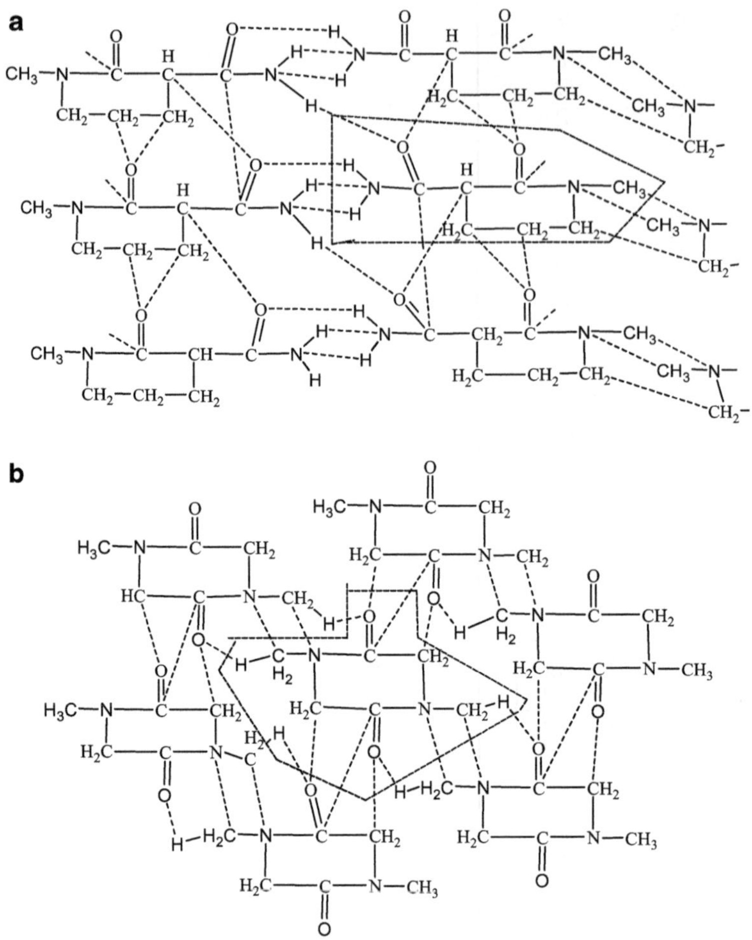

Fig. 10.4 Schematic picture of the crystalline structure of *N*-acetyl-L-prolinamide (**a**) and 4-dimethyl-2,5-piperazine dione(cyclo-glycilglycine) (c-SARSAR) (**b**) with the network of the hydrogen bonds and the specific interactions

The relatively high melting temperature points to the special depths of the crystal structure transformation in the melting process of 1,4-dimethyl-2,5-piperazine dione (cyclo-glycilglycine), (c-SARSAR), which molecule forms the crystalline structure of ten specific interactions 4D=C=O → CH_3, 4D–N → CH_3–N, 2DO=C → C=O, formed by the essentially unshared $2s^2$ electron pair of the carbon atom with reduced stability (Table 10.4). Simultaneously, *N*-acetyl-L-prolinamide with stable hydrogen bond 2D–N•••H–N, 2D=C=O•••H–N (Fig. 10.4b), and four hydrogen bonds D=C=O•••H–CH_2 of low stability are melting at slightly different temperatures and with higher degrees of the transformation of the crystal structure. This implies that at the melting of this compound there are gaps for specific interactions with reduced hydrogen bonds of low stability. As a result the relatively high values of enthalpies and entropies of melting have values comparable with the values of the same characteristics of *N*-acetyl-L-alanineamide (L-NAAA). However, in the netted structure of the crystal of this compound peptide, hydrogen bonds with high energy values are practically not affected (Table 10.2) and the network is preserved. The sharp increase of the melting temperature of *N*-acetyl-L-valinamide (509.0 K) and the enthalpies (39.10 ± 0.23 kJ mol^{-1}) of melting point to the approximation of the properties of the ethyl group, with the isostructural methyl group located at the first carbon atom in the function of the ligand and therefore, the stability of the specific interaction formed is increased. In this connection the degree of transformation of the crystal structure in the temperature sequence is relatively reduced, but its temperature dependence is certainly increased, the enthalpy of melting is increased and the entropy factor (76.82 ± 0.45 JK^{-1}mol^{-1}) in the system increases. The even higher parameters for the transformation of the structure of acetyl-L-isoleucineamide, the temperature (T = 529 K), enthalpy (41.81 kJ mol^{-1}), and entropy (78.93 JK^{-1} mol^{-1}) of melting point to the stabilization of the interaction, formed by the alkyl fragment with increasing number of carbon atoms in the chain with the isostructural methyl group located at the first carbon atom in the isopropyl ligand (Table 10.4). Thus, on the basis of the analysis one can conclude that maintaining practically unchanged all realized specific interactions and hydrogen bonds in crystalline *N*-acethyl-glycineamide and its derivatives, these compounds are divided into two groups. The first group consists of compounds with the propyl isostructural group with the isomeric methyl group at the second carbon atom of the chain *N*-acetyl-D-leucineamide and *N*-Acetyl-L-leucineamide with inherent reduced temperature of melting and transformation of the crystal structure. The second group of compounds of *N*-acetyl-L-alanineamide, *N*-acetyl-L-valinamide, and *N*-acetyl-L-isoleucineamide is characterized by the stabilizing influence of an increasing number of carbon atoms of the chain from the methyl to the propyl fragment with the isomeric methyl group located at the first carbon atom of the chain.

The definite correlation of the melting temperatures and energy parameters of the intermolecular interactions indicates the fact that the least low melting temperature of *N*-acetyl-L-leucineamide points to their formation by the propyl fragment with the isomeric methyl group located at the second carbon atom of the chain

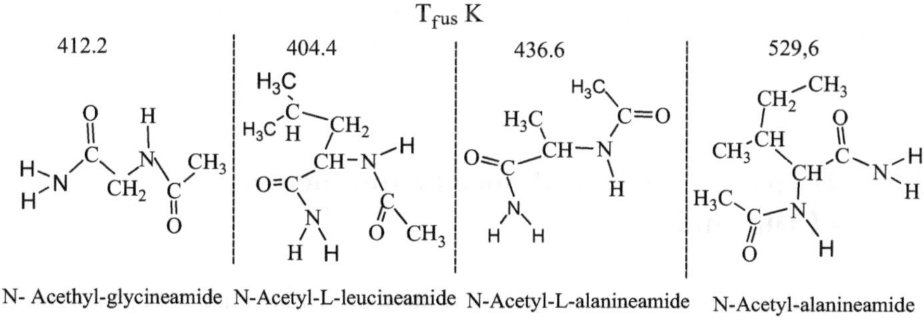

that is less stable compared with the isomeric isostructural methyl group formed of *N*-acetyl-L-alanineamide and especially with the propyl fragment with the isomeric

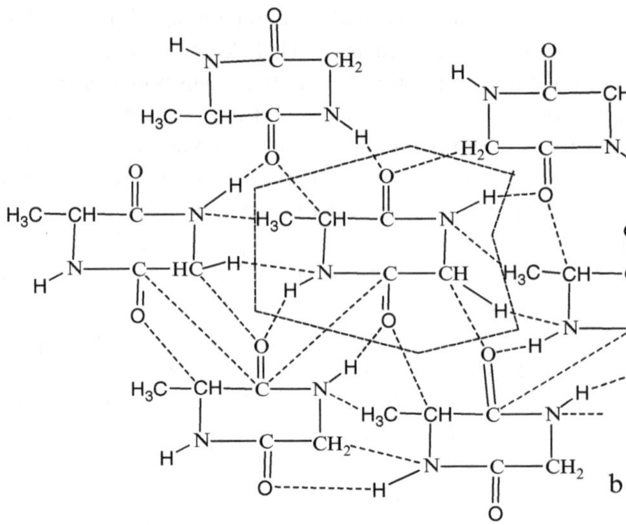

Fig. 10.5 Schematic picture of the crystalline structure of 3-methyl-2,5-piperazine dione(cyclo-glycilglycine), (c-GA), with the network of the hydrogen bonds and the specific interactions

methyl group located at the first carbon atom of the chain of the *N*-acetyl-L-isoleucineamide molecule. Increasing the length of the chain of this ligand is accompanied by stabilization of the specific interaction formed and the energy of the peptide hydrogen bonds will practically be kept unchanged.

The compounds of the series of 2,5-piperazine dione(cyclo-glycilglycine), 3-methyl-2,5-piperazine dione(cyclo-glycilglycine) (c-GA), and 3,6-dimethyl-2,5-piperazine dione(cyclo-glycilglycine) (c-AA) with four peptide hydrogen bonds D=C=O•••H–N–CH_2– and eight specific interactions 4D=C=O → CH_2, 2D–N → CH_3 and 2DO=C → C=O, forming the network crystal structure (Fig. 10.5), have a high temperature of destruction (560 K), melting (556.1 and 543.9 K), and relatively reduced degrees of the transformation of crystal structure on melting (Table 10.4) compared with *N*-acetyl-L-isoleucineamide, reflecting the high stability of the specific interactions formed. The presence of the two isostructural methyl groups with location at the hexamerous hetero cycle in *trans*-position points to the increased stability of the specific interactions formed.

10.3 Hydrogen Bonds and Specific Interactions of Dipeptides

The dipeptide molecules with an increased number of functional groups and bond vacancies compared with simplest peptides and with three oxygen atoms of the carbonyl groups and three nitrogen atoms with strong acceptor properties are deficient in donor functional groups, representing the electron density. Dipeptides

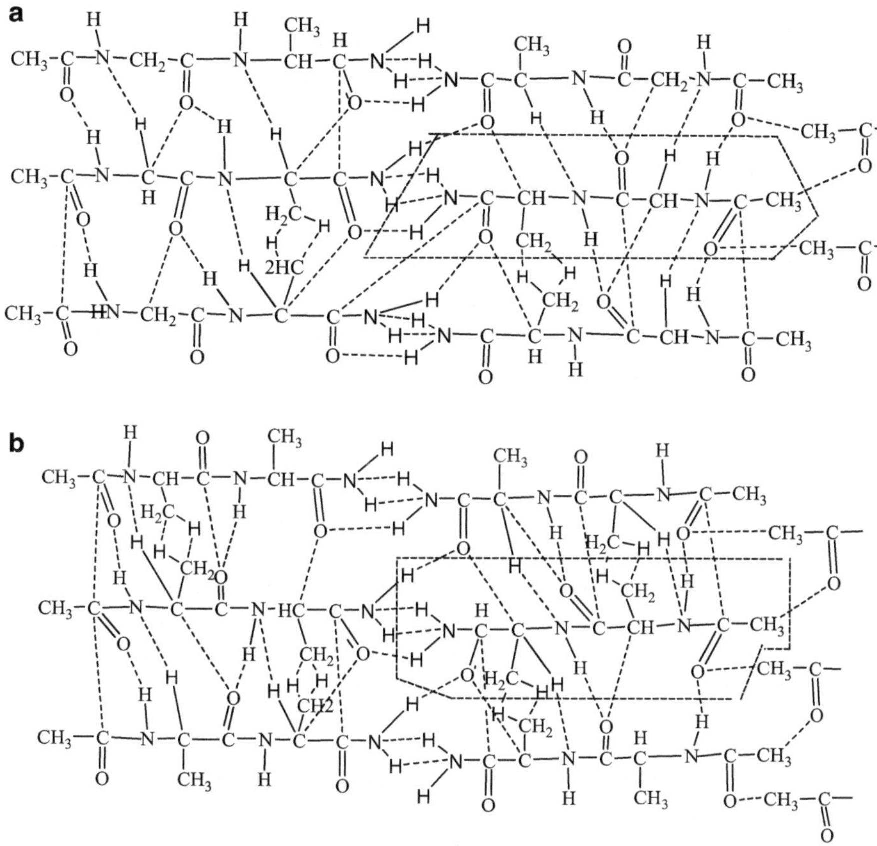

Fig. 10.6 Schematic picture of the crystalline structure of *N*-acetyl-glycil-L-alaninamide (NAGAA) (**a**) and *N*-acetyl-L-alanyl-L-alaninamide (**b**) with the network of the hydrogen bonds and the specific interactions

N-acetyl-glycil-L-alaninamide (NAGAA) and *N*-acetyl-L-alanyl-L-alaninamide (NAA_2A) with an equal number of bond vacancies and one and two isostructural methyl groups, respectively, form eight hydrogen bonds of three types, DN•••H–N (two), D=C=O•••H–N (two), and D=C=O•••H–N–CH_2 (four) and 12 specific interactions (Fig. 10.6). The crystalline structure of these compounds with the network of hydrogen bonds and specific interactions contains 4D=C=O → CH and 2D=C=O → CH_3 of increased stability, formed with the terminal methyl groups, and three interactions DO=C → C=O with an energy of 5.7 kJ mol^{-1}. In addition to the mentioned interactions, nitrogen atoms of the amide groups with free electron pairs form with the hydrogen atoms of the methylene groups the hydrogen bonds 3D–N•••H–CH, of low stability (1.0 kJ mol^{-1}).

A distinctive feature of these dipeptides is the presence the isostructural methyl group in the *N*-acetyl-glycil-L-alaninamide molecule and two of the same groups in the molecule chain of *N*-acetyl-L-alanyl-L-alaninamide, forming two and four specific interactions $DH_3C \rightarrow H–CH_2$, of low stability. Nevertheless, the principal difference of these compounds is interconnected with the same isostructural methyl groups, one in the *N*-acetyl-glycil-L-alaninamide molecule and two in the *N*-acetyl-L-alanyl-L-alaninamide molecule. The molecules of these compounds form two and four peptide hydrogen bonds, D=C=O•••H–N–CH_2–*iso* (16.0 kJ mol^{-1}), of reduced stability (Fig. 10.6). On this basis one can conclude that carbonyl and amide groups, fringed by the isostructural methyl groups, form hydrogen bonds with significantly reduced stability, rather than similar terminal fragments of the same molecule and therefore its contribution to the vaporization enthalpy has a relatively small value. Thus, the reduced melting temperature (Table 10.5) and melting enthalpy of *N*-acetyl-L-alanyl-L-alaninamide is caused by the gap of a small number of specific interactions with reduced stability at the gap of twice the number of hydrogen bonds of low stability, significantly increasing the sublimation entropy of this compound. The increased melting temperature of *N*-acetyl-glycil-L-alaninamide points to the gap of the larger number of specific interactions with relatively increased stability compared with the one formed by the same center with substituted hydrogen atom, formed at the chiral center (methylene group) and hydrogen bonds D=C=O•••H–N–CH_2–*iso*. The even lower melting temperature of *N*-acetyl-L-propyl-glycinamide illustrates the gap of the specific interactions, formed by the CH_2 group of the propyl cycle with the carbonyl oxygen atom and the nitrogen atom of the amide group due to the deficiency of acceptors of electron density.

We note that the terminal fragment of *N*-acetyl-glycil-L-alaninamide and *N*-acetyl-L-alanyl-L-alaninamide molecules with the amino group forms two hydrogen bonds D–N•••H–N, the energies of which depend on the charge at the nitrogen and the hydrogen atoms of this group, and therefore the energy value of this interaction is little affected by the influence of the isostructural methyl group. As a result the energy value of this type of hydrogen bond remains unchanged and is equal to 9.7 kJ mol^{-1}. The energy of the second type of hydrogen bond, D=C=O•••H–N, formed by the second hydrogen atom of the amino group with the oxygen atom of the carbonyl group,

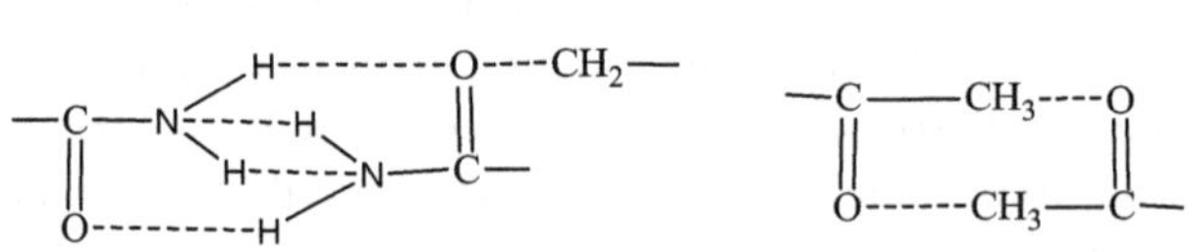

directly connects with the methylene group and gives it less electron density. However, it is partly restored by the shifting of the electron density from the nitrogen atom, leading to the increased difference with the hydrogen atom of the same amino group and formation of the hydrogen bond D=C=O•••H–N with an energy of 17.5 kJ mol^{-1}. In turn the same carbonyl oxygen atom forms a specific interaction with the carbon atom of the methylene group with an energy value of 3.6 kJ mol^{-1}.

Table 10.5 Temperatures, enthalpies, and entropies of fusion of dipeptides [21]

Compounds	Formula	Structure	T_{fus}, K	$\Delta_{fus}H°$ (T) kJ mol^{-1}	$\Delta_{fus}S°$ (T) J K^{-1}mol^{-1}
N-acetyl-glycil-L-alaninamide, NAGAA	$C_7H_{13}N_3O_3$	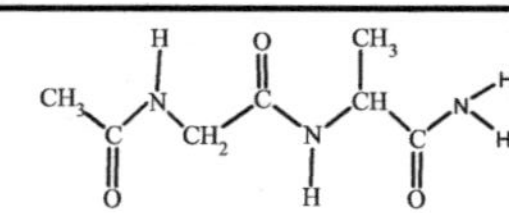	522.7 ± 0.2	50.6 ± 0.5	96.8 ± 1.0
N-acetyl-L-alanyl-L-alaninamide, NAA$_2$A	$C_8H_{15}N_3O_3$	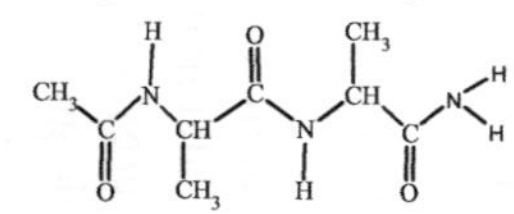	511.2 ± 0.4	38.1 ± 0.8	74.5 ± 1.6
N-acetyl-L-propyl-glycinamide, NAPGA	$C_9H_{15}N_3O_3$	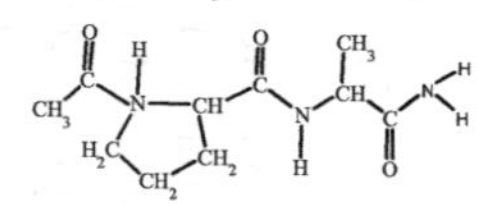	434.1 ± 0.1	32.2 ± 0.4	73.7 ± 0.5

The oxygen atoms of the carbonyl group and the nitrogen atom –N–H of the peptide group of the central fragment of *N*-acetyl-glycil-L-alaninamide and *N*-acetyl-L-alanyl-L-alaninamide molecules receive less electron density from the methylene groups connected with the isostructural methyl group as a result of the partial shifting to the carbon atom of the same group and, therefore, have

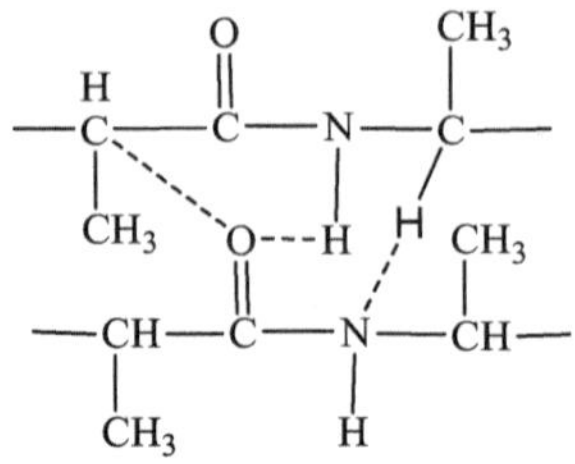

reduced values of negative charges and reduced differences in the charges of the oxygen atom and hydrogen atoms of the amide group. As a result, the peptide hydrogen bond D=C=O•••H–N–CH_2–*iso* should be less stable compared with the hydrogen bond D=C=O•••H–N formed and the energy of the second type of peptide hydrogen bond D=C=O•••H–N–CH_2 (20.1 kJ mol^{-1}). Thus, one obtains values of energies of these types of hydrogen bonds that are

$$D{=}C{=}O\cdots H\text{-}N\ (17.5) > D{=}C{=}O\cdots H\text{-}N\text{-}CH_2\text{-}iso\ (16.0\ \text{kJ mol}^{-1})$$

accurate and reflect the specific nature of the interaction. The last or the second peptide group *N*-acetyl-L-alanyl-L-alaninamide forms a similar type of peptide hydrogen bond with a reduced energy value of 16.0 kJ mol^{-1}. The oxygen atom of the carbonyl group of this peptide group for *N*-acetyl-L-alanyl-L-alaninamide and *N*-acethyl-glycineamide shifts the electron density from the carbon atom of the terminal methyl group and provides it with an increased charge compared with the methylene group. Therefore, the carbonyl atom of oxygen gives more electron density to this carbon atom of the methyl group at the formation of the specific interaction D=C=O → CH_3, providing it with increased stability and leading to the inequality in the energies of interactions of D=C=O → CH_3 > D=C=O → CH_2 with the specific interaction of the methylene group.

The *N*-acethyl-glycineamide molecule with the fringing methylene group, with strong acceptors of the electron density in the oxygen atom of the carbonyl group and nitrogen atom of the –N–H group in the absence of the isostructural methyl group, forms the peptide hydrogen bond D=C=O•••H–N–CH_2– with an energy of 20.1 kJ mol^{-1}. It points to the similarity of formation of this peptide hydrogen bond with the peptide fragments of other dipeptide molecules and this allows us to take the energy value of the same type of interaction as equal to the value for *N*-acetyl-glycil-L-alaninamide and, consequently, it reflects the accuracy of the inequality in the energies of the hydrogen bonds formed,

D-N•••H-N (9.7) < D=C=O•••H-N-CH_2-*iso* (16.0) < D=C=O•••H-N (17.5) > D=C=O•••H-N-CH_2- (20.1 kJ mol^{-1})

The thermodynamic analysis allows us to state that the netted crystalline structure of *N*-acetyl-L-alanyl-L-alaninamide is formed by hydrogen bonds D–N•••H–N (9.7), D=C=O•••H–N (17.5), D=C=O•••H–N–CH_2–*iso* (16.0), and D–N•••H–CH (1.0 kJ mol^{-1}) and the specific interactions D=C=O → CH_2 (3.6), DO=C → C=O (5.7) and DC=O → CH_3 with energy value of 5.80 kJ mol^{-1} and it has analogs with the network of liquid methyl 2-(*N*,*N*-dimethylamino). The sublimation enthalpy of crystalline *N*-acetyl-L-alanyl-L-alaninamide (NAA_2A) was calculated with the help of Eq. 10.4, including the contributions of the energies of all hydrogen bonds and the specific interactions and the contribution of two isostructural methyl groups (1.5 × 2 kJ mol^{-1}),

$$\Delta_{vap}H^0(298K)_{NAA_2A} = 2D\text{-}N{\cdots}H\text{-}N + 2D{=}C{=}O{\cdots}H\text{-}N + 4D{=}C{=}O{\cdots}H\text{-}N\text{-}CH_2\text{-} + 4D{=}C{=}O{\rightarrow}CH_2 + 2D{=}C{=}O{\rightarrow}CH_3 + 3D\text{-}N{\cdots}H\text{-}CH + 3DO{=}C{\rightarrow}C{=}O + 2D iso CH_3) \quad (10.4)$$

The value of sublimation enthalpy of *N*-Acetyl-L-alanyl-L-alaninamide obtained is equal to 157.8 kJ mol^{-1}. Taking into account the energy value of the peptide hydrogen bond (20.1 kJ mol^{-1})

$$\Delta_{vap}H^0(298\ K)_{NAGAA} = 2D\text{-}N{\cdots}H\text{-}N + 2D{=}C{=}O{\cdots}H\text{-}N + 2D{=}C{=}O{\cdots}H\text{-}N\text{-}CH_2\text{-} + 2D{=}C{=}O{\cdots}H\text{-}N\text{-}CH_2\text{-iso} + 4D{=}C{=}O{\rightarrow}CH_2 + 2D{=}C{=}O{\rightarrow}CH_3 + 3D\text{-}N{\cdots}H\text{-}CH + 3DO{=}C{\rightarrow}C{=}O + D iso CH_3) \quad (10.4a)$$

of crystalline *N*-acetyl-glycil-L-alaninamide (NAGAA) and Eq. 10.4a, we obtained the sublimation enthalpy of this compound to be equal to 164.3 kJ mol^{-1}. All energy values of the hydrogen bonds of the peptides, discussed in the previous chapters and given in Table 10.6, illustrate its consistency and adequate reflection of the nature of the interactions.

10.4 Energies of Hydrogen Bonds, Specific Interactions, and the Configurations of Dipeptide Chains

Maintaining the stability of the energies values of the formed hydrogen bonds and the specific interactions of the repeating sequence of carbonyl, amide =N–H, and methylene groups allows us to obtain the values of the energy contributed by the polypeptide chain configuration and information on the energy values of the terminal functional groups. The available information on the energy contribution of the isostructural methyl group, connected with the methylene group, and

Table 10.6 Energies of the hydrogen bonds of the peptide series

Compounds	Formula	Structure	D=C=O•••H–N	Compounds	Formula	Structure	D=C=O•••H–N
Liquid				Cristal			
2-Methoxyethylamine	C_3H_9NO	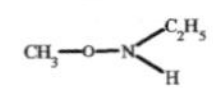	–	6-Caprolactam	$C_6H_{11}NO$	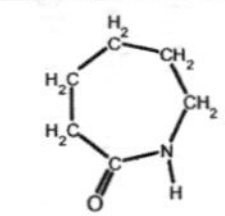	16.5
Methyl carbamate	$C_2H_5NO_2$	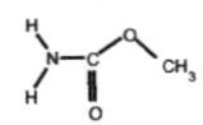	7.8	2-Aminotropone[a]	C_7H_7NO	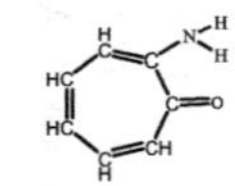	8.4
N-methyl carbamic acid, ethyl ester	$C_4H_9NO_2$	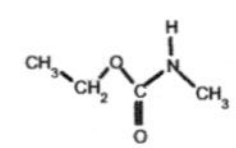	9.1	3-Pyridinecarboxamide *N*-oxide	$C_6H_6N_2O_2$	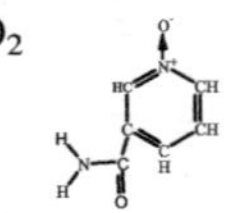	9.2
Ethyl carbamate	$C_3H_7NO_2$	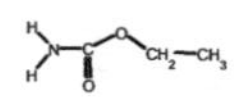	12.3	4-Pyridinecarboxamide *N*-oxide	$C_6H_6N_2O_2$	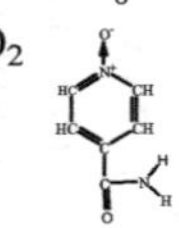	13.0
Propyl carbamate	$C_4H_9NO_2$	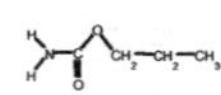	13.1	Acetamide	C_2H_5NO	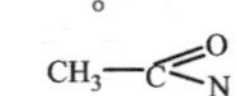	17.5
2-Pyrrolidone	C_4H_7NO	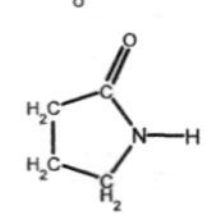	14.45	*N*-methylacetamide	C_3H_7NO	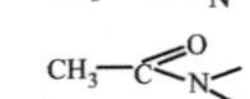	17.6
Acetamide	C_2H_5NO	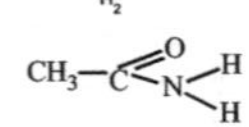	15.3	*tert*-Butylacetamide	$C_6H_{13}NO$	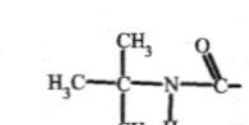	17.6
N-methyl-acetamide	C_3H_7NO	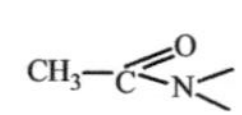	15.3	Acetanilide, Phenylacetamide	C_8H_9NO	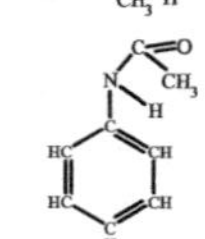	17.1
N-ethyl-acetamide	C_4H_9NO	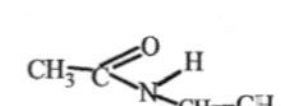	15.3	*N*-(2-Methylphenyl) acetamide	$C_9H_{11}NO$	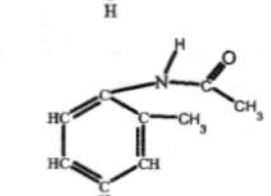	17.1

N-propyl-acetamide	$C_5H_{11}NO$	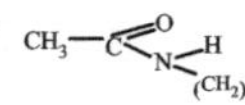	15.3	*N*-(4-Methylphenyl) Acetamide	$C_9H_{11}NO$	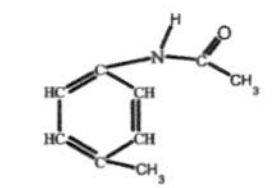	17.1
N-methyl-2- methyl-propionamide	$C_5H_{11}NO$	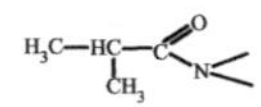	15.3	3-Pyridinecarboxamide	$C_6H_6N_2O$	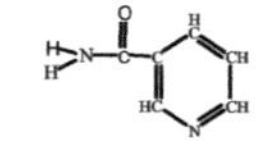	21.3 ± 3.0
–	–	–	–	4-Pyridinecarboxamide	$C_6H_6N_2O$	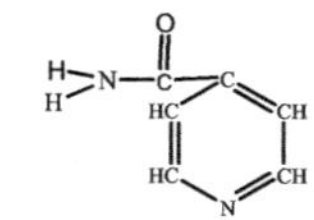	18.8 ± 3.0
–	–	–	–	Aminoformic acid	CH_3NO_2	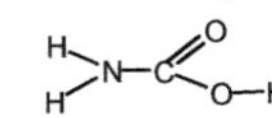	17.5
–	–	–	–	Glicine, aminoacetic acid	$C_2H_5NO_2$	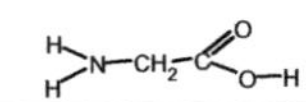	17.5

[a]2,4,6-Cycloheptatriene – 1-one, 2-amine

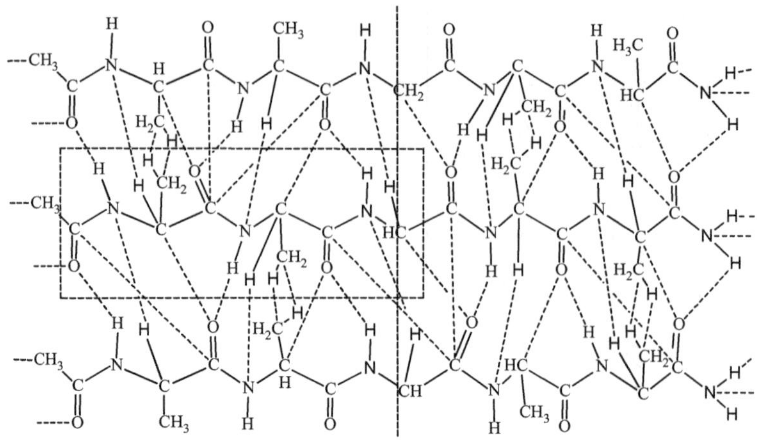

Fig. 10.7 Schematic picture of the crystalline structure of tetrapeptede of the derivative *N*-acetyl-glycil-L-alaninamide, NAGAA, with the network of the hydrogen bonds and the specific interactions

relative to other alkyl groups allows us to significantly broaden the possibilities of thermodynamic analysis for obtaining information on the configuration of the polypeptide chain of a number of proteins. The information obtained on the energies of the peptide hydrogen bonds and the specific interactions of dipeptide *N*-acetyl-L-alanyl-L-alaninamide, NAA_2A, should be used for obtaining the energy contribution of the peptide chain configuration and sublimation enthalpy of the tetrapeptide. The schematic picture presented in Fig. 10.7 of the crystalline structure of *N*-acetyl-glycil-L-alaninamide (NAGAA) with the network of peptide hydrogen bonds and the specific interactions allow us to make a choice of its types and number, the component figuration of the peptide chain, the terminal functional groups, and to determine its energies. The dipeptide configuration of *N*-acetyl-glycil-L-alaninamide (NAGAA) includes two types of peptide hydrogen bonds $D{=}C{=}O{\cdots}H{-}N{-}CH_2{-}$ and $D{=}C{=}O{\cdots}H{-}N{-}CH_2{-}iso$, differing in the values of their energies, caused by the influence of the isostrucutural methyl group on the reduction of the electron density from the carbon atom of the methylene group to the oxygen atoms of the carbonyl group and nitrogen atom, the fringing CH_2 group of the peptide chain, and reducing the stability of the formed hydrogen bond.

Therefore at three atoms of nitrogen and oxygen of three carbonyl groups, *N*-acetyl-glycil-L-alaninamide molecule forms four peptide hydrogen bonds $D{=}C{=}O{\cdots}H{-}N{-}CH_2{-}$ with an energy of 20.1 and two $D{=}C{=}O{\cdots}H{-}N{-}CH_2{-}iso$ with reduced stability and an energy value of 16.0 kJ mol^{-1}. Each nitrogen atom is in a tetra-coordinated condition and forms a hydrogen bond $D{-}N{\cdots}H{-}CH$ of reduced stability with the hydrogen atom of the methylene group with an energy of 1.0 kJ mol^{-1} at a total of six in the peptide chain configuration. This configuration of the peptide chain of the compound includes four specific interactions $DO{=}C \rightarrow C{=}O$ with energies of 5.7 kJ mol^{-1} and six interactions $D{=}C{=}O \rightarrow$

CH_2, formed by methylene groups with energy values of 3.6 kJ mol^{-1}. An additional contribution of energy introduces two isostructural methyl groups with a value of 1.5 mol^{-1} or specific interactions formed with low stability and energies of 0.75 kJ mol^{-1}. The terminal functional fragments of the peptide chain with the ethyl and amide group form two specific interactions andtwo hydrogen bonds

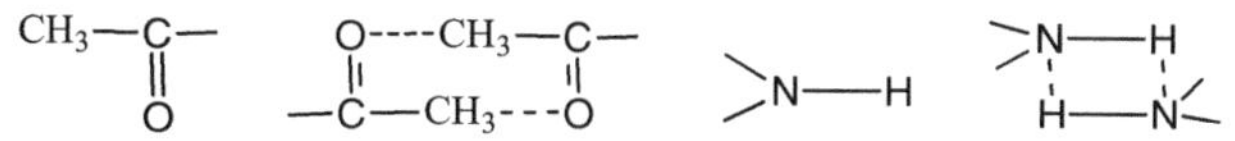

with the energies of 5.8 and 9.7 kJ mol^{-1}, respectively. Therefore, the energies contributed by the peptide hydrogen bonds is 112.4 kJ mol^{-1} or the total contribution of all hydrogen bonds, including the energies of the hydrogen bonds D–N•••H–CH, is of 118.4 kJ mol^{-1}. The total energy of the realized stable and low stable specific interactions $DH_2C \rightarrow H–CH$ is equal to 47.4 kJ mol^{-1} and the energy of the peptide chain configuration is 165.8 kJ mol^{-1}. Contributions of the terminal functional groups are 11.8 and 19.4 kJ mol^{-1} ,respectively. The energies of the peptide chain configuration and terminal functional groups obtained allow us to calculate the sublimation enthalpy of the tetrapeptede with this configuration (362.8 kJ mol^{-1}) and other peptide forms of compounds.

The crystalline structure with a network of hydrogen bonds and the specific interactions of dipeptide *N*-acetyl-L-alanyl-L-alaninamide, NAA_2A, differ from the considered one by the influence of an additional one, the isostructural methyl group, which results in increasing the number from two to four peptide hydrogen bonds DC=O•••H–N–CH_2–*iso* with reduced stability (16.0 kJ mol^{-1}) and an increase to eight of the number of specific hydrogen bonds, formed by the isostructural methyl groups. As a result, the total energy value of contribution of the peptide hydrogen bonds (104.2 kJ mol^{-1}) and the total contribution of the hydrogen bonds (110.2 kJ mol^{-1}) is reduced. The contribution of the energies of the specific interactions to the configuration of the peptide chain (50.4 kJ mol^{-1}) is increased as a result of increasing of number of specific interactions of low stability and the total energy contribution of the peptide chain configuration, formed by dipeptide *N*-acetyl-L-alanyl-L-alaninamide, is equal to 160.6 kJ mol^{-1}. The calculated value of the sublimation enthalpy of the formed tetrapeptede with a configuration of the peptide chain *N*-acetyl-L-alanyl-L-alaninamide is equal to 352.2 kJ mol^{-1}.

10.5 Energies of Hydrogen Bonds, Specific Interactions, and the Configurations of Dipeptide Chains of Same Proteins

There is an increasing interest in amino acids and in the derivatives of *N*-acethylglycineamides and peptides, which are the simplest structural elements of proteins, affecting different aspects of their properties. The the amino acid and peptide molecules in solutions and crystalline condition are in zwitter-ionic form [29–31], and we consider this condition for the molecular crystal [32] with stable peptide hydrogen bonds. In this connection we notice that they possess high stability and the process of its degradation begins in the crystal form and proceeds with the sublimation process [8]. It complicates the experiments on the determination of the vapor pressure and accurate thermodynamic parameters of the vaporization process. As a result, the data in the literature [8, 10–16] on sublimation enthalpies of the amino acids have high errors, and particular values of these parameters are significantly dependent on the method used –the direct determination of the vapor pressure or the effusion method of Knudsen including registration of ions by the mass spectrum [33, 34]. In this connection, by the use of correlation dependencies, refined experimental data, estimated sublimation enthalpies for a number of compounds, and thermodynamic analysis, we obtained the energies of the hydrogen bonds and the specific interactions of the amino acids of the normal structure and with isostructural groups at the methylene group of the chain (Chap. 9). In this regard, we note the fact of increased energy values of the hydrogen bonds of the amino acids with the normal structure compared with the values of similar compounds with isostructural groups. In particular, the energy of the hydrogen bond D–O•••H–O of glycine for the aminoacetic acid was obtained to be equal to 30.2 kJ mol^{-1} and the energy of the same bond of its analog with the isostructural methyl group L-(D)- α-alanine, aminopropionic acid equal to 29.7 kJ mol^{-1}. The value of the energy of the hydrogen bond D–O•••H–O of compounds with the normal structure can be specified if one is prepared to use the contribution of the increasing number of methylene groups at compounds of the aminoformic acid (28.45) – aminoacetic acid (30.2) – aminopropionic acid (31.2 kJ mol^{-1}), reflecting the tendency of the growth of this bond, by the value of 30.4 kJ mol^{-1}. From the difference of this value and the obtained energy value of the hydrogen bond (29.7 kJ mol^{-1}) of compounds with the isostructural methyl group at the center we obtain a value of 0.65 kJ mol^{-1}. This value, reflecting the difference in energies of the peptide hydrogen bonds DC=O•••H–N–CH_2– of the amino acids with methylene and D=C=O•••H–N–CH_2–*iso* with the isostructural methyl group, fringed by strong acceptors of the electron density, is within the experimental error limits and clarifies the sublimation enthalpies of the amino acids (Tables 9.2 and 9.3). Note that for compounds of *N*-acethylglycineamide (NAGA), *N*-acetyl-L-alanineamide (L-NAAA), *N*-acetyl-L-leucineamide (L-NALA), and *N*-acetyl-D-leucineamide (D-NALA) the experimentally obtained sublimation enthalpies had high accuracy because of a high degree of purification of the starting compounds of normal structure. The nature of the formation of the peptide hydrogen bonds D=C=O•••H–N–CH_2– (20.1 kJ mol^{-1})was reveaied as more stable than compounds with the isostructural methyl group D=C=O•••H–N–CH_2–*iso* (16.0 kJ mol^{-1}), located between the strong acceptors of the electron density already mentioned. Consequently, the energy of the hydrogen bond D=C=O•••H–N–CH_2–*iso* in crystalline aminopropionic acid corresponds to the value of 17.5 kJ mol^{-1}, lying within the error of the energy value (16.0 kJ mol^{-1}) of the same type of peptide hydrogen bond of crystalline peptides *N*-acetyl-L-

a

b

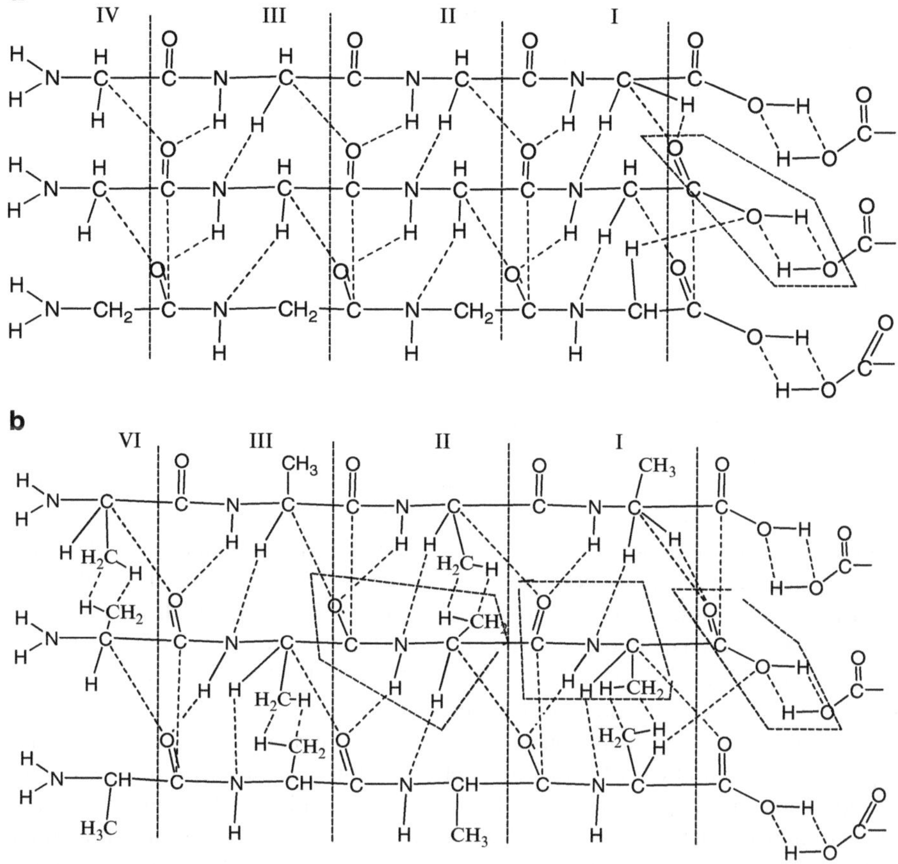

Fig. 10.8 Schematic picture of the crystalline structure of tetrapeptede of the derivative aminoacetic acid, Glicine (**a**) and L-(D)-α-Alanine, aminopropionic acid (**b**) with the network of the hydrogen bonds and the specific interactions

alanineamide and other compounds of this series. In turn, the aminoacetic acid molecule forms the peptide hydrogen bond D=C=O•••H–N–CH_2– with an energy value of 18.35 kJ mol^{-1}. The value of the energy contribution of the isostructural methyl group is used in calculations of the energies of the sequentially repeating configuration of the peptide chain at the base of *N*-acetyl-L-alanineamide (L-NAAA); it is 1.5 kJ mol^{-1}.

The results of the thermodynamic analyzes of glycine, aminoacetic acid, and the values of the energies of the specific interactions given in Table 9.2 in conjunction with the energy of the peptide hydrogen bond D=C=O•••H–N–CH_2– (18.45 kJ mol^{-1}) of glycine proposed above, aminoacetic acid allows us to go to a structural-energy review of the protein peptide chain and consider the repeating sequence of the peptide chain to obtain the values of the sublimation enthalpy of dipeptide, tripeptide with a larger number of configurations (confermers) of this chain.

First, it is necessary to note that for the schematic picture presented in Fig. 10.8a of the repeating sequence of the configuration of the polypeptide chain of the aminoacetic acid derivative, glycine, includes the functional terminal fragment and the terminal group. The terminal fragment of this chain includes the amino group with the methylene

65.4 kJ mol^{-1} 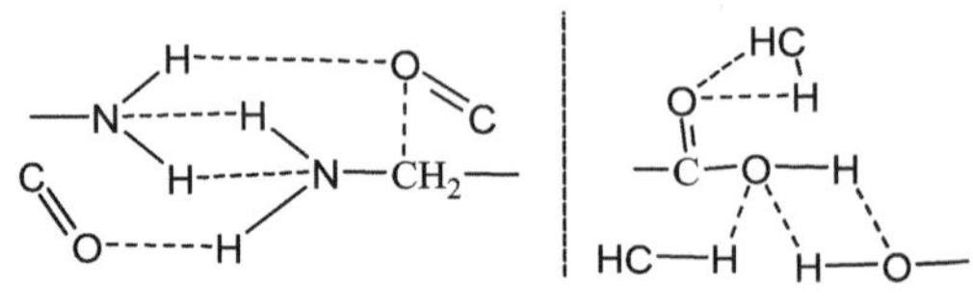 68.3 kJ mol^{-1}

group, forming two hydrogen bonds and two peptide hydrogen bonds and the specific interactions, which contribute a total energy of 65.4 kJ mol^{-1}. The terminal functional carboxyl group forms two hydrogen bonds, one of low stability and two specific interactions DO=C → C=O (5.7), DO → CH_2 (3.6 kJ mol^{-1}) with an energy contribution of 68.3 kJ mol^{-1}. Still the presented fragments are identical and have sequentially repeating configurations of the polypeptide chain, formed by six intermolecular interactions, presenting two peptide hydrogen bonds D=C=O•••H–N–CH_2, two hydrogen bonds of low stability 2D–N•••H–CH (1.0), two specific interactions of two types D=C=O → CH_2 (3.6) and DO=C → C=O (5.7) with total energy contribution of 46.3 kJ mol^{-1}.

Consequently, the chain consists of an acid fragment and a sequence of a repeating configuration of the polypeptide chain and the terminal functional group. Thus, the sublimation enthalpy of the dipeptide is obtained as the total energy contribution of the terminal functional fragment and the terminal functional carboxyl group and one configuration of the peptide chain and is equal to 180.0 kJ mol^{-1}. In the case of a tripeptide the energy of configuration of the chain is added and in the case of the tetrapeptide two identical configurations and the contributions of their energies are added and we obtain the values of the sublimation enthalpy of 226.3 kJ mol^{-1} and 272.6 kJ mol^{-1}, respectively.

The L-(D)-α-alanine, aminopropionic acid molecules with the isostructural methyl group at the methylene group have hydrogen bonds, and the specific interactions with energy values of similar aminoacetic acids contain, in addition, two specific interactions of low stability DH_3C → H–CH_2 on introducing an energy value of 1.5 kJ mol^{-1} and a configuration of sequentially repeating peptide chain (Fig. 10.8b). Thus, we obtain the value of the energy contribution of the terminal fragment as 64.7 kJ mol^{-1}; the terminal functional carboxyl group is kept at 68.3 kJ mol^{-1} and the sequentially repeating contributions of the peptide chain configuration increase up to 45.6 kJ mol^{-1}. The sublimation enthalpies of dipeptide, tripeptide, and tetrapeptide have values of 178.6, 224.2, and 269.8 kJ mol^{-1} respectively.

10.6 Extrastabilization Effect of Isostructural Methyl Group at Biochemical Compounds and Proteins

Understanding the role of the extra stabilizing effect of the isostructural methyl group requires the consideration of the charges accounting for the atoms in the molecule of the corresponding compounds. Necessary information for the targeted accuracy of the important role played by the isostructural methyl group could be obtained from existing limited results of research on the distribution of the charges in the molecules and the small number of biochemically active compounds. The identity of the repeating structures, the $–CH–(CH_2)_3–CH(CH_3)–$ fragments, in acetylene alcohol C_{15} and C_{20} [28], in saturated, and unsaturated ketones C_{18}, for which the sublimation enthalpies [35] were determined, and vitamin A_1-aldehyde C_{20}, β-retinene, retinene C_{26}-analogy, vitamin A_1-retineneoxime, vitamin A_2-carboxylic acids form $C_{20}H_{26}O_2$ and vitamin A_2-dihydropetinole $C_{20}H_{28}O$ with known values of the charges at the atoms of the alkyl chain, allow us to obtain the specificity of the distribution of the electron density at the atoms and to get information on the role of the isostructural methyl group. The principal difference of acetylene alcohol, saturated, and unsaturated ketones C_{18} from compounds of vitamin A is that in the alkyl chains of ketones

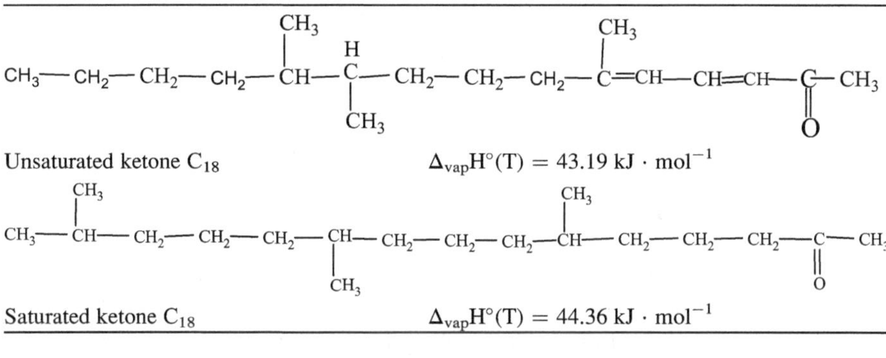

Unsaturated ketone C_{18} $\Delta_{vap}H°(T) = 43.19\ kJ \cdot mol^{-1}$

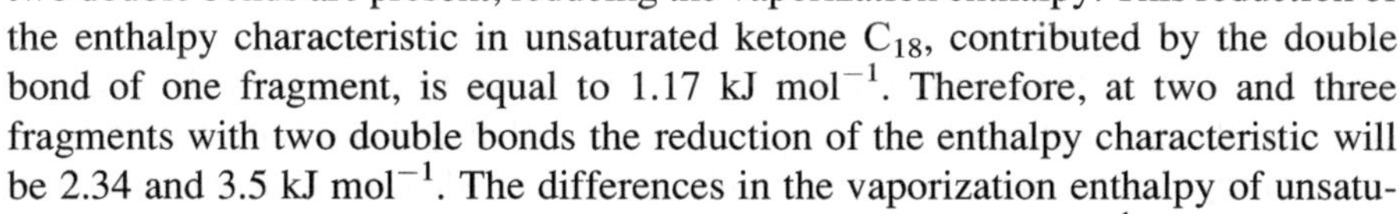

Saturated ketone C_{18} $\Delta_{vap}H°(T) = 44.36\ kJ \cdot mol^{-1}$

two double bonds are present, reducing the vaporization enthalpy. This reduction of the enthalpy characteristic in unsaturated ketone C_{18}, contributed by the double bond of one fragment, is equal to 1.17 kJ mol^{-1}. Therefore, at two and three fragments with two double bonds the reduction of the enthalpy characteristic will be 2.34 and 3.5 kJ mol^{-1}. The differences in the vaporization enthalpy of unsaturated ketones of pseudoionone and ketones C_{18} (6.35 kJ mol^{-1}) and saturated ketones C_{13} and C_{18} (6.43 kJ mol^{-1}) is within experimental error limits[35].

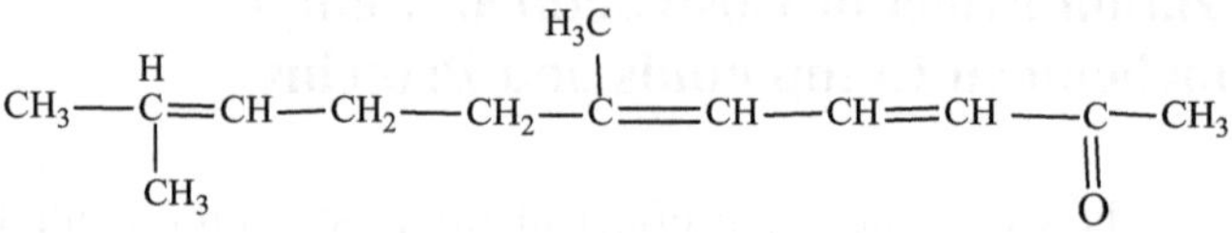

6,10-dimethyl-3,5,9-undecatrien-2-one

Pseudoionone $\Delta_{vap}H^{\circ}(T) = 36.84\ kJ \cdot mol^{-1}$ $\Delta_{vap}S^{\circ}(T) = 83.9 \pm 1.0\ J \cdot mol^{-1}T^{-1}$

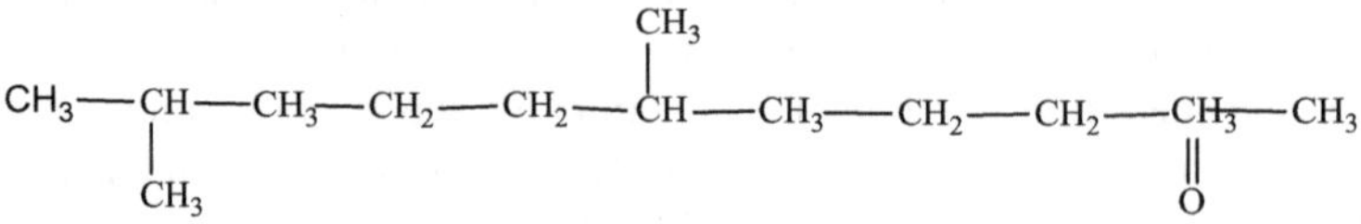

6,10-dimethyl-2-undecanone

Hexahydropseudoionone $\Delta_{vap}H^{\circ}(T) = 37.93\ kJ \cdot mol^{-1}$ $\Delta_{vap}S^{\circ}(T) = 83.9 \pm 1.0\ J \cdot mol^{-1}T^{-1}$

A fundamentally important feature is the extremely low energy contribution, introduced by fragments of five methylene groups with the average value of 6.40 $kJ \cdot mol^{-1}$, with 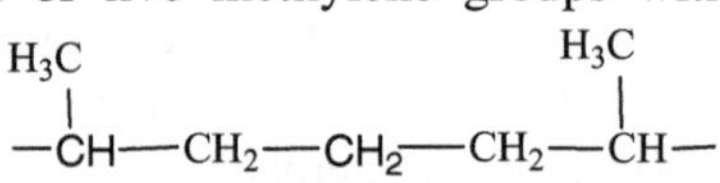 fringing isostructural methyl groups, at the energy contribution of one methylene group of the alkyl chain of ketone, with the normal structure, of 5.0 kJ mol^{-1}.

First, we consider the distribution of the charges at the atoms of vitamin A_1-aldehyde C_{20}, β-retinene molecule [36]. It is interesting that there exists an equality of negative charges at the carbon atoms of isostructural methyl groups in the vitamin A_1-aldehyde C_{20}, β-retinene molecule (β-retinene $C_{20}H_{28}O$) (1.047–1.048e) and retinene C_{26}-analogy (retinene C_{25} –analog) (1.047–1.048e^-) and an equality of positive charges at the hydrogen atoms (0.937–0.940e) and (0.938–0.941e), respectively, in the molecules of the same compounds (Fig. 10.9). The increased value of the charges ia always a characteristic for the carbon atoms (1.048e^-) and hydrogen atoms (0.940) of the isostructural methyl groups, located in the average fragments. It follows that each hydrogen atom transmits 0.060e^- of its total value of 0.180e to the carbon atom of the same methyl group, which obtains the charge (1.047–1.048e) and the remaining part (0.132) it gives to carbon atoms of the unsaturated alkyl chain of fragment, fringed by the isostructural methyl groups. In this case, the distribution of the electron density is such that even a carbon atom of the chain always has an increased negative charge, and an odd carbon atom of the chain loses part of its electron density and remains positively charged. The sequence in the electron population is that carbon atoms at the terminal of the isostructural methyl groups of the chain are characterized by a reduced value in the β-retinene (0.889–0.890e) molecule or in the retinen C-25-analog (0.890–0.892e) molecule. In this case the carbon atom of the chain of the average methyl group has a slightly increased charge (0.910) and (0.915e), respectively, in the molecules of the compounds considered. This implies that the carbon atom of the chain, located at the isostructural group, always has a reduced positive charge. An important feature of

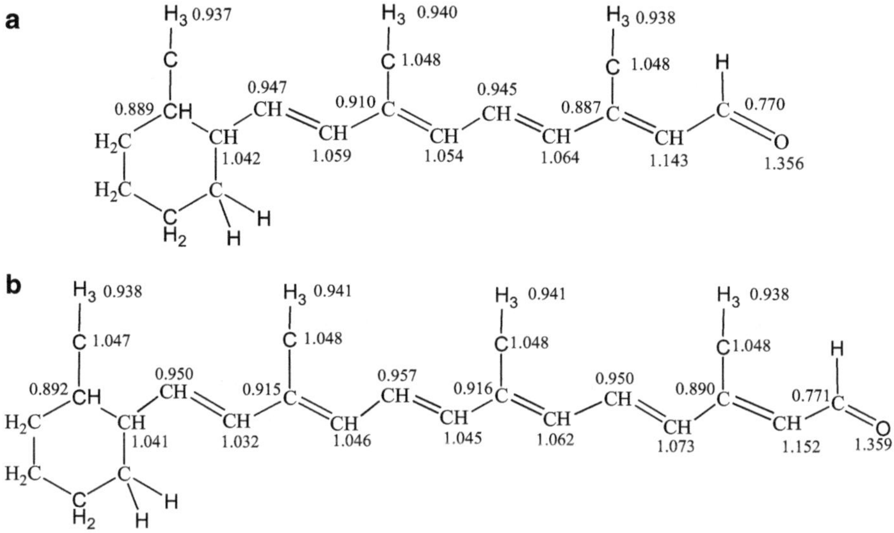

Fig. 10.9 Electron structure of molecule of vitamin A_1-aldehyde C_{20}, β-retinene (**a**) and retinene C_{26}-analogy (**b**) [36]

the distribution of the electron density in the carbon chain is that the odd carbon atom is positively charged, while the even atom has a negative charge (Fig. 10.9). It points to the special role of the isostructural methyl group, which, while maintaining its negative charge at the carbon atom, acts as a guarantee of the distribution of the electron density inside the carbon chain. This role is saved within the fragment fringed by the CH_3-groups. The influence of the oxygen of the aldehyde group is expressed in the increase of positive charge of the contacting carbon atom (0.770 and 0.771e). Therefore, this carbon atom transmits not only the electron density obtained from the hydrogen atoms (0.180e), but also transmits part of its own density (0.230e), providing the negative charge of the aldehyde oxygen atom (1.359e) and in addition transmits at least 0.070e to the second (even) carbon atom of the chain. The next odd one, the third carbon atom contacting with the isostructural methyl group, at β-retinene and retinene analog $C_{20}H_{28}O$, shifts electron density to the fringing second and fourth carbon atoms of the first fragment, providing it with donor properties on the formation of the specific interaction H–C → C–H in the molecules that are close in the environment in crystalline and liquid condition. In this case the second, even carbon atom of the chain of the considered aldehydes obtains the biggest charge 1.143 and 1.152e^- preserving the negative charge at the oxygen atom (1.356 and 1.359e^-). It should be noted that the increase of the number of fragments up to four at the retinene analog $C_{20}H_{28}O$ contributes to the increase of the charge at the second carbon atom of the chain. On the basis of this analysis we can conclude that the oxygen atom of the aldehyde group obtains 0.230e^- directly from the carbon atom in contact and the latter ones obtain 0.129e^- as a result of the shifting of the electron density on the chain fragment due to the increase in positive charge of the odd carbon atoms, mainly, due to contact with the carbon atom of the isostructural methyl groups. This implies that the following holds:

First, at three and larger numbers of fragments the aldehyde group forms a hydrogen bond =C=O•••H–C with significant difference of charges at the oxygen atoms

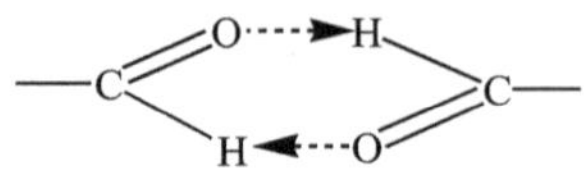

and hydrogen atoms of this group with the permanent energy value independent of the chain fragments.

Second, all further fragments regardless of their number introduce an identical contribution of energy to the vaporization enthalpy, similar to the methylene groups of ketone with the normal molecule structure [35].

This conclusion is proved by the energies established for the specific interactions and hydrogen bonds of saturated C_{18} and unsaturated C_{18} ketene and acetylene alcohol C_{15} and C_{20} [28]. Thus, the value of the vaporization enthalpy of this series of compounds is determined by the energy contribution of the cycle reduced by the energy contribution of the substituted hydrogen atoms, the total value of the energy contribution of fragments, and the energy of the hydrogen bond and the specific interactions, formed by two terminal CH groups.

Replacement of the aldehyde group by the terminal oxime group >N—O—H in retinenoxime (Fig. 10.10a) does not change the charge of the carbon atom (1.048e) of the isostuctural methyl group and does not contribute visible changes to the hydrogen atom of this group (0.941–0.945e). The presence of the terminal group of oxime leads to some reduction of the negative charge of the even carbon atoms of the chain and to the reduction of positive charge of the odd carbon atoms. These insignificant changes of the charges do not result in significant changes of energies of the H–C → C–H specific interactions, formed by the CH-groups of fragments of vitamin A_1-retineneoxime. The basic differences of contributions of the energies to the enthalpy characteristics of vaporization contributed by the carbonyl oxygen and hydrogen atom of the aldehyde group of vitamin A_1-aldehyde C_{20} and the alcohol hydroxyl group of the vitamin A_1-retineneoxime molecule are caused by the difference of its charges 1.358 and 1.908e, respectively. The high acceptor properties of oxygen atoms of alcohol and carboxyl groups lead to the mentioned reduction of positive charge of the odd and negatively charged even carbon atoms of the chain. The difference of the charges on contacting with these groups carbon atoms (1.022 and 0.768e) reflects the significant difference in acceptor properties of the oxygen atoms of the alcohol and carboxyl groups. Precisely stated, the carboxyl group provides the contacting carbon atom with a charge of 0.768e in the vitamin A_2-carboxylic acid molecule of the form $C_{20}H_{26}O_2$ (Fig. 10.10b), which determines its high donor properties and the formation of a

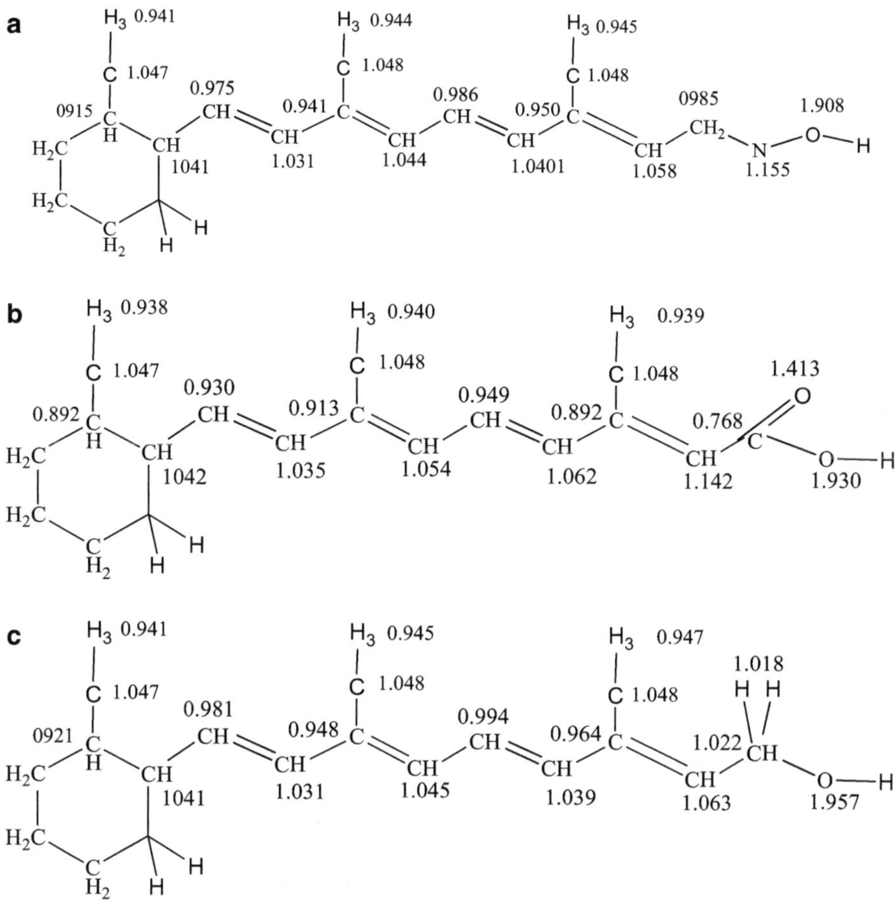

Fig. 10.10 Electron structure of molecule vitamin A_1-retineneoxime (**a**), vitamin A_2-carboxylic acids form $C_{20}H_{26}O_2$ (**b**), and vitamin A_2- dihydropetinole $C_{20}H_{28}O$, acid form (**c**) [36]

stable specific interaction. It is concluded that this sequence in the change of the charge at oxygen atoms of the hydroxyl groups in the considered compounds holds:

Vitamin A_2-reteneoxime		Vitamin A_2-carboxylic acids $C_{20}H_{26}O_2$	Vitamin A_2-hydropentenole acid form
>N—O—H		—C(=O)—O—H	O–H
H–O	1.908	1.930	1.957e
CH_2	0.983	1.022	0.768e

and is accompanied by stabilization of the hydrogen bonds. The carbonyl oxygen atom also has a high value of electron density (1.413e) and expresses high donor properties and the ability to form hydrogen bonds and specific intermolecular interactions. The values of the energies of the specific interactions, formed by the

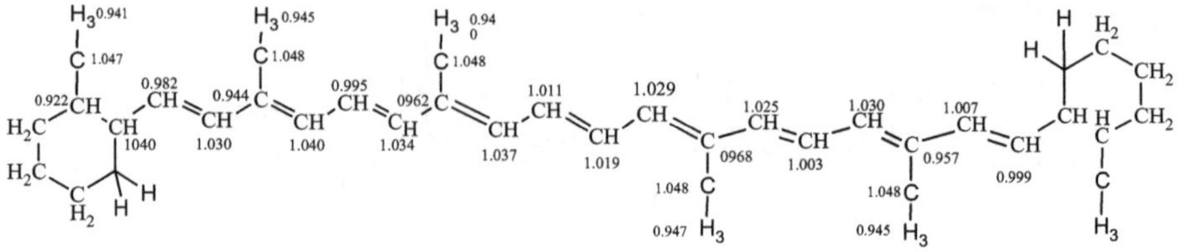

Fig. 10.11 Electron structure of molecule δ-corotine [36]

isostructural methyl group H_2–C–H → CH_3 of low stability, remains practically unchanged and it is estimated from the thermodynamic analysis of similar compounds. On the presence of the nitrogen atom in the vitamin A_2-reteneoxime molecule of higher negative charge ($1.153e^-$), than for the carbon atom (1.022) in vitamin A_2-hydropentenole (vitamin A_2–dehydropentenole) (Fig. 10.10c), we note that it is in the same position in the chain, pointing to the possibility of the formation of more stable specific interactions compared with the interaction H–C → C–H, formed by the essentially unshared $2s^2$ electron pair of the carbon atoms.

A similar distribution of the charges at the atoms is preserved in symmetric compounds (Fig. 10.11) despite the increasing number of carbon atoms in the average fragment, fringed by isostuctural methyl groups. The observed distribution of the structurally-charged parameters inside the molecular interaction in aldehydes, alcohol, and carbon acids, in compounds with nonsymmetrical and symmetrical structure reflects the stabilizing effect of the isostructural methyl group, participating in the distribution of electron density in the chain and influencing the intermolecular interactions. A slight fluctuation of the charges of the carbon and hydrogen atoms of the isostructural methyl groups, accompanying more significant changes of the charges of the odd and even carbon atoms of the chain fragments, reflects its role in the regulation of the electron density in the molecules of aldehyde, oxymes, alcohol, and carbon acids, providing an equal energy contribution by each chain fragment. From the results of the thermodynamic analysis and the structural-charge condition of atoms of molecules of organic, biochemicals, including amino acids, peptides, and proteins one can draw the following conclusions:

First, the isostructural methyl group participates in the distribution of electron density in the molecule and the formation of fragments being more beneficial for the molecule's distribution of the charges at the group atoms.

Second, participation of the methyl groups in the distribution of electron density in the molecule is a general feature and is a phenomenon that occurs in the chemistry of organic and biochemical compounds and is inseparable from the chemistry of the peptides and proteins.

Third, being formed by the isostructural methyl groups fragments of 4–5 atom groups with similar energy contributions to the enthalpy characteristics of vaporization, they provide more information on the thermodynamic character for the series of compounds and less on experimental research.

10.7 Specific Interactions at the Secondary Structure of Protein

The hydrogen bond is the only interaction and consequently the principal important feature in the structure of the peptide and protein at the C=O group in the sequence of intermolecular bonding with the N–H group. In both parallel and anti-parallel β-sheets the N–H group of one peptide chain participates in the formation of a stable hydrogen bond with the C=O group of another peptide chain. The theoretical research above and the results of the calculations led to an accurate value of the energy of the hydrogen bond >N—H◄···O=C<, formed by the N–H group with the oxygen atom of the carbonyl group C=O with an energy value of 17.5 kJ mol^{-1} for the crystalline condition of the compound. The higher energy values of the hydrogen bonds obtained, formed by the hydroxyl groups of amino alcohol and amino acids, are ascribed to the unique peptide hydrogen bond. Even "soft scorn" to the isostructural methyl group has to change for extra stabilizing effect in the fragment of the chain of organic and biochemical compounds. It testifies to the fact that it is necessary to consider the mutual influence of all participants of the chain fragments. Even in the simplest protein chain considered in the formed fragment of secondary structure there is

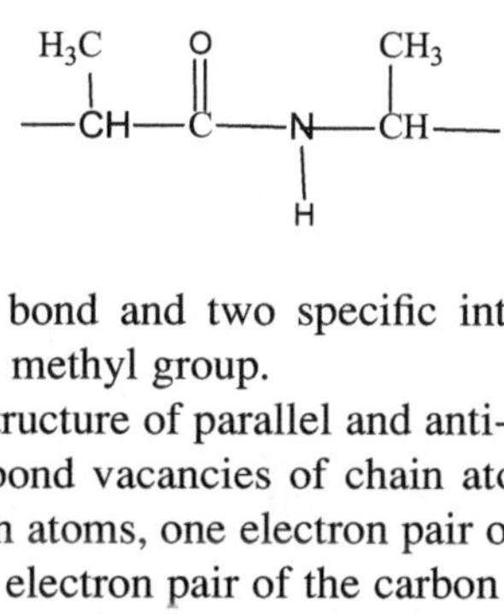

alternation of the peptide bond and two specific interactions H_3–C → H–CH_2, formed by the isostructural methyl group.

When considering the structure of parallel and anti-parallel β-sheets it is necessary to remember all the bond vacancies of chain atoms, including the two free electron pairs of the oxygen atoms, one electron pair of the nitrogen atom, and the essentially unshared $2s^2$(c) electron pair of the carbon atoms. Due to the deficit of bond vacancies of the hydrogen atom in the methylene group, it acts as acceptor of the electron density from the nitrogen donor atom in the formation of the hydrogen bond of low stability N•••H–C(CH_3). The isostructural methyl group also forms two specific interactions of low stability, H_3C → H–CH_2 (1.0 kJ mol^{-1}). The two peptide hydrogen bonds =C=O•••H–N are stable specific interactions with an energy of 17.5 kJ mol^{-1}, the two specific interactions =C=O → CH(CH_3) are formed by the oxygen atom with carbon atom of the methylene group and one carbon atom of the carbonyl group O=C → C=O with the energy of 5.7 kJ mol^{-1}. Thus, each protein fragment forms four hydrogen bonds and five specific interactions with a similar fragment of contacting chains that are close in the environment (Fig. 10.12).

The networks of hydrogen bonds and the specific interactions in crystal in various protein secondary structures with parallel β-sheet are given in Fig. 10.13.

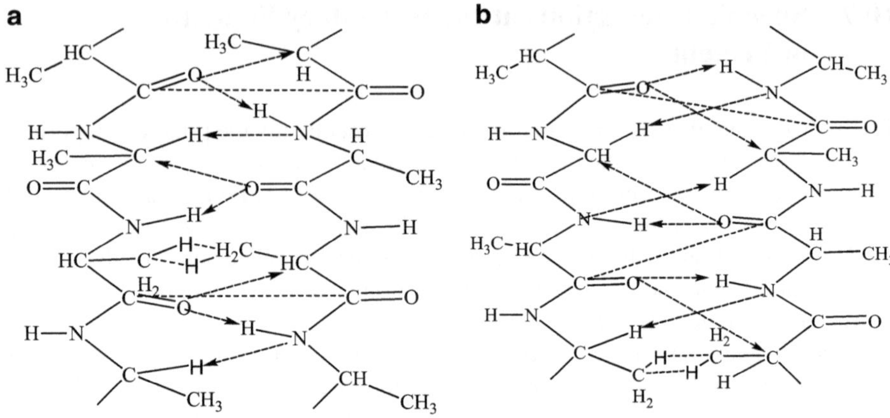

Fig. 10.12 Schematic picture of hydrogen bonds and the specific interactions in various protein secondary structures: parallel β-sheet (**a**) and antiparallel β-sheet (**b**)

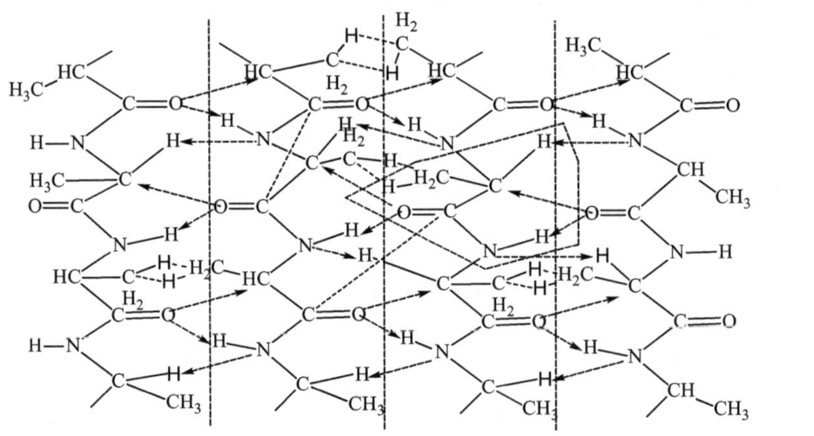

Fig. 10.13 Network of hydrogen bonds and the specific interactions in various protein secondary structures of parallel β-sheet

The specific interactions formed and the hydrogen bonds include its mutual influence and with the enthalpy characteristic of vaporization available allow us to obtain the values of its energies.

In this case the energy value of the specific interaction, formed by the methylene group $=C=O \rightarrow CH(CH_3)$, is accepted to be equal to 3.6 kJ mol^{-1} and the unknown energy value of the hydrogen bond of low stability N•••H–C(CH_3) as equal to 1.0 kJ mol^{-1} (Sect. 10.4). The value of the total energy contribution of all types of hydrogen bonds and the specific interactions obtained, formed by a fragment of the chain protein, is equal to 51.4 kJ mol^{-1}.

References

1. Arnautova EA, Zacharovas MV, Pivina TC (1996) The methods of calculation of sublimation enthalpies of organically molecular crystalline. Izvectia RAS Ser Chem 12:2872–2881
2. Lilley TH (1988) In: Jones M (ed) Biochemical thermodynamics, 2nd edn. Elsevier, Amsterdam, Chap. I
3. Barone G, Glancola C, Lelley TH, Mattia CA, Puliti R (1992) Enthalpies and entropies of fusion of same substituted deputizes-comparison with crystal data. J Therm Anal 38(12):2771–2778
4. Chickos JS, William EA (2002) Enthalpies of vaporization of organic and organometallic compounds, 1880–2002. J Phys Chem Ref Data 31:519–879
5. Chickos JS, William EA (2003) Enthalpies of Sublimation of organic and organometallic compounds, 1910–2001. J Phys Chem Ref Data 32:537–698
6. Karlson TA (1981) Photoelectron and OGE- spectroscopy. Mech. Eng. Russia, Leningrad. 431 p
7. Nefedov VI (1984) Infra red – spectroscopy of chemical compounds. Chemistry, Moscow, p 255
8. De Kruif CG, Voogd J, Offringa JCA (1979) Enthalpies of sublimation and vapor pressures of 14 amino acids and peptides. J Chem Thermodyn 11:651–656
9. Cottrel T, Коттерел T (1956) Stability chemical bond. Inostr. Liter, Moscow
10. Takaki S, Chihara H, Seki S (1959) Vapor pressure of molecular cristals. XIII. Vapor pressure of –glycine cristal. The energy of proton transfer. Bull Chem Soc Jpn 32:84–88
11. Svec HJ, Clyde DD (1965) Vapor pressures of some – α-amino acids. J Chem Eng Data 10:151–152
12. Ngauv SN, Chihara H, Seki S (1977) Thermodynamiqul de composes azotes. III. Etude termochimiqul de la L-α-Alanine. Thermochim Acta 20:371–374
13. Sabban R, Laffiti M (1978) Thermodynamiqul de composes azotes. III. Etude termochimiqul de la L-α-Alanine. Bull Soc Chim Fr 1:50–52
14. Lebedev YA, Miroschnichenko EA (1981) Thermochemistry of the vaporization of organic compounds. The vaporization and sublimation enthalpy and saturated pressure. Nauka, Moscow
15. Arnautova EA, Zakharova MV, Pivina TS, Smolenskii EA, Sukhachev DV, Shcherbukhin VV (1996) Methods for calculating the enthalpies of sublimation of organic molecular crystals. Russ Chem Bull 45(12):2723–2732
16. Ivanov EV, Puchovskie Yu P, Abrosimov VK (2004) About formation of weak hydrogen bonds CH•••O in liquid methanol. Abstract of the IX international conference, The problem of solvation and complex formation in solutions, Plyos, p 345
17. Ferro D, Barone G, Della Gatta G, Piacente V (1987) Vapor pressures and sublimation enthalpies of urea and some of the derivatives. J Chem Thermodyn 9:915–923
18. Ferro D, Della Gatta G, Barone GJ (1988) Enthalpies of sublimation and fusion for *N*-acetyl substituted glycine, L-alanine, and D-leucine amides. Therm Anal 34:835–841
19. Puliti R, Mattia CA, Barone G, Della Gatta G, Ferro D (1990) Crystal-structure and thermodynamics of phase-transitions of *N*-acetyl-L-valinamide. Thermochim Acta 162:229–240
20. Politi R, Mattia CA, Barone G, Giancola C (1991) Structures of N2-acetyl-DL-alaninamide and N2-acetyl-DL-leucinamide. Acta Crystallogr C 47:1658–1662
21. Barone G, Giancola C, Lilley TH, Mattia CA, Politi R (1992) Enthalpies and entropies of fusion of some substituted peptides. Thermochim J Therm Anal 38:2771–2778
22. Abate L, Della Gatta G, Somsen G (1994) Heat-capacities of 16 small peptides (*N*-acetyl-*N*-methylamino acid-aides) measured by differential scanning calorimetric. Thermochim Acta 239:7–16
23. Abate L, Palecz B, Giancola C, Della Gatta G (1997) Heat capacities, and enthalpies and entropies of fusion of some uncharged small peptides (*N*-acetylamino acid amides and 2,5-diketopiperazines). J Chem Thermodyn 29:359–368

24. Del Vecehio P, Barone G, Sabbah R, Della Gatta G, Abate L (1999) Thermodenamics of sublimation of *N*-acetyl amides of some amino acids. J Chem Thermodyn 31:1001–1008
25. Puliti R, Mattia CA, Barone G, Giancola C (1989) Structure of some *N*-acetylamides of the amino acides. Acta Crystallogr C 45:1554–1557
26. Baev AK (1969) Phase condition and complex formation ability of halogenide metals. In: General and applied chemistry, vol 1. Vysheischaia Educating, Minsk, pp 207–218
27. Baev AK (1972) Complex formation ability halogenides of second – six groups periodical system. In: General and applied chemistry, vol 5. Vysheischaia Educating, Minsk, pp 35–51
28. Baev AK (2012) Specific intermolecular interactions of organic compounds. Springer, Hiedelberg/Dordrecht/London/New York, p 432
29. Kvic A, Karaghoule AB, Koetzle TF (1977) Deformation of the electron density of –glycilglyciat 82 K. The neutron diffraction study. Acta Crystallogr B 33:3796–3801
30. Destro R, Bianchi R, Morosi G (1989) Electrostatic properties of L-alanamine from X-Ray diffraction an 23K and ab initio calculations. J Phys Chem 93:4447–4457
31. Coppens PH, Abramov Y, Carducci M et al (1999) Experimental charge densities and intermolecular interactions: electrostatic and topological analysis of DL-Histidine. J Am Chem Soc 121:2585–2593
32. Paraichic G, Orvill-Tomas U (eds) (1984) Molecular interactions. Mir, Moscow, pp C184–C232
33. Vatagin VC, Badelin VG (1993) Derivatografical investigation of glycol containing peptides. Russ J Chem Thermodyn Thermochem 2(1):101–105
34. Badelin VG, Tyunina YB, Girichev GV, Giricheva NI, Pelinec OV (2007) Interaction between structure of molecule of the amino acids and dipeptides and heat effect of their sublimation. Russ J Struct Chem 48:698–704
35. Baev AA, Baev AK (2011) Thermodynamics of saturated and unsaturated retones C13 and C18 and the energies of specific intermolecular interactions. Russ J Phys Chem 85(5):823–832
36. Kalinun FL, Lobov VP, Zidkov VA (1971) Reference book on biology. Naukiva Dumka, Kiev, p 1012

Chapter 11 Specific Intermolecular Interactions of Urea, Uracile, and Their Derivatives

11.1 Structural-Energetic Properties of Urea

There are two tautomeric forms of urea:

$$H_2N\text{-}C(=O)\text{-}NH_2 \leftrightarrow HN{=}\ C(\text{-}OH)\text{-}NH_2$$

α-form β-form

The derivatives of the β-tautomer (isourea) are characterized by strong basic properties, forming ions $H_2NCON_3^+$ and $H_2NC(NH)O^-$ in a water environment [1]. "Pure" crystalline urea is a mixture of mono crystals with a tetragonal system, the lattice constants of which belong to the structural class $P42_1m$, $Z = 2(mm^2)$, and have the values a = 0.5645, b = 0.5645, and c = 0.4704 nm [2]. The crystalline structures of urea have the shapes of needles and stratiforms, the layers of which consist of plane prisms, formed by peptide (amide) bonds >C=O•••H–N< [1]. The urea molecule has a planar structure with a sufficiently strong electric moment of dipole $\mu = 4.6$ D or $\approx 15.3{\bullet}10^{-30}$ cm at 298 K [3] owing to the asymmetric distribution of the density of n- and π-bonding electrons.

According to X-ray data [4], the length of the molecule $(NH_2)_2CO$ in crystalline lattice is close to 0.24 nm. The planarity of the urea molecule is proved by RE- and neutron microscopic research, and by the results of IR-spectroscopy and other methods. The planar structure of urea amides is caused by the presence of the carbonyl group >C=O. However, in [4] it is noted that planarity of the molecules with a carbonyl group is a peculiarity of its composition only in the basic quantum condition. IR-frequencies of stretching vibrations >C=O are influenced by inductive and resonance electronic effects of the molecules close in the environment. To the urea molecule is ascribed a pyramidal structure with several degrees. This condition belongs to the excited state of the molecule, expressed in the result of $n \rightarrow \pi^*$-electron transfer and by interaction of a nucleophile with a carbonyl group with the participation of stereo electron effects [5]. Provided the location of the nucleophile group (Nu=O or N) of a neighboring molecule is near the carbon atom,

A.K. Baev, *Specific Intermolecular Interactions of Nitrogenated and Bioorganic Compounds*, DOI 10.1007/978-3-642-37472-2_11,

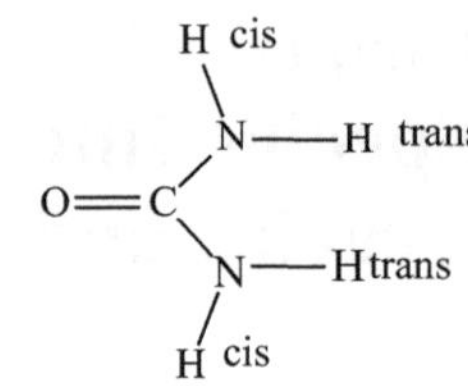

Fig. 11.1 Structural formula of the molecule urea at crystalline state

but lacks a direct bond with it, the elongation of the bond =C=O occurs and the system requires a pyramidal structure. The amine group possesses a high basicity and, consequently, it acts as a strong electron donor. In this connection, the hydrogen atom of the NH_2-group performs the role of electron acceptor. The exactly mixed electron-donor function of the urea molecule causes a significantly expressed association in crystalline urea [6]. In the molecule $(NH_2)_2CO$, there is a small difference in the interatomic distances of C–O (0.1262) and C–N (0.3335 nm). In the author's opinion [5], the hydrogen atoms of the amino groups are in *cis*- and *trans*-positions (Fig. 11.1).

The crystalline structure of urea is characterized by the presence of various specific contacts, and different H-bonds, including ramified ones. Given the data presented in [7], the interatomic distance N•••O for hydrogen bonds with an implementation of the role of urea as donor and acceptor is characterized by the average value of 0.298 nm with limits 0.281 and 0.318 nm. According to Vinogradov [8] and Muidinov et al. [9], [8, 9] when using the unit graph in the urea crystalline analysis, each molecule of the chain forms two bonds OH•••N and N–H•••O with an interatomic distance H•••O of all four bonds of 0.2329 nm. The chains are connected directly by bonds N–H•••O. Each molecule forms 11 similar bonds with the H•••O distance equal to 0.2136 nm. Two bonds are in the plane and two bonds perpendicular to the first plane. These bonds unite molecules to the layers $L^4_4(4)$. In the structure is present a tetramerous cycle, which is part of any chain $C_2^4(2)$ and layer $L^4_4(4)$. This cycle is formed by two bonds N–H•••O with an interatomic distance equal to 0.2136 nm, and two other H•••O bonds with a distance of 0.2326 nm between the amide atom of hydrogen and the oxygen atom of the carbonyl group. By X-ray and neuron diffraction [8, 9] it was shown that the formation of one bond N–H•••O=C< in amide crystal is accompanied by a reduction in the intermolecular distance of C–H of $2.5 \cdot 10^{-3}$ nm, and the >C=O bond is extended approximately at $1.4 \cdot 10^{-3}$ nm. As a result, a shifting of the electron density from the orbital with an unshared electron pair of the nitrogen atom to the carbonyl group is induced . It results in a reduction in the magnetic shielding of the amine proton in accordance with the shifting of its resonance in the direction of the weak field in NMR spectra [7]. In other words, a reverse dative bond forms between the nitrogen atom of the amino group and the carbon atom

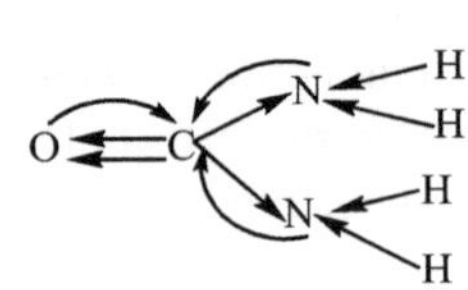

of the carbonyl fragment, stabilizing the intermolecular bond C–N and destabilizing the nitrogen atom–hydrogen bond, or we have the specific intermolecular interaction $CH_3(-CH_2)n$–N → $CH_3(-CH_2)n$–N of the alkyl derivative compound of urea. Reduction of the C–N bond and, consequently, strengthening of its π–character in urea crystal is accompanied by an increase in the net rigidity of the H-bond. A similar reverse dative bond is realized between the oxygen atom of carbonyl and carbon. However, the high acceptor ability of the oxygen atom causes an increased shift of the electron density from the carbon atom and even the nitrogen atom, providing it with an increased positive charge. A partial shift of the electron density from the nitrogen atom causes an increased shifting of the electron density from the hydrogen atoms of the amino groups, which contributes to the increase in the difference of the negative charge of the oxygen atoms and positive charge of the hydrogen atoms. As a result, the oxygen atom possesses a high ability to form a hydrogen bond with high stability. Exactly stated, the presence of the nitrogen atom with an amino group at the carbonyl's carbon atom leads to the oxygen having the ability to form a much more stable hydrogen bond in comparison with the similar oxygen atom in the ketone molecule. At the same time, the alkyl derivatives of urea are able to have specific interactions with the participation of a pentacoordinated carbon atom, also with increased stability. According to the data presented [10] by Gartland and Craven, the formation of urea bonds =C=O•••H–N points to the more effective acceptor properties of this crystalline compound compared with proton donors. It means that hydrofilicity of the peptide bond is necessary for the electron properties of the carbonyl's oxygen atom [6] and hydrogen atom and, in the case of the alkyl derivatives, of the carbon atom of the terminal methyl group.

This conclusion is consistent with the results of tenzometric research [11] and quantum chemical calculations [7, 12]. Thermodynamic analysis of the heat capacities of urea, its alkyl-substituted derivatives and dimethylformamide solutions in water [13], and IR and RE spectra in the amide – salt – water systems [14] led Arnett [6] to the conclusion regarding the increased proton donor ability of water in comparison with NH-groups and domination of the urea crystalline structure of rather weak intermolecular bonds. Kucherjavei and Lebedev [15] noted that the observed monotonic changes of the excess ratio of thermal broadening of urea generally do not reflect the occurrence of a thermal influence on the structural changes in urea crystal. However, in the range of 298–308 K, very high values are expressed of the isobaric coefficients of thermal broadening in the local orders of the crystal structure, associated via the hydrogen bond. In terms of polymorphic changes, the presence of these anomalies is considered to be of a high order, without manifestation of the latent change in the volume of the system on the condition of the constancy of its enthalpy [15].

Arnett [6] supposes that, given the condition that the equality of the hydrogen bond H–O•••H–O of liquid water is 15.5 kJ mol^{-1} [16], the energy of the H-bond >C=O•••H–N in crystalline urea should not exceed 15 kJ mol^{-1}. According to the author [6], this assumption finds experimental confirmation in the estimated energy value =C=O•••H–N in the urea crystalline structure, established from sublimation enthalpy with the use of the isoelectronic isobutene model. The obtained value is close to ~14.5 kJ mol^{-1} [17]. However, this value also includes the energy of detraction of hydrocarbon dipoles [6]. Therefore, the authors determined the energy of the H-bond in urea crystal to be equal to ~10.5 kJ mol^{-1} with the use of data presented in [18] on the vaporization enthalpy of tetra-dimethylurine, 45.6 kJ mol^{-1}. The enthalpy of melting of crystalline urea is given in two works as 14.55 [1] and 13.6 kJ mol^{-1} [19], and the melting temperatures are 405.86 [1], 405.75 [19], and 405.85 K [15].

Research on urea water solutions led the authors to [6] the unconvincing conclusion that the direct bond >C=O•••NH in hydrate complexes is not realized, and that the possibility of the formation of amine dimers is unlikely. The fact of the presence of urea clusters in water solutions contradicts this conclusion. Fundamental research on urea as the major product and participant in the metabolic process of living organisms significantly broadened our knowledge of its crystalline condition, and of structural rearrangements of urea derivatives in water. However, attempts at establishing hydrogen bonds in crystalline urea have not been successful and the problem of the hydrogen bonds remains unsolved.

A significant moment in this question was the assignment of two free electron pairs of the oxygen atom of carbonyl to the bond vacancies, two additional hydrogen bonds being formed with one of the hydrogen atoms in each of the two amine groups of the contacting urea molecule. The two remaining hydrogen atoms in each amino group interact with the oxygen atoms of chains agglomeration, that is, taking into account two bond vacancies carrying negative charges, and four with positive charges of the urea molecule. Consequently, the urea molecule does not fully realize its bond vacancies and continues to be nonparticipating with contacting molecules close in the environment. Thus, the significant lack of representations with the position of graphs [8, 9] is an unnatural deprivation of the two bond vacancies of the urea molecule to participate in the formation of specific intermolecular interactions. At the same time nitrogen atoms of two of the groups supporting free electron pairs, not taken into account in the theory of graphs [8, 9] and other ideas, while exactly free electron pairs of the nitrogen atoms of ammonia and dimethyltetraacetic acid participate in the formation of the hydrogen bond and coordination bond with metal kations, respectively. Here we draw attention to the apparent ability of the nitrogen atoms of the amino groups of the alkylamine complexes in [20–23] of urea derivatives NMF and *N*,*N*-DMF to participate in the formation of intermolecular interactions. In all the compounds mentioned, the nitrogen atom acts as a donor of electrons and gives its electron pair for general use.

As mentioned earlier, the urea molecule structure reveals the presence of different specific contacts for the formation of H-bonds with the molecule close in the environment. On the one side, two free electron pairs of two nitrogen atoms of the amino groups and two free electron pairs of the oxygen atoms of the carbonyl group tend to manifest donor properties and, on the other side,

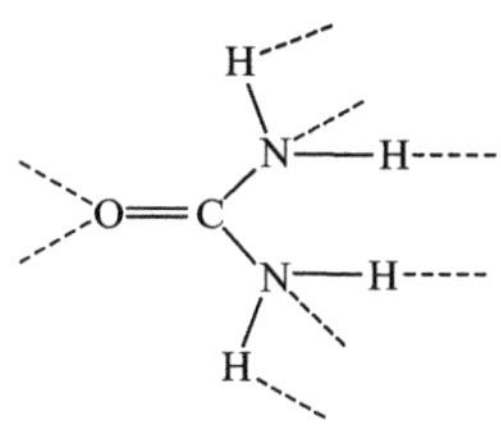

the positively charged hydrogen atoms with acceptor properties of the two amino groups of the urea molecule express acceptor properties. Therefore, the urea molecule possesses eight free bond vacancies, each of which is realized in the formation of the hydrogen bonds. Two of the least stable hydrogen bonds N•••H–N are formed by each nitrogen atom with one hydrogen atom of the amino group, with the location of the similar group of the urea molecule close in the environment . In turn, two other H-bonds are formed by the hydrogen atoms of the two amino groups of the urea atoms with nitrogen atoms of the molecules close in the environment. The remaining four free vacancies of the oxygen atoms of the carbonyl group and two hydrogen atoms of the two amino groups form more stable

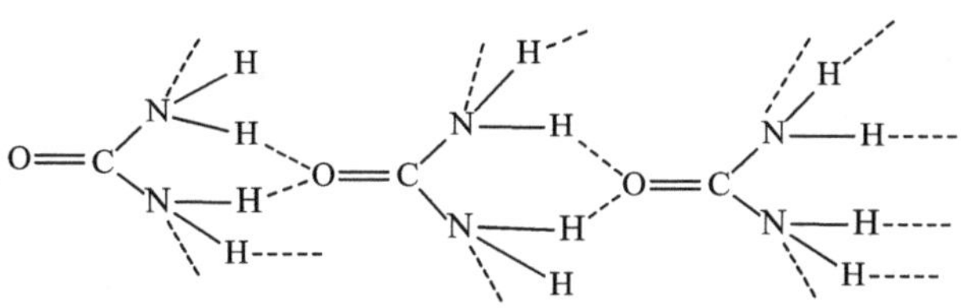

hydrogen bonds of the second type, >C=O•••H–N, contacting similar groups of the urea molecules close in the environment.

The results of the study of water solutions prove the presence of eight bond vacancies in the urea molecule and its participation in the formation of the hydrogen bonds.

In the range of near hydration, r = 0.4 nm, around the urea molecule, there are located eight water molecules, forming stable hydrogen bonds with it [24–28]. The difference in the geometrical parameters of the urea and the water molecules is not an obstacle for the small difference in the interaction enthalpy between them for the molecules of solvents in bulk. It leads to the appearance of the polypeptide structure, similar to the local tetrahedric formations of solvent molecules. This conclusion is consistent with the information on the influence of stereoelectronic effects on the process of urea hydration [6]. In accordance with the widely known [28] data on the carbonyl's oxygen atom and amine's hydrogen atoms, there are 2.5 and 3.2 molecules of H_2O, respectively. Its total number of coordination is equal to 5.7 (n_h) of water molecules in the first hydrate shell of the urea molecules, which is in good compliance with the number of strong hydrogen bonds, =C=O•••H–O, formed by four bond vacancies with the participation of two hydrogen atoms and the oxygen atom. Modeling of the system by the method of molecular dynamics and thermo chemical ($n_h = 5$) [18] research leads to a near value. Neuron graphic research ($n_h = 7.1 \pm 0.5$) [29] reflects an intermediate number of coordinating water molecules for eight bond vacancies of the urea molecules. It is a direct indication of the existence of cluster forms of urea in water solution at the given concentrations for the experiment and of the formation of a peptide group of the urea molecule with more stable hydrogen bonds in comparison with the ones realized in liquid and crystalline water.

11.2 Energies of Hydrogen Bonds of Urea

The results of the research conducted using different methods allow us to state that in the environment close to the urea molecule, there are six molecules, two of which are coordinated because of the four stable hydrogen bonds =C=O•••H–N and the four less stable hydrogen bonds N•••H–N (Fig. 11.2). The analogy of the hydrogen bond N•••H–N in liquid and solid urea is the hydrogen bond in liquid and crystalline conditions of ammonia. The energies of the second more stable type of hydrogen bond,

=C=O•••H–N, can be determined by the difference between the vaporization enthalpies of urea and the total value of the less stable four hydrogen bonds N•••H–N with the help of Eq. 11.1,

$$D{=}C{=}O\cdots H\text{-}N = (\Delta_{evap}H^0(T) - 4D\text{-}N\cdots H\text{-}N)/4 \qquad (11.1)$$

where $\Delta_{evap}H°(T)$ is the vaporization enthalpy or urea sublimation. This makes sense, because the enthalpy and entropy characteristics of phase changes are interrelated with the number and energy of bursting intermolecular interactions, as revealed in [30, 31].

However, it should be noted that the correctness of the establishment of the energies of the hydrogen bonds in the given approach requires additional explanation. The strong acceptor, which is the oxygen atom of the carbonyl group, shifts the electron density from the nitrogen atom and carbon atom, leading to a positive charge at these atoms. As a result, the hydrogen atom of the amino group acquires an increased positive charge in comparison with the charge of the hydrogen atom of ammonia. In this connection, we clarify the possible error of the calculation of the energy of the stable hydrogen bonds. For this purpose, we use tetramethylurea, the specific interactions of which are identical to the ones realized in liquid methylamines. As mentioned earlier, replacement of the hydrogen atom in ammonia by the methyl group is accompanied by a reduction in the electron density at the nitrogen atom and there is a destabilized n-σ – interaction [32]. Therefore, the hydrogen bond of the ammonia formed is more stable than the specific interaction by the methyl group, D–N•••H–N > DN → CH_3–N. A similar effect of the increase in positive charge at the nitrogen atom is observed at the shifting of the electron density to the oxygen atom of the carbonyl group in the urea molecule and

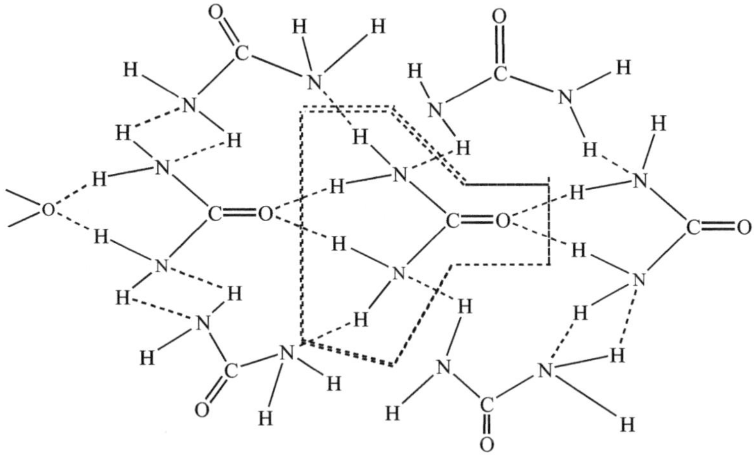

Fig. 11.2 Schematic picture of the liquid and crystalline structure of urea

the molecule of tetramethylurea. One can follow the possible manifestation of the correlation in the shifting of the electron density at the nitrogen atoms of trimethylamine and tetramethylurea at the additive energy contribution of the specific interaction of liquid dimethyl ether and trimethylamine and, on the other side, at the energies of the specific interactions of liquid tetramethylurea. The three-coordinated condition of the nitrogen atom in the molecule of tetramethylurea changes to the tetra-coordinated condition in its liquid and crystalline conditions because of the formation of four specific interactions, N → CH_3–N, with molecules close in the environment (Fig. 11.2). In this connection, the energy of the specific interaction for the same coordination, 4, taking into account the value of the positive charge at the nitrogen atom $N(CH_3)_3$, should be determined by the vaporization enthalpy (25.5 kJ mol^{-1}) [33] equal to 6.4 kJ mol^{-1}. Coordination of the oxygen atom in dimethyl ether and tetramethylurea remains equal to four; we obtain the energy of the formed specific interaction CH_3–O → CH_3 in liquid dimethyl ether from the vaporization enthalpy (22.54 kJ mol^{-1}) [33] equal to 5.63 kJ mol^{-1}. Taking into account that vaporization enthalpy is interrelated with the number and energy of the realized specific intermolecular interactions, we obtain a vaporization enthalpy of tetramethylurea of 48.0 kJ mol^{-1}, which exceeds the experimentally established value (45.6 kJ mol^{-1}) by 2.4 kJ mol^{-1} [18]. It follows that the molecular structure of tetramethylurea includes N → CH_3–N bonds, less stable than in the molecule of trimethylamine on the condition of formation of four specific interactions in liquid $N(CH_3)_3$. The error introduced to the vaporization enthalpy of each bond is equal to 0.6 kJ mol^{-1}. It is obvious that this difference between the experimentally obtained values of energies of the specific interactions N → CH_3–N and the values calculated on the basis of the principle of the continuous contribution to the enthalpy characteristic is caused by its increase. It is connected with the fact that the real coordinating number of the methyl group around the nitrogen atom in liquid trimethylamine is equal to six instead of four in the molecule of tetramethylurea. Naturally, the calculation of the vaporization enthalpy of tetramethylurea with the use of the energy of the specific interaction N $\rightarrow$ CH_3–N with six coordinating methyl groups, provided there are equal positive charges at the nitrogen atoms in molecules of tetramethylurea, contributes to the error. It is connected with the fact that with the change of the coordination number, the energy value of the specific interaction is changed significantly, and the contribution to the positive charge at the nitrogen atom contributed by each methyl group is sharply reduced. Avoiding the influence of the endo-effect of coordination of the methyl groups is possibly subject to change to the hydrogen atom in methylamine, which is proved by the change of the energy of the specific interaction N $\rightarrow$ CH_3–N: trimathylamine (4.25) < dimethylamine (4.80) < methylamine (5.33 kJ mol^{-1}). The equal numbers of methyl groups in the dimehtylamine and in the amine group of tetramethylurea allow us to use the energy value of the specific interaction DN $\rightarrow$ CH_3–N equal to 4.8 kJ mol^{-1}. Such an assumption is quite correct for amine-containing compounds, as was illustrated in Chap. 2. The calculation of energy of the specific interaction implemented

$$\text{D-O}\rightarrow\text{CH}_3\text{- N} = (\Delta_{\text{vap}}H^\circ(298\text{ K}) - 4\text{D-N}\rightarrow\text{CH}_3\text{-N})/4 \qquad (11.2)$$

leads to the establishment of the correct energy value of this type of interaction (6.6 kJ mol^{-1}), which is more stable in comparison with the specific interaction of the same type realized in liquid methyl ether (5.63 kJ mol^{-1}). This makes sense because the shifting of the electron density from the nitrogen atom to the oxygen atom in the urea molecule leads to the shifting of the electron density from the carbon atom of the methyl group, which enhances its acceptor properties with the increased charge at the oxygen atom and stabilizes the specific interaction D–O $\rightarrow$ CH_3–N in comparison with the bond DO $\rightarrow$ CH_3–O of dimethyl ether.

The presence of obvious contradictions in the charges of the nitrogen atoms in the molecules of ammonia and urea and the difference in the number of coordinating molecules in liquid ammonia (six) and in urea nitrogen atoms (four) inevitably introduces errors in the thermodynamic calculations of the energies of the stable hydrogen bond D=C=O•••H–N, and similarly in the calculation implemented of the vaporization enthalpy of tetramethylurea.

The formation of four stable hydrogen bonds =C=O•••H–N, and four hydrogen bonds N•••H–N of reduced stability by the urea molecule determines the properties of the liquid and solid state of urea and the high enthalpy characteristics of the sublimation and vaporization processes. Therefore, the value of sublimation enthalpy given in [19, 34] without error indication (88 kJ mol^{-1}) is a reduced one, and 98.6 kJ mol^{-1}, given in [35], is increased. The correct experimentally obtained value of the sublimation enthalpy [36] and the melting enthalpy (14.55 kJ mol^{-1}) of urea [1] allow us to determine the energies of the hydrogen bonds, formed in its liquid and crystalline conditions.

The urea vaporization enthalpy, as the difference of the enthalpy of sublimation (95.5 $\pm$ 0.3 kJ mol^{-1}) [36] and of melting (14.55 kJ mol^{-1}) [1], is equal to

80.95 kJ mol^{-1}. Taking the energy of the hydrogen bond D–N•••H–N as equal to the energy value of the hydrogen bond in liquid ammonia (3.86 kJ mol^{-1}) at the coordination of six hydrogen atoms at the nitrogen atom of liquid ammonia, we obtained the increased energy value of the hydrogen bond in liquid urea for D=C=O•••H–N: 16.4 kJ mol^{-1}. At the coordination of four hydrogen atoms at the nitrogen atom in liquid ammonia and, correspondingly, four hydrogen atoms at the nitrogen atom in liquid urea (Fig. 9.2), the energies of the hydrogen bond D–N–H•••N–H (5.91 kJ mol^{-1}) are significantly stabilized. As a result, we obtained the correct energy value of the hydrogen bond; it is equal to 14.3 kJ mol^{-1}. Using the conformity of the four-coordinated condition of the nitrogen atom in the crystalline condition of urea and ammonia, the energy of the hydrogen bond D–N•••H–N of 6.34 kJ mol^{-1} and equality (11.1), we determined the energy of the stable hydrogen bond D=C=O•••H–N to be equal to 17.5 kJ mol^{-1}. It should be mentioned that the estimated energy value of the hydrogen bond of 10.5 kJ mol^{-1} in liquid urea [37], calculated according to the scheme proposed in [18] with the vaporization enthalpy of tetramethylurea (45.6 kJ mol^{-1}), is significantly reduced and, therefore, it cannot explain the existence of urea in its cluster form in water solutions .

The established energies of the hydrogen bonds in solid and liquid urea definitely provide clarity to questions of water solutions of urea. The average value of urea hydration enthalpy of 15.34 ± 0.15 kJ mol^{-1} from the most reliable research results from 12 studies, drawn from [37], illustrates the endo-thermicity of the urea dissolution. Energy consumptions are caused by the destruction of the crystalline lattice and the creation of a salvation cavity in a solvent, not compensated by the effect of the formation of the hydrogen bonds of water – urea, D=C=O•••H–O. The results of this research and the appearance of the maximum in the dependency of isotopic effects (about 0.6 kJ mol^{-1}) in the field of temperature of 310 K to the urea melting enthalpy in water from the temperature are caused by the higher energy of the hydrogen bond in solid urea (17.5 kJ mol^{-1}) in comparison with the energy of the same bond in water 10.99 kJ mol^{-1}, calculated from the water vaporization enthalpy (43.96 kI mol^{-1}) with two bond vacancies of two free electron pairs of the oxygen atom and two positively charged hydrogen atoms, forming tetrahedric structure of liquid and crystalline water and four hydrogen bonds H–O•••H–O.

The natural change in the energies of the hydrogen bonds of urea and water explains

D(sol.)=C=O•••H-N (17.5) > D(liq.)=C=O•••H-N (14.3) > D(sol)H-O•••H-O (12.89) > D(liq.)H-O•••H-O (10.99 kJ mol^{-1})

the existence of urea clusters in diluted water solutions, predicted in the work [28] and experimentally proved by the results of research on ultra sound speed [38–42] and proton dissociation of alcohol [43] of the water-urea system, by spectra of NMR [44] and IR absorption [45].

An important fact for an understanding of urea water solutions is the generally accepted phenomenon of a reduction in the role of urea as a destroyer of water structure with temperature increase. It created serious difficulties in the understanding of, and formation of consensus on, the temperature border of the effect of negative hydration of the urea molecules. In this regard, the sound position is that at low temperature, water is more structured than molecules near urea. On the contrary, with temperature increase the structure becomes less structured in the bulk and more similar to the structure near urea. From this we can conclude that the temperature increase is expressed significantly at the destabilization of hydrogen bonds realized in water, the rupture of which needs less energy than that in urea. At the same time, the entropy of the structural component of the endothermic process of urea dissolution remains unchanged up to 310 K and it is only higher, at 313.15 K, that it becomes visible and its reduction more significant [37]. It means that the hydration process is accompanied by the formation of less stable hydrogen bonds, realized with solid urea. Consequently, breaking an equal number of hydrogen bonds at the reduced energy requires less of the enthalpy component of the process of urea hydration. It is important to draw attention to the fact that with the increase in temperature, not only the energy of the hydrogen bond but also the coordination number, and, consequently, the number of molecules of the near sphere are reduced, prmarily because of the rupture of the more weak hydrogen bonds H–N•••H–N in solid urea. The reduction of the coordination number causes some destabilization of the bonds; nevertheless, the enthalpy component of the hydration process is significantly underestimated in comparison with the original condition, which points to the prevailing effect of reduction of the coordinating molecules to stabilize the remaining bonds. Thus, the different character of the temperature dependency of the destabilization of the hydrogen bonds in urea and water leads to a feathered area of temperature transition of negative salvation and therefore is not precisely known, which is proved by the data presented in [37]. It should be noted that the theoretical interpretation of the process of negative salvation is a complex problem, and in order to understand this, reliable thermodynamic parameters, obtained from the detailed thermodynamic analysis of complex vaporization and sublimation processes and dependent on the number and energies of various types of bursting hydrogen bonds and specific interactions, are necessary.

11.3 Energies of Hydrogen Bonds and Specific Interactions of Derivative Urea

The interest of researchers in urea as a product of, and participant in, the metabolism of living organisms broadened the number of derivatives of this compound. The established thermodynamic properties of this class of compounds differ from the standard conditions in a number of cases; however, for a larger number, there is the possibility to reveal the nature of the formed intermolecular interactions and to determine the energies with certainty. Taking this into account, it is reasonable to consider the expression of the influence of the reverse dative bond at the

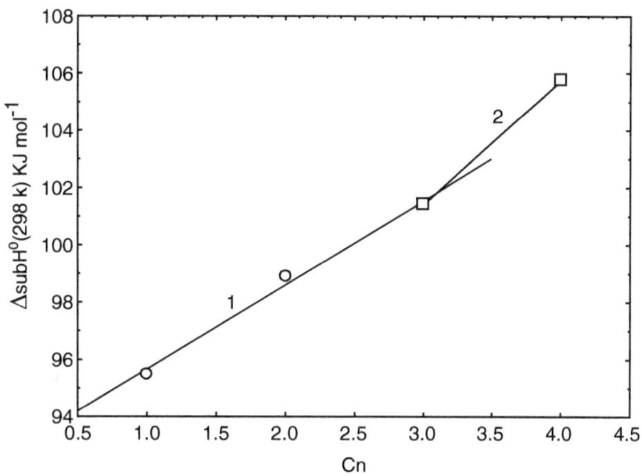

Fig. 11.3 Dependence of the sublimation enthalpy of 1-Alkylureas on the number of carbon atom in the alkyl ligand [36]

sublimation enthalpies of the alkyl urea derivatives. For the most reliable interpretation of the enthalpy characteristics ,we conducted thermodynamic analysis of the experimental values of the sublimation enthalpies of alkyl urea derivatives [36, 46–52] with the normal structure of the alkyl ligand. The dependence of sublimation enthalpy on the number of carbon atoms in the alkyl chain presented in Fig. 11.3 illustrates the location of the values of the thermodynamic properties, obtained with low error, on two lines practically with additive energy contribution with an increasing number of methylene groups in the alkyl fragment. Surely, such a sequence of its change reflects the equal manifestation of the reduction in the influence of the reverse dative bond on the specific interaction formed, clearly reflecting its completion at 1-n-propylurea. Further increase in the energy contribution of the methylene group to the enthalpy characteristic is caused by its participation in specific interactions with similar groups of the alkyl ligand of the molecules close in the environment, with a general energy contribution of 4.6 kJ mol^{-1}. The equal contribution of the other compounds with five or six carbon atoms in the ligand allow us to estimate its sublimation enthalpies with the accuracy of the experiment implemented (Table 11.1).

A similar character of dependence of vaporization enthalpy on the number of carbon atoms in the alkyl ligand is realized at alkylamines, dialkylamines, and trialkylamines with a lower manifestation of the stabilizing effect by the weakening of the influence of the reverse dative bond with increasing number of methylene groups in the alkyl chain (Fig. 11.4). The increasing number of alkyl ligands in the molecule is accompanied in the same sequence by an increasing effect, undermined by the influence of the reverse dative bond, ending at the propyl amines. This is characteristic for tryamines, which is proved by [20–23] by the insolvency of ideas explaining the steric effect. The role of the latter is limited by Coulomb interactions between carbon atoms connected directly with the nitrogen atom. The dependence of vaporization enthalpies of the alkylamines (1–3) and dialkylamines (4–6) on the sublimation enthalpy of 1-alkylureas (1–3) given in Fig. 11.5 reflects the inadequacy

Table 11.1 Energies of the hydrogen bonds and specific interactions (kJ mol^{-1}) in crystalline derivative urea (298.15 K)

Compounds	Formula	$\Delta_{vap}H°(T)$	References	D–N•••H–N	D–N → CH_3–(CH_2)n–N	D=C=O → CH_3–N	D=C=O•••H–N
Urea	CH_4N_2O	80.05	[1, 36]	5.91	–	–	14.1
Tetramethylurea	$C_5H_{12}N_2O$	45.6	[18]	–	4.8	6.6	–
		$\Delta_{sub}H°(T)$					
Urea	CH_4N_2O	95.5 ± 0.3	[36]	6.34	–	–	17.5
1-Methylurea	$C_2H_6N_2O$	95.5 ± 0.5	[36]	6.34	6.34	–	17.5
1-Ethylurea	$C_3H_8N_2O$	98.9 ± 1.1	[46, 51]	6.34	n = 1, 8.05	–	17.5
1-n-Propylurea	$C_4H_{10}N_2O$	101.4 ± 0.6	[36]	6.34	n = 2, 9.4	–	17.5
1-n-Buthylurea	$C_5H_{12}N_2O$	105.8 ± 0.7	[36]	6.34	9.4	$\Sigma DCH_2 = 4.6$	17.5
1-n-Pentylurea	$C_6H_{14}N_2O$	110.4	(4.6)	6.34	9.4	$\Sigma DCH_2 = 9.2$	17.5
1-n-Hexylurea	$C_7H_{16}N_2O$	115.0	(4.6 × 2)	6.34	9.4	$\Sigma DCH_2 = 13.6$	17.5
1,1-Dimethylurea	$C_3H_8N_2O$	93.5 ± 0.3	[36]	6.34	5.7	–	17.5
1,3-Dimethylurea	$C_3H_8N_2O$	89.3 ± 0.4	[36, 49]	6.34	4.8	–	17.5
1,1-Diethylurea	$C_5H_{12}N_2O$	95.7 ± 0.7	[36]	6.34	5.9	–	17.5
1,3-Diethylurea	$C_5H_{12}N_2O$	95.4 ± 0.3	[36, 46, 48]	6.34	5.9	–	17.5
1-*iso* Propylurea	$C_4H_{10}N_2O$	100.6 ± 1.3	[49]	6.34	8.05	D*iso*CH_3 = 1.7:2 = 0.85	17.5
1-*iso*-Buthylurea	$C_5H_{12}N_2O$	103,6 ± 1.1	[49]	6.34	9.3	2.2:2 = 1.1	17.5
1-*sec*-Buthylurea		101.9 ± 0.5	[36]	6.34			17.5
1-*tert*-Buthylurea	$C_5H_{12}N_2O$	103.40.7	[36]	6.34	8.05	2.5:4 = 0.62	17.5
1,1-di-*tert*-Buthylurea	$C_9H_{20}N_2O$	91.1 ± 1.0	[36, 46, 48]	6.34	4.5	2.5:4 = 0.62	17.5
		92,9 ± 0.7					
Acethylurea	$C_3H_6N_2O_2$	103.1 ± 0.7	[35]	6.34	8.05	DC=O•••H–CH_2 = 1.05	17.5

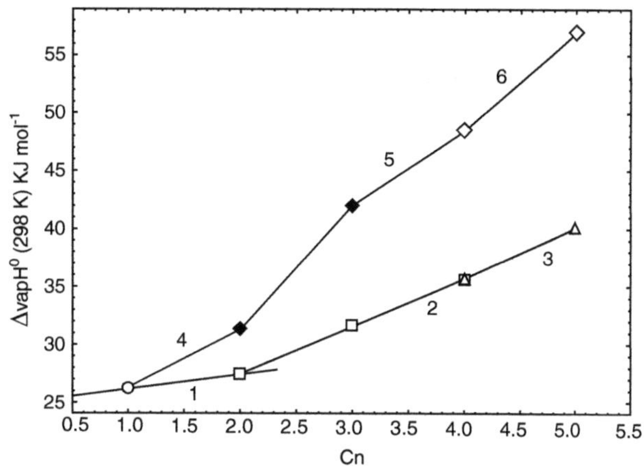

Fig. 11.4 Dependence of the sublimation enthalpy Alkylamines (*1–3*) and Dialkylamines (*4–6*) on the number of carbon atom in the alkyl chain (*1*, *2*) and (*4–5*) influence of reverse dative bond, (*2*, *3*) and (*5*, *6*) influence of CH_2-groups

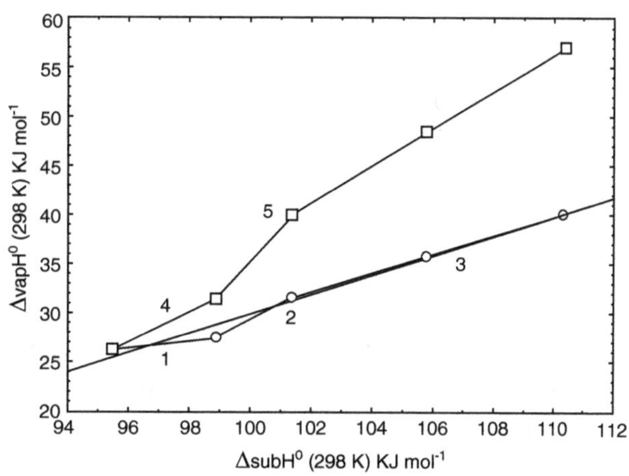

Fig. 11.5 Correlation between the sublimation enthalpy Alkylamines (*1–3*), Dialkylamines (*4–6*), and the sublimation enthalpy 1-Alkylureas: nonidentity influence of reverse dative bond on the stability of the specific interactions DN → R–N Alkylamines and Alkylureas (*1*, *2*) and (*4–5*), identical influence of CH_2-groups

of the manifestation of the reverse dative bond in the stability of the specific interactions of DN → R–N for alkylamines and alkylureas, caused by the specificity of shifting of the electron density in a molecule of alkylurea, CH_3–N–C=O.

Replacement of a hydrogen atom in urea by the methyl group reduces the number of hydrogen bonds, formed by amine up to two, and results in the formation of two specific DN → CH_3–N interactions and preserving of the crystal structure with the network of specific interactions and hydrogen bonds (Fig. 11.6). This replacement of the hydrogen atom is not accompanied by a reduction in the sublimation enthalpy of methylurea, which reflects the practically insignificant difference in the energies of the hydrogen bond and the specific interaction for the replacement of the hydrogen atom. As mentioned earlier, this fact is well in agreement with the research results of RE spectroscopy, which pointed to the small difference in charges at the nitrogen atoms on the replacement of hydrogen by the methyl group [53, 54]. Taking this into account and considering that the difference in the energies of the specific interactions will not exceed the experiment error,

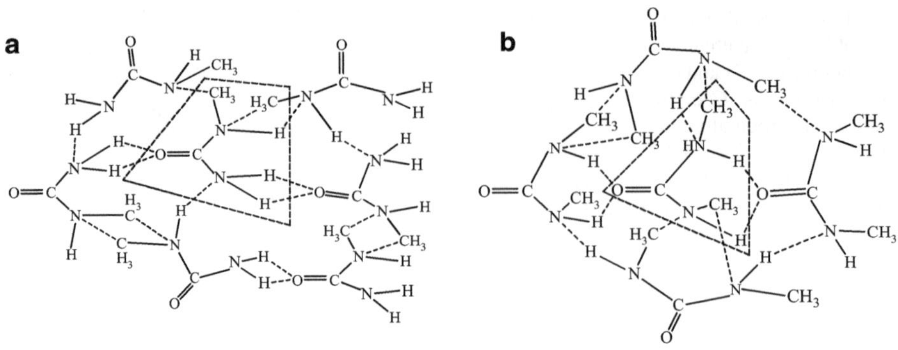

Fig. 11.6 Schematic picture of the crystalline structure of 1,1-Dimethylurea (**a**) and 1,3–Dimethylurea (**b**) with the network of the specific interactions

we should accept the equality of energies of the hydrogen bond and the specific interaction D–N•••H–N=D–N–CH_3 ← N = 6.34 kJ mol^{-1}.

The increase in the sublimation enthalpy of the alkyl derivatives of urea at the series of compounds

$$\text{1-Ethylurea } (98.9) < \text{1-n-Propylurea } (101.4) < \text{1-n-Buthylurea } (105.8)$$
$$< \text{1-n-Pentylurea } (110.4) < \text{1-n-Hexylurea } \left(115.0 \text{ kJ mol}^{-1}\right)$$

reflects the stabilization of the specific interactions at the replacement of methyl by the ethyl and propyl ligand. The completion of the influence of the reverse dative bond at the propyl ligand of urea causes the stability of the specific interaction DC=O•••H–N for all further compounds of this series, the energies of which should be determined with the help of Eq. 11.3,

1-Methylurea – 1-n-Propylurea

$$\text{DN-CH}_3\leftarrow\text{N} = (\Delta_{sub}H^{\circ}(T) \text{ - 4 D>C=O}\bullet\bullet\bullet\text{H-N - 2DN-H}\bullet\bullet\bullet\text{NH})/2 \quad (11.3)$$

From the data given in Table 11.1, it follows that the energies of the specific interactions of D–N → R–N increase in the order

$$\text{1-Methylurea } (6.34) < \text{1-Ethylurea } (8.05) < \text{1-n-Propylurea } \left(9.4 \text{ kJ mol}^{-1}\right).$$

Replacement of the second hydrogen atom at the location of 1,1-dimethylurea by the methyl group is accompanied by a reduction in the enthalpy characteristic at 2.0 kJ mol^{-1} in comparison with the sublimation enthalpy of urea or 1.0 kJ mol^{-1} at one methyl group. In other words, at each D–N → CH_3–N specific interaction, we have an energy of 0.5 kJ mol^{-1}, which value corresponds to the experimental error. The location of the methyl groups at 1,3-dimethylurea *cis* is accompanied by an additional reduction of 4.2 kJ mol^{-1} in the sublimation enthalpy and 2.1 kJ mol^{-1} in the energy of the specific interaction. A similar destabilizing effect is observed at the location of the ethyl ligand of 1,1-diethylurea (Fig. 11.6b), which reduces the total energy value of the specific interaction D–N → CH_2–CH_3–N by 2.1 kJ mol^{-1}. At the same time, the location of the ethyl ligand at locations 1,3-diethylurea has a destabilizing effect that is not observed. The energy of the specific interaction of DN → R–N of the compounds 1,1-alkylurea, 1,3-alkylurea is determined with the help of Eq. 11.4:

$$\text{D-N}\rightarrow\text{R -N} = (\Delta\text{vapH}^{\circ}(298\ \text{K}) - 4\text{D>C=O}\bullet\bullet\bullet\text{H-N})/4 \qquad (11.4)$$

where D–N → R–N–D–N → CH_3–N or D–N → CH_3–CH_2–N.

From the results given in Table 11.1 of the calculations implemented, it follows that the energies of the specific interactions of D–N → CH_3–N and D–N → CH_3–CH_2–N are destabilized in the order

D–N → CH_3–N: 1-Methylurea (6.34) > 1,1-Dimethylurea (5.7) > 1,3-Dimethylurea (4.8 kJ mol^{-1})

DN → CH_3–CH_2–N: 1-Ethylurea (8.05) > 1,1-Diethylurea (5.9) ≈ 1,3-Diethylurea (5.9 kJ mol^{-1})

Urea alkyl derivatives with the isostructural methyl group form a crystalline structure with the network of hydrogen bonds, stable specific interactions, and interactions of low stability, formed by the CH_3 group (Fig. 11.7a). The energy contribution of the isostructural methyl group to the sublimation enthalpy is determined by the difference between the sublimation enthalpies of the urea derivative with the isostructural methyl group 1-*iso* propylurea, 1-*iso*-buthylurea, and 1-*tert*-buthylurea and the enthalpy characteristics of the urea derivative with the normal structure and an equal number of carbon atoms in the chain of the alkyl fragment of 1-ethylurea, 1-n-propylurea, and 1-ethylurea:

iso-Propylurea:

$$\text{DisoCH}_3 = \Delta_{\text{sub}}\text{H}^{\circ}(\text{T})\text{is.pr} - \Delta_{\text{sub}}\text{H}^{\circ}(\text{T})\text{et} \qquad (11.5)$$

Tertbuthylurea:

$$\text{DisoCH}_3 = \Delta_{\text{sub}}\text{H}^{\circ}(\text{T})\text{tert bu} - \Delta_{\text{sub}}\text{H}^{\circ}(\text{T})\text{et} \qquad (11.6)$$

iso-Buthylurea:

$$\text{DisoCH}_3 = \Delta_{\text{sub}}\text{H}^{\circ}(\text{T})\text{is bu} - \Delta_{\text{sub}}\text{H}^{\circ}(\text{T})\text{pr} \qquad (11.7)$$

(Table 11.1). The values obtained for the energy of the specific interactions of low stability, DH_3C → H–CH_2, reflect the natural tendency of its stabilization,

D-H_2C–H–H–CH_2–:1-*tert*-Buthylurea (0.60) < 1-*iso* Propylurea (0.85) < 1-*iso*-Buthylurea (1.1 kJ mol^{-1})

with a reduction in its number, the bonded carbon atom fringing with the nitrogen atom in a molecule of 1-*tert*-buthylurea, and an increase in the number of carbon atoms of the chain in a molecule of 1-*iso*-buthylurea.

The acetylurea molecule with an acetyl fragment forms a crystal structure with the network of specific interactions similar to those of ethylurea in addition to four

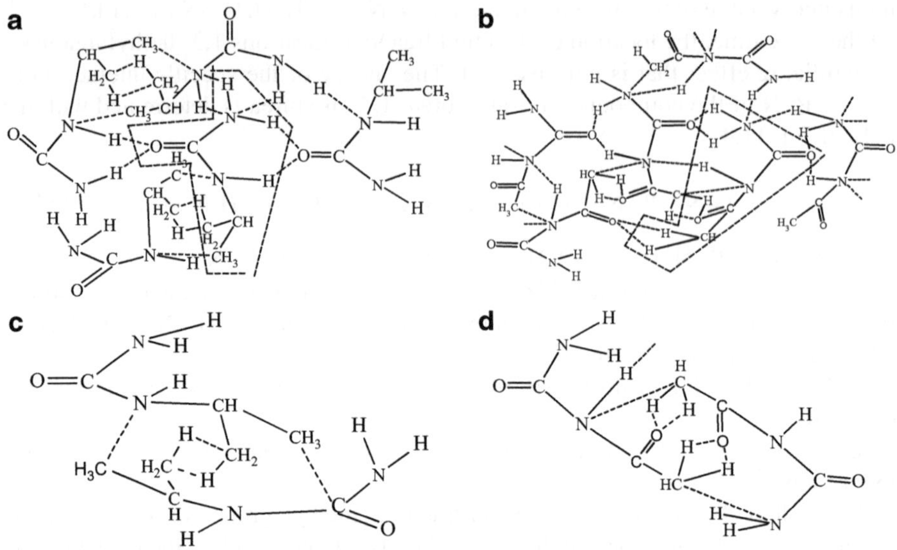

Fig. 11.7 Schematic picture of the crystalline structure of *iso*-Propylurea (**a**), Acethylurea (**b**) with the network of the specific interactions, interactions of the isostructural CH_3-group of iso-propyl (**c**) and acethyl ligand (**d**)

hydrogen bonds of low stability, formed by the hydrogen atoms of the methyl group of the acetyl ligand with the carbonyl oxygen atoms of the same ligands of contacting molecules that are close in the environment (Fig. 7b). It allows us to draw the conclusion that energies of the hydrogen bonds of two types DN•••H–N, D=C=S•••H–N and specific interactions, formed by the ethyl group in the crystal of these compounds, are equal to each other and the difference between the sublimation enthalpies of acethylurea and ethylurea is the total value (4.2 kJ mol^{-1}) of the energy contribution of four hydrogen bonds of low stability. Thus, we obtain the energy value of the hydrogen bond, formed by the carbonyl atom and hydrogen atoms of the terminal group of

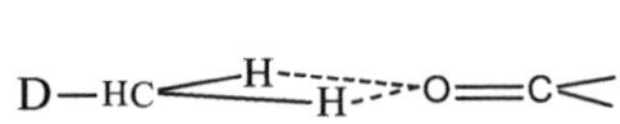

the acetyl fragment (Fig. 11.7d), equal to the value 1.05 kJ mol^{-1}.

11.4 Energies of Hydrogen Bonds and Specific Interactions of Thiourea and its Derivative

Replacement of the oxygen atom in the urea molecule (95.5 kJ mol^{-1}) by the sulfur atom is accompanied by a significant increase in the sublimation enthalpy of thiourea (112.0 kJ mol^{-1}). A replacement of the hydrogen atom similar to urea by the methyl

Table 11.2 Energies of the specific interactions (kJ mol^{-1}) of ethers [56]

	DO → CH_3–O	DO → CH_3–CH_2–O	DO → CH_3–O	DO → CH_3–CH_2–O
Compounds	Cristal	Cristal	Liquid	Liquid
$O(CH_3)_2$	6.59	–	5.63	–
$O(C_2H_5)_2$	–	8.83	–	6.78
	DS → CH_3–O	DS → CH_3–CH_2–O	DS → CH_3–O	DS → CH_3–CH_2–O
$S(CH_3)_2$	–	–	6.98	–
$S(C_2H_5)_2$	–	–	–	8.95

group in its analogue thiourea is accompanied by a small change in sublimation enthalpy, whose numerical value is within experimental error (Table 11.1 and 11.2). This fact is in good compliance with results of research by the method of RE spectroscopy [53, 54], pointing to an insignificant reduction in the electron density for the nitrogen atom on the replacement of the hydrogen atom by the methyl group. It follows that the practically equal value of the sublimation enthalpy of thiourea and 1-methylurea points at least to a small difference in the energies of the hydrogen bond, D–N•••H–N, and the specific interaction D–N → CH_3–N, formed by the methyl group. In other words, the values of the energies of these interactions should lie within experimental error. Since both interactions are formed by the amino group with two hydrogen atoms, forming four hydrogen bonds, or with two methyl groups, forming four specific interactions with the alkyl ligand in crystal, the energy of the hydrogen bond should be established from the sublimation enthalpy of ammonia at the four-coordinated condition of the nitrogen atom (6.34 kJ mol^{-1}). Taking into account that the obtained experimental values of the sublimation enthalpies of thiourea and its derivatives are determined with errors of 2–5 kJ mol^{-1}, we accepted the energy of the specific interaction formed by the methyl group to be equal to 6.34 kJ mol^{-1}. It is symptomatic that the sublimation enthalpies of thiourea and its derivatives exceed at 14–15 kJ mol^{-1} the values of the same urea properties and, in particular, 1-methylurea. It follows that the sulfur atom, with two electron pairs located at a remote energy level, is able to obtain a large electron density of the contacting carbon atom. This inevitably causes a further shifting of the electron density from the nitrogen atom to the carbon atom and, in turn, from the hydrogen atoms to the nitrogen atom of the same amino group and further to the carbon atom and sulfur atom, =C=S, of this fragment. This sequence in the shifting of the electron density leads to the biggest difference in the negative charge of the sulfur atom and positive charge of the hydrogen atom of the amino group. At the same time, the difference in the charge of the hydrogen and nitrogen atoms of the amino group remains practically the same as in ammonia. As a result of the changes in the nature of the charges in the thiourea molecule, there is a significantly stabilized hydrogen bond D=C=S•••H–N, and we practically kept the stability of the hydrogen bond of the second type, D–N•••H–N. A similar character in the distribution of the electron density is kept in the molecule of 1-methylthiourea, 1,3-dimethylthiourea, trimethylthiourea between the sulfur atom and carbon atom of the methyl group, nitrogen, and the carbon atom of the same methyl group, and thus the corresponding differences in the charges of marked atom pairs are preserved. While the main

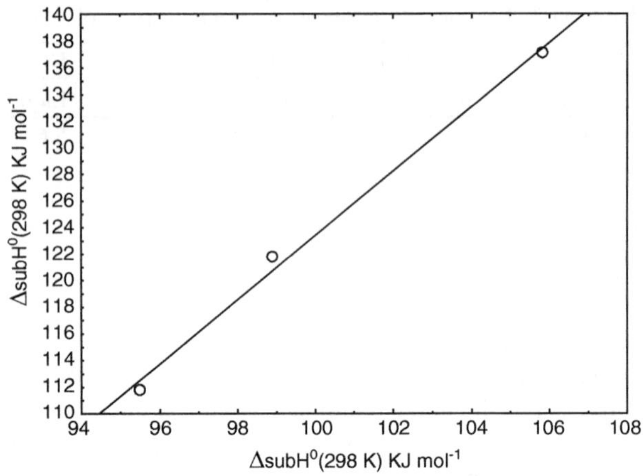

Fig. 11.8 Correlation between the sublimation enthalpy 1,3-alkylthioureas and the sublimation enthalpy 1-alkylureas: influence of reverse dative bond on stability of the specific interactions (*1*) and influence of CH_2-groups (*2*)

electron density of the nitrogen atom is given by the contacting carbon atom, partly returning as a result of the reverse dative bond, the increase in the number of carbon atoms in the alkyl ligand from one to two and maximum to three is practically not expressed at the charge of the nitrogen atom. Therefore the energy of the formed specific interaction is determined mainly by the charge of the carbon atom of the terminal methyl group of the alkyl ligand. We can therefore conclude that the relative strength of the hydrogen bond, D–N•••H–N, and the specific interaction D–N → CH_3–N or D–N → CH_3–CH_2–N, are preserved in the crystalline and liquid conditions of methylthiourea, 1,3-dimethylthiourea, trimethylthiourea, and derivatives with ethyl, propyl ligands. The correct examples proving these ideas can serve the sublimation enthalpies of compounds, within experimental errors. In this regard, it is reasonable to pay attention to the energies of the specific interactions, realized in liquid and solid methyl, ethyl ethers, and dimethyl and diethyl sulfide (Table 11.2). The data given in the table point to a bigger stability of the specific interactions S → CH_3–S and S → CH_3–CH_2–S, formed by sulfur in comparison with the energies of the same interactions formed by oxygen.

The limited data on the enthalpy characteristics of urea derivatives drive us to consider the correlation of the sublimation enthalpies. The linear character of the dependence of the sublimation enthalpies of 1-alkylureas and 1,3-alkylthioureas (Fig. 11.8) illustrates the mutual compensation of energies of the specific interactions at the first representatives of compounds with methyl and ethyl ligands and the bigger number of methylene groups in the latter. One should draw attention to the fact that urea derivatives in crystals form two hydrogen bonds, two specific interactions at four hydrogen bonds of high stability, while thiourea derivatives form four specific interactions by alkyl ligands. The linear character of the correlation dependence of the thermodynamic properties allows us to use the method of interpolation and extrapolation for the determination of the sublimation enthalpies of experimentally unstudied propylthiourea, pentylthiourea, and hexylthiourea (Table 11.3). The dependence presented in Fig. 11.9 reflects its dual nature, on

Table 11.3 Energies of the hydrogen bonds and specific interactions (kJ mol^{-1}) in solid derivative thiourea (298.15 K)

Compounds	Formula	$\Delta_{sub}H^{\circ}(T)$ [55]	T K	D–N–CH_3 ← N	D–N–CH_3 ← S=C	D–N–H•••S=C
Thiourea	CH_4N_2S	112.0 ± 2	298	–	–	21.7
1-Methylthiourea	$C_2H_6N_2S$	112.9 ± 3	298	6.34	–	21.7
1,3-Dimethylthiourea	$C_3H_8N_2S$	111.8 ± 3	298	6.25	–	21.7
		107.3 ± 4.0				21.2
1-Ethylthiourea	$C_3H_8N_2S$	118.8 ± 5	298	8.65	–	21.7
1,3-Diethylthiourea	$C_5H_{12}N_2S$	121.7 ± 3	298	8.70	–	21.7
1,3-Di-n-propylthiourea	$C_7H_{16}N_2S$	132.5 ± 3.0[a]	298	11.4[a]	–	21.7
1,3-Propylthiourea	$C_7H_{16}N_2S$	129.3[b]	298	10.6	–	21.7
1,3-Dibutylthyourea	$C_9H_{20}N_2S$	137 ± 3.0	298	10.6	$\Sigma DCH_2 = 7.7$	21.7
1,3-Dipentylthyourea	$C_{11}H_{24}N_2S$	145.3[b]	298	10.6	$\Sigma DCH_2 = 16.0$	21.7
1,3-Dihexylthyourea	$C_{13}H_{28}N_2S$	153.6[b]	298	10.6	$\Sigma DCH_2 = 24.3$	21.7
Trimethylthiourea	$C_4H_{10}N_2S$	83 ± 3.0	333	6.34 × 4	7.15	21.7[a]
Tetramethylthiourea	$C_5H_{12}N_2S$	83.0 ± 0.2	298	6.34	14.4	–
		$\Delta_{vap}H^{\circ}(T)$				
1,3-Dimethylthiourea	$C_3H_8N_2S$	93 ± 4.0	359	5.91	–	17.3[a]

[a]It is not correct

[b]Elucidated by extrapolation and interpolation

Fig. 11.9 Dependence of the sublimation enthalpy 1,3-alkylthioureas on the number of carbon atom in the alkyl chain

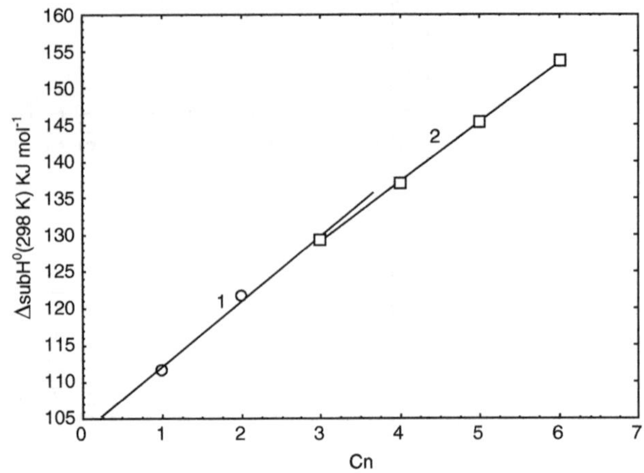

the one side at the first section of the manifestation of the influence of the reverse dative bond at the stabilization of the specific interactions and, on the other side, after reaching the maximum stability of N → CH_3–CH_2–CH_2–N of the interaction. For propylurea, the further increase in the enthalpy characteristic is caused by the energy contribution of the increasing number of methylene groups in the alkyl chain. The energy contribution of the latter is relatively small at one methylene group in comparison with the manifestation of the dative bond.

Similar to urea, the energy of the hydrogen bond of D=C=S•••H–N for thiourea is determined with the help of Eq. 11.1, taking into account the equality of energies of the hydrogen bond D–N•••H–N (6.34 kJ mol^{-1}) realized in liquid ammonia at the four-coordinated nitrogen atom. The increased energy value, formed by the sulfur atom fragment =C=S of 4.2 kJ $mol^{-1,}$ in comparison with the energy of the same type of hydrogen bond, D=C=O•••H–N (17.5 kJ mol^{-1}), formed by urea, reflects its higher electron donor properties. As noted earlier, the insignificant reduction in the charge of the nitrogen atom on the replacement of the hydrogen atom of the amino group by the methyl group allows us to follow the manifestation of its influence on the change of stability of the specific interactions for the different compounds of thiourea derivatives. In this regard, the increased value of the sublimation enthalpy of thiourea in comparison with that of 1-methylthiourea (Table 11.3) allows us to attribute the difference of its values to the increased stabilization of the hydrogen bond, D–N•••H–N.

The increased value of the sublimation enthalpy of 1-ethylthiourea, rather than *N*-methylthiourea, points to the increased energy of the specific interaction D–N → CH_3–CH_2–N, formed by the ethyl ligand. At the same time, the reduced value of the sublimation enthalpy of 1,3-dimethylthiourea, rather than 1-methylthiourea, points to the fact that replacement of two hydrogen bonds in the crystal of the latter at the same number of specific interactions of D–N → CH_3–N in a crystal of 1,3-dimethylthiourea is accompanied by a reduction of the enthalpy characteristic at 1.1 kJ mol^{-1} or an energy of the specific interaction D–N → CH_3–N of approximately 0.5 kJ mol^{-1}. Using the established energy of the hydrogen bond D=C=S•••H–N of crystalline thiourea, we determined the energies of the specific interactions of D–N → CH_3–N and D–N → CH_3–CH_2–N, respectively, 1-methylthiourea, 1,3-dimethylthiourea, and 1-ethylthiourea and D–N → CH_3–CH_2–N with the help of Eqs. 11.3 and 11.4, reflecting the mentioned tendencies of its changes at methyl derivatives of the thiourea compounds (Table 11.3). At the same time, unlike 1,3-diethylurea, in its analogues 1-ethylthiourea (8.65) and 1,3-diethylthiourea (8.70 kJ mol^{-1}) there is no observed reduction in the energies of the specific interactions D–N → CH_3–N, caused by *cis*- and *trans*-isomers, possibly connected with the experimental errors of the determination of the sublimation enthalpies of the derivatives of thiourea.

The results of the calculations of energies of the D–N → CH_3–CH_2–CH_2–N specific interactions for 1,3-dipropylthiourea with the help of Eq. 11.4 and compounds with a bigger number of methylene groups of the alkyl chain, taking into account the total energy contribution to enthalpy characteristics

$$\text{D-N}\rightarrow\text{CH}_3\text{-CH}_2\text{-CH}_2\text{-N} = (\Delta\text{vapH}^\circ(298\ \text{K}) - 4\text{D=C=S}\cdots\text{H-N} - \Sigma\text{DCH}_2)/2 \qquad (11.8)$$

(Eq. 11.8), are given in Table 11.3.

The obtained energies of the specific interactions illustrate its stabilization in the order of

1-Methylthiourea (6.34) < 1-Ethylthiourea (8.65) (10.6 kJ mol^{-1})

DN → CH_3–N 1,3-Methylthiourea (6.34) < DN → CH_3–CH_2–N 1,3-Diethylthiourea (8.70) < DN → CH_3–CH_2–CH_2–N 1,3-Dipropylthiourea (10.6 kJ mol^{-1})

The methyl groups of tetramethylthiourea form two types of specific interactions by four bonds each. The bonds DN–CH_3 ← N formed of the methyl groups at one nitrogen atom with the energy of 6.34 kJ mol^{-1} allow us to determine the energy of the second type of specific interaction with the help of Eq. 11.2. The obtained high value of the energy of the specific interaction of D=C=S → CH_3–N (14.4 kJ mol^{-1}) is caused by the shifting of a significant electron density from the carbon atom of fragment =C=S, and by the increased shifting of the electron density from the carbon atom of the methyl group to the nitrogen atom and further to the nitrogen of the same fragment. Thus, the high acceptor ability of the sulfur atom leads to an increased difference in the charges of the sulfur atom and carbon atom of the methyl group and a high ability of the hydrogen bond and specific interaction D=C=S → CH_3–N.

Three methyl groups and the hydrogen atom of the amine groups of trimethylthiourea form two hydrogen bonds D=C=S•••H–N, two specific interactions D=C=S → CH_3–N and four specific interactions DN → CH_3–N, of which the energy of the specific interaction remains unknown, formed by the sulfur atom of the fragment =C=S. The energy of this type of interaction, equal to the difference between the sublimation enthalpy and the total value of energies of the hydrogen bonds and specific interactions, is determined with the help of Eq. 11.9,

$$\text{D=C=S}\rightarrow\text{CH}_3\text{-N} = (\Delta\text{vapH}^\circ(298\ \text{K}) - 2\text{D=C=S}\cdots\text{H-N} - 2\text{D-N}\rightarrow\text{CH}_3\text{-N})/2 \qquad (11.9)$$

The value obtained for the energy of the specific interaction D=C=S → CH_3–N (7.15 kJ mol^{-1}) of crystalline trimethylthiourea is twice reduced in comparison with the specific interaction realized in solid tetramethylthiourea. Note that this value, obtained for 333 K, also remains reduced and at the standard conditions. Nevertheless its increased value could be realistic on the condition of this bond being located in *cis*-position with a stable hydrogen bond D=C=S•••H–N. The reduced value is also the energy of the hydrogen bond D=C=S•••H–N of liquid 1,3-dimethylthiourea, of which the vaporization enthalpy is obtained at 359 K and which is calculated by Eq. 11.4 without corrections to the temperature dependence (Table 11.3). In the calculations the energy of the hydrogen bond D–N•••H–N (5.91 kJ mol^{-1}) of liquid ammonia with four-coordinated nitrogen atom was used.

11.5 Energies of Hydrogen Bonds and Specific Interactions Uracile and Its Derivatives

11.5.1 Uracile and Its Derivatives

Urea in the shape of the compound of malonylurea is an important derivative of pyrimidine, considered in Chap. 5. In turn, in the process of hydrolyzing nucleic acids present in the nucleus of a cell, important derivatives of pyrimidine uracyl, cytosine, and timine, existing in three tautomeric forms are obtained. For the first two compounds and derivatives there exist sublimation enthalpies, leading to thermodynamic analysis and obtaining of energy parameters, formed by the hydrogen bonds and specific interactions.

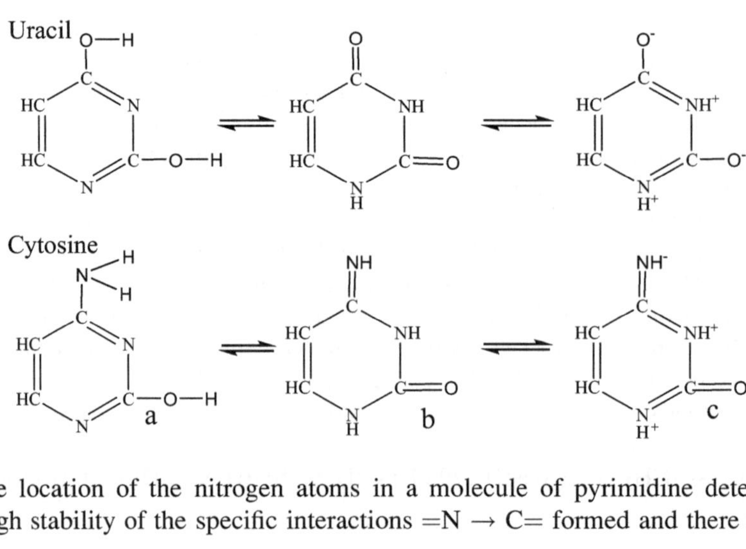

The location of the nitrogen atoms in a molecule of pyrimidine determines the high stability of the specific interactions =N → C= formed and there remain

Table 11.4 Energies of the hydrogen bonds (kJ mol) of solid derivative benzoic alcohols [56]

Compounds	DH–O•••H–O	Compounds	DH–O•••H–O
Phenol	12.7	–	–
1.2-Dihydroxybenzene	10.4	1.2.3-Trihydroxybenzene	11.6
1.3-Dihydroxybenzene	10.5	1.2.4-Trihydroxybenzene	12.7
1.4-Dihydroxybenzene	12.1	1.3.5-Trihydroxybenzene	14.8
2-Methylphenol	8.7	3-Methylphenol	2.4
4-Methylphenol	9.5	–	–

two unoccupied locations of these bonds for cycle carbon atoms that form interactions =HC → CH= with energy values of the same type of bonds, realized in liquid and crystalline benzene. According to the data of the work of [55] the energies of the hydrogen bonds of crystalline phenol and dihydroxybenzenes are changed in the interval 12.7–10.4 kJ mol^{-1} depending on the location of the hydroxyl group in the benzene ring with the values of 12.7 and 12.1 kJfor mol phenol and 1.4-dihydroxybenzene, respectively (Table 11.4). It is obvious that the two nitrogen atoms present in the ring of the uracyl molecule with a rigid six-membered cycle have a definite influence on the energies of the hydrogen bonds, formed by the hydroxyl groups at locations C(1) and C(3), providing it with increased stability compared with the phenol derivative.

The transition of the hydrogen atoms of the hydroxyl groups to the cycle's nitrogen atoms in tautameric forms caused by low energy consumptions is easy enough, and this presupposes the increase in the differences in the charges of the hydrogen atom and oxygen atom and stabilization of the hydrogen bond H–O•••H–O formed by the hydroxyl groups of uracyl at the locations C(1) and C(3).

The planarity of the urea molecule caused by the presence of the =C=O fragment with two double bonds and with practically insignificant pyramidality, formed by the hydrogen atoms of the amino groups, and planarity of the molecules of uracyl and its derivatives could not lead to a significant difference in the energies of the hydrogen bonds formed by the hydroxyl groups of these compounds and urea. The principal difference in the energies of its hydrogen bonds is in the rigidity of the uracyl cycle and its derivatives. The open structure of urea leads to the increased difference in the charges of the oxygen atom of the =C=O fragment and the hydrogen atom of the amino group, caused by the shifting of the electron density.

The approach to thermodynamic analysis and the determination of the energy of the hydrogen bond mentioned above is based on the known value of energy of the specific interaction HC → CH (7.40 kJ mol), formed by the CH group of the benzene ring and its identity with the energy of the same bond, formed by the pyrimidine cycle and its derivatives. The fact that we have the known value of energy of the specific interaction, formed by the metoxy group of 1,3-dimethyluracil, equal to the energy of the same type of interaction (6.59 kJ mol) of crystalline dimethyl ether (Chap. 5) makes this task easier. The molecule of 1,3-dimethyluracil with ten free bond vacancies forms the crystal structure with the network of specific interactions (Fig. 11.10), where two specific interactions D=HC → CH, four D=N → CH= and two interactions DO → CH_3–O participate. From this one

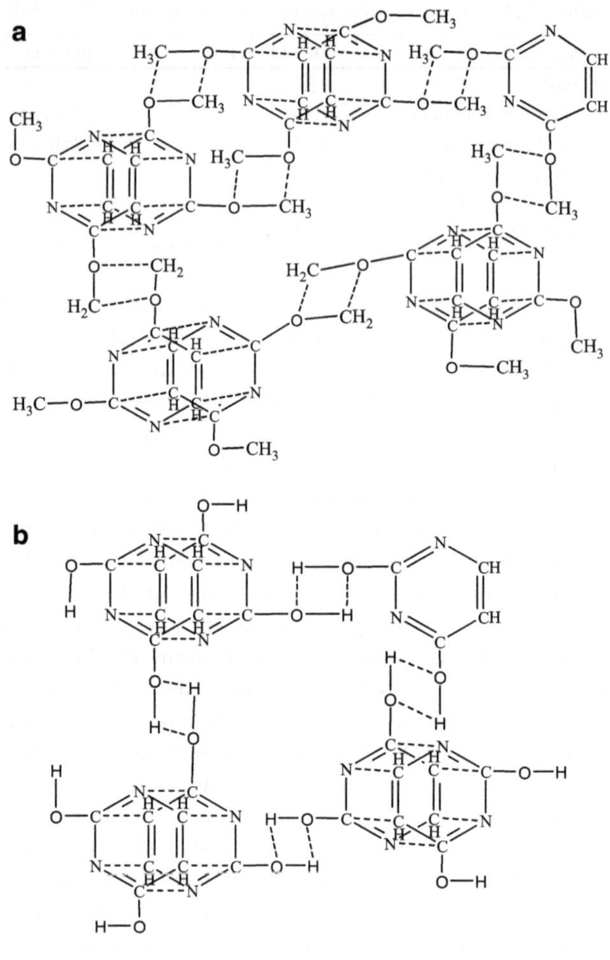

Fig. 11.10 Schematic picture of the crystalline structure of 1,3-Dimethyluracil (**a**), Uracyl (**b**), with network of the hydrogen bonds and specific interactions

can conclude that the energy of the specific interaction of D=N → CH= can be determined with the help of Eq. 11.10,

$$D{=}N{\rightarrow}CH{=} \;=\; (\,\Delta subH^0(298\ K)dmu - 2D{=}HC{\rightarrow}C\ H\text{-}\ 4D\text{-}O{\rightarrow}CH_3\text{-}O)\,/\,4 \qquad (11.10)$$

The energy value of this interaction leads to the possibility of determining the energy of the hydrogen bond of uracyl, forming the crystal structure with the network of specific interactions (Fig. 11.10a) with participation of two types of six interactions of 4D=N → CH=, 2D=HC → CH and four hydrogen bonds (Fig. 11.10b),

$$D\text{-}O\cdots H\text{-}O = (\Delta_{sub}H^0(298\ K)ur\text{-}\ 2D{=}HC{\rightarrow}CH\text{-}\ 4D{=}N{\rightarrow}CH{=}\,)\,/\,4 \qquad (11.11)$$

The results of the calculations are given in Table 11.5, which shows that the energy of the hydrogen bond of uracyl (15.1 kJ mol) exceeds the energies of

Table 11.5 Energies of the hydrogen bonds and specific interactions (kJ mol^{-1}) of the crystalline structure of uracyl and its derivatives

Compounds	Formula	Structure	$\Delta_{sub}H^{\circ}(T)$ [55]	T K	D–O → CH_3–O	D=HC → C–H	D=N → C=	D–O•••H–O
1,3-Dimethyluracil	$C_6H_8N_2O_2$	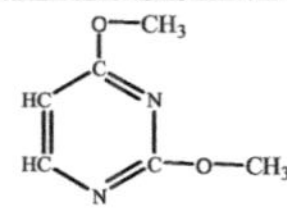	96.4 ± 1.4 96.9 ± 1.2	298	6.59	7.40	13.9	–
Uracyl	$C_4H_4N_2O_2$	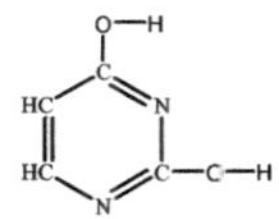	131. 5	298	–	7.40	13.9	15.1
5-Methyl uracyl (Thimine)	$C_5H_6N_2O_2$	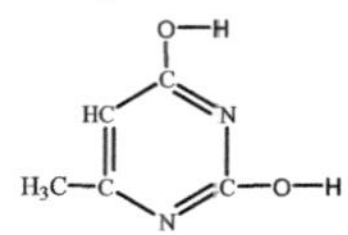	131.3 ± 4 134.1 ± 4.2 C	298	–	7.40	13.9	15.1 $DCH_3 = 0.60$
6-Methyl uracyl	$C_5H_6N_2O_2$	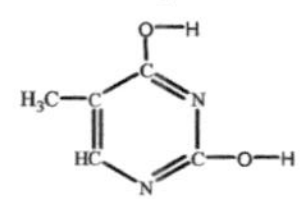	131 ± 5	298	–	7.40	13.9	15.1
1-Methyl uracyl	$C_5H_6N_2O_2$	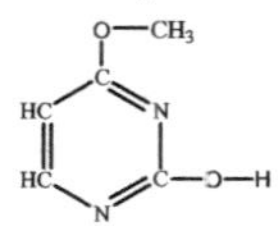	112.5 ± 2.5	358	6.59	7.40	13.9	14.4
3-Methyl uracyl	$C_5H_6N_2O_2$	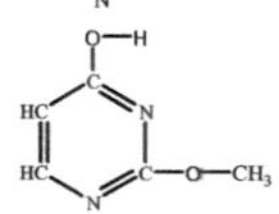	118,8 ± 3.0	382	6.59	7.40	13.9	17.55
1,3-Dimethyl-5-ethyluracil	$C_8H_{12}N_2O_2$	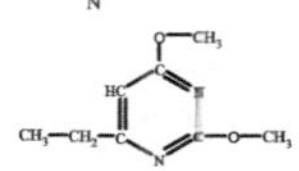	99.3 ± 0.2	308	6.59 × 2	7.40 × 2	13.9 × 4	$DC_2H_5 = 1.2$
1,3-Dimethyl-5-propyluracil	$C_9H_{14}N_2O_2$	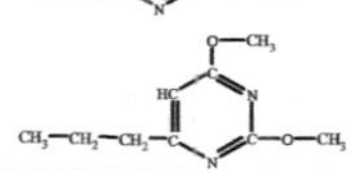	1111.0 ± 1.6[a]	322	6.59 × 4	7.40 × 2	13.9 × 4	$DC_3H_7 = 7.0$

(continued)

Table 11.5 (continued)

Compounds	Formula	Structure	$\Delta_{sub}H^{\circ}(T)$ [55]	T K	D–O → CH_3–O	D=HC → C–H	D=N → C=	D–O•••H–O
1,3-Dimethyl-5-isopropyluracil	$C_9H_{14}N_2O_2$	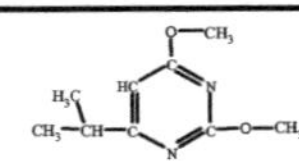	102.9 ± 1.6	322	6.59 × 4	7.40 × 2	13.9 × 4	DC_3H_7 DC_2H_5 + DCH_3 = 2.4 + 0.6
1,3-Dimethyl-5-butyluracil	$C_{10}H_{16}N_2O_2$	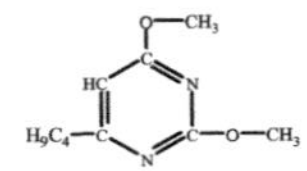	106.3 ± 1.3	298	6.59 × 4	7.40 × 2	13.9 × 4	DC_4H_9 = 4.7
1,3,5-Trimethyluracil	$C_7H_{10}N_2O$	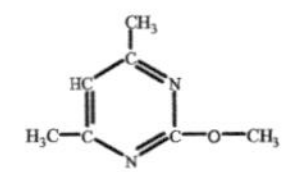	103.5 ± 1.5	328	6.59 × 4	7.40 × 2	13.9 × 4	ΣDCH_3 = 19.9
1,3,6-Trimethyluracil	$C_7H_{10}N_2O$	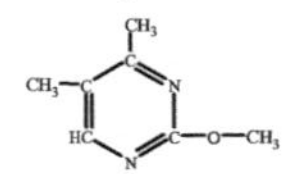	106.7 ± 2.5	320	6.59 × 2	7.40 × 2	13.9 × 4	ΣDCH_3 = 23.1
5-Aminouracil	$C_4H_5N_3O_2$	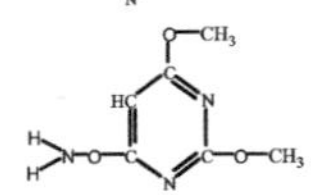	145.6	298	DN•••H–N = 15.1:4 = 3.8	7.40 × 2	13.9 × 4	15.1

the same type of bond of phenole (12.7 kJ mol), 1.4-dihydroxybenzene (12.1 kJ mol), and 1.4-dihydroxybenzene (10.5 kJ mol). It also exceeds the energy of the specific interactions D=N → CH= (13.9 kJ mol) and DHC → CH (7.40 kJ mol), realized in solid uracyl and 1,3-dimethyluracil. The absence of the difference in sublimation enthalpies of uracyl and 5-methyl uracyl, and 6-methyl uracyl points to the low contribution of the isostructural methyl group at locations C(5) and C(6) (1.5 kJ mol^{-1}) to the enthalpy characteristic exceeding to some extent the energy contribution of the substituted hydrogen atom of the CH group of the cycle (1.0 kJ mol^{-1}). Thus, the replacement of the methyl group at location C(5) practically does not feature in the calculations for the energies of the hydrogen bonds of crystalline uracyl, 5-methyl uracyl, and 6-methyl uracyl. The energy of the hydrogen bond of crystalline 1-methyl uracyl and 3-methyl uracyl is determined by the difference in the sublimation enthalpy of these compounds reduced by the total value of energies of the three types of specific interactions divided by the number of hydrogen bonds. We have

$$\text{D-O}\cdots\text{H-O} = (\Delta\text{subH}^0(298\ \text{K})\text{amur} - 2\text{D=HC}\rightarrow\text{CH} - 4\text{D=N}\rightarrow\text{C=} - 2\ \text{D-O}\rightarrow\text{CH}_3\text{-O})\ /\ 2 \qquad (11.12)$$

The results of the calculations (Table 11.5) are implemented without temperature dependence of the energy parameter; nevertheless they reflect the stabilization of the hydrogen bond at the replacement of the metoxy group's location at C(1) at C(3) and the hydroxyl group in the reverse sequence from C(3) to C(1) accurately. Therefore, replacement of one hydroxyl group in a molecule of uracyl by the metoxy group in 1-methyl uracyl and further at the replacement of location of the metoxy and hydroxyl group at C(1) to C(3) is accompanied by a change of values in the energies of the hydrogen bonds and by the character of change of its stability at the compounds of the series of

DO•••H–O: Uracyl (15.1) $<$ 1-Methyl uracyl (14.4) $<$ 3-Methyl uracyl (17.55 kJ mol^{-1})

The largest energy value of the hydrogen bond at the location of the hydroxyl group at C(3) between two nitrogen atoms of the hetero cycle points to the higher positive charge of the carbon atom in this location. As a result, the carbon atom takes the main excess electron density from the hydrogen atom, leading to the increased difference in negative self-charge and positive charge of the hydrogen atom, forming hydrogen bonds with the similar hydroxyl group of the molecule in contact. In this connection we should mention that the hydroxyl groups at locations C(1) and C(3) of the uracyl molecule have a mutual influence and thus, the energies of the formed hydrogen bonds in crystalline condition are the average values of two hydrogen bonds, formed by 1-methyl uracyl and 3-methyl uracyl.

The nature of the specific interactions of the alkyl fragments is determined by the essentially unshared 2 s^2 electron pair of the carbon atom with its excess or lack of electron density, with negative or positive charge, and the donor or acceptor properties allow us to conclude that the obtained value of energy and enthalpy characteristic of the vaporization process is accurate. In this regard the determination of the energy contribution to the sublimation enthalpy of compounds of the series 1,3-dimethyl-5-ethyluracil – 1,3-dimethyl-5-butyluracil (Fig. 11.11) located at the C(5) fragment, is interesting. The value of this energy contribution is equal to the difference of sublimation enthalpies of each compound of this series with enthalpy characteristic of 1,3-dimethyluracil. The high value of the energy contribution of the propyl fragment with three chain carbon atoms,

Fragment	Ethyl	Isopropyl	Propyl	Buthyl
	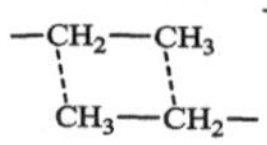	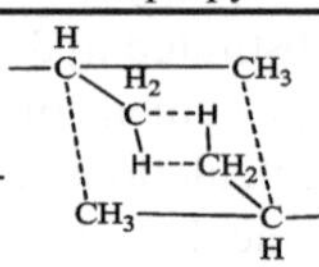	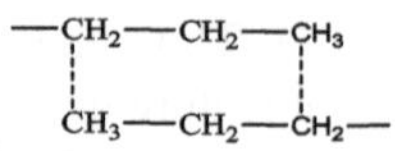	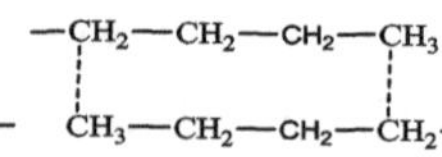
$\Sigma DCH_3 \rightarrow CH_3\text{–}R$	2.4	6.0	14.1?	9.4
$DCH_3 \rightarrow CH_3\text{–}R$	1.2	1.8 / 1.2	–	4.7

exceeding the contribution of butyl and twice the ethyl fragment is the reason for the inaccurate value of the sublimation enthalpy of 1,3-dimethyl-5-propyluracil, obtained with significant difference from standard conditions (322 K). The most realistic value of the energy contribution of the butyl fragment is where the influence of the reverse dative bond ends. These data in conjunction with the contribution of the energies of the two methyl groups of 1,3,5-trimethyluracil and 1,3,6-trimethyluracil allow us to obtain more detailed information on the influence of the location of the methyl group at the carbon atoms at locations C(5) and C(6). The energies of the contributions of the two methyl groups considered, and, consequently, the specific interactions formed, can be determined with the help of Eq. 11.13,

$$DCH_3 = (\Delta_{sub}H^0(298\ K)tmu - 4D{=}N{\rightarrow}C{=} - 2DO{\rightarrow}CH_3\text{-}O) / 2 \qquad (11.13)$$

From the data given in Table 11.5 it follows that the location of the alkyl fragment at the carbon atom C(5) near the nitrogen atom N(4) is accompanied by insignificant stabilization of the specific interaction, formed by the ethyl group (1.2 kJ mol^{-1}) 1,3-dimethyl-5-ethyluracil and the same group of the isopropyl fragment (1.8 kJ mol^{-1}) 1,3-dimethyl-5-isopropyluracil and further by the butyl fragment (4.7 kJ mol^{-1}) of the molecule of 1,3-dimethyl-5-butyluracil. The isostructural methyl group of the isopropyl fragment forms a specific interaction of less stability (0.60 kJ mol^{-1}). It follows that the presence of the two metoxy groups in a molecule of 1,3-dimethyl-5-alkyl at C(1) and C(3) significantly reduces the electron density of these carbon atoms and, on the other side, the carbon atom located at C(5) with the nitrogen atom N(4) has a reduced difference in the charges compared with the carbon atom of the ethyl, isopropyl, and butyl groups.

Replacement of the O-CH_3 fragment at the methyl group at the carbon atom C(1) and the location of the second methyl group at C(5) is accompanied by significant energy contribution of these groups (19.9 kJ mol^{-1}) to the sublimation enthalpy of 1,3,5-trimethyluracil (Table 11.5). At the same time, the location of the two methyl groups at the carbon atoms C(1) and C(6) is accompanied by an increase of the energy contribution up to 23.1 kJ mol^{-1} and additional stabilization of the specific interaction $H_3C \rightarrow H\text{–}CH_2$ with 4.92 up to 5.75 kJ mol^{-1}. Based on this,

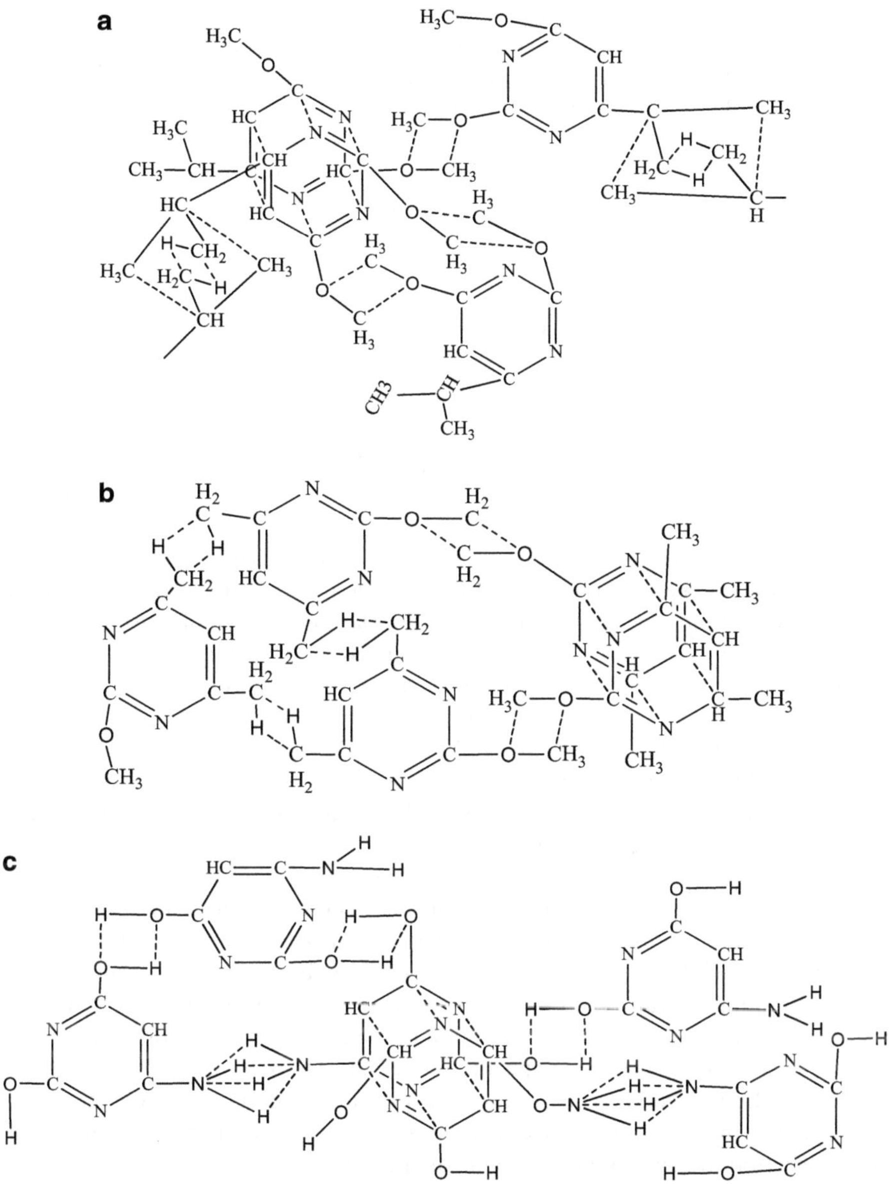

Fig. 11.11 Schematic picture of the crystalline structure of 1,3-Dimethyl-5-isopropyluracil (**a**), 1,3,5-Trimethyluracil (**b**), and 5-Aminouracil (**c**), with network of the hydrogen bonds and specific interactions

one can conclude that the CH group at location C(6) has an increased electron density compared with the closely located C(1) and C(5), in contact with the nitrogen atoms, and on the other side, the largest difference of charges of the carbon and hydrogen atoms. It allows us to conclude that the obtained energies of the specific interactions validate our conclusion on the increased donor properties of the carbon atom C(6) in the pyrimidine molecule.

The molecule of 5-aminouracil with the amino group leads to the formation of the second type of hydrogen bond, DN•••H–N, in the crystal structure with the network of specific interactions, forming definite pyramidal fragments (Fig. 11.11) in it. The location of the amino group at the carbon atom C(5) with its low electron density compared with C(6) leads to hydrogen bonds N•••H–N with reduced stability. The energy of this type of hydrogen bond is determined by the difference of sublimation enthalpies for 5-aminouracil and uracyl reduced by the contribution of the substituted hydrogen (1.0 kJ mol^{-1}) of the amino group or its difference from the total value of energies of two types of specific interactions and hydrogen bonds of crystalline uracyl with the help of Eq. 11.14

$$D\text{-}N\cdots H\text{-}N = (\Delta_{sub}H^0(298\ K)amur - 4D{=}N{\rightarrow}\c{C}{=}\ - 2D{=}HC{\rightarrow}CH - 4D\text{-}O\cdots H\text{-}O\text{-}\ DH)/4 \quad (11.14)$$

The obtained energy value of this type of hydrogen bond of 3.8 kJ mol^{-1} with the use of the known energy parameters of the specific interactions and hydrogen bond D–O•••H–O, forming the network of intermolecular interactions of crystal, points to its reduced stability compared with the one realized in crystalline aminopyridines.

On the basis of the thermodynamic analysis one can state that the energy contribution of the isostructural methyl group with location at C(5) and C(6)

5-Methyl uracyl (131.3 ± 4) 5-Methyl uracyl (131 ± 5)

to the stable hydrogen bonds (15.1 kJ mol^{-1}) is equal to 1.2 kJ mol^{-1} and increases by five times (6.59 kJ mol^{-1}) for the compound with less stable specific interactions, DO → CH_3–O, formed by

1,3,5-Trimethyluracil (103.5 ± 1.5) 1,3,6-Trimethyluracil (106.7 ± 2.5 kJ mol^{-1})

the metoxy group at location C(3) and the methyl group with location at C(1) and C(5). The location of the amino group at C(5) contributing energy to the sublimation enthalpy (15.1 kJ mol^{-1}) is similar to the hydrogen bond, formed by the hydroxyl groups of uracyl, and the energy of D–N•••H–N is equal to 3.8 kJ mol^{-1}. It is significant that the energy contribution of the amino group is comparable with the energies of the more stable specific interaction D=N → C= (13.9 kJ mol^{-1}) and the hydrogen bond. The location of the amino group at C(6) should be accompanied by a significant stabilization of the hydrogen bond DN•••H–N. In this connection the cytosine molecule is of special interest.

11.5.2 Cytosine and Its Derivatives

The cytosine molecule with the hydroxyl and amino group and pyrimidine cycle forms the structure of the crystal with the network of four D–N•••H–N, two

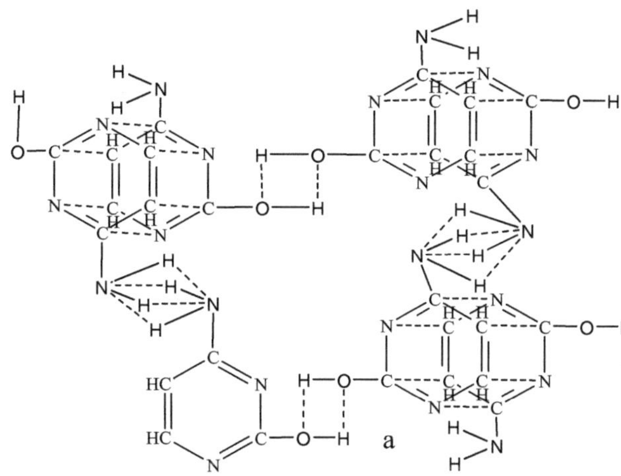

Fig. 11.12 Schematic picture of the crystalline structure of cytosine with network of the hydrogen bonds and specific interactions

D–O•••H–O hydrogen bonds of increased stability, and six specific interactions of two types, D=N → C= and D=HC → C–H (Fig. 11.12), including a tetragonal bipyramidal component. The known values of the energies of the specific interactions formed and the hydrogen bond, formed by the hydroxyl group, allow us to determine the total energy contribution to the sublimation enthalpy of the amino group and, consequently, the energies of the formed hydrogen bonds are to be obtained from the difference of the enthalpy characteristic and the sum of the energies of the remaining interactions with the help of Eq. 11.14. The value of the energy contribution of the amino group is 54.4 kJ mol^{-1} and the energy of the hydrogen bond D–N•••H–N is 13.6 kJ mol^{-1}). Note that the unusually high value of the energy of this bond is consistent with the above mentioned assumption and its reliability is based on the experiment with the mentioned error. A small difference in the energies values of the hydrogen bonds and specific interactions, lying within the experimental error limits

D–N•••H–N (13.6) $\approx$ D=N → C= (13.9) $\leq$ D–O•••H–O (15.1 kJ mol^{-1})

explain the tautomeric transformations of forms "a, b," and "c" cytosine (Sect. 11.5) with content of the first for small quantities [55], which proceed at room temperature.

The value of the energy contribution of the methyl group to the sublimation enthalpy of 1-methylcytosine and the specific interaction formed should be determined with the help of Eq. 11.15.

$$DN{\rightarrow}CH_3\text{-}N = (\Delta_{sub}H^0(298\ K)mcy - 4D{=}N{\rightarrow}C{=} - 2DHC{\rightarrow}CH - 4DO{\cdots}H\text{-}O - 2DN{\cdots}H\text{-}N)/2 \quad (11.15)$$

The obtained energy value of this type of specific interaction of 10.6 kJ mol^{-1} is approximately twice as large and should exceed the energy of the same type of interaction, realized in methylamine The observed natural transformation of energies of the specific interactions and hydrogen bonds of crystalline 1-methylcytosine

Table 11.6 Energies of the hydrogen bonds and specific interactions (kJ mol^{-1}) of the crystalline structure of cytosine and its derivatives

Compounds	Formula	Structure	$\Delta_{sub}H^{\circ}(T)$ [55]	T K	D–N•••H–N	D–N → CH_3–N=	D–O → H_3–O/ DCH$_3$	D–O•••H–O
D=HC → CH = 7.40; D=N → C = 13.9								
Cytosine	$C_4H_5N_3O$	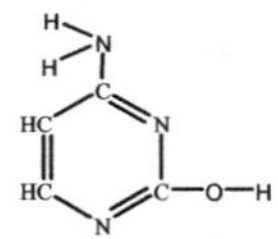	155 ± 3.0	298	= 13.6	–	–	15.1
1-Methylcytosine	$C_5H_7N_3O$	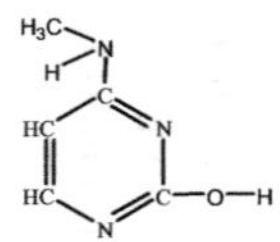	149.1 ± 9.0	298	= 13.6	10.6	–	15.1
1,5-Dimethylcytosine	$C_5H_7N_3O$	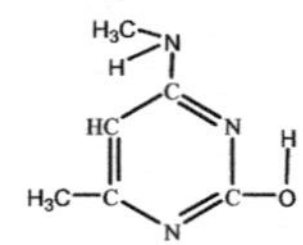	132.8 ± 0.6	390–437	7.8	7.8	/DCH$_3$ = 0.6	15.1
1,5-Dimethyl-*N*-methoxycytosine	$C_7H_{11}N_3O$	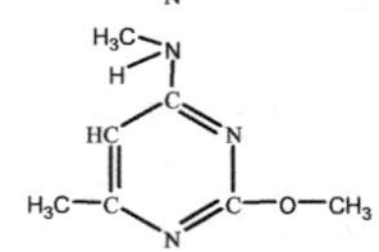	95.6 ± 0.7?	327–363	2.9.	2.6	6.59 × 2 / DCH$_3$ = 0.6	–
1-Methyl-*N*-hydroxycytosine	$C_5H_7N_3O_2$	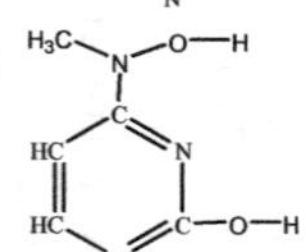	126.7 ± 1.5	–	–	–	–	–
1,5-Methyl-4N-metoxycytosine	$C_6H_9N_3O_2$	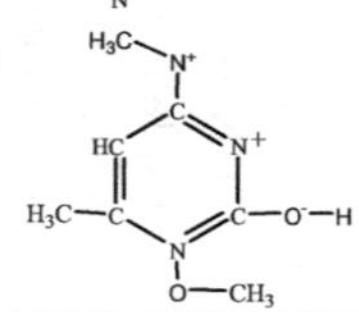	106.9 ± 0.4	320–352	–	7.40 × 2	13.9 × 4	15.1

D–N → CH_3–N (10.6) < D–N•••H–N (13.6) ≈ D=N → C= (13.9) ≤ D–O•••H–O (15.1 kJ mol^{-1})

points to the preservation of the ability of this compound of tautomerization.

The sharp reduction in the total energy contribution of $HNCH_3$ and CH_3 group (12.6 kJ mol^{-1}) to the sublimation enthalpy is observed at 1,5-dimethyl-*N*-methoxycytosine (Table 11.6). A similar influence of the methyl group is expressed in the redistribution of the electron density at derivatives of 5-methyl uracyl, 6-methyl uracyl, and 1,3-dimethyl-5-ethyluracil with an energy contribution of 0.60 kJ mol^{-1}. Thus, the isostructural methyl group expresses the significant influence on the redistribution of the electron density in a molecule of derivatives of cytosine at its location from C(5) and hydroxyl or metoxy group with C(3) or C(2) and C(3) and influences of the hydrogen bond, formed by the amino group. It testifies to the fact that the change of energy of this hydrogen bond greatly influences the process of tautomerization and the temperature conditions of its realization. On the basis of these data arranging the considered compounds in the order of the natural reduction of the tendency to tautomerization appears possible,

Cytosine > 1-Methylcytosine > 1,5-Dimethylcytosine > 1,5-Dimethyl-*N*-methoxycytosine > 1,*N*,*N*-Trimethylcytosine > Uracyl

References

1. Knunyanec ILM (1964) Chemical entomology. Sov Entomol 3:639
2. Heger G, Klein S et al (1980) A comparative analysis of the crystal structures of carbamide and thiocarbamide. Zetschr Kristallogr 153:237–254
3. Clellan AI (1963) Tables of experimental dipole moments. Freeman, San Francisco
4. Katz M, Lobo PW, Minano AS (1969) Polarizability and apparent radius of urea. J Chem Phys at Phys Chem Biol 66(6):1045–1048
5. Dominikova F, Chargitty I (eds) (1997) Molecular structure: precise methods of investigation. Mir, Moscow, p 671
6. Arnett EM (1967) Modern problems of physic organic chemistry. Mir, Moscow, pp 195–341
7. Mullen D, Hellner EA (1978) Simple refinement of the density distribution of the bonding electrons. V. Bond electron density distribution in urea, $CO(NH_2)_2$ at 123 K. Acta Cristallogr 34(5):1624–1627
8. Vinogradov C (1984) In: Rataichik G, Orvill-Tomas WJ (eds) Molecular interactions. Mir, Moscow, pp 184–232
9. Muidinov RY, Zorkii PM (1999) A comparative analysis of the crystal structures of carbamide and thiocarbamide. J Struct Chem 40(6):931–939
10. Gartland GL, Craven BM (1974) Hydrogen bonding NHO=C of barbiturates: the (1:1) crystal complex of urea and 5,5-diethylbarbituric acid (barbital). Acta Cryst 30(4):980–987
11. Wolfenden R (1978) Interaction of the peptide bond with solvent water: a vapor phase analysis. Biochemistry 17(2):201
12. Orita Y, Pullman A (1977) Quantum mechanical studies of environmental effect on biomolecules, VII hydration of urea. Theor Chim Acta 45(4):257–258
13. Bonner OD, Bednarek JM, Arisman RK (1977) Heat capacities of ureas and water in water and dimethylformamides. J Am Chem Soc 99(9):2898–2902
14. Bonner OD, Jordan CF (1976) The interaction of salts, amides, and water. Physiol Chem Phys 8(4):293–301

15. Kucherjavei V, Lebedev VV (1970) Synthesis and applied carbamides. Chemistry, Leningrad, p 441
16. Belousov VP, Morachevakei AG, Panov MY (1981) Heat properties of non-electrolyte solutions. Chemistry, Leningrad, p 264
17. Pimentel V, Mak-Klellan O (1964) Hydrogen bonds. Mir, Moscow, p 462
18. Spencer JN, Hovick JW (1970) Solvation of urea and methyl-substituted urea by water and DMF. Can J Chem 66(3):562–565
19. Gy J, Van Hook WA (1981) Effect in aqueous system. Thermodynamics of urea-h_4/H_2O and urea-d_4/D_2O solutions. J Phys Chem 85(25):3480–3493
20. Baev AK (2003) Novel approaches to the structure of functional solvents and elementorganic compounds with saturated hydrocarbon ligands and their solvates in binary systems. Book of abstract of 27th international conference on solution chemistry, Debrecen, p 79
21. Baev AK (2004) Insolvency of sp^3-gibrydixation model of carbon atom and chemistry of non-electrolytes solutions. Book of abstract of 9th international conference on the problems of solvation and complex formation in solution, Ivanovo, pp 62–64
22. Baev AK (2004) Lack of co-ordination of thermodynamic properties of elementorganic compounds with sp^3-hibrydization model of carbon atom and its insolvency. Book of abstract. Russian research symposium on thermochemistry and calorimeters, Niznie Novgorod, pp 40–41
23. Baev AK (2009) Phenomenological significance of no identical sp^3-hibridization model and nature of interaction in chemistry of liquid non-electrolytes and chemical thermodynamics. Book of abstract of the 10th international conference on physical and coordination chemistry of porphyrines and them analogous, Ivanovo, p 86
24. Tanaka H, Touhara H, Nakanishi K, Watanabe N (1984) Computer experiment on aqueous solution. IV. Molecular dynamics calculation on the hydration of urea in an infinitely dilune aqueous solution with a new urea – water pair potential. J Chem Phys 80(10):5170–5186
25. Cristinziano P, Leli F, Amodeo P, Barone V (1987) Molecular dynamics studies of associations in solutions an NPT simulations of the urea dimer in water. Chem Phys Lett 140(4):401–405
26. Cristinziano P, Leli F, Amodeo P et al (1989) Stability and structure of formamide and urea dimera in aqueous solutions. J Chem Soc Faraday Trans Part 1 85(3):621–632
27. Klrssinger M, Rademacher P (1979) Konformationsanalyse durch Photoelektronen-Spektroskopie. Angev. Chem. Bd 91(11):885–896
28. Astrand PO, Wallqvist A, Karlstrom G, Linse F (1991) Properties of urea-water solvation calculated from a new ab initio polarizable intermolecular potential. J Chem Phys 95(11): 8419–8429
29. Finney JL, Turner J (1988) Neutron scattering studies of molecular hydration in solution. Electrochim Acta 106(20):1183–1190
30. Baev AK (1969) Problems of chemical nature of phase transformation. In: General and applied chemistry, vol 1. Vysheischaia Educating, Minsk, pp 197–206
31. Baev AK (1969) Phase condition and complex formation ability of halogenide metals. In: General and applied chemistry, vol 1. Vysheischaia Educating, Minsk, pp 207–218
32. Nefedov VI, Vovna VI (1989) Electronic structure of organic and elementorganic compound. Nauka, Moscow, 176p
33. Baev AK (2001) Thermodynamic of vaporization of alkyl compounds elements second – six groups. Russian J Izvestiya Vischich Ucheb Zaved Chem Chem Technol 44(1):3–13
34. Suzuki K, Onshi S, Koide T, Seki S (1956) Vapor pressure of molecular crystals. XI. Vapor pressure of molecular crystalline urea and diformylhydrazine, energies of hydrogen djnds in these crystals. Bull Chem Soc Jpn 29:127–131
35. Chickos JS, Acree WE (2002) Enthalpies of vaporization of organic and organometallic compounds, 1880–2002. J Phys Chem Rev 31(2):537

36. Emel'yanenko VN, Kabo GJ, Verevkin SP (2006) Measurement and prediction of thermochemical properties: improved increments of sublimation and standard enthalpies of formation of the alkyl derivatives of urea. J Chem Data 51:79–87
37. Ivanov EV, Abrosimov VK (2001) Biological activity compounds in solutions. In: Kutepov AM (ed) Structure, thermodynamics, reaction ability. Nauka, Moscow, p 402
38. Agakova K, Takenak AN, Sasaki K (1970) Ultrasonic study of dilute aqueous solutins of urea guanidine hydrochloride and dioxane. Bull Chem Soc 49(3):636–641
39. Endo H (1973) The adiabatic compressibility of nonelectrolyte aqueous solutions in relation to the structures of water and solutions. Bull Chem Soc Jpn 46(4):1106–1111
40. Mathieson JG, Conway BE (1974) H_2O-D_2O solvent isotope effect in the apparent molal volume and compressibility of urea. J Solut Chem 3(10):782–788
41. Ogawa T, Yasufa M, Mizutami K (1984) Volume and adiabatic compressibility of amino acids in urea water mixture. Bull Chem Soc Jap 57(3):662–669
42. Juszkiewieiz A (1988) Study on hydration of urea and amides by use of the ultrasonic method. Zitchr Phys Chem (BRD) 158(1):87–96
43. Puliti R, Mattia CA, Barone G, Giancola C (1989) Structure of some N-acetylamides of amino acides. Acta Crystallogr C 45:1554–1557
44. James DW, Armishaw RF, Frost RL (1976) Structure of aqueous solutions, libration band studies of hydrophobic and hydrophilic effects in solutins of electrolytes and nonelectrolytes. J Phys Chem 80(12):1346–1350
45. Finer EG, Franks F, Tail M (1972) Nuclear magnetic resonance studies aqueous urea solutions. J Am Chem Soc 94(13):4424–4429
46. Zaitsav DV, Kabo GJ, Kazuro AA, Sevruk VM (2003) The effect of the failure of isotropy of a gas in an effusion cell on the vapor pressure and enthalpy of sublimation for alkyl derivatives carbamide. Thermochim Acta 406:17–28
47. Kabo GJ, Kazuro AA, Sevruk VM (1995) Thermodynamic properties of organic compounds in crystalline state 2 heat capacities and enthalpies of phase transition of the alkyl derivatives of urea in the crystalline state. J Chem Eng Data 40:371–393
48. Zaitsav DV, Verevkin SP, Paulechka VU, Kabo GJ, Sevruk VM (2003) Comprehensive study of vapor pressure and enthalpies of vaporization of cyclohexyl esters. J Chem Eng Data 48: 1399–1400
49. Piacente V, Ferro D, Della Gatta G (1990) Sublimation enthalpy of eleven urea. Thermochim Acta 158:79–85
50. Ferro D, Barone G, Della Gatta G, Piacente V (1987) Vapor pressure and sublimation enthalpies of urea and some of its derivatives. J Chem Thermodyn 19:915–923
51. Fiorari P, Ferro D (1987) Sublimation enthalpy of monobutyl urea. Thermochim Acta 112: 387–389
52. De Wit HG, Van Miltenburg JC, De Kruif CG (1987) Thermodynamic properties of molecular organic crystals containing nitrogen, oxygen, and sulfur. 1. Vapor pressure and enthalpies of sublimation. J Chem Thermodyn 15:657–663
53. Karlson, TA (1981) Photo electronic and OGE-spectrosopy. Mechanical Engineering, Leningrad, 431p
54. Nefedov VI (1984) X-ray spectroscopy of chemical compounds. Chemistry, Moscow, 255p
55. Chickos JS, Acree WE (2003) Enthalpies of sublimation of organic and organometallic compounds, 1910–2001. J Phys Chem Rev Data 32:537–698
56. Baev AR (2012) Specific intermolecular interactions of organic compounds. Springer, Hiedelberg/Dordrecht/London/New York, 434p

Index

A.K. Baev, *Specific Intermolecular Interactions of Nitrogenated and Bioorganic Compounds*, DOI 10.1007/978-3-642-37472-2,

C

D

E

MIX
Papier aus verantwortungsvollen Quellen
Paper from responsible sources
FSC® C105338

If you have any concerns about our products,
you can contact us on
ProductSafety@springernature.com

In case Publisher is established outside the EU,
the EU authorized representative is:
Springer Nature Customer Service Center GmbH
Europaplatz 3, 69115 Heidelberg, Germany

Printed by Libri Plureos GmbH
in Hamburg, Germany